2nd Edition

EUROPEAN WRECKS & RELICS

Otger van der Kooij

MIDLAND PUBLISHING LIMITED

This book
is dedicated to the memory of
Cilia van der Kooij

European Wrecks & Relics
An illustrated survey of preserved,
instructional and derelict airframes
in 22 countries.

Otger van der Kooij © 1998

ISBN 1 85780 085 0

This fully revised second edition
published 1998 by
Midland Publishing Limited
24 The Hollow, Earl Shilton,
Leicester, LE9 7NA, England

Printed and bound in England

Worldwide distribution (except North America):
Midland Counties Publications (Aerophile) Limited
Unit 3 Maizefield, Hinckley Fields,
Hinckley, Leics, LE10 1YF, England

All rights reserved. No part of this publication
may be reproduced, stored in a retrieval system,
transmitted in any form or by any means,
electronic, mechanical or photo-copied, recorded
or otherwise, without the written permission
of the publishers and copyright holder.

Contents

Introduction	4
Notes	6
Acknowledgements	7
Abbreviations	8
Map of Europe	9
Austria	10
Belgium	14
Cyprus	31
Czech Republic	33
Denmark	54
Finland	64
France	71
Photograph Section: One	129
Germany	161
Greece	232
Hungary	246
Iceland	256
Italy	257
Photograph Section: Two	289
Italy, concluded	321
Malta	329
Netherlands	331
Norway	349
Poland	356
Portugal	383
Slovak Republic	397
Spain	403
Sweden	418
Switzerland	432
Turkey	446
Stop Press	458
Type Index	460
Location Index	473

Cover photographs:

Front: Amazing line of eight Piaggio P.148s at the Ditellandia park at Castel Volturno near Mondragone in Italy. All aircraft are pole mounted above the parking lot.

Rear, top: The collection at Zruč in the Czech Republic started life as a number of stored aircraft at the nearby village of Druztova. Illustrated are L-29R 2608, MiG-15SBs 1170 and 1836, MiG-19S 0219 and PZL M18 OK-TGA.

Rear, bottom: All three highway exits to Linköping have a pole-mounted SAAB aircraft on display. At the most western exit is J29F 29441 '52'.

Title page photograph:

The famous collection at Savigny lès Beaune in France has two CM170s on display, both unmarked. The Château itself houses a large collection of motorbikes and a single Mignet HM-14 'Flying Flea'.
All Otger van der Kooij

Introduction

I still remember my visit to Mike Bursell in early 1994. We talked about his continuing efforts in compiling *Wrecks & Relics* data, the ideas for a second edition of *European Wrecks & Relics* and the ways I could help. Little did I know that a few months later Mike would inform me that he was no longer able to continue his efforts on the second edition due to his change in job and his family life taking up more and more of his free time. Shortly afterwards, Ken Ellis of Midland Publishing telephoned to ask if I would continue the process. To this day I still don't know why I said yes...!

Some four years and a whole lot of typing later the second edition lies in front of you. Although on the outside it may look like the previous edition, the inside has changed considerably. Not only does the book contain 33% more pages but 'new' countries have been added. After the fall of the 'Iron Curtain', the former Eastern Bloc countries became the new 'spotters' holiday destinations and in this edition the Czech Republic, Hungary, Poland and Slovak Republic have been added. A total of 22 countries are covered in this new edition.

Two comments from the first edition bear repetition. Firstly, some readers of this volume may wonder why a book purporting to cover 'European Wrecks & Relics' omits the UK and Eire. The answer lies in the fact that this volume is intended to complement the original and long-standing *Wrecks & Relics* which covers the UK and Eire and has so far run to sixteen volumes, and has a well earned place as the standard biennial reference work on the subject. Secondly, the reader may wonder why there are no detailed maps. It is felt that the book is already sufficiently weighty and instead, where available, directions have been included to the lesser known locations, where a particular relic is still in residence. No attempt has been made to give directions to large towns and aerodromes for which references are readily available elsewhere. Where directions are given, they are intended to give the reader an idea of where to look on a map, rather than to provide comprehensive instructions on how to get there. There are many suitable maps on the market for potential 'tourists' and the compiler has found the Michelin series to be of particular help over the years – they do at least try to mark all the airfields!

Another change from the first edition is the absence of locations where only the smaller light civilian *Wrecks & Relics* aircraft could be found. This has been caused partially due to the lack of sufficient reports and the belief that most readers are more interested in military aviation, museums, airliners and the larger general aviation types. Only civil aircraft from roughly the size of commuter types (eg Jetstream, An-2, etc) and larger are included as a matter of course, unless they are part of larger collections or museums.

The layout has been adapted to be more in line with the UK edition of *Wrecks & Relics*, the 16th edition of which is currently available. All aircraft which should still be current at a location are easily identified with a ❏ in front of the serial column. With the large numbers of types being disposed of, or scrapped, several locations give listings for clarity and ease of presentation and in these cases, no box is shown. Readers should have no problems with the format adopted and the notes given on page six should help in this.

Since the first edition in early 1989 a lot has changed in the aviation enthusiast's world. Even before the fall of the 'Iron Curtain' visits to Eastern European countries were possible. My visit to the open day at Mošnov in the former Czechoslovakian Republic in September 1989 showed that there was not a lot of difference between an airshow in the West and the East. Besides the aircraft, there were also the familiar local 'hot dog' and book stands at Mošnov. Taking pictures during that day was no problem. Later it became obvious (certainly after the fall of the 'Iron Curtain' and the ending of the Warsaw Pact) that there was also a lot of aviation enthusiasm, not only in Czechoslovakia but also in Poland and Hungary, and that local people were also collecting historical data on aircraft. This was, and in some places still is, not always accepted by the local authorities and enthusiasts always need to be alert to the potential hazards of their hobby. These countries have, in the same manner as western European countries, annual airshows, aviation museums and aircraft collections. The Czech Republic, Hungary and Poland are even getting membership of the NATO alliance.

Another major event in the early 1990s was the fall of the Berlin wall, which led to the reunion of East and West Germany on 3rd October 1990. The 'West Germans' received a whole lot of 'new' aircraft from the 'East'. Only a handful of aircraft were used by the new Germany. Most were put into storage in October 1990 and have found their way slowly into museums and private collections.

From September 1973, talks were held between all European countries, Canada and the United States over arms reduction in Europe. Most were held in Helsinki and one result was the CSBM (Confidence and Security Building Measures) treaty. Further talks at Vienna about the MBFR (Mutual Balanced Force Reductions) measurements did not result directly in a treaty, but lead to further talks between only 22 European countries. They signed the CFE (Conventional Armed Forces in Europe) document. This document set out to reduce and equalise the military force in the ATTU (Atlantic to the Ural) zone and came in effect on 1st January 1991. This CFE treaty gave maximum limits to how many armoured fighting vehicles, artillery, combat aircraft and assault helicopters the NATO and Warsaw Pact countries should have. Despite the collapse of the Warsaw Pact itself, the CFE treaty was still in effect.

Reduction of combat aircraft could be established in three different manners; severing (cutting the fuselage in three parts), deformation (a minimum of 30% deformation of the fuselage) or the usage as a target drone. The first option was the most commonly used. Combat helicopter forces could be reduced by either severing (cutting the tail boom from the cabin), deformation (a minimum of 30% deformation of the cabin) or the use of explosives (blowing up the helicopter). Later additions to the treaty gave also the possibility for the usage of CFE aircraft for instructional purposes, static display or a reduction because of accidents (crashes).

Some bases contained many dismantled combat aircraft that could be seen over the years – the validation process allowing for inspections on the ground and by satellite or later through the 'Open Skies' agreement. Germany dismantled (and finally scrapped) 141 MiG-21s at Dresden. Hungary dismantled ten MiG-21s at Kecskemét, Italy a large number of G91s at Améndola, Brindisi and Cérvia and F-104s at Villafranca (Italy reduced more aircraft than they had to by the treaty), Poland scrapped 91 aircraft (including 22 MiG-17s and 32 MiG-21s at Mierzecice) and Czechoslovakia 64 MiG-21s (from Vodochody, some went later to the Slovak Republic). Note that some of these aircraft, with the exception of the German MiG-21s, found their way to museums and collections. Only the Russian Federation had to reduce combat helicopters (115 in total).

While this book was in the completion phase, the future was considered. It is planned that a third edition will follow within a shorter time span than between this edition and the first. For this new edition I'm hoping to expand coverage still further into Eastern Europe, to countries like Romania, the former Yugoslavian states and other countries, when there is enough information at hand. For this I depend on readers inputs, including prints and slides. A book like this can only be possible with such contributions. All comments, corrections, sightings, updates, etc for this and the future editions are more then welcome.

Finally, special thanks to a number of people who helped to made this edition a reality. Mike Bursell and Ken Ellis checked the text and helped smooth over my English. Thanks to Aad van der Voet for turning the script from my ancient PC-editor to the modern Words version and thanks to Eddy Wierenga for taking care of one of the countries. Also special thanks to Jurgen van Toor for final proof reading and Andy Marden, Dimitri Schmidt, Berry Vissers, Hans van der Vlist for their researches and checking.

Otger van der Kooij
July 1998

E-mail: wrecks@wxs.nl
Fax: 31 (0)70 31 75 692

Postbus 1017
2260 BA Leidschendam
Netherlands

Warning!
In some of the countries covered by this book, photography, taking notes and even just looking at civil or military aircraft and airfields are not as commonly accepted as it is in countries such as the UK and the Netherlands. Although it may sometimes look easy, a hobby like aircraft spotting is liable to misinterpretation and may lead to arrest and imprisonment. When executing your hobby always be cautious and always obey all local laws. If in doubt, make enquiries with the local authorities, or better still enquire with the relevant country's embassy *before* setting out.

European Wrecks & Relics is put together using the best information available to the author. Every effort is made to be as accurate as possible. However, niether the author nor Midland Publishing Limited can be held responsible for any errors or changes that may occur in the location or status of aircraft or places listed.

Notes

After a short narrative introduction, each country's locations are given in alphabetic order. With each location, the text will give (where known) information about current or previous aircraft. If aircraft have left a certain location, their new resting place (if this is within the coverage of the book) will be <u>underlined</u>. Aircraft that have moved outside the coverage of this book do not have their destination underlined. After the serial number, the construction number, if known, is given within brackets. Aircraft within each location are given in a specific order. First come foreign military aircraft, listed in order of their country. All these aircraft have their (armed forces) country of origin mentioned, where appropriate. Within each country of origin the aircraft are listed in serial order (if there is a serial system) or by type. Civil aircraft then follow in order of registration, alphabetically.

The entry for each aircraft is arranged in five columns, as follows:

Column 1 - Registration/Serial
Lists the current or last known registration or serial of an aircraft. If an aircraft is painted in incorrect markings or has been painted in previously worn markings this will be displayed in the remarks column. There are two points (..) in this column if the serial of registration is not known or has not been allocated. In similar manner, full scale models or replicas are marked with two points as they normally have no registration allocated. Aircraft from these last two options will also have the markings they wear mentioned in the remarks column.

Column 2 - Aircraft Type
Due to the limited space the aircraft type will be displayed in its most commonly used variant. A Svenska Aeroplan Aktiebolaget Sk37 Viggen will be listed as just as Sk37. Licence built aircraft which received a new designation in the process will be listed with that new designation. A French built Ju52 is listed as an AAC.1 and the Czech built Il-14 is known as an Avia 14. Local military designations are used. The Danish called their Drakens F-35s, TF-35s and RF-35s, while the Swedish referred to the Alouette 2 as Hpk2.

Column 3 - Construction Number
This column is self evident. Occasionally, eg Belgian SF-260s, the line number will be given in this column in place of a c/n. If that is the case, this will be mentioned in the text.

Column 4 - Status and Remarks
The status of the aircraft can be split into seven different types: decoy, dumped, instructional, preserved, stored, target and under restoration. This can be followed by several different remarks, including further explanations of the status (eg preserved, at gate), the marking carried on the aircraft (if these are false or incorrect it will be mentioned as; marked as '12345') or a previous location, serial or markings.

Column 5 - 'Last Noted' Date
This last column should give a good indication what can expected at the mentioned location. A last noted date from many years ago may indicate that the aircraft may have gone or that it is a not very often visited location. If only a year (ie with no month) appears in this column it indicates that according to the museum or official sources the aircraft was still there in that year, but there were no actual sightings of the aircraft during that year. A long dash (–) indicates that the aircraft *should* be there but this had not been confirmed. Also, when an aircraft is expected to move to the given location the long dash will be used.

Finally, at locations where a large number of aircraft were located in the past, these will also be listed in the columns, but the box at the beginning of the line is deleted and in Column 5 can be found their destination, civil registration or the word 'gone' if the aircraft have left for a unknown location or 'scrapped' if the aircraft is known to have been scrapped on site.

Type and Location Index
The last two sections of the book provide a way of finding entries other than by country. Because types mentioned in column 2 of the tables are necessarily abbreviated, this information is expanded in the type index where possible. Aircraft names are given with aircraft designations (eg, by SAAB 32 the name Lansen is added). For the more uncommon types, again where possible, the factory name is given in brackets; G46 will be listed as G46 (Fiat). The location index is self evident. Please note the additional location names of the 'Stop Press' pages on the last page of the locations index.

Acknowledgements

A book like *European Wrecks & Relics* relies on the inputs, sightings and reports from a large group of people. Without their help this book would not be as it is today. Thank you all very much.

Alan Allen, Peter R Arnold, Martin Bach, Frank Bernard, Jack Bosma, Martin Brasley, Pascal Brugier, Julian Bloomfield, Tom Bouwmeester, Daniel Brackx, Heinrich Breuer, Pascal Brugier, Rod Burden, Dean Charnley, Ingvar Claesson, David Con-ciotti, Enrique Cortes, Aiden Curley, Howard Curtis, John Davis, Alf-Rico Denck, Bruno Despret, Jaap Dijkstra, Alistair Duncan, Ron Duurland, Simon Ellwood, Erik Jan Elgelen, Oluf Eriksen, Tieme Festner, Rui Ferreira, Pete Foster, Stefan Goosens, Rainer Göpfert, Paul Gross, Wayne Groves, Ján Jando, Clive Harden, Jan Hendriksen, Walter Heukersfeld, Peter Hillman, Lars Hoebers, Hans Hoogers, Hans Holzer, Anton Homma, Leo Hoogerbrugge, Bertrand Hugot, Paul van den Hurk, John Hutchins, Paul A Jackson, Magnus Johansson, Marcel de Jong, Martin Kaye, Elmar Keetman, Ralf Keil, Miroslav Khol, Ron Kraan, Pavel Kucera, S Kudela, Christoph Kugler, Stewart Lanham, Birger Larsen, Dave Lee, Mike Lee, David Legg, Erik Leijdens, Hans Lichtenhof, Roberto Lima, Josef Loidl, Piotr Lopalewski, Alan Macey, Nikolaus Marx, Daniela Messina, Frank Mink, Tony Morris, Johan Mulder, Frits von Münching, Richard Nels, Sten-Olaf Nieminen, Guus Ottenhof, Arnold ten Pas, Andy Patsalides, Gerard Post, M A Powell, Magnus Pedersen, Per Thorup Pedersen, Bo Bang Petersen, Bob Prescott, Harry Prins, Ian Roberts, Jozef Rodák, Sven Scheiderbauer, Saviour Schembri, Chris Schmidt, Tony Seeley, Roger Seroo, Graham Sheppard, Stanislav Skala, Jaroslaw Sobocinski, Peter Spooner, Mike Staines, Peter Stevens, Jeffrey van Summeren, Bob Symes, Karel Tarantik, Dave Thompson, Reiner Tippman, Hannu Valtonen, Paul Veenboer, Holger Veh, Rick Versteeg, Patrick Vinot Prefontaine, Gert-Jan Vis, Brian Waclawek, Alan Warnes, Fabrice Wassong, Renate Westermann, Graham Wickens, Paul Wigley, Stephan Williams, Jürgen Wolter, David Woods, Henry Wydler and Wijgert Ijlst.

The compiler wishes to acknowledge and recommend the following, used in the preparation of this book.

Magazines and periodicals: *Aeronews* of Belgium (Aviation Society of Antwerp), *AIM* (Spotting Group Volkel), *Air-Britain News* (Air-Britain Historians), *Airforces Monthly* and *Airforces Research* (Key Publishing), *Airnieuws* (Stichting Airnieuws Nederland), *Aviation Letter* (Lundkvist Aviation Research), *British Aviation Review* (British Aviation Research Group), *CAPC Magazin*, Centerline (Stichting Valkenburgse Vliegtuigspotters), *Full Stop* (Aviation Group Leeuwarden), *Letecké Listy*, Luftfahrt Journal (Coincat), *Military Aviation Review* (formely *Strobe*), *On Finals* (Luctvaartvereniging Twenthe), *Panoravia*, *Scramble* (Dutch Aviation Society) and *Zipper* (International F-104 Society).

Books: *Aircraft Museums and Collections of the World*, parts 3, 4 and 9: Bob Ogden; Bob Ogden Publications.
Belgian Military Wrecks & Relics: Daniel Brackx; Flash Aviation, 1991.
Die andere Deutsche Luftwaffe: Wilfried Kopenhagen; Motor Buch Verlag, 1994.
European Air Forces Directory: Ian Carroll; Mach III plus, 1997.
European Wrecks & Relics: Mike Bursell; Midland Counties Publications, 1989.
Flygplan på Museum: Sölve Fasth; Allt om Hobby, 1993.
Ilustrovana Historie Lecectvi: various; Edice Triada.
Militair 1982: John Andrade; Aviation Press, 1982.
Soviet Transports: Peter Hillman, Stuart Jessup, Guus Ottenhof; The Aviation Hobby Shop, 1996.
Survivors 97: Vince Horan; Gatwick Aviation Society Publication, 1997.
The Complete RDAF: Per Thorup Pedersen; Per Thorup Pedersen, 1996.
United States Air Force and Army Serial Batches: A J McGregor, D Wellden; Mach III plus, 1993.
United States Navy and Marine Corps Serial Batches A J McGregor, D Wellden; Mach III plus, 1995.
US Military Aircraft Designations and Serials since 1909: John M Andrade; Midland Counties,1979 rptd 1997.
Warbirds Directory: John Chapman, Geoff Goodall; Warbirds Media Company, 1996.
4+ Publikace: various; Nakladatelstvĺ 4+.

In this high tech age the internet also proved very useful, especially the military spotters list of Mil-Spotters (mil-spotters-forum-subscribe@makelist.com) and the *Scramble* (http://www.scramble.nl) and Italian database on-line (http://www.spotters.it). Also the *Scramble Bulletin Board System* (BBS) was very helpful.

Abbreviations

Use of abbreviations is limited as much as possible and will normally only appear in column 5.

AMARC	Aircraft Maintenance and Regeneration Center	RSweAF	Royal Swedish Air Force
arr	arrived	SAAF	South African Air Force
a/c	aircraft	soc	struck off charge
CAF	Canadian Armed Forces	SpaAF	Spanish Air Force
camo	camouflage	SwiAF	Swiss Air Force
canx	cancelled	temp	temporarily
CASA	Construcciones Aeronauticas SA	TurkAF	Turkish Air Force
c/n	construction number	UN	United Nations
c/s	colour scheme	USAAF	United States Army Air Force
BDR	Battle Damage Repair	USAF	United States Air Force
Belg AF	Belgian Air Force	USN	United States Navy
CFE	Combat Forces Europe	WGAF	West German Air Force
DASA	Deutsche Aerospace	WGArmy	West German Army
dep	departed	WGNavy	West German Navy
DGMFA	Depósito General de Material de Força Aérea	wfu	withdrawn from use
		yel	yellow
EWR-1	European Wrecks and Relics – first edition		
FAF	French Air Force		
FMV	Försvarets Materielverk		
f/n	first noted		
GAF	German Air Force		
HQ	Headquarters		
IIAF	Islamic Iranian Air Force		
ItAF	Italian Air Force		
KLM	Koninklijk Luchtvaart Maatschappij		
km	kilometres		
LH	Left hand		
l/n	last noted		
MBB	Messerschmitt Bölkow Blohm		
ntu	not taken up		
NVA	Nationalen Volksarmee		
OGMA	Oficinas Gerias de Material Aeronauticao		
PdF	Patrouille de France		
PolAF	Polish Air Force		
PortAF	Portuguese Air Force		
RAAF	Royal Australian Air Force		
RAF	Royal Air Force		
RCAF	Royal Canadian Air Force		
RDanAF	Royal Danish Air Force		
RJordAF	Royal Jordanian Air Force		
RNethAF	Royal Netherlands Air Force		
RNethNavy	Royal Netherlands Navy		
RNN	Royal Netherlands Navy		
RNoAF	Royal Norwegian Air Force		
RSV	Reparto Sperimantale Volo		

Countries covered

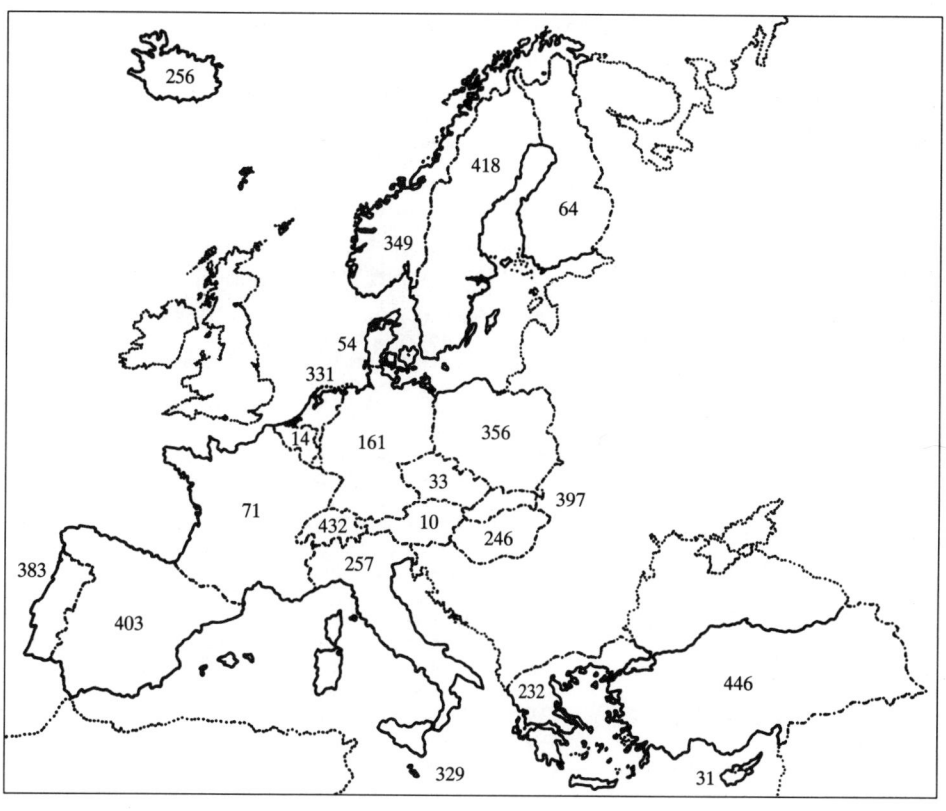

Austria	10	Italy	257
Belgium	14	Malta	329
Cyprus	31	Netherlands	331
Czech Republic	33	Norway	349
Denmark	54	Poland	356
Finland	64	Portugal	383
France	71	Slovak Republic	397
Germany	161	Spain	403
Greece	232	Sweden	418
Hungary	246	Switzerland	432
Iceland	256	Turkey	446

AUSTRIA

The major *Wrecks & Relics* event in Austria has been the opening of the new museum at Bad Ischl. The owner is currently building a new shelter for the outdoor display aircraft and hopes to acquire all sub-types of the MiG-21. The museum at Graz has also received a number of new aircraft over the past years. The older museums at Wien seem to have lost their aircraft on display, placing them in storage.

AIGEN IN ENNSTALL
This main Austrian military Alouette base has placed one of the helicopters in storage.
☐ 3E-KC   Alouette 3   1387   stored   8-95

BAD ISCHL
The Museum Fahrzeug-Technik-Luftfahrt displays a large collection of former Warsaw Pact type aircraft, although one has already gone. The Mi-8PS 9344 has left for St Johann im Pongau. The Czech MiG-21PFM arrived in Austria in 1994 and was used for an exhibition in Wien. It arrived at Bad Ischl in 1996. The museum can be found along road 145 near Lauffen and is open from 1st April to 30th October, between 09:00 and 18:00.

☐ 350		MiG-15bis	..	preserved, ex Hungarian AF	6-97
☐ 2610		L-29RS	792610	preserved, ex Sobeslav, Czech AF	6-97
☐ 4143		Mi-4	04143	stored, ex Sobeslav, Czech AF	6-97
☐ 0201		MiG-19S	050201	stored, ex Zruč, ex Druztová, ex Czech AF	6-97
☐ 0310		MiG-21F-13	560310	preserved, ex Sobeslav, ex Czech AF	6-97
☐ 0603		MiG-21F-13	760603	stored, marked as '1002', ex Zruč, ex Czech AF	6-97
☐ 4405		MiG-21PFM	764405	preserved, ex Letňany, ex Czech AF	6-97
☐ 2014		MiG-23MF	0390324627	preserved, ex Laage, ex 336/NVA	6-97
☐ 2232		MiG-21SPS	94A4506	preserved, ex Rothenburg, ex 833/NVA	6-97
☐ 2378		MiG-21UM	02695163	preserved, ex Rothenburg, ex 204/NVA	6-97
☐ 2386		MiG-21U-400	661119	preserved, ex Rothenburg, ex 237/NVA	6-97
☐ 2523		Su-22M-4	26510	preserved, ex Laage, ex 723/NVA	6-97
☐ 2547		Su-22M-3UK	17532369809	preserved, ex Laage, ex 113/NVA	6-97
☐ 2814		L-39ZO	731017	preserved, ex Rothenburg, ex 154/NVA	6-97
☐ 9308		Mi-8T	10538	preserved, ex Briest, ex 922/NVA	6-97
☐ 9338		Mi-8PS	10550	preserved, ex Briest, ex 962/NVA	6-97
☐ 29466	I red	J29F	29466	preserved, inside	6-97
☐ 3A-BK		L-19E	23937	stored, fuselage only, ex 57-6023	5-97
☐ 3A-BL		L-19E	23938	stored, fuselage only, ex 57-6024	5-97
☐ D-HOLD		Enstrom F.28A	185	preserved, inside	5-97
☐ OK-NYA		An-2	113902	preserved, crashed 9-7-94	6-97

GERASDORF
Just north east of Wien is the small town of Gerasdorf. Displayed here on some poles with a Mercedes parts dealer is the former Polish MiG-21.
☐ 1706   MiG-21PF   761706   preserved, ex Wiener Neudorf, ex Polish AF   97

GRAZ THALERHOF
The military gate is guarded by a J29, which can be seen from the lake at the Schwarzl Freizeitzentrum on the west side of the airfield. It has been reported that also an unconfirmed Vampire trainer maby preserved here.

☐ 29588	D red	J29F	29588	preserved, at gate	6-97
☐ 25		J35D	35339	instructional, ex 35339/RSweAF	6-97
☐ OE-KSS		SAAB 91D	91456	preserved, ex 3F-SS	10-96

AUSTRIA - 11

The Osterreichisches Luftfahrt Museum on the civil side of the field was first opened to the public on 28th May 1987. The museum is open daily from May to September between 11:00 and 20:00. The Bf110 has been deleted from the list as the 'substantial wreckage' has now degenerated to the port engine, propeller and a few small parts. Recent arrivals are a Swiss Hunter, Hungarian MiG-21 and Swedish Draken.

❏ 4406		MiG-21MF	964406	stored, ex Kecskemét, ex Hungarian AF	10-96
❏ 2919		An-2TP	1G29-19	stored, ex Polish AF	3-98
❏ 1326		Lim-2	1B-1326	stored, ex Polish AF	6-97
❏ 0221		TS-11-100	1H-0221	stored, ex Polish AF	3-98
❏ 32510	02	J32E	32510	preserved, ex Swedish AF	7-97
❏ 35804		Sk35C	35804	preserved, ex Swedish AF	6-97
❏ J-1733		Venom FB.54	903	preserved, ex Swiss AF	10-93
❏ J-4094		Hunter F.58	..	stored, ex Swiss AF	7-97
❏ 29541	H yel	J29F	29541	stored	3-98
❏ 4C-AH		Yak-11	171229	preserved, wfu 9-65, ex 4A-AH	10-93
❏ OE-APD		Luscombe 8F	6012	preserved, also reported as HB-DUO	—
❏ OE-ATP		CeF150L	1093	stored	6-97
❏ OE-CCW		Beech A45T	G-2	preserved	10-96
❏ OE-DBS		Meteor FL55BM	1132	stored	6-97
❏ OE-LSA		SA226TC	TC-315	stored, crashed 17-9-84, ex N1014X	3-98

HUBHOF

The Märchenland amusement park is in Hubhof near Aggsbach-Markt, some 30km east of Linz. Preserved out here in the open is a 'Tunnan'.

❏ 29560	F yel	J29F	29560	preserved, white/blue c/s	5-97

LINZ

Town: The Vampire at the Höheren Technischen Bundeslehranstalt at the Paul Hahnstrasse 4 has been swopped with a J29 from the airfield.

❏ 29447	B yel	J29F	29447	preserved, ex Linz -Hörsching	10-96

Airfield – Hörsching: As correctly surmised in the first issue, there are in fact two J29s preserved here. J29 29447 has moved to the town of Linz. The Austrian Air Force has placed a number of AB204s in storage since 1982. Some have been sold to the Swedish Armed Forces, including; 4D-BI (3118), 4D-BJ (3139, which became 03139), 4D-BN (3156, became 03156), 4D-BO (3157, became 03316) and 4D-BQ (3189, became 03189). Stored here since 1993 are SAAB 91s. Some of these have also found new homes; 3F-SI moved to Tulln, as did 3F-SM, 3F-SP and 3F-SR.

❏ 29443	M yel	J29F	29443	preserved, on pole	6-97
❏ 3C-OF		OH-58B	42244	dumped, crashed 28-2-92	10-93
❏ 3F-SB		SAAB 91D	91453	stored	10-93
❏ 3F-SF		SAAB 91D	91463	stored	10-93
❏ 3F-SO		SAAB 91D	91450	stored, for museum in Wien	10-93
❏ 3F-SW		SAAB 91D	91462	stored, dismantled	10-93
❏ 4D-BE		AB204B	3066	stored	4-92
❏ 4D-BF		AB204B	3072	stored, dismantled	10-96
❏ 4D-BH		AB204B	3107	stored	10-93
❏ 4D-BM		AB204B	3155	stored, dismantled	10-96
❏ 4D-BV		AB204B	3201	stored, dismantled	10-96
❏ 4D-BY		AB204B	3204	stored, dismantled	10-96
❏ 5C-VF		Vampire T.55	15694	preserved, ex town, ex XH320	10-96
❏ ..		AB204B	..	dumped	10-93
❏ ..		AB204B	..	dumped	10-93
❏ ..		SAAB 105ÖE	..	dumped	10-93

| AUSTRIA - 12 |

PUNITZ
An An-2 is preserved in the small town of Punitz along the Austrian border with Hungary.
❑ CCCP-44998 An-2 .. preserved, ex Wiener Neustadt Ost —

RAMSAU AM DACHSTEIN
Unconfirmed reports suggest that a 'Tunnan' can still be found in this small village.
❑ 29587 C red J29F 29587 preserved —

SALZBURG
The Skytrain formerly stored at Wiener Neustadt was first noted here at the airfield in June 1996.
❑ OE-LBC C-47A 13073 stored, ex Wiener Neustadt Ost, ex N86U 6-96

SCHWAZ
The abendlokal (pub) Giovanni in the Stausre Au 1 (the name of the street) has taken delivery of a former Czech Air Force MiG-15. Schwaz is some 15km north east of Innsbruck.
❑ 1836 MiG-15 231836 preserved, ex Druztová 96

ST JOHANN IM PONGAU
Between St Johann and Bischofshofen (some 50km south of Salzburg) a Mi-8 is in use as a clubhouse at a private helicopter landing strip.
❑ 9344 Mi-8PS 10585 preserved, ex Bad Ischl, ex Briest, ex 990/NVA 97

ST MARTIN
At the back of some military barracks near the airfield of Zeltweg is an incompletely marked AB204 dumped. The camp is also close to the highway.
❑ .. AB204B .. dumped, only marked as 4D-B 6-97

TULLN LANGENLEBARN
The instructional L-19A 3A-CF has become OE-CFF. L-19E 3A-BF has also departed from storage.
❑ 3A-BH	L-19E	23935	stored, ex 57-6021	6-97
❑ 3A-BI	L-19E	23936	stored, ex 57-6022	6-97
❑ 3A-CB	L-19A	21510	stored, ex 51-4625	6-97
❑ 3A-CD	L-19A	21930	stored, ex 51-5045	6-97
❑ 3A-CE	L-19A	21990	instructional, ex 51-5105	5-95
❑ 3A-CH	L-19A	23079	stored, ex 51-12624	6-97
❑ 3A-CM	L-19A	22085	stored, ex 51-7531	6-97
❑ 3A-CU	L-19A	22827	stored, ex 51-16955	6-97
❑ 3B-HA	OH-13H	2479	stored, crashed 21-2-66, ex Wien, ex 58-6984	6-97
❑ 3F-SI	SAAB 91D	91467	instructional, parts only, ex Linz	5-95
❑ 3F-SJ	SAAB 91D	91468	preserved	10-96
❑ 3F-SM	SAAB 91D	91448	stored, ex Linz	6-97
❑ 4D-BS	AB204B	3191	instructional	5-95

Flying from here is a historic flight with military types. All aircraft are owned and flown by the military.
❑ 29449	F J29F	29449	preserved, flyable	6-97
❑ 3A-CG	L-19A	22589	preserved, flyable, ex 51-12275	6-97
❑ 3B-HD	OH-13H	2482	preserved, flyable, ex instructional, ex 58-6987	5-95
❑ 3F-SP	SAAB 91D	91451	preserved, flyable, ex Linz	6-97

AUSTRIA - 13

WIEN
Town: The Heeresgeschichtliches Museum can be found at the Arsenal, Object 1, in the eastern suburbs of Wien. A visit in 1996 revealed that the J29F 29556, Yak-18 3A-AA, Yak-11 4C-AF and LT-6G 4C-TE have been placed in temporary storage. Their OH-13H 3B-HA moved to Tulln.

☐ 20.01	Albatros B.I	01	preserved	10-96

The Technisches Museum für Industrie und Gewerbe at the Mariahilferstrasse 212 in Wien should have a number of aircraft on display, but there are no recent reports from this museum.

☐ 3B-HO	OH-13H	2493	preserved, wfu 6-76, ex 58-6998	8-89
☐ D-ENPE	Fi156C-3	110253	preserved, ex OE-ADO, ex 3818/Swedish AF	—
☐ OE-BLF	Job 15-150/2	069	preserved	—
☐ OE-BVM	Dove 5	04488	preserved	—

Airfield – Schwechat: The aircraft at the viewing gallery at the international airport have seen some changes. The L-29 was swopped for the Harvard and N1002. Harvard F-BMJO (88-42005) and N1002 D-EOAR (163) have both left for an unknown location.

☐ 2614	L-29R	792614	preserved, unmarked, ex Czech AF	11-97
☐ 5427	Pembroke C.54	1019	preserved, nose only, ex WGAF	7-97
☐ 7809	Sycamore Mk.52	13461	preserved, orange/black c/s, ex WGArmy	11-97
☐ 29392 I yel	J29F	29392	preserved, marked as 'H yellow'	11-97
☐ G-AGRW	Viking 1A	115	preserved	11-97
☐ ..	Zlin 37A	..	preserved, unmarked	11-97

WIENER NEUDORF
A furniture centre in the Industriezentrum NO-Süd acquired ex Polish Air Force MiG-21PF 1706 during December 1990. The engineless MiG was displayed in original colours on a container. It had moved to Gerasdorf by 1997.

WIENER NEUSTADT OST
The stored An-2 CCCP-44998 moved on to Punitz during 1989. The C-47A N86U was bought by the First Austrian DC-3 Dakota Club and has been in storage here for some time. It moved to Salzburg in 1996.

ZELTWEG
Preserved at the gate of this military airfield is a Fiat G46.

☐ 3A-BB	G46-4B	157	preserved, at gate, ex MM53397/Italian AF	6-97
☐ 3F-SR	SAAB 91D	91455	stored	6-97

BELGIUM

The population of *Wrecks & Relics* in Belgium is largely controlled by the Koninklijk Leger Museum/Musée Royal de l'Armée. In this Belgian section the term Brussels museum refers to their aircraft. This is a healthy situation, as compared to countries like the Netherlands, where a lot of former military aircraft have recently been scrapped. In Belgium military aircraft preserved at the gates of military bases are owned by the Brussels museum. The big changes over the last few years in Belgium have been the end of operational flying of the Mirage 5s and the storage of more than 30 F-16s. Some of these 'modern' F-16s have found their way to museums.

ANTWERPEN
A former Brustem Thunderstreak, as well as some smaller aircraft, are used as instructional airframes at the Stedelijk Instituut voor Middelbaar en Technisch Onderwijs at the Emile Verhaerenlaan 24 (on the Antwerpen left bank). The F-84F is on loan from the Brussels museum.

❏ FU-36	F-84F	..	instructional, ex 52-7157	7-91
❏ OO-AJK	N1203	261	instructional	—
❏ OO-WIF	AA-1B Trainer	0264	instructional	—

ARLON
Mounted on a pole at the Victory museum is a Skytrain painted in USAAF colours. The museum at Arlon is easy to find as it is directly at one of the exits of the motorway to Luxembourg.

❏ F-BAIF	C-47B	16371/33119	preserved, marked as '2100847', ex 76787/FAF 7-97

ATH
Stored for many years in a yard at the Rue de Liessies are three Hunters. Two were withdrawn from use by the Air Force after making belly-landings, while the third was damaged during maintenance by gunfire from an other Hunter. During 1996 the owner of the yard died, which could mean the end for the aircraft here. Ath can be found some 20km north west of Mons.

❏ ID-102	Hunter F.4	8651	stored, crashed 19-9-58, soc 30-10-58	6-96
❏ IF-29	Hunter F.6	T119	stored, crashed 28-5-58, soc 17-11-58	6-96
❏ IF-92	Hunter F.6	T164	stored, soc 5-8-59	6-96

BALEN
A scrapdealer in the area bought two Magisters from the Belgian Air Force in 1996. One of them, MT-29 (286), is submerged in an unknown lake, while the other, MT-31, left for a new museum in Meerhout.

BEERVELDE
A haulage company along the A14/E3 Antwerpen-Gent motorway near Beervelde has a Starfighter. The aircraft is stored behind some containers.

❏ FX-79	F-104G	9137	stored, crashed 14-12-79	4-98

BEVEKOM
This is the only Belgian Air Force base which has a dual name (in both languages, Beauvechain is the French name). The F-16s left the base early in 1996, their place being taken by the SF-260s formerly based at Goetsenhoven as well as Alpha Jets and CM170s from Brustem. Not only have the operational fighters left, most of the *Wrecks & Relics* have also left the base. Spitfire SG-31/YL-B (marked as 'SG-3/MN-A') was exchanged by the Brussels museum for a Bristol F.2b. SG-31 left on 3rd May 1990 and is now under restoration

at Audley End, UK, and will become G-BSKP. The remains of the dumped F-16s FA-42 (6H-42, crashed 9-10-86) and FA-113 (6H-113, crashed 12-5-95, ex Evere) have been scrapped. The former decoy F-84Fs FU-123, FU-191 and FU-192 have all gone to <u>Weelde</u>. The dumped FU-143 and FU-194 have been scrapped. Preserved Starfighter FX-04 (ex Evere) went to Brustem in September 1996 for the airshow and then returned here. FX-76 (ex Koksijde) moved on to <u>Kampenhout</u>, while FX-58, FX-78 and FX-89 (all ex Koksijde) all went to Weelde. Their place will be taken by two former Brustem aircraft and one Hunter from the Weelde store (the two stored Hunters will be rebuild to one complete airframe). All three are Brussels museum owned and will become part of the new The Golden Falcon museum here which is hoping to open its gates in 1998.

❏ CF-03		SA226T	T-262	dumped, remains, crashed 16-4-80	1-98
❏ EG-79	S2-R	Meteor F.8	6521	stored, arr 4-11-97, ex Brustem	2-98
❏ FA-10		F-16A	6H-10	expected, ex Weelde	—
❏ FR-32		RF-84F	..	stored, arr 8-7-96, ex Weelde, ex Bierset	2-98
❏ FT-24		T-33A	7788	preserved, ex Brustem, 52-9892	2-98
❏ FX-04		F-104G	9019	preserved, ex stored, ex Evere	2-98
❏ ID-16		Hunter F.4	..	stored, marked as 'IF-83', ex Weelde, ex Bierset	2-98
❏ ID-26	JE-P	Hunter F.4	..	stored, ex Weelde, ex Koksijde, ex Zellik	2-98
❏ MT-3		CM170	260	stored, inside shelter S13, ex Weelde	1-98
❏ MT-46		CM170	145	stored, inside shelter S13, ex Weelde	1-98
❏ ST-14		SF-260M	10-14	stored, ex Goetsenhoven, ex Saffraanberg	1-98

BEVEREN AAN DE IJZER
A scrap dealer here bought Mirage 5 BA-35 during the early 1980s. The aircraft was damaged by an engine explosion and ensuing fire on 25th January 1979. The Mirage (less its tail) moved to <u>Savigny lès Beaune</u>, where it was used to complete a composite Mirage. The tail went to Peuti.

BEVERLO
The former Laage MiG-23ML 2021 was used here as an instructional airframe before moving to <u>Weelde</u>.

BEVINGEN
Only half a mile or so from Brustem is a military camp which served as an administrative/instructional site for its neighbouring base. Their preserved Hunter was removed from its pole in 1996 and will go to Weelde.

❏ IF-65		Hunter F.6	T146	stored, will go to Weelde	5-97

BIERSET
The 'ownership' of the military side of Bierset has changed hands, the Air Force having left and the Army Aviation (from their former bases in Germany) taking over in 1994. All Air Force related W*recks & Relics* material has also left. Mirage 5s BA-02 and BA-15 went to <u>Weelde</u>, BA-45 and BA-51 to <u>Gosselies</u> and BA-55 to <u>Florennes</u>. Fate of the dumped BD-06 is unknown. RF-84F FR-33 went to <u>Florennes</u>, while FR-32 and FR-34 have gone to Weelde. The museum at <u>Savigny lès Beaune</u> bought of a number of F-84Fs; FU-10, FU-97, FU-106 and FU-116. FU-74 is now at <u>Esneux</u>. FU-156 moved to <u>Weelde</u>. FU-29/BA-02 was destroyed in September 1993 during the airshow and FU-159/BA-066 was scrapped in the same year. The fate of FU-93/BA-03 is unknown. The Army will position the Islander B-05/LE and Do27 D-02 at/near the gate in the near future.

❏ A-34		Alouette 2	1791	preserved, at officers mess	5-96
❏ A-37		Alouette 2	1803	stored, white UN c/s	9-96
❏ A-38		Alouette 2	1810	stored, dismantled	9-95
❏ A-55		Alouette 2	1998	stored	11-97
❏ A-76		Alouette 2	2110	stored, wreck, crashed 23-2-95	5-96
❏ B-05	LE	BN-2B-21	501	stored, ex Saffraanberg	11-95
❏ D-02		Do27J-1	2058	stored, ex Brasschaat, ex Saffraanberg	11-95

BELGIUM - 16

The old airfield, some kilometres to the north of the main field, is no longer in use. Portuguese Corsair 5501 was scrapped a long time ago, while the F-84F FU-26 was moved to the main airfield and went later to Savigny lès Beaune, together with the stored FU-31. Preserved Hunter F.4 ID-16 (marked as 'IF-83/JE-P') is now at Weelde.

BRASSCHAAT
Do27 D-02 came from Saffraanberg where it served as an instructional airframe and arrived here for restoration. It has left for Bierset in 1995. Cessna 182R G-03 was blown over by a Dutch Hercules at Brussels and was in store here. It moved to Evere in December 1996. The stored ex Laage MiG-23MF 2021 went to Weelde.

❏ OO-FDH   Auster AOP.6   2834   preserved, marked as 'A-16', ex A-15, ex VT995   9-96

Also located on the base (in the military hangars and consequently not accessible to the casual visitor) is the Eric Voormezeele aircraft collection, several of which are airworthy with others undergoing restoration. Gone are the Lysander (4617?) and Bolingbroke OO-BLH (ex 9947/RCAF, both to Le Bourget). N1101 OO-RLR went to a museum in Haine St Pierre (France) and P2.05 OO-PTO (31) became F-AZPK. N1002 F-BCAS went to Valence and also the Harvard 2A H-39 (88-9828) has gone. Current non flyers should included:

❏ 34	DJ	N1101	34	stored, ex 34/French AF	2-95
❏ FG-244		T-28D	171-50	stored, ex Oostende, ex Zaire AF, ex 50-0244	95
❏ A-18		Alouette 2	1624	stored	9-96
❏ F-BDCX		SV-4C	552	under restoration, to become OO-HYW	2-95
❏ I-AEHO		G46-4A	143	stored, ex MM53093/Italian AF	9-96
❏ OO-MSA		MS733	178	under restoration, ex 178/French	2-95
❏ OO-STO		MS505	269/6	stored, ex F-BBUK	2-95
❏ OO-SVC		SV-4C	1086	under restoration, ex 1086/French	2-95
❏ OO-VAF		N1101	119	stored, ex F-BHER, ex 119/French	2-95
❏ OO-VOR		Piel CP301A	238	stored, at Antwerpen ?	2-95

BREE
In the back garden of a house in this town is a MiG-23UB. The aircraft will be used as travelling exhibit.

❏ 2056   MiG-23UB   A1078504   preserved, marked as '23', ex Nieuw Loosdrecht   7-97

BRUSSELS
Town: The large Koninklijk Leger Museum/Musée Royal de l'Armée is at the Jubelpark (or Parc du Cinquantenaire in French) and is still expanding. Over the past years a number of newer aircraft have arrived such as a Mi-24, Mirage 5, F-16, some MiGs and other ex Eastern block aircraft. Bolingbroke 9895 moved on to Saffraanberg for restoration and returned here via Brustem. Alouette 2 A-13 (1566) went to Ota. S-58C B-13 OT-ZKM (58-395) was scrapped and SV-4B V-33 (ex Saffraanberg) is now preserved in the terminal at Brussels-Zaventem. AL-60B2 I-MACL (6218/38) was sold, while DH.82A OO-EVA (85873) and L-4H OO-RVA (12221) have moved to France. DH.82A OO-SOE (3858) returned to flying status (from Antwerpen). During 1996 the SF-260M ST-16 was on display for the 50 years of the Belgian Air Force exhibition, the aircraft has since returned to operational flying. CM170 MT-49 was on temporary display, while the museum's MT-24 was under restoration at Evere. Auster AOP.6 OO-FDI (ex A-16) went to Kbely in 1992. MS500 F-BFCD has moved to museum store at Kapellen.

❏ 1558		Battle I	..	preserved, as 'T-70', ex 'R3950/HA-L', ex RCAF	7-97
❏ 9895		Bolingbroke IVT	..	preserved, marked as '10038/XD-A', ex RCAF	7-97
❏ 18534		CF-100 Mk.5	434	preserved, ex RCAF	7-97
❏ 2808		L-29R	892808	preserved, arr 18-12-91, ex Kbely, ex Czech AF	7-97
❏ 1317		LET C-11	171317	preserved, marked as '1706', ex Kbely, Czech AF	7-97
❏ 3911		MiG-15bis	623911	preserved, ex Kbely, ex Czech AF	7-97
❏ P-130		Chipmunk Mk.20	C1/0109	preserved, ex Geldenaken, ex Danish AF	7-97
❏ 4421		MiG-23BN	4421	preserved, marked as 'red 23', ex Egypt AF	7-97
❏ 320	UQ	MD450	320	preserved, ex French AF	7-97

BELGIUM - 17

☐ C.2531		Caudron G.3	..	preserved, ex French	7-97
☐ NF11-3		Meteor NF.11	..	preserved, ex Koksijde, ex French AF	7-97
☐ 3085		G91R/3	348	preserved, ex WGAF	7-97
☐ 9633		Mi-24D	340273	preserved, ex Basepohl, ex 528/NVA	7-97
☐ C.227/16		Aviatik C.I	832	under restoration, ex German	7-97
☐ 3471/18		Halberstadt C.V	1541	preserved, marked as '3441/18', ex German	7-97
☐ 5141/18		LVG C.VI	..	preserved, ex German	7-97
☐ 2157		MiG-21F-13	..	preserved, marked as '77 red', ex Indonesia AF	7-97
☐ 5316		F-86F	191-938	preserved, ex Portuguese AF, ex 52-5242	7-97
☐ XH292		Vampire T.11	15160	preserved, ex RAF	7-97
☐ 68-0590	SW	RF-4C	3579	preserved, ex Zweibrücken	7-97
☐ 35067	34	J35A	35067	preserved, ex Swedish AF	7-97
☐ BA-15		Mirage 5BA	15	preserved, ex Weelde, ex Bierset	7-97
☐ CP-46	OT-CEH	C-119G	11254	preserved, ex 53-8151	7-97
☐ EG-224	K5-K	Meteor F.8	M1	preserved	7-97
☐ FA-01		F-16A	6H-1	preserved, ex Weelde	7-97
☐ FR-28		RF-84F	..	preserved, ex 51-1945	7-97
☐ FT-34		T-33A	9584	preserved, ex 55-3043	7-97
☐ FU-30		F-84F	..	preserved, ex 52-7169	7-97
☐ FX-12		F-104G	9029	preserved	7-97
☐ FZ-107		F-84G	..	preserved, as 'FZ-153/3R-E', ex 'FZ-132', ex 51-10667, with tail from FZ-71/51-10195	7-97
☐ HD-78		Hanriot HD.1	VII.5153	preserved	7-97
☐ ID-46		Hunter F.4	AF-HOF59	preserved, marked as 'IF-70'	7-97
☐ MB-24	ND-N	Mosquito NF.30	984597	preserved, ex RK952	7-97
☐ PL-21		Grunau SG-38	..	stored, glider	3-97
☐ PL-33		Grunau Baby II	09	stored, glider	3-97
☐ PL-36		Grunau Baby II	..	stored, glider	3-97
☐ PL-37		Grunau Baby III	82/55	stored, glider	3-97
☐ RM-4	OT-ZAD	Pembroke C.51	P66/21	preserved	7-97
☐ SG-55	GE-R	Spitfire FR.XIVe	6S.648170	preserved, with parts of SG-37, ex MV246	7-97
☐ SM-15		Spitfire LF.IXc	CBAF-IX-1301	preserved, as 'MJ360/GE-B', ex MJ783	7-97
☐ SP-49		SPAD XIII C.1	..	preserved	7-97
☐ A-11		Alouette 2	1535	preserved	7-97
☐ B-6	OT-ZKF	HSS-1	SA181	preserved	7-97
☐ B-06	LF	BN-2A-21	510	preserved, ex Butzweilerhof, ex G-BDPU	7-97
☐ D-04		Do27J-1	2101	preserved	7-97
☐ H-21		AT-6D	88-15950	preserved, ex 7630/SAAF, ex EZ256	7-97
☐ K-16	OT-CWG	C-47B	20823	preserved, ex 43-16357	7-97
☐ OL-L87		L-18C	18-3149	preserved, ex 53-4749	7-97
☐ MT-24		CM170	281	preserved, ex restoration at Evere	7-97
☐ O-16		Oxford I	PAC/W/936	preserved, ex MP455	7-97
☐ P-4		P31 Proctor IV	H-578	preserved, crashed 10-5-49, ex NP171	7-97
☐ V-28		SV-4B	1170	preserved	7-97
☐ V-56		SV-4B	1198	preserved	3-97
☐ V-57		SV-4B	1199	preserved, marked as 'OO-ATD', ex Zellik	3-97
☐ V-64		SV-4B	1206	preserved	7-97
☐ 8		RAF RE.8	326	preserved, possibly ex A4719	7-97
☐ S85		Sopwith 1½ Strutter	..	preserved	7-97
☐ 5.160		Schreck FBA	55	preserved	7-97
☐ B5747		Camel F.1	..	preserved	7-97
☐ N5024		Nieuport 23C.1	..	under restoration	3-97
☐ T9800 ?		M14A Magister	1992	preserved, ex OO-NIC, ex G-1	7-97
☐ LF658		Hurricane IIc	..	preserved, marked as 'LF345/ZA-P'	7-97
☐ F-BCNT		MS315	350	preserved	3-97

BELGIUM - 18

	Registration	Type	c/n	Notes	Date
☐	F-BEJO	MS230	1066	preserved, marked as '1066'	7-97
☐	F-BERF	N1002	184	preserved, marked as '184'	7-97
☐	F-BFZC	SV-4C	28	preserved, marked as '102/5.S.25', ex OO-CLH	7-97
☐	G-ACGR	Gull Four II	D29	preserved	3-97
☐	N67160	A-26B	28044	preserved, ex D-CAFY	7-97
☐	OB-ADT	Fi156C-3	5503	preserved, ex code KR-QX and marked as such	3-97
☐	OO-AFH	SABCA Vivette	2	preserved, ex O-BAFH and marked as such	3-97
☐	OO-ANP	Kreit & Lambrickx KL2	..	preserved	3-97
☐	OO-ASB	Tipsy S.2	29	preserved, marked as 'OO-TIP', ex G-AFVH	3-97
☐	OO-CNP	DH.89A	6458	under restoration, ex OO-AFG, ex G-AKNV	3-97
☐	OO-EVD	DH.82A	86517	stored, ex T-22, ex NM209	—
☐	OO-EVS	DH.82A	3272	stored, ex G-AOJX, ex G-AOJX	3-97
☐	OO-FDA	Auster AOP.6	2818	stored, ex A-3, ex VT979 and marked as such	3-97
☐	OO-FDE	Auster AOP.6	2826	preserved, ex A-11, ex VT990	7-97
☐	OO-LUT	UC-61K	951	preserved, ex F-BAMB, ex 43-14987	7-97
☐	OO-MAA	KZ III	72	under restoration	3-97
☐	OO-SJA	B707-329	17623	preserved, front fuselage	3-97
☐	OO-SNE	Bü181B	021969	preserved, ex TP-CP and marked as such	7-97
☐	OO-SOI	DH.82A	83728	preserved, marked as 'T-24/UR-!', ex G-AMJD	7-97
☐	OO-SRA	SE210-6N	64	preserved, wfu 8-8-74, ex F-WJAK	3-97
☐	OO-SRS	SV-4D	1208	preserved	3-97
☐	OO-SRZ	SR-7B	1003	preserved	3-97
☐	OO-SZP	Ka6CR	196	preserved, glider	3-97
☐	OO-SUD	DHC-3	297	preserved, ex 144669/USN	7-97
☐	OO-TAO	MS892A	10457	preserved	3-97
☐	OO-ZBA	Grunau Baby II	..	preserved, glider	—
☐	OO-ZPJ	Gö IV Goevier	413	preserved, glider	3-97
☐	OO-ZSA	SZD-8bis	255	preserved, glider	3-97
☐	OO-11	HM-293	..	preserved	3-97
☐	OO-15	Jodel D.9	..	stored	3-97
☐	OO-33	HM-293	..	preserved, marked as 'OO-BAM'.	3-97
☐	OO-40	Jodel D.9	558	stored	3-97
☐	..	Bataille Triplane	..	preserved, oldest Belgian built aircraft	3-97
☐	..	Blériot XI	..	preserved, Olieslagers built	3-97
☐	..	Bristol F.2B	..	preserved, marked as 'B4/66'	7-97
☐	..	Caudron C.800	9816/256	preserved, glider	3-97
☐	..	Farman F.11A-2	..	preserved	3-97
☐	..	Farman-Voisin	..	preserved	3-97
☐	..	Fokker Dr.I	..	preserved, replica with original parts, as '425/17'	3-97
☐	..	HM-290	..	stored	3-97
☐	..	Hunter F.6	41H-691136	preserved, nose only (purpose-built instr airfr)	3-97
☐	..	Kassel 12	..	preserved	3-97
☐	..	MS317		under restoration	3-97
☐	..	Rumpler C.IV	..	stored, fuselage only	—
☐	..	SABCA Junior	2	stored	—
☐	..	SABCA Junior	3	stored	—
☐	..	SABCA Junior	10	stored	—
☐	..	SABCA Junior	11	preserved	3-97
☐	..	SV-4C	180 ?	preserved	3-97
☐	..	Tipsy Nipper	49 ?	preserved, composite marked as 'OO-NIP'	3-97
☐	..	Voisin LA5B	..	preserved, fuselage	3-97
☐	..	Z-12 Zögling	..	stored, glider	—

Donated to the Koninklijke Militaire School (KMS) on 12th May 1997 is a former Air Force SF-260 which had a landing accident on 26th June 1995. Note that the number 10-33 is not the construction number, but a line

BELGIUM - 19

number (33rd aircraft for 10th customer, being Belgium).
❏ ST-33	SF-260M	10-33	instructional	5-97

At the Congress Kolom part of Brussels an Anson C.19 is stored in a garage. The aircraft will be restored in the near future.
❏ TX192	Anson C.19	..	stored, ex Le Bourget	—

Airfield - Zaventem/Melsbroek: The new terminal building of Zaventem airport received a Stampe from the Brussels museum, while C-119G CP-10 was under restoration at the Melsbroek (military) side and has become a gate guard there. Parts of the scrapped CP-11 (ex Neuville) were used for this aircraft. The stored Pembroke will be restored and placed next to the Flying Boxcar at the gate. The dumped F-104G FX-18 (ex Saffraanberg) was transported from here to Evere during May 1995. Stored Niger Air Force C-130H 5U-MBH (4831) is restored to flying status. Over the years a good many airliners have passed through the airport for storage and/or scrapping. Of these B707-328B 4X-ATE (18456), DC-8-53 9Q-CBF (45629), DC-8-55 N915R (45916) and Convair 580s OO-VGH (28) and N73165 (368) have been scrapped. Stored B707-328B OO-TYC flew out as N2909B, as did B720-051B OY-APV which became TF-AYD. Also at Zaventem are the SABENA Old Timers, well known with their flying Lysander OO-SOT (marked as '2442/MA-D', Brussels museum owned). This association restores aircraft for the Brussels museum and currently has two DH.82s and two Tipsy Trainers in storage. The Ju52 OO-AGU was expected to move to the Brustem for further restorations during March 1998.

❏ 54	NC856A	54	stored, ex French AF	5-97
❏ CP-10	C-119G	10690	preserved, ex Neuville, ex Koksijde, ex 52-2701	4-98
❏ RM-7 OT-ZAG	Pembroke C.51	P22/27	preserved, ex Molenheide, arr 16-1-98	4-98
❏ V-33	SV-4B	1175	preserved, arr 3-10-95, ex Brussels	4-98
❏ 5N-AUG	A310-222	329	stored	3-98
❏ F-WGYR	A310-222	331	stored, all white, ex OO-SCI	2-96
❏ G-AFJR	Tipsy Trainer 1	2	stored, Brussels museum owned	—
❏ G-AFJV	Tipsy Trainer 1	10	stored, Brussels museum owned	—
❏ OO-AGU	Ju52/3mg7e	501196	stored, ex 6309/Port AF, parts from 6310	7-96
❏ OO-EVT	DH.82A	84875	stored, ex G-AMTP, ex T6534, Brussels owned	—
❏ OO-SOF	DH.82A	82592	stored, ex G-AMTL, ex W6420, Brussels owned	—
❏ OO-TEF	A300B1	2	stored, ex F-WUAC	4-98

BRUSTEM

Until 18th November 1996 the base housed the training units of the Belgian Air Force, flying with Alpha Jets and the last of the CM170s. The T-33 and Meteor are long term residents, while the Magisters, having come to the end of their flying life, were new arrivals. All these have moved on. CM170 MT-30 went to Herstal, while MT-29 and MT-31 went to Weelde. Meteor EG-79 and T-33 FT-24 have gone to Bevekom. Some of the F-84Fs have also gone; FU-51 is now at Temploux, FU-66 went to Kleine Brogel, FU-92 moved to Weelde and FU-105 went to Marche. Bolingbroke '10038' which was under restoration here moved to Brussels in December 1996.

❏ FU-28	F-84F	..	stored, ex 52-7166, will go to Kapellen	11-96
❏ FU-82	082 F-84F	..	stored, ex 53-6587, will go to Kapellen	11-96

CHARLEROI

Stored at this airfield is a former Shabair B707.
❏ 9Q-CSZ	B707-323C	19577	stored, ex ZS-LSH	12-96

CHIÈVRES

Nothing has changed at the base of HQ SHAPE. The Thunderstreak is still on the base, while the Meteor is preserved at barracks close to the airfield. Both aircraft carry false codes.

❏ EG-18	'VT-R'	Meteor F.8	6339	preserved, at gate, soc 1-58, arr 9-7-58	6-96
❏ FU-33	'3R-B'	F-84F	..	preserved, at gate, arr 10-6-86, ex 52-7192	6-96

DINANT
High above the town of Dinant, at the citadel, the Meteor is still present. This aircraft was civil registered as OO-ARU between 27th August 1958 and 2nd March 1961. Nowadays it is marked with the code 'K5-K' and is owned by the Brussels museum. The aircraft will go to Koksijde and become a gate guard there.
☐ OO-ARU   Meteor F.8   6496   preserved, ex EG-162 and marked as such   5-97

ESNEUX
On 5th February 1994 a Thunderstreak was submerged into the water-filled former quarry Le Gombe. It now serves as a practice object for the local scuba-diving club and was noted as such several times in 1997.
☐ FU-74   IS-E   F-84F   ..   submerged, ex Bierset, ex 53-6716   4-97

EVERE
Evere is the home of the air accident investigation branch. Many crashed airframes pass through here, often not more than bits and pieces, and normally not staying for a long period. Exceptions are the preserved Thunderflash, Starfighter and the MiG-21bis and some airframes of the Belgian AF recruiting unit (21 Logistic Wing). Preserved SV-4B V-33 went to Saffraanberg on 8th May 1990. F-104 FX-04 moved on to Bevekom, while recruiting exhibit CM170 MT-9 is now at Weelde. The remains of the crashed F-16A FA-113 were transported to Bevekom on 17th June 1996. Fouga MT-24 was under restoration here from Brussels and was temporarily swopped with MT-49.

☐ 2429	MiG-21bis	N75033205	preserved, ex Geel, ex Dresden, ex 874/NVA	12-96
☐ FR-29	RF-84F	..	preserved, ex 51-11279	11-96
☐ FX-18	F-104G	9040	dumped, ex Melsbroek, ex Saffraanberg	5-95
☐ FX-21	F-104G	9046	preserved, marked as 'FX-23', ex Saffraanberg	10-94
☐ MT-17	CM170	274	preserved, cockpit only (recruiting unit)	—
☐ MT-49	CM170	222	preserved, recruiting unit	6-97
☐ ST-33	SF-260M	10-33	wreck, crashed 24-6-96	96
☐ A-68	Alouette 2	2068	stored, wreck	4-96
☐ G-03	Ce182R	67850	stored, wreck, damaged 22-9-94, ex Brasschaat	12-96

FLORENNES
Florennes has become a well known base in Europe due to the Tactical Leadership Programmes (TLPs) which are held here. A recent arrival on the *Wrecks & Relics* side is F-84E K-6 from the Netherlands. It arrived here for the local Lallemand Museum (on the western side of the base and open to the public) to complete the line of former Belgian Air Force aircraft which operated from Florennes. The wreck of F-16A FA-08 moved to Peutie. RF-84F FR-27 left on 29th January 1996 for Spa. F-84F FU-152 was completely burnt by the mid 1980s and has been scrapped. F-84 FU-154 went to Brustem. The stored F-104s FX-83 and FX-85 have moved to Weelde.

☐ K-6	F-84E	1358	preserved, marked as 'FS-7/3R-B', ex RNethAF	5-98
☐ BA-51	Mirage 5BA	51	stored, ex Gosselies, ex Bierset	11-97
☐ BA-55	Mirage 5BA	55	dumped, ex Bierset	3-97
☐ BR-04	Mirage 5BR	304	preserved, arr 6-1-97, ex Weelde	5-98
☐ FA-04	F-16A	6H-4	preserved, arr 13-1-97, ex Weelde	5-98
☐ FR-33	RF-84F	..	preserved, ex Bierset, ex EA-305/WGAF	12-97
☐ FU-50	F-84F	..	decoy, ex 52-7011	9-97
☐ FU-91	F-84F	..	dumped, burnt out, ex 53-6693	8-93
☐ FU-103	F-84F	..	preserved, marked as 'FU-66', ex 53-6597	5-98
☐ FU-108	F-84F	..	preserved, at gate, ex 53-6610	5-98
☐ FU-144	F-84F	..	preserved, marked as 'FU-52/UR-H'	5-98
☐ FU-154	F-84F	..	stored, 'FU-66' on right side, ex 53-6806	5-98
☐ FU-179	F-84F	..	decoy, ex 53-6941	11-97
☐ FX-70	F-104G	9119	stored, ex Koksijde	3-97
☐ SG-57	'RL-D'	Spitfire FR.XIV	6S.432331	preserved, ex RM921	11-97

GEEL
Former NVA MiG-21 2429 was temporarily in use here as instructional airframe. It was used to train NATO observers who were present at the destruction of former Eastern bloc aircraft. The aircraft moved to Evere.

GELDENAKEN
Chipmunk P-130, ex Danish AF, which was under restoration here moved back to Brussels.

GITS
The Skytrain K-1/OT-CWA, which was just off the N63 between Roeselare and Torhout, was offered for sale in August 1992. It was bought by Dakota Unlimited, who have moved the aircraft to Schaffen Diest.

GLONS
A Starfighter is preserved at a military barracks in this village. Glons is 10km east of Tongeren.

❏ FX-15	F-104G	9034	preserved, ex Saffraanberg, arr 2-9-96	12-97

GOETSENHOVEN
SF-260 ST-14 was declared unrepairable after an emergency landing in August 1988. After this it became a gate guard and has moved on to Bevekom. In September 1996 the EFS moved to Bevekom. Three SF-260s rebuilt as 'link' trainers (ST-10, ST-13 and ST-28) have not been noted for a long time and may have been disposed of. The dumped ST-29 may have also been scrapped.

❏ ST-29	SF-260M	10-29	dumped, crashed 7-6-93	6-95

GOSSELIES
Some ex Air Force Mirage 5s (BA-45, BA-51 and BD-13) arrived here as a spare parts source for the Mirages which have been sold to Chile. All should have left by late 1996. Of these only BA-51 was seen again, it went to Florennes. The Mirages which were in store at Koksijde and Weelde also passed through here before going to Chile. Going back to the first issue, the mortal remains of Britannia 9U-BAD (13454) were last noted in April 1981 and were scrapped soon after.

❏ BA-45	Mirage 5BA	45	stored, ex Bierset	10-96
❏ BD-13	Mirage 5BD	213	stored, tail at Balen	10-96

HELCHTEREN
Although the ranges are on military property, access is permitted at weekends and it is still possible to find these aircraft after a long walk. They remain complete and are relatively unharmed.

❏ FU-63	BA-56	F-84F	..	target, ex 52-10519	10-88
❏ FU-181	BA-02	F-84F	..	target, ex 52-6371	10-88
❏ FU-183	BA-06	F-84F	..	target, ex 52-6403	10-88

HERSTAL (Milmort)
The former FN-moteurs (now called Techspace Aero SA) which built F-16 and Transall engines among others, has erected a former Brustem Magister as gate guard. The aircraft still wears its 1995 demo colours.

❏ MT-30	CM170	287	preserved, ex Brustem, arr 27-5-96	9-96

KALKEN
A scrap and surplus dealer located on the north side of the town (Kalken itself is south of the E3 motorway be-

tween Gent en Lokeren) has dealt with aircraft scrap. Nothing has been seen in recent years except some 15 F-16 canopies.

KALMHOUT
Douglas C-47B K-40/OT-CWS (33244) here was last noted early 1979 and is said to have been be scrapped soon afterwards.

KAMPENHOUT
Some 15km south east of Mechelen is the village of Kampenhout. Close by the Taveerne De Wilg at the Langedonckstraat in Boortmeerbeek a private collector has two Starfighters.

❏ FX-76	F-104G		9131	preserved, ex Bevekom, ex Koksijde	4-97
❏ FX-100	F-104G		9176	preserved, ex Kleine Brogel, ex Koksijde	4-97

KAPELLEN
This Belgian Army complex is used by the Brussels museum as its main storage and restoration facility and could become a future satellite to the Brussels exhibition halls. Besides numerous tanks and artillery pieces, more and more aircraft, coming from Westerloo, Tielen and Weelde are finding their way to this depot.

❏ FR-34	RF-84F		..	expected, ex Weelde, ex Bierset	—
❏ FU-28	F-84F		..	expected, ex Brustem, ex 52-7166	—
❏ FU-82	082	F-84F	..	expected, ex Brustem, ex 53-6587	—
❏ FU-92	F-84F		..	expected, ex Weelde, ex Brustem, ex 53-6702	—
❏ FU-105	F-84F		..	stored, arr 10-2-97, ex Marche, ex 53-6707	2-97
❏ FU-156	F-84F		..	expected, ex Weelde, ex Bierset, ex 53-6834	—
❏ FU-184	F-84F		..	stored, front fuselage only, ex Westerloo	12-96
❏ FU-186	F-84F		..	stored, aft fuselage only, ex Westerloo	12-96
❏ FU-191	F-84F		..	expected, ex Weelde, ex Bevekom, ex 52-6374	—
❏ FU-192	F-84F		..	expected, ex Weelde, ex Bevekom, ex 52-6605	—
❏ FX-02	F-104G		9017	stored, arr 26-10-94, ex Koksijde	12-96
❏ FX-47	F-104G		9090	stored, arr 7-11-94, ex Koksijde	12-96
❏ FX-52	F-104G		9095	stored, arr 11-10-94, ex Koksijde	12-96
❏ FX-69	F-104G		9115	stored, arr 11-10-94, ex Koksijde	12-96
❏ PL-13	Ka2		77/55	stored, glider, ex Tielen	12-96
❏ PL-50	Weihe 50		62.07	stored, glider, ex Tielen	12-96
❏ F-BFCD	MS500		374	stored, ex Brussels	12-96
❏ F-BSAM	MS880B		1555	stored, ex Tielen	12-96
❏ G-AKIS	M38 Messenger 2A		6725	stored, ex Westerloo	12-96
❏ OO-ABN	Auster J/1		2047	stored, ex Tielen	12-96
❏ OO-ARM	P44 Proctor V		Ae84	stored, ex Westerloo, ex G-AHZY	12-96
❏ OO-AVX	L-4J		13128	stored, ex Tielen, ex 44-80832	12-96
❏ OO-BLJ	Bü181B		0216168/FR14	stored, ex Westerloo, ex OO-RVA	12-96
❏ OO-EVG	DH.82A		86338	stored, ex Tielen, ex T-15, ex NL981	12-96
❏ OO-EVH	DH.82A		85832	stored, ex Tielen, ex T-27, ex DE972	12-96
❏ OO-FDB	Auster AOP.6		2820	stored, ex Tielen, ex A-7, ex VT981	12-96
❏ OO-FDC	Auster AOP.6		2824	stored, ex Tielen, ex A-8, ex VT988	12-96
❏ OO-FDD	Auster AOP.6		2817	stored, ex Tielen, ex A-9, ex VT978	12-96
❏ OO-FDJ	Auster AOP.6		2832	stored, ex Tielen, ex A-17, ex VT993	12-96
❏ OO-FDL	Auster AOP.6		2836	stored, ex Tielen, ex A-22, ex VT997	12-96
❏ OO-GEG	L-4H		11694	stored, ex Tielen, ex 43-30403	12-96
❏ OO-KBL	Mooney M.20		1293	stored, ex Tielen, ex F-BKBL	12-96
❏ OO-LGA	L-18C		18-1597	stored, ex Tielen, ex F-BLMM	12-96
❏ OO-LMV	UC-61K		1069	stored, ex Westerloo, ex N9759F	12-96

☐ OO-OPO	P40 Prentice T.1	PAC.F215	stored, ex Westerloo, ex G-AOPO, ex VS613	12-96
☐ OO-RAK	Ercoupe 415CD	4789	stored	12-96
☐ OO-SEH	Ce310B	35630	stored, ex Westerloo, ex 9Q-CES, ex OO-CES	12-96
☐ OO-SEL	Ce310B	35524	stored, ex Westerloo, ex OO-CUC	12-96
☐ OO-SND	Aeronca 7AC	7AC-4047	stored, ex Tielen, crashed 18-8-81	12-96
☐ OO-SOW	DH.82A	84567	stored, ex Tielen, ex G-APPT, ex T6100	12-96
☐ OO-SOX	DH.82A	83830	stored, ex Tielen, ex T7303	12-96
☐ OO-ZJN	Br905 Fauvette	45	stored, glider, ex Tielen, ex F-CCJO	12-96
☐ ..	Auster AOP.6	..	stored	12-96
☐ ..	Chandelon Helicopter	..	stored, ex Brussels	12-96

KEIHEUVEL
A Thunderstreak is still displayed by the local aero club at Keiheuvel.

☐ FU-197		F-84F	..	preserved, ex 52-6584	8-96

KLEINE BROGEL
The Voodoo is kept in storage for the Brussels museum and eventually will move there. Two F-16 wrecks (FA-35 and FA-85) were noted in September 1995. Both were only short lived and have moved on (presumably scrapped). Surely gone is F-84F FU-49/BA-05 which was completely burnt by September 1993 and has been scrapped. The FU-177/BA-04 has moved to Leopoldsburg. Starfighter FX-100 is now at Kampenhout, while the dumped fuselage of FX-96 (9166), last noted in November 1995, has gone.

☐ 58-0322	ZR	F-101B	690	stored, ex Zweibrücken, Brussels owned	4-98
☐ FS-17	'DMB'	F-84E	..	preserved, ex 'FS-2/KB-1', ex 'FS-2/8S-W'	6-96
☐ FU-66	RA-T	F-84F	..	preserved, at gate, ex Brustem, ex 53-6677	10-97
☐ FU-67	BA-03	F-84F	..	decoy, ex 53-6681, soc 19-1-65	—
☐ FU-145	Z6-E	F-84F	..	stored, ex 53-6613	6-96
☐ FU-188	8S-N	F-84F	..	preserved, ex 'FU131/8S-H', ex 52-6369	10-97
☐ FX-41		F-104G	9084	dumped, ex Koksijde	9-96
☐ FX-86		F-104G	9147	preserved	11-95
☐ FX-94		F-104G	9164	preserved, ex Koksijde	9-95

KOKSIJDE
In 1997 it was considered that the base of Koksijde should be closed for fixed wing aircraft. These plans have not gone through and Koksijde will remain open for all types. The famous storage facilities here has closed and its task has been taken over by Weelde. Currently only three *Wrecks & Relics* aircraft remain at the airfield.

☐ B-4	OT-ZKD	HSS-1	SA145	preserved, on airfield, Brussels museum owned	10-97
☐ B-8	OT-ZKH	HSS-1	SA185	preserved, at barracks, Brussels museum owned	10-97
☐ ID-123	7J-P	Hunter F.4	8679	preserved, at barracks	10-97
☐ OO-ARU		Meteor F.8	6496	expected, still at Dinant, ex EG-162	—

Weelde has taken over the task of storage airfield. All airframes stored here over the past eight years have left. The Brussels museum aircraft went to Weelde or Kapellen, the cocooned Mirages were sold to Chile and were overhauled at Gosselies. All the Starfighters have also been disposed of.

NF11-3		Meteor NF.11	..	stored, l/n 7-92	to Brussels
191	8-NK	Mystère 4A	191	stored, ex Savigny, l/n 7-94	to Weelde
145		Super Mystère B2	145	stored, faded code 12-ZR, l/n 7-94	to Weelde
ID-26	OV-K	Hunter F.4	..	stored, arr 10-9-90, ex Zellik, l/n 7-94	to Weelde
BA-01		Mirage 5BA	01	stored, cocooned, l/n 7-92	to Gosselies
BA-04		Mirage 5BA	04	stored, cocooned, l/n 7-92	to Gosselies
BA-23		Mirage 5BA	23	stored, cocooned, l/n 7-92	to Gosselies
BA-37		Mirage 5BA	37	stored, l/n 7-91	to Gosselies

BELGIUM - 24

BA-39	Mirage 5BA	39	stored, cocooned, l/n 7-92	to Gosselies
BA-46	Mirage 5BA	46	stored, cocooned, l/n 7-92	to Gosselies
BA-48	Mirage 5BA	48	stored, cocooned, l/n 7-92	to Gosselies
BA-50	Mirage 5BA	50	stored, cocooned, l/n 7-92	to Gosselies
BA-52	Mirage 5BA	52	stored, l/n 7-91	to Gosselies
BA-56	Mirage 5BA	56	stored, cocooned, l/n 7-92	to Gosselies
BA-57	Mirage 5BA	57	stored, cocooned, l/n 7-93	to Gosselies
BA-59	Mirage 5BA	59	stored, cocooned, l/n 7-93	to Gosselies
BA-62	Mirage 5BA	62	stored, cocooned, l/n 7-92	to Gosselies
BD-14	Mirage 5BD	214	stored, l/n 7-91	to Gosselies
BD-15	Mirage 5BD	215	stored, l/n 7-92	to Gosselies
CP-10	C-119G	10690	dumped, 3C-ABB ntu, dep 25-5-90	to Neuville
CP-11	C-119G	10685	dumped, ex 51-2696, dep 25-5-90	to Neuville
CP-21	C-119G	10952	dumped, ex 52-6022	scrapped, parts to St Niklaas
FC-01	TF-104G	5786	stored, l/n 11-90	to United States
FC-02	TF-104G	5787	stored, l/n 11-90	to United States
FC-03	TF-104G	5788	stored, l/n 11-90	to United States
FC-04	TF-104G	5101	stored, l/n 11-90	to United States
FC-06	TF-104G	5103	stored, l/n 11-90	to United States
FC-07	TF-104G	5104	stored, l/n 11-90	to United States
FC-08	TF-104G	5105	stored, dep 9-88	to Savigny lès Beaune
FC-10	TF-104G	5107	stored, l/n 11-90	to United States
FC-11	TF-104G	5108	stored, l/n 11-90	to United States
FC-12	TF-104G	5109	stored, l/n 11-90	to United States
FX-02	F-104G	9017	stored, l/n 7-94	to Kapellen
FX-03	F-104G	9018	stored, l/n 11-90	to United States
FX-07	F-104G	9022	stored, l/n 11-90	to United States
FX-10	F-104G	9027	stored, l/n 11-90	to United States
FX-41	F-104G	9084	stored, l/n 7-91	to Kleine Brogel
FX-44	F-104G	9087	stored, l/n 11-90	to United States
FX-45	F-104G	9088	stored, l/n 3-89	to Ledeberg
FX-47	F-104G	9090	stored, l/n 7-94	to Kapellen
FX-48	F-104G	9091	stored, l/n 11-90	to United States
FX-51	F-104G	9094	stored, l/n 11-90	to Kissimmee, USA
FX-52	F-104G	9095	stored, l/n 7-94	to Kapellen
FX-57	F-104G	9100	stored, l/n 11-90	to United States
FX-58	F-104G	9101	stored, l/n 7-91	to Bevekom
FX-59	F-104G	9102	stored, l/n 11-90	to United States
FX-60	F-104G	9103	stored, dep 2-89	to Hermeskeil
FX-62	F-104G	9105	stored, l/n 11-90	to United States
FX-64	F-104G	9107	stored, dep 6-89	to Mojave, USA
FX-65	F-104G	9108	stored, l/n 7-91	to Burbank, USA
FX-67	F-104G	9113	stored, l/n 11-90	to United States
FX-69	F-104G	9115	stored, l/n 9-96	to Kapellen
FX-70	F-104G	9119	stored, l/n 7-91	to Florennes
FX-72	F-104G	9121	stored, l/n 11-90	to United States
FX-74	F-104G	9126	stored, l/n 11-90	to United States
FX-76	F-104G	9131	stored, l/n 6-90	to Bevekom
FX-78	F-104G	9133	stored, l/n 7-91	to Bevekom
FX-80	F-104G	9138	stored, l/n 11-90	to United States
FX-81	F-104G	9139	stored, dep 6-89	to Pennsaukin, USA
FX-82	F-104G	9140	stored, dep 6-89	to Chino, USA
FX-83	F-104G	9141	stored, l/n 7-91	to Florennes
FX-84	F-104G	9142	stored, dep 6-89	to Mojave, USA
FX-85	F-104G	9146	stored, l/n 7-91	to Florennes

BELGIUM - 25

FX-89	F-104G	9153	stored, l/n 6-90	to Bevekom
FX-90	F-104G	9154	stored, dep 11-88	to Savigny lès Beaune
FX-93	F-104G	9160	stored, dep 11-89	to Aix Les Milles
FX-94	F-104G	9164	stored, l/n 6-90	to Kleine Brogel
FX-99	F-104G	9172	stored, l/n 6-90	to Bangor, USA
FX-100	F-104G	9176	stored, l/n 6-90	to Kleine Brogel
OO-GEO	Do28D-1	4023	stored, ex Oostende	to Weelde

KORTRIJK - WEVELGEM
Besides the locally based Landuyt aircraft collection which has a number of vintage aircraft under restoration or flying, three ex French Air Force Caravelles arrived here for storage pending sale. The first one to go was F-WQCV (264) which became EL-AWY, followed by the F-WQCT (240) as EL-WNA in December 1996. Still one Caravelle should be current here.

❏ F-WQCU	SE210-11R	251	stored, arr 16-6-96, ex 251/French AF	8-97
❏ OO-SXA	EMB121	121038	stored, ex PT-MBF	7-97

LEDEBERG
Ledeberg is a part of Gent. The local Volvo dealer at the Burgermeester Crommenlaan had Starfighter FX-45 (ex Koksijde), which arrived here on 25th March 1989, in use as eyecatcher. The F-104 has moved on to <u>Baarlo</u>.

LEOPOLDSBURG
Town: Preserved at the Thematisch Militair Openluchtmuseum at the Ruiterijkamp barracks is a former Kleine Brogel decoy Thunderstreak mounted on a pole.

❏ FU-177	BA-04 F-84F	..	preserved, ex Kleine Brogel, ex 53-6888	11-96

Airfield: Stored in a hangar is the fuselage of an SF-260M which was declared wfu after a accident.

❏ ST-07	SF-260M	10-07	stored, fuselage, crashed 13-2-88	2-98

LOMMEL
The legerstock at the Kolonie part of Lommel (along the border, just south of the Dutch town of Eindhoven) has two aircraft. The MiG-21PFM is for sale, while the Jet Provost is on permanent display.

❏ 4407	MiG-21PFM	94A4407	stored, ex Czech AF	3-98
❏ XN512	Jet Provost T.3	..	preserved, ex RAF	3-98

LIBIN (Transinne)
The Euro Space Centrum, a museum/attraction park along the Brussels-Luxemburg highway (exit 24) had for many years a skeleton Space Shuttle look-alike as an eyecatcher. This was joined by a Mirage and SF-260.

❏ BA-26	Mirage 5BA	26	preserved, ex Weelde	7-97
❏ ST-11	SF-260M	10-11	preserved, crashed 7-4-92	—

MARCHE
Thunderstreak FU-105 (ex Brustem) was used by the local para-commando unit as instructional airframe. The aircraft moved to <u>Kapellen</u> in early 1997.

MEERHOUT
The Wings of War museum is a small new Belgian museum. The only complete aircraft is a Magister painted in

Rode Duivels colours. There are also some parts of a Fairey Battle
- ❏ MT-31   CM170   288   preserved, ex Balen, ex Weelde, ex Brustem   12-96

MOLENHEIDE
The Molenheide leisure and camping complex had former BAF Pembroke RM-7. The aircraft was mounted on poles and could be found just off the N15 from Kleine Brogel to the motorway, some 2km north of Helchteren. The aircraft moved to Brussels on 16th January 1998.

NEUVILLE
A small microlight airfield at the south side of this town received two Flying Boxcars from Koksijde on 25th May 1990. CP-10 was used as a clubhouse, while the fuselage from CP-11 was dumped at the airfield. In the mid 1990s the airfield was closed and the C-119s moved on. CP-10 appeared at Brussels Zaventem, while CP-11 (10685) was scrapped. The MiG is still stored at a local Toyota garage. Neuville is just south of Philippeville (along the N5), not far from Florennes.
- ❏ 613   MiG-15UTI   ..   stored, ex Polish AF   9-97

OOSTENDE
The Vrij Technisch Instituut received a Skyservant in 1989 on loan from the Brussels museum. The aircraft had made an emergency landing at Oostende and was disposed off. It went to Koksijde and later to Weelde, only to return to Oostende for use as instructional airframe for the school in the town. The stored B707 TY-BBW moved to Wetteren in July 1997.
- ❏ 5T-BOB   Ce421B   421B-0039   dumped   2-97
- ❏ LX-DKT   C-47A   10253   stored, ex F-BEIG   2-97
- ❏ OO-GEO   Do28D-1   4023   instructional, in town   7-95

OVERBOELARE
In use as a clubhouse and bar at the airfield here is a Douglas DC-4. The aircraft arrived at Brussels with the false registration 'N2893C' and was impounded. After being sold at a public auction, it made its last flight to here on 5th July 1967 and is still resident after 30 years.
- ❏ N90443   C-54A   10352   preserved, marked as 'N2893C', ex HP-298   6-96

PEER
Preserved just outside the airfield of Kleine Brogel are two former Air Force aircraft at the Tiger Air Forces museum. The Starfighter has the tail of FX-86. Also here is the tail of F-16A FA-79 (crashed 1986). The location of the museum is at the Steenweg (Wijchmaal).
- ❏ FU-185   BA-07   F-84F   ..   preserved, ex 52-6569   5-97
- ❏ FX-61   F-104G   9104   preserved, with tail from FX-86   5-97

PEUTIE
Peutie is a small village situated just west of Brussels airport, near the town of Vilvoorde. A scrapdealer here, near the local police station, has acquired some ex Polish fighters which are for sale/lease. The Belgian aircraft consist of an empty Magister, pieces of a crashed F-16 and the tail of Mirage 5BA BA-35 (ex Beveren). The CM170 has the left tailplane from MT-38, the right tailplane from MT-30 and the right wing from MT-46.
- ❏ 412 ?   CM170   412 ?   stored ex Frecnch AF   4-98
- ❏ 427   4-WA   CM170   427   stored, ex Luxeuil, ex French AF   4-98
- ❏ 418   Lim-5   1C-0418   stored, ex Hoornsterzwaag, ex Polish AF   4-98
- ❏ 1213   Lim-5   1C-1213   stored, ex Tilburg, ex Polish AF   4-98
- ❏ MT-43   CM170   322   stored, crashed 26-9-85, see notes   4-98

BELGIUM - 27

SAFFRAANBERG
The Bolingbroke 9895 arrived here for restoration. It moved on after partial completion to Brustem and later to Brussels. The dumped remains of F-16A FA-52 (6H-52) were scrapped during the early 1990s, while RF-84F FR-30 has moved on to Temploux. Starfighter FX-18 went to Melsbroek in 1991. FX-21 to Evere marked as 'FX-23' and FX-15 is now at Glons. Parts of CM170 MT-23 and MT-37 were noted here in the 1990s, all moved to Weelde. Instructional SF-260M ST-14 moved on to Goetsenhoven. The school has over the years been used for some restoration projects, including Islander B-05 (to Bierset), Do27J D-02 (to Bierset) and SV-4B V-33 (ex Evere, to Brussels). The Alouette 2 A-10 was not noted on the open day in 1997 and may also have gone. The Air Force instructional school here holds its annual open days on Ascension Day.

❏ BA-03	Mirage 5BA	03	instructional	5-98
❏ BA-16	Mirage 5BA	16	instructional	5-98
❏ BA-17	Mirage 5BA	17	instructional	5-98
❏ FA-03	F-16A	6H-3	instructional	5-98
❏ FB-03	F-16B	6J-3	instructional, ex Weelde	5-98
❏ FX-39	F-104G	9079	instructional	5-98
❏ FX-53	F-104G	9096	preserved, silver c/s	5-97
❏ ST-13	SF-260M	10-13	instructional	5-98
❏ A-10	Alouette 2	1534	instructional	9-96
❏ A-27	Alouette 2	1711	instructional	5-97

SCHAFFEN - DIEST
This airfield is used by the para-commando troops of the Belgian Army and houses a very active civil aero club. The Skytrain is under restoration here.

❏ K-1	OT-CWA	C-47B	15056/26501 stored, ex Gits, ex OO-SMA, ex 43-49240	—

SPA
In early 1996 a RF-84F Thunderflash arrived in Spa from Florennes. Its now preserved at the airfield.

❏ FR-27	RF-84F	..	preserved, arr 29-1-96, ex Florennes, ex 51-1922	8-97

TEMPLOUX
The local civil airfield of Namur received two fighters on loan from the Brussels museum in the mid 1990s. Both were restored to perfect condition and were freshly resprayed.

❏ FR-30	RF-84F	..	preserved, ex Saffraanberg, ex 51-17015	5-97
❏ FU-51	F-84F	..	preserved, ex Brustem, ex 52-7215	5-97

TIELEN
The Brussels museum had a storage facility here at a military compound. Of the stored aircraft DH.82A OO-EVM (86507, ex NM199) moved to France, while all the others listed in the first edition of this book went to Kapellen (which see).

WAARSCHOOT
Waarschot can be found some 25km west of Brussels. Since 15th September 1992 a Viscount can be found here. It is mounted on poles along road number N9 to Gent.

❏ G-AZNA	Viscount 813	350	preserved, ex Southend, ex ZS-CDX	9-96

WEELDE
From the mid 1990s part of this reserve base has been used as storage facility. Most of the F-16s arrived by air, as did most of the Mirages. The cocooned F-16s are in operational reserve, while the uncocooned ones are used

as a spare parts source. The first F-16 to return to operational service was FA-46 (6H-46, arrived 28-3-95) which left by August 1997. The Brussels museum took delivery of the stored F-16A FA-01 in 1996, while FA-04 is now at Florennes and FA-10 will go to Bevekom. F-16B FB-03 moved on to Saffraanberg. Of the stored Mirages BA-11 (dep 26-9-94), BA-60 (dep 21-9-94), BD-12 (dep 17-5-95), BR-13 (dep 2-1-96), BR-25 (dep 8-1-96), BR-26 (dep 19-2-96) and BR-27 (dep 26-2-96) went to Gosselies to be overhauled for the Chilean Air Force. BA-15 (ex Bierset) moved to Brussels, BA-21 went to Deblin (on 12-12-94), BA-26 to Libin and BR-04 to Florennes (on 6-1-97). Besides the Air Force, the Brussels museum is also using the storage facilities here. The latter had a number of older aircraft collected from other airfields and stored them here. Unfortunately these aircraft sustained a lot of damage during the different moves. In future the museum will only keep aircraft on wheels stored here due to operational needs which dictates flexibility. RF-84F FR-32 (ex Bierset) went on 8th July 1996 to Bevekom. CM170s MT-29 and MT-31 (ex Brustem) have moved to Balen, MT-33 is now at Savigny lès Beaune and MT-3 and MT-46 went to Bevekom. Both dumped Hunter F.4s, ID-16 and ID-26, have moved to Bevekom. The civil Skyservant OO-GEO arrived for storage here from Koksijde, the aircraft is now at Oostende. Aircraft marked * belong to the Brussels museum.

❑ 191	8-NK	*	Mystère 4A	191	stored, ex Koksijde, ex Savigny	3-98
❑ 145		*	Super Mystère B2	145	stored, ex Koksijde, ex Savigny	3-98
❑ 2021		*	MiG-23ML	0390324638	stored, arr 12-2-96, ex Beverlo/Brasschaat	3-98
❑ 9635		*	Mi-24D	340274	stored, ex 530/NVA and marked as such	3-98
❑ 1014		*	TS-11-100 bisB	1H-1014	stored, arr 8-7-96, ex Polish AF	3-98
❑ J-4077		*	Hunter F.58	..	stored, arr 4-10-95, ex Swiss AF	3-98
❑ BA-08			Mirage 5BA	08	stored, arr 13-1-94	3-98
❑ BA-10			Mirage 5BA	10	stored, arr 6-94	3-98
❑ BA-18			Mirage 5BA	18	stored, arr 6-94	3-98
❑ BA-20			Mirage 5BA	20	stored, arr 14-1-94	3-98
❑ BA-22			Mirage 5BA	22	stored, arr 6-94	3-98
❑ BA-27			Mirage 5BA	27	stored, arr 13-1-94	3-98
❑ BA-30			Mirage 5BA	30	stored, arr 6-94	3-98
❑ BA-31			Mirage 5BA	31	stored, arr 10-1-94	3-98
❑ BA-33			Mirage 5BA	33	stored, arr 20-1-94	3-98
❑ BA-42			Mirage 5BA	42	stored, arr 10-1-94	3-98
❑ BA-43			Mirage 5BA	43	stored, arr 6-94	3-98
❑ BA-44			Mirage 5BA	44	stored, arr 10-1-94	3-98
❑ BA-53			Mirage 5BA	53	stored, arr 12-1-94	3-98
❑ BA-54			Mirage 5BA	54	stored, arr 10-1-94	3-98
❑ BD-09			Mirage 5BD	209	stored, arr 13-1-94	3-98
❑ BD-10			Mirage 5BD	210	stored, arr 20-1-94	3-98
❑ BD-11			Mirage 5BD	211	stored, arr 11-1-94	3-98
❑ BR-03			Mirage 5BR	303	stored, arr 13-1-94	3-98
❑ BR-07			Mirage 5BR	307	stored, arr 12-1-94	3-98
❑ BR-08			Mirage 5BR	308	stored, arr 10-1-94	3-98
❑ BR-09			Mirage 5BR	309	stored, arr 12-1-94	3-98
❑ BR-10		*	Mirage 5BR	310	stored, arr 6-94	3-98
❑ BR-12			Mirage 5BR	312	stored, arr 10-1-94	3-98
❑ BR-14			Mirage 5BR	314	stored, arr 10-1-94	3-98
❑ BR-15			Mirage 5BR	315	stored, arr 10-1-94	3-98
❑ BR-17			Mirage 5BR	317	stored, arr 12-1-94	3-98
❑ BR-21			Mirage 5BR	321	stored, arr 12-1-94	3-98
❑ BR-22			Mirage 5BR	322	stored, arr 13-1-94	3-98
❑ BR-23			Mirage 5BR	323	stored, arr 12-1-94	3-98
❑ BR-24			Mirage 5BR	324	stored, arr 10-1-94	3-98
❑ FA-02			F-16A	6H-2	stored, arr 25-10-95	3-98
❑ FA-05			F-16A	6H-5	stored, arr 16-3-94	3-98
❑ FA-09			F-16A	6H-9	stored, arr 6-4-94	3-98
❑ FA-10			F-16A	6H-10	stored, arr 6-4-94, will go to Bevekom	3-98
❑ FA-16			F-16A	6H-16	stored, arr 4-5-94, in cocoon	3-98

❏ FA-17	F-16A	6H-17		stored, arr 26-8-94, in cocoon	3-98
❏ FA-18	F-16A	6H-18		stored, arr 26-10-94, in cocoon	3-98
❏ FA-19	F-16A	6H-19		stored, arr 10-5-95, in cocoon	3-98
❏ FA-20	F-16A	6H-20		stored, arr 28-3-95, in cocoon	3-98
❏ FA-21	F-16A	6H-21		stored, arr 6-4-94, in cocoon	3-98
❏ FA-22	F-16A	6H-22		stored, arr 16-3-94	3-98
❏ FA-23	F-16A	6H-23		stored, arr 5-5-94, in cocoon	3-98
❏ FA-25	F-16A	6H-25		stored, arr 4-5-94, special c/s	3-98
❏ FA-26	F-16A	6H-26		stored, arr 27-7-94, in cocoon	3-98
❏ FA-28	F-16A	6H-28		stored, arr 16-3-94	3-98
❏ FA-30	F-16A	6H-30		stored, arr 4-5-94, in cocoon	3-98
❏ FA-31	F-16A	6H-31		stored, arr 17-5-94, in cocoon	3-98
❏ FA-32	F-16A	6H-32		stored, arr 17-5-94, in cocoon	3-98
❏ FA-34	F-16A	6H-34		stored, arr 26-10-94, in cocoon	3-98
❏ FA-36	F-16A	6H-36		stored, arr 19-5-94, in cocoon	3-98
❏ FA-37	F-16A	6H-37		stored, arr 28-3-95, in cocoon	3-98
❏ FA-38	F-16A	6H-38		stored, arr 5-5-94	3-98
❏ FA-40	F-16A	6H-40		stored, arr 24-8-94	3-98
❏ FA-43	F-16A	6H-43		stored, arr 28-7-94	3-98
❏ FA-44	F-16A	6H-44		stored, arr 27-10-94	3-98
❏ FA-45	F-16A	6H-45		stored, arr 16-3-94, in cocoon	3-98
❏ FA-51	F-16A	6H-51		stored, arr 16-3-94, in cocoon	3-98
❏ FR-34	* RF-84F	..		stored, ex Bierset, ex EA-334/WGAF	3-98
❏ FU-92	* F-84F	..		stored, arr 13-12-95, ex Brustem, ex 53-6702	3-98
❏ FU-123	* F-84F	..		stored, ex Bevekom, ex 53-6768	11-96
❏ FU-156	* F-84F	..		stored, ex Bierset, ex 53-6834	3-98
❏ FU-191	* F-84F	..		stored, ex Bevekom, ex 52-6374	5-97
❏ FU-192	* F-84F	..		stored, ex Bevekom, ex 52-6605	3-98
❏ FX-58	* F-104G	9101		stored, arr 9-95, ex Bevekom, ex Koksijde	5-97
❏ FX-78	* F-104G	9133		stored, arr 9-95, ex Bevekom, ex Koksijde	5-97
❏ FX-83	* F-104G	9141		stored, arr 1-96, ex Florennes, ex Koksijde	5-97
❏ FX-85	* F-104G	9146		stored, arr 9-95, ex Florennes, ex Koksijde	5-97
❏ FX-89	* F-104G	9153		stored, arr 9-95, ex Bevekom, ex Koksijde	5-97
❏ MT-9	CM170	266		stored, Rode Duivels c/s, ex Evere	5-97
❏ MT-23	CM170	280		stored, fuselage only, ex Saffraanberg	5-97
❏ MT-37	* CM170	312		stored, wings/tail only, ex Saffraanberg	5-97

WESTERLOO
The Brussels museum had a storage facility in a military area here which used to hold some of the museum aircraft, Meteor F.8 EG-247 (marked as 'EG-244') went to the Czech museum at Kbely, together with the ex Royal Netherlands Air Force Harvard B-67 (ex Brussels). All the other stored aircraft here (see first edition) moved to Kapellen.

WESTOUTER
The Skytrain is still stored at Westouter near the Fench border.

❏ K-31	OT-CNR	C-47B	16064	stored, fuselage only	7-95

WETTEREN
In the night of 10th July 1997, a B707-321 arrived at this village (just east off Gent). Here it will be converted into a restaurant.

❏ TY-BBW	B707-321	18084	preserved, ex Oostende, ex TY-AAM	7-97

BELGIUM - 30

ZELE
A private owner here (some 10km east of Kalken) has the fuselage of a Lockheed T-33A. It has not been seen recently, but should still be current.
❑ FT-25   T-33A   9063   stored, crashed 26-3-76, ex 53-5742   9-92

ZELLIK
There was a large Belgian Air Force depot at Eckstein Kaserne (along the Zellik-Relegem/Wemmel road). Stampe V-57 moved on to Brussels (in June 1964!), while their Hunter ID-26 was placed in storage at Koksijde.

CYPRUS

Reports from this island are quite rare. However the situation in Cyprus has changed a bit since the first issue of this book. Most notable is the arrival of the aircraft at Paphos.

AKROTIRI
Based at this airfield is a detachment of the US Army with UH-60As. Of the W*recks and Relics* here, Vulcan XL317 and Canberra WJ768 have both gone during the late 1980s. In return a Phantom and a second Whirlwind have arrived.

❏ XS929	L	Lightning F.6	95262	preserved, ex Binbrook	2-96
❏ XV470	BD	Phantom FGR.2	3288	stored	2-96
❏ XD184		Whirlwind HAR.10	WA27	preserved, at gate, ex Larnaca	2-96
❏ XJ437		Whirlwind HAR.10	WA53	instructional	—

DHEKELI
The British Army has her 16 Flight based here with Gazelles AH.1s. The flight used two Scout AH.1s for battle damage repair training. Both choppers, XR602 (F9524) and XT648 (F9654), were last noted in april 1993 and have gone.

ERCAN
Derelict here by 1989, having been picked clean of spare parts, was Caravelle 10B1R TC-ARI (235). The aircraft made its last flight on 27th June 1988 from Istanbul and it was scrapped a year later. In the early 1990s Caravelle 10R TC-AKA (239) was noted here in storage. This aircraft was scrapped during 1993.

KALO LAKATAMIA
The DC-6 was noted at this airfield in a dismantled state during 1996.

❏ N19CA	DC-6B	43744	stored, ex Larnaca, ex TF-ISC	10-96

LARNACA
Whirlwind HAR.10 XD184 at the British military gate here moved to Akrotiri. The stored DC-6 N19CA went to Kalo Lakatamia.

❏ 5B-DAK	B707-123B	17632	stored, derelict	10-89

NICOSIA
An ex CSA Tu-104 is dumped near the western threshold. The aircraft is in bad condition after its landing accident on 29th August 1973.

❏ 5B-DAB	Trident 2E	2155	stored, hangared	2-95
❏ 5B-DAE	Trident 2E	2134	dumped, burned, ex G-AZND	6-91
❏ OK-MDE	Tu-104A	86601202	dumped	6-95

PAPHOS
After Mr Savvas Constantinides had bought two complete former RAF Shackletons in the early 1990s, the deal of acquisition of Canberra T.17 WJ981 (R3/AE3/6606), Sea Hawk WV795 (6056), Whirlwind HAS.7 XN385 (WA315) and Lightning T.5 XS458 (95018) all fell through and all aircraft remained in the UK. Three stored Mi-26s (RA-06263, RA-06264 and RA-06265) were first noted in October 1993. The Helicopters were in good

condition and flew on 8th November 1995 to Cyprus Port on their way to Peru.
❑ WL747	Shackleton AEW.2	..	stored, ex Coventry	1-98
❑ WL757	Shackleton AEW.2	..	stored, ex Coventry	1-98
❑ WR967	Shackleton AEW.2	..	stored, nose only	8-92
❑ F-AZEN	MD312	250	stored, ex Aix, N250DF ntu	1-98
❑ G-JETP	Jet Provost T.52A	PAC/W/17634	expected, ex 355/Singapore AF	—

CZECH REPUBLIC

The most prominent *Wrecks and Relics* site in the Czech republic is the large museum at the Kbely airfield. More then 150 aircraft are part of the collection. Since the dismise of the 'Iron Curtain' the museum at Kunovice was officially opened and the new collections at Vyškov and Zruč were established. Unfortunately a large number of preserved and stored MiG-15s and L-29s, which could be found all around the country, have gone over the last few years.

BAKOV
Parked right on the highway from Praha to Liberec is an Il-18 which is in use as a restaurant.
❑ OK-WAJ   Il-18D   187010101   preserved   10-96

BECHYNĚ
During 1993 the operational MiG-21s of the based 9SBOLP left Bechyně. Although the base is still held by the military, flying activity is rare nowadays. The MiG-15 should still be preserved inside the base, but the fate of the MiG-21s is unknown. A MiG-21F-13 was noted dumped alongside the runway in September 1991 and only the figures '01' could be read on the aircraft. This aircraft could be one of the two stored airframes which were not seen in 1991.
❑ 3689   MiG-15bisSB   613689   preserved   94
❑ 0103   MiG-21F-13   460103   stored, wfu 6-87   4-93
❑ 1001   MiG-21F-13   061001   stored, wfu 12-89   4-93
❑ ..   MiG-21F-13   ..   decoy, see note   9-91

BILÉ POLIČANY
North of Hradec Králové the village of Bilé Poličany can be found. In the village centre MiG-15bisSB 3775 was discovered in April 1993. Within a year it has moved to Kunčina Ves.

BOŘETICE
Bořetice is located some 20km north of Břeclav, close to the Brno-Bratislava highway. On the southside of the village a Delfin is preserved. The aircraft is painted in a non standard colour scheme.
❑ 0804   L-29   390804   preserved, unmarked   5-95

BREZOVÁ
The Czech Republic has more that a dozen villages named Brezová! *This* Brezová can be found halfway along the Plzeň-Praha highway, just north of Zebrák. Their MiG-15 3904 moved to Roudnice in 1997.

BRNO
Town: In the town centre the Techniké Muzeum Brno is located on the corner of Orli and Josefka streets. Displayed inside the museum yard were a civil Zlin 226B, a MiG-19S and an autogiro. It has been reported that the museum was closed for rebuilding by late 1995, but the aircraft were still on site and could be seen from outside. The former CSA Tu-134A has became a restaurant here. Its located near a lake in the western suburbs of Brno.
❑ 0511   MiG-19S   150511   preserved, arr 4-74   9-96
❑ OK-80   XA-66 Aeron   ..   preserved, autogiro   8-95
❑ OK-CFH   Tu-134A   2351801   preserved, wfu 4-7-91   9-97
❑ OK-MPR   Zlin 226B   ..   preserved   2-97

Airfield – Slatina: MiG-21s were operational from Brno Tuřany until the mid 1980s. Brno Slatina, where the maintenance of these MiGs was carried out, was connected to Tuřany via a long taxitrack. After the departure of the operational MiG-21s and the completion of the highway to Prostějov (which separates Tuřany from Slatina), the only aircraft left were those of the technical school. A number of these airframes were noted stored outside. Some have not been seen for some time and may have moved inside as the repair of one of the school's hangars was finished in the mid 1990s. A number of former airframes have found new homes. L-39C 0002 went to Kbely. The museum at Kunovice acquired Mi-4 1874 and MiG-19PM 1040. MiG-21F-13 0520 went to the transport museum in Líšeň and Su-7UM 1014 was bought by Mr Tarantik from Zruč. The aero club had Avia 14S 1108, but it moved to Líšeň in late 1992. The other Avia 14S 3133, which was preserved between some hangars at the military side, moved to Vyškov.

❑ 2404	Il-28RT	52404	instructional	9-97
❑ 6017	Mi-1	506017	instructional	9-96
❑ 3512	MiG-15bis	613512	instructional	9-97
❑ 0107	MiG-21F-13	460107	instructional	7-97
❑ 0512	MiG-21F-13	660512	instructional	9-96
❑ 0704	MiG-21F-13	760704	instructional	7-97
❑ 1308	MiG-21PF	761308	instructional	7-97
❑ 2701	MiG-21MF	962701	instructional	9-93
❑ B-2929	Mi-2	5310929069	stored, wreck in hangar	9-97

BROUMOV BYLNICE

The main road from Zlín to Trenčín passes through Broumov Bylnice. L-29 0107 (290107) was in the yard of the local school and could (only just) be seen from the main road. Unfortunately it was scrapped in 1995.

BUBOVICE

A yard in the village centre in Bubovice, some 25km south west of Praha held, besides a Hungarian glider, dismantled MiG-15bis 3814 (623814) . This was still there in 1996, but was not noted in 1997.

ČÁSLAV

After a number of air base closures, Čáslav is still one of the few which will be held open. MiG-15 5206 should still be preserved inside the base, while MiG-15bisSB 3738 (623738), which was dumped for many years on the north side of the airfield, was removed and scrapped by late 1992. Stored MiG-15bisSB 3677 (613677) was transported to East Fortune, UK in 1993. Three MiG-23BNs (5733, 9819 and 9831) were declared wfu on 18th October 1994. Of these, 9819 (0393219819) was scrapped soon afterwards, while 9831 was transported to Kbely on 10th September 1995. 5733 was still noted at Čáslav in September 1996 and moved a few kilometres north to the airfield of Kolín later that year. On 10th June 1993 MiG-23BN 9829 (0393219829) crashed and its remains were noted at Čáslav in October 1993. By 1994 this MiG was also scrapped. Two Su-7s are preserved among the buildings on the north side of the base.

❑ 5206	MiG-15	225206	preserved	94
❑ 5516	Su-7BM	5516	preserved, ex Přerov, wfu 12-89	11-97
❑ 5526	Su-7BM	5526	preserved, ex Přerov, wfu 12-89	11-97

ČERVENE JANOVICE

In this south east of Čáslav MiG-15bis 3839 used to be preserved. By 1992 it had moved to Ceska Olešná.

ČESKÁ OLEŠNÁ

The owner of the MiGs at Ceska Olešná, some 15km north east of Jindrichův Hradec, is contracted by the government if non operational aircraft have to be relocated. For example, he brought the Czech MiG-15s to Duxford and East Fortune. Short term inmate here was MiG-15bis 3839 which arrived from Červene Janovice

CZECH REPUBLIC - 35

in 1993 and moved on to Roudnice in late 1994. Note that the two sites here, one along a main road and one at a farm, have been combined into one. All aircraft can now be found along the main road through the village.

❏ 2462	MiG-15UTI	722462	stored, forward fuselage only		5-97
❏ 2514	MiG-15UTI	922514	stored		5-97
❏ 8810	MiG-15bis	528810	stored		5-97
❏ 0876	MiG-19S	650876	stored		5-97
❏ 1041	MiG-19PM	651041	stored		5-97

ČESKÁ SKALICE
The preserved Il-62 OK-YBA (90602) was last noted dismantled in July 1992. The aircraft was gone by 1994 and reported to have left for Germany.

ČESKÁ TŘEBOVÁ
The museum at Zruč knows where to find its MiG-15s. They bought MiG-15bisSB 3943 from this village.

ČESKÉ BUDĚJOVICE
Operational flying from this MiG-23BN and MiG-29 base was stopped in 1994, leaving the airfield only with the task of heavy maintenance of Čáslav's MiG-23MLs. Since the mid 1990s a large number of aircraft have been stored here, none of which are really visible from outside the airfield. Operational aircraft returned here in 1997 in the form of the Czech test unit LZO with its MiG-21s and L-39s.

❏ 3043	MiG-15bisSB	713043	preserved		6-97
❏ 2409	MiG-23ML	0390322409	stored		6-97
❏ 2410	MiG-23ML	0390322410	stored		6-97
❏ 3641	MiG-23MF	0390213641	dumped, wfu 1994		6-97
❏ 3645	MiG-23MF	0390213645	dumped, wfu 1994, nose to Vyškov		6-97
❏ 3880	MiG-23MF	0390213880	stored, wfu 1994		95
❏ 3920	MiG-23MF	0390213920	stored, wfu 1994		95
❏ 3924	MiG-23MF	0390213924	stored, wfu 1994		95
❏ 7182	MiG-23MF	0390217182	dumped, wfu 1994		6-97
❏ 7184	MiG-23MF	0390217184	stored, wfu 1994		6-97
❏ 9817	MiG-23BN	0393209817	stored, wfu 22-9-94		6-97
❏ 9830	MiG-23BN	0393209830	dumped, wfu 29-6-94		6-97
❏ 9861	MiG-23BN	0393209861	dumped, wfu 15-6-94		6-97
❏ 9862	MiG-23BN	0393209862	stored, wfu 15-6-94		6-97
❏ 9863	MiG-23BN	0393209863	dumped, wfu 18-10-94, nose to Kbely		6-97
❏ 9866	MiG-23BN	0393209866	dumped, wfu 5-10-94		6-97
❏ 9868	MiG-23BN	0393209868	stored, wfu 8-6-94		6-94

A number of the stored MiG-23MFs have found new homes. They have gone to local museums and two have left for the USA. Other MiG-23s have been scrapped and are also listed below. More may have been scrapped and may include some of the above mentioned aircraft. All Czech MiG-29s have been sold to Poland.

1080	MiG-23BN	0393211080	stored, wfu 29-6-94	scrapped during 1995
1083	MiG-23BN	0393211083	stored, wfu 1-6-94	scrapped during 1995
3646	MiG-23MF	0390213646	stored, wfu 1994	to Kbely
3887	MiG-23MF	0390213887	stored, wfu 1994	to Zruč
3888	MiG-23MF	0390213888	stored, wfu 1994	scrapped during 1995
3922	MiG-23MF	0390213922	stored, wfu 1994	to Kbely
5735	MiG-23BN	0393215735	stored, wfu 8-6-94	to Kbely
5739	MiG-23BN	0393215739	stored, wfu 8-6-94	to Roudnice
5741	MiG-23BN	0393215741	stored, wfu 8-6-94	scrapped during 1995
5742	MiG-23BN	0393215742	stored, wfu 1-6-94	scrapped during 1995

CZECH REPUBLIC - 36

5744	MiG-23BN	0393215744	stored, wfu 8-6-94	to Wilmington, USA in 1997
7183	MiG-23MF	0390217183	stored, wfu 1994	to Vyškov
7805	MiG-23UB	A1037805	stored, wfu 1994	to Wilmington, USA
9139	MiG-23BN	0393209139	stored, wfu 1-6-94	scrapped during 1995
9142	MiG-23BN	0393209142	stored, wfu 8-6-94	scrapped during 1995
9545	MiG-23BN	0393209545	stored, wfu 1-6-94	scrapped during 1995
9548	MiG-23BN	0393209548	stored, wfu 1-6-94	scrapped during 1995
9549	MiG-23BN	0393209549	stored, wfu 15-6-94	scrapped during 1995
9550	MiG-23BN	0393209550	stored, wfu 5-10-94	scrapped during 1995
9814	MiG-23BN	0393209814	stored, wfu 8-6-94	to Zruč
9820	MiG-23BN	0393209820	stored, wfu 8-6-94	to Vyškov
3810	MiG-29	32038	stored, wfu 1994, dep 12-95	to Polish AF
4012	MiG-29	32040	stored, wfu 1994, dep 1-96	to Polish AF
4402	MiG-29UB	N50903014528	stored, wfu 1994, dep 12-95	to Polish AF
5414	MiG-29	32354	stored, wfu 1994, dep 1-96	to Polish AF
5616	MiG-29	32356	stored, wfu 1994, dep 1-96	to Polish AF
5918	MiG-29	32359	stored, wfu 1994, dep 12-95	to Polish AF
7702	MiG-29	26377	stored, wfu 1994, dep 12-95	to Polish AF
8304	MiG-29	26383	stored, wfu 1994, dep 1-96	to Polish AF
8906	MiG-29	26389	stored, wfu 1994, dep 12-95	to Polish AF
9207	MiG-29	26392	stored, wfu 1994, dep 12-95	to Polish AF

CHOTUSICE
The main gate of Čáslav air base is in the village of Chotusice. In this village a dismantled MiG-15 can be found. The aircraft arrived in September 1994 from Josefodol. The serial number is for some unknown reason presented in reverse order on the aircraft.

❏ 3934　　　MiG-15bis　　　623934　　　stored, marked as '4393', ex Josefodol　　　6-97

CHRUDIM
The civil airfield of Chrudim is located on the south west side, not to be mixed up with the former Slovair helipad on the south side. On the far side of the field is a small military camp where a Mi-4 is preserved.

❏ 0542　　　Mi-4　　　..　　　preserved　　　9-97

DEHTÍN
The Delfin formally stored at Klatovy moved a few kilometres north west to its hew home at Dehtín. It is parked in the courtyard of a private house.

❏ 2810　　　L-29R　　　892810　　　stored, ex Klatovy　　　6-97

DOLNI POUSTEVNA
In the extreme northern part of the republic, along the German border, is Dolni Poustevna. The aircraft is preserved only a few hundred meters from the border crossing, in front of a flat.

❏ 3797　　　MiG-15bisSB　　　623797　　　preserved　　　5-95

DRUZTOVÁ
The museum collection at Zruč started life as a number of stored aircraft in Druztová. Some 15 aircraft were stored in a yard near the museum owner's house and since 1992 they have moved gradually to the museum location at Zruč (see which). Zruč is only 5km from Druztová. Exceptions are L-29RS 2612 (marked as '2601') which went to Praha, MiG-15SB 1836 moved to Schwaz and 1170 which went to Bremerhaven. Aircraft still in the yard are two unidentified Il-28 noses (both ex Malacky) and a number of smaller civil aircraft.

☐ OK-KGC	Aero Super 45	..	stored		9-97
☐ OK-TGA	PZL M18	1Z018-10	stored		9-97

HAVLÍČKUV BROD
Until 1992 the 52vrlt was based here with its Mi-2s and Mi-8s. Even before the unit's disbandment the only wreck on the airfield, Mi-4 4153 (04153), was scrapped.

HOŘICE
The aero club of Hořice has, as have many other aero clubs, a former Czechoslovakian military aircraft in their possession. This time not a MiG-15, but the reconnaissance version of the L-29 Delfin.
☐ 2811	L-29R	892811	preserved	9-93

HRABYNĚ
East of Mosnov is a large monument to the Soviet soldiers who liberated the Czechoslovakian Republic. The small museum beside this monument has recently sold two of its aircraft, Avia 14 3114 and Mi-1 4033, to Zruč.
☐ 2514	Mi-4	12114	preserved	5-95
☐ 3952	MiG-15bisSB	713952	preserved	5-95
☐ 0711	MiG-21F-13	760711	preserved, wfu 2-85	5-95

HRADEC KRÁLOVÉ
The last operational aircraft left the base in 1993 and only the MiG-15 is still preserved. Both Avia 14s which were stored here, were scrapped. These were the glass nosed Avia 14FG 3111 (703111) and sole Avia 14SRS 3160 (013160, SRS standing for Speciální Radarovy System).
☐ 3947	MiG-15bis	713947	preserved	9-97

HRBOV
The village of Hrbov, just off the Praha-Brno highway at the Velke Meziříčí zapad (zapad meaning west) exit, had its own MiG-15SB, 0562 (220562). This was last noted in May 1995 and was no longer there in June 1997.

JIHLAVA
Another aero club and another MiG-15. This time at the Jihlava airfield which is close to the village of Svábka.
☐ 0543	MiG-15	220543	preserved	6-94

JILEM
An autobazar (cardealer) owner has taken delivery of the former Veľká Losenice MiG-15. Jilem is located some 25km east of Jindřichův Hradec.
☐ 3935	MiG-15bis	623935	preserved, ex Veľká Losenice	6-97

JIMLÍN ZEMĚCHY
Some 20km east of the sleeping airfield of Žatec is the little village of Jimlín Zeměchy.
☐ 3801	MiG-15bisSB	623801	preserved	9-97

JOSEFODOL
MiG-15bis 3934 which used to be preserved here at a local factory moved to Chotusice in July 1994.

KBELY
Near the helicopter maintenance hangars, on the extreme west side of the airfield, a small dump was discovered in 1991. Of the two dumped helicopters the Mi-24 should still be present, while the dumped Mi-8P 0815 moved during early 1993 to the museum at Zruč. During 1994 L-410M 0502 ran out of airframe hours and was used by the airfield's fire brigade for a few months before moving to Zruč. Also one of the VZLU (state test unit) aircraft (L-410UVP OK-028) which was stored here in 1995 went to Zruč, together with the fuselage of MiG-21MF 1207 (l/n 6-96). The stored An-12BP 2105 (4342105, last flight 26th May 1994) was not part of the museum and during late 1997 is was sold as LZ-SFJ. Stored Tu-134A 1407 (1351407, last flight 26th November 1996) was sold in November 1997 as EW-65861. Stored with the VZLU since 1992 was L-410UVP CCCP-67500 (851404), which became N63020 in early 1997.

❏ 7109	An-24	17307109	stored, last flight 25-9-96	9-97
❏ 0711	Mi-2	5110711088	stored	9-97
❏ 0910	Mi-8	0910	stored	9-97
❏ 0143	Mi-24D	M340143	dumped	10-93
❏ CCCP-67674	L-410UVP	912604	stored, became OK-WDU ?	9-97
❏ OK-018	Il-18	180002202	stored, ex DDR-STC	1-96
❏ OK-020	Yak-40	9431436	stored, ex VZLU	9-97

On 1st January 1995 the museum at the airfield was placed in the hands of a new management. The first result of this new leadership was the removal of most of the older aircraft which were parked in the inner courtyard of the museum. Basically, only one aircraft of each type is retained in the inner court, the rest being placed in storage next to the museum. The L-39 prototype 3905 (X05) carries a correct serial number and has always flown as such, although there is also L-39ZA 3905 (633905). The real story behind the Korean MiG-15 will probably never be known. Some ten Soviet built North Korean Air Force MiG-15s were delivered to Czechoslovakia and replaced in North Korea by the same number of Czechoslovakian built MiG-15s. It is not known if 530738 was one of those which came from Korea. The camouflage colours were applied by the museum many years ago. L-39V 0740, which was preserved on the inner courtyard, was noted at Letňany in 1996, together with the MiG-15bis 3949. HC-3A OK-06 which was resorted to Trenčin returned there and will become part of the new Slovak military museum. NTZ denotes Narodni Technické Muzuem, the museum in Praha town. The museum at Kbely is open daily except on Mondays and Fridays.

Outside (inner courtyard):

❏ EG-247	B2-R	Meteor F.8	M24/7021	preserved, ex Brussels, ex Belgian AF	9-96
❏ 33 white		L-39C	330207	preserved, ex Soviet AF	9-97
❏ XT899	B	Phantom FGR.2	2507	preserved, ex RAF	9-97
❏ 00878		F-5E	R1059	preserved, ex Vietnam AF, ex 73-0878	9-97
❏ 6926		Il-28RTR	56926	preserved, ex BA-11, ex DE-51, arr 16-9-69	9-97
❏ 2827		L-29R	892827	preserved, arr 19-12-89	9-97
❏ 4003		Mi-1M	404003	preserved	9-97
❏ 0538		Mi-4	20138	preserved	9-97
❏ 0313		Mi-8T	0313	preserved, ex Letňany	9-97
❏ 2611		MiG-15UTI	142611	preserved	9-97
❏ 2626		MiG-15UTI-P	722626	preserved, ex OK-010	9-97
❏ 3255		MiG-15bisSB	613255	preserved	9-97
❏ 0872		MiG-17F	0872	preserved	9-97
❏ 0414		MiG-19S	150414	preserved, arr 20-4-70	9-97
❏ 0813		MiG-19P	650813	preserved, arr 13-3-66	9-97
❏ 0308		MiG-21PF	760308	preserved, arr 29-5-89	9-97
❏ 0948		MiG-21US	09685148	preserved, ex 5148	9-97
❏ 2410		MiG-21MF	96002410	preserved	9-97
❏ 4411		MiG-21PFM	94A4411	preserved, wfu 8-90	9-97
❏ 3646		MiG-23MF	0390213646	preserved, special c/s, ex České Budějovice	9-97
❏ 5735		MiG-23BN	0393215735	preserved, ex České Budějovice	9-97
❏ 9825		MiG-23BN	0393219825	preserved, special c/s, arr 1-7-94	9-97
❏ 9831		MiG-23BN	0393219831	preserved, special c/s, arr 10-9-95, ex Čáslav	9-97

CZECH REPUBLIC - 39

☐ 7905		MiG-23UB	A1037905	preserved	9-97
☐ 1017		Su-7UM	1017	preserved, arr 20-4-89	9-97
☐ 6513		Su-7BKL	6513	preserved, wfu 9-90	9-97

Hall A:

☐ 21		Aero Ae-10	21	preserved, locally built Brandendurg 76	5-97
☐ E 10		Letov S-20	50	preserved	5-97
☐ L-BALB		Aero A-10.3	3	preserved	5-97
☐ L-BIZL		SPAD VII C-1	11583	preserved, marked as '2'	5-97
☐ OK-ATW		Taylor E2 Cub	147	preserved	5-97
☐ OK-IZZ		Avia BH-11C	18	preserved, ex L-BONK and marked as such	5-97
☐ OK-TBX		Zlin XII	170	preserved	5-97
☐ OK-ZOB		Letov S-218	18	preserved, ex C-49 and marked as such	5-97
☐ ..		Aero A-12	..	preserved, replica, marked as '4'	5-97
☐ ..		Aero A-18	..	preserved, replica, marked as '5'	5-97
☐ ..		Aero A-18C	..	preserved, replica, marked as '5', original wings	5-97
☐ ..		Aero Ab-11	..	preserved, replica, marked as 'L-BUCD'	5-97
☐ ..		Aero Ap-32	..	preserved, replica, marked as 'E 2'	5-97
☐ ..		Avia B-10	..	preserved, replica, marked as 'C 155'	5-97
☐ ..		Avia BH-11	..	preserved, replica, marked as 'OK-LIQ'	5-97
☐ ..		Avia BH-11k	..	preserved, replica, marked as '17'	5-97
☐ ..		HM-14	..	preserved	5-97

Hall B:

☐ B-67		AT-16ND	14A-1462	preserved, ex FT422 and marked as such	5-97
☐ 38 red		Il-2m3	12438	preserved, ex Soviet Air Force	5-97
☐ 77 white		La-7	..	preserved, on loan from NTZ	5-97
☐ TE565	NN-N	Spitfire LF.IXe	..	preserved, on loan from NTZ	5-97
☐ D-EBIG		DH.82A	86536	preserved, marked as 'R5148/35', ex PH-UAM	5-97
☐ OO-FDI		Auster AOP.6	2835	preserved, ex A-16 and marked as such	5-97
☐ SP-BHA		CCS-13	420891	preserved, Polish built Polikarpov Po-2	5-97

Hall C:

☐ 5502		Avia B-33	5502	preserved, locally built Il-10, wfu 5-60	5-97
☐ 0010		L-29	190010	preserved	9-97
☐ 3905		L-39	X05	preserved, ex OK-184, ex OK-25	9-97
☐ 1125		L-200D	171125	preserved	5-97
☐ 1720		MiG-15	231720	preserved	9-97
☐ 1015		MiG-17PF	1015	preserved	9-97
☐ 1043		MiG-19PM	651043	preserved, arr 20-4-70	9-97
☐ 0613		MiG-21F-13	760613	preserved, arr 8-5-79	9-97
☐ 1501		MiG-21R	94R01501	preserved	9-97
☐ 3922		MiG-23MF	0390213922	preserved, ex České Budějovice	9-97
☐ 9863		MiG-23BN	0393209863	preserved, cockpit only, ex České Budějovice	9-97
☐ 5616		Su-7BM	5616	preserved, arr 20-4-89	9-97
☐ 1727		LET C-11	171727	preserved, locally built Yak-11	5-97
☐ 30		Yak-17	IS-1001	preserved	5-97
☐ H 6		Avia B-534	226	preserved	5-97
☐ HX-51		Yak-23	10101	preserved	9-97
☐ UC-26		Avia CS-199	565	preserved, locally built Bf109G-12	5-97
☐ UF-25		Avia S-199	178	preserved, locally built Bf109K	5-97
☐ V-34		Avia S-92	4	preserved, locally built Me262A-1	9-97
☐ B-2530		Mi-2	539430105	preserved	9-97
☐ OK-08		TOM-8	4	preserved	5-97
☐ OK-045		Zlin 135	1	preserved, ex Zlin Z-35	5-97

CZECH REPUBLIC - 40

☐ DM-WKB	Zlin 326AS	596		preserved	5-97
☐ OO-FRE	Zlin 22	82		preserved, ex OO-GUY, ex LX-MAI	5-97
☐ OK-AIP	K-65 Cap	475228		preserved, ex HO-20 and marked as such	5-97
☐ OK-AQO	Aero C-104A	227		preserved, ex A-27 and marked as such	5-97
☐ OK-BGL	Praga E-114M	125		preserved	5-97
☐ OK-BHM	M-1C Sokol	127		preserved	5-97
☐ OK-DJR	Zlin 381	64		preserved, marked as 'UA-264', locally Bü181D	5-97
☐ OK-DMO	Aero 45	4911		preserved, ex 4911	5-97
☐ OK-EXF	Zlin 26	36		preserved, ex UC-36 and marked as such	5-97
☐ OK-FRI	Zlin 26	601		preserved	5-97
☐ OK-IRG	Zlin 50LS	0017		preserved, ex OK-HRE (Zlin Z-50LA)	5-97
☐ OK-IVA	HC-2	1		preserved, ex OK-09 and marked as such	5-97
☐ OK-KHN	L-40	150002		preserved	5-97
☐ OK-KMA	Zlin 226A	0108		preserved	5-97
☐ OK-KOS	L-60	001		preserved, ex OK-01	5-97
☐ OK-RXA	HC-102	0436		preserved	5-97
☐ SP-AHB	L-4J	12406		preserved, marked as '10', ex 44-80110	5-97
☐ ..	Avia Ba-122	..		preserved, replica marked as 'OK-AVE'	5-97
☐ ..	BAK-01	1		preserved	5-97
☐ ..	HC-4	..		preserved, unfinished, frame only	5-97
☐ ..	L-39M	..		preserved, mock-up, marked as '79'	5-97

Hall D (not yet open to public):

☐ OK-BAK	A19 Cloud	5		preserved, ex Trenčín, ex G-ACGO	9-96
☐ OK-EFJ	Tu-134A	4323128		preserved, cockpit only, wfu 23-1-95	9-96
☐ ..	Bensen B-8	..		preserved, unmarked	9-96

Storage: The Kbely museum has a large collection of stored aircraft. Some are awaiting their turn in the museum, while others are only stored as exchange objects. Known to have left over the years are L-29RS 2405 (792405) which was sold as N29AD to the USA. L-29R 2808, LET C-11 1317 and MiG-15 3911 were all exchanged for the Meteor and AT-16 with the museum at Brussels, while 0906 moved to Orange. MiG-21PF 1309 (761309) was exported to the USA and has been noted at Wilmington, DE, since. HC-102 OK-RVY was restored at Trenčín before arriving at Kbely. It moved back to Trenčín in 1995. Formerly stored at Letnany, Mi-8T 0133 was scrapped by the museum during late 1995. MiG-21U-400 0132 (660132) was last noted in September 1991 and has gone since. The stored Delfin 2818 went to Praha, while Tatra T.131 OK-TAB is now on display at Kopřivnice. The serial of MiG-15bisT 3131 may be false.

☐ J-1161	Vampire FB.6	670	stored, ex Wohlen, ex Swiss AF	9-96
☐ ..	L-29A	893027	stored, single seat version, ex Soviet AF	9-96
☐ 7006	An-2	117006	stored	9-96
☐ 2904	An-24V	77302904	stored	9-97
☐ 3108	Avia 14T	813108	stored, arr 12-3-85	9-97
☐ 6102	Avia 14FG	806102	stored, on airfield, ex OK-MCB	9-97
☐ 1087	Avia B-33	1087	stored, locally built Il-10, will go to France	9-96
☐ 5271	Avia CB-33	271	stored, locally built Il-10U	9-96
☐ 0603	Il-14FG	4340603	stored, arr 24-11-76	9-97
☐ 0501	Il-28U	650100501	stored, ex CD-10, arr 15-9-69	9-97
☐ 2107	Il-28B	52107	stored	9-97
☐ 2303	Il-28RT	52303	stored	9-97
☐ 0003	L-29	190003	stored	9-97
☐ 0108	L-29	290108	stored, camo colours	9-96
☐ 2606	L-29RS	792606	stored	5-97
☐ 0002	L-39C	130002	stored, ex Brno Slatina	9-97
☐ 0725	L-39V	630725	stored	9-96
☐ 3908	L-39V	X08	stored, ex Vodochody	9-97
☐ 0705	L-200A	170705	stored	9-96

CZECH REPUBLIC - 41

☐ 0414	XL-160	150414	stored	9-96
☐ 1706	LET C-11	171706	stored, locally built Yak-11, ex OK-JIM	9-96
☐ 2710	Li-2D	23442710	stored, arr 27-9-67, also marked as 'OK-WDI'	9-97
☐ 3002	Li-2F	23443002	stored, arr 27-7-67, ex D-24, OK-GAH	9-97
☐ 1005	Mi-1	11005	stored	9-96
☐ 6014	Mi-1	506014	stored	9-96
☐ 0751	Mi-4	0751	stored, arr 20-3-79	9-97
☐ 2143	Mi-4A	02143	stored	9-96
☐ 0819	Mi-8T	10819	stored, ex Letňany	9-97
☐ 0830	Mi-8PS11	10830	stored, ex B-8130	9-97
☐ 1232	Mi-8T	041232	stored, ex Letňany	9-97
☐ 1532	Mi-8T	..1532	stored, ex Letňany	9-97
☐ 0738	MiG-15bis	530738	stored, Korean marks, see notes, Soviet built	9-97
☐ 1186	MiG-15SB	141186	stored, arr 3-1-68	9-97
☐ 1585	MiG-15SB	141585	stored, arr 2-4-68	9-96
☐ 1713	MiG-15SB	231713	stored, arr 2-4-68	9-97
☐ 2512	MiG-15UTI	922512	stored, last flight 18-3-83	9-97
☐ 3131	MiG-15bisT	713131	stored, ex Zruč, see note	9-97
☐ 3671	MiG-15bisR	613671	stored	9-97
☐ 3841	MiG-15bis	623841	stored	9-97
☐ 0101	MiG-17PF	0101	stored	9-96
☐ 0201	MiG-17PF	0201	stored	9-96
☐ 0742	MiG-19P	650742	stored, arr 27-11-67	9-96
☐ 1006	MiG-19S	851006	stored	9-96
☐ 0133	MiG-21US	01685133	stored, will go to Zruč	9-97
☐ 0202	MiG-21F-13	460202	stored, for Narodni Technické Muzeum	9-97
☐ 0210	MiG-21F-13	460210	stored	9-97
☐ 0212	MiG-21F-13	460212	stored	9-97
☐ 0241	MiG-21US	02685141	stored	9-96
☐ 0302	MiG-21PF	760302	stored, arr 29-5-89	9-97
☐ 0304	MiG-21F-13	560304	stored	9-97
☐ 0305	MiG-21F-13	560305	stored, arr 17-7-90	9-97
☐ 0313	MiG-21F-13	560313	stored	9-97
☐ 0712	MiG-21F-13	760712	stored, fuselage only	9-97
☐ 1013	MiG-21F-13	161013	stored, camouflaged, arr 24-5-89	9-97
☐ 1014	MiG-21F-13	161014	stored, wfu 4-90	9-91
☐ 1305	MiG-21PF	761305	stored	8-93
☐ 1903	MiG-21R	94R01903	stored	9-97
☐ 2820	MiG-21U-600	662820	stored	9-97
☐ 3166	MiG-21UM	516931066	stored	9-97
☐ 3181	MiG-21UM	516931081	stored	9-97
☐ 4609	MiG-21PFM	94A4609	stored, wfu 8-90	9-97
☐ 4916	MiG-21U-600	664916	stored, ex Vodochody, arr 17-8-94	9-97
☐ 7705	MiG-21MF	967705	stored, camouflaged	9-97
☐ 5919	Su-7BKL	5919	stored, wfu 9-90	9-97
☐ 6428	Su-7BKL	6428	stored, arr 16-1-90	9-97
☐ 5001	XM-12 Makrol	001	stored	9-96
☐ 0723	Yak-40	9230723	stored, ex OK-BYG, arr 22-6-83	9-97
☐ DD-39	Avia B-33	..	stored, locally built Il-10, will go to France	9-96
☐ V-35	Avia CS-92	5	stored, locally Me262B-1A, ex V-31 (51104)	9-97
☐ B-2047	Mi-2	5311147060	stored	9-97
☐ C-FBAM	Stinson 108-2	..	stored	9-96
☐ N143J	DC-3-229	1995	stored, ex N143JR	9-97
☐ OK-06	Aero 145	02002	stored, ex OK-KDA (Aero 45S)	9-96
☐ OK-10	MiG-15UTI	142744	stored, ex 2744	9-97

☐ OK-12	Zlin 37	0006	stored, ex OK-UJE	9-96
☐ OK-20	XL-410	690003	stored, ex OK-YKF, OK-063	9-96
☐ OK-70	XL-29	..	stored, ex OK-14, 3rd prototype	9-97
☐ OK-010	MiG-15UTI	..	stored	9-97
☐ OK-078	Zlin 43S	0001	stored, ex OK-YKN	9-97
☐ OK-182	XL-39	002	stored	9-97
☐ OK-ADR	Aero C-103A	224	stored, fuselage only, locally built Siebel Si-204	9-96
☐ OK-LDA	Tu-104A	76600503	stored	9-97
☐ OK-EGN	Aero 45	5079	stored, parts only	9-97
☐ OK-EPC	Aero 45	..	stored	9-96
☐ OK-HLJ	Zlin 126	721	stored, flyable	9-96
☐ OK-JEN	Yak-12R	14425	stored	9-96
☐ OK-JZE	LET C11	171511	stored, locally built Yak-11, ex OK-JIL, flyable	9-96
☐ OK-KGE	Aero 45S	170419	stored, on loan to CSA (Czech National Airline)	9-96
☐ OK-KIS	An-2R	1G190-13	stored	9-97
☐ OK-KMB	Zlin 226AS	0208	stored	9-96
☐ OK-MCI	Avia 14/32	805119	stored	9-97
☐ OK-MPM	Zlin 226B	258	stored	9-96
☐ OK-NAA	Il-18V	189001604	stored, ex CCCP-75703	9-97
☐ OK-RUV	SM-2	165007	stored	9-97
☐ OK-RVE	HC-102	0108	stored	9-96
☐ OK-SJB	Zlin XZ-37	0001	stored	9-96
☐ OK-XSB	Zlin 42	0003	stored	9-96
☐ OK-ZUD-15	Nieuport 11CR	00943	stored, 7/8 scale replica, flyable	9-96
☐ SP-CRU	PZL 104-35	59049	stored	9-97
☐ ..	LET C-11	171525	stored, wreck	9-96

KLATOVY

The owner of the civil airfield of Klatovy has his own MiG-21. The MiG is kept on the field and is in flyable condition. Engine runs are carried out a few times a year. The co-located L-29 2810 moved to Dehtín.

☐ 4943	Mi-1	..	expected, ex Líně	—
☐ 0607	MiG-21F-13	760607	stored, wfu 7-90	5-95

KLECANY

Between Praha and Vodochody, south west of to the airfield of Vodochody, are some military barracks. Inside, a MiG-15bis is preserved.

☐ 3941	MiG-15bis	713941	preserved	9-95

KOLÍN

Preserved at the local airfield is a former Čáslav storage MiG-23.

☐ 5733	MiG-23BN	0393215733	preserved, ex Čáslav	9-97

KOPŘIVNICE

The Technické Muzeum Tatra has a former Kbely Tatra T.131 (locally built version of a Bü131) on display. Kopřivnice is a few kilometres south of Příbor.

☐ OK-TAB	Tatra T.131.1	167	stored, on loan, ex Kbely, ex OK-AXM	97

KRALUPY NAD VLTAVA

MiG-21F-13 1005 in this town, north of Praha, moved to Roudnice.

CZECH REPUBLIC - 43

KUNČINA VES
This small village norh west of Rychnov nad Kněž has a MiG-15. The aircraft came from Bile Policany.

❏ 3775	MiG-15bisSB	623775	preserved, ex Bile Policany		98

KUNOVICE
The airfield of Kunovice is well known as the LET factory is situated here. During the early 1990s the stored aircraft near the LET factory were moved to the aero club side of the airfield and became the Slovacke Letecke Muzeum. L-29 0009 (290009) which was already stored near the aero club was sold to the USA and left during 1991. XL-410 OK-YKE was preserved on a schoolyard close to the airfield and was last noted there in September 1993. LET C-11 OK-NYA (172630) was stored in the museum for many years, but was scrapped in 1995. The museum has the following aircraft:

❏ 1103	Avia 14FG	601103	preserved	7-97
❏ 3157	Avia 14T	013157	preserved	7-97
❏ 0113	L-29	290113	preserved	7-97
❏ 0517	L-29A	390517	preserved, single seat version, ex OK-SZA	7-97
❏ 2613	L-29R	792613	preserved	7-97
❏ 1874	Mi-4A	1874	preserved, ex Brno Slatina	7-97
❏ 3005	MiG-15bisSB	713005	preserved	7-97
❏ 1040	MiG-19PM	651040	preserved, ex Brno Slatina	7-97
❏ 0514	MiG-21F-13	660514	preserved, wfu 10-83	7-97
❏ 0510	Su-7UB	0510	preserved	7-97
❏ 5530	Su-7BM	5530	preserved	7-97
❏ PK-35	LET C-11	171721	preserved, locally built Yak-11	7-97
❏ DDR-SNK	Zlin 37A	06-15	preserved, ex DM-SNK	7-97
❏ OK-022	L-410M	750401	preserved	7-97
❏ OK-030	L-410UVP	770001	preserved, ex OK-166, c/n 760604 ntu	5-97
❏ OK-ADO	L-410A	710005	preserved	7-97
❏ OK-ADP	L-410A	710101	preserved	7-97
❏ OK-ADQ	L-410A	700003	preserved, ex OK-AZA, ex OK-AKG	7-97
❏ OK-EKB	L-410A	740309	preserved, ex OK-EDB, ex OK-EKB	7-97
❏ OK-FHA	Aero 45	51163	preserved	7-97
❏ OK-IFD	Zlin 326	739	preserved	7-97
❏ OK-MPW	Zlin 226B	269	preserved	7-97
❏ OK-PHB	L-200D	170814	preserved	7-97
❏ OK-RFS	L-200D	171116	preserved	7-97
❏ OK-YKE	XL-410	690001	preserved, ex Kunovice village, ex OK-060	7-97
❏ OK-ZKA	XL-410	700004	preserved	7-97

LETŇANY
The aero club here has a MiG-21PFM. The aircraft came from the Vodochody store and is owned by the same person who owned the MiG-15s and L-29s in Praha Holešovice. An-2 OK-RYA which was stored here for many years moved during the summer of 1995 to Zruč.

❏ 5409	MiG-21PFM	94A5409	preserved, arr 30-5-94	9-97

On the opposite side of the road alongside Kbely airfield, the aircraft maintenance facilities of the Letecke Opravny Kbely (LOK) are located. Guarding the gate is a MiG-21F with false markings. Stored near the hangars since 1993 were five Mi-8Ts, all moved to the Kbely museum store. Mi-8Ts 0313 and 1532 were moved by late 1994, while Mi-8Ts 0133, 0819 and 1232 during the summer of 1995. The dumped MiG-21F-13 0603 was relocated to Druztová in 1992. The L-39V arrived here in 1996 for further storage.

❏ 0740	L-39V	630740	stored, ex Kbely	10-96
❏ 3949	MiG-15bis	713949	stored, ex Kbely	10-96
❏ 0619	MiG-21F-13	760619	preserved, marked as '8519'	9-97

Some of the stored MiG-21s at Vodochody arrived here during 1993 and 1994. At Letnany they were demilitarised and prepared for onward sale. All have left in by 1995. The are:

0906	MiG-21F-13	960906	ex Vodochody, arr 12-7-94, dep early 1995	to Kbely
1011	MiG-21F-13	161011	ex Vodochody, arr 11-7-94, dep 11-11-94	to Clakston
1101	MiG-21F-13	161101	ex Vodochody, arr 13-7-94	to N1011E
1104	MiG-21F-13	161104	ex Vodochody, arr 13-7-94, dep 18-11-94	to USA
1108	MiG-21F-13	261108	ex Vodochody, arr 11-7-94	to N6285D
1109	MiG-21F-13	261109	ex Vodochody, arr 11-7-94	to N6285L
1112	MiG-21F-13	261112	ex Vodochody, arr 13-7-94	gone
1114	MiG-21F-13	261114	ex Vodochody, arr 12-7-94	to N221YA
2916	MiG-21U-600	662916	ex Vodochody, arr 22-7-93	gone
4315	MiG-21PFM	94A4315	ex Vodochody, arr 11-7-94, dep 17-10-94	to USA
4402	MiG-21PFM	94A4402	ex Vodochody, arr 12-7-94, dep 22-11-94	to Oregon
4405	MiG-21PFM	94A4405	ex Vodochody, arr 4-1-94, dep 16-2-94	to Austria
4406	MiG-21PFM	94A4406	ex Vodochody, arr 29-3-94, dep 25-11-94	to Orange

LÍNĚ
Town: On the road to the main gate of the airfield a former Líně MiG-19 is preserved.

❑ 1045	MiG-19PM	651045	preserved	8-97

Airfield: Líně was still in use by helicopters from Přerov and the MiG-21s of the Czech test unit LZO in 1997. By December 1997 the military had abandoned the airfield. The MiG-15 and MiG-21 are preserved near the officers' mess and the MiG-19 is parked in the middle of the buildings area. All these will leave the base. The Mi-1, which was stored near the hangars, will go to Klatovy.

❑ 4943	Mi-1	..	preserved, will go to Klatovy	8-96
❑ 3779	MiG-15bis	623779	preserved	8-96
❑ 0508	MiG-19S	150508	preserved	8-96
❑ 0312	MiG-21F-13	560312	preserved, wfu 9-90	8-96

LÍŠEŇ
One of the suburbs of Brno is Líšeň. In this part of the town the Technické Muzeum Brno has a second building, named Expozice Mestske Hromadne Dopravy. Inside are mostly buses and other means of transport, outside are a L-29R and a MiG-21. The Avia 14S 1108 (911108) arrived in Líšeň during 1992 (from Brno Slatina aero club), but was not part of the museum. It was last noted during February 1993 and was scrapped shortly afterwards.

❑ 2821	L-29R	892821	preserved, ex Trenčín, marked as '8291'	6-97
❑ 0520	MiG-21F-13	660520	preserved, ex Brno Slatina	6-97

LUŽANY
Some 10km to the west of Jičín the village of Lužany can be found. Opposite the church MiG-15bisSB 3133 was preserved. The aircraft moved to Zruč in 1996.

MOŘCOV
About 20km south of the below mentioned Mošnov is Mořcov. At the swimming pool MiG-15bisR 3023 (713023) was preserved in a very bad shape. It was last noted in October 1994 and had gone by 1997.

MOŠNOV
The MiG-21 unit here was disbanded late 1993, while the transport aircraft and helicopters had already gone earlier in that year. Former instructional airframes Mi-4 (4139) and Su-7BM (5521) moved to Vyškov. The Avia

14 moved from the military side to the civil terminal.

❑ 3145	Avia 14T	913145	preserved	10-94
❑ 2404	L-29RS	792404	stored, ex instructional	6-94
❑ 0418	MiG-21F-13	660418	instructional, wfu 8-87	10-92
❑ 5618	Su-7BM	5618	stored, ex instructional, wfu 11-87	6-94

NÁMĚŠŤ NAD OSLAVOU
Airfield: On the Czech Sukhoi base are a number of preserved aircraft which are painted with year numbers. All these are related to special occasions on the airfield. The MiG-15 at the gate is still anonymous, while the MiG-15 at the dump cannot be seen from outside. It was still current during 1994.

❑ 2807	L-29RS	892807	preserved	11-97
❑ 3128	MiG-15bisSB	713128	dumped	94
❑ ..	MiG-15	..	preserved, at gate, marked as '1956/1986'	11-97
❑ 5317	Su-7BM	5317	preserved, marked as '1990', ex '1964'	11-97
❑ 5320	Su-7BM	5320	preserved, marked as '1965', ex '1956'	11-97

Barracks: Close to the airfields gate is a military barrack with an unmarked Su-7BM. It was thought that this was 5315, but that is incorrect as 5315 was at Prešov at the same time. It had gone after September 1992.

NECHVALIN
North east of Kyjov (30km west of Uherské Hradiště) the village of Nechvalin has its own MiG-15.

❑ 3945	MiG-15bis	713945	preserved	6-97

NEDAŠOV
Close to the border with Slovakia a local school at Nedešov (just east of Brumov Bylnice) also had their own MiG-15. As with many others, this MiG-15bisSB 3687 (613687) was scrapped recently.

NEUBUZ
A local restaurant here had two preserved MiG-21s (0403 and 0404) outside. The restaurant closed in 1993 and the MiGs were moved to <u>Slušovice</u> by 1994.

NOVÉ MĚSTO NAD METUJE
An Avia 14, together with a Mi-4, is preserved with the aero club at Nové Město nad Metuje.

❑ 3173	Avia 14T	013173	preserved	6-95
❑ 4142	Mi-4	04142	preserved	6-94

OLOMOUC
The town Olomouc has in its centre, near the ice stadium, a former CSA Tu-104 in use as a restaurant.

❑ OK-NDF	Tu-104A	8350801	preserved	9-97

PARDUBICE
Town: On the south western side of the town, in a housing area just outside the airfield, a local playground had a MiG-19PM 1101 (651101) until it was scrapped during 1992.

Airfield: Three aircraft used to be preserved on the air base. L-29R 2828 (892828) was sold as N29SV. The other two are normally kept in or around some of the shelters. MiG-15SB 3004 (713004) on the dump was last noted during the airshow in May 1993 and has since been scrapped.

❑ 0309	MiG-21F-13	560309	stored, wfu 8-89	5-97
❑ 6509	Su-7BKL	6509	stored, wfu 9-90, Kbely museum owned	5-97

PETŘÍKOVICE
There are several villages with the name Petříkovice in the Czech republic. The one with its MiG-15 is located along road number 123, some 20km west of Tábor.

❑ 3916	MiG-15bisSB	623916	preserved	5-95

PLZEŇ
The airfield of Plzeň Bory was abandoned in 1992 with the operational helicopters moving to the nearby base of Líně, while the preserved Mi-4 2543 at the gate went to Zruč.

PRAHA
Town: At a secondary school at an unknown location in Praha, a former Kbely Delfin is preserved. Also somewhere in town should be a film studio with the former Zruč Mi-1.

❑ 4033	Mi-1	404033	preserved, ex Zruč, ex Hrabyně	97
❑ 2818	L-29RS	892818	preserved, ex Kbely	97

Town - Holešovice: In the centre of the town, near the river Vltava, the residential area of Holešovice has a used car dealer with the name of Okase. During June 1993 MiG-15bis 3794 (623794) moved to Duxford, UK. MiG-15bis 3922 (623922) was last noted in September 1995 and left for an unknown location. Mi-4 9147 (09147) arrived mid 1991 from Tábor-Všechov, but moved to Weston-super-Mare, UK, late in 1993.

❑ 2612	L-29RS	792612	preserved, marked as '2601', ex Druztová	10-96
❑ 9903	L-29R	599903	preserved	9-97
❑ 1397	MiG-15SB	231397	preserved, faded old serial S-57	10-96
❑ 6502	Su-7BKL	6502	preserved, ex Motol, ex Přerov	9-97

In the south western side of the Holešovice area the Narodni Technické Muzeum is located. It can be found near of the Sparta Praha football stadium. The museum is open daily from 09:00 to 17:00 (except Mondays).

❑ RA-05	HC-102	..	preserved, ex 0002, OK-10	9-96
❑ 010.091	Anatra Anasalja	3979	preserved, marked as '11120'	9-96
❑ 119/15	Knoller C II	15	preserved	9-96
❑ 4	LWF V Scout	..	preserved	9-96
❑ L-BILG	VBS-1 Kunkadlo	..	preserved	9-96
❑ OK-AHN	M-1C Sokol	118	preserved	9-96
❑ OK-AVO	Avia BH-10	14	preserved	9-96
❑ OK-AXY	Aero C-104	254	preserved, locally built Bü131D	9-96
❑ OK-IPF	Avia BH-9	9	preserved, unmarked	9-96
❑ OK-IRF	Zlin 50LA	0016	preserved	9-96
❑ OK-TBZ	Zlin XIII	1	preserved	9-96
❑ ..	Blériot XI	..	preserved	9-96
❑ ..	HM-14	..	preserved	9-96
❑ ..	Praha PB.3	..	preserved	9-96

Town - Malesice: The LOM Praha facility has a preserved MiG-21F-13 in their factory grounds.

❑ 0506	MiG-21F-13	660506	preserved, wfu 10-83	9-97

Town - Motol: The area at Holešovice was becoming too small and the Okase dealer set up a second yard in the western part of Motol town (near the end of the highway from Plzeň). The yard is simply marked Okase 2 and for some time Su-7BKL 6502 was stored here. The aircraft moved to the first Okase yard in 1997.

CZECH REPUBLIC - 47

PŘEDMĚŘICE NAD LABEM
Glass nosed Avia 14FG 6103 was preserved outside a factory at this village close to Hradec Králové. From here it moved to Zruč during August 1992.

PŘEROV
The current *Wrecks & Relics* situation at Přerov is unclear since the departure of the L-29s, L-39s and MiG-21s to Čáslav and Pardubice. Their place has been taken by the helicopters from Líne and Prostejov. During May 1995 only two Su-7s were noted outside, the rest is maybe scrapped or have left: Su-7UM 1015 and 1016 to Vyškov, Su-7Ms 5516 and 5526 to Čáslav, Su-7BKL 6427 and 6511 to Vyškov and Su-7BKL 6502 to Praha. Sliač received Su-7BKLs 6430, 6501, 6503, 6504, 6505, 6506 and 6508. Also stored MiG-19 1102, MiG-21F-13 1111 and MiG-21PFs 1212 and 1313 left for Vyškov. In 1995 some seven covered MiG-21s and one MiG-23 were also noted as stored. By September 1996 no more fighters were noted in (outside) storage.

❏ 0110	L-39C	230110	instructional, fuselage only	7-93
❏ 0316	MiG-21F-13	560316	stored	9-90
❏ 0407	MiG-21F-13	660407	stored, cockpit only, wfu 12-90	4-91
❏ 0513	MiG-21F-13	660513	stored	9-90
❏ 0517	MiG-21F-13	660517	preserved	10-93
❏ 0912	MiG-21PF	760912	stored	9-90
❏ 1012	MiG-21F-13	161012	stored, wfu 12-89	4-91
❏ 1015	MiG-21F-13	161015	stored	8-91
❏ 1110	MiG-21F-13	261110	dumped, wfu 12-89	9-91
❏ 1213	MiG-21PF	761213	stored	9-90
❏ 1214	MiG-21PF	761214	stored	9-90
❏ 1303	MiG-21PF	761303	stored	9-90
❏ 6002	Su-7BKL	6002	stored	8-92
❏ 6006	Su-7BKL	6006	stored	9-90
❏ 6009	Su-7BKL	6009	stored	9-90
❏ 6011	Su-7BKL	6011	stored	5-93
❏ 6023	Su-7BKL	6023	stored	5-93
❏ 6426	Su-7BKL	6426	stored	8-92
❏ 6429	Su-7BKL	6429	stored	10-93
❏ 6514	Su-7BKL	6514	stored	7-93

PŘIBOR
Follow the signs to the local Hotel Letka (which means Squadron Hotel), near the town centre, there is a MiG-15 preserved in a small yard alongside the road.

❏ 3669	MiG-15bis	613669	preserved	10-94

PROSTĚJOV
The paratrainer Il-14 is a bit of a mystery. The serial number quoted for it, 2407, does not fit in any known serial number batch. As the aircraft is currently in bare metal and has no tail it cannot be checked easily. Although the helicopters have left Prostějov, this Il-14 still remains near the tower. During the early 1990s three aircraft from here moved to Vyškov, these were the Avia 14T 3144, Mi-1 4005 and Mi-4 0599.

❏ 3109	Avia 14T	813109	preserved, with aero club	9-96
❏ 2407	Il-14	..	instructional, see note	9-96

RANÁ
When the MiG-15 at Raná (north of Louny) was discovered in 1992 the aircraft was already dismantled. During 1994 the aircraft was pushed off the local airfield into a small ravine and probably still lies there.

❏ 3213	MiG-15bisSB	613213	dumped, ex stored	6-94

RONOV NAD DOUBRAVKA
South east of the Čáslav base lies Ronov nad Doubravka in which a MiG-15 is preserved. It can be found at the back of the local post office.
❑ 3832 MiG-15bis 623832 preserved 6-94

ROUDNICE
This Roudnice is not where the famous Memorial Air Displays are held every year, but it is a small village some 10km west of Hradec Králové. A surplus yard has a number of aircraft.
❑ 3839 MiG-15bis 623839 preserved, ex Česká Olešná, ex Červené Janovice 9-97
❑ 3904 MiG-15bisSB 623904 stored, ex Brezová 9-97
❑ 0503 MiG-19S 150503 preserved, ex Studenec 9-97
❑ 1005 MiG-21F-13 061005 stored, ex Kralupy nad Vltava, wfu 12-87 9-97
❑ 5739 MiG-23BN 0393215739 stored, ex České Budějovice, wfu 8-6-94 9-97
❑ OK-DFI Tu-134A 3351908 stored, cockpit only, wfu 17-1-92 9-97

SEC
The holiday town of Sec (south west of Čáslav) has a number of camping sites. One of them, along the lake, has a former CSA Il-18 as gate guard.
❑ OK-PAE Il-18V 181002902 preserved 8-95

SEZIMOVO ÚSTÍ
The town of Sezimovo Ústí (south of Tábor) is split in two halves by road number 3. In the part named Sezimovo Ustí II a MiG-19 is preserved at a local playground.
❑ 1042 MiG-19PM 651042 preserved, wfu 6-72 6-94

SLUŠOVICE
For many years two MiG-21PFs were preserved at a restaurant at Neubuz. Since the closure of the restaurant the two MiG-21s moved a few kilometres south to Slušovice for storage. Both aircraft moved on; 0403 went to Zorgvlied and 0404 to Doetinchem. In town, near the local airfield, an Il-18 is in use as a restaurant.
❑ OK-PAG Il-18V 181004201 preserved 9-95

SOBĚSLAV
The aero club at Soběslav has no more aircraft left as L-29RS 2610, Mi-4 4143 and MiG-21F-13 0310 were bought by the museum at Bad Ischl and MiG-21PF 0812 moved to Baak.

STARA BOLESLAV
Outside the headquarters buildings of the Czech Air Force at Stara Boleslav a MiG-21PFM is preserved.
❑ 4401 MiG-21PFM 94A4401 preserved, wfu 5-90 10-96

ŠTĚPÁNOV
Preserved at the gate of the military barracks in Štěpánov (some 10km north of Olomouc) is a Delfin.
❑ 2616 L-29R 792616 preserved 6-94

STOD
At an orphanage MiG-15bisT 3906 (623906) was preserved. Unfortunately it was scrapped in 1994.

STRAČOV
Just south of the main road from Jičín to Hradec Králové the little village of Stračov can be found. The local football field in the village centre held MiG-15bis 1822 (231822) which was discovered in April 1993. It was scrapped two years later.

STUDENEC
MiG-19S 0503 could be found uphill near a school in the village of Studenec, which is located some 10km north of Nova Paka. The aircraft moved to Roudnice in 1996.

STUDENKÁ
The collector at Zruč bought the Avia 14T 3146, which used to be preserved here. The aircraft moved to Zruč in mid 1993.

TÁBOR
On the civil field of Tábor (on the east side of town) Mi-4 9147 was discovered in April 1991. By September 1991 the helicopter had moved to the airfield of Tábor Všechov on the west side of Tábor. It was there until October 1992 when it moved again, this time to Praha.

TOUŽIM
Until 1994 former CSA Tu-104A OK-LDC (76600602) was in use as a restaurant in Toužim (south west of Karlovy Vary). In 1995 the aircraft was relocated to Ústí nad Ladem.

ÚSTÍ NAD LADEM
Halfway between Dresden and Praha is the town of Ustí nad Ladem. The former Touzin Tu-104 is still in use here as a restaurant. The more exact location of the Tu-104 has been quoted as Bahratal.

❏ OK-LDC	Tu-104A	76600602	preserved, ex Toužim	6-95

VEĽKÁ LOSENISE
Another mover is MiG-15bisSB 3935 from this village (east of Havlíckův Brod), to Jilem.

VELKE CHVALKOVICE
Just east of Česká Skalice one of the country's famous MiG-15s can be found. The aircraft is preserved in a school garden alongside the main road through the village.

❏ 3653	MiG-15bis	613653	preserved	6-94

VLACHOVICE
The MiG-15 at Vlachovice didn't survive long. Within a year after the discovery of MiG-15bisSB 3924 (623924) in May 1993 it was scrapped locally.

VODOCHODY
Besides building new L-39/L-59s, the Aero factory at Vodochody is also renovating the older Delfins for export. The former preserved Su-7BM 5308 and stored L-410 OK-FDC moved to Cerbaiola, while stored L-39V 3908 is now part of the museum collection at Kbely. Details of every stored L-29 and MiG-21 (of which 64 were CFE aircraft) are known and are listed below. None of them remain at Vodochody.

CZECH REPUBLIC - 50

1234	L-29	591234	stored, l/n 4-93	to N62187
1235	L-29	591235	stored, l/n 7-93	to OK-TYP
1326	L-29	591326	stored, l/n 3-94, N70750 ntu	to N9204X
1327	L-29	591327	stored, l/n 5-95	to N70751
1330	L-29	591330	stored, l/n 3-94, N62188 ntu	to N9196N
1408	L-29	591408	stored, l/n 8-94, N7082K ntu	to N9196X
1416	L-29	591416	stored, l/n 10-94, N7082P ntu	to N63DV
1706	L-29	591706	stored, l/n 8-94, N70752 ntu	to F-AZSD
2813	L-29R	892813	stored, l/n 7-93	to N31088
2814	L-29R	892814	stored, l/n 10-94, N7076N ntu	to N700PB
2815	L-29R	892815	stored, l/n 3-94, N7076R ntu	to N129JC
2819	L-29R	892819	stored, l/n 8-94, N7075Z ntu	to N19CZ
2820	L-29R	892820	stored, l/n 5-95, N7076J ntu	to N9137W
2824	L-29R	892824	stored, l/n 10-94, N7076G ntu	to N9105V
2825	L-29R	892825	stored, l/n 6-92	to N2825Q
2829	L-29R	892829	stored, l/n 10-94, N7076D ntu	to N27SR
2830	L-29R	892830	stored, l/n 8-93, N70759 ntu	to N82674
0307	MiG-21F-13	560307	stored, scrapped on 19-5-93	gone
0317	MiG-21F-13	560317	stored, scrapped on 26-5-93	gone
0406	MiG-21F-13	660406	stored, scrapped on 24-8-93	gone
0408	MiG-21F-13	660408	stored, scrapped on 25-5-93	gone
0410	MiG-21F-13	660410	stored, scrapped on 25-5-94	gone
0411	MiG-21F-13	660411	stored, scrapped on 25-8-93	gone
0416	MiG-21F-13	660416	stored, dep 9-6-93	to Sliač
0417	MiG-21F-13	660417	stored, scrapped on 24-8-93	gone
0420	MiG-21F-13	660420	stored, dep 25-6-93	to Sliač
0502	MiG-21F-13	660502	stored, dep 9-6-93	to Sliač
0516	MiG-21F-13	660516	stored, dep 25-6-93	to Sliač
0518	MiG-21F-13	660518	stored, dep 9-6-93	to Sliač
0604	MiG-21F-13	760604	stored, dep 9-6-93	to Sliač
0605	MiG-21F-13	760605	stored, dep 25-6-93	to Sliač
0606	MiG-21F-13	760606	stored, scrapped on 26-5-94	gone
0609	MiG-21F-13	760609	stored, dep 25-6-93	to Sliač
0614	MiG-21F-13	760614	stored, dep 9-6-93	to Sliač
0617	MiG-21F-13	760617	stored, dep 9-6-93	to Sliač
0701	MiG-21F-13	760701	stored, dep 25-6-93	to Sliač
0710	MiG-21F-13	760710	stored, dep 25-6-93	to Sliač
0817	MiG-21U-400	660817	stored, dep 17-8-94	to Vyškov
0906	MiG-21F-13	960906	stored, dep 12-7-94	to Letňany
1011	MiG-21F-13	161011	stored, dep 11-7-94	to Letňany
1101	MiG-21F-13	161101	stored, dep 13-7-94	to Letňany
1104	MiG-21F-13	161104	stored, dep 13-7-94	to Letňany
1108	MiG-21F-13	261108	stored, dep 11-7-94	to Letňany
1109	MiG-21F-13	261109	stored, dep 11-7-94	to Letňany
1112	MiG-21F-13	261112	stored, dep 13-7-94	to Letňany
1114	MiG-21F-13	261114	stored, dep 12-7-94	to Letňany
1117	MiG-21U-400	661117	stored, scrapped on 30-5-94	gone
2419	MiG-21U-600	662419	stored, dep 23-7-93	to Sliač
2916	MiG-21U-600	662916	stored, dep 22-7-93	to Letňany
4315	MiG-21PFM	94A4315	stored, dep 11-7-94	to Letňany
4402	MiG-21PFM	94A4402	stored, dep 12-7-94	to Letňany
4405	MiG-21PFM	94A4405	stored, dep 4-1-94	to Letňany
4406	MiG-21PFM	94A4406	stored, dep 29-3-94	to Letňany
4409	MiG-21PFM	94A4409	stored, dep 3-7-93	to Sliač
4410	MiG-21PFM	94A4410	stored, scrapped on 16-8-94	gone

CZECH REPUBLIC - 51

4413	MiG-21PFM	94A4413	stored, dep 23-7-93	to Sliač
4414	MiG-21PFM	94A4414	stored, dep 23-7-93	to Sliač
4415	MiG-21PFM	94A4415	stored, dep 23-7-93	to Sliač
4916	MiG-21U-600	664916	stored, dep 17-8-94	to Kbely
4917	MiG-21U-600	664917	stored, scrapped on 16-8-94	gone
4918	MiG-21U-600	664918	stored, scrapped on 16-8-94	gone
5406	MiG-21PFM	94A5406	stored, scrapped on 20-10-93	gone
5407	MiG-21PFM	94A5407	stored, scrapped on 20-10-93	gone
5409	MiG-21PFM	94A5409	stored, dep 30-5-94	to Letňany
5410	MiG-21PFM	94A5410	stored, dep 3-7-93	to Sliač
5411	MiG-21PFM	94A5411	stored, dep 29-3-94	to Seattle, USA
7113	MiG-21PFM	94A7113	stored, scrapped on 17-5-94	gone
7114	MiG-21PFM	94A7114	stored, scrapped on 11-5-94	gone
7201	MiG-21PFM	94A7201	stored, scrapped on 24-5-94	gone
7202	MiG-21PFM	94A7202	stored, scrapped on 4-5-94	gone
7203	MiG-21PFM	94A7203	stored, scrapped on 11-5-94	gone
7204	MiG-21PFM	94A7204	stored, scrapped on 16-8-94	gone
7206	MiG-21PFM	94A7206	stored, dep 3-7-93	to Sliač
7207	MiG-21PFM	94A7207	stored, dep 3-7-93	to Sliač
7208	MiG-21PFM	94A7208	stored, scrapped on 4-5-94	gone
7209	MiG-21PFM	94A7209	stored, dep 3-7-93	to Sliač
7908	MiG-21PFM	94N7908	stored, dep 23-7-93	to Sliač
7909	MiG-21PFM	94N7909	stored, scrapped on 21-7-94	gone
7910	MiG-21PFM	94N7910	stored, dep 23-7-93	to Sliač
7912	MiG-21PFM	94N7912	stored, scrapped on 28-4-94	gone
7913	MiG-21PFM	94N7913	stored, dep 3-7-93	to Sliač
7914	MiG-21PFM	94N7914	stored, scrapped on 28-4-94	gone
8001	MiG-21PFM	94N8001	stored, dep 17-8-94	to Vyškov
8002	MiG-21PFM	94N8002	stored, scrapped on 11-5-94	gone

VOJKOVICE
For more then 20 years a MiG-15 has been preserved in the village of Vojkovice (along road 48 to the Polish border, near Ceský Těšín). The serial number of the aircraft has deteriorated over the years. The owner of the aircraft, a former MiG-15 pilot, has confirmed its serial number and construction number, although there are reports that the aircraft is 3508 (243508).

❏ 3006	MiG-15bisSB	713006	preserved	6-94

VYŠKOV
Town: In the town of Vyškov the Czech Air Force has a large complex. In there should be a number of instructional airframes. Only one aircraft is confirmed.

❏ 1003	Su-25K	25508110003	instructional	7-96

Airfield: The museum at the airfield of Vyškov has expanded rapidly over the last few years. All aircraft are in excellent condition. The Mi-1 is currently at Brno for restoration. The name of the museum is Nadece Letecké Historické Spolecnosti Vyškov and it is open on Saturdays and Sundays between 09:00/12:00 and 14:00/18:00.

❏ 3133	Avia 14R	913133	preserved, ex Brno Slatina	5-98
❏ 3144	Avia 14T	913144	preserved, arr 29-6-91, ex Prostejov	5-98
❏ 0720	L-39V	630720	preserved	5-98
❏ 0735	L-39V	630735	preserved	5-98
❏ 4005	Mi-1	404005	preserved, ex Prostejov, see note	7-97
❏ 0599	Mi-4	0599	preserved, ex Prostejov	5-98
❏ 4139	Mi-4	04139	preserved, ex Mošnov	5-98

CZECH REPUBLIC - 52

❑ 0818	Mi-8T	10818	preserved	5-98
❑ 3912	MiG-15bisSB	623912	preserved, arr 16-2-91	5-98
❑ 0412	MiG-19S	150412	stored	5-98
❑ 1102	MiG-19PM	651102	preserved, ex Přerov	5-98
❑ 0817	MiG-21U-400	660817	preserved, arr 17-8-94, ex Vodochody	5-98
❑ 1003	MiG-21F-13	061003	stored, sectioned	9-97
❑ 1106	MiG-21F-13	261106	preserved, arr 3-91, wfu 7-90	5-98
❑ 1111	MiG-21F-13	261111	preserved, ex Přerov	5-98
❑ 1212	MiG-21PF	761212	preserved, ex Přerov	5-98
❑ 1311	MiG-21PF	761311	preserved	5-98
❑ 1313	MiG-21PF	761313	preserved, ex Přerov	5-98
❑ 2101	MiG-21R	94R02101	preserved	5-98
❑ 2703	MiG-21MF	962703	preserved	5-98
❑ 8001	MiG-21PFM	94N8001	preserved, arr 17-8-94, ex Vodochody	5-98
❑ 3645	MiG-23MF	0390213645	preserved, nose only	5-98
❑ 5734	MiG-23BN	0393215734	preserved, arr 14-6-94	5-98
❑ 7183	MiG-23MF	0390217183	preserved, ex České Budějovice	5-98
❑ 9820	MiG-23BN	0393209820	preserved, ex České Budějovice	5-98
❑ 1015	Su-7UM	1015	preserved, ex Přerov	5-98
❑ 1016	Su-7UM	1016	preserved, ex Přerov	5-98
❑ 5521	Su-7BM	5521	preserved, marked as '21', ex Mošnov	5-98
❑ 6427	Su-7BKL	6427	preserved, ex Přerov	5-98
❑ 6511	Su-7BKL	6511	preserved, ex Přerov	5-98
❑ OK-PLU	L-200A	170812	preserved	7-98
❑ OK-WJT	Zlin 37A	03-14	preserved	7-98

ZÁBŘEH
The airfield of Zábřeh, between Opava and Ostrava, has recently offered their Avia 14 for sale.

❑ 3159	Avia 14T	013159	preserved	5-94

ZBRASLAVICE
The airfield of Zbraslavice has one of the few original, instead of the locally built Avia 14, Ilyushin 14s. Stored MiG-19 0918 went in 1997 to Zruč.

❑ 0507	Il-14P	4340507	preserved, to go to Zruč	3-97

ZLIN
Near the tram station in the eastern part of town MiG-15bis 3845 (623845) could be found in a local playground. It was scrapped in 1995.

ZRUČ
Since its start in 1992 the Air Park at Zruč has become one of the largest collections in the country. A number of aircraft still have to arrive from storage at Druztová, while other aircraft, like an Il-10, will arrive in the near future. Some of the preserved aircraft were sold, MiG-19S 0201 and MiG-21F-13 0603 (marked as '1002', ex Letnany) both moved on to Bad Ischl, while the MiG-15bisT 3131 went to Kbely and An-2 OK-RYA to Kolbermoor. Mi-1 4033 (ex Hrabyne) and the tailboom of Police Mi-2 B-2743 (5310343097) both went to a film studio somewhere in Praha. MiG-15bis 3242, of which only the cockpit was present, was scrapped. MiG-21MFs 1207 and 4307 will be assembled as one complete aircraft. The museum is open daily in the 'summer' between 10:00 and 19:00.

❑ 3114	Avia 14T	913114	preserved, ex Hrabyně	1-98
❑ 3146	Avia 14T	913146	preserved, ex Studenká	1-98

CZECH REPUBLIC - 53

❏ 6103	Avia 14FG	806103	preserved, ex Přeměřice nad Labem	1-98
❏ 0507	Il-14P	4340507	expected, still at Zbraslavice	—
❏ 0101	L-29	290101	preserved, ex Druztová	1-98
❏ 2608	L-29R	792608	preserved, ex Druztová	1-98
❏ 2611	L-29RS	792611	preserved, ex Druztová	1-98
❏ 0502	L-410M	750502	preserved, ex Kbely	1-98
❏ ..	Mi-1M	..	preserved	1-98
❏ 8745	Mi-2	518745074	stored, ex Druztová	1-98
❏ 2543	Mi-4	12543	preserved, ex Plzeň Bory	1-98
❏ 0815	Mi-8P	10815	stored, ex Kbely	1-98
❏ 0551	MiG-15	220551	preserved	1-98
❏ 2501	MiG-15UTI	722501	preserved	1-98
❏ 3001	MiG-15bisSB	713001	preserved	1-98
❏ 3133	MiG-15bisSB	713133	preserved, ex Lužany	1-98
❏ 3943	MiG-15bis	713943	preserved, ex Česká Třebová	1-98
❏ 5237	MiG-15	225237	preserved	1-98
❏ 0219	MiG-19S	050219	preserved, ex Druztová	1-98
❏ 0918	MiG-19PM	650918	preserved, ex Zbraslavice	1-98
❏ 0602	MiG-21F-13	760602	preserved, wfu 8-85	1-98
❏ 1009	MiG-21F-13	161009	preserved, wfu 6-89	1-98
❏ 1207	MiG-21MF	961207	preserved, fuselage only, ex Kbely	1-98
❏ 4307	MiG-21MF	964307	preserved, rear fuselage and wings	1-98
❏ 9409	MiG-21MF	969409	preserved, cockpit only	1-98
❏ 3887	MiG-23MF	0390213887	preserved, ex České Budějovice	1-98
❏ 9814	MiG-23BN	0393209814	preserved, ex České Budějovice	1-98
❏ 1014	Su-7UM	1014	preserved, ex Druztová, ex Brno Slatina	1-98
❏ OK-028	L-410UVP	810625	preserved, ex Kbely	1-98
❏ OK-AKF	Zlin 37A	13-26	preserved, blue colours	1-98
❏ OK-AYA	Aero C-3A	313 ?	preserved, fuselage, locally built Si-204D-1	1-98
❏ OK-FIU	Mi-2	534542125	preserved, ex B-2542	1-98
❏ OK-IYB	L-410UVP	770102	preserved, ex Druztová, ex OK-162	1-98
❏ OK-YKG	Zlin 37	08-23	preserved, with wings of OK-VJM	1-98
❏ OK-YXB	Mi-8T	041032	preserved, with boom of OK-YXC, ex 1032	1-98
❏ ..	Il-12	..	stored, cockpit only	1-98
❏ ..	L-410UVP	770003	stored, static test airframe, never flown	1-98

DENMARK

Dispite the happy notes in the Danish introduction of the first edition, large numbers of F-84s, F-86s and F-100s have sadly been scrapped since the early 1990s. Another sad note is that the museum at Billund closed in late 1997. All aircraft had to be off the museum site by April 1998. Additionally the two museums in København no longer have aircraft. On a happier note, the new collection at Helsingør is hoping to open its doors (starting with a limited display) in 1998.

ÅLBORG

Town: In a local park (the Fjordparken), a Thunderjet has been in use as a children's plaything since 1971. The Alouette 3 can be found in Ålborg Marinemuseum. F-104G R-888 could be seen in the Ålborg Techniske museum, but this was closed in May 1997 and it moved to Stauning in August 1997.

❏ 51-10622	KU-U	F-84G	..	preserved, ex FZ-118/Belgian AF	2-98
❏ M-070		Alouette 3	1070	preserved, arr 1-9-95, ex Ålborg airfield	2-98

Airfield: The F-84 at the gate carries the incorrect code KR-A, it never wore this code in its operational life. The F-86D on the dump was used as a spares source for the Sabre at the gate. Stored Alouette 3 M-070 moved to the Marinemuseum in town. CF-104 R-851 is planned to be preserved at the main gate, while R-896 moved to Helsingør, together with two-seater RT-667. Lynx S-196 is missing its cockpit (which can be found at Frederikshavn).

❏ 35420	49	J35F	35420	stored, ex RSweAF	2-98
❏ F-470		F-86D	173-603	dumped, ex 51-8470	2-98
❏ F-947		F-86D	173-091	under restoration, marked as 'F-326', ex 51-5947, with tail from F-043/51-6043	2-98
❏ F-971		F-86D	173-115	instructional, ex 51-5971	2-98
❏ R-771		CF-104	1071	preserved, at gate, ex Vandel	5-98
❏ R-825		CF-104	1125	preserved, with Esk726, ex 12825/RCAF	2-98
❏ R-832		CF-104	1132	preserved, with Esk723, ex 12832/RCAF	2-98
❏ R-851		CF-104	1151	instructional, ex 104851/CAF	2-98
❏ RT-664		CF-104D	5334	instructional, ex 104664/CAF	2-98
❏ 51-504		Meteor NF.11	5545	preserved, at gate, ex Billund, ex WM387/RAF	2-98
❏ 51-9681	AT-D	F-84G	..	dumped, parts only, used to restore 51-9966	10-97
❏ 51-9966	'KR-A'	F-84G	..	preserved, at gate, ex SI-Q	2-98
❏ S-196		Lynx Mk.80	WA196	instructional, unmarked, crashed 14-9-85	1-98
❏ OY-XFA		Polyt III	1	stored, ex Z-931, ex 92-931	11-93

ÅLHOLM

Ålholm Automobiel museum at the Ålholm castle had two aircraft. Currently the museum is closed and some parts are being sold off. Is has been reported that the Draken is currently preserved in a local playground.

❏ A-007		F-35	351007	preserved, arr 3-5-94	5-94
❏ ..		Wright Flyer	..	preserved, replica	5-94

BILLUND

The Danmarks Flyvemuseum was part of the Mobilium which also included the car- and the falck-museum. The Mobilium was sadly closed in late 1997. The buildings were bought by the Lego company and all aircraft had to be out of the buildings by April 1998. The Starfighter RT-657 will move to Ålborg and RF-84F C-264 to Skrydstrup. No decisions are yet made over the other aircraft. Spitfire 41-401 is not the real 401, this came to light during restoration at Billund, but is ex RAF MA298. The original 401 was scrapped. MA298 was given Danish instructional serial number F.M.S.m.4. F-86D F-028 (ex Skrydstrup) has moved to Helsingør, together with Meteor 43-469 and F-100 GT-961 (both ex Skrydstrup). The Egeskov museum received Alouette 3 M-338,

DENMARK - 55

Chipmunk P-127, Hutter OY-AXH and Monospar OY-DAZ (ex København) from here. Meteor 51-504 went to Ålborg. Note that the aircraft currently in storage at Vandel have all been seen here in the museum in the past. Sabre F-421 (l/n June 1994) has moved on to Skrydstrup. First aircraft to move after the closure in 1998 are F-84G 51-10777, C-47A K-681, S-55C S-883, Meteor 43-461, DH.89A OY-AAO and Proctor OY-ACP. All went to Helsingør.

❑ 29487	07	J29F	29487	preserved, ex F3/RSweAF	5-97
❑ A-001		F-35	351001	preserved, arr 14-4-92, l/f 2-11-90	5-97
❑ DT-289		T-33A	7073	preserved, on pole, ex Værløse, ex 51-9289	2-98
❑ DT-491		T-33A	5786	preserved, arr 11-11-88, ex Værløse, ex 51-4491	5-97
❑ DT-497		T-33A	5829	preserved, on pole, ex Værløse, ex 51-6497	2-98
❑ DT-923		T-33A	6707	preserved, on pole, ex 51-8923	2-98
❑ GT-927		TF-100F	243-203	preserved, arr 22-6-92, ex Værløse, ex 56-3927	5-97
❑ RT-657		CF-104D	5327	preserved, arr 28-4-86, ex 12657/RCAF	5-97
❑ 11-113		KZ IIT	121	preserved, hands on display, ex 121	5-97
❑ 31-309		Harvard IIB	14A-966	preserved, arr 14-10-86	5-97
❑ 110		Avro 504N	OV.50	preserved, arr 23-7-89, ex København	5-97
❑ 158		Hawker Dankock	OV.54	preserved, arr 11-7-89, ex København	5-97
❑ F.M.S.m.4		Spitfire LF.IX	..	preserved, marked as '41-401', ex MA298	5-97
❑ LN-BIF		PT-26 Cornel	T43-4640	preserved, marked as '179', ex 253/RNoAF	5-97
❑ OY-ACT		KZ III	50	preserved, in car museum, ex 61-611	5-97
❑ OY-ADS		RC560E	612-40	preserved	5-97
❑ OY-ATX		H/T-O 2G	..	preserved, glider, ex Kongelunden, ex OY-100	5-97
❑ OY-DAX		Grunau Baby IIa	2	preserved, glider, ex Kongelunden, ex OY-29	5-97
❑ OY-DRH		Cessna 310	35407	preserved, ex G-ASSZ, ex N5207A	5-97
❑ OY-DZU		KZ IV	70	preserved	5-97
❑ OY-FAV		PA23-250	27-2154	preserved	5-97
❑ OY-KRD		SE210-3	47	preserved, ex Kastrup	5-97
❑ OY-98		Polyt I	..	preserved, glider, ex Kongelunden, ex OY-BEX	5-97
❑ SE-BWC		SAAB L17A	17320	preserved, marked as 'E', ex17320/RSweAF	5-97
❑ BS		Berg & Storm BS.III	3	preserved, arr 12-9-89, ex København	5-97
❑ ..		Astir CS77	..	preserved, glider, nose/forward fuselage only	5-97
❑ ..		Ellehammer	..	preserved, replica (1906 model)	5-97
❑ ..		F-100	..	preserved, procedure trainer, cockpit only	96
❑ ..		Gyrocopter VS17	..	preserved	5-97
❑ ..		Nielsen & Winther Aa	..	preserved, replica with original parts	5-97

A number of the museum's aircraft are stored or under restoration. Most are in a hangar at the nearby airfield. Now the museum is closed the future of these aircraft is uncertain. The nose of DC-7 OY-KNB used to be a restaurant, the remainder of the aircraft was scrapped at Kastrup. F-84G A-047 and TF-100F GT-961 moved on to Helsingør. The Meteor B-499 moved to Videbæk, while SG-38 OY-86 went to Vejle. The gliders which were not seen in May 1997 were all stored off site.

❑ A-803	F-84G	..	stored, ex Skrydstrup, ex 51-9803	5-97
❑ C-264	RF-84F	145	stored, ex 51-11264	5-97
❑ E-401	Hunter F.51	41H-680260	stored, ex 47-401, ex WW591/RAF	5-97
❑ J-49	Fokker D.XXI	109	stored, fuselage frame only	5-97
❑ K-687	C-47A	19200	stored, ex BW-P/RNoAF, ex 42-100737	5-97
❑ 12-131	Chipmunk T.20	C1/0695	under restoration, with HO Aero	10-97
❑ G-AKDK	M65 Gemini 1A	6469	stored, ex Kongelunden	5-97
❑ OY-AVA	P66 President	P66/79	stored, ex København, ex 69-697, ex G-AOJG	5-97
❑ OY-ATR	Chipmunk T.20	C1/0802	under restoration to flying condition, ex P-140	10-97
❑ OY-CBB	CeFTB337	..	stored	5-97
❑ OY-DHZ	Dove 6	04476	stored, ex D-IBYW	5-97
❑ OY-DXH	SZD-30	W397	stored, glider	97
❑ OY-EAZ	UC-61A	962	stored, ex Kongelunden, ex SE-CPA	5-97
❑ OY-KNB	DC-7C	44929	stored, nose only	97

DENMARK - 56

☐ OY-MUX	Mü13D	01536	stored, off site		97
☐ OY-VOX	Weihe	216	stored, glider, ex Kongelunden, ex 8301/RSweAF		97
☐ OY-XIE	L-13	026924	stored, glider		97
☐ SE-AKN	Klemm KL.35D	1873	stored, ex København, ex 5087/RSweAF		5-97
☐ ..	Lund HL-1	..	stored, ex Kongelunden		97

DC-8 5N-ATS (45817) mentioned in the first issue was repaired and flown out again.

EGESKOV CASTLE
The Egeskov Veteranmuseum possesses a varied collection of aircraft, several of them on loan from other organisations. The museum can be found some 2km west of Kværndrup on the A8 (about 30km south of Odense). SAAB B17 SE-BWC went to Billund, as did Spitfire '41-401' and PT-26 LN-BIF. Fa330 60127 was only temporarily at Billund. The nose of DC-7 OY-KND moved to Helsingøre.

☐ 60127	Fa330A-1	60127	preserved, ex Billund, ex Egeskov	5-97
☐ A-012	F-35	351012	stored, arr 26-5-94	5-97
☐ E-426	Hunter F.51	41H-680285	preserved, nose section only, ex 47-426	5-97
☐ M-388	Alouette 3	1388	preserved, ex Billund	5-97
☐ P-127	Chipmunk T.20	C1/0106	preserved, ex Billund	5-97
☐ R-814	CF-104	1114	preserved, arr 6-4-87, ex 104814/CAF	5-97
☐ 51-9792	F-84G	..	stored, ex KP-K, tail from 51-10094	5-97
☐ OY-AXH	Hütter H17A	186	preserved, glider, ex Billund, ex OY-61	5-97
☐ OY-BAK	DH.82A	86356	preserved, ex NL913, ex 'S-11'	5-97
☐ OY-DAZ	Monospar ST-25	95	preserved, ex Billund, ex G-AEYF	5-97
☐ ..	F-104	..	preserved, simulator (cockpit only)	5-97
☐ ..	HM-14	..	preserved, replica	5-97

ESBJERG
Stored at this airfield are a number of early Jetstreams.

☐ OY-CRP	HP137	209	stored, ex N2009	5-97
☐ OY-CRR	HP137	217	stored, ex N12217	5-97
☐ OY-CRS	HP137	225	stored, ex N12225	5-97
☐ OY-CRT	HP137	233	stored, ex N2S	5-97

FARUM
The NBC School at Farum, which is a few kilometres north of Værløse, lost Thunderjet 51-9978 to Helsingør. Its place was taken by a Starfighter.

☐ R-855	CF-104	1155	preserved, arr 4-4-86, ex 104855/CAF	5-97

FREDENSBORG
A private collection here, the Spangkuk Air Museum, has closed. T-33A DT-404 moved to Tirstrup.

FREDERIKSHAVN
The naval base has the cockpit of a Lynx as an instructional airframe with the Action and Information School, it came from Værløse. The rest of it is at Ålborg.

☐ S-196	Lynx Mk.80	WA196	instructional, marked as 'AIS', cockpit only	10-95

GRENÅ
The Per Udsen factory, a well known manufacturer of military hardware, has a Draken. After restoration it will

DENMARK - 57

be placed at the gate.
☐ A-019   F-35   351019   stored, arr 7-6-94   6-94

GUDSØ
A visit to this civil defence fire exercise area in 1997 revealed that Sabre F-307 (163-440) and Thunderjet A-143 had both been scrapped locally. The aircraft were last noted in August 1992.

HADSUND
Several ex Polish Lims and MiGs went through here in 1991, complete with engine, instruments and machine guns. They were reduced to tiny bits of metal, there is nothing left.

HELSINGE
A private collector here acquired some original parts of a Bf110 from Russia. These are under restoration.

HELSINGØR
The Danmarks Tekniske Museum is located at two sites in the town. Future plans are to move the museum to one new site in København, with the aircraft and some cars going to the Flyvemuseet Ellerhammer. The museum also has some original wings from the Donnet-Leveque A and these will be matched to the aircraft. For the replica Maagen 2, wings and a new fuselage will be built.

☐ OY-ABX	Grunau Baby IIb	..	preserved, glider, ex OY-87	5-97
☐ OY-AFX	Polyt II	PFG2	preserved, glider, crashed 27-3-59, ex OY-55	5-97
☐ ..	Ellehammer	..	preserved	5-97
☐ ..	Ellehammer helicopter		preserved, experimental, 1/3 scale	5-97
☐ ..	Ellehammer monoplane		preserved	5-97
☐ ..	Glenten (Svendsen)	..	preserved, replica	5-97
☐ ..	Larsen		preserved, glider	5-97
☐ Maagen 3	Maagen 3	OV.5	preserved, replica	5-97
☐ ..	Ternen (D L Type A)		preserved, with wings from Maagen 2	5-97

A new museum here, the Flyvemuseet Ellerhammer, expected to open its gates in late 1998. It is housed in a former shipbuilder's building, currently under renovation. The museum has already taken delivery of a number of aircraft, some of which were originally planned for the museum at Slangerup.

☐ 32571	09	J32E	32571	expected, F16M markings, ex RSweAF	—
☐ 51-10777		F-84G	..	preserved, marked as 'A-777/SY-H', ex Billund	6-98
☐ A-047		F-84G	..	stored, cockpit section only, ex 52-3047	6-98
☐ A-005		F-35	351005	expected, still at Karup	—
☐ AR-108		RF-35	351108	expected, still at Karup	—
☐ DT-102		T-33A	6886	stored, ex Værløse, ex 51-9102	6-98
☐ DT-847		T-33A	6179	stored, ex Værløse ex 51-6847	6-98
☐ F-028	'AL-E'	F-86D	173-172	stored, ex Billund	6-98
☐ GT-961		TF-100F	243-237	stored, ex Billund, ex Skrydstrup, ex 56-3961	6-98
☐ K-681		C-47A	9664	stored, marked as 'OY-DDA', ex Billund	6-98
☐ L-861		PBY-6A	2105	stored, ex Værløse, ex 82-861	6-98
☐ R-896		CF-104	1196	stored, with tail from R-707, ex Ålborg	6-98
☐ RT-667		CF-104D	5337	stored, ex Ålborg, ex 12667/RCAF	6-98
☐ S-883		S-55C	55-1031	stored, ex Billund, ex Værløse	6-98
☐ 43-461		Meteor F.4	G5-294	stored, ex Billund	6-98
☐ 43-469		Meteor F.4	G5-302	stored, ex Billund, ex Skrydstrup	6-98
☐ 51-9978	FS-987	F-84G	..	stored, ex Farum, ex SI-G	6-98
☐ OY-AAO		DH.89A	6775	stored, ex Billund, ex Kongelunden	6-98

DENMARK - 58

☐ OY-ACP	P34 Proctor III	H274	stored, ex 62-605 and marked as such, ex Billund	6-98
☐ OY-ALL	Chipmunk T.20	C1/0881	expected, ex P-142	—
☐ OY-KND	DC-7C	45211	stored, nose only, ex Egeskov	5-97

HERNING
In the city are civil defence barracks, where a Super Sabre is in use as an instructional airframe.

| ☐ GT-874 | TF-100F | 243-150 | instructional, ex Karup, ex 56-3874 | 5-97 |

HOLSTEBRO
With the civil defence in this town, Thunderjet A-652 was scrapped during 1990/1991.

JAGERSPRIS
Still extant here in 1987 were the mortal remains of three F-84G Thunderjets (A-114, 51-9944/SI-S and 51-10753/SE-K) which had been expended as range targets. The aircraft did not survive into the 1990s.

JONSTRUP
The military Flyvevabnets Officersskole in Jonstrup, close to Værløse, has taken delivery of a Karup Draken.

| ☐ A-002 | F-35 | 351002 | preserved, at barracks, ex Karup | 97 |

KARUP
Town: A playground in Karup village received a F-84G Thunderjet in the late 1970s.

| ☐ 51-9838 | SI-A | F-84G | .. | preserved, ex 51-9838, with tail from 51-10752 | 5-97 |

Airfield: The airfield is located halfway between Viborg and Herning. Instructional ex Swedish Air Force S35E Drakens 35925 and 35929 were scrapped by March 1992, while 35992 went to Skrydstrup. Some of the stored Danish Drakens flew out under civil markings; A-020 became N20XD, AR-106 became N106XD, AT-155 became N155XD and AT-156 became N156XD. A-002 went to Jonstrup. Meteor 22-265 and 44-499 have moved on to Billund, while by the late 1980s F-86Ds F-018, F-118, F-123, F-303, F-427, F-474, F-504, F-952, F-953 and F-985 were scrapped. RF-84Fs C-248, C-281, C-473 and C-670 were also all scrapped (by March 1992). Three Sabres (F-500, F-977 and F-984) went to Kjevik on 17th September 1992 for battle damage training by the Norwegian Air Force. Scrapped by March 1992 were the remains of F-100Ds G-279 and G-773. Stored TF-100F GT-874 moved to Herning. F-84G 51-10477 (with tail from 51-16665) was burned by 1974.

☐ 35552	J35F	35552	instructional, cockpit only, ex RSweAF	6-97
☐ 35905	S35E	35905	instructional, ex RSweAF	8-94
☐ 35931	S35E	35931	instructional, marked as 'BOR-931', ex RSweAF	6-97
☐ A-005	F-35	351005	stored, allocated to Helsingøre	5-97
☐ A-006	F-35	351006	instructional	8-94
☐ A-008	F-35	351008	instructional	6-97
☐ A-018	F-35	351018	instructional	6-97
☐ AR-104	RF-35	351106	instructional	6-97
☐ AR-108	RF-35	351108	stored, allocated to Helsingøre	6-97
☐ AR-109	RF-35	351109	stored, allocated to København	8-94
☐ AR-112	RF-35	351112	preserved, at main gate	6-97
☐ AR-113	RF-35	351113	preserved, flyable	11-96
☐ AR-118	RF-35	351118	instructional	6-97
☐ AT-158	TF-35	351158	stored, ex OY-SKA	10-97
☐ C-054	RF-84F	364	stored, ex preserved, ex 51-17054	6-97
☐ C-253	RF-84F	209	stored, ex instructional, ex FAF, ex 52-7253	6-97
☐ C-324	RF-84F	315	stored, ex FAF, ex 52-7324	6-97

DENMARK - 59

☐ C-581	RF-84F	633	preserved, at flying school, ex 53-7581	6-97
☐ C-649	RF-84F	688	stored, ex 53-7549	6-97
☐ C-651	RF-84F	670	instructional, ex stored, ex 53-7651	6-97
☐ C-865	RF-84F	037	stored, ex preserved, ex 51-1865	6-97
☐ DT-450	T-33A	5745	stored, forward fuselage only, ex Værløse	6-97
☐ DT-884	T-33A	6668	under restoration to airworthy state, ex Stauning	6-97
☐ R-757	CF-104	1057	instructional, ex Tirstrup, ex 12757/RCAF	5-97
☐ R-758	CF-104	1058	instructional, ex Tirstrup, ex 104812/CAF	9-92
☐ R-812	CF-104	1112	instructional, ex Vandel, ex 104812/CAF	8-95
☐ RT-655	CF-104D	5325	instructional, ex 12655/RCAF	8-95
☐ RT-660	CF-104D	5330	instructional, ex Tirstrup, ex 12660/RCAF	8-95
☐ S-884	S-55C	55-1032	stored, ex New Waltham (UK)	6-97
☐ OY-SKA	TF-35	351158	stored, ex AT-158	12-97

A museum has been set up on the base and is open every Wednesday afternoon, between 13:00 and 16:00.

☐ A-014	F-35	351014	preserved, arr 18-5-94	1-98
☐ A-665	F-84G	..	preserved, with tail from 51-10477, ex 51-16665	1-98
☐ C-274	RF-84F	155	preserved, ex 51-11274	1-98
☐ DT-905	T-33A	6689	preserved, arr 14-3-81, ex 51-8905	1-98
☐ GT-949	TF-100F	243-225	preserved, ex 56-3949	1-98
☐ 31-324	Harvard IIB	14A-1420	preserved, arr 6-5-82, ex FT380	1-98
☐ 44-491	Meteor F.8	G5/365	preserved, arr 14-3-81	1-98
☐ OY-ATK	KZ VII	182	preserved, ex O-620, ex Værløse	1-98

KØBENHAVN

Town: The Tojhusmuseet was empty in 1990. All aircraft mentioned in EWR-1 moved to Billund. Also empty is the Osterbrogades Kaserne. Their last aircraft, Klemm SE-AKN, went to Billund. A new building for the military Tojhusmuseet will be built here in the near future. For this new Tojhusmuseum, RF-35 AR-109 is allocated, is currently stored at Karup.

Airfield – Kastrup: The stored President OY-AVA and SE-210 OY-KRD moved to Billund. The airfield has a number of dumped and derelict airframes. Of these B720-025 OY-DSP (18241) was scrapped in the 1990s, while the nose of DC-7C OY-KNB moved to Billund (the remainder being scrapped here).

☐ OY-BAV	DC-6B	45198	dumped, poor condition, ex N574	8-96
☐ OY-BVH	F27-200	10200	instructional, ex A2-ADG	8-96
☐ OY-BZW	SA226TC	TC-328	dumped, ex ZS-LHJ	8-96
☐ OY-DMT	DC-7C	44136	dumped, ex CF-PWD	9-97
☐ OY-STD	SE210-10B3	238	dumped, unmarked, ex EC-CMS	9-96
☐ LN-PIP	DC-8-32	45256	instructional, tractor handling, ex N8038A	9-97

Airfield – Roskilde: Three LET410s were impounded here in July 1993. They were sold and OK-TDB (882039) became S9-TAU, OK-TDG (882040) became S9-TAV and SP-FTK (912530) became OK-LET.

☐ TF-ODN	HP137	214	stored, ex N17RJ	5-97

KONGELUNDEN

The Danmarks Flyvemuseum had a store on a military site here but with the establishment of a permanent museum at Billund all these airframes gravitated to the new location.

ODENSE

The Air Pub in Kongesgade in the centre of town has a Hughes helicopter inside.

☐ OY-HAD	H269A-1	25-0031	preserved, crashed 24-7-68, N8718F ntu	9-95

DENMARK - 60

OKSBØL
New range targets arrived here during 1988 in the form of half a dozen Sabres from Ålborg and Skrydstrup.

❑ A-181	F-84G	..	target, ex K-84/RNethAF, ex 51-10181	6-94
❑ F-016	F-86D	173-160	target, ex AB-N, ex 51-6016	6-94
❑ F-034	F-86D	173-178	target, ex AL-H, ex 51-6034	6-94
❑ F-346	F-86D	173-479	target, ex 51-8346	6-94
❑ F-431	F-86D	173-564	target, ex 51-8431	6-94
❑ F-473	F-86D	173-606	target, ex 51-8473	6-94
❑ F-994	F-86D	173-138	target, ex AB-O, ex 51-5994	6-94

PLEJERUP
PBY-5A Catalina L-857 stored here moved on to the museum at Stavanger on 11th November 1989.

RANDERS
The civil defence in the city have an F-84G in poor condition.

❑ 51-10482	SE-A	F-84G	..	instructional, with tail from 51-10209	5-97

RYVANGEN
In the end the F-84G 52-2981 did not go anywhere and was scrapped locally.

SKRYDSTRUP
This southerly Danish Air Force base is situated some 15km west of highway 47. Some of the preserved aircraft here moved to Billund; F-84G A-803, F-86D F-028, TF-100F GT-961, Harvard T.2B 42-309 and Meteor F.4 43-469. F-86D F-421 (ex Billund) will return to the museum after restoration.

❑ 35922	52	S35E	35922	instructional, ex Karup	6-96
❑ 51-10603		F-84G	..	preserved, at main gate as 'SKP', ex SY-C	5-97
❑ A-004		F-35	351004	instructional, l/f 2-9-93	6-96
❑ F-421		F-86D	173-554	under restoration, ex Værløse, will go to Billund	6-96
❑ F-946		F-86D	173-090	dumped, ex 51-5946	8-96
❑ GT-870		TF-100F	243-146	preserved, in Esk730 area, ex 56-3870	8-96
❑ GT-908		TF-100F	243-184	preserved, in Esk727 area, ex 56-3908	6-96
❑ RT-662		CF-104D	5332	instructional, ex 12662/RCAF	6-96
❑ 51-10487	SE-V	F-84G	..	dumped, parts only, ex FZ-77/Belgian AF	8-96

SLANGERUP
The proposed Nordsjællands Flyvemuseum here was cancelled. All the stored aircraft have gone to the new Flyvemuseet Ellehammer at Helsingør.

STAUNING
This airfield is the home of the Dansk Veteranflysamlung whose museum displays a large number of aircraft and maintains even more flying aircraft. T-33A DT-884 returned to Karup. KZ IIT OY-FAM (111) was scrapped. KZ VII D-EBTO (149) and OY-AAN (148) were sold, as was Champion OY-ALA (7AC-1024), Chipmunk OY-ALZ (C1/0067, ex P-121) and Maule OY-ANZ (6027C). KZ VII OY-AVR (176) has also gone. Note that OY-ATK and OY-AZZ are on loan from the Foreningen af Nuværende og tidligere Artilleriflyvere from Billund. KZ IIT OY-FAN is still in the museum, but will move back to the owner at Svendborg. Until that time the wings of this aircraft are fitted to 11-105.

❑ A-009		F-35	351009	preserved, arr 27-4-93, l/f 18-12-91	5-97
❑ A-057	KP-X	F-84G	..	preserved, ex 52-3057	5-97

DENMARK - 61

☐ A-515	F-84G	..	preserved, cockpit only, ex 51-10515		5-97
☐ O-622	KZ VII	184	preserved, ex Billund		5-97
☐ R-888	CF-104	1188	preserved, ex Ålborg, ex 104888/CAF		8-97
☐ OY-AAU	KZ VII	158	preserved		5-97
☐ OY-ABT	J-3F-50 Cub	2475	preserved		5-97
☐ OY-ACE	KZ Ellehammer	204	preserved, replica		5-97
☐ OY-ADM	KZ IIT	113	preserved, ex 11-105, with parts of 119 and wing of OY-FAN		5-97
☐ OY-AEA	KZ IIK	27	preserved, ex OH-KZT		5-97
☐ OY-AKM	PA16	16-101	preserved, ex D-EKOW		5-97
☐ OY-ALF	J-3C-65 Cub	12591	preserved, marked as '480295', ex D-EGAR		5-97
☐ OY-AMG	Druine Turbulent	278	preserved		5-97
☐ OY-AOL	KZ X-2	205	preserved, ex OY-ACL		5-97
☐ OY-ASX	KZ G1	44	preserved, glider, ex OY-54		5-97
☐ OY-AVZ	Bü181B-1	273	preserved, ex 25073/RSweAF		5-97
☐ OY-AUX	Grunau Baby IIb	PFG8	preserved, glider		5-97
☐ OY-AXS	Bergfalke II	01	preserved, glider		5-97
☐ OY-AXU	Spatz B	524	preserved, glider		5-97
☐ OY-AZZ	L-18C	18-3165	preserved, unmarked, ex Y-654, ex 66-654		5-97
☐ OY-BLX	H/T-O 2G	5	preserved, glider		5-97
☐ OY-DBC	SV-4B	1204	preserved, ex V-62/Belgian AF		5-97
☐ OY-DEZ	DH.87B	8040	preserved, ex G-AMZO, ex SE-ALD		5-97
☐ OY-DIZ	KZ IV	43	preserved		5-97
☐ OY-DOU	KZ IIS	13	under restoration, ex Sonderborg, ex SE-ANM		5-97
☐ OY-DRR	KZ VIII	203	preserved, ex D-EBIZ		5-97
☐ OY-DSH	Taylorcraft Plus D	228	preserved, ex D-ECOD, ex 'LB381'		5-97
☐ OY-DVZ	PL-12 Airtruk	1238	preserved, ex VH-ETZ		5-97
☐ OY-DZA	KZ III	66	preserved		5-97
☐ OY-ECH	DH.82A	85234	preserved, ex OO-DLA		5-97
☐ OY-FAE	KZ IIT	119	preserved, frame from c/n 113, ex 11-111		11-97
☐ OY-FAI	Hollænder HT.1	2	preserved		5-97
☐ OY-FAN	KZ IIT	110	stored, ex 11-102		5-97
☐ OY-KZI	KZ I	..	preserved, replica		5-97
☐ OY-REX	Bergfalke II	102	preserved, glider, ex D-0040, ex D-0002		5-97
☐ OY-XIT	Doppelraab IV	03	preserved, glider, ex D-6351		5-97
☐ ..	Ellehammer	..	preserved, replica		5-97

The museum also has a number of aircraft under restoration or in store, some of which are not at Stauning.

☐ 16126	83	Harvard IIB	14-426	under restoration, off site, ex RSweAF	12-90
☐ O-624		KZ VII	187	stored, off site	—
☐ 31-306		Harvard IIB	14A-748	stored, off site, parts only, ex FH114	8-94
☐ D-EKOM		Auster J/1	124	stored, frame only, ex G-AFWN	7-95
☐ OY-AKX		Grunau SG-38	..	stored, glider, ex OY-94	7-95
☐ OY-ALW		M28 Mercury 6	6268	stored, ex D-EHAB	7-95
☐ OY-AVJ		Rearwin 9000L	567D	stored, off site, ex Malmö, ex SE-AGB	7-95
☐ OY-AVR		KZ VII	176	stored, ex O-615	97
☐ OY-DHJ		Chipmunk 22	C1/0470	stored, off site, ex G-AMMA	8-96
☐ OY-DRL		PL-12 Airtruk	1135	preserved, temp in museum in Erslev	—
☐ OY-DVP		DH.82A	85506	stored, off site, wreck, ex N38013	7-95
☐ OY-EFV		Stinson 108-2	353	stored, wreck, crashed 6-4-96, ex D-EFAD	97

At the airfield side, the ex Thisted An-2R CCCP-55721 (1G49-13) was broken up by early 1997. The two Cessna's were not seen in 1997 and may have gone.

☐ LN-PBC	Ce208B	208B-0310	stored, wreck, ex OY-TCC, ex N1015E	4-95
☐ OY-ECJ	Ce421B	421B-0508	stored, wreck, ex N69855, D-ICOA ntu	4-95

| DENMARK - 62 |

On the road between the airfield and the museum is a Dove.
❏ OY-BHZ	Dove 8	04270	preserved, ex G-BLIF, ex WB534	5-97

THISTED
The wrecked and stored An-2 CCCP-55721 moved on to Stauning.

TIRSTRUP
Three of the aircraft at this reserve airfield moved to Karup: CF-104s R-757 and R-758 and dual RT-660. The two F-86Ds, the dumped F-361 (173-494) and stored F-451 (173-584), were scrapped by 1995. The T-33 arrived for a new museum, which hopefully will open its gate in the near future. Also here are parts of Liberator GR.VIII KK259 (shot down 9th April 1940) and Blenheim IV R3821 (shot down 13th August 1940). Parts of the latter are also at the terminal at Ålborg, at Billund and in England.
❏ DT-404	T-33A	5689	stored, ex Fredensborg, ex Værløse, ex 51-4404	97
❏ OY-ANO	F24W41A	834	under restoration, ex D-EHIB	4-95

TØNDER
Preserved at the local aero club on the airfield is a Draken.
❏ AR-102	F-35	351102	preserved, arr 19-5-94	6-97

VÆRLØSE
Most of the *Wrecks & Relics* aircraft of this airfield have moved on. F-86D F-421 went to Skrydstrup. T-33A DT-102 and DT-847 moved to Helsingør. Billund received T-33s DT-289, DT-491 and DT-497, TF-100F GT-927, KZ VII O-621 (which is now OY-AVH) and S-55C S-883. T-33A DT-404 went to Fredensborg, DT-571 to Sola, while DT-450 turned up at Karup. Lynx S-196 is now an instructional airframe at Frederikshavn, KZ VII O-620 went to Karup, Catalina L-861 to Helsingør and KZ IIT OY-FAE went to Stauning. Finally, Meteor 44-481 was scrapped in the late 1980s.
❏ 0734	3-H-141	Lynx Mk.23	WA035	stored, unmarked, ex Argentine Navy	5-98
❏ A-010		F-35	351010	instructional, arr 20-2-95, l/f 25-3-92	5-98
❏ P-143		Chipmunk T.20	C1/0878	stored, ex instructional	6-94
❏ RT-654		CF-104D	5324	instructional, ex 12654/RCAF	11-94
❏ 51-10731	KP-R	F-84G	..	instructional, ex FZ-100/Belgian AF	6-94
❏ OY-CFE		Ce150L	0712	instructional, ex D-ECHX	97

VANDEL
Town: The Egnsmuseet has a F-84G. The museum is in the town of Vandel along the Billund-Vejle road.
❏ A-708	F-84G	..	preserved, ex 51-10708	2-98

Airfield: The older *Wrecks & Relics* aircraft have all gone: F-84Gs A-024 and A-043 were scrapped by the late 1980s as were F-86Ds F-026, F-060, F-062, F-096, F-119, F-429, F-469 and F-960. The rebuild of the H500M H-210 was finished and it is flying again. The Starfighter R-812 went to Karup. Ålborg received F-104 R-771 and F-84G 51-9681. Decoy F-84G 51-10651 was scrapped in September 1989. All aircraft currently in store will return to the Air Force now the museum at Billund is closed. Note that the construction number and former serial for R-704 is correct. There was already a R-703 (6055, ex 62-12703) flying in Denmark.
❏ A-017	F-35	351017	stored, ex Billund, l/f 25-5-93	97
❏ AR-105	RF-35	351105	stored, l/f 14-5-92	97
❏ AR-115	RF-35	351115	stored, l/f 28-6-92	97
❏ R-704	CF-104	1003	stored, ex Billund, ex 12703/RCAF	97
❏ R-846	CF-104	1146	stored, ex Billund, ex 104846/CAF	97
❏ 22-265	Meteor T.7	G5-354	stored, ex Billund, ex Karup	97

DENMARK - 63

☐ R-846	CF-104	1146	stored, ex Billund, ex 104846/CAF	97
☐ 22-265	Meteor T.7	G5-354	stored, ex Billund, ex Karup	97

VEJLE
The Danish Transport Centre here has SG-38 Schulgleiter OY-86 on display. The glider came from Billund.

VIDEBÆK
Under restoration at a local technical school, in the Opsund part of Videbæk, is a Meteor. After completion the aircraft will go to the museum at Stauning.

☐ B-499	Meteor F.8	G5-373	under restoration, ex Billund, ex Karup	5-97

VOJENS
Preserved in a playground in Ostergarde Street (east of the town centre) next to a church is an F-84G.

☐ A-525	F-84G	..	preserved, ex 51-10525	7-97

FINLAND

The *Wrecks & Relics* situation in Finland has remained quite stable over the past ten years. Some of the aircraft have moved location within the country and only one new collection, at Kormu, has been established. The military have withdrawn some of their Drakens and all MiG-21s from service and some of these can now be found at different locations.

ÄHTÄRI
CM170 FM-71 was displayed here in the Mini Suomi park (little Finland park), Ähtäri is some 100km north of Tampere. The park itself is some 7km east of Ähtäri. The Draken DK-206 replaced CM170 FM-71 which went to Halli.

| ❏ DK-206 | SAAB 35BS | 35266 | preserved, marked as 'DK-206A', ex Halli | 6-97 |

HALLI
Draken DK-206 (35266) was damaged during a ground fire on 18th January 1974. The aircraft moved here as 'DK-942' (942 was the Halli telephone area code) as an instructional airframe. Later reserialled as 'DK-247' for display duties and finally moving to Ähtäri as 'DK-206A'. Other instructional Drakens leaving here were DK-202 to Tampere and DK-212 to Rovaniemi. The former Ähtäri CM170 FM-71 was last noted here in June 1994 and is now at Vesivehmaa. CM170 FM-43 and MiG-21F-13 MG-78 moved to Kymi. Airworthy Fw44J OH-SZO went to Tikkakoski.

❏ 35252	J35B	35252	preserved, ex RSweAF	6-96
❏ DK-208	SAAB 35BS	35214	instructional, ex 35214/RSweAF	6-97
❏ DK-261	SAAB 35F	35460	instructional, ex 35460/RSweAF	6-96
❏ MG-34	MiG-21F-13	0404	instructional, marked as 'MG-124'	6-96
❏ MG-121	MiG-21bis	N75098132?	stored	6-97
❏ MG-123	MiG-21bis	N75098140	stored	6-96
❏ MK-143	MiG-21UM	516999411	stored	6-95

Of the long list of stored Magisters in the first edition a few more fates are known. FM-14 and FM-45 turned up at Tikkakoski, while FM-67 is now at Kauhava. Still preserved on a town square is a Folland Gnat.

| ❏ GN-103 | Gnat F.1 | FL16 | preserved | 6-97 |

The Hallinportti Ilmailumuseo at the airfield is open during June, July and August between 13:00 and 17:00, during weekends the opening times are between 12:00 and 18:00. The IVL C24 8F.4 moved to Helsinki.

❏ AE-47	Aero A-11	..	preserved, ex Vesivehmaa	6-97
❏ BU-59	Bulldog IVA	7810	preserved	6-97
❏ HA-41	Haukka II	2	preserved	6-97
❏ LK-1	VL Sääski II	2	preserved	6-97
❏ MU-1	MiG-15UTI	922221	preserved, outside	6-97
❏ 1E.18	Caudron G.3	..	preserved	6-97
❏ 5A.1	Rumpler 6B	..	preserved	6-97
❏ OH-VKK	Karhu 48	5	preserved, fuselage frame only	..

HELSINKI
Town: At the south side of the Helsinki harbour, the Sotamuseo is set up in an old fortress on an island. The C.24 was actually built at the island.

| ❏ 8F.4 | IVL C.24 | 1 | preserved, ex Halli | 7-90 |

Noted in another place in the harbour near Jätkäsaari was a former Soviet MiG-23MLD. This MiG moved on to the airfield at Malmi.

FINLAND - 65

Airfield - Malmi: The ex Soviet MiG-23 has been decorated with 21 Afghan mission markings. It has no wings. Tilli OH-XTL moved to <u>Vantaa</u>. The spares source C-47 DO-8 (19309), which keeps DC-3A OH-LCH flying, is still stored at the airfield.

❏ 35 red	MiG-23MLD	0390320549	preserved, ex Soviet	6-97
❏ DO-8	C-47A	19309	stored, ex OH-LCD, ex 42-100846	8-95
❏ OH-EFC	Aero 45	03-007	stored, ex OY-EFC	6-95
❏ OH-BLK	Be65-B80	LD-298	stored, ex SE-EUU	7-90

Airfield – Vantaa: The Suomen Ilmailumuseo is still at the airport (some 20km north of Helsinki) and has expanded its collection slowly. The museum is open daily between 12:00 and 18:00. The fuselage of the stored Bandierante OH-EBA (which had nothing to do with the museum) is now at a scrapyard at <u>Riihimäki</u>.

❏ BL-180		Blenheim V	V/20	stored, rear fuselage only	6-97
❏ CA-84		Caudron C.60	24	stored, frame only	6-97
❏ DK-206		SAAB 35BS	35245	preserved, outside, ex 35245/RSweAF	6-97
❏ DK-262		SAAB 35CS	35823	expected	—
❏ DO-3		DC-2-115	1562	preserved, fuselage only, ex OH-LDB	6-97
❏ FM-42	X	CM170	FM-42	preserved	6-97
❏ GA-58		Gamecock II	15	preserved, rear fuselage only	6-97
❏ GN-105		Gnat F.1	FL23	stored	6-97
❏ GN-106		Gnat F.1	FL28	preserved	6-97
❏ HK-1		SM-1Sz	A07029	preserved	6-97
❏ HR-3		Mi-4	09114	preserved	6-97
❏ HS-1		Mi-8T	13301	preserved, outside	6-97
❏ IL-2		IVL A.22 Hansa	2	preserved, ex 4D.2	6-97
❏ MG-111		MiG-21bis	N75064540	preserved, outside	6-97
❏ MK-105		MiG-21UM	..	expected	—
❏ PY-5		VL Pyry II	I/4	stored, fuselage only	6-97
❏ PY-16		VL Pyry II	15	stored	6-97
❏ PY-27		VL Pyry II	26	preserved	6-97
❏ PY-30		VL Pyry II	29	stored	6-97
❏ SÄ-122		VL Sääski II	II/5	preserved, ex 'SÄ-131'	6-97
❏ SF-9		SAAB 91D	91355	preserved	6-97
❏ SG-1		Grunau SG-38	79	preserved, glider	3-97
❏ TU-169		VL Tuisku	20	stored, frame	97
❏ TU-178		VL Tuisku	29	preserved	6-97
❏ UT-1		UTI-4	..	preserved, dual Polikarpov I-16, ex VH-22	6-97
❏ VA-2		Vampire FB.52	V0692	preserved	6-97
❏ VH-25		VL Vihuri II	..	stored, forward fuselage only	97
❏ VT-9		Vampire T.55	15720/VT0699	preserved	6-97
❏ G-36		Grunau ESG-9	22	preserved, glider	3-97
❏ H-5		Harakka I	5	preserved, glider	10-95
❏ H-56		Harakka II	24	preserved, glider	3-97
❏ OH-ABB		Ju A50ce	3530	preserved, inside terminal, ex D-1915	3-97
❏ OH-BAA		Grunau Baby IIb	12/43	preserved, glider	8-95
❏ OH-BAR		Grunau Baby IIb	27	preserved, glider	8-95
❏ OH-BBA		Be95-A55	TC-261	preserved, ex OY-DPK, ex OH-APU	3-97
❏ OH-CBQ		CeF150J	0526	preserved	—
❏ OH-CMB		Zlin 37	04-09	stored, wreck	6-95
❏ OH-FSA		Fi 156K-1	4230	preserved, ex OH-VSF, ex ST-112	6-97
❏ OH-HIA		Bell 47D-1	646	preserved	3-97
❏ OH-HKA		Heinonen HK-1	1	preserved	8-95
❏ OH-KCC		SZD-10bis	W-51	preserved, glider	8-95
❏ OH-KLA		Klemm L-25	137	stored, ex OH-ABA	6-95
❏ OH-LRB		Convair 440	73	preserved	3-97
❏ OH-MVL		DHC-2	141	preserved, ex Border Police	6-97

FINLAND - 66

❏ OH-OAA	Meise	1/45	preserved, ex OH-134		8-95
❏ OH-PJN	PA28R-180	28R-30885	preserved, SE-FDY ntu		8-95
❏ OH-PXA	Pik-10	1	preserved, glider		8-95
❏ OH-SAA	WWS-1	147	preserved, glider, ex OH-PIK6		8-95
❏ OH-SME	Letov S-218A	26/VL	preserved, marked as 'SM-153', ex SM-162		6-97
❏ OH-TEA	Eklund TE-1b	1	preserved		8-95
❏ OH-VII	VL Viima II	VI-21	preserved, ex VI-21		8-95
❏ OH-VKB	DC-3A	1975	preserved, ex SE-BAC		3-97
❏ OH-VKL	Karhu 48B	6	preserved, ex OH-KUA		8-95
❏ OH-VKU	Lockheed L18-07	18-2006	preserved, ex F-ARTF		8-95
❏ OH-WAC	WA54	138	preserved		97
❏ OH-YKA	Pik-3a	1	preserved, glider, ex OH-PCA		8-95
❏ OH-YMA	Pik-11	1	preserved, glider		8-95
❏ OH-XQA	Quickie 1	1049	preserved		97
❏ OH-XTL	VL Tuuli TL-III	1	stored, ex TL-1		6-97
❏ OH-XYY	Kokkola Ko-04	01	preserved, autogiro		8-95
❏ OH-152	Pik-5b	16	preserved, ex OH-PAR and marked as such		8-95
❏ OH-177	SZD-9bis	P-284	preserved, glider, ex OH-KBZ		8-95
❏ OH-201	Pik-3c	3	preserved, glider, ex OH-YKY		8-95
❏ OH-318	Pik-12	3	preserved, ex OH-KYC and marked as such		8-95
❏ OH-355	Fibera KK-1e	13	preserved, glider, ex OH-LKE		8-95
❏ OH-368	Fibera KK-1e	21	preserved, glider, ex OH-LKI		8-95
❏ OH-425X	Pik-20	001	preserved, glider		8-95
❏ OH-450X	Bryan HP-16	01	preserved, glider, fuselage only		8-95
❏ OH-571	Pik-16c	38	preserved, glider, airworthy		8-95
❏ ..	Adaridi	1	preserved		8-95
❏ ..	Blomqvist & Nyberg	1	stored		8-95
❏ ..	HM-14	..	preserved, replica, marked as 'OH-BFA'		—
❏ ..	Kassel 12A	1	stored, at Vesivehmaa (?), marked as '13'		7-91
❏ ..	L-13N	..	preserved, glider, marked as 'OH-VLK'		8-95

HYRYLÄ
The Ilmatorjuntamuseo, or in plain English the Anti Aircraft Museum was opened here in 1969. Hyrylä is a few kilometres north of Helsinki Vantaa.

❏ FM-50	CM170	FM-50	preserved, outside	6-97

IMATRA
The Rajamuseo here has been open since 1993 and traces the history of the Border Police. One of the force's Beavers was on display here (OH-MVM), but has moved to <u>Tikkakoski</u> for storage.

KARSTULA
Stored in a local industrial estate near the Teras-Astra AB company is a stripped Lim-5, which came from the scrapyard at Riihimäki.

❏ 1709	Lim-5	1C-1709	stored, ex Kormu, ex Polish AF	6-97

KAUHAVA
Kauhava is the Finnish Air Force main training base. Some of the former equipment of the KoulLLv is still present on the air base. The rest of CM170 FM-67 should be at Paimio.

❏ FM-21	K	CM170	225	preserved, at gate	7-97
❏ FM-67		CM170	FM-67	instructional, cockpit only, ex Halli	6-95

FINLAND - 67

❏ FM-82	M	CM170	FM-82	preserved, at gate	7-97
❏ SF-7		SAAB 91D	91353	instructional	6-97

KORMU
The Kormui Romuttamo scrapyard near Riihimäki has dealt with a lot of aircraft in the past. At the yard are also unidentified parts of a DC-3, DC-8, AB206, Yak-52s and Yak-55s. Robinson R22 OH-HAI moved from here to a bar (Fat Larrys on the Kirkkokatu 10) in Tampere. Lim-5 1709 moved on to Karstula. Recently the owner has decided to build a museum with the aircraft. All aircraft seen in June 1997 (and hopefully some other new ones) will be part of this new museum. The museum carries the name Sotilas ja Lentoteknikkan Museo and hopes to open its gates to the public in the near future.

❏ GN-113	Gnat FR.1	FL24	stored, ex Tikkakoski, ex XN326	6-97
❏ FM-14	CM170	246	stored, ex Tikkakoski, ex Halli	—
❏ FM-61	CM170	FM-61	stored, composite	6-97
❏ SF-10	SAAB 91D	91356	stored, crashed 16-6-72	6-97
❏ CCCP-15667	Mi-2	530520097	stored, ex Aeroflot	92
❏ CCCP-15687	Mi-2	510547127	stored, ex Aeroflot	6-97
❏ CCCP-25267	Mi-8	..	stored, ex Aeroflot	92
❏ N26RT	SA226T	T-216	dumped, crashed 24-2-89	92
❏ OH-CAR	Ce500	0144	dumped, crashed 19-11-87	92
❏ OH-COS	CeF172M	0909	dumped, ex LN-BWX	—
❏ OH-CSI	CeF172F	0105	dumped	92
❏ OH-EBA	EMB110P1	110226	dumped, ex Helsinki	92
❏ OH-HPY	Enstrom F.28A	136	dumped	6-95
❏ OH-SSB	SC7-3-301	1838	dumped, crashed 1-11-89	92
❏ RA-19536	Ka-26	7404610	stored	6-97

KUOPIO RISSALA
At the civil side of the military MiG-21 and Hawk base is one of the old MiG-21Fs. Restored Fokker D.XXI FR-110 moved to the museum at Tikkakoski.

❏ MG-61	MiG-21F-13	1117	preserved, at civil side	7-97

KYMI
The local airfield houses four former Air Force jets. All are preserved outside around the local aero club, while the gliders are all inside.

❏ DK-259	SAAB 35F	35499	preserved, ex 35499/RSweAF	9-97
❏ FM-43	CM170	FM-43	preserved, ex Halli	9-97
❏ GN-107	Gnat F.1	FL31	preserved	9-97
❏ MG-78	MiG-21F-13	1205	preserved, ex Halli	9-97
❏ MG-116	MiG-21bis	N75083895	expected	—
❏ H-12	Harakka I	11	preserved, glider	—
❏ H-34	Harakka III	1	preserved, glider	—
❏ H-57	Harakka II	25	preserved, glider	—
❏ OH-348	K8b	14	preserved, glider, ex OH-RTU	—

PAIMIO
Some 23km east of Turku is the town of Paimio. Near the Shell Hevonpää company (not in Hevonpää itself) the car and motorbike museum called the Paimion Paroni Automuseo can be found. The museum has a Polish MiG-17 and should also have a Magister.

❏ 1505	Lim-5	1C-1505	preserved, ex Polish AF	6-97
❏ FM-67	CM170	FM67	preserved, with parts of FM-61	—

FINLAND - 68

PORI
The technical college in this town has a former Tikkakoski museum SAAB Safir.
| ☐ SF-3 | SAAB 91D | 91349 | stored, ex Tikkakoski | 97 |

ROVANIEMI
HävLLv 11 is based at Rovaniemi and flies SAAB 35s and Hawks. The gate here is guarded by a Gnat.
☐ DK-200	SAAB 35A	35026	preserved, ex 35026/RSweAF	6-97
☐ DK-212	SAAB 35BS	35257	stored, ex Halli, ex 35257/RSweAF	6-95
☐ GN-110	Gnat F.1	FL44	preserved, at gate	6-97
☐ SF-36	SAAB 91B	91246	instructional, crashed 6-3-56, ex OH-SFA	6-97

TAMPERE
Town: In the town of Tampere were many factories with the name Valmet, but outside the main Valmet factory (nowadays called Finavitec) in the west side of the town is a Viima. The name of the road alongside the factory is probably Tarmontaku.
| ☐ VI-1 | VL Viima | VI-1 | preserved, in a glass pagoda | 7-97 |

The Tampereen Teknillinen Museo was closed in 1989 after damage by a fire. All the aircraft have moved to Tikkakoski, except for the Letov, the fate of which is still unknown.
| ☐ SM-141 | Letov S-21B | 5/VL | under restoration | — |

Airfield - Pirkkala: At the military airfield an unmarked MiG-25 is preserved. It is ex Soviet, but that's all that is known about the aircraft. Draken DK-202 is probably a travelling recruiting exhibit as it was noted at Halli in 1994 and 1995, at Turku in April 1996 and at the airshow at Pirkkala in June 1996
☐ ..	MiG-25RBS	02050740	preserved, ex Soviet	7-97
☐ DK-202	SAAB 35BS	35265	stored, ex Halli, ex Turku, ex 35265/RSweAF	6-96
☐ ..	SAAB 35XS	..	dump	6-97
☐ PA-12	PA-28RT-201	8018091	dump, crashed 1-3-96	6-97

TEMMES
Temmes is some 50km south of Oulu. Preserved outside an ice cream shop in an ex Aeroflot An-2.
| ☐ CCCP-70133 | An-2TP | 1G137-07 | preserved | 6-95 |

TIKKAKOSKI
North of Jyväskylä is the airfield of Tikkakoski. To the south of the airfield is the Keski-Suomen Ilmailumuseo. The museum is open between 10:00 and 20:00 (1st June to 15th August) and 11:00 and 17:00 (16th August to 31st May).
☐ AV-57		Avro 504K	..	preserved, ex 1H.49, ex G-EBNU, ex E448	1-98
☐ DO-4		C-47A	14070/25515	preserved, ex OH-LCF, ex 43-48254	1-98
☐ FM-45	A	CM170	FM-45	preserved, ex Halli	1-98
☐ FR-110	7	Fokker D.XXI	III/11	preserved, ex Kuopio Rissala	1-98
☐ GL-12		Gourdou Lesseure GL22B3	60	preserved, ex 8F.12 and marked as such	1-98
☐ GN-101		Gnat F.1	FL8	preserved, ex G-39-6	1-98
☐ GN-104		Gnat F.1	FL19	preserved, outside on pole	1-98
☐ HM-671		VL Humu	..	preserved	1-98
☐ HR-1		Mi-4	07114	preserved	1-98
☐ MA-24		Martinsyde F.4	D4326	preserved	1-98
☐ MG-92		MiG-21F-13	1721	preserved	1-98
☐ MK-103		MiG-21U	1416	preserved	1-98
☐ MS-52		Morare MS50C	..	preserved, ex 2G.7	1-98

FINLAND - 69

☐ MT-507	O	Bf109G-6Y	167271	preserved	1-98
☐ MU-4		MiG-15UTI	722375	preserved	1-98
☐ NH-4		Il-28R	1106	preserved	1-98
☐ PM-1		VL Pyörremyrsky	1	preserved	1-98
☐ SF-2		SAAB 91D	91348	preserved	1-98
☐ SF-8		SAAB 91D	91354	preserved, forward fuselage only	1-98
☐ SZ-4		Fw44J	2895	preserved, ex Halli, ex SZ-4, ex OH-SZO	1-98
☐ VH-18		VL Vihuri II	II/17	preserved	1-98
☐ VT-8		Vampire T.55	15719	preserved	1-98
☐ 1		Polikarpov U-2	11429	preserved, Soviet colours	1-98
☐ OH-EJA		DH.60X	..	preserved, ex OH-MAH, ex OH-ILC, ex MO-105	1-98
☐ OH-HRC		Mi-1M	1811	preserved, ex Frontier Guard	1-98
☐ ..		Thulin D	..	preserved, replica, marked as 'F1'	1-98
☐ ..		HM-14	..	preserved, incomplete	1-98
☐ ..		Päätalo Tiira	..	preserved	1-98

In a nearby shed, usually not open to the public, is the museum's store. Two aircraft are known to have left, Gnat GN-113 and CM170 FM-14 both are now at Kormu.

☐ 26		P-39Q	..	preserved, ex Soviet, ex 44-2664	1-98
☐ 7108	NE-ML	Bf109F	7108	stored, aft fuselage, wings and engine	1-98
☐ ..		Blenheim I	..	stored, cockpit only	98
☐ ..		Blenheim I	..	stored, cockpit only	98
☐ BL-200		Blenheim IV	VI/3	stored, marked as 'BL-147'	98
☐ DO-1		DC-2-115E	1354	stored, ex DC-1, ex SE-AKE, ex PH-AKH	98
☐ FK-113		Fokker C.X	..	stored, fuselage frame, ex Soesterberg	98
☐ ..		Fokker C.X	..	stored, derelict fuselage frame	98
☐ HC-452		Hurricane I	41H/11096	stored, ex HU-452, ex N2394/RAF	98
☐ HK-2		SM-1/600Sz	A07030	stored, ex Tampere	98
☐ MY-5		Valmet Myrsky II	II/4	stored, frame only	98
☐ MY-9		Valmet Myrsky II	II/13	stored, frame only	98
☐ MY-10		Valmet Myrsky II	II/2	stored, frame only	98
☐ PR-2		Pembroke C.53	P66/70	stored	98
☐ PY-1		VL Pyry I	1	stored, ex Tampere	98
☐ PY-35		VL Pyry II	I/34	stored	98
☐ SÄ-95		VL Sääski I	1	stored, frame only, ex Tampere, ex K-SASA	98
☐ SZ-35		Fw44J	2928	stored	98
☐ VA-6		Vampire FB.52	VO696	stored	98
☐ VT-6 ?		Vampire T.55	..	stored, cockpit only	98
☐ 8E.3		Fokker D.X	..	stored, frame only, ex Tampere	98
☐ OH-BAD		Grunau Baby II	15/36	stored, glider	98
☐ OH-CNH		CeF172H	0519	stored	98
☐ OH-ILA		DH.60X	447	stored, ex Tampere, ex K-SILA	98
☐ OH-KAA		HM-14	1	stored, ex Tampere	98
☐ OH-MVM		DHC-2	790	stored, ex Imatra	98
☐ OH-NOA		UC-64A	646	stored, fuselage only, ex HB-UIK, ex 44-70381	98
☐ OH-OAB		Meise	..	stored, glider	98
☐ OH-PAX		Pik-5b	21	stored, glide	98
☐ OH-PBA		Pik-5c	23	stored, glider	98
☐ OH-SFB		SAAB 91D	91440	stored, ex SF-31	98
☐ OH-SFD		SAAB 91D	91443	stored, ex SF-34	98
☐ OH-SFE		SAAB 91D	91444	stored, ex SF-35	98
☐ OH-VTP		Valmet L.90	001	stored, aft fuselage	98
☐ OH-WAB		Weihe	201	stored, glider, ex OH-JÄMI3, ex OH-135	98
☐ OH-199		Pik-3b	5	stored, glider, ex OH-YKE	98
☐ OH-209		SZD-10bis	..	stored, glider, ex OH-KCD	98

FINLAND - 70

☐ OH-258	SZD-22c	..	stored, glider	98
☐ ..	Arado 66	..	stored, burned wreck, fuselage frame only	98
☐ ..	Harakka H-17	..	stored, glider	98
☐ ..	Harakka H-54	..	stored, glider	98
☐ ..	Valmet LEKO 70	..	stored, prototype	98
☐ ..	MiG-3	..	stored, rear fuselage only	98
☐ ..	Il-2M3	..	stored, cockpit only	98
☐ ..	Tu SB-2bis	..	stored, rear fuselage only	98
☐ ..	VL Tuisku	..	stored, derelict, fuselage frame	98

During 1997 two MiG-21F-13s were noted in store. They wore the same colours as the MiG-21bis' and even had fake pitot tubes installed. The airfield is also known as Jyväskylä Luonetjärvi.

☐ ..	MiG-21F-13	..	stored, decoy	7-97
☐ ..	MiG-21F-13	..	stored, decoy	7-97

UTTI
The C-47 is still in use as para trainer. The preserved Messerschmitt has been placed under a glass pagoda.

☐ DO-5	C-47A	19795	instructional, fuselage only, ex 43-15329	6-97
☐ HS-2	Mi-8T	13302	instructional	6-95
☐ HS-12	Mi-8T	13308	stored, crashed 13-4-82	6-97
☐ MG-35	MiG-21F-13	0405	stored, in hangar	6-94
☐ MT-452	Bf109G-6	165277	preserved, at gate	6-97

UUSIKAUPUNKI
SAAB SF-5 may still be displayed at the local SAAB car factory here. Uusikaupunki is some 50km north west of Turku. Also at Uusikaupunki is a car museum which should have Pik-3b glider OH-175 on display.

☐ SF-5	SAAB 91D	91351	preserved	90

VESIVEHMAA
The Finnish Aviation Museum Society has a storage hangar here which can be visited during the weekends between 1st May and 31st October (between 10:00 and 18:00, admission free). The MiG-15, Gnat and CM170 are on display, the remainder are stored or under restoration. Stored Aero A11 AE-47 moved on during the mid 1990s to Halli, while Caudron CA-84 went to Vantaa.

☐ AEj-59	Aero A-32	..	stored	6-97
☐ CA-50	Caudron C.59	..	stored, ex 2E.5	6-97
☐ CA-556	Caudron C.714	8583/6	stored	6-97
☐ FO-75	Fokker C.VE	..	stored, fuselage only	6-97
☐ FM-71	CM170	FM-71	stored, ex Halli, ex Ähtäri	6-97
☐ GN-112	Gnat FR.1	FL47	stored	6-97
☐ HA-39	Haukka I	..	stored	6-97
☐ KA-147	VL Kotka II	I/3	stored	6-97
☐ MU-2	MiG-15UTI	822028	stored	6-97
☐ PY-26	VL Pyry II	I/27	stored, fuselage only	6-97
☐ RI-140	Blackburn Ripon IIF	II/12	stored	6-97
☐ SZ-35	Fw44J	..	stored, fuselage only	5-93
☐ 3C.30	Breguet 14.A.2	..	stored	6-97
☐ ..	Gamecock	..	stored, fuselage only	6-97
☐ ..	IVL K.1 Kurki	..	stored, marked as '9205'	6-97
☐ ..	Tu ANT-40	..	stored, fuselage only as '16/250', ex Soviet AF	6-97

FRANCE

Since the early 1990s, the French military disposed of a large number of the older aircraft from their inventory. Operational flying with aircraft like Mirage 3s, Mirage 5s, CM170s, CM175s and Br1150s stopped. Flying with types like the Jaguar, Mirage 4 and MS760 has also been reduced over the past years. The base at Châteaudun has played a key role in the disposal of these types, nearly 400 aircraft are known to have ended their operational life here. Large numbers of aircraft from here have passed on to collections and museums, while others were brought back to flying status with civil owners. France has always been a country with a large aviation heritage. Many of the civil airfields still have their own preservation or historic group, ranging from a single preserved aircraft to a large number of flying vintage and 'warbird' aircraft. The number in brackets after the location name are those of the France departementes.

ABBEVILLE - DRUCAT (80)
The Aero Club de la Somme's derelict MD450 Ouragan 215 continued in residence at the airfield until late 1990. By early 1991 she had gone to <u>Savigny lès Beaune</u> in exchange for Mystère 4A 287 which is painted up as '316/8-NM'. The Mystère was in place by June 1991.

❏ 287		Mystère 4A	287	preserved, marked as '316/8-NM', ex Savigny	6-97

AIRE SUR L'ADOUR (40)
The Magister reported in the previous edition of EWR-1 is not with the CNES, but with Potez Aéronautique at the local airfield. Its old code 11-OF is just visible.

❏ 167	11-OF	CM170	167	preserved	6-97

AIX - LES MILLES (13)
The gate of the military side of the airfield still has the real Mystère 4A 289, while preserved at offices is a unknown Magister in Patrouille de France colours. The aircraft from the fire dump, H-34A SA57, MD312 238 and N2501F 87, had all been scrapped in the late 1980s. Magister 454 arrived here from Châteaudun, but has moved on to <u>Ambérieu</u>.

❏ ..		CM170	..	preserved, in PdF colours	7-95
❏ 289	2-EY	Mystère 4A	289	preserved, at gate	5-98

Of the large Escadrille Pegase collection only a few aircraft have been noted here over the last few years. Aircraft known to have departed are F-104G FX-93 (ex Koksijde) which went to <u>Orange</u>, Super Mystère B2 46 moved on to <u>Brive</u>. CM170 45 is now civil as G-FUGA, T-33 21121 went to <u>Cuers</u>, Flamant 250/F-AZEN is now at <u>Papos</u>, and the nose section of the Canso F-ZBAR went to <u>Rochefort</u>. Stored N2501F 105/62-SJ flew again as F-AZVM. Made its first civil flight on 20th May 1995.

❏ 197		N1101	197	stored, at La Mole ?	—
❏ 338	30-MN	Vautour 2N	338	stored	—
❏ F-AZNA		N2504	01	stored, civil side	5-98

ALBERT (80)
A company manufacturing aircraft components has set up a small collection. The building is easy to find as the Flamant is parked outside and a 1:1 scale Airbus A320 is painted on the outside wall.

❏ IB427		Vampire Mk.52	..	stored, ex Brienne le Château, ex Indian AF	3-98
❏ 275	316-KC	MD311	275	preserved, ex L'Aigle	3-98
❏ 20		N3400	20	stored	3-98
❏ ..		N1101	109 ?	stored, with wings from F-BLQY	3-98
❏ ..		NC865	..	stored	3-98
❏ F-GIFZ		MH1521M	315	stored	3-98

FRANCE - 72

ALBERTVILLE (73)
Stored here in 1995 was an ex ALAT Alouette 3 is good condition It was only missing its rotor and tailboom. This SA316B 1088 should have become F-GKHM by now.

ALÈS - DEAUX (30)
The Magister came from Montpellier airport. It may have returned there.

❏ 126		CM170	126	preserved, ex F-WZLQ, ex Montpellier	5-94

AMBÉRIEU EN BUGEY (01)
T-6 114688 which has been at the gate since 1982 has recently been joined by a CM170 and a Super Mystère. Outside the base at the roundabout is a Mirage 3R painted as '356/33-TE'. Its correct identity is unknown as the real 356 is still at Rochefort. Breguet 941 3 (l/n June 1989) went to Aubenas.

❏ 454		CM170	454	preserved, ex Aix Les Milles, ex Châteaudun	6-96
❏ ..		Mirage 3RD	..	preserved, outside gate, marked as '356/33-TE'	8-96
❏ 240	12-YP	Mirage F1C-200	..	instructional, cockpit only, ex Châteaudun	6-96
❏ 136	12-YH	Super Mystère B2	136	preserved	8-96
❏ 114688	RC	T-6G	182-373	preserved, ex 51-14688	8-96

ANGERS (49)
Town: In a barracks (Caserne Berthezene) in the town, the Armée de Terre's 21 Regiment du Genie have a Vertol H-21C preserved. It has the legend 'Gaubourgs' on the fuselage sides and was still current during November 1988.

❏ ..		H-21C	..	preserved	11-88

At the cross roads of the RN32 and N260 on the south west side of Angers is the Ecole la Baronnerie (formerly the Ecole Technique Privée St Julian). Outside is a CM170 on a pole.

❏ 78		CM170	78	preserved, on pole, ex Châteaudun	6-97

Airfield - Avrillé: The Musée de l'Air Regional of the Ailes Anciennes Anjou have two hangars at this airfield in which preserved and flyable aircraft are on view. In 1997 only two 'military' aircraft were recorded. The group has some two dozen aircraft. This airfield will close in the near future, a new one along the D766 near Narce is being built.

❏ 335		CM170	335	preserved	6-97
❏ 184	328-EO	N2501F	184	preserved	6-97

ANNECY - MEYTHET (74)
The Museum des Deux Guerres at the local airfield has not really grown over the last ten years. H-34A SA92 moved on to an unknown location and a new arrival here is a former Air France B727.

❏ 85		CM170	85	preserved	4-98
❏ F-BPJK		B727-228	20202	preserved	4-98

ANNEMASSE (74)
At the small airfield of Annemasse a Broussard is preserved. The town of Annemasse is along the border with Switzerland, a few kilometers from Geneva.

❏ 316	315-SN	MH1521M	316	preserved, ex F-WGKR	3-95

APT - ST CHRISTOL (84)
The gate at Base Aérienne 200 should still be guarded by H-34 SA84, while H-34A SA72 has not been reported

| FRANCE - 73 |

here since 1983 and has been deleted.
| ❏ SA84 | 67-OB | H-34A | SA84 | preserved, at gate | — |

AUBENAS (07)
In 1991 an amusement park (Aéro City Parc) was built here next to the airfield and by 1994 three aircraft and a plastic AS565 were noted in the park. The CM170 is a more recent addition and comes from the local aero club.
❏ 98		Alouette 2	1134	preserved	8-97
❏ 3		Br941S	03	preserved, marked as '305', ex Ambérieu	8-97
❏ 124		CM170	124	preserved	8-97
❏ F-BTGL		Be65-A80	LD-227	preserved, ex D-ILKE	8-97
❏ ..		AS565	..	preserved, mock-up	8-97

AULNAY SOUS BOIS (93)
The Parc des Sports de la Rose des Vents had N2501 50 on display. This machine returned to Le Bourget.

AVIGNON - CAUMONT (84)
Preserved by the tower is a Caravelle in Air France colours. The registration was modified to the F-P... homebuilt range on the occasion of a regional RSA rally. Preserved MH1521 124 has become F-GRES.
| ❏ F-BOHA | | SE210-3 | 242 | preserved, near tower, marked as 'F-POHA' | 6-96 |

AVORD (18)
Inside the gate is MD312 229, while preserved nearby is a Mirage 4. The Cap 10s are left overs from the disbanded EFIPN.307 and all may have gone by now. Gone for sure are stored Cap 10s 27 (to F-GKAJ) and 116 (to F-GRRA). In 1989 Ouragan 31 was rediscovered still in use with the fire section for rescue training.
❏ 2	VT	Cap 10	2	instructional, with wings from c/n 19	5-92
❏ 20 ?		Cap 10	20	stored	5-96
❏ 119		Cap 10B	119	stored	5-96
❏ 126	307-SL	Cap 10B	126	stored, fuselage only in hangar	5-92
❏ 129		Cap 10B	129	stored	5-96
❏ 130		Cap 10B	130	stored	5-96
❏ 146	319-GE	MD312	146	preserved	3-98
❏ 229	319-DW	MD312	229	preserved, near main gate	3-98
❏ 31	314-TA	MD450	31	instructional, at fire station	9-89
❏ 29	BB	Mirage 4A	29	preserved, near main gate, in silver c/s	3-98

BARCELONNETTE (04)
Barcelonnette is located close to the Italian border. On the airfield, along the D900, a Flamant is preserved
| ❏ 200 | 13-TA | MD312 | 200 | preserved | 6-97 |

BAYEUX (14)
The Bayeux Memorial Museum here has the wreckage of a Spitfire on display.
| ❏ NH341 | | Spitfire IXb | .. | preserved, wreck | 96 |

BEAUNE (21)
Cap 10 19/313-SQ which arrived here for display in the prospective Jodel museum, was not noted during 1994 and 1996. The museum itself never materialised.

FRANCE - 74

BEAUVAIS - TILLÉ (60)
Stored here by Mr Blondel should still be a N1101 Noralpha.
❏ 74	LZ	N1101	74	stored	—

BERGERAC - ROUMANIÈRE (24)
The C-53 TZ-AJW was first noted here in storage in 1994. By early 1996 it had flown out as N49AG and was shortly afterwards painted in the colours of the Belgian Air Force as K-36/OT-CWG. Flamant 236 has not been noted for some years and may have been scrapped. It was reported preserved north of the airfield.

BEYNES (78)
The restaurant at this glider field has a former Châteaudun Mirage 3 outside.
❏ 578	3-JM	Mirage 3E	578	preserved, ex Châteaudun	6-97

BEZIERS - VIAS (34)
Stored at the civil airfield in 1991 was Broussard 118. It has become F-GMMH.

BIARRITZ (64)
Mirage 3A 06, which was preserved at the AMD-BA (Dassault) factory, moved to <u>Savigny lès Beaune</u>.

BORDEAUX - MÉRIGNAC (33)
The airfield of Bordeaux can be split up into several parts. Preserved at the military gate are a Mystère 4 and a Mirage 3R (with an incomplete code). The Broussard is parked beside one of the hangars. Mystère 4A 48 moved on to <u>Montélimar</u>.

❏ 346	33-C	Mirage 3R	346	preserved, ex stored	3-98
❏ 306	8-MW	Mystère 4A	306	preserved	3-98
❏ 285		MH1521M	285	stored	3-98
❏ 124	21-LC	Super Mystère B2	124	preserved	87
❏ 636	92-AW	Vautour 2B	636	preserved	7-94
❏ 44-35859	328-EY	A-26C	29138	preserved	7-94

A museum has been set up in a hangar on the airfield. It is currently not open to the public, but if funds become available the collection hopes to move and open to the public. Stored CM170 117 has not been seen since June 1991 and has gone. The nose section of N2501SNB 200 has also gone. In addition the collection has a number of smaller aircraft off site, including an MS733 and N1203.

❏ 1370	EQ	CM170	370	stored, intended for Cameroon AF, ntu	3-98
❏ 40		Etendard 4M	40	stored, ex Hyères	3-98
❏ 204	DG	Mirage 3B	204	stored	3-98
❏ 01		Mirage 2000B	01	stored, inside Dassault area	98
❏ 299		Mystère 4A	299	stored, ex Le Bourget	3-98
❏ 01		Mystère 4N	01	stored	3-98
❏ 202	319-CS	MD312	202	stored	3-98
❏ 228	43-BU	MH1521M	228	stored	3-98
❏ 99	JBW	N3400	99	stored, ex Le Bourget	3-98
❏ 55		NC856	55	stored	3-98
❏ 105		NC856	105	stored	3-98
❏ 234	CQ	SE210-6R	234	stored, ex F-BRGX	3-98
❏ 158	12-YP	Super Mystère B2	158	stored, ex Taverny	3-98
❏ ..		T-6	168-23001?	stored	3-98
❏ 21049		T-33AN	49	stored	3-98

FRANCE - 75

☐ ..	Vautour 2B	..	expected	—	
☐ D-ACVK	SE210-10B1R	176	stored, ex EC-CAE	3-98	
☐ F-AMPA	Mauboussin M.121/35	110	stored	3-98	
☐ F-BTTF	Mercure 100	6	stored, arr 19-9-94	3-98	
☐ F-CBRO	Fauvel AV36	115	stored, glider	3-98	
☐ F-WAQM	..	..	stored, glider	3-98	
☐ F-WMSH	Falcon 20C	1/401	stored	3-98	

Opposite the terminal is a large complex, where companies like SOGERMA and Airbus carry out maintenance on several types, and Dassault building aircraft. Stored here for several years were six CM170s which were destined for Cameroun, but never delivered. Of these, 1370/EQ moved to the museum. The CEV Caravelle was used for Zero-G tests and has been stored here since mid 1996. This aircraft also belongs to the museum now. Its task was taken over by the previously stored A300B2-1C F-BUAD (003). The dumped SE210-10B3 SE-DEB (247) was last noted here in August 1991 and was scrapped. The two stored ex Zaire AF C-130s, 9T-TCC (4588) and 9T-TCF (4589), both last noted in 1995, have been sold to the French Air Force. The Mercure belongs to a local engineering school.

☐ 1330	EL	CM170	330	stored, for Cameroon AF, ntu	3-98
☐ 1341	EM	CM170	341	stored, for Cameroon AF, ntu	3-98
☐ 1347	EN	CM170	347	stored, for Cameroon AF, ntu	3-98
☐ 1351	EO	CM170	351	stored, for Cameroon AF, ntu	3-98
☐ 1371	ER	CM170	371	stored, for Cameroon AF, ntu	3-98
☐ 9T-TCA		C-130H	4411	stored, ex Zaire Air Force, ex 71-1067	3-98
☐ D-AAST		SE210-10B1R	230	stored, ex EC-BIE	8-96
☐ F-BJTP		SE210-3	152	stored, ex preserved, marked as 'F-CCIB'	8-96
☐ F-BTTI		Mercure 100	9	instructional, arr 6-1-95	1-96
☐ PJ-TAC		L100-30	5225	stored	7-95

A quartet of CM170 Magisters were in open store by the SOGERMA hangars by the end of 1985 after having trained foreign pilots. All had gone by the early 1990s; 69/F-ZVLF became F-GELI (later N669FM) and 99/F-ZVLG became N99FR, The fates of 51/F-ZVLE and 102/F-ZVLL are unknown.

BOUILLY (10)

Six Djinns (FR103, FR117, FR128, FR130, FR137 and FR145) were stored in December 1987 at the premises of a former agricultural firm named Phitagri. A search ten years later revealed that the choppers were no longer present. Two earlier examples have left, FR111 and FR132 both going to Jacou.

BOURG EN BRESSE - CEYZÉRIAT (01)

The CM170 preserved at this airfield along the highway still carries the ferry registration F-WFUQ.

☐ 523	CM170	523	preserved, ex F-WFUQ	8-96

BOURGES (18)

The Etablissement Technique du Bourges (ETBS), in the south east part of town, between the D976 and RN76, held a number of dismantled airframes, mostly Super Mystère B2s. Of these, most found new homes, 02 went to Savigny lès Beaune, together with 91 and 118. 73 is now at Orange and 90 at Chartres.

☐ 1220	GTA	Alouette 2	1220	stored	3-98
☐ 01		Super Mystère B2	01	stored, silver c/s	5-92
☐ ..		T-33A	..	stored	9-89

The ALAT facility at Bourges lost Nord 3202 coded MDZ to Montauban by 1989, it was replaced outside the Ecole Superieure du Materiel by an O-1E.

☐ 24523	BGS	O-1E	24523	preserved	5-92

BRÉTIGNY SUR ORGE (91)
The unknown T-33 which was displayed at the mess has not been seen there for a long time and may be the aircraft which is near the Eurocontrol buildings. The fuselage of former Air France Boeing 707-328 F-BHSR (18245), which was on the dump is also long gone. Ce411 6/AD is stored at the main apron, while the other two may have left.

☐ 248		AB	Ce411	0248	stored	10-96
☐ 185		AC	Ce411	0185	stored	10-96
☐ 6		AD	Ce411	0250	stored	4-98
☐ SKY421		68-OF	H-34A	..	preserved, Eurocontrol area	6-95
☐ 02			Mirage 3R	02	preserved, at gate	3-98
☐ 6		AE	Mirage 4A	6	stored	4-97
☐ 116		CE	SE210-3	116	stored, ex OH-LED	5-97
☐ 355			Vautour 2N	355	stored	4-97
☐ ..			T-33A	..	preserved, Eurocontrol area	—

BRIARE (45)
The Musée Automobile de Briare should still have their Bird Dog. The aircraft is parked outside in the yard of the museum.

☐ 24710		O-1E	24710	preserved, pale blue c/s	3-90

BRIENNE LE CHÂTEAU (10)
The Musée Aéronautique de Champagne was set up in the early 1980s in one of the hangars at the airfield. The Flamant is owned by the aero club but is parked near the museum. CM170 31 should also belong to the aero club. Indian Vampire IB427, which was stored off site, moved to <u>Albert</u>. It seems that all aircraft not noted in July 1994 have gone.

☐ 6250		C-45	..	preserved, ex 1383/RCAF	4-90
☐ 7		CM170	7	preserved	7-94
☐ 31		CM170	31	stored, with aero club	6-93
☐ 235		MD312	235	preserved	6-93
☐ NF11-1		Meteor NF.11	..	preserved, Le Bourget owned	7-94
☐ 91	8-OZ	MH1521M	91	preserved	6-89
☐ 28	314-TY	Mystère 4A	28	preserved	7-94
☐ 31	B	N2501F	31	preserved	7-94
☐ 57		NC856N	57	preserved, also quoted as serial 93	87
☐ 147563		SP-2H	..	preserved, Le Bourget owned	7-94
☐ 14115	314-TM	T-33A	5409	preserved, ex 51-4115	6-93

BRIGNOLES (83)
Stored here at the Sécurité Civile premises is an MD311. This can be found on the north side of Brignoles between the B35 towards Bras and the highway A8.

☐ 266	MD312	266	preserved	12-93

BRIVE - LA GAILLARDE (19)
On the airfield, which is at the western side of Brive, ASPAA has set up an collection of former French Air Force aircraft. Super Mystère B2 53/12-YO went to <u>Orange</u> town. Also here are some CM170s and CM175s in civil markings.

☐ 44		CM170	44	preserved, ex Salon	8-95
☐ 16		Etendard 4M	16	preserved, ex Hyères	8-95
☐ 44	3.10-LC	Mirage 3C	44	preserved	8-96
☐ 359	33-TH	Mirage 3RD	359	preserved, ex Châteaudun	2-97

FRANCE - 77

☐ 288		MD311	288	preserved	8-95
☐ 46		Super Mystère B2	46	preserved, nose only, ex Aix	8-96

BRUZ (35)
A visit in 1986 failed to locate the Mirage with the CELAR but it is still thought to be here.

☐ 8	2-FB	Mirage 3C	8	preserved	83

CAEN (14)
Town: At the Second World War Memorial should be a Hawker Typhoon IB. It is uncertain whether this is a replica or perhaps a rebuild based on the wreck recovered in the Bayeux region a few years ago.

Airfield – Carpiquet: The airport entrance had Mystère 2C 52 on display. It is owned by the Musée de l'Air and moved on to Orange. ACE/Transvalair obtained a number of N2501s from Châteaudun. Some have returned to civil service (113 became HC-BPU, 135/F-GEXH became 9Q-CCD, 142/F-GEXS became 9Q-CNE and 169 became 9Q-CKO), while 100/F-WFYF has been scrapped. Others are still stored here. Also based here is Aviation Aviantik with a number of flying warbirds, including an ex Swiss Hunter. At the military side the fuselage of Noratlas 38 is stored between some buildings and can be seen from the rear of the camp.

☐ 38		N2501F	38	stored, fuselage only, ex Châteaudun	2-97
☐ 180		N2501SNB	180	stored, F-WFYH ntu	6-97
☐ F-WECE		N2501F	96	stored	7-96
☐ F-WFYG		N2501F	179	stored, ex F-WGLZ, ex Châteaudun	6-94

CALAIS - MARCK (62)
Preserved at the airfield at the town where the Eurotunnel begins is a former Châteaudun Mirage 3.

☐ 515		Mirage 3E	515	preserved, ex Châteaudun	3-97

CAMBRAI - EPINOY (59)
The puzzle of the Super Mystère with three different serials has been resolved. Long term resident Super Mystère 88 has been repainted as '148/12-YP' and is still at the gate. The real 148 is at Lyon. A second Super Mystère was here in the late 1980s. This was 139 was painted as '177/12-YA' and has since been relocated to Creil. Mystère 4A 186/12-UT which was noted here in 1989, belongs to the museum at Le Bourget and returned there. The Mirage 3 does not carry a serial, but during the open day in 1996 its construction number was checked. The stored CM170 539/103-CX went to Châteaudun in 1997.

☐ 213	103-XN	CM170	213	preserved	5-98
☐ 554	103-CZ	CM170	554	stored	5-98
☐ SA162	68-DJ	H-34A	SA162	preserved	5-98
☐ 88		Super Mystère B2	88	preserved, at main gate, marked as '148/12-YP'	5-98
☐ 471	45-103	Mirage 3E	471	stored, ex code 3-IK	5-98
☐ 250	12-KQ	Mirage F1C-200	..	dumped, in scrap compound, crashed 21-3-84	6-93
☐ 172		N2501F	172	stored, fuselage only, ex Évreux	5-98

CANNES - MANDELIEU (06)
Preserved at the entrance of this airfield is a Beech. The stored Mi-2 went to Villefranche sur Saône.

☐ F-BUOP		Beech E18S	BA-184	preserved, marked as 'FBCCI'	8-96

CAPTIEUX (33)
Preserved at the radar site gate here is a former Châteaudun Mirage 3.

☐ 519	3-XC	Mirage 3E	519	preserved, ex Châteaudun	6-96

CASTELNAU MAGNOAC (65)
The Aero Club de Castres still have their Flamant, which was SOC on 2nd February 1974.

☐ 220	319-CK	MD312	220	preserved	7-91

CAZAUX (33)
The remote Base Aérienne 120 has a large number of preserved and stored aircraft. Vautour 615/92-AK moved to Châteaudun. The other airframes noted in 1989 (see the first edition) may have been relocated to the large live weapons range at the south and north side of the base. Not much of the range is visible from outside and airframes may come and go without being noticed. Noted in 1997 were also an unidentified Alize, Canberra, Flamant, Noratlas and Vautour.

☐ 2041	341-ED	Alouette 3	2041	stored, in compound	7-97
☐ 203		MD312	203	stored, in compound	7-97
☐ SA86	67-OY	H-34A	SA86	preserved	6-95
☐ ..		Jaguar A	..	stored, in compound	7-97
☐ 209		Mirage 3B	209	stored, in compound, ex Châteaudun	7-97
☐ 2		Mirage 3C	2	stored, in compound	6-93
☐ 32	10-RE	Mirage 3C	32	stored, ex Rochefort	7-97
☐ 309	33-CV	Mirage 3R	309	stored, in compound	7-97
☐ 433	13-QD	Mirage 3E	433	stored, in compound, ex Châteaudun	7-97
☐ 04		Mirage F1	04	preserved, by fire station	7-97
☐ 100	NG	MS760	100	preserved, behind fire station	7-97
☐ 114		MS760	114	stored, in compound	7-97
☐ 120	8-ME/8-NE	Mystère 4A	120	preserved, on pole	7-97
☐ 28		N2501F	28	stored, in compound	7-97

CHAMBON SUR VOUEIZE (23)
Some 4km south of this village, up a hill at a private strip is the unique HD-321.

☐ 01		HD-321	01	stored	3-98

CHARTRES (28)
Base Aérienne 122 has three aircraft preserved near its main gate. The Mirage 3 is new here and has taken the place of the Flamant and H-34, which were mentioned as departed in the first edition. All three aircraft should be visible at the base from the town's cathedral spire.

☐ 62		CM170	62	preserved, at main gate	4-97
☐ 469		Mirage 3E	469	preserved, at main gate, ex Châteaudun	6-97
☐ 90	12-YQ	Super Mystère B2	90	preserved, near main gate, ex Bourges	8-96

CHÂTEAUDUN (28)
The airfield of Châteaudun needs no further introduction. To start with the aircraft which are preserved on and around the base. The Mirage 3R is outside the main gate, while the Super Mystère is also outside the base at the officers mess on the north east side of the airfield. Preserved Mirage 3B-RV 245 moved to Orly.

☐ 476	312-AG	CM170	476	preserved, ex stored	11-96
☐ 303	33-TM	Mirage 3R	303	preserved, ex instructional	6-97
☐ 358		Mirage 3RD	358	preserved, ex stored	11-97
☐ 367	33-TP	Mirage 3RD	367	preserved, at main gate	5-98
☐ 584	4-AN	Mirage 3E	584	preserved, ex stored, near tower	5-98
☐ 1	AP	Mirage 4A	1	preserved	5-98
☐ 24	13-SP	Mirage 5F	24	preserved, near tower	5-98
☐ 9		Mirage F1C	..	preserved	6-97
☐ 89		Mystère 4A	89	preserved, marked as '68/7-AM'	3-96

FRANCE - 79

☐ 278	8-MB	Mystère 4A	278	preserved	6-97
☐ 79	10-RH	Super Mystère B2	79	preserved, outside base at officers' mess	5-98
☐ 124		MD315	124	preserved	5-98
☐ 231		MD450	231	preserved	6-97
☐ 171		N2501F	171	preserved	5-98
☐ 29061		F-84F	..	preserved, ex 52-9061	5-98
☐ 615	92-AK	Vautour 2B	615	preserved, ex Cazaux	7-97

The instructional airframes, are (mostly) gathered inside and outside small hangars at the north west side of the base, just behind the main gate. A scrap compound can also be found on this side. From this compound the cockpit of the dumped Mirage F1C 240 is now in use as recruiting aid. The Super Sabre is the odd one out. It is not known if the serail given is its real one.

☐ 53-1580		F-100A	192-75	stored, Thunderbird c/s, see text	11-97
☐ 2327	67-JF	Alouette 3	2327	instructional	8-96
☐ ..	7-PE	Jaguar A	..	instructional	8-96
☐ ..	3.10-LL	Mirage 3C	..	dumped, fuselage only, mayby 29 or 35	6-95
☐ 340	2-ZH	Mirage 3R	340	instructional, in shelter, ex stored	6-95
☐ 402	13-QL	Mirage 3E	402	instructional, but see Savigny	8-96
☐ 484		Mirage 3E	484	instructional, ex stored	11-97
☐ 494		Mirage 3E	494	instructional, ex stored	6-97
☐ 547	3-IQ	Mirage 3E	547	instructional, ex stored	11-97
☐ 517	3-IS	Mirage 3E	517	instructional, ex stored	6-97
☐ 02		Mirage F1	02	instructional	8-96

The resident Entrepot de l'Armée de l'Air 601 (EAA.601) is responsible for the overhaul, storage and disposal of many AdlA types. In recent years CM170s, Mirage 3s and Mirage 5s have been retired from operational service. Also large numbers of Jaguars, Mirage 4s and MS760s have arrived for storage.

☐ 203		CM170	203	stored	6-95
☐ 208	315-IJ	CM170	208	stored	11-96
☐ 210		CM170	210	stored	5-98
☐ 223	11-OD	CM170	223	stored	11-96
☐ 226		CM170	226	stored	6-95
☐ 227		CM170	227	stored	11-97
☐ 232	33-QG	CM170	232	stored	8-96
☐ 233		CM170	233	stored, main ramp	11-97
☐ 237	315-PY	CM170	237	stored	6-95
☐ 242		CM170	242	stored, main ramp	11-96
☐ 316		CM170	316	stored	7-97
☐ 333		CM170	333	stored	6-94
☐ 340		CM170	340	stored, main ramp	11-96
☐ 342		CM170	342	stored, marked with 'GE315 - 580000 heures'	2-97
☐ 352	315-QR	CM170	352	stored	6-87
☐ 378	315-..	CM170	378	stored	5-98
☐ 393	33-XB	CM170	393	stored	6-95
☐ 397		CM170	397	stored	6-95
☐ 405	312-TF	CM170	405	stored	6-95
☐ 408		CM170	408	stored	6-95
☐ 411		CM170	411	stored, main ramp	11-96
☐ 414		CM170	414	stored	11-97
☐ 421		CM170	421	stored	6-95
☐ 423	12-XM	CM170	423	stored	6-95
☐ 425		CM170	425	stored	5-98
☐ 426	126-VP	CM170	426	stored	4-97
☐ 429		CM170	429	stored	5-98
☐ 430	125-HC	CM170	430	stored	5-96

FRANCE – 80

☐ 433		CM170	433	stored	7-97
☐ 442	312-TV	CM170	442	stored, in hangar	6-95
☐ 443		CM170	443	stored	9-87
☐ 445	312-UH	CM170	445	stored	5-98
☐ 446	102-CK	CM170	446	stored	7-97
☐ 447	312-TO/312-TN	CM170	447	stored	5-98
☐ 452	TK	CM170	452	stored, PdF c/s	2-97
☐ 455		CM170	455	stored	4-97
☐ 458	312-AR	CM170	458	stored	5-98
☐ 461	312-TE/312-AE	CM170	461	stored	11-96
☐ 464		CM170	464	stored, PdF c/s	6-95
☐ 465	312-AK	CM170	465	stored	5-98
☐ 480	312-TH	CM170	480	stored	6-95
☐ 481		CM170	481	stored	8-96
☐ 483	312-AO	CM170	483	stored	5-96
☐ 484	312-AI	CM170	484	stored	5-98
☐ 485	312-AJ	CM170	485	stored	6-95
☐ 486	312-TX	CM170	486	stored	6-95
☐ 492		CM170	492	stored	11-95
☐ 496	312-TX	CM170	496	stored	6-95
☐ 500		CM170	500	stored	5-98
☐ 502		CM170	502	stored	11-97
☐ 509		CM170	509	stored	8-96
☐ 512	312-AU	CM170	512	stored, Tiger c/s	8-96
☐ 516		CM170	516	stored	7-97
☐ 517		CM170	517	stored	8-96
☐ 519		CM170	519	stored	5-98
☐ 524		CM170	524	stored	88
☐ 527		CM170	527	stored, PdF c/s	3-94
☐ 530		CM170	530	stored, PdF c/s	6-95
☐ 537	312-AP	CM170	537	stored	5-98
☐ 539		CM170	539	stored, ex Cambrai	11-97
☐ 546	312-..	CM170	546	stored	6-95
☐ 547	113-CL	CM170	547	stored	5-98
☐ 553		CM170	553	stored	4-97
☐ 563	312-	CM170	563	stored	11-97
☐ 568		CM170	568	dumped, ex stored	11-96
☐ 569	12-XK	CM170	569	stored	6-95
☐ 573		CM170	573	stored	8-96
☐ 575	312-TO	CM170	575	stored	2-97
☐ A2	11-RA	Jaguar A	..	stored	5-98
☐ A5		Jaguar A	..	stored	2-97
☐ A7		Jaguar A	..	dumped	11-97
☐ A9	7-PB	Jaguar A	..	stored	11-95
☐ A11	7-PA	Jaguar A	..	dumped, ex stored	7-97
☐ A13		Jaguar A	..	stored	5-98
☐ A16		Jaguar A	..	stored	6-95
☐ A19	7-HE	Jaguar A	..	dumped, ex stored	7-97
☐ A21		Jaguar A	..	dumped, desert c/s	5-96
☐ A23		Jaguar A	..	stored	5-98
☐ A27	11-MQ	Jaguar A	..	dumped, ex instructional	5-98
☐ A32	7-IC	Jaguar A	..	dumped, ex stored, desert c/s	5-98
☐ A33		Jaguar A	..	stored	6-95
☐ A37		Jaguar A	..	stored	5-98
☐ A40		Jaguar A	..	stored	5-98

FRANCE - 81

☐ A46		Jaguar A	..	stored	5-98
☐ A48		Jaguar A	..	stored	11-97
☐ A49		Jaguar A	..	stored	11-97
☐ A54	11-MW	Jaguar A	..	stored	5-98
☐ A70		Jaguar A	..	stored	5-98
☐ A73		Jaguar A	..	stored	5-98
☐ A74		Jaguar A	..	stored	5-98
☐ A76		Jaguar A	..	stored	5-98
☐ A79		Jaguar A	..	stored	11-97
☐ A80	N	Jaguar A	..	dumped, no undercarriage, ex stored	7-97
☐ A89		Jaguar A	..	stored	5-97
☐ A91	11-YG	Jaguar A	..	stored	6-95
☐ A101	11-RK	Jaguar A	..	stored, in hangar	6-95
☐ A118		Jaguar A	..	stored	6-95
☐ A122		Jaguar A	..	stored	5-97
☐ A126		Jaguar A	..	stored	6-95
☐ A129		Jaguar A	..	stored	5-97
☐ E7	7-H.	Jaguar E	..	stored, in hangar	6-95
☐ E40		Jaguar E	..	stored	6-95
☐ 41		Mirage 3C	41	stored	9-87
☐ 206	DF	Mirage 3B	206	stored	5-97
☐ 212	13-FK	Mirage 3B	212	stored	10-93
☐ 213	13-SR	Mirage 3B	213	stored, in hangar	6-95
☐ 214	13-FI	Mirage 3B	214	stored	6-93
☐ 220	DG	Mirage 3B	220	stored, gone	6-94
☐ 243	DO	Mirage 3B-RV	243	stored	11-97
☐ 246	DB	Mirage 3B-RV	246	stored	3-94
☐ 248	DE	Mirage 3B-RV	248	stored	3-94
☐ 249	13-FE	Mirage 3B-RV	249	stored	3-94
☐ 250	DD	Mirage 3B-RV	250	stored, marked 'Dernier Vol Mirage IIIB'	5-98
☐ 266	13-PP	Mirage 3BE	266	stored	6-97
☐ 269	13-PB	Mirage 3BE	269	stored	6-97
☐ 270		Mirage 3BE	270	stored	7-97
☐ 301		Mirage 3R	301	stored, dismantled	3-96
☐ 325		Mirage 3R	325	stored	5-93
☐ 327		Mirage 3R	327	stored, ex instructional, ex 33-CO	5-98
☐ 348	33-CM	Mirage 3R	348	stored	10-93
☐ 349		Mirage 3R	349	stored	9-88
☐ 350		Mirage 3R	350	stored, ex 2-ZG	9-87
☐ 360	33-TI	Mirage 3RD	360	stored	11-97
☐ 364		Mirage 3RD	364	stored	9-91
☐ 368		Mirage 3RD	368	stored	9-91
☐ 449	13-QL	Mirage 3E	449	stored	6-88
☐ 470	3-XJ	Mirage 3E	470	stored	5-98
☐ 479	3-XG	Mirage 3E	479	stored	5-98
☐ 491	3-XT	Mirage 3E	491	stored, special c/s	8-96
☐ 495	3-XL	Mirage 3E	495	stored	8-96
☐ 502		Mirage 3E	502	stored	10-93
☐ 504		Mirage 3E	504	stored, gone	6-94
☐ 506	3-XA	Mirage 3E	506	stored	5-98
☐ 508		Mirage 3E	508	stored	8-96
☐ 513	3-JG	Mirage 3E	513	dumped, ex instructional	5-98
☐ 520	3-IT	Mirage 3E	520	stored	5-98
☐ 521	3-..	Mirage 3E	521	stored	5-98
☐ 524	3-XH	Mirage 3E	524	stored	5-97

FRANCE - 82

❏ 527		Mirage 3E	527	stored	11-97
❏ 529		Mirage 3E	529	stored	5-98
❏ 530		Mirage 3E	530	stored	2-97
❏ 534	3-XB	Mirage 3E	534	stored	5-98
❏ 535	3-IK	Mirage 3E	535	stored	11-96
❏ 539		Mirage 3E	539	stored, special c/s	5-98
❏ 550		Mirage 3E	550	stored	5-98
❏ 565		Mirage 3E	565	stored, gone	6-94
❏ 566	3-ID	Mirage 3E	566	stored	5-98
❏ 567		Mirage 3E	567	stored	5-98
❏ 570		Mirage 3E	570	stored	11-97
❏ 577	3-X.	Mirage 3E	577	stored	5-98
❏ 579	3-XJ	Mirage 3E	579	stored	5-98
❏ 587	13-QL	Mirage 3E	587	stored	5-98
❏ 610		Mirage 3E	610	stored	5-98
❏ 611	3-XS	Mirage 3E	611	stored	11-97
❏ 615	4-BM	Mirage 3E	615	stored	5-98
❏ 617	3-XD	Mirage 3E	617	stored, ex instructional	11-97
❏ 619	4-BJ	Mirage 3E	619	stored	11-97
❏ 620	3-XK	Mirage 3E	620	stored	5-98
❏ 621		Mirage 3E	621	stored	5-98
❏ 624	3-IZ	Mirage 3E	624	stored	5-98
❏ 625		Mirage 3E	625	stored	5-98
❏ 7		Mirage 4A	7	stored, marked as '01', white c/s	6-95
❏ 14	AM	Mirage 4A	14	stored	6-97
❏ 19	AR	Mirage 4A	19	stored	6-95
❏ 21	AT	Mirage 4A	21	stored	6-95
❏ 24	AW	Mirage 4A	24	stored, white c/s	6-95
❏ 26	AY	Mirage 4A	26	stored	7-97
❏ 27	AZ	Mirage 4A	27	stored	11-97
❏ 28	BA	Mirage 4A	28	stored	6-97
❏ 32	BE	Mirage 4A	32	stored	11-97
❏ 34	BG	Mirage 4A	34	stored	11-97
❏ 44	BQ	Mirage 4A	44	stored	6-95
❏ 46	BS	Mirage 4A	46	stored	8-96
❏ 47	BT	Mirage 4A	47	stored	6-95
❏ 49	BV	Mirage 4A	49	stored	6-97
❏ 54	CA	Mirage 4A	54	stored	11-97
❏ 55	CB	Mirage 4A	55	stored	11-97
❏ 57	CD	Mirage 4A	57	stored, special c/s	11-97
❏ 2	13-PG	Mirage 5F	2	stored	8-96
❏ 4		Mirage 5F	4	stored	11-97
❏ 6	13-PS	Mirage 5F	6	stored	11-97
❏ 7	13-PM	Mirage 5F	7	stored	11-97
❏ 10		Mirage 5F	10	stored	11-97
❏ 11	13-PH	Mirage 5F	11	stored	11-97
❏ 13	13-PB	Mirage 5F	13	stored	5-96
❏ 17	13-PK	Mirage 5F	17	stored	11-97
❏ 21	13-PA	Mirage 5F	21	stored	11-97
❏ 22	13-PK	Mirage 5F	22	stored	11-97
❏ 29	13-PH	Mirage 5F	29	stored	6-95
❏ 32	13-PR	Mirage 5F	32	stored	11-97
❏ 34	13-PG	Mirage 5F	34	stored	11-97
❏ 35	13-PM	Mirage 5F	35	stored	7-97
❏ 36	13-PN	Mirage 5F	36	stored	11-97

FRANCE - 83

☐ 38	13-PT	Mirage 5F	38	stored	11-97
☐ 39		Mirage 5F	39	stored	11-97
☐ 41		Mirage 5F	41	stored	11-97
☐ 43	13-SM	Mirage 5F	43	stored	11-97
☐ 44	13-SB	Mirage 5F	44	stored	11-97
☐ 46	PM	Mirage 5F	46	stored, marked 'Dernier Vol Mirage V'	11-97
☐ 47	13-SN	Mirage 5F	47	stored	11-97
☐ 48		Mirage 5F	48	stored	11-97
☐ 49	13-SL	Mirage 5F	49	stored	11-97
☐ 50	13-PC	Mirage 5F	50	stored	11-97
☐ 53	13-PO	Mirage 5F	53	stored	8-96
☐ 55		Mirage 5F	55	stored	8-96
☐ 56	13-PL	Mirage 5F	56	stored	8-96
☐ 16		Mirage F1C	..	stored	3-96
☐ 19		Mirage F1C	..	stored	5-98
☐ 20		Mirage F1C	..	stored	3-96
☐ 22		Mirage F1C	..	stored	3-96
☐ 25	12-ZP	Mirage F1C	..	stored	5-98
☐ 30		Mirage F1C	..	stored	3-96
☐ 33		Mirage F1C	..	stored	3-96
☐ 37		Mirage F1C	..	stored	3-96
☐ 38		Mirage F1C	..	stored	3-96
☐ 40	30-FM	Mirage F1C	..	stored	3-96
☐ 41		Mirage F1C	..	stored	3-96
☐ 42		Mirage F1C	..	stored	3-96
☐ 43		Mirage F1C	..	stored	3-96
☐ 49		Mirage F1C	..	stored	3-96
☐ 55		Mirage F1C	..	stored	3-96
☐ 62	12-ZR	Mirage F1C	..	stored	3-96
☐ 68		Mirage F1C	..	stored	3-96
☐ 69	30-SF	Mirage F1C	..	stored	3-96
☐ 74	33-FO	Mirage F1C	..	stored	3-96
☐ 75		Mirage F1C	..	stored	3-96
☐ 81	12-ZA	Mirage F1C	..	stored, desert c/s	6-95
☐ 84	30-FJ	Mirage F1C	..	stored	6-95
☐ 201	33-FI	Mirage F1C-200	..	stored	3-96
☐ 205	33-FX	Mirage F1C-200	..	stored	8-95
☐ 211		Mirage F1C-200	..	stored	3-96
☐ 213	30-SR	Mirage F1C-200	..	stored	6-95
☐ 511		Mirage F1B	..	stored	5-95
☐ 659	33-CA	Mirage F1CR	..	stored	6-95
☐ 88	5-NL	Mirage 2000C	..	stored, in shelter	6-95
☐ 19		MS760	19	stored	5-98
☐ 23	133-KA	MS760	23	stored	7-97
☐ 24		MS760	24	stored	8-96
☐ 25	314-DF	MS760	25	stored	8-96
☐ 29	314-..	MS760	29	stored	5-98
☐ 44	314-D.	MS760	44	stored	8-96
☐ 60	314-DE	MS760	60	stored	5-96
☐ 62	314-DD	MS760	62	stored	5-98
☐ 77		MS760	77	stored	8-94
☐ 78	5-HG	MS760	78	stored	5-98
☐ 80	DE	MS760	80	dumped, ex instructional	5-98
☐ 81	314-DD	MS760	81	stored	5-96
☐ 97		MS760	97	stored	2-97

FRANCE - 84

☐ 109		N2501F	109	stored, fuselage only	5-96
☐ 141		N2501F	141	dumped, fuselage only, desert c/s	5-96

Of the above mentioned types a large number should have left Châteaudun but no details are known. The aircraft below are confirmed as having left the base.

78		CM170	78	stored, l/n 1987	to Angers
168	GF	CM170	168	stored, l/n 6-95	to Toulouse
171		CM170	171	stored, l/n 6-94	to Cherance
172		CM170	172	stored, l/n 5-93	to Rennes as F-WKYD
173		CM170	173	stored, l/n 6-93	to N338DM
179		CM170	179	stored, l/n 5-97	to N179PS
197		CM170	197	stored, l/n 6-93	to N315VB
205	13-TC	CM170	205	stored, l/n 3-94	to Rochefort
209		CM170	209	stored, l/n 6-93	to N332DM
217	33-XA	CM170	217	stored, l/n 6-95	to Dôle as F-GLRE
224	307-KY	CM170	224	stored, l/n 5-97	to N224PS
228		CM170	228	stored, l/n 6-93	to N336DM
230		CM170	230	stored, l/n 6-95	to Le Havre as F-GLMO
315		CM170	315	stored, l/n 7-91	to Rennes as F-WKYF
326		CM170	326	stored, l/n 6-94	to N326F
334	91-GB	CM170	334	stored, l/n 4-90	to N6222N
336	11-OC	CM170	336	stored, l/n 3-94	to N326DM
343		CM170	343	stored, l/n 3-94	to Rochefort
344	315-...	CM170	344	stored, l/n 6-93	to N344FM
346		CM170	346	stored, l/n 6-93	to N346DM
353		CM170	353	stored, l/n 6-93	to N335DM
356	GH	CM170	356	stored, l/n 6-93	to N325DM
361		CM170	361	stored, l/n 6-93	to N330DM
362		CM170	362	stored, l/n 6-94	to N362F, later to N8MT
363	7-JH	CM170	363	stored, l/n 6-94	to N363F
369	315-	CM170	369	stored, l/n 6-95	to F-GPCJ
377		CM170	377	stored, l/n 6-93	to Rochefort
381		CM170	381	stored, l/n 6-94	to N381F, later to N381JK
383	315-PV	CM170	383	stored, l/n 6-95	to N383FM
385		CM170	385	stored, l/n 6-93	to N385F
387		CM170	387	stored, l/n 1989	to Speyer
388	315-II	CM170	388	stored, ex preserved, l/n 4-97	to Nordholz
394		CM170	394	stored, l/n 6-95	to Touchay
398	315-QV	CM170	398	stored, l/n 6-94	to N398F
399	4-CT	CM170	399	stored, l/n 6-93	to Rochefort
402		CM170	402	stored, l/n 10-93	to Rochefort
403	312-HC	CM170	403	stored, l/n 5-97	to N403PS
404	330-DH	CM170	404	stored, l/n 3-94	to Rochefort
409		CM170	409	stored, l/n 6-93	to Orange
413	128-VM	CM170	413	stored, l/n 6-93	to Cuers as F-GNBG
415		CM170	415	stored, l/n 10-93	to N415FM
417		CM170	417	stored, l/n 5-97	to N417PS
418		CM170	418	stored, l/n 5-97	to N418PS
420		CM170	420	stored, l/n 4-97	to USA
422		CM170	422	stored, l/n 11-95	to USA
424		CM170	424	stored, l/n 6-95	to La Baule as F-GJMN
431		CM170	431	stored, l/n 5-97	to N431PS
432		CM170	432	stored, l/n 5-97	to N432PS
434		CM170	434	stored, l/n 6-93	to N434F
435		CM170	435	stored, l/n 6-95	to Rennes as F-WKYH

438	312-TW	CM170	438	stored, l/n 6-93	to F-WIMQ
440	312-...	CM170	440	stored, l/n 6-93	to F-WIMN
450	312-AV	CM170	450	stored, l/n 4-97	to N450PS
454		CM170	454	stored, l/n 1988	to Aix
460	312-TB	CM170	460	stored, l/n 1989	to Lyon
463	312-UD	CM170	463	stored, l/n 6-95	to N531PA
470	312-AS	CM170	470	stored, l/n 6-95	to N570PS
472	312-UB	CM170	472	stored, l/n 11-96	to N572PS
478		CM170	478	stored, l/n 1989	gone
479		CM170	479	stored, l/n 11-97	to F-GLEZ
482	312-TL	CM170	482	stored, l/n 11-97	to F-GJIJ
487	312-AS	CM170	487	stored, l/n 4-97	to N487FM
495		CM170	495	stored, PdF c/s, l/n 5-96	to N495F
497		CM170	497	stored, l/n 5-96	to USA
508	312-TI	CM170	508	stored, Tiger tail, l/n 11-97	to N508F
511		CM170	511	stored, l/n 4-97	to USA
513	312-AU	CM170	513	stored, l/n 4-97	to N513FM
520	312-UF	CM170	520	stored, l/n 6-95	to N532PA, later to N520F
525	312-UC	CM170	525	stored, l/n 6-95	to N525FJ
526	312-TS	CM170	526	stored, l/n 11-96	to USA
528	102-HB	CM170	528	stored, l/n 4-97	to N528F
532		CM170	532	stored, l/n 1989	to Lyon as F-GNYN
534		CM170	534	stored, l/n 11-95	to N534FM
555		CM170	555	stored, l/n 11-96	to N555FA
561		CM170	561	stored, l/n 1989	to Beauvais as F-WIGJ
564	VO	CM170	564	stored, l/n 6-93	to N323DM
566	312-TB	CM170	566	stored, l/n 6-95	to N533PA
579		CM170	579	stored, l/n 11-95	to N319DM
A8	11-EB	Jaguar A	..	dumped, l/n 3-96	to Savigny lès Beaune
A36	7-IP	Jaguar A	..	stored, desert c/s, l/n 3-96	to Savigny lès Beaune
A72		Jaguar A	..	stored, l/n 6-95	to Savigny lès Beaune
203	DR	Mirage 3B	203	stored, FG on nosewheel door, l/n 10-93	to Istres
205	13-FC	Mirage 3B	205	stored, l/n 10-93	to Rochefort
207	13-FH	Mirage 3B	207	stored, l/n 6-95	to Évreux
209		Mirage 3B	209	stored, l/n 5-93	to Cazaux
219	DK	Mirage 3B	219	stored, l/n 6-94	to Tours
222	13-FF	Mirage 3B	222	stored, l/n 6-94	to Orange
224	13-FA	Mirage 3B	224	stored, l/n 10-93	to Nantes
241	DK	Mirage 3B-RV	241	stored, l/n 7-97	to Orange
247	DQ	Mirage 3B-RV	247	stored, l/n 6-93	to Varennes sur Allier
260	13-PT	Mirage 3BE	260	stored, l/n 8-96	to Pakistan AF
261		Mirage 3BE	261	stored	to Pakistan AF
267	13-PU	Mirage 3BE	267	stored, l/n 8-96	to Pakistan AF
268	13-PV	Mirage 3BE	268	stored, l/n 8-96	to Pakistan AF
274		Mirage 3BE	274	stored	to Pakistan AF
275		Mirage 3BE	275	stored, l/n 8-96	to Pakistan AF
304	33-TN	Mirage 3R	304	stored, l/n 5-93	to Savigny lès Beaune
313		Mirage 3R	313	stored, l/n 1991	scrapped
324	2-ZM	Mirage 3R	324	dumped, ex instructional, l/n 3-96	to Savigny
329	33-CM	Mirage 3R	329	stored, l/n 1988	to CEV (operational)
330	33-CC	Mirage 3R	330	stored, l/n 1988	to CEV (operational)
333	33-CB	Mirage 3R	333	stored, l/n 9-87	to Nancy
335	33-TK	Mirage 3R	335	stored, l/n 1987	to CEV (operational)
351		Mirage 3RD	351	stored	to Rennes
352		Mirage 3RD	352	stored, l/n 1991	to Paris Orly

FRANCE - 86

355		Mirage 3RD	355	stored, l/n 1991	to Speyer
356	33-TE	Mirage 3RD	356	stored, l/n 3-89	to Rochefort
359	33-TH	Mirage 3RD	359	stored, l/n 1988	to Brive
363		Mirage 3RD	363	stored, l/n 8-96	to Orange
432		Mirage 3E	432	stored, l/n 9-91	to Speyer
433	13-QD	Mirage 3E	433	stored, l/n 7-91	to Cazaux
460	13-QR	Mirage 3E	460	stored, l/n 1991	to Le Bourget
469	3-IF	Mirage 3E	469	stored, l/n 6-94	to Chartres
499	3-IG	Mirage 3E	499	stored, l/n 11-95	to Savigny
501		Mirage 3E	501	instructional, ex stored, l/n 6-93	to Savigny
514		Mirage 3E	514	stored, l/n 10-93	to Rennes
515		Mirage 3E	515	stored, l/n 6-94	to Calais
519	3-XC	Mirage 3E	519	stored, l/n 5-97	to Captieux
538	3-XF	Mirage 3E	538	stored, l/n 6-95	to Elvington, UK
549	4-BF	Mirage 3E	549	stored, l/n 11-95	to Contrexéville
554		Mirage 3E	554	stored, l/n 8-96	to Brazil AF
560		Mirage 3E	560	stored, l/n 6-88	to CEV (operational)
564	3-XN	Mirage 3E	564	stored, l/n 8-96	to Brazil AF
572	3-IU	Mirage 3E	572	stored, l/n 4-97	to Luxeuil
578	3-JM	Mirage 3E	578	stored, l/n 6-95	to Beynes
586	3-XF	Mirage 3E	586	stored, l/n 6-95	to St. Ramber t d'Albon
616		Mirage 3E	616	stored, l/n 6-95	to Le Havre
4	AC	Mirage 4A	4	stored, l/n 6-95	to Rochefort
45	BR	Mirage 4A	45	stored, l/n 3-94	to Paris
18		Mirage 5F	18	stored, l/n 5-96	to Pakistan AF
19		Mirage 5F	19	stored, l/n 8-96	to Pakistan AF
25	13-PN	Mirage 5F	25	stored, l/n 8-96	to Pakistan AF
31	13-SK	Mirage 5F	31	stored, l/n 8-96	to Pakistan AF
51		Mirage 5F	51	stored, l/n 8-96	to Pakistan AF
52		Mirage 5F	52	stored, l/n 7-97	to Pakistan AF
57		Mirage 5F	57	stored, l/n 8-96	to Pakistan AF
58		Mirage 5F	58	stored, l/n 8-96	to Pakistan AF
2	30-MG	Mirage F1C	..	stored, l/n 10-95	to Rochefort
3		Mirage F1C	..	stored, l/n 10-95	to Rochefort
6		Mirage F1C	..	stored, l/n 10-95	to Rochefort
14		Mirage F1C	..	stored, l/n 10-95	to Rochefort
17		Mirage F1C	..	stored, l/n 3-96	to Rochefort
18	12-KF	Mirage F1C	..	stored, ex preserved, l/n 4-97	to Rochefort
29		Mirage F1C	..	stored, l/n 10-95	to Rochefort
36		Mirage F1C	..	stored, l/n 10-95	to Rochefort
39		Mirage F1C	..	stored, l/n 10-95	to Rochefort
47	12-ZG	Mirage F1C	..	stored, l/n 10-95	to Rochefort
50		Mirage F1C	..	stored, l/n 10-95	to Rochefort
60		Mirage F1C	..	stored, l/n 10-95	to Rochefort
63		Mirage F1C	..	stored, l/n 10-95	to Rochefort
70		Mirage F1C	..	stored, l/n 10-95	to Rochefort
14		MS760	14	stored, l/n 8-96	to Long Beach, USA
51		MS760	51	stored, l/n 2-97	to N751PJ
53		MS760	53	stored, l/n 5-96	to N53PJ
F-ZBAC	61	DC-6B	44898	stored, l/n 6-91	to N4390F
F-ZBAD	62	DC-6B	45066	stored, l/n 6-91	to N4390X

As a follow-on on to the long list of stored aircraft from the first edition, all Mystère 4As have gone. Of these, parts from 116 and 142 ended up at Savigny lès Beaune. The second Super Mystère, 53/12-YO, left for Brive. All 2501s have gone from Châteaudun, except for 109 and 141 mentioned above. Most were scrapped, but some survived and of these 92, 149 and 151 are at Savigny lès Beaune, 125 went to Orléans, 146 went to Viuz en Sallez, 179 went to Caen and 194 ended up at Le Bourget. Noratlas 38 was described as scrapped in the first edition, but the fuselage is now at Caen.

CHÂTEAUROUX - DÉOLS (36)
The Caravelle is still preserved here in spite of the effort in the first edition to remove it! Broussard 20, which was stored at the back of an old shed, became F-GGKG. A number of Airbuses can be found in temporary store or under conversion to freighters congiguration.

| ❏ F-BYCD | | SE210-6N | 67 | preserved, ex OO-SRE | 7-97 |

CHÂTELLERAULT (86)
Two CM170s are preserved in this town. On the north side on the road to Dangé is 338. It is mounted on a pole outside the SOCATA/SNECMA factory. 133 is preserved on the airfield at the south side of the town.

| ❏ 133 | | CM170 | 133 | preserved, on pole, with aero club | 4-97 |
| ❏ 338 | 315-IB | CM170 | 338 | preserved, on pole | 4-97 |

In the middle of nowhere, along the D9 towards Monthoiron, some 9km south of Châtellerault, two aircraft stored. They are on the right hand side of the road (coming from Châtellerault) behind a large metal gate and thick hedge.

| ❏ 199 | 41-GE | MD312 | 199 | stored | 8-96 |
| ❏ F-GDQV | | MH1521M | 294 | stored | 8-96 |

CHÂTILLON EN DIOIS (26)
The camp site at the western end of town was closed in the early 1990s and F-100D 52734/11-YP moved to Nansy-Essey by June 1993.

CHERANCE (72)
Preserved at the local airfield is an ex Châteaudun Magister.

| ❏ 171 | | CM170 | 171 | preserved, ex Châteaudun | 11-95 |

CHERBOURG - MAUPERTUS (50)
The MD312 Flamant was handed over to the airfield firemen in 1985 and was still there in early 1992.

| ❏ 170 | 319-CR | MD312 | 170 | instructional, fuselage only | 3-92 |

CLÈRES (76)
The Musée Automobile de Clères (at the Château de Clères) still has the T-6G Texan outside. Inside the museum is a Potez 53 in camouflage colours. It is not known what has happened to the museum's Stampe SV4 F-BMME (109), which was last seen here in 1987.

| ❏ 115102 | D-44 | T-6G | 182-789 | preserved, outside, ex 51-15102 | 10-96 |
| ❏ F-AMJP | | Potez 53 | 3322 | preserved, marked as 'E2705' | 7-95 |

CLERMONT FERRAND - AULNAT (63)
On this base a Mirage 3C acts as gate guard at military Atelier Industriel de l'Aéronautique. It replaces the former gate guarding Mirage F1 03, which may have moved somewhere inside. The Mirage F1 was last noted in late 1989.

| ❏ 70 | | Mirage 3C | 70 | preserved, at gate, on pole | 5-96 |

FRANCE - 88

CLOYES SUR LE LOIR (28)
The scrapyard here was checked in August 1991 and no aircraft were found. Both Br941s, 1/62-NA and 2/62-NB, were scrapped.

COGNAC - CHÂTEAUBERNARD (16)
The training unit here has replaced the CM170s with the newer Epsilons. Some CM170s can still be found on the base, together with a prototype Epsilon which should be in use as an instructional airframe. Magister 196 went to Le Bourget.

❑ 198	315-GC	CM170	198	preserved, near hangars	5-97
❑ 309	315-QA	CM170	309	stored, in hangar	10-95
❑ 324	315-QD	CM170	324	preserved, on pole near flightline	3-98
❑ 376	315-QA	CM170	376	stored, ex Saintes	5-97
❑ 404	13-QN	Mirage 3E	404	preserved	3-98
❑ 02	VJ	TB30	02	instructional	—

COLMAR (68)
Town: Preserved some 500 meters from the Colmar's civil airfield, Houssen, near the Parc des Expositions, along the N83, is a former Meyenheim Mirage 5.

❑ 15	132-BA	Mirage 5F	15	preserved, ex Colmar Meyenheim	6-97

Airfield – Meyenheim: The base of Colmar is currently the home of two Mirage F1CT units. Formerly flying from here were several types of Mirages, of which some examples can still be found on the base.

❑ 13	13-EC	Mirage 3C	13	preserved, on pole at gate	6-97
❑ 403	13-QB	Mirage 3E	403	preserved	6-97
❑ 483		Mirage 3E	483	stored, only tail preserved by 6-97	2-98
❑ 607	3-XE	Mirage 3E	607	dumped, only tail preserved by 6-97	6-95
❑ 37	13-PB	Mirage 5F	37	dumped	6-92
❑ 602	92-AB	Vautour 2B	602	dumped, north east shelter area	5-89

COMPIÈGNE (60)
Preserved on a pole at the French Air Force barracks in the town of Compiègne (not on the ALAT airfield) on the south side of town is a Mirage 3R. It can be seen from outside.

❑ 322	33-TB	Mirage 3R	322	preserved, on a pole	5-98

CONTREXÉVILLE (88)
Preserved at the military airfield at Contrexéville (BA942) is a former Châteaudun Mirage 3E.

❑ 549		Mirage 3E	549	preserved, ex Châteaudun	6-97

COULANGES SUR L'AUTIZE (79)
A night club here has an ex Air France Caravelle. It was last noted in 1986 and may have gone.

❑ F-BHRI		SE210-3	17	preserved	86

CREIL (60)
Town: The CM170 which used to be at the aero club of Creil Senlis is said to have moved to a college in Creil.

❑ 1	AO	CM170	1	instructional	—

Airfield – Senlis: After EC.10 was disbanded in the mid 1980s the base was reduced to care and maintenance status and the preserved Super Mystère relocated to Cambrai. With the establishment of ETL01.062 here in the early 1990s the Super Mystère came back from Cambrai and was again placed at the gate. The HD-34 is preserved at the civil side of the airfield.

❏ 139	10-RE/10-SE	Super Mystère B2	139	preserved, at gate, ex marked as '177/12-YA'	6-97
❏ ..		N2501	..	stored, fuselage only	5-96
❏ F-BICV		HD-34	8	preserved, with aero club	5-96

CRESSANGES (03)
A Broussard may still be kept here at a private location.

❏ 15		MH1521M	15	stored	89

CUERS - PIERREFEU (83)
The French Navy maintenance facility here has a N1101 as gate guard. The Etendard 4MP is preserved on the base. Acquired in 1990s were eleven ex US Navy F-8 Crusaders. These were used for spares and are nowadays stored at Cuers.

❏ 11x		F-8J	..	stored, ex US Navy, see note	8-96
❏ 80		Br1050	80	dumped, fuselage only	7-93
❏ 17		Br1150	17	dumped, fuselage only	8-96
❏ 166		Etendard 4MP	166	preserved	6-96
❏ 6	MA	N260	6	dumped, ex LN-LMG	5-92
❏ 181	CAN-12	N1101	181	preserved, near main gate	7-93

On the civil side is a small collection of aircraft, including some flying types. Stored CM175 05 has become F-AZPI and CM170 413 became F-GNBG.

❏ J-4095		Hunter F.58	..	preserved, ex Emmen	8-97
❏ 30		Etendard 4M	30	preserved	8-97
❏ 251	44C	MH1521M	251	preserved	8-97
❏ 440	13-QF	Mirage 3E	440	preserved	8-97
❏ 21121		T-33AN	121	preserved, marked '14045/TR-045', N12424 ntu	8-97

DAX (40)
The French Army has a large fleet of training helicopters based at this southern airfield..

❏ A-414		Alouette 3	1414	stored, in hangar, ex Marseille, ex RNethAF	7-97
❏ 1094	DAX	Alouette 2	1094	preserved, at gate, ex code BBN	9-97
❏ 266		MH1521M	266	preserved	7-97
❏ 24585	CUW	O-1E	24585	stored, in hangar	7-97
❏ ..		SA341	..	stored, fuselage only	7-97

The Musée de l'Aviation Légère de l'Armée de Terre is located on the northern edge of the airfield and is open daily between 14:30 and 17:30, except for Sundays and holidays. The museum's AB47G 056/CAN-7 went to the museum at Rochefort and the stored H-21C FR106 to Toulouse.

❏ 9408		Mi-8TB	10568	preserved, ex Parow, ex 814/NVA	2-98
❏ 1076		Alouette 2	1076	preserved	2-98
❏ 1162		Alouette 2	1162	preserved	2-98
❏ 1634	BEJ	Alouette 2	1634	preserved	2-98
❏ 001	AE	Alouette 3	001	preserved, marked as '1001'	2-98
❏ 160	ZP	AB47G-2	160	preserved	2-98
❏ 1314	BDM	B47G-1	1314	preserved	2-98
❏ 55-864	AXC	H-19D	55-864	preserved, ex Le Luc, Le Bourget owned	2-98
❏ 55-1086	AVW	H-19D	55-1086	preserved	2-98
❏ FR94	BEJ	H-21C	FR94	preserved	2-98
❏ SA143		HSS-1	SA143	preserved	2-98

☐ ..		L-18C	18-1363 ?	preserved	2-98
☐ 269	MIA	MH1521M	269	preserved	2-98
☐ 66	AJR	N3202	66	preserved	2-98
☐ 99		N3202	99	preserved, unmarked	2-98
☐ 75	MFA	N3400	75	preserved	2-98
☐ ..	SU	NC856A	..	preserved	2-98
☐ 24530	CEA	O-1E	24530	preserved	2-98
☐ 24725	BUA	O-1E	24725	preserved	2-98
☐ 1201		SA349-2	1201	preserved	2-98
☐ 1003	BSP	SA361	1003	preserved	2-98
☐ FR14	ATH	SO1221	FR14	preserved	2-98
☐ 133		UH-12A	133	preserved, ex F-OAHB	2-98
☐ F-BDIZ		SV-4C	496	preserved, ex 496 and marked as such	2-98
☐ F-BDQQ	UA	MS505	656	preserved	2-98
☐ F-BFYP		PA22	22-256212	preserved, ex 22-256212 and marked as such	2-98

The museum should have some more aircraft, the current status of which is not always known. O-1E 24521 departed for an unknown destination in exchange for the PA-22. Stored Alouette 2 1046 became F-GIJK, while 1234/JAR now flies as F-GKGC.

☐ ..		L-18C	..	stored, incomplete, ex Belgian Army	7-95
☐ FR69		H-21C	FR69	stored, Le Bourget owned	7-95
☐ 78		N3400	78	stored	—
☐ FR116	CBB	SO1221	FR116	stored, wreck	87
☐ FR149	BNG	SO1221	FR149	stored	—
☐ ..	BVV	SO1221	..	stored	87

DIJON - LONGVIC (21)

This French base houses two squadrons of Mirage 2000C fighters. Beside these operational aircraft, a number of EWR-1 can also be found on this airfield.

☐ SKY479		H-34A	..	preserved, ex Nîmes	4-97
☐ 187		MD312	187	dumped, fire dump	6-89
☐ 221		MD312	221	dumped, scrap compound	6-89
☐ 30	2-EH	Mirage 3C	30	preserved	6-97
☐ 425	2-LG	Mirage 3E	425	preserved	6-97
☐ 438		Mirage 3E	438	instructional	7-94
☐ 14	13-PW	Mirage 5F	14	preserved	6-97
☐ 6	2-FP	Mirage 2000C	..	dumped	7-94
☐ 503	2-FO	Mirage 2000B	..	dumped	7-94
☐ 290		Mystère 4A	290	preserved, PdF c/s	6-97

DINAN - TRÉLIVAN (22)

The CM170, which arrived at the Aero Club de Dinan on 14th October 1986, is still current.

☐ 80		CM170	80	preserved	8-97
☐ F-SEBF		F27-200	10320	stored, no engines or undercarriage	8-97

DINARD - PLEURTUIT (35)

The maintenance facility here are split into two parts. At one side of the airfield is TAT which has a large fleet of stored Fokkers and Fairchilds. As more of the older aircraft in the fleet are replaced, the number of stored aircraft here grows. Some have been only temporarily here, including F27-500 9Q-CBI (10459) which became G-JEAP and 9Q-CBU (10425) which became G-BVZW.

☐ 5V-MAH		DHC-5D	71	stored, ex Togo AF	8-97

FRANCE - 91

☐ F-GBRU	F-27J	43	stored, ex N2771R	2-97
☐ F-GBRV	F-27J	44	stored, arr 8-93, ex N2772R	2-97
☐ F-GCPN	FH-227B	533	stored, ex Toulouse, F-ODMR	9-96
☐ F-GCPT	FH-227B	547	stored, ex N4227	2-97
☐ F-GCPU	FH-227B	548	stored, ex N4228	2-97
☐ F-GCPV	FH-227B	526	stored, ex 5N-BMO	8-96
☐ F-GCPX	FH-227B	535	stored, ex N4222	2-97
☐ F-GCPY	FH-227B	538	stored, ex N4223	9-96
☐ F-GCPZ	FH-227B	561	stored, fuselage only, in forest, ex N4234	8-93
☐ F-GDYU	F28-4000	11142	stored, ex 5N-ANU	2-97
☐ F-GIAI	F28-1000	11013	stored, ex C-GQBS	2-97
☐ F-GNFD	C-47B	32561	stored, ex G-BVRB	6-97
☐ F-GNZB	F28-1000	11073	stored, ex F-ODZB	2-97
☐ I-ALGM	N262A-14	8	stored, ex N420SA	2-97
☐ I-TIBB	F28-1000	11010	stored, ex LN-SUX	2-97
☐ PH-JHG	F28-6000	11001	stored, in forest	7-94

At the other side of the airfield a number of smaller companies have maintenance hangars, resulting in a number of stored aircraft. Not all of which end up as scrap. N262 F-BVFI (14) flew as 9Q-CVC, DHC-6 F-GCVR (200) became N753AF, CASA 212 F-GHOX (302) became T9-ABA, while F27-100 (10121) went to Rotterdam as F-WKPX. F27-500 TR-LCW (10687) became N1005L.

☐ F-BVFG	N262	10	stored, ex 7T-VSR	10-96
☐ F-GKPY	F27-100	10224	stored, arr 5-93, ex LX-LGA	12-95
☐ F-GIHR	F-27J	117	stored, ex Toulouse, F-ODBY	6-97
☐ F-OGJB	F-27J	103	stored, F-GGZP ntu, ex F-BGRR	6-97
☐ OO-SVL	F27-100	10113	stored, unmarked, arr 2-92, ex F-BYAP	6-97
☐ 9U-BHE	F27-100	10166	stored, arr 2-92, ex OO-SVN	6-97

DOMME (24)
The Aero Club du Sarladais lost their MD315R 39/30-QR (which was fitted with the fins of No. 41) by June 1993. It has been reported that the aircraft have moved to a private location.

ECROUVES - BLATZEN (54)
On the D400 just west of Toul, a military barracks (BA551) has Vautour 2N on display at the main gate.
☐ 308	30-MC	Vautour 2N	308	preserved	6-98

ÉPINAL (88)
It is not known if the plans to convert a Caravelle into a bar here succeeded. The last time the SE210 was noted it was still intact.
☐ F-BYCY	SE210-6N	233	stored, ex YU-ANG	8-89

ÉTAIN - ROUVRES (55)
There are no further reports of the preserved H-19 (or Whirlwind) here since EWR-1. The digits 7615 on the helicopter suggest that it is 52-7615, a known ALAT example.
☐ 52-7615	H-19	..	preserved, ex code CYX	6-89

ÉTAMPES (91)
The Escadrille du Souvenir has left the airfield here and their Noratlas (98/64-BC) and Mistral (50/7-BM) have gone. The tailbooms of the Noratlas were noted at Montmirault. Only a few *Wrecks & Relics* still remain. Stored

FRANCE - 92

Broussard 64 became F-AZEZ.

❏ 140		C-45	..	stored, unmarked, ex HB140	6-97
❏ ..		C-45	..	stored, marked as 'MT10/'MT122'	6-97
❏ 87	BR	N1101	87	stored	6-97
❏ 106		N1101	106	stored	6-90
❏ 30		N3202	30	stored	6-90

EU SUR MER - LE TRÉPORT (76)
Some 20km north east of Dieppe is the coastal town of Eu. Preserved at the local airfield is a Flamant.

❏ 218	MD312	218	preserved, with aero club	10-96

ÉVREUX - FAUVILLE (27)
Preserved at the main gate of Base Aérienne 105 is a Noratlas. Preserved at a gate on the north side is an H-34, while preserved between buildings on the west side is a Broussard. The local aero club has their own relic in the form of a Sahara, which is been reported to have left by September 1997. The fuselages of N2501s 7 and 17 have both gone, while 172/VR moved to Cambrai.

❏ 501	64-PE	Br765	501	preserved, with aero club	4-97
❏ F227	64-GP	C-160NG	230	dumped, burned out 17-3-93	4-97
❏ SA167	68-OC	H-34A	SA167	preserved	12-97
❏ 253		MH1521M	253	preserved	5-98
❏ 207	13-FH	Mirage 3B	207	stored, ex Châteaudun	4-97
❏ 156	64-BJ	N2501F	156	preserved, at main gate	5-98

FALAISE (14)
Monsieur du Valleroy received a Flamant in 1984. Its current whereabouts are unknown, but it is thought that the aircraft is at his private house.

❏ 161	MD312	161	preserved	—

FÉNIERS (23)
Some 100km west of Clermont Ferrand is the small village of Feniers. The local military barracks should have a Magister preserved.

❏ 477	AJ	CM170	477	preserved, ex Limoges	5-90

FONTENAY - TRÉSIGNY (77)
The last Br763 Provence in France can be found at the airfield of Fontenay. The airfield, some 40km east of Paris, is along the N4 route.

❏ 306	Br763	306	preserved, marked as 'F-BACC'	6-97

FRÉJUS - ST RAPHAËL (83)
The French Navy left the airfield in 1995. Only the Flamant should still be here.

❏ 149		HSS-1	SA149	preserved, at gate	9-94
❏ 04		Lynx	..	instructional, ex XX911	—
❏ 113	30-QY	MD315R	113	preserved, near civil side	8-97

GAP - TALLARD (05)
Noted here was the fuselage of a Neptune. It was mounted on a rig with wheels as some sort of traveling exhibit.

FRANCE - 93

On the airfield is a Magister.

❏ 195		CM170	195	preserved	6-95
❏ ..		SP-2H	..	preserved, marked as '205', fuselage only	9-92

GAVRES (56)
Standing on the ranges here (over the river from Lorient) since at least 1982 and probably for a lot longer, were some unknown helicopters (including a Bell 47G and HSS-1, both l/n 1988). A v isit in May 1997 confirmed that they had both been removed from their previous resting places.

GRENOBLE - LE VERSAUD (38)
The Centre d'Etude et de Loisirs Aerospatiaux de Grenoble (CELAG) has a number of helicopters in store or as instructional airframes on the airfield. Some of the helicopters do not carry original markings. The ex ALAT H-34 is painted as a Navy example. Alouette 2 1252 is also ex ALAT, but will be restored as a wheeled version in Aeronavale colours. Alouette 3 2084 is ex Armée de l'Air but is in Sécurité Civile markings. Likewise 2112 is ex Air Force (crashed 1992) and painted in Gendarmerie colours. Crashed Alouette 3 1968 was used as a spares source for the other two. Nowadays only small parts remain. The Gazelle (SA341 or SA342) frame was taken from the production line for fitting tests and has never flown. O-1E 24720 was sold to the USA in June 1995.

❏ 129		Alouette 2	1219	instructional	8-97
❏ 57	57-JL	Alouette 2	1071	instructional	8-97
❏ 236		Alouette 2	1440	instructional	8-97
❏ 1252		Alouette 2	1252	instructional	8-97
❏ 1730		Alouette 2	1730	instructional	8-97
❏ 2084	67-IK	Alouette 3	2084	instructional, crashed 16-4-84	8-97
❏ 2112		Alouette 3	2112	instructional, as 'GMS', crashed 8-9-89	8-97
❏ FR41	ARB	H-21C	FR41	instructional	8-97
❏ SA177		H-34A	SA177	instructional, marked as '177/F'	8-97
❏ ..		Gazelle	..	stored, see note	8-97
❏ ..		H-19D	4525	instructional, marked as '3RHC'	8-97
❏ ..	S	SA330	..	stored, cockpit only	8-96

GUISCRIFF - SCAËR (56)
Since the CM170 arrived by air on 18th October 1988, no reports have been received that it is still current here. It is certainly not outside but may reside in one of the newly built hangars here.

❏ 143	CM170	143	preserved	10-88

HÉNIN BEAUMONT (62)
Along the main road through town is the Parc Municipal. Here a Mystère 2C was preserved and was last noted in June 1991. By July 1995 it had moved to Touchay.

HOURTIN (33)
An unknown HSS-1 may still be preserved in the naval base.

❏ ..	HSS-1	..	preserved	4-86

HYÈRES - LE PALYVESTRE (83)
After the withdrawal from operational service of the CM175 Zéphyrs and Etendard 4Ms, a number could be found at Hyères. Etendard 4M 3 went to La Ferté Alais, 16 to Brive, 37 to Le Castellet, 40 to Bordeaux, 41 to Montmirault and 56 to Le Bourget. Some of the CM175s may have gone as well.

❏ 16	Br1050	16	stored, ex preserved	8-96

FRANCE - 94

☐ 48		Br1050	48	preserved	8-97
☐ 4		CM175	4	dumped	7-95
☐ 6		CM175	6	dumped, wreck, crashed 22-10-92	8-94
☐ 12		CM175	12	dumped	8-94
☐ 18		CM175	18	dumped, ex preserved	6-92
☐ 21		CM175	21	stored	8-97
☐ 7		Etendard 4M	7	stored	9-92
☐ 11		Etendard 4M	11	dumped, fuselage only	8-96
☐ 13		Etendard 4M	13	preserved, at gate	1-98
☐ 36		Etendard 4M	36	preserved	8-97
☐ 04		SA3210	04	stored	1-98
☐ 05		SA3210	05	stored	1-97

ISTRES (13)

Town: In the industrial estate, the yard of Etablissements Fondi had H-34A SA95. It moved to <u>La Valette</u>.

Airfield – Le Tubé: Stored Mirage 3C 3 moved to <u>Salon</u>, while N2501 145 from the fire dump has been scrapped. The Vautour and Mirage 3E can be seen from the road to the Dassault factory.

☐ ..	1	CM170	..	preserved, PdF c/s	6-96
☐ 202	FG	Mirage 3B	203	stored, ex Châteaudun	6-96
☐ 434	13-QK	Mirage 3E	434	dumped	1-98
☐ 147	CM	Mystère 2C	147	preserved, at gate	6-97
☐ 621	92-AN	Vautour 2B	621	dumped	1-98

JACOU (34)

The Aero 34 Group here had intended to restore an Alouette 3 and a Djinn. With no reports from here since December 1988, so it is not known if they succeeded. Note that this group also has a base at Le Cies.

☐ FR111	CAA	SO1221	FR111	stored, ex Bouilly	12-88
☐ FR132		SO1221	FR132	stored, ex Bouilly	12-88
☐ ..		SO1221	..	stored, ex Bouilly	12-88
☐ F-BRAQ		Alouette 3	1447	stored	12-88

LA BAULE - ESCOUBLAC (44)

The expected Super Mystère 179 from Captieux never arrived here, but went to <u>Vannes</u>. T-33 53091 of Ailes Anciennes - Le Baule also went to <u>Vannes</u>. N1101 177 was restored to flying condition and became F-GJBQ. Les Ailes Anciennes de la Baule-Pays de Loire have a number of 'warbirds' here.

☐ 133		MH1521M	133	stored, spares source for F-BMJO	—

LA FERTÉ ALAIS (91)

The Salis and other collections here comprise an enormous number of vintage aircraft. As most aircraft are airworthy or under restoration its been decided that for this edition a listing will be given of aircraft which do not fall into these two categories. While this is a shortcoming, it is no use just to copy the list from the first edition and some other publications as the exact status of all these aircraft is not known. Better to visit the airfield and the museum (with a tour through the restoration hangars) yourself. The aircraft noted as dumped are all to the west side of the airfield, away from the Salis Collection. Stored A-37B 68-7958 moved to <u>Le Castellet</u>.

☐ C.6-188	421-68	T-6G	168-160	stored, ex Cuatro Vientos, ex SpaAF, ex 49-3056	5-98
☐ 83		CM170	83	stored, ex Montmirault	7-93
☐ 3		Etendard 4M	3	preserved, ex Hyères	3-98
☐ 3	OH	N262A	3	preserved, ex F-BLHE	3-98
☐ 120	NR	N1101	120	dumped	3-98

FRANCE - 95

☐ 134	DZ	N1101	134	dumped	3-98
☐ 87		N3400	87	dumped	8-96
☐ 98		N3400	98	dumped	3-98
☐ 331	PN	NC702	331	dumped	3-98
☐ 114374	DD	T-6G	182-61	dumped, ex 51-14374	3-98
☐ 114707	KN	T-6G	182-394	dumped, ex 51-14707	3-98
☐ 115017	RD	T-6G	182-704	dumped, ex 51-15017	3-98
☐ F-AZBN		UC-64A	774	dumped, marked as 'D-ANBZ', ex 44-70509	6-95
☐ F-BLOZ		DC-3A	13142	stored, marked as '315101', ex F-BAXG	3-98

L'AIGLE (61)
The Aero Club de l'Aigle had their Flamant preserved at the airfield along the N26, west of l'Aigle. The aircraft was last noted in July 1995 and is now at <u>Albert</u>.

LAGNY SUR MARNE (77)
On the far eastern side op Paris, near the town of Lagny, is the Parc Euro Disney. In here a former Air Atlantic Skytrain is preserved.

☐ G-AMPP	C-47B	26717	preserved, marked as 'G-AMSU', ex KK136	93

LANDIVISIAU (29)
In a simulator building, the prototype Super Etendard 01 (rebuilt from Etendard 4M 68) is in use as an instructional airframe.

☐ 52	Etendard 4M	52	preserved, at gate	7-97
☐ 60	Etendard 4M	60	stored	6-97
☐ 01	Super Etendard	01	instructional, forward fuselage, ex Etendard 68	6-97
☐ ..	F-8E(FN)	..	instructional, forward fuselage only	6-97

LANEUVEVILLE EN SAULNOIS (57)
Stored in the back of one of the hangars is a Bird Dog. This small airfield is along the D955.

☐ 24504	AMK	O-1E	24504	stored, ex Nancy Essey	7-95

LANNION - SERVEL (22)
In the Eurocopter hangar a Broussard coded 'KAO' was noted, it probably came from a sale at Montauban. The aero club has a CM170 in their grounds.

☐ 60		CM170	60	preserved, PdF c/s	8-97
☐ ..	KAO	MH1521M	..	stored	6-92

LANVÉOC - POULMIC (29)
Although it has been reported that Etendard 50 should be preserved at the base, it has not been seen here. It may well be inside somewhere. The long standing HSS-1 is clearly visible.

☐ 51	Etendard 4M	51	preserved, on pole at dockyard	6-97
☐ 7	HSS-1	SA50	preserved, at gate, ex 58-1007	8-97
☐ 122	SA321G	122	stored, in hangar	6-96

LASCLAVERIES - THÈZE (64)
The clubroom at this microlight field consists of a Noratlas. The airfield is located north of Pau, on the east side

of the N134 road.
| ☐ 118 | | N2501F | 118 | preserved, with booms of 42 | 7-95 |

LATRESNE (33)
Just south east of Bordeaux is the Centre de Formation et de Perfectionnement de l'Aéronatique de Bordeaux (CFPAB). This school has received two aircraft from its closing partner, CFPAP at Villebon sur Yvette.
☐ 329		Alouette 2	1703	instructional	3-98
☐ 8		CM170	8	instructional, ex Villebon sur Yvette	3-98
☐ 96		CM170	96	instructional	3-98
☐ 01		Mirage F1E	01	instructional	3-98
☐ 68	NB	MS760	68	instructional	3-98
☐ F-ZWRU		Alpha Jet	02	instructional, ex Villebon sur Yvette	3-98

LA VALETTE (83)
The H-34 from the Istres estate should be preserved at a shop called 'Cargo'.
| ☐ SA95 | H-34A | SA95 | preserved | — |

LE CASTELLET (83)
Beside a number of flyable vintage aircraft, such as CM175 F-AZPF (28) and Broussard F-BVSS (55), there are also some stored and preserved aircraft at this racetrack airfield.
☐ 68-7958	A-37B	..	stored, ex La Ferté Alais, ex Vietnam AF	8-96
☐ 124541	F4U-5NL	..	under restoration, ex 0433/2-A-202/Arg Navy	8-96
☐ 37	Etendard 4M	37	preserved, ex Hyères	8-96
☐ 350	CM170	350	stored	8-96
☐ 379	CM170	379	stored	8-96

LE HAVRE (76)
Preserved at the local airfield on the north west side of town is a former Châteaudun Mirage 3E. Also here are some flyable warbirds including an ex RAF Jet Provost, a Swiss Alouette 2 and a Vampire.
| ☐ 616 | Mirage 3E | 616 | preserved, ex Châteaudun | 2-97 |

LE LUC - LE CANNET (83)
Preserved within the ALAT base is Vertol H-21C FR107. An H-19 with construction number 55-864 was hangared here at the open day in June 1989 and since moved to Dax.
| ☐ FR107 | LRZ | H-21C | FR107 | preserved, ex gate guard | 9-95 |

LEMPDES (63)
High above some buildings in a village on the east side of the airfield a colourful Viscount. It is preserved in Lempdes at the Concorde supermarket along the N89.
| ☐ F-BOEA | Viscount 708 | 12 | preserved, crashed 28-12-71, ex G-ARER | 9-97 |

LES ANDELYS (27)
The Musée Normandie-Niemen at the rue Raymond Phelip in Les Andelys has set up a small museum to recall the history of the unit in the second world war. It is about the 96 pilots who flew 5240 missions with 273 confirmed victories and 36 possible ones at the Eastern front. Preserved outside is the Mirage F1.
| ☐ 101 | 30-LI | Mirage F1C | .. | preserved | 7-95 |

FRANCE - 97

LES MUREAUX (78)
Noratlas 202 is still preserved at the Aerospatiale works here. The airfield is also home to a number of warbirds and a ALAT unit.

| ❏ 202 | N2501F | 202 | preserved | 4-97 |

LIMOGES - ROMANET (87)
Not much of this base (Base Aérienne 274, near the Zone Industrial Romanet) can be seen due to a high wall surrounding it. Their CM170 477 went to Feniers, but they should still have:

| ❏ 7 | 2-FD | Mirage 3C | 7 | preserved | 89 |
| ❏ 63 | | Mystère 4A | 63 | preserved | 89 |

The Musée de la Resistance du Centre has completed restoration of ex Luftwaffe Reggiane Re2002 OV-BI whose identity could possibly be No.15.

| ❏ .. | | Re2002 | 15 | preserved, marked as 'OV-BI' | 7-92 |

LOHEAC
Preserved with a local automobile museum is a Magister.

| ❏ 169 | | CM170 | 169 | preserved | 97 |

LONS - LE SAUNIER (39)
The preservation group here maintains a number of aircraft in flying condition. Their last stored *Wrecks & Relics* aircraft, Flamant 217 became F-AZFS and moved to Lyon.

| ❏ F-EURO | | N2501 | 148 | preserved | 96 |

LORIENT (56)
Town: Preserved within the Ecole des Fusilliers-Marins (within the naval arsenal alongside the River Blavet) is Sikorsky HSS-1 144.

| ❏ 144 | | HSS-1 | .. | preserved | 89 |

Airfield - Lann Bihoué: With the replacement of the older Atlantics by the Nouvelle Génération Atlantiques, a number of the former type were stored here and scrapped. Sold for scrap in 1996 were 65, 66, 67 and 68. One Atlantic is preserved and may have replaced the Texan which has not been seen for some years. The two dumped SP-2Hs 147562 and 147568 have not been reported since 1990 and must have been scrapped.

❏ 86		Br1050	86	preserved, at gate	8-97
❏ 7		Br1150	7	stored	8-97
❏ 85104		JRB-4	..	preserved, ex Rochefort	6-96
❏ 60		Etendard 4M	60	stored	6-93
❏ 43981		SNJ-5	..	preserved	89
❏ 147567		SP-2H	..	preserved, at gate	8-97
❏ 147569		SP-2H(ECM)	..	stored, to go to Plobannalec	6-96

LUCHEUX (80)
Still preserved at the radar station (CDC05.925) at the top of a hill (near Doullens) here is Super Mystère 154.

| ❏ 154 | 12-YA | Super Mystère B2 | 154 | preserved | 9-97 |

LUNÉVILLE (54)
Some 30km east of Nancy is the town of Lunéville. Inside some hangars at the local airfield are two *Wrecks & Relics* aircraft. The Magister used to be preserved at the French base at Berlin Tegel.

FRANCE - 98

☐ C-555		C3603	..	stored	7-97
☐ 545		CM170	545	stored, PdF colours, ex Berlin Tegel	7-97

LUXEUIL (70)
Town: A local museum's Vautour 358 was stored at the Barreau Combustibles Carburants in the Rue Colonel Thienault (ZI des Achelots). The aircraft was seen their in April 1992 and it has moved to Orange in late 1997

Airfield - St Sauveur: The dumped Mirage 2000s here were no more then just some recognisable parts. Stored CM170 427/4-WA moved to Peutie.

☐ 202	339-WW	CM170	202	preserved	5-98
☐ 166		MD312	166	dumped	6-96
☐ 568	4-BI	Mirage 3E	568	preserved, on pole	5-98
☐ 572	BR	Mirage 3E	572	preserved	5-98
☐ ..		Mirage 3	..	dumped	5-92
☐ 321	4-BJ	Mirage 2000N	..	dumped, remains only, crashed 17-5-90	5-92
☐ 328	4-CN	Mirage 2000N	..	dumped, remains only, crashed 9-4-91	5-92

LYON (69)
Town – Mont Verdun: On one of the hills on the north side of Lyon, along the A7 highway is Base Aérienne 942. Preserved at this military ATC station is the real Super Mystère 148 (see Cambrai).

☐ 148	12-YH	Super Mystère B2	148	preserved	9-95

Airfield – Bron: At this civil airfield at the south side of Lyon an ex Gendarmerie Cessna is stored.

☐ 2147	JAB	CeU206F	2147	stored	3-97

Airfield – Corbas: The airfield here is split into civil and military areas. At the ALAT gate is a an H-19 on which only the code AVV is readable. At the civil side it was intended to set up a small museum, but that has failed. Collection'saircraft are all stored outside

☐ 460		CM170	460	stored, ex Châteaudun	4-90
☐ ..	AVV	H-19	..	preserved	8-97
☐ 189		MD312	189	stored	5-98
☐ 96		MS733	96	stored	5-98
☐ F-AZFS		MD312	217	stored	5-98
☐ F-GNYN		CM170	532	stored, ex Châteaudun	6-96

Airfield – Satolas: The Caravelle at the international airfield does not carry its original colours and has no registration applied. It is marked Aeroport de Satolas.

☐ F-BHRM		SE210-3	37	preserved, unmarked	7-96

MARMANDE (47)
On display at the aerodrome was Noratlas 111/63-VF, but it was not noted here in 1995. It had moved on by then to Villeneuve sur Lot. The only 'aircraft' noted here in 1995 was the mock-up fuselage of a Falcon 50.

MARSEILLE - MARIGNANE (13)
Town: Preserved on a roundabout, along the D9 road, in the village of Marignane is an unmarked Alouette 2.

☐ ..		Alouette 2	..	preserved, unmarked, white c/s	8-97

Airfield: The damaged Cameroon Air Force C-130 TJX-AC (4747) was first noted in the summer of 1990. The C-130H was shipped from rebuilt Bordeaux by boot in 1997. As part of a sale of Cougars to the Netherlands Air Force, Aerospatiale took some Alouette 3s in return. Of these A-267 went to for Orange and A-414 went to Dax.

The were CL-215s recently withdrawn from use and replaced by CL-415s. Going back in time, the two ex AMI Trackers mentioned in the first edition have long been scrapped.

☐ TJX-AC		C-130H	4747	stored, ex Cameroon AF	12-96
☐ A-226		Alouette 3	1226	stored, ex RNethAF	12-96
☐ A-281		Alouette 3	1281	stored, ex RNethAF	12-96
☐ A-336		Alouette 3	1336	stored, ex RNethAF	12-96
☐ A-383		Alouette 3	1383	stored, ex RNethAF	12-96
☐ 2x		SA321	..	stored, unmarked	5-98
☐ F-BNKE		SE210-3	224	preserved, near terminal	8-94
☐ F-BTOA		SE210-12	274	stored, fuselage only, wfu 6-4-91	96
☐ F-BTTH		Mercure 100	8	stored	5-98
☐ F-GCVM		SE210-12	270	stored, ex OY-SAA	5-98
☐ F-GJGC		G159	111	stored	7-93
☐ F-ZBBE	05	CL-215	1005	stored, ex Sécurité Civile	5-98
☐ F-ZBAR	21	CL-215	1021	stored, ex Sécurité Civile	5-98
☐ F-ZBAY	23	CL-215	1023	stored, ex Sécurité Civile	5-98
☐ F-ZBBD	29	CL-215	1029	stored, ex Sécurité Civile	5-98
☐ F-ZBBH	26	CL-215	1026	stored, ex Sécurité Civile	5-98
☐ F-ZBBJ	28	CL-215	1028	stored, ex Sécurité Civile	5-98
☐ F-ZBBT	40	CL-215	1040	stored, ex Sécurité Civile	5-98
☐ F-ZBBV	46	CL-215	1046	stored, ex Sécurité Civile	5-98
☐ F-ZBBW	47	CL-215	1047	stored, ex Sécurité Civile	5-98
☐ F-ZBDD	24	CL-215	1024	stored, ex Sécurité Civile	5-98

Noted in the late 1980s were a number of helicopters stored in and around the maintenance hangars. Of these SA360 F-WSQL went to <u>Rochefort</u>. All the others have gone (all scrapped or have found new homes).

☐ 944	HSS-1	..	stored	5-90
☐ FC-22	SO1221-PS	013	stored	87
☐ F-BTRP	SA321F	01	stored	87
☐ F-WXFI	SA341	1515	stored	6-89
☐ F-WZAK	SA361H	1012	stored	87
☐ ..	SO1221	001	stored	89

MEAUX - ESBLY (77)
Magister 31 left the local airfield for the aero club at <u>Brienne le Château</u>.

MELUN - VILLAROCHE (77)
The SNECMA factory here has a private museum displaying many engines and some aircraft. These are a Vautour and a Mirage 3C. The museum is only open on Wednesdays from 09:00. On the other side of the airfield, a Mirage 3R is stored in the CEV hangar together with some civil CM170s and warbirds.

☐ 21	Mirage 3C	21	preserved	3-98
☐ 318	Mirage 3R	318	stored	5-98
☐ 337	Vautour 2N	337	preserved	3-98

MENGAM (29)
It is still not known if the Mirage 2000 at the Thomson-CSF facility here is a genuine airframe or a test rig. Definitely a genuine article was the ex Sembach F-4 64-0922 which was also here. It moved on to <u>Plobannalec</u>.

MERVILLE - CALONNE (59)
The IAAG (Institut Aéronautique Amaury de la Grange) posesses a number of instructional airframes for

teaching engineering skills. The engines of some of the aircraft are still installed and engine runs are carried out once in a while. The Caravelle and Super Mystère are parked inside, the rest outside. As all hangars were checked in 1995, some of the aircraft mentioned in the first edition have left; AB47G 028 was scrapped and Super Mystère 173 went to Pau. Gone, with their fate unknown, are Be65s F-BKBU and F-BRPL and Cessna 310 F-BKBS.

❑ 129		N2501F	129	preserved	6-96
❑ 167		Super Mystère B2	167	instructional	6-95
❑ F-BHRT		SE210-3	55	instructional	6-95
❑ F-BHSF		B707-328	17618	preserved	6-96
❑ F-BMCF		Viscount 724	54	preserved, ex CF-TGQ	6-96
❑ F-BOJA		B727-228	19543	instructional	6-95
❑ F-GCVK		SE210-12	276	instructional, arr 6-93, ex OY-SAG	6-96

METZ - FRESCATY (57)
The F-84 on the cover of the first edition could been seen from outside. It has not been seen since the early 1990s and may have left (or are the trees getting too big ?). There is still one N2501 at the airfield which has also incorrectly been reported as 42.

❑ 28879	9-AU	F-84F	..	preserved, ex 52-8879	3-90
❑ SA170	67-EH	H-34A	SA170	preserved	4-91
❑ 41		N2501GABRIEL	41	preserved, silver c/s	7-97
❑ 114314	VM	T-6G	182-1	preserved, ex 51-14314	3-90

MOISSAC (82)
A Broussard lies derelict on this small airfield.

❑ 127		MH1521M	127	stored	8-96

MOISSELLES
This airfield has the stored fuselage of a Nord 3400. The exact location of Moisselles is unknown.

❑ 45		N3400	45	stored	6-97

MONTAUBAN (82)
The main ALAT unit responsible for the overhaul, storage and disposal of their aircraft is at Montauban. The preserved Djinn is not visible from outside, while the Nord 3202 was not noted in 1993.

❑ 281	MJC	MH1521M	281	stored, wfu 1-7-93	7-93
❑ ..	MDZ	N3202	..	preserved, ex Bourges	2-89
❑ 24572	BVW	O-1E	24572	stored, ex Berlin	6-93
❑ ..		SO1221	..	preserved	6-93

MONTCEAU LES MINES (71)
The l'Association de Restauration d'Aeronefs et de Collectionneurs de Materiel Aéronautique du Musée SV-4 Aero obtained a new MD312 after their own Flamant F-AZEP (286) crashed. This Flamant, 237, found its way onto the civil register and becoming F-AZFE. Strong rumour has it that two Super Mystère B2s should also be here.

MONT DE MARSAN (40)
Town: A Noratlas is preserved in the town at the Army barracks of the 6RCP. The barracks are alongside the D932 which runs from the town centre to the east.

❑ 186	328-EJ	N2501SNB	186	preserved, with tail from 105	6-97

FRANCE - 101

Airfield: A Mystère and Mirage 4 are parked behind the main gate. The Vautour and Super Mystère are preserved on the west side of the base. Nearby is a dump which held a Mirage 3B and an unknown Mirage 3 (upside down, maybe 48/10-RK). Dumped in a car scrapyard at the airfield was a Mirage without a tail.

❑ 542	330	CM170	542	preserved, near flighline	7-95
❑ 48	10-RK	Mirage 3C	48	dumped	6-91
❑ 215	13-FC	Mirage 3B	215	dumped	6-97
❑ 405		Mirage 3E	405	dumped	7-95
❑ 436		Mirage 3E	436	dumped	7-95
❑ ..		Mirage 3	..	dumped, no tail	6-97
❑ 43	BP	Mirage 4A	43	preserved, behind main gate	7-95
❑ 601		Mirage F1CR	..	preserved	3-98
❑ 234		Mystère 4A	234	preserved, behind main gate	7-95
❑ 153	12-YY	Super Mystère B2	153	preserved	6-97
❑ 364	92-AY	Vautour 2N	364	preserved	6-97

MONTELIMAR - ANCÔNE (26)

The Musée Europeen de l'Aviation de Chasse at the local airfield exchanged some aircraft in 1995. Magister 38 left for <u>Savigny lès Beaune</u> in a swop for an ex Belgian Thunderstreak, while Magister 52 went to <u>Berlin Gatow</u> in exchange for the Bronco, 9927. This will be used for spare parts to keep the former 9924, now F-AZKM airworthy. The Mirage 3EX was an experiment from Dassault, fitted with canards and more modern equipment.

❑ FU-26		F-84F	..	preserved, ex Savigny, ex Belgian AF	5-98
❑ 2003	6	MiG-23MF	0390313097	preserved, ex Manching, ex 582/NVA	5-98
❑ 2394		MiG-21U-600	663820	preserved, ex Rothenburg, ex 281/NVA	5-98
❑ 9454		T-33A	9117	preserved, ex Manching, ex 53-5778	5-98
❑ 9858		G91T/3	0021	preserved, ex Manching, ex 3419/WGAF	5-98
❑ 9927		OV-10B	338-12	preserved, ex WGAF, ex D-9956	5-98
❑ XD613		Vampire T.11	..	preserved, ex RAF	5-98
❑ J-4067		Hunter F.58	..	preserved, ex Swiss AF	5-98
❑ 57		CM170	57	stored	5-98
❑ 101		CM170	101	preserved	5-98
❑ 150		CM170	150	stored	5-98
❑ 172		MD312	172	preserved	5-98
❑ 214		MD450	214	preserved, PdF c/s	5-98
❑ 124		MH1521M	124	preserved	5-98
❑ 55		Mirage 3C	55	preserved, ex Rochefort	5-98
❑ 467		Mirage 3EX	467	preserved, unmarked	5-98
❑ 48		Mystère 4A	48	preserved, ex Bordeaux	5-98
❑ 21113		T-33AN	113	preserved, in USAF marks, ex Saintes	5-98
❑ ..		Mirage G8	..	preserved, replica with original parts	3-96
❑ F-AZAI		MD312	228	preserved	5-98
❑ F-GGKS		MH1521M	23	preserved, ex Pau	5-98
❑ N56NA		C-53C	4979	preserved, ex N400RS	5-98

MONTLAUR EN DIOIS (26)

At a farm in this small village along the D93 (Gap-Valence) the mortal remains of a number of HSS1/S58s can be found. Bits and pieces were found in August 1996 at the site which could maybe amount to one complete airframe. Serials noted were 139 and 954 (both ex Aeronavale).

MONTLUÇON (03)

The Zeland-Gazuit factory had an Ouragan on display by January 1987. Unfortunately, neither the factory itself nor the Ouragan were found when searched for in May 1996.

MONTMIRAULT (91)
The Lycee d'Enseignement Alexandre Denis is located along the main road through the village which goes from Etampes to La Ferte Alais (how convenient!). The school uses a nice collection of instructional airframes which are all located in one building. The aircraft can be seen (with some effort) from outside the building. Also at the school are the tailbooms of a Noratlas (98/64-BC, ex Etampes). Magister 83 moved to the nearby airfield of La Ferté Alais. Wa421 F-BOYV (408) and HR200/100 F-WSQP (01) have not been seen for some years and have been deleted.

❏ 88		Alouette 2	1106	instructional	3-98
❏ 41		Etendard 4M	41	instructional, ex Hyères	3-98
❏ 1376		HSS-1	..	instructional, ex 58-1376	3-98
❏ 164		MD312	164	preserved	3-98
❏ 179		MH1521M	179	instructional	3-98
❏ 191		MH1521M	191	instructional	3-98
❏ 418	QQ	Mirage 3E	418	instructional	3-98
❏ F-BJBG		T-6G	88-16151	instructional, ex 41-33966	3-98
❏ F-BJUP		RC680F	1064-61	instructional	3-98

MONTPELLIER (34)
Town: The Lycee J Mermoz at Montpellier has no known instructional airframes left. Super Mystère 72 moved to Berlin. CM170 126/F-WZLQ, which arrived at the school in 1988, moved on to the airfield below and moved to Alès Deaux a few years later.

Airfield – Fréjorgues: This airfield is the home base of Air Littoral. Some aircraft at their ESMA training school are used as instructional airframes. The airfield also has some stored aircraft.

❏ F-BTTE	Mercure 100	5	instructional, arr 24-11-94	97
❏ F-BUAN	A300B2-1C	132	stored	7-93
❏ F-GBEJ	N262A-11	7	instructional, ex HB-ABC	97
❏ F-GGVP	B737-2K2C	20943	stored, ex VT-EKC	7-93
❏ F-WZLQ	CM170	126	stored, dismantled, see Alès Deaux	97

MORLAIX (29)
After they replaced their Mercures with Fokker 70s, Air Inter did their best to preserve their Mercures as far from each other as possible. The above mentioned F-BTTE at Montpellier was placed in the very south of France, F-BTTG is located at Morlaix, some 50km from Brest, in the western side of France.

❏ F-BTTG	Mercure 100	7	stored, arr 15-11-95	6-97
❏ OY-TOV	N262	54	instructional, ex N91205	10-95

MORTAGNE AU PERCHE (61)
At this airfield is yet another Magister.

❏ 136	CM170	136	preserved	6-97

MOURMELON LE GRAND (51)
The hulk of prototype SA330 Puma F-ZWWR lies derelict in a military camp here.

❏ F-ZWWR	SA330	05	dumped	9-88

MULHOUSE (68)
The Musée de l'Automobile is established in a mill near the airfield of Mulhouse and the Swiss town of Basel. Beside hundreds of motor cars, a Vickers Viking is also displayed.

❏ G-AIVG	Viking 1B	220	preserved	97

NANCY (54)

Airfield – Essey: A Bird Dog is preserved at the ALAT side of the airfield. It made a heavy landing while owned by the aero club and has since been restored to display status.

☐ F-GCSB		O-1E	24508	preserved, ex 24508/AOI and marked as such	12-93

The Musée de L'Aéronautique opened its purpose-built building on the airfield of Nancy Essey in 1992. The collection at Nancy's Parc de Haye and the stored aircraft at the ALAT airfield all found a new home here. Note that the Mirage 3C 27 is the original aircraft. The aircraft at Toulouse and Koblenz carry false markings

☐ FU-76	'11-RQ'	F-84F	..	preserved, ex '28946/3-IV', ex Belgian AF	6-98
☐ J-4107		Hunter F.58A	..	preserved, ex Meiringen, ex Swiss AF	6-98
☐ 2196		F-104G	7065	preserved, ex WGAF	6-98
☐ 3093		G91R/3	357	preserved, ex WGAF	6-98
☐ 882		MiG-21SPS	94A5207	preserved, ex NVA	6-98
☐ 046	BDS	AB47G	046	preserved	6-98
☐ 24	3-KB	CM170	24	preserved	6-98
☐ 19600	3C-Q	F-84E	..	preserved, ex Nîmes, ex 51-9600	6-98
☐ 52734		F-100D	224-1	preserved, as '42272/11-EF', ex Châtillon	6-98
☐ 55-3181	APY	H-19D	..	preserved	6-98
☐ SA59		H-34A	..	preserved, tail marked 80	6-98
☐ K33		Ka6	333	preserved, glider, Le Bourget owned	6-98
☐ 260		MD311	260	preserved, wfu 5-12-84	6-98
☐ 8		MD450	8	preserved	6-98
☐ NF11-9	BF	Meteor NF.11	..	preserved, ex WM304/RAF, Le Bourget owned	6-98
☐ 27	3.10-LE	Mirage 3C	27	preserved	6-96
☐ 571	3-XB	Mirage 3E	571	preserved, ex Nancy Ochey	6-98
☐ 173		MS733	173	preserved, with wings of 186	6-98
☐ 23	314-ZK	Mystère 4A	23	preserved, ex code 7-CP	6-98
☐ 189	316-FN	N2501F	189	preserved, wfu 30-11-87	6-98
☐ 100	FIA	N3400	100	preserved, ex code 'MIA'	6-98
☐ 148334		SP-2H	7262 ?	preserved, Le Bourget owned	6-98
☐ 113	12-YQ	Super Mystère B2	113	preserved, ex 12-ZK, wfu 12-7-77	6-98
☐ 35061		T-33A	8400	preserved, ex 53-5061, with tail from 21127	6-98
☐ 41553	33-XU	RT-33A	9184	preserved, ex 54-1553	6-98
☐ 307		Vautour 2N	307	preserved, Le Bourget owned	6-98
☐ F-BHPH		N1203	355	stored, fuselage only	6-98
☐ F-BHRY		SE210-3	61	preserved	6-98
☐ F-BKYD		MS880B	289	preserved	6-98
☐ F-BLQV		N1101	143	preserved	6-98
☐ F-GEOA		C-47B	16004/32752	preserved, marked as '2108979/H2'	6-98
☐ F-GGGD		MH1521M	290	preserved, ex 290/11-OB and marked as such	6-98
☐ F-ZWWL		CM173/Potez 94A	01	preserved, Le Bourget owned	6-98
☐ ..		HM-14	..	preserved	

Outside, at the back of the museum, a number of aircraft are stored. Some are waiting to get in, others have had their time inside and moved outside recently.

☐ 459		CM170	459	stored, ex F-WGGC	6-95
☐ ..	AOP	H-21C	..	stored	6-90
☐ 148		MD312	148	stored, ex F-WZXA	6-98
☐ 54	X	N2501F	54	stored	6-98
☐ 97	63-WB	N2501F	97	stored	6-98
☐ N49549		KC-97L	17062	stored, at airfield, ex Le Bourget, 53-0280	6-98

With the closure of the store at Parc de Haye some aircraft have gone missing. They may have moved on or still be held in storage. O-1E 24504 went to Laneuveville, while N3202 64 became F-AZJK in 1996. T-33A 21255 and CM170 30 are now at Reims.

FRANCE - 104						
☐ 65	67-JL	Alouette 2	1070	preserved		3-96
☐ 198		MH1521M	198	preserved, Parc de Haye		6-89
☐ 59		N3202	59	preserved		6-89
☐ 100	AJE	N3202	100	preserved		—
☐ 51	AYM	NC856	51	preserved		—
☐ 108		NC856	108	preserved		5-89
☐ FR144		SO1221	FR144	preserved, Parc de Haye		6-89
☐ F-BDQT		MS505	700/36	preserved, Parc de Haye		—
☐ F-BEXU		F24R	1063	preserved, Parc de Haye, ex KK445		—
☐ F-CBEL		Castel C311P	300	stored, glider		—
☐ F-CCFS		Br904S	9	preserved, glider, Parc de Haye		—
☐ F-CRMI		Castel C25S	169	preserved, glider, Parc de Haye		—
☐ F-PCRL		Bü181	112/121	preserved, glider, Parc de Haye		—

Airfield – Ochey: Preserved at the entrance to the base is Mystère 4A 185, this aircraft was reported in 1992 with the serial 183. Stored Mirage 3E 571/3-XB went to Essey. Mirage 3E 617/3-XA has been reported here as stored, but this aircraft should still be at Châteaudun.

☐ 217	2-FE	Mirage 3B	217	stored	6-98
☐ 333		Mirage 3R	333	stored, marked as '469/3-II', ex Châteaudun	6-98
☐ 498	'3-EC'	Mirage 3E	498	preserved, ex code 3-XE	6-98
☐ 608	3-XE	Mirage 3E	608	stored	6-98
☐ 185		Mystère 4A	185	preserved, at gate	6-98

NANTES - CHÂTEAU BOUGON (44)
The Super Constellation is preserved at the airfield's entrance. Fairchild F-27J F-GDXY (85) was broken up here for spares and its fuselage was noted at Le Bourget in May 1993.

☐ 224		Mirage 3B	224	stored, ex Châteaudun	6-97
☐ 18		N2501F	18	stored, special c/s	6-97
☐ F-BRAD		L1049	4519	preserved, ex F-BGNJ	6-97

NARBONNE (11)
A Super Mystère is preserved at the gate of radar site of Detachment Air 90.115. The site is on a hill near to Narbonne Plage, which is clearly marked from the Narbonne Centre exit of the highway. An ex Rochefort Magister should also be here.

☐ 377		CM170	377	preserved, ex Rochefort, ex Châteaudun	—
☐ 54	5-NO	Super Mystère B2	54	preserved, at gate	2-97
☐ ..		MH1521M		preserved	2-97

NICE (06)
Town: The Musée du Train Miniature had moved here from nearby St Jacques de Grasse by early 1990.

☐ 135		CM170	135	preserved, ex St Jacques de Grasse	—

Airfield – Côte d'Azur: Preserved just outside the airport is Caravelle F-BJTH, which arrived on 7th July 1981. She currently wears an aluminum/yellow/black colour scheme.

☐ F-BJTH		SE210-3	124	preserved, ex F-WJTH	6-97

NÎMES (30)
Airfield - Courbessac: Nîmes Courbessac can be split into two parts. Base Aérienne 726 is on the west side of the N86, the road from the A9 highway. Preserved here are some aircraft which are very difficult to see from outside. The Magister can be seen from the road between the airfield and the base, just before the roundabout.

F-84E 19600 moved on to Nancy and H-34A SKY479 to Dijon. In 1995 a fighter could be seen within the barracks, but it could not be determined whther it was the Mystère or Ouragan.

❏ 17		CM170	17	preserved, ex instructional	5-96
❏ SA163		H-34A	SA163	instructional	9-87
❏ 232	SC	MD450	232	preserved, between buildings	8-94
❏ 143	5-OP	Mystère 2C	143	preserved, near gate	5-90

At the other side of the road is the airfield. Although it is mainly a civil airfield, the military have a small section here where the Noratlas is preserved. In the past the Ouragan could also be found here.

❏ 78	312-BE	N2501F	78	stored	5-96
❏ F-AZEH		MD311	274	stored	8-95
❏ F-AZGE		MD312	158	stored, unmarked	7-95

Airfield – Garons: With the introduction of the Atlantique NG (or ATL2 as they are sometimes called), many Br1150 Atlantics became surplus. A large number of them were stored and later scrapped at Nîmes Garons. In 1996 only some dozen aircraft were still here. Noted here before 1995 were 1 (l/n 8-93), 9 (to Plaisir), 11 (l/n 9-94, '100.000 heures' marks), 13 (l/n 9-94), 15 (l/n 6-93), 23 (l/n 8-94, '100.000 heures' marks), 27 (l/n 10-92, to Geneve), 37 (l/n 9-94), 38 (l/n 6-93) 47 (l/n 10-92), 44 and 50 were both sold for scrap in late 1995, 53 (l/n 5-96) and 57 (l/n 5-96) went to Pakistan. During 1992 some former Swiss Vampires were temporarily stored here before the moved on or became civil. All were here too briefly to be included. Besides the Atlantics, some civil airliners also found their last resting place at Nîmes Garons. Scrapped were SE210 F-BUZC (94), F-GATZ (175) and F-GAPA (99). Caravele 11 F-BRGU (237) was destroyed here for a movie in 1994, while stored F-GGKD (255) returned to live again as HK-3913X. DC-8-33 F-GDRN (46091) was scrapped here in late 1988.

❏ 01	Atlantique NG	01	stored, ex Br1150 42	5-96
❏ 5	Br1050	5	preserved	5-98
❏ 11	Br1050	11	dumped	6-97
❏ 51	Br1050	51	dumped	6-97
❏ 76	Br1050	76	dumped	6-97
❏ 04	Br1150	04	stored, ex Atlantique NG 001	5-98
❏ 3	Br1150	3	stored	5-96
❏ 5	Br1150	5	stored	5-96
❏ 25	Br1150	25	stored, marked as '400.000 heures de Vol'	1-98
❏ 31	Br1150	31	preserved	5-98
❏ 35	Br1150	35	stored	5-98
❏ 41	Br1150	41	stored	3-95
❏ 48	Br1150	48	stored	5-95
❏ 49	Br1150	49	stored	8-96
❏ 51	Br1150	51	stored	1-98
❏ 54	Br1150	54	stored	8-96
❏ ..	Br1150		stored, marked 'Br1150 Atlantic'	3-96
❏ 87	C-47	4579	preserved, ex EI-ALR, ex 41-18487	5-98

NOYANT SUR ALLIER (03)
Broussard 156 is still here. It is in a fenced compound on a road 200 meters from the village square.

❏ 156	MH1521M	156	preserved	6-96

ORANGE (84)
Town: The Aviomodelli France Company at an industrial estate on the west side of Orange Centre is setting up a collection of aircraft. Currently there are no plans to make it into a museum, the owner currently only wants to save aircraft from scrapping.

❏ 0906	MiG-21F-13	960609	stored, ex Kbely, ex Letnany, ex Czech AF	5-98
❏ A-267	Alouette 3	1267	stored, unmarked, ex Marseille, ex RNethAF	5-98

FRANCE - 106

❏ J-1183		Vampire FB.6	692	stored, ex Sion, ex Swiss AF	5-98
❏ U-1227		Vampire T.55	987	stored, ex Swiss AF	5-98
❏ 409	5-MD	CM170	409	stored, ex Châteaudun	5-98
❏ 5	44-CA	MH1521M	5	stored	5-98
❏ 222	13-FF	Mirage 3B	222	stored, ex Châteaudun	5-98
❏ 241		Mirage 3B-RV	241	stored, ex Châteaudun	5-98
❏ 363		Mirage 3RD	363	stored, inside, ex Châteaudun	5-98
❏ 526	13-QU	Mirage 3E	526	stored, ex Orange Caritat	5-98
❏ 52	10-LF	Mystère 2C	52	stored, inside, ex Caen	5-98
❏ 53	12-YO	Super Mystère B2	53	stored, ex Brive, ex Châteaudun	5-98
❏ 73	12-ZI	Super Mystère B2	73	stored, ex Bourges	5-98
❏ 358		Vautour 2N	358	stored, ex Luxeuil, Le Bourget owned	5-98

On the west side of Orange, near Martignan (on the other side of the motorway), an other collection can be found with the name Lafayette War Museum. The aircraft are all stored outside.

❏ FX-93		F-104G	9160	stored, ex Aix, ex Belgian AF	5-98
❏ 4406		MiG-21PFM	94A4406	stored, ex Letnany, ex Vodochody, ex Czech AF	5-98
❏ 193		N1101	193	stored, fuselage only	2-97
❏ F-BSEQ		SIAI S205/20R	4-237	stored, ex LN-VYY	2-97

Airfield – Caritat: The Mirage 2000 unit here has a large number of preserved aircraft. Four aircraft are gathered near the main gate of the base. Some serials look doubtful as they are from crashed aircraft.

❏ 68	5-NI	Mystère 4A	68	preserved	5-96
❏ 143	5-WT	Mystère 2C	143	preserved	5-96
❏ 121	5-OL	Super Mystère B2	121	preserved, crashed 18-5-71	5-96
❏ 38	5-OC	Mirage 3C	38	preserved, ex Rochefort	5-96
❏ 326	OU	Mirage 3R	326	preserved, crashed 20-3-65	5-96
❏ 361	33-TJ	Mirage 3RD	361	stored, ex Orange town, ex Strasbourg	98
❏ 424	13-Q.	Mirage 3E	424	dumped, crashed 8-10-70	5-96
❏ 426	13-QG	Mirage 3E	426	preserved	5-96
❏ ..		Vampire	..	preserved, as '10035/5-NT', ex Swiss AF ?	5-96

ORLÉANS (45)
Town: The Musée de Para holds the fuselage of a N2501 and parts of a second example. The museum can be found in the northern side of Orléans in a garden in the residential area.

❏ 125		N2501F	125	preserved, ex Châteaudun	3-98

Airfield – Bricy: The Super Mystère which was presumed to be gone, was rediscovered on a secondary gate. Noratlas 155 on the dump has been scrapped.

❏ 29	61-QP	N2501F	29	preserved	5-98
❏ 9		Super Mystère B2	9	preserved	9-94
❏ F-WESE		C-160V	V3	stored, for spares, ex TR-LWE, ex 5002/WGAF	6-97

PARIS (75)
Town: During 1992 there was an exhibition at the Centre Georges Pompidou named 'Manifesta'. The exhibition was about design and a Mirage 3E was included. It is not known if the Mirage is still here.

❏ 423	QQ	Mirage 3E	423	preserved	10-92

Town - Porte de la Villette: The Cité des Sciences et de l'Industrie museum, near the Porte de la Villette peripherique exit, has a Mirage 4. The aircraft is on the second floor and has been displayed since 1992. Opening hours are between 10.00 and 18.00 (Sundays to 17.00).

❏ 45	BR	Mirage 4A	45	preserved, ex Châteaudun	10-96

FRANCE - 107

Town - Vilgenis (91): The Centre d'Instruction de Vilgenis (Air France Technical School) is located west of Orly airport on the D120 between Massy and Vilgenis. A complete tour of the facility in 1995 found a reduced number of aircraft, T-33s 16520 (5852), 16524, 16525/OF (5857) and 35339 (8678) were not noted. Of these 16524 went to St Dizier. Unfortunately not all the civil aircraft were recorded in 1995.

❑ 18693	13-TF	T-33A	6477	instructional, ex 51-8693	6-95
❑ 29867	314-WL	T-33A	8173	instructional, ex 52-9867	6-95
❑ F-BHAH		SAAB 91C	91296	instructional, under restoration, ex SE-XAF	85
❑ F-BHAJ		SAAB 91C	91298	instructional, spares source, ex SE-XAH	85
❑ F-BHRA		SE210-3	1	instructional, wfu 19-12-75, ex F-WHRA	5-96
❑ F-BHSL		B707-328	17919	instructional	5-96
❑ F-BJBH		AT-6D	88-15778	instructional	6-95
❑ F-BJBJ		AT-6C	88-12046	instructional	6-95
❑ F-BJBS		AT-6F	121-42497	instructional, ex 44-81775	6-95
❑ F-BJTO		SE210-3	148	instructional, nose and tail only	85

Airfield – Charles de Gaulle (95): During cabin cleaning work on a newly delivered Air France A340, the aircraft caught fire and burned out. The wreck was still to be found near the maintenance hangars in 1994

❑ F-GNIA	A340-211	10	dumped, damaged 20-1-94	3-94

Airfield – Le Bourget (93): The airfield of Le Bourget needs no further introduction. Stored C-47As F-GEFU (19074) and F-GEFY (14152/25597) were scrapped (noses to Cranfield, UK). Stored F-27J F-OGJC (107) was last noted in 1989 and has gone. Stored (non museum aircraft) are:

❑ 6W-SAV	Rallye 235A	3380	stored, fuselage only, ex Senegal AF, ex F-ZVLI	6-97
❑ F-BIUK	F27-100	10247	stored	5-98
❑ F-GCJO	FH-227B	530	stored, ex N703AU	4-97
❑ F-GDXY	F-27J	85	stored, fuselage only, ex Nantes	5-93
❑ F-GKGA	SN601	11	stored	6-97
❑ F-GKGD	SN601	24	stored	6-97
❑ F-OGJC	F-27J	107	stored, ex F-GBRT	5-98
❑ F-OGOQ	Do228-201	8056	stored	6-97
❑ TJ-AGH	CASA 212-200	159	stored	5-98
❑ TN-311	Alouette 3	1918	stored, wreck, ex Congo AF	6-97

Le Bourget houses the famous Musée de l'Air et de l'Espace. A large fire on 11th May 1990 in the Dugny storage of the museum destroyed the following aircraft: Amiot 351 117, Beech E18S F-SCBF (BA-68), Bristol 156 Bolingbroke 9947 (11-880-264), Maurice 50 F-PHFI (1), Dassault 320 F-WPXB (01), DH.82A R5238 (83097, marked as 'K2570'), Dewoitine VII 16, Dumorald Autoplan F-WFOQ (01), F-5G Lightning 44-53247 (8502), Mignet HM-8, MS149 F-AJFJ (29), MS1500 F-ZJND (01), MS472 283, MS505 F-BCMD (599/34), MS XXIX AI 1598, N3202 42/AJC (42) and 67/AIA (67), N3400 70/MJA (70), and 108/JBY (108), B-25J N7681C (108-37495, ex 44-86701), NA64 Yale N3415C (164-2224), Robin ATL F-WFNA (01), Santos Dumont 14bis replica, SNCASO.30P F-ZABI (18), SV-4C F-BIZY (82), Spitfire IX BS464/GW-S, Lysander 1589, T-33A 21064 (064), AB47G F-YEAA (040), UH-12A 157 and Perrin Helicion III (01). Luckily the museum still has a large number of aircraft left. In June 1995 ex Cognac CM170 196 was noted here, it should have left for Brazil. N3400 99 moved to Bordeaux. The museum is open daily (but closed on Mondays) between 1st May and 31st October between 10:00 and 18:00. During the rest of the year it closes at 17:00. Preserved at the entrance are:

❑ 23	CM170	23	preserved, on pole, PdF c/s	3-98
❑ 26	CM170	26	preserved, on pole, PdF c/s	3-98
❑ 29	CM170	29	preserved, on pole, PdF c/s	3-98

Gallery:

❑ ..	Astra Wright BB	..	preserved	3-98
❑ ..	Blériot XI	..	preserved	3-98
❑ ..	Blériot XI-2	686	preserved, named 'Pegoud'	3-98
❑ F-WAHR	Breguet 14A2	2016	preserved	3-98

FRANCE - 108

☐	C.324	Caudron G.3	..	preserved	3-98
☐	C.1720	Caudron G.4	..	preserved	3-98
☐	..	Deperdussin B	334	preserved, marked as 'F1'	3-98
☐	F1258	DH.9	..	preserved	3-98
☐	..	Donnet Leveque A	01	preserved, replica	3-98
☐	..	Farman HF-20	..	preserved, marked as '275'	3-98
☐	15	Farman MF-7	446	preserved, ex Belgian AF	3-98
☐	..	Ferber 6bis	1	preserved, replica	3-98
☐	6796/18	Fokker D.VII	..	preserved, Albatros built, ex Luftwaffe	3-98
☐	..	Hydravion Fabre	10001	preserved, replica	3-98
☐	5929/18	Junkers J9	..	preserved, ex Luftwaffe	3-98
☐	..	Levasseur Antoinette	..	preserved	3-98
☐	..	Morane H	156	preserved	3-98
☐	N.556	Nieuport 11	..	preserved	3-98
☐	..	Nieuport 11N	..	preserved	3-98
☐	..	Paumier Biplan	1	preserved	3-98
☐	2690/18	Pfalz D.XII	3240	preserved, ex Luftwaffe	3-98
☐	..	REP type K	58	preserved	3-98
☐	..	Santos Dumont XX	..	preserved	3-98
☐	556	Sopwith 1½	556	preserved, built by Liore et Olivier	3-98
☐	S.254	SPAD VII	254	preserved	3-98
☐	S.5295	SPAD XIII C1	15295	preserved	3-98
☐	V-955	Voisin LA5	..	preserved	3-98
☐	..	Voisin Farman 1bis	..	preserved	3-98
☐	..	Vuia 1	1	preserved	3-98
☐	..	Wright Flyer	..	preserved, replica	3-98

Hall A (permanently closed):

☐	F-ABAO	MS30A1	2283	preserved	7-95
☐	F-AEDD	SPAD 52	3125/8	preserved	7-95
☐	F-AJOR	Schreck FBA 17HT4	195	preserved	7-95
☐	F-AJTE	Dewoitine D530	06	preserved	7-95
☐	F-ALQT	Potez 36/13	2620	preserved	7-95
☐	F-APOZ	Gourdou Leseurre B7	3	preserved	7-95
☐	F-BAOP	Farman F.192	7248/4	preserved, marked as 'F-AJJB', ex F-AQCP	7-95
☐	F-PFLN	Caudron C.109/1	6192/6	preserved, ex F-AIQI	7-95
☐	N12845	Nieuport Delage 29C.1	010	preserved, ex R251	7-95
☐	1685	Breguet 19GR	1685	preserved	7-95
☐	..	Breguet 19	3	preserved	7-95
☐	..	Caudron C.366	6808/4	preserved	7-95
☐	..	Caudron C.714R	01	preserved	7-95
☐	..	Junkers F13	609	preserved	7-95
☐	..	Kellner Bechereau E60	01	preserved	7-95
☐	..	HM-14	..	preserved	7-95

Hall B (permanently closed):

☐	4	Yak-3	..	preserved, ex Soviet AF	2-96
☐	C.4K-156 471-28	HA1112M-1L	156	preserved, ex Spanish AF	2-96
☐	62	NC900A-8	62	preserved, marked as '7298', locally Fw190A-8	2-96
☐	..	Dewoitine D520	862	preserved, marked as '277'	2-96
☐	..	MS500	1034	preserved, marked as 'D-EMAW'	2-96

Hall C (Hall des Prototypes):

☐	01	Leduc 0.16	01	preserved, marked as Leduc 0.10	3-98
☐	01	Leduc 0.22	01	preserved	3-98

FRANCE - 109

☐ 01		Mirage 3A	01	preserved		3-98
☐ 01		Mirage 3V	01	preserved		3-98
☐ 01		Mirage G8	01	preserved		3-98
☐ 01		Mystère 4A	01	preserved		3-98
☐ 02		N1500	02	preserved, F-WZUI		3-98
☐ 01	Y	SO9000	01	preserved, F-ZWRY		3-98
☐ F-WFRQ		SO1110	01	preserved		3-98
☐ F-WFKC		Breguet G111	01	preserved		3-98
☐ F-WFKY		SO6000	03	preserved, composite of c/n 03 and 05		3-98
☐ F-WGVD		SO1220	002	preserved, ex F-BGVD		3-98
☐ F-ZWWE		SA3210	01	preserved		3-98
☐ F-ZW.Z		SNECMA C.400 P2	01	preserved		3-98
☐ ..		SA610A Ludion	001	preserved		3-98

Hall D (Hall de la Cocade):

☐ 02		Alouette 2	02	preserved		3-98
☐ 28		CM170	28	preserved, nose only		3-98
☐ 28875	4-VA	F-84F	..	preserved, ex 52-8875		3-98
☐ 54841	13-PI	F-86K	221-81	preserved, ex 55-4841		3-98
☐ 52736	11-EF	F-100D	224-3	preserved, ex 55-2736		3-98
☐ 154	4-LT	MD450	154	preserved		3-98
☐ 01		Mirage 2000	01	preserved, F-ZWRS		3-98
☐ 122		MS472	122	preserved, nose only		3-98
☐ 105		Mystère 4A	105	preserved, marked as '289/2-EY'		3-98
☐ 210		Mystère 4A	210	preserved, nose only		3-98
☐ 4	7-CE	SE535 Mistral	4	preserved		3-98
☐ 11		Super Mystère B2	11	preserved, marked as '153/12-YY'		3-98
☐ 114522		T-6G	182-209	preserved, marked as '14915/RM', ex 51-14522		3-98
☐ F-BFAN		HD-10	01	preserved, ex F-WFAN		3-98
☐ F-BHHI		SE210	02	preserved, nose only, ex F-WHHI		3-98
☐ F-OGFI		C-47A	15010/26455	preserved, nose only, ex 43-49194		3-98
☐ F-WGVA		Payen Pa49B	01	preserved		3-98
☐ F-WGVC		Hirsch C100	01	preserved		3-98

Hall E (L'entre deux guerres et l'Aviation Légère):

☐ 1	VU	Cap 20	01	preserved		3-98
☐ 57		N3202	57	preserved		3-98
☐ 334		Caudron C.800	334	preserved, glider		3-98
☐ CS-ADG		Caudron C.635M	8519/428	preserved, marked as 'F-ANRO'		3-98
☐ F-AINX		Caudron C.60	6184/49	preserved		3-98
☐ F-AJGP		Bernard 191GR	2908/02	preserved		3-98
☐ F-AHBE		SPAD 54/1	8	preserved		3-98
☐ F-AYOL		Farman F.455	01	preserved		3-98
☐ F-AOFX		Caudron C.277R	7156/14	preserved		3-98
☐ F-APXO		Potez 43/7	3588/11	preserved		3-98
☐ F-BBQL		SV-4C	149	preserved		3-98
☐ F-BCNM		MS317	6582/328	preserved		3-98
☐ F-BDAD		LeO C.302	3	preserved, autogyro		3-98
☐ F-BEKH		Secan SUC10	31	preserved, marked as 'F-BBXY'		3-98
☐ F-BGMQ		MS230	1048	preserved, in military markings		3-98
☐ F-BHCD		DH.89A	6706	preserved, ex HG721		3-98
☐ F-BICY		N1203	373	preserved		3-98
☐ F-BJSY		RF2	02	preserved, ex F-WJSY		3-98
☐ F-BMJJ		Piel CP1310	938	preserved		3-98
☐ F-BORT		Zlin 326	923	preserved		3-98

FRANCE - 110

☐ F-CAEX		Habicht II	1	preserved, glider	3-98
☐ F-CAJA		Br901	01	preserved, glider	3-98
☐ F-CBGR		Weihe	30	preserved, glider	3-98
☐ F-CBYN		Castel C301S	1054	preserved, glider, fuselage only	3-98
☐ F-CCAY		Siren Bertin C34	02	preserved, glider	3-98
☐ F-CCHX		SZD-24	W-150	preserved, glider, fuselage only	3-98
☐ F-CRBT		Meise	12/259	preserved, glider	3-98
☐ F-CRRB		Fauvel AV36	102.36	preserved, glider	3-98
☐ F-CRLL		SA-104	253	preserved, glider	3-98
☐ F-HMFU		Farman F.60	3	preserved	3-98
☐ F-PEPF		Jodel D.9	01	preserved	3-98
☐ F-PTXJ		Colomban MC.10	01	preserved	3-98
☐ F-PVQI		Croses EC.6	01	preserved	3-98
☐ F-WFDQ		SE3101	01	preserved	3-98
☐ F-WYDD		Gary R.01	01	preserved, autogiro	3-98
☐ F-ZABY		Arsenal Air 100	..	preserved, glider	3-98
☐ ..		Cierva C.8L-II	..	preserved, replica, marked as 'G-EBYY'	3-98
☐ ..		Potez 53	3402	preserved, marked as '10'	3-98
☐ ..		Castel C242	104	preserved, glider, fuselage only	3-98
☐ ..		Grunau SG-38	31	preserved, glider	3-98
☐ ..		Grunau SG-38	173	preserved, glider	3-98
☐ ..		HM-8	..	preserved	3-98

Hall Concorde:

☐ 076		AB47G	076	preserved	3-98
☐ 9	AH	Mirage 4A	9	preserved, arr 28-4-93	3-98
☐ 1		He162A	120223	preserved	3-98
☐ 9		Polikarpov I-153	..	preserved, ex Soviet AF	3-98
☐ 44-20371		P-47D	..	preserved	3-98
☐ RR263		Spitfire LF.XVIe	CBAF-IX-3310	preserved, marked as 'TB597/GW-B'	3-98
☐ F-WTSS		Concorde 100	001	preserved	3-98
☐ N9772F		P-51D	122-31597	preserved, as '466318/MO-C', ex 44-53871	3-98
☐ N61053		PT-17	75-2419	preserved, ex 41-8860/USAAF	3-98

Outside:

☐ 9939		G91R/3	515	preserved, ex 3246/WGAF	3-98
☐ 35069	36	J35A	35069	presrerved, ex RSweAF	3-98
☐ J-4099		Hunter F.58	..	preserved, ex Swiss AF	3-98
☐ 63-8300	WW	F-105G	..	preserved, ex USAF	3-98
☐ 61		Br1150	61	preserved	3-98
☐ 763	AM	Canberra B.6	71310	preserved, ex WJ763/RAF	3-98
☐ 27		CM175	27	preserved	3-98
☐ 42	EF	Mirage 3C	42	preserved	3-98
☐ 226	DI	Mirage 3BE	226	preserved	3-98
☐ 334	33-CC	Mirage 3R	334	preserved	3-98
☐ 460	13-QR	Mirage 3E	460	preserved, ex Châteaudun	3-98
☐ 148335		SP-2H	7264	preserved	3-98
☐ 01		Super Mirage 4000	01	preserved	3-98
☐ F-BLCD		B707-328B	18941	preserved	3-98
☐ F-BTTD		Mercure 100	4	preserved, arr 3-5-95	3-98
☐ F-GCVL		SE210-12	273	preserved, ex OY-SAE	3-98

Dugny Store:
KC-97L N49549 moved on to <u>Nancy Essey</u>, while T-6G 11351 is now flying as F-AZIU. Anson TX192 went to <u>Brussels</u>. Mystère 4A 299 went to <u>Bordeaux</u>. The museum at <u>Rochefort</u> received Aquilon 203 53 and Dewoitine

FRANCE - 111

D520 650. Some of the stored aircraft can be found under restoration with Ailes Anciennes hangar at Dugny. There are also a large number of gliders in storage. The storage site holds its open day every September.

☐ 1607		HFB320	1024	stored, ex F-WZIH, ex WGAF		9-97
☐ 2240		F-104G	7118	stored, ex WGAF		9-97
☐ BR.2I-129		CASA 2111B	..	stored, ex Spanish AF		9-97
☐ 29665	20	J29B	29665	stored, ex RSweAF		9-97
☐ 32515	04	J32E	32515	stored, ex RSweAF		5-98
☐ J-277		D-3801	15	stored, marked as 'C1', ex Swiss AF		9-97
☐ J-1155		Vampire FB.6	664	stored, ex Swiss AF		9-97
☐ J-1636		Venom FB.54	846	stored, ex Swiss AF		9-97
☐ 216		AAC.1	216	stored, marked as '334'		9-97
☐ 126979	MK	A-1D	7779	stored		9-97
☐ 01	AS	Alouette 2	01	stored		9-97
☐ 41-39162		A-26B	6875	stored, marked as '44-68219'		9-97
☐ 44-34773		A-26B	28052	stored, ex F-ZLAA, ex N67944		9-97
☐ 4	62-ND	Br941S	04	stored		5-98
☐ 02		Br1001	02	stored, F-ZWRE		9-97
☐ 10		Br1050	10	stored		9-97
☐ 71		C-47A	12471	stored, ex N12471		9-97
☐ 92449	FA	C-47A	12251	stored, ex F-BEFB, ex 42-92449		5-98
☐ 1		CM170	1	stored		97
☐ 17		CM175	17	stored		9-97
☐ 45061	CA	DC-7C	45061	stored, ex LN-MOG		5-98
☐ 45819	FC	DC-8-55F	45819	stored		5-98
☐ 603		Dewoitine D520	603	stored		97
☐ 56		Etendard 4M	56	preserved, ex Hyères		9-97
☐ 121	Y	HSS-1	SA121	stored, with tail from SKY992 ?		9-97
☐ SA101		H-34A	SA101	stored		9-97
☐ 130077	6	HUP-2	..	stored, with Ailes Anciennes		5-98
☐ A04	E	Jaguar A	..	stored		9-97
☐ N1957C		L-18C	18-1430	stored, ex ALAT, ex 51-15430		7-95
☐ 2503	F-ZVMV	L-749A	2503	stored, ex F-BAZR		5-98
☐ WU-21	4	Lancaster B.7	..	stored, with Ailes Anciennes, ex NX664		6-97
☐ 919		Mauboussin M.123	..	stored		6-92
☐ 130	OF	MD315	130	stored		9-97
☐ 280		MD311	280	stored		97
☐ NF11-5		Meteor NF.11	..	stored		5-98
☐ NF14-747		Meteor NF.14	..	stored, ex WS747/RAF		9-97
☐ 18	314-DA	MH1521M	18	stored		9-96
☐ 74		Mirage 3C	74	stored		9-97
☐ 02		Mirage 4A	02	stored, F-ZADS		9-97
☐ 186	12-UT	Mystère 4A	186	stored		9-97
☐ 16		N262A	16	stored		5-98
☐ 135	LX	N1101	135	stored		97
☐ 50	64-BH	N2501F	50	stored, ex Aulnay sous Bois		6-97
☐ 162	F	N2501F	162	stored		9-97
☐ 194	316-FP	N2501F	194	stored, ex Châteaudun		9-97
☐ 67	AIA	N3202	67	stored, damaged by fire		6-92
☐ 131	CUC	N3400	131	stored		9-96
☐ 111	YT	NC856	111	stored. With Ailes Anciennes		5-98
☐ 141	FG	SE210-3	141	stored, ex F-BJTK		5-98
☐ 3	V	SE5003	3	stored		6-97
☐ 59	12-ZS	Super Mystère B2	59	stored		9-97
☐ 90	12-ZQ	Super Mystère B2	90	stored		9-97
☐ 121		T-28A	174-344	stored, ex 51-7491		9-97

☐ 055	GC	T-33A	8394	stored, ex 53-5055	5-98
☐ 35211		T-33A	8550	stored, ex 53-5211	9-97
☐ 330	30-ML	Vautour 2N	330	stored	9-97
☐ 634	92-JD	Vautour 2B	634	stored	5-98
☐ F-AZCL		Aero 45-1	4927	stored	9-97
☐ F-AZDY		MD312	156	stored, with Ailes Anciennes	5-98
☐ F-AZPT		PT-17	..	stored	6-92
☐ F-BPTT		WA51	01	stored, ex F-WVKT	7-95
☐ F-BBNA		Bü181C-3	FR15	stored, marked as 'SV-NJ'	7-95
☐ F-BGSO		B-17G	8289	stored, ex 44-8889	9-97
☐ F-BICR		HD-34	4	stored	5-98
☐ F-BJTR		SE210-3	22	stored, ex OH-LEB	2-97
☐ F-BJSF		MS880B	3	stored	97
☐ F-BNAN		Potez 842	3	stored	9-97
☐ F-BZCK		N2501F	44	stored, with Ailes Anciennes, ex code 64-BI	5-98
☐ F-BRQE		RC680FL	1716-137	stored, with Ailes Anciennes	5-98
☐ F-CAAA		Glider	1	stored, glider	97
☐ F-CAYN		Castel C311P	19	stored, glider	97
☐ F-CCCD		Weihe	32	stored, glider	97
☐ F-CCGV		Br905	..	stored, glider	97
☐ F-CCHR		Fauvel AV45	2	stored, glider	97
☐ F-CDIY		WA30	207	stored, glider	97
☐ F-CRBS		Castel C310P	131	stored, glider	97
☐ F-CHRB		SA-103	69	stored, glider	97
☐ F-CRFX		Castel C25S	203	stored, glider	97
☐ F-CRPP		SA-103	106	stored, glider	97
☐ F-OBIP		Sandringham VII	SH-57C	stored, ex VH-APG	9-97
☐ F-PHZI		HM320	01M	stored	97
☐ F-PINS		Jodel D.119	613	stored	97
☐ F-WLKB		Falcon 20	01	stored, with Dassault	3-98
☐ F-ZBBF		Do28A-1	3032	stored	6-92
☐ N61909		JRB-4	..	stored, with Ailes Anciennes, ex 44676/USN	5-98
☐ ..		Caudron C.714	8357/5	stored, Ailes Anciennes	5-98
☐ ..		Heinkel He46D	846	stored	97
☐ ..		REP type D	24	stored, bare frame	3-97

Airfield – Orly (94): Omitted from EWR-1 was Caravelle 4N F-BYAI (139) which was in use since 1985 for fire/security training. It was completely scrapped by late 1997. The preserved Concorde, Caravelle F-BVPZ, Mercure and two Mirage 3's are here with Athis Paray Aviation Aero Exposition, on the south side of the airfield. Scrapped during 1991 was SE210-12 F-BNOH (269), while in 1993 F-BNOG (271) and F-BTOC (278) were scrapped. The nose of the latter went to Toussus le Noble.

☐ 245	Mirage 3B-RV	245	preserved, ex Châteaudun	5-98
☐ 352	Mirage 3RD	352	preserved, ex Châteaudun	5-98
☐ F-BPJU	B727-214	19683	dumped, ex N528PS	4-98
☐ F-BTTJ	Mercure 100	10	preserved, ex stored	5-98
☐ F-BTOD	SE210-12	279	dumped, wfu 5-6-91	4-98
☐ F-BVPZ	SE210-6N	218	preserved, ex YU-AHF	5-98
☐ F-WTSA	Concorde 100	02	preserved	5-98

PAU - UZEIN (63)

The preservation group Squadron 64 disbanded many years ago. A number of aircraft of the former collection are still at the airfield. CM170 14, Mirage 3R 310/33-TF and Super Mystère 4A 73/12-YP (ex Merville) all left for Savigny lès Beaune. Stored MH1521 23/MCA became F-GGKS and can now be found at the museum at Montelimar. The military ALAT gate, on the other side of the airfield, is guarded by an Alouette 3.

☐ 258	341-RL	Alouette 2	1505	preserved		89
☐ 1185		Alouette 3	1185	preserved, at gate		5-96
☐ 193	328-EU	N2501F	193	preserved		5-96
☐ F-AZEL		MD312	177	stored, ex F-WZEL		6-97

Just down the road, on the southern side of the airfield, the military ETAP barracks have a Noratlas preserved.

☐ 161	63-BP	N2501F	161	preserved		5-96

PERPIGNAN (66)
Town: The Noetinger family keeps an interesting collection of aircraft (the Musée d'Aviation du Mas Palegry) at their vineyard. The museum is signposted from the Perpignan south highway exit. Only the fin of F-86K MM55-4815 was ever here. RC680V-TU F-BSTM (1540-6) was scrapped. There are also some smaller aircraft, including a Blériot IX on loan from Le Bourget.

☐ MM51-1928	3-43	RF-84F	..	preserved, with tail from 52-7457, ex Italian AF	2-97
☐ 51-1077	FS-077	F-84G	..	preserved, nose only, ex USAF	2-97
☐ VX950		Vampire FB.5	..	preserved	2-97
☐ 05		CM170	05	preserved, on pole, outside	2-97
☐ 015	721-EP	MH1521M	015	preserved, outside	2-97
☐ ..	8-MD	Mystère 4A	..	preserved, nose only	2-97
☐ F-BMGQ		MS733	101	preserved	2-97
☐ ..		Mirage F1	..	preserved, mock up, marked as '01'	2-97

Airfield – Llabanere: A large amount of *Wrecks & Relics* airliners have been stored here over the years. Of these Vanguards F-BVRZ (741), F-BXAJ (725), F-BXOG (739), G-AXNT (737), G-AZRE (729) and PK-MVH (746) have been scrapped. Caravelles F-BXOO (76) and F-GBMJ (149) have also been scrapped. More lucky were SE210-10B1R EC-CPI (236) which became F-GFBI and EC-DCN (199) which became F-GFBH.

☐ F-BJEN	SE210-10B3	185	stored, ex OH-LSC	2-97
☐ F-BMKS	SE210-10B3	181	stored, ex Toulouse, ex OH-LSA	2-97
☐ F-BTOB	SE210-12	277	stored	8-94
☐ F-GFBA	SE210-10B1R	243	stored, ex D-ABAV	2-97
☐ G-APEK	Vanguard 963C	714	stored	2-97
☐ 5N-AOY	SE210-6N	180	stored, ex F-GBMK	8-89

PERSAN BEAUMONT (95)
Although the Mirage 5 is a recent arrival here, it does not look too good. The construction plate of this aircraft was checked and still revealed it as a Mirage 5J, the designation for its original customer, the Israeli Air Force. Stored Broussard 255/5-ML became F-GGHL. Persan is some 25/30km north of Paris.

☐ 33		Mirage 5F	33	preserved	6-97

PÉZENAS - NIZAS (34)
The MD315R Flamant radar trainer, 66/30-QW, at this small airfield was last noted in January 1987 and was scrapped by June 1994.

PHALSBOURG (67)
Not noted at the recent open days were a H-19 (or Whirlwind) and Broussard. MH1521 38/MJC moved to Angers, while the fate of the anonymous H-19 coded MST is unknown.

PLAISIR (78)
At the buildings of the Thomson firm, the fuselage of a Atlantic is in use as a mock-up for avionics development

For the Atlantique. It can clearly be seen from outside.

☐ 9	Br1150	9	instructional, ex Nîmes	4-97

PLASSAC (17)
Monsieur Francis Bouyer's car dismantlers yard (on the RN137 south of Saintes) still has one H-34. There were no traces of the Djinn and H-34A SA91 (marked as 'P92').

☐ SKY615	H-34A	22	stored, ex 'P-96'	8-96

PLOBANNALEC (29)
Somewhere south of Brest is the Aero-Sub Multi Services. Besides a Phantom in fair condition there are currently six HSS-1/H-34s in a poor state. The Etendard arrived after May 1997.

☐ 64-0922		F-4C	1401	stored, ex Mengam, ex Sembach, ex USAF	5-97
☐ 14		Etendard 4M	14	stored	97
☐ 147569		SP-2H	..	expected, still at Lann Bihoué	—
☐ 135	N	HSS-1	SA135	stored, ex Rochefort	5-97
☐ 148		HSS-1	SA148	stored, ex Rochefort	5-97
☐ 512	V	HSS-1	58-512	stored, ex Rochefort	5-97
☐ 640		HSS-1	58-640	stored, with tail from SA150, ex Rochefort	5-97
☐ 688		HSS-1	58-688	stored, with tail from SA119, ex Rochefort	5-97
☐ SKY705		H-34A	58-705	stored, ex Toussus	5-97

PLONEIS (29)
The discotheque 'Le Moulin' has a Super Constellation which is missing its engines and outer fins. The colours are that of Air France, for which it never flew. The aircraft can be found along the D765, west of Quimper.

☐ F-BHBG	L-1049	4626	preserved	8-97

POITIERS (86)
The Musée de l'Art Populaire 'Chez Manuel' can be found along Route National 10 between the Futuroscope and Poitiers. There is no sign of the anonymous Djinn and the museum owners said it had gone a long time ago. The Armée de Terre caserne is in the town centre along the river and easy to find. They have their N2501 fuselage stored on a grass field outside the main barracks.

☐ 23	312-BE	N2501F	23	preserved	3-98
☐ SA142		HSS-1	..	preserved	3-98
☐ F-BBKX		N1203	57	preserved	3-98

PONTARLIER (25)
Preserved on a pole at the local aero club is a former Rochefort instructional airframe.

☐ 162	CM170	162	preserved, ex Rochefort	5-98

PONT D'AIN (01)
On top of a wash-house on a local campsite is an MS733, which wears military colours.

☐ F-BLYC	MS733	..	preserved, marked as '96'	6-91

POURVILLE SUR MER (76)
Due to the lack of visitors the Musée du Aout 19 was closed in the early 1990s. Its NC856 94/AGM was placed in storage somewhere in the Dieppe area.

REIMS (51)

Town: Displayed at the Centre de Controle Aérienne is MD312 Flamant 215. It is located off a roundabout (with a Farman monument) on the RN44 on the southern outskirts of Reims. The Musée de l'Ancien College des Jesuites in the city along the Rue Gambetta has an MH1521M Broussard on display in a courtyard.

❏ 215		MD312	215	preserved	9-97
❏ 305		MH1521M	305	preserved	9-97

Airfield – Champagne: The French moved their Mirage F1CR unit from Strasbourg to Reims in the mid 1990s. They also took their Thunderflash to their new home. The Vautour was mounted at the gate in 1974.

❏ 37577	33-CK	RF-84F	..	preserved, ex Strasbourg	9-97
❏ 347	30-FB	Vautour 2N	347	preserved, at gate	9-97

Airfield – Prunay: Stored here at the back of the Reims Aviation factory is Broussard 164. At the aero club site the group ACRAA maintain some airworthy aircraft.

❏ 164	YC	MH1521M	164	stored	9-95
❏ 21255	314-VH	T-33A	255	preserved, ex Nancy	9-97
❏ 30		CM170	30	preserved, ex Nancy, ex F-WDHG	9-97

RENNES - ST JACQUES (35)

The fuselage of N2501F Noratlas 187 was used by the ALAT here for paratroop training. It has not been noted since June 1992 and has probably been scrapped. Also to be found here is the Yankee Delta group with some vintage aircraft, including CM170s F-WKYD and F-WKYE.

❏ U-1123		Vampire T.55	983	stored, F-AZHV ntu, ex Sion, ex Swiss AF	6-97
❏ 351		Mirage 3RD	351	preserved, on airfield, ex Châteaudun	8-97
❏ 514		Mirage 3E	514	stored, all white, with aero club, ex Châteaudun	8-97
❏ 252	AGA	MH1521M	252	preserved, at military gate	8-97
❏ F-GCVJ		SE210-12	275	preserved, on airfield, ex OY-SAF	7-97

ROANNE (42)

Air France SE210 Caravelle F-BHRF arrived here on 6th June 1981 for preservation and is still present.

❏ F-BHRF	SE210-3	12	preserved	6-95

ROCAMADOUR (46)

The French Army took over this base in 1985. After they had disposed of the MD311 and Mystère from the gate, only the CM170 remained. This may have gone as well, as it has not been noted since 1986.

❏ 21	CM170	21	preserved	3-86

ROCHEFORT (17)

Town: The scrapyard near the railway station was checked in 1995. All those aircraft mentioned in the first edition have gone.

Airfield - Soubise: The Musée de Tradition de l'Aéronautique Navale is still not open to the general public. The aircraft can be viewed by prior arrangement, normally only on Mondays, when volunteers carry out maintenance.

❏ 056	CAN-7	AB47G	056	preserved, ex Dax	6-97
❏ 138		Alouette 2	..	stored	6-97
❏ 53		Aquilon 203	53	preserved, ex Le Bourget	6-97
❏ 15		Br1050	15	preserved, outside	6-97
❏ 716		C-47B	16700/33448	preserved, ex K-36/Belgium AF, ex KP229	6-97
❏ 1		CM175	1	preserved	6-97

FRANCE - 116

☐ 20		CM175	20	stored	6-97
☐ 650		Dewoitine D520	650	preserved, ex Le Bourget	6-97
☐ 05		Etendard 4M	05	preserved	6-97
☐ ..		H-21	..	preserved, marked as 'H20'	6-97
☐ 150		HSS-1	SA150	preserved	6-97
☐ 183		HSS-1	SA183	stored	6-97
☐ M05	J	Jaguar M	M05	preserved, ex instructional	6-97
☐ 25		JRB-4	..	preserved, ex 66425/USN	6-97
☐ 294		MD312	294	preserved	6-97
☐ 258	CAN-16	MH1521M	258	preserved	6-97
☐ 33		MS760	33	preserved, ex instructional, outside	6-97
☐ 925		PA31-350	31-7300925	preserved, ex instructional	6-97
☐ 709		SNB-5	..	preserved, ex 134709/USN	6-97
☐ 144688		SP-2H	..	preserved	6-97
☐ 02		Super Etendard	02	preserved, ex instructional	6-97
☐ 7		SV-4	..	preserved	6-97
☐ 3820		T-6D	..	preserved	6-97
☐ F-CAFB		Caudron C.800	9865.205	preserved	6-97
☐ F-WSQL		SA360	001	preserved, ex Marseille, may have gone	95
☐ F-ZBAR		PBV-1A	CV449	stored, nose only, ex Aix, ex CF-NJP	6-97

Reports from the Aeronavale technical training school have been few. Recent sightings are of those aircraft which have been seen outside their hangars. Known departures are the five HSS-1s, 135, 148, 512, 640 and 688, which moved on to Plobannalec. Etendard 4M 06 went to Vannes, together with Crusader 01. Savigny lès Beaune bought Alize 04. Around the year 2000 it is planned that the Marine and Air Force school will reform as one large military school. The school at Soubise will then be closed.

☐ 3		Br1050	3	instructional	88
☐ 4		Br1050	4	instructional	5-96
☐ 8		Br1050	8	preserved, outside main gate	8-97
☐ 21		Br1050	21	stored	7-95
☐ 40		Br1050	40	stored	5-96
☐ 02		CM170M Esquif	02	instructional	87
☐ 10		CM175	10	instructional	88
☐ 24		CM175	23	instructional	88
☐ 01		Etendard 4M	01	preserved, main gate	8-97
☐ 03		Etendard 4M	03	dumped	87
☐ 1		Etendard 4M	1	instructional	5-96
☐ 34		Etendard 4M	34	stored, in hangar	4-97
☐ 16		F-8E(FN)	..	instructional	87
☐ 03		Lynx	..	instructional, ex XX904	5-96
☐ 31		MS760	31	instructional	88
☐ 7	MB	N260	7	instructional, ex F-BLHN	5-96
☐ 9	ME	N260	9	instructional, ex F-WLHP	5-96

Airfield - St Agnant: The Air Force instructional school, Escadron Avion 21.317, still has a large number of airframes. Recent arrivals are some Mirage F1Cs, while the first Mirage 2000 has also arrived. A number of Magisters left in the 1990s, these included 162 to Pontarlier, 165 which became N165F, 166 moved to Val Doise, 222 went to Cap Feyrefite, 332 is now at Forez, while 377 (ex Châteaudun) should be at Narbonne. Also 400/315-IH has been sold. Mirage 3B-RV 242/13-SG was reported here in July 1994, but this seems to be an error. Known departures of Mirage 3Cs are; 1 to Saintes, 32/10-RE to Cazaux and 55 to Montelimar. The last Mirage 3C, 38/10-RF (last noted July 1994) left for Orange The fates of the other aircraft mentioned in the first edition are unknown. The Gruman Widgeon belongs to the Musee Historique de l'Hydraviation at Biscarosse and is here only for restoration.

☐ ..		Alouette 2	..	instructional, frame only	4-97
☐ 205	13-TC	CM170	205	instructional, ex Châteaudun	4-97

☐ 231		CM170	231	instructional	4-97
☐ 343		CM170	343	instructional, ex Châteaudun	4-97
☐ 345		CM170	345	instructional	4-97
☐ 349		CM170	349	instructional	4-97
☐ 399		CM170	399	instructional, ex Châteaudun	4-97
☐ 402		CM170	402	instructional, ex Châteaudun	4-97
☐ 404		CM170	404	instructional, ex Châteaudun	4-97
☐ 466		CM170	466	instructional	4-97
☐ 467		CM170	467	instructional	4-97
☐ 196		MD312	196	preserved, ex instructional	4-97
☐ 205	13-FC	Mirage 3B	205	instructional, ex Châteaudun	4-97
☐ 315	33-GL	Mirage 3R	315	instructional	4-97
☐ 317	33-TF	Mirage 3R	317	instructional	4-97
☐ 319	33-NR	Mirage 3R	319	instructional	4-97
☐ 321	33-CJ	Mirage 3R	321	instructional	4-97
☐ 343	33-CP	Mirage 3R	343	instructional	4-97
☐ 345	33-CN	Mirage 3R	345	instructional	4-97
☐ 356		Mirage 3RD	356	instructional, ex Châteaudun	4-97
☐ 369	33-TR	Mirage 3RD	369	instructional	4-97
☐ 370	33-TS	Mirage 3RD	370	instructional, to go to Touchay	4-97
☐ 429	2-LO	Mirage 3E	429	instructional	4-97
☐ 454	13-QD	Mirage 3E	454	instructional	4-97
☐ 497	H	Mirage 3E	497	instructional	7-96
☐ 500	3-IH	Mirage 3E	500	instructional	4-97
☐ 516	3-JR	Mirage 3E	516	instructional	4-97
☐ 555	4-BE	Mirage 3E	555	instructional	4-97
☐ 563	21-317	Mirage 3E	563	preserved	4-97
☐ 573	13-QJ	Mirage 3E	573	instructional	4-97
☐ 574	13-QF	Mirage 3E	574	instructional	7-96
☐ 01		Mirage 3T	01	preserved, at gate	7-97
☐ 4	AC	Mirage 4A	4	preserved, ex Châteaudun	7-97
☐ 2		Mirage F1C	..	instructional, ex Châteaudun	7-96
☐ 3	12-ZO	Mirage F1C	..	instructional, ex Châteaudun	7-96
☐ 6	30-SJ	Mirage F1C	..	instructional, ex Châteaudun	4-97
☐ 14		Mirage F1C	..	instructional, ex Châteaudun	4-97
☐ 15	30-FD	Mirage F1C	..	instructional	7-96
☐ 17		Mirage F1C	..	instructional, ex Châteaudun	7-96
☐ 18	12-KF	Mirage F1C	..	instructional, ex Châteaudun	6-97
☐ 29	12-ZM	Mirage F1C	..	instructional, ex Châteaudun	4-97
☐ 36		Mirage F1C	..	instructional, ex Châteaudun	7-96
☐ 39	12-ZM	Mirage F1C	..	instructional, ex Châteaudun	4-97
☐ 47	12-ZG	Mirage F1C	..	instructional, ex Châteaudun	4-97
☐ 50	12-YA	Mirage F1C	..	instructional, tiger c/s, ex Châteaudun	4-97
☐ 60	12-ZH	Mirage F1C	..	instructional, ex Châteaudun	4-97
☐ 63	30-FJ	Mirage F1C	..	instructional, ex Châteaudun	4-97
☐ 78		Mirage F1C	..	instructional, ex Châteaudun	7-96
☐ 03		Mirage 2000	03	instructional	4-97
☐ 122	63-WD	N2501F	122	preserved	7-97
☐ N750M		G44A Widgeon	1341	under restoration, ex 37711/US Navy	4-97

RODEZ (12)
The Flamant which arrived here from Châteaudun by August 1987 is still preserved at the airfield. The airfield is some 10km north west of the town.

☐ 232		MD312	232	preserved	2-97

FRANCE - 118

ROMANS SUR ISÈRE (26)
Preservation group Aero Phoenix are based at the aerodrome. The only *Wrecks & Relics* aircraft should be the Martinet, all the others are now in flying condition, including the recently restored MD311 290 (to F-AZGX).

| ☐ 350 | PR | NC702 | 350 | preserved, ex St Martin d'Hostun | .. |

ROMORANTIN - PRUNIERS (41)
This airfield is split into two parts. Preserved at the gate on the airfield is a Mirage 3, while preserved at the barracks on the other side of the road along the airfield is an Ouragan.

☐ 297	UG	MD450	297	preserved, on other side of the road	3-98
☐ 331	33-TN	Mirage 3R	331	preserved, at gate	3-98
☐ ..		MH1521M	..	stored	5-92

ROUEN (76)
The Noratlas is still outside some military barracks in the Rue Louis Blanc.

| ☐ 24 | | N2501F | 24 | preserved | 7-95 |

SAINTES - THÉNAC (17)
There ought to be a number of instructional aircraft at this AdlA airfield. Not much could be identified from outside. Only the Mirage 3C was readable. The Mystère 4A has no serial, as has the black Super Mystère. T-33A 21113 went to Montelimar, N2501F 154 went to Speyer and CM170 376 returned to Cognac. Notatlas 170 has being reported as scrapped.

☐ 384		CM170	384	instructional	10-86
☐ 1		Mirage 3C	1	stored, ex instructional, ex Rochefort	7-97
☐ 64		Mirage 3C	64	instructional	87
☐ 11		Mystère 4A	11	instructional, faded code 314-TQ	6-97
☐ 124		N2501F	124	preserved	88
☐ 156		Super Mystère B2	156	stored, black c/s	7-97

SALON DE PROVENCE (13)
Airfield – Eyguières: This second airfield of Salon, on the west side of the town, has only a small aero club which used to have a Magister. This aircraft, No. 44, moved on to Brive by late 1995.

Airfield – Salon de Provence: In early 1996 the last operational Magisters left the airfield of Salon de Provence. Most were send, after brief storage here, to Châteaudun.

☐ 499	AF	CM170	499	preserved, on pole near flightlines, ex 312-AF	5-96
☐ 531		CM170	531	preserved	5-95
☐ 572		CM170	572	preserved, tiger c/s	5-96
☐ 3	2-FA	Mirage 3C	3	preserved	9-94

SARRE UNION (67)
Preserved with the Aero Club de Sarre Union is MD311 255/316-KN.

| ☐ 255 | 316-KN | MD311 | 255 | preserved | 7-95 |

SAVIGNY EN SEPTAINE (18)
The French Air Force storage facility still have an ex Rochefort Super Mystère within their grounds. The aircraft can be seen from the public road and although the barracks are very close to the airfield of Avord, they are still a separate location.

| ☐ 99 | 12-YJ | Super Mystère B2 | 99 | preserved | 3-98 |

SAVIGNY LÈS BEAUNE (21)

The largest private collection of aircraft in France can be found with the Association des Amis du Musée du Château, at Savigny lès Beaune. This museum is open daily between 09:00/12:00 and 14:00/18:00. In summer the last entry is extended to 18:30. A number of the aircraft are completed with parts from other aircraft, the most common practice being tail swopping between aircraft, although the Belgian Mirage 5BA has been made up of four different types of aircraft (see later). Some of the recent arrivals have been devoid of any marks.

❏ FC-08		TF-104G	5105	preserved, ex Koksijde, ex Belgian AF	5-98
❏ FR-26		RF-84F	..	preserved, ex 51-1886, ex Belgian AF	5-98
❏ FU-45		F-84F	..	preserved, ex 52-7210, ex Belgian AF	5-98
❏ FX-90		F-104G	9154	preserved, ex Koksijde, ex Belgian AF	5-98
❏ ID-44		Hunter F.4	102-8138	preserved, ex Belgium AF	5-98
❏ MT-33		CM170	290	preserved, ex Weelde, ex Belgium AF	5-98
❏ 104799		CF-104	1099	preserved, ex Sollingen, with tail from 104750	5-98
❏ 2343		MiG-21MF	96001091	preserved, ex Rothenburg, ex 774/NVA	5-98
❏ 3243		G91R/3	512	preserved, ex Hermeskeil, ex Frankfurt	5-98
❏ JA-339		CL-13B Sabre 6	1651	preserved, ex Hermeskeil, ex WGAF	5-98
❏ 185		Vampire T.55	15775	preserved, ex Irish AF, Le Bourget owned	5-98
❏ 720		SBLim-2	1A07020	preserved, unmarked, ex Polish AF	5-98
❏ 306		Lim-6bis	1F-0306	preserved, ex Polish AF	5-98
❏ 2718		MiG-21U-600	662718	preserved, ex Polish AF	5-98
❏ 813		Su-7BKL	7813	preserved, ex Łódż, ex Polish AF	5-98
❏ 1734		T-6	..	under restoration, unmarked, ex Port AF	5-98
❏ 1801		G91T/3	0003	preserved, ex Port AF, ex 3403/WGAF	5-98
❏ 5216		F-84G	..	preserved, ex Alverca, ex Port AF, ex 51-10838	5-98
❏ XM178		Lightning F.1A	95065	preserved, ex RAF	5-98
❏ 63-8357		F-105F	..	preserved, ex Hermeskeil, ex Hahn, ex USAF	5-98
❏ J-1178		Vampire FB.6	687	preserved, ex Sion, ex Swiss AF	5-98
❏ J-1545		Venom FB.54	755	stored, ex St Julien de Cassagnas, ex Morges	5-98
❏ 04		Br1050	04	preserved, ex Rochefort	5-98
❏ 14		CM170	14	preserved, unmarked, ex Pau	5-98
❏ 38		CM170	38	preserved, unmarked, ex Montelimar	5-98
❏ 2		CM175	2	preserved	5-98
❏ ..		Etendard 4M	..	preserved, nose only	2-98
❏ 29003	4-SA	F-84F	..	preserved, with aft fuselage and tail from FU-21	5-98
❏ 42130	11-YF	F-100D	223-10	preserved, ex 54-2130, with tail from 54-2235	5-98
❏ 42295		F-100D	223-175	preserved, nose only, USAF marks, ex 54-2295	5-98
❏ 63937	11-YH	F-100F	243-213	preserved, ex 56-3937, with tail from 54-2293	5-98
❏ WAD130	CUV	H-19D	..	under restoration	5-98
❏ SA114	68-OF	H-34A	SA114	preserved	5-98
❏ A8	11-EB	Jaguar A	..	preserved, ex Châteaudun	5-98
❏ ..	11-YA	Jaguar A	..	preserved, ex Châteaudun	5-98
❏ 230	4-US	MD450	230	preserved	5-98
❏ F6		Meteor T.7	..	preserved	5-98
❏ NF11-24		Meteor NF.11	..	preserved, ex WM301/RAF, Le Bourget owned	5-98
❏ 001		Mirage 3O	O01	preserved, unmarked, Australian prototype	5-98
❏ 06		Mirage 3A	06	preserved, ex Biarritz	5-98
❏ 50	3.10-LD	Mirage 3C	50	preserved, desert c/s	5-98
❏ 216		Mirage 3B	216	preserved, with tail from 218	5-98
❏ 323	33-TB	Mirage 3R	323	preserved	5-98
❏ 354		Mirage 3RD	354	preserved, ex code 33-TG	5-98
❏ 501		Mirage 3E	501	stored, fuselage only, ex Châteaudun	5-98
❏ ..		Mirage 3E	..	preserved, unmarked	5-98
❏ 18	AQ	Mirage 4A	18	preserved	5-98
❏ 9	13-SH	Mirage 5F	9	preserved, L on nosewheel door	5-98
❏ 013		Mystère 2C	013	preserved	5-98

FRANCE - 120

❏ 24	8-NQ	Mystère 4A	24	preserved, nose only	5-98
❏ 37		Mystère 4A	37	preserved, with tail from 47	5-98
❏ 69	G	Super Mystère B2	69	preserved, ex code 12-YG, ex Captieux	5-98
❏ ..	12-ZC	Super Mystère B2	..	preserved, nose only	5-98
❏ 21029		T-33AN	029	preserved, Le Bourget owned	5-98
❏ 2	92-AB	Vautour 2A	2	preserved, ex Cazaux	5-98
❏ 304		Vautour 2N	304	preserved, ex Brétigny, Le Bourget museum	2-98
❏ SE-DCF		Meteor NF.11	..	preserved, unmarked	5-98
❏ ..		HM-14	..	preserved, inside, unmarked	5-98

There are some other aircraft here, apart from the museum and storage area. The fuselage of Noratlas 151 is welded to that from an unidentified example.

❏ ..		Mirage F1	..	stored, fuselage only, desert c/s	5-98
❏ 92		N2501F	92	stored, fuselage only, ex Châteaudun	5-98
❏ 151	F	N2501F	151	stored, fuselage only, ex Châteaudun	5-98
❏ ..		N2501	..	stored, fuselage only, see note	8-97

In storage area a large number of aircraft are lined up for use as spare parts and for future exchange. All aircraft were confirmed here in the summer of 1997.

❏ BR-19		Mirage 5BR	19	stored, tail only, ex Belgian AF	6-97
❏ FU-10	BA-01	F-84F	..	stored, ex Bierset, ex Belgian AF, ex 52-7115	8-97
❏ FU-21	021	F-84F	..	stored, ex Belgian AF	8-97
❏ FU-29		F-84F	..	stored, tail only, ex Bierset, ex Belgian AF	6-97
❏ FU-31	BA-08	F-84F	..	stored, ex Bierset, ex Belgian AF, ex 52-7178	8-97
❏ FU-97	BA-09	F-84F	..	stored, ex Bierset, ex Belgian AF, ex 53-6539	8-97
❏ FU-106	BA-05	F-84F	..	stored, ex Bierset, ex Belgian AF, ex 53-6722	8-97
❏ FU-116	BA-07	F-84F	..	stored, ex Bierset, ex Belgian AF, arr 23-5-89	8-97
❏ FU-186		F-84F	..	stored, tail only, ex Belgian AF	6-97
❏ 104648	857B	CF-104D	5318	stored, ex Sollingen, ex CAF	8-97
❏ 19572		F-84E	..	stored, ex 51-19572	6-97
❏ 52739	11-YL	F-100D	224-6	stored, ex 55-2739, with tail from 54-2130	8-97
❏ 64017	11-YB	F-100F	243-293	stored, ex 56-4017	8-97
❏ A36	7-IP	Jaguar A	..	stored	5-98
❏ A72		Jaguar A	..	stored, ex Châteaudun	8-97
❏ 215		MD450	215	stored, ex Abbeville	8-97
❏ 35	3.10-LB	Mirage 3C	35	stored, desert c/s	8-97
❏ 37	3.10-LI	Mirage 3C	37	stored, desert c/s	8-97
❏ 304	33-TN	Mirage 3R	304	stored, ex Châteaudun	8-97
❏ 324	2-ZM	Mirage 3R	324	stored, ex Châteaudun	5-98
❏ 402	13-QL	Mirage 3E	402	stored, ex Châteaudun	5-98
❏ 456		Mirage 3E	456	stored, tail only	8-97
❏ 499	3-IG	Mirage 3E	499	stored, ex Châteaudun	5-98
❏ ...		Mirage 3	..	stored, fuselage part only, desert c/s	6-97
❏ 39	8-MT	Mystère 4A	39	stored	8-97
❏ 100	8-NP	Mystère 4A	100	stored, ex preserved	8-97
❏ 116	8-NL	Mystère 4A	116	stored, ex Châteaudun	8-97
❏ 149		N2501F	149	stored, fuselage only, ex Châteaudun	5-98
❏ 02	CV	Super Mystère B2	02	stored, ex Bourges	8-97
❏ 50	12-ZE	Super Mystère B2	50	stored, ex Captieux	8-97
❏ 91	12-ZV	Super Mystère B2	91	stored, ex Bourges	8-97
❏ 118	12-YS	Super Mystère B2	118	stored, ex Bourges	8-97
❏ 21127		T-33A	127	stored, ex code 314-UZ	8-97
❏ SE-DCH		Meteor NF.14	5549	stored, nose and tail only	8-97

The fuselage of Belgian Mirage 5BA BA-35 is fitted with the tail from Mirage 5F 54, nose cone from a Mirage

3E and wings of a Mirage 3C. This composite is now at Hermeskeil. Ex Belgian F-84F (most likely FU-26) moved to Montelimar. Other (parts of) aircraft seen over the years include:

BA-35		Mirage 5BA	35	stored, fuselage only, l/n 5-90	to Hermeskiel
FU-26	BA-06	F-84F	..	stored, ex Belgian AF, l/n 4-95	to Montelimar
104750		CF-104	..	stored, ex Sollingen, l/n 5-96	to Ota
1527		Harvard IIA	88-10673	stored, ex Ota, ex PortAF, ex 41-33553, l/n 5-96	gone
36	10-RB	Mirage 3C	36	stored, blue c/s, l/n 8-97	to Łódź
67		Mirage 3C	67	stored, rear fuselage only, l/n 3-96	gone
91	10-RG	Mirage 3C	91	stored, l/n 10-90	gone
310	33-TF	Mirage 3R	310	stored, ex Pau, l/n 6-94	to Hermeskeil
313	33-TU	Mirage 3R	313	stored, l/n 12-93	to Ota
54		Mirage 5F	54	stored, aft fuselage only	to Hermeskeil
24		Mystère 4A	24	stored, tail only, l/n 5-96	gone
142	8-NU	Mystère 4A	142	stored, ex Châteaudun, l/n 1990	gone
178	8-NK	Mystère 4A	178	stored	to Koksijde
191	8-NS	Mystère 4A	142	stored, l/n 4-88	gone
237		Mystère 4A	237	stored, tail only, l/n 6-94	gone
287	8-NM	Mystère 4A	287	stored, l/n 6-89	to Abbeville
293		Mystère 4A	293	stored, l/n 6-89	to Touchay
83	12-ZT	Super Mystère B2	83	stored, l/n 6-89	gone
145	12-ZR	Super Mystère B2	145	stored	to Koksijde
173	12-YP	Super Mystère B2	173	stored, ex Pau	to Hermeskeil

SÉZANNE - ST REMY (51)
During 1994 preserved Flamant 153 moved to Til Chatel to help restore their Flamant 226.

SOLENZARA (20A)
Preserved at BA126 on the French Island of Corsica are a Vautour 2N and Mirage 3E.

❑ 557	2-LJ	Mirage 3E	557	preserved	3-96
❑ 370	30-QY	Vautour 2N	370	preserved	3-96

ST BRIEUC (22)
C-47A Skytrain F-GESB (13835, ex Aeronavale 35) was stored at the new airfield, west of St Brieuc, whilst F-BHKX (11995) was dumped on the old disused airfield. No reports from here since August 1986.

ST CYR L'ÉCOLE (78)
All helicopters mentioned, including the preserved H-34A SA85, in the first edition have gone from this airfield. The last ones were noted here in June 1989.

ST DALMAS DE TENDE (06)
A bit difficult to find, but the Magister is still in this little village along the RN204 near the Italian border. In the centre of the village follow the sign to the firebriage station and from their to the estate 'Le Miniere'. Here, inside a modern looking garden, (or perhaps an unfinished building with no roof) the Magister can be found.

❑ 125		CM170	125	preserved	9-95

ST DIZIER - ROBINSON (52)
With the disbandment of the Jaguar units at Toul, St Dizier houses the last operational Jaguar fighter units. Preserved at their gate are a Mirage 4 and a Thunderstreak. The T-33 and CM170 were inside the airbase.

❏ 46		CM170	46	preserved	6-92
❏ 11		Mirage 3C	11	stored, traveling recruiting aid	6-93
❏ 16	AO	Mirage 4A	16	preserved, at gate	7-97
❏ 16524	338-H	T-33A	5856	preserved, ex 51-6524	6-92
❏ 28897	1-ET	F-84F	..	preserved, at gate, ex 52-8897	7-97

ST GATIEN DES BOIS (14)
MD315R Flamant 51 was found in a bad state in 1988 at the parking lot of the 'Top Gun' disco. It may not have survived into the 1990s.

❏ 51	30-QS	MD315R	51	stored	7-88

ST GEOIRS (38)
Preserved at this new Grenoble airfield is a CM170.

❏ 119	CM170	119	preserved	9-95

ST JACQUES DE GRASSE (06)
A search during 1995 did not locate the CM170 135. The site of the Musée du Train Miniature was vacant and the locals reported that the museum (and its CM170) had moved to the nearby town of Nice. Unfortunately there are no reports from Nice either.

ST JULIEN DE CASSAGNAS (30)
Former Swiss Venom J-1545 (ex Morges) was first noted here in the Parc Ornothologique des Isles in 1995. The aircraft moved to Savigny lès Beaune in late 1996.

ST MALO - ST SERVAN (35)
The aero club's Magister, 160, was last noted in 1990. A search for the airfield here in 1996 failed. A Michelin road map from 1985 still showed the airfield, but on the map from 1995 the airfield was no longer there.

ST MÈRE ÈGLISE (50)
The town of Sainte Mère Èglise is located along the N13 towards Valognes. In the centre is the Airborne Troops Museum, with two aeronautical exhibits.

❏ 25	C-47A	19288	preserved, marked as '315159', ex 42-100825	7-97
❏ 45-17241	CG-4A	..	preserved, glider	5-93

ST NAZAIRE - MONTOIR (44)
Outside an Airbus factory at the local airfield is a Vautour. Inside should still be a Bretagne and Fouga 90.

❏ 37	SO30P	37	preserved, marked as 'F-BANZ'	3-89
❏ 632	Vautour 2B	632	preserved, at factory gate	6-97
❏ F-WZJB	Fouga 90	01	preserved	3-89

ST PÉRAVY LA COLOMBE (45)
Some 3km west of the Orléans Bricy airbase is the small village of St Péravy La Colombe. On the eastern side at the petrol station, the fuselage of a Noratlas still remains. The village is along the main road from the airfield of Orléans to Châteaudun.

❏ 12	N2501F	12	stored, fuselage only	5-98

ST PHILIBERT DES CHAMPS (14)
Nothing has been heard of the H-34 which should be with the private collector Monsieur Joseph.
❏ SA55   68-OA   H-34A   SA55   preserved   87

ST RAMBERT D'ALBON (26)
The Aero Retro preservation society here has a large number of flying vintage and 'warbird' aircraft including a Skyraider and ex RAF Chipmunks. Stored in their hangars are a CM170, HSS-1 and Mirage 3. The HS748 does not belong to Aero Retro and is parked outside on another part of the airfield and has been stripped of most useable parts.

❏ 206	CM170	206	stored, ex F-WMDM	3-97
❏ 641	H-34A	58-641	stored	3-97
❏ 586	Mirage 3E	586	stored, ex Châteaudun	3-97
❏ F-GHKL	HS748-2A	1677	stored	3-97

ST VALERY EN CAUX (76)
The aero club's Magister is in excellent condition as it is always kept in a hangar.
❏ 13   30-QI   CM170   13   preserved   2-97

STRASBOURG (67)
Town: Under restoration with Monsieur Le Reverend at a private location is SO1221 Djinn FR146.
❏ FR146   SO1221   FR146   under restoration   9-88

Airfield – Entzheim: The last operational Mirage F1CRs left the base in the mid-1990s. Mirage 3RD 361 moved to museum at <u>Orange</u> and RF-84F 37577 is now at <u>Reims</u>.
❏ 314   33-TC   Mirage 3R   314   preserved, up a pole near terminal   6-95

Airfield - Neuhof: The group Ailes Anciennes - Alsace are based at the airfield here (also known as Polygone). Despite their ambitions as reported in EWR-1, as yet only the Mystère 4A is preserved here.
❏ 33   Mystère 4A   33   preserved, painted as '57/10-RE'   9-97

TARBES - OSSUN (65)
In July 1995 an unknown MS760 Paris was displayed at the Aerospatiale factory, while a TB30 Epsilon was derelict on the field (reported as 42/VO). In October 1995 this aircraft was noted flying as 42/315-VO. Was the derelict TB30 restored to flying condition or is there another Epsilon here?
❏ ..   MS760   ..   preserved   7-95

TARNOS (40)
The CM170 painted in Patrouille de France colours and mounted on a pole within the Turbomeca factory has been joined by an Alouette 2 in the colours of the Gendarmerie. Both remain unidentified.
❏ ..   Alouette 2   ..   preserved, unmarked, in Gendarmerie c/s   5-96
❏ ..   CM170   ..   preserved, unmarked, in PdF c/s   5-96

TAVERNY (95)
Displayed outside the Centre de Commandement des Forces Strategiques at Base Aérienne 921 in the suburbs of Paris was Super Mystère 158. It was no longer noted in June 1991 and went to <u>Bordeaux</u>. It was replaced by a silver Mirage 2000 which was first noted during that month.
❏ ..   Mirage 2000   ..   preserved   6-91

TIL CHÂTEL (21)
On this airfield, some 20km north of Dijon, the aero club has restored Flamant 226 to flying condition as F-AZES. For this they used spares from 153 (ex Sezanne). Neither aircraft was noted at the airfield in 1997.

TOUCHAY (18)
At this location, some 40km south of Bourges, a new Les Cocardiers du Ciel museum has been set up. Currently visits are only possible during the weekends.

❏ 394		CM170	394	preserved, ex Châteaudun	3-98
❏ 370	33-TS	Mirage 3RD	370	expected, still at Rochefort	—
❏ 104	10-SQ	Mystère 2C	104	preserved, ex Henin Beaumont	3-98
❏ 293		Mystère 4A	293	preserved, ex Savigny lès Beaune	3-98

TOUL - ROSIÈRES (54)
1998 is last year that the airfield of Toul Rosières is open. After the disbandment of EC03.011 in mid 1997, the last operational unit here, the base will close. It is not known what has happened to the Mirage 3R and F-100D.

❏ 42131	11-MJ	F-100D	223-11	preserved, ex 54-2131	6-93
❏ A26	11-RO	Jaguar A	A26	preserved	97
❏ 316		Mirage 3R	316	stored, in tiger c/s	5-91

TOULON - ST MANDRIER (83)
Preserved at the gate of this Aeronavale helicopter base is an HSS-1.

❏ 163	Alouette 2	1163	stored, dismantled	6-96
❏ 213	Alouette 3	1213	stored, dismantled	6-96
❏ 182	HSS-1N	SA182	preserved	6-96
❏ 102	SA321G	102	stored	6-96

TOULOUSE (31)
Town: The Ecole Nationale Superieure de l'Aéronautique (ENSAE, or SupAero) is restoring their Mirage 3A 02. It will be mounted on a pylon. CM170 178 moved to Toulouse Blagnac.

❏ 02	C	Mirage 3C	02	under restoration	97

The Ecole d'Essais Aéronautique de Toulouse has a Mirage F2 prototype. It was last been noted at an airshow at Toulouse Francazal, but should normally live here.

❏ 01	Mirage F2	01	instructional	6-93

An institute with the name UIUT Toulouse is using a Magister, which used to be preserved at Blagnac.

❏ 103	CM170	103	instructional, ex Blagnac	97

Town - Lhers: On the gate at Base Aérienne 292 was Mystère 4A 182 which arrived from Rocamadour around 1982/1983. It is believed that the base has closed and the Mystère moved on (or was scrapped).

❏ 182	8-MZ	Mystère 4A	182	preserved	1-87

Airfield – Blagnac: Ailes Anciennes - Toulouse have a large collection of aircraft preserved and stored on a site within the Airbus factory grounds. The collection can only be visited on Saturday mornings via one of the gates of the Airbus factory. Their CM170 103 moved to UIUT in Toulouse town. Broussard 139 (ex Montauban) became civil and N1101 81 moved to Germany.

❏ FU-125		F-84F	..	preserved, ex 53-6760, ex Belgian AF	8-96
❏ 770		MiG-21SPS	94A4509	preserved, ex NVA	8-96
❏ 2191		F-104G	7060	preserved, ex WGAF	10-97
❏ 52-6789	789	F-84F	..	preserved, ex Ramstein, ex Greek AF	8-96

FRANCE - 125

☐ J-4065		Hunter F.58	..	stored, arr 8-95, ex Swiss AF	8-96
☐ XE950		Vampire T.11	..	stored, arr 14-11-86, ex RAF	8-96
☐ 54-2239	FW-239	F-100D	223-119	stored, ex Bitburg, ex FAF, ex USAF	8-96
☐ 58-0282		F-101B	..	preserved, ex USAF	8-96
☐ 504		Br765	504	stored	2-97
☐ 085		C-45	..	preserved, ex F-ZJAD	8-96
☐ 168		CM170	168	preserved, ex Châteaudun	8-96
☐ 178		CM170	178	preserved, PdF c/s	8-96
☐ 19572		F-84G	..	stored, arr 6-82, ex 51-9725	8-96
☐ 52-7603	APZ	H-19D	..	stored	8-96
☐ FR106		H-21C	FR106	preserved, marked as 'FR26', ex Dax	8-96
☐ SA116	116-RB	H-34A	SA116	preserved, ex '58-002/GR'	8-96
☐ 227		MD312	227	preserved, arr 2-11-83	8-96
☐ NF11-8	BG	Meteor NF.11	..	preserved, arr 2-1-84, Le Bourget owned	8-96
☐ 78	MAB	MH1521	78	stored	8-96
☐ 86		Mirage 3C	86	preserved, marked as '27/3.10-LE'	8-96
☐ 1		Mystère 4A	1	stored, arr 29-10-83, Le Bourget owned	8-96
☐ 44		Mystère 4A	44	preserved, PdF c/s, arr 27-3-83	8-96
☐ 88	BZ	N1101	88	preserved	8-96
☐ 191		N2501F	191	preserved, arr 25-4-84	2-97
☐ 130		N3400	130	stored	8-96
☐ 23		RF-3	..	stored	8-96
☐ FR101		SO1221	..	stored, Le Bourget owned	8-96
☐ 01	C	SO6025 Espadon	01	preserved, ex St Dié	8-96
☐ 48	12-UA	Super Mystère B2	48	preserved, ex 21-BT	8-96
☐ ..	OJ	T-6G	..	preserved, arr 11-11-80, maybe s/n 50-8092	8-96
☐ 14230		T-33A	5524	preserved, marked 58-0468, ex 51-4230	8-96
☐ 640		Vautour 2B	640	preserved, arr 4-2-81, Le Bourget owned	8-96
☐ F-AMKT		Caudron C.282/8	6770/26	preserved, Le Bourget owned	8-96
☐ F-BETX		J-3C Cub	12763	stored, marked as 'L-39', ex D-EBOR	8-96
☐ F-BMMT		MS733	106	preserved, arr 9-5-81, marked as '106/VO'	8-96
☐ F-BPMF		MS893A	10761	stored	8-96
☐ F-BPPA		Boeing 377SG	002	preserved, Super Guppy, ex N212AS	8-96
☐ F-CCHN		SM30	01	stored, glider, nose only	8-96
☐ F-CROE		N1300	9	stored, glider	8-96
☐ F-CRQH		N1300	200	stored, glider	8-96
☐ F-GHMU		SE210-10B3	249	preserved, ex HB-IKD	2-97
☐ F-WBHA		Deltaviex	01	stored, arr 30-9-84	8-96
☐ F-WZJL		AS350	..	stored	8-96
☐ G-ALWC		C-47A	13590	preserved, F-GBOL ntu, ex KG723	2-97
☐ ..		Alpha Jet	..	preserved, mock-up	8-96
☐ ..		Gardan GY80 Horizon	..	stored	8-96
☐ ..		HM-8	..	stored	8-96
☐ ..		HM-14	..	stored	8-96
☐ ..		Jodel D.119	..	stored	8-96
☐ ..		Miles ..	..	stored, cockpit only	8-96

Other aircraft at the airfield are a Concorde which is preserved near one of the Airbus gates. Noted dumped near one of the hangars in June 1990 was a silver Transall, this is presumably a test airframe that has never flown. F-27s F-GCPN (533) and F-GIHR (117) were noted in storage here in 1991 and both moved on to Dinard, while F-GIPD (92) was scrapped on site in 1997. Stored Caravelle F-BMKS flew out to Perpignan.

☐ F-BAIF	C-47B	16371/33119	stored, ex 44-76787	6-90
☐ F-BTOE	SE210-12	280	stored, on pole	6-95
☐ F-GELP	SE210-10B3	187	stored, ex I-GISO	1-97
☐ F-WTSB	Concorde 100	201	preserved, Airbus factory	8-96

Airfield – Francazal: During the open house in 1996 only the preserved aircraft were noted, there was no sign of the other two aircraft and they may all have left. Formerly preserved MD311 282/316-KY became F-AZFX.

❑ 114	316-FQ	N2501F	114	dumped	6-93
❑ 115	63-VA	N2501F	115	stored	6-95
❑ 208	340-HB	N2501F	208	preserved, at main gate	6-97
❑ SA154	68-DI	H-34A	SA154	preserved	6-96

TOURS - ST SYMPHORIEN (37)

EAC00.314, as the local AdlA unit is called today, still flies the Alpha Jet. Some aircraft which served with the unit are preserved at the base. The Beech is preserved on a roundabout just north west of the airfield.

❑ 187	UI	MD450	187	preserved	6-97
❑ 227	UN	MD450	227	stored	9-91
❑ 250	TO	MD450	250	preserved	6-95
❑ 219	DK	Mirage 3B	219	dumped, ex Châteaudun	6-95
❑ 401	13-QH	Mirage 3E	401	dumped	6-95
❑ 22	314-TG	Mystère 4A	22	preserved	6-97
❑ 21330	314-UM	T-33AN	330	preserved	6-95
❑ F-BTMA		Beech 99	U-90	preserved, ex N921GP	3-98

TOUSSUS LE NOBLE (78)

In the hangar of the Sodeteg Formation (opposite the large Farman hangar) are a number of instructional airframes. Stored H-34A SKY705 (l/n 6-91) moved on to Plobannalec.

❑ A-05		Alouette 2	1379	stored	5-98
❑ A-23		Alouette 2	1666	stored, ex Belgian Army	6-97
❑ 100		CM170	100	instructional	3-98
❑ 240		MD312	240	instructional	3-98
❑ 1654	JAG	Alouette 2	1654	instructional	3-98
❑ D-HOBU		Alouette 2	1890	instructional	3-98
❑ F-BGOA		Dove 6B	04344	instructional	3-98
❑ F-BTOC		SE210-12	278	instructional, cockpit only, ex Paris Orly	3-98
❑ F-BVAF		PA31P	43	instructional, ex SE-GAE	3-98
❑ F-GBQC		CeF150	1577	instructional	3-98
❑ ..		Alouette 2	..	instructional, purpose built rig	3-98
❑ ..		Alouette 3	..	instructional, purpose built rig marked SF III 1	3-98
❑ ..		Alouette 3	..	instructional, purpose built rig marked SF III 3	3-98
❑ ..		Gazelle	..	instructional, purpose built rig	3-98

VALENCE - CHABEUIL (26)

Still preserved at the civil side of this airfield is a Broussard. Its former code GG is no longer readable.

❑ 24		MH1521M	24	preserved	5-96

VANNES - MEUCON (56)

The Ailes Anciennes - Armorique group has a hangar at this former military field in which also their airworthy Broussard is kept. The museum's hangar is open on Sunday afternoons.

❑ 06		Etendard 4M	06	preserved, ex Rochefort	3-98
❑ 01		F-8A	..	stored, ex Rochefort, ex 143719/USN	3-98
❑ 70	10-KE	MH1521M	70	preserved	3-98
❑ 01	G	N2200	01	stored, in hangar	3-98
❑ 160		N2501F	160	preserved	3-98
❑ 179	12-YN	Super Mystère B2	179	preserved, ex Captieux	3-98

FRANCE - 127

☐ 21009	T-33AN	009		stored, in hangar	3-98
☐ 53091	T-33A	9632		preserved, ex La Baule, ex 55-3091	3-98
☐ F-CAJB	Grunau SG-38	35		stored, in hangar, glider	3-98
☐ F-CBGB	N2000	10409/79		stored, in hangar, glider	3-98

VARENNES SUR ALLIER (03)
Varennes Sur Allier is some 25km north of Vichy and preserved with a military maintenance centre (EAA606) is a former Châteaudun Mirage 3.

☐ 247	Mirage 3B-RV	247	preserved, ex Châteaudun	94

VIERZON (18)
Les Ailes Vierzonnaises continue to look after Flamant 191 which was still current in March 1998.

☐ 191	XA	MD312	191	preserved	3-98

VILLACOUBLAY (78)
That close to Paris, yet no reports of the preserved H-34. This is largely due to the fact that not much can be seen of the airfield from outside.

☐ SKY506	67-SV	H-34A	..	preserved	88

VILLEBON SUR YVETTE (91)
The Centre de Formation et de Perfectionnement de l'Aéronutique de Paris (CFPAP) will close down much of its activities in the near future. Alpha Jet F-ZWRU and CM170 8 moved to the CFPAB in <u>Latresne</u>.

☐ 16	P	Mirage 3C	16	instructional	3-98
☐ 21	12-ZX	Super Mystère B2	21	instructional	3-98

VILLEFRANCHE SUR SAÔNE (69)
Stored at this airfield are two Mi-2s with many parts missing.

☐ YU-HDE	Mi-2	541129069	stored, ex Cannes, ex Yugoslav AF	9-97
☐ YU-...	Mi-2	..	stored, UN markings	9-97

VILLENEUVE SUR LOT (47)
Preserved at this small airfield is a Noratlas in civil markings. The aircraft came from Marmande which is some 60km to the west.

☐ F-WFYF	N2501F	111	preserved, ex Marmande	8-96

VILLEPERDUE (37)
Wrecked Broussard 182 (l/n 1986) was not noted in 1993 and may have been scrapped by the yard. Villeperdue is some 30km south of Tours along the D21.

☐ FR60	H-21C	FR60	stored, with tail from FR94, ex Dax	4-97
☐ 21028	T-33A	028	stored	5-93

VISAN - VALRÉAS (84)
Still preserved at the airfield are two aircraft in good condition. The Fouga is painted in white and blue colours.

☐ 522	CM170	522	preserved, ex F-WFTX	8-96
☐ 253	MD312	253	preserved	8-96

VITROLLES (13)
The Lycee P Mendes-France has a number of instructional airframes. Of these only the Noratlas is parked outside. Vitrolles in on the east side of the A55 highway alongside Marseille airport.

❏ 372		CM170	372	instructional	1-98
❏ 114		MH1521M	114	instructional	10-93
❏ 296		MH1521M	296	instructional, F-GFJY ntu	1-98
❏ 453	13-QA	Mirage 3E	453	instructional	1-98
❏ 205	A	N2501F	205	instructional, outside	1-98
❏ F-WVKD		SA365	..	instructional	—

VIUZ EN SALLEZ (74)
Along the D907 from Genève towards Taninges is the small village of Viuz en Sallez (some 20km east of Genève). Preserved next to the local school is a N2501, which is in use as a chapel.

❏ 146	MC	N2501F	146	preserved, ex Châteaudun	6-96

BUREAU CENTRAL DES RELATIONS EXTERIEURES
This Air Force unit has a number of airframes which are used as recruiting aids and can be seen airshows and other events in France. Maybe the Mirage 3A and 3C nose are from the same aircraft.

❏ 171	341-HV	Alouette 2	1306	preserved	6-97
❏ 55	312-HB	CM170	55	preserved	—
❏ 111		CM170	111	preserved, PdF colours	6-97
❏ 401	315-XV	CM170	401	preserved	—
❏ E1	7-JH	Jaguar E	2	preserved	6-97
❏ 08	DZ	Mirage 3A	08	preserved, nose only, see note	—
❏ 83	RO	Mirage 3C	83	preserved, nose only	91
❏ ..	30-SB	Mirage 3C	..	preserved, nose only, see note	6-92
❏ ..	10-RC	Mirage 3C	..	preserved, nose only, see note	6-93
❏ ..	2-EA	Mirage 3B	..	preserved, nose only	7-95
❏ 79	30-LA	Mirage F1C	..	preserved, cockpit only	6-97
❏ 240	12-YP	Mirage F1C-200	..	preserved, cockpit only, ex Châteaudun	6-97
❏ ..	30-AD	Mirage F1C	..	preserved, cockpit only	5-98

UNKNOWN
Three Magisters are known to have left Rochfort - St Agnant. Their exact location is unknown as the location names mentioned for the aircraft could not be found in the index of the common Michelin roadmap of France.

❏ 166	CM170	166	l/n 7-94 at Rochefort, to <u>Val Doise</u>	—
❏ 222	CM170	222	l/n 7-94 at Rochefort, to <u>Cap Feyrefite</u>	—
❏ 332	CM170	332	l/n 7-94 at Rochefort, to <u>Forez</u>	—

AUSTRIA

MiG-15bis 350 at Bad Ischl.
Otger van der Kooij

L-19 Bird Dogs 3A-BK and 3A-BL at Bad Ischl.
Otger van der Kooij

C-47A Skytrain N86U at Wiener Neustadt Ost (now at Salzburg).
Otger van der Kooij

BELGIUM

F-84F Thunderstreak FU-51 at Temploux.
Otger van der Kooij

RF-84F Thunderflash FR-27 at Spa.
Dimitri Schmidt

F-84F Thunderstreak FU-66 'RA-T' at Kleine Brogel.
Julian Bloomfield

TS-11-100bisB Iskra 1014 at Weelde.
Richard Nels

BELGIUM

Ouragan 320 'UQ'
at the Brussels museum.
Gerard Post

Super Mystère B2 145 '12-ZR'
at Weelde.
Julian Bloomfield

F-104G Starfighter FX-61
at the Tiger Air Forces Museum,
Peer.
Berry Vissers

BELGIUM

HSS-1 Seabat B-4 'OT-ZKD' on temporary display at Brustem (normally at Koksijde).
Dean Charnley

Airbus A300B1 OO-TEF at Brussels.
Johan Mulder

C-47B Skytrain '2100847' (F-BAIF) at Arlon.
Gerard Post

CZECH REPUBLIC

MiG-15bisSB 3912 at Vyskov.
Otger van der Kooij

MiG-19PM 1040 at Kunovice.
Paul Gross

MiG-21US 0241 in store at Kbely.
Johan Mulder

CZECH REPUBLIC

L-29R 2811 (note under fuselage fairing) at Horice. *Otger van der Kooij*

Il-28RT 2404 at Brno-Slatina. *Paul Gross*

Su-7BKL 6011 stored at Prerov. *Otger van der Kooij*

MiG-23BN 5733 at Kolin. *Otger van der Kooij*

Avia 14T 3173 at Nove Mesto nad Metuje. *Otger van der Kooij*

Mi-4 4142 at Nove Mesto nad Metuje. *Otger van der Kooij*

L-410UVP Turbolet OK-IYB at Zruc. *Otger van der Kooij*

DENMARK
F-84G Thunderjet 51-10622 'KU-U' in the Fjordparken at Alborg. *Otger van der Kooij*
RF-84F Thunderflash C-581 at the FLSK, Alborg. *Otger van der Kooij*
F-86D Sabre F-421 at Værlose (now at Skrydstrup). *Berry Vissers*

DENMARK
RF-35 Draken AR-112 stored at Karup and now on the main gate. *Berry Vissers*
Meteor NF.11 51-504 on the gate at Alborg; behind is F-86D F-947 marked as 'F-326'. *Otger van der Kooij*
TF-100F Super Sabre GT-949 at Karup. *Otger van der Kooij*

DENMARK

C-47A Skytrain K-687 at Billund.
Stephan de Bruijn

KZ VII O-622 (and youngster!) at Stauning.
Otger van der Kooij

PBY-6A Catalina L-861 under restoration at Helsingor.
Otger van der Kooij

FINLAND

Gnat F.1 GN-103 at Halli.
Alan Warnes/Airforces Monthly

Lim-5 1505 at Paimio.
Berry Vissers

MiG-25RBS c/n 02050740 at Tampere Pirkkala.
Berry Vissers

FINLAND

Bf 109G-6Y MT-507 'O' in the foreground of the Keski-Suomen Ilmailumuseo at Tikkakoski.
Keski-Suomen Ilmailumuseo

SAAB 91B Safir SF-36 at the technical school at Rovaniemi.
Berry Vissers

CM170 Magisters FM-21 'K' and FM-82 'M' at Kauhava.
Tieme Festner

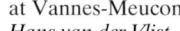

FRANCE

Mystère IVA 287 marked as '316' '8-NM' at Abbeville-Drucat.
Otger van der Kooij

Mystère IVA 278 '8-MB' at Châteaudun.
Arnold ten Pas

Super Mystère B2 179 '12-YN' with Ailes Anciennes-Armorique at Vannes-Meucon.
Hans van der Vlist

FRANCE

Mirage IVA 4 'AC' on the gate at Rochefort - St Agnant.
Otger van der Kooij

Ouragan 214 on display at Montelimar.
Otger van der Kooij

Mirage F1C 14, instructional at Rochefort - St Agnant.
Otger van der Kooij

FRANCE

Mirage IIIB 205 '13-FC', instructional at Rochefort - St Agnant.
Otger van der Kooij

Mirage IIIE 584 '4-AN' at Châteaudun.
Otger van der Kooij

Mirage IIIRD 361 '33-TJ' with Aviomodelli at Orange.
Otger van der Kooij

FRANCE

CM170 Magister 522
at Visan - Valréas.
Otger van der Kooij

F-8A Crusader 01
at Vannes - Meucon.
Otger van der Kooij

CM170 Magister 523 with ferry
registration F-WFUQ still applied,
at Bresse - Creyzériat.
Otger van der Kooij

FRANCE

Hunter F.4 ID-44 '044'
at Savigny lès Beaune.
Otger van der Kooij

Deltaviex 01 F-WBHA
at Toulouse-Blagnac.
Otger van der Kooij

H-34A Choctaw SA114 '68-OF'
at Savigny lès Beaune.
Otger van der Kooij

FRANCE

MD312 Flamant 196 at Rochefort-St Agnant.
Otger van der Kooij

HD-321 01 stored on a farm at Chamon sur Voueize.
Otger van der Kooij

N2501F Noratlas 184 '328-EO' at Angers-Avrillé.
Otger van der Kooij

FRANCE

Nord 262 3 'OH' at La Ferté Alais.
Gerard Post

HSS-1 Seabat 135 'N' plus three others at Plobannalec.
Otger van der Kooij

F27-200 Friendship 10320 F-SEBF stored at Dinan.
Julian Bloomfield

FRANCE

Caravelle 3 F-BOHA, marked as 'F-POHA', at Avignon-Caumont.
Alan Warnes/Airforces Monthly

HS748-2A F-GHKL stripped for spares at St Rambert d'Albon.
Otger van der Kooij

Fairchild F27J F-OGJB stored at Dinard.
Otger van der Kooij

V953C Merchantman (Vanguard) G-APEK at Perpignan.
Otger van der Kooij

GERMANY

F-86F '130136' 'FU-036'
in the domestic area at Bitburg.
Paul Gross

RF-4C Phantoms 68-0577 and
69-0360 'GA' scrapped at Bitburg.
Otger van der Kooij

CL-13B Sabre 6 D-9539
at Berlin-Gatow.
Otger van der Kooij

GERMANY
RF-84F Thunderflash EA-315, marked as 'EA-105', at Roth. *Dean Charnley*
F-104G Starfighter 2167, marked as '4033', at Buchel. *Julian Bloomfield*
G91T/3 3402 in Bavarian colours at Fürstenfeldbruck. *Elmar Keetman*

GERMANY

MiG-21F-13 '705' (true identity unknown) at Dresden.
Arnold ten Pas

MiG-21MF 2319, withdrawn at Drewitz in the mid-1990s.
Paul Gross

MiG-21UM 2380 at Baschutz, later moved to Wurzburg.
Otger van der Kooij

GERMANY

MiG-17 22 in the school at See, near Rothenburg.
Otger van der Kooij

Mi-2 '17 white' left behind at Gut Kummersdorf.
Hans van der Vlist

Su-22M-4 2504, complete with armament and over-run barrier, at Cottbus. *Otger van der Kooij*

GERMANY

Mirage 5BA BA-35 (Belgian) in French markings as '13-PL' at Hermeskeil.
Gerard Post

Alouette II 'PQ-131' (correct identity unknown) at Dessau.
Otger van der Kooij

Su-7BM 01 on show at Beelitz.
Berry Vissers

GERMANY

Mi-2 380 at the Cottbus museum.
Otger van der Kooij

Mi-8PS 9380 stored at Bückeburg.
Brian Waclawek

Mi-24P 9649 under restoration for display at Bückeburg.
Julian Bloomfield

GERMANY

Pembroke C.54 5421
at Hermeskeil.
Otger van der Kooij

Do28D2 Skyservant 5911
stored at Nordholz.
Otger van der Kooij

OV-10B Bronco 9918
at Schwenningen.
Gerard Post

GERMANY

FWP149D 9094 on display inside Fürstenfeldbruck. *Brian Waclawek*

Do31 D-9530 preserved at its maker's airfield at Oberpfaffenhofen. *Jaap Dijkstra*

Do24T-3 HD.5-3 preserved at Oberpfaffenhofen. *Jaap Dijkstra*

C-54G Skymaster 45-0557 on display at Berlin-Tempelhof and now in store there. *Johan Mulder*

C-47A Skytrain N62443 in its former Jordanian colours at Hermeskeil. *Berry Vissers*

Devon C.2/2 G-DVEN (formerly WB533 'DA') at Frankfurt. Now at Meresburg. *Aad van der Voet*

CASA 2111 BR.2I-14 in Luftwaffe colours as 'G1FL' at Hermeskeil. *Gerard Post*

GERMANY
VEB-14P 482 shortly after arrival at Finow. *Otger van der Kooij*
VEB-14P DM-SAF at Halle; it moved to Pulsforde shortly afterwards. *Aad van der Voet*
Il-62 DDR-SEH preserved at Alach. *Aad van der Voet*

GREECE

F-5A Freedom Fighter '063' (identity unknown) at Stilisa. *Chris Schmidt*

F-84G Thunderjet 19752 'FS-752B' in the memorial display at Lárissa. *Chris Schmidt*

F-84F Thunderstreak 53-7216 '216' at Athine. *Chris Schmidt*

GREECE

RF-84F Thunderflash 17011 displayed at the Helliniki Aerporia Moussio at Athine, Tatoi.
Alan Warnes/Airforces Monthly

TF-102A Delta Dagger 62327 at Elefsis.
Stephan de Bruijn

TF-104G 5901 'TF-5901' in use as a decoy at Tanágra.
Chris Schmidt

(Greece continues on page 289)

GERMANY

Germany has seen a dramatically in its military aviation over the past ten years. All Starfighters, G91s and Alpha Jets have been withdrawn from service. An even greater impact was the reunion of East and West Germany in October 1990. Hundreds of former East German aircraft were given new serials in what was the former West German serial system, although only a few types remained in service. Today only two eastern types are still flying, the MiG-29s and some of the L-410s. Due to the arms reduction treaty 141 MiG-21s were scrapped and the remainder were sold to museums, collections and private individuals all over the world. A number of aircraft was sold to other air armes. Where known, all former East German aircraft are listed at their initial storage airfield, although some remained there only for a short time. Note that the NVA called their MiG-21PFs (type 76) modified with a new radar MiG-21PFM, which are not the same as the factory built MiG-21PFMs. MiG-21PFMs (type 94) were named in NVA service MiG-21SPS or MiG-21SPS-K.

AHAUS
A privately owned Starfighter is preserved in this village.
☐ 2126   F-104G   6687   preserved, ex Fassberg   3-98

AHLHORN
After the departure of the USAF A-10A detachment in the late 1980s, Hubschrauber Transport Geschwader 64 with its UH-1Ds also left the base in the mid-1990s. Consequently nearly all the *Wrecks & Relics* aircraft have left. Instructional F-4C 63-7446 and F-101B 58-0265 went to Speyer. Dumped UH-1D 7084 moved to Fassberg, while 7126 (8186) was scrapped. Sycamore 7823 (13481) was last noted in 1987 and 7837 moved to Landsberg. RF-84F EB-354 was last noted on the dump in April 1993. Sabre JB-112 (1775) went to Wittmund.
☐ 5356   N2501D   186   preserved, at gate   4-93

ALACH
After being stored at Erfurt airfield, the Il-62 moved a few kilometres west to Alach were nowadays it can be found outside a furniture shop.
☐ DDR-SEH   Il-62   31405   preserved, wfu 9-11-89, ex Erfurt, ex DM-SEH   8-91

ALLSTEDT
This former Soviet airfield, south west of Lutherstadt Eisleben, had a MiG-15 on the dump. As all operational aircraft have left here the MiG has presumably also been removed.
☐ 93   MiG-15UTI   812759   dumped, ex Soviet AF   7-91

ALTENBURG
The gate guard at this former Soviet military airfield, Yak-28R 91 blue (8961310), is said to have left for Hannover in 1992, but where in Hannover is still not clear.

ALTENSTADT
Altenstadt is located near Schongau, some 60km south west of München. The Luftlande und Lufttransport Schule should still have Transall 5003, which includes parts of 5005 (A05). The remains of 5005 were burned in the mid-1970s at Trauen. Three UH-1Ds (7037 (4368), 7206 (8326) and an unknown example) were last noted in 1986 and have all left.
☐ 5003   C-160A   A01   instructional, sections only, ex YA-053   4-91
☐ 5337   N2501D   160   preserved   4-91

ALTES LAGER
This former VVS airfield is also known as Jüterbog. Gate guard MiG-23ML 70 red only moved some 20km to the north to the village of Beelitz when the Soviets pulled out of this base.

ANKLAM
A number of aircraft are preserved at this airfield in the former DDR. They are part of the Lilienthal museum.

❏ D-FONG	An-2	112208	preserved, ex DDR-SKG, ex 863/NVA		3-97
❏ D-HOAG	Ka-26	7705908	preserved, ex DDR-SPG, ex DM-SPG		3-97
❏ DDR-TED	PZL 106BR	10880212	preserved		3-97

ANKUM
A new museum will be set up near this town, for which aircraft are currently stored (most in containers) off site.

❏ 2033	MiG-23ML	0390324038	stored, ex Laage, ex 569/NVA		—
❏ 2223	MiG-21SPS	94A4209	stored, ex Rothenburg, ex 717/NVA		—
❏ 2390	MiG-21U-600	662617	stored, ex Rothenburg, ex 265/NVA		—
❏ 2401	MiG-21US	07685145	stored, ex Rothenburg, ex 217/NVA		5-98
❏ 2409	MiG-21US	04685139	stored, ex Rothenburg, ex 242/NVA		—
❏ 2625	F-104G	9177	stored, ex Diepholz		5-98
❏ 2831	L-39ZO	232304	stored, ex Rothenburg, ex 182/NVA		—
❏ 3100	G91R/3	366	expected, still at Oldenburg ?		—
❏ 9164	FWP149D	186	stored		5-98
❏ 9902	G91R/3	0068	stored, ex Fürstenfeldbruck		5-98
❏ 9520	T-33A	1653	stored, ex Rotenburg Wümme, ex 58-0684		—
❏ 679	MiG-21F-13	740901	stored, marked as '696', ex Uetersen, ex Neuhard.		—
❏ 969	MiG-21SPS	94A6501	stored, ex Uetersen, ex Bautzen		—
❏ D-EFWP	FWP149D	118	stored, fuselage only, ex 9098		5-98

ANSBACH
The US Army airfield east of Ansbach is better known by the Americans as Katterbach. On the north west side of the base, a 'Huey' with a false serial number was dumped for many years.

❏ 67-17241	UH-1H	9439	dumped, marked as '20001', ex US Army		5-92

ASCHERSLEBEN
The Polizei Historische Samlung Sachsen Anhalt has a small museum at the Polizei training centre on the south east side of town. The collection has three aircraft and these can be viewed on weekdays. The museum is not sign-posted, ask for the museum at the main gate of the Polizei centre.

❏ D-HZPC	Mi-2	538839114	preserved, ex Schönefeld, ex DDR-VPG		11-97
❏ D-HZPT	Ka-26	7404108	preserved, marked as 'DDR-VPK', ex Dessau		11-97
❏ DDR-WBY	PZL 104-35	62163	preserved, marked as 'DDR-VPS', ex Dessau		11-97

AUENHAUSEN
The radar station on a hilltop near Auenhausen was visited in early 1992 by a HEER CH-53. The helicopter carried a Starfighter for preservation at the site.

❏ 2908	F-104F	5058	preserved, marked as '2705', ex BB-371		92

AUGSBURG
The Fliegendes Museum Augsburg was forced to move to a new location as a Beechcraft overhaul plant was to be built at Augsburg. All aircraft had left by late 1992. Most of the non-flying aircraft were moved to Tannheim

near Memmingen. These included MS733 71/YA, F-104G 2046, G91R/3 3074 (ex Ganderkesee), MiG-15UTI 140, Lim-5 413 and DH.82A Tiger Moth D-EFTF. Bf108B-2 A-208 (marked as 'D-IOIO') and HA1112K-1L D-FMBB came from the MBB factory in town and were last noted in the museum in April 1991. As the name of the museum suggests it also had a large number of flyable aircraft (which are not covered in this book). Some of the aircraft were later noted at Sandown on the Isle of Wight. These included Harvard 4 D-FABE (CCF4-499) and An-2T D-FOFM (12802). Fates for L-4H Cub D-EFIX (12233) and DH.89A Dragon Rapide D-IGUN (6437), both last noted here in April 1991, are not known. Il-18D DDR-STH arrived on 6th May 1990 and moved on in 1993 to Hermeskeil, together with Tu-134A DDR-SCK which arrived on 30th September 1990.

AURICH
The gate of Blucherkaserne in Aurich (75km north west of Oldenburg) is guarded by an F-104F carrying an F-104G serial number. It is the former 2916 and not 2908 as was mentioned in the first edition.

| ❏ 2916 | F-104F | 5069 | preserved, marked as '2086', ex 59-5016 | 6-97 |

BAD DÜBEN
Bad Düben was the major East German Air Force instructional school. Although there is only one report from here (Mi-8T 636, noted in February 1993 and which went Butzweilerhof) a large number of aircraft are known to have lived here. Aircraft from Bad Düben are now in collections and museums like Berlin, Butzweilerhof, Dessau and Finow.

BAD OEYNHAUSEN
The Motor Technica Museum (formerly known as the Norddeutsches Auto Motorrad und Technik Museum) of Bad Oeynhausen is located in Rehme, along autobahn 2 from Bielefeld to Hannover. MiG-21 2244 and Mi-8S 9319 are both painted in bright colours to attract passers-by to the museum. From Soesterberg, the museum received a Starfighter and the crashed remains of a Ju88 and P-47. The museum is open Monday to Friday between 10:00 and 17:00. At the weekend it is open till 18:00.

❏ 42-7924	QP-F	P-47D	..	preserved, inside, wreck	4-97
❏ D-5804		TF-104G	5804	preserved, ex Soesterberg, ex RNethAF	5-98
❏ 2058		MiG-23UB	A1038034	preserved, ex Laage, ex 106/NVA	5-98
❏ 2244		MiG-21SPS	94A6408	preserved, marked as '353', ex 953/NVA	5-98
❏ 2348		MiG-21MF	96002112	preserved, marked as '2228', ex 782/NVA	5-98
❏ 2383		MiG-21UM	07695156	preserved, ex Rothenburg, ex 266/NVA	5-98
❏ 2391		MiG-21U-600	662619	preserved, ex Rothenburg, ex 272/NVA	5-98
❏ 2550		Su-22UM-3	17532371002	preserved, ex Laage, ex 146/NVA	5-98
❏ 2812		L-39ZO	731015	preserved, ex Rothenburg, ex 152/NVA	9-97
❏ 2833		L-39ZO	831123	preserved, ex Rothenburg, ex 189/NVA	5-98
❏ 7666		Alouette 2	1732	preserved, inside	4-97
❏ 9319		Mi-8PS	10552	preserved, ex Briest, ex 970/NVA	5-98
❏ 9339		Mi-8PS	10551	preserved, ex Basepohl, ex 966/NVA	5-98
❏ ..		Ju88	..	preserved, inside, wreck	4-97
❏ D-EHIT		P149D	257	preserved, ex D-EJCA, ex 9175	5-98
❏ D-EIKW		Do27A	289	preserved, ex 5621	8-97
❏ D-HAHN		Sycamore Mk.52	13445	preserved, inside, ex D-HOPF, ex 7806	4-97
❏ D-IBIB		RC680E	805-68	preserved	4-97
❏ SP-WWF		An-2R	1G173-57	preserved, wfu 17-12-91	5-98
❏ ..		Mi-2	001/9/76	preserved, unmarked, inside, incomplete	4-97

BAD SOODEN - ALLENDORF
High above this village (40km east of Kassel) on the former German east-west frontier a border post has been converted into a museum. Displayed are two helicopters.

❏ 01 yellow	Mi-24V-1	3532422810014	preserved, ex Dessau, ex Soviet AF		97
❏ 9384	Mi-8TB	10578	preserved, ex Basepohl, ex 752/NVA		97

BASCHUTZ

A small private collection here was set up the early 1990s (the former NVA base of Bautzen is only a few kilometres away from Baschutz). Some aircraft have moved on. MiG-21UM 2380 went to a museum in Wurzburg and L-39ZO 2804 (731104, ex Rothenburg) went to Breighton, Yorkshire in the UK. PZL106 DDR-TCA moved to Cottbus.

❏ 2243	MiG-21SPS	94A5606	preserved, ex Rothenburg, ex 948/NVA		9-97
❏ D-EWMJ	Zlin 42MU	0020	preserved		9-97
❏ DDR-SUD	Zlin 37A	18-22	preserved		9-97

BASEPOHL

This airfield used to be the homebase of Kampfhubschraubergeschwader 5 (KGH-5) with Mi-8s and Mi-24s and Hubschrauberstaffel zur Fürung und Aufklarung 105 (HSFA-105) with Mi-2s and Mi-9s. After the reunion of East and West Germany the aircraft were placed under the command of HFS-80 and the last military flights were made on 28th November 1994. All aircraft were placed in storage and from here and Cottbus Mi-24s left in May 1995 for Hungary. Later that year a further number of Mi-24s were sold, this time to Poland. Other helicopters have left for various museums and other military sites. The airfield was empty in late 1996.

9339	Mi-8PS	10551	stored, ex 966/NVA, l/n 3-95	to Bad Oeynhausen
9380	Mi-8PS	10597	stored, ex 732/NVA, l/n 4-93	to Bückeburg
9381	Mi-8TB	10561	stored, ex 133/NVA, l/n 4-94	gone
9382	Mi-8TB	10563	stored, ex 135/NVA, l/n 4-93	to Fassberg
9383	Mi-8TB	10576	stored, ex 750/NVA, l/n 6-94	gone
9384	Mi-8TB	10578	stored, ex 752/NVA, l/n 3-95	to Bad Sooden
9385	Mi-8TB	10580	stored, ex 763/NVA, l/n 3-95	to Friedrichsthal
9386	Mi-8TB	10581	stored, ex 764/NVA, l/n 4-93	to Schönefeld
9387	Mi-8TB	10582	stored, ex 768/NVA, l/n 3-95	to Friedrichsthal
9389	Mi-8TB	10591	stored, ex 937/NVA, l/n 4-94	gone
9390	Mi-8TB	10593	stored, ex 939/NVA, l/n 5-94	gone
9391	Mi-9	340005	stored, ex 409/NVA, l/n 3-95	gone
9392	Mi-9	340006	stored, ex 411/NVA, l/n 3-95	to Berlin
9393	Mi-9	340007	stored, ex 416/NVA, l/n 3-95	gone
9394	Mi-9	340008	stored, ex 426/NVA, l/n 3-95	to Uberlingen
9620	Mi-24D	B4001	stored, ex 403/NVA, l/n 3-95	to Hungary
9621	Mi-24D	B4002	stored, ex 406/NVA, l/n 3-95	to Bentlage
9622	Mi-24D	110158	stored, ex 408/NVA, l/n 5-92	to Holzdorf
9623	Mi-24D	B4004	stored, ex 412/NVA, l/n 3-95	to Poland
9624	Mi-24D	B4071	stored, ex 417/NVA, l/n 3-95	to Hungary
9625	Mi-24D	B4072	stored, ex 418/NVA, l/n 3-95	to Hungary
9626	Mi-24D	110159	stored, ex 421/NVA, l/n 5-94	to Weston-super-Mare
9627	Mi-24D	110162	stored, ex 434/NVA, l/n 3-95	to Hungary
9628	Mi-24D	110164	stored, ex 447/NVA, l/n 3-95	to Hungary
9629	Mi-24D	340272	stored, ex 485/NVA, l/n 3-95	to Poland
9630	Mi-24D	110166	stored, ex 494/NVA, l/n 10-90	to United States
9631	Mi-24D	110167	stored, ex 495/NVA, l/n 3-95	to Poland
9632	Mi-24D	110168	stored, ex 496/NVA, l/n 3-95	to Hungary
9633	Mi-24D	340273	stored, ex 528/NVA, l/n 7-95	to Brussels
9634	Mi-24D	340271	stored, ex 529/NVA, l/n 3-95	to Poland
9635	Mi-24D	340274	stored, ex 530/NVA, l/n 3-95	to Weelde
9636	Mi-24D	340275	stored, ex 532/NVA, l/n 3-95	to Hungary
9637	Mi-24D	340276	stored, ex 533/NVA, l/n 3-95	to Poland

GERMANY - 165

9638	Mi-24D	340277	stored, ex 544/NVA, l/n 7-95	to Poland
9639	Mi-24D	340278	stored, ex 547/NVA, l/n 10-90	to 9831
9640	Mi-24P	340330	stored, ex 357/NVA, l/n 10-90	to 9833
9641	Mi-24P	340331	stored, ex 358/NVA, l/n 3-95	to Hungary
9642	Mi-24P	340332	stored, ex 361/NVA, l/n 3-95	to Hungary
9643	Mi-24P	340333	stored, ex 387/NVA, l/n 3-95	to Berlin
9644	Mi-24P	340334	stored, ex 415/NVA, l/n 3-95	to Hungary
9645	Mi-24P	340335	stored, ex 422/NVA, l/n 3-95	to Hungary
9646	Mi-24P	340336	stored, ex 439/NVA, l/n 3-95	to Hungary
9647	Mi-24P	340337	stored, ex 442/NVA, l/n 10-90	to 9834
9648	Mi-24P	340338	stored, ex 444/NVA, l/n 3-95	to Hungary
9649	Mi-24P	340339	stored, ex 464/NVA, l/n 3-95	to Bückeburg
9650	Mi-24P	340340	stored, ex 480/NVA, l/n 3-95	to Hermeskeil
9651	Mi-24P	340341	stored, ex 512/NVA, l/n 10-90	to United States

BAUTZEN
Town: The Offiziershochschule für Militarflieger had two aircraft preserved on their parade grounds. Both (L-29 311 and MiG-21SPS 829) went to the Luftwaffe Museum collection and were transported to Uetersen.

Airfield: During the last years of the NVA, the airfield was closed due to major runway repairs. The based L-39s were relocated to Rothenburg. The aircraft preserved at Bautzen were part of the Historische Vorstartlinie (historic flightline) near the gate. All of these aircraft found their way to museums in the early 1990s. The museum at Uetersen took delivery of An-14A 995, Il-28B 208, L-29 338, Mi-4 569, MiG-17F 083, Lim-5P 615 (marked as '091'), MiG-19PM 335, MiG-21F-13 645 and MiG-21SPS 969. Finow received VEB-14P 482, while SM-1U 001 and Yak-11 225 both were noted at Kamenz in 1991. The dumped L-29s 345 (591528) and 359 (191531) were both scrapped in the early 1990s, the remains went to Rothenburg.

BAYREUTH
Sold from Rothenburg to a so far unknown location within the town of Bayreuth is a MiG-21US.

❏ 2405	–	MiG-21US	02685134	preserved, ex Rothenburg, ex 225/NVA	—

BEELITZ
Mobeldiscount Optimal along road 2 from Beelitz to Buchholz (both south west of Berlin) has acquired two former Soviet aircraft. The oft quoted construction number of 7805 for the Su-7 is not correct. 7805 can be found on a plate in the left wheelbay, while a similar plate in the right wheelbay shows the number 7806.

❏ 70 red	MiG-23ML	10389	preserved, ex Altes Lager, ex Soviet AF	5-96
❏ 01 red	Su-7BM	..	preserved, ex Grossenhain, ex Soviet AF	5-96

BENTLAGE
The gate of the Theodor Blank kaserne has been guarded for more than 20 years by an H-34, while the Lehrwerkstatt also has some ex air force aircraft. The ex Basepohl Mi-24D 9621 (B4001) was temporarily stored here before passing on, with the aid of a CH-53, to the museum at Duxford, UK.

❏ 5686		Do27B-5	394	dumped	6-97
❏ 7532		Alouette 2	1324	preserved, on pole near tower	4-98
❏ 7611		Alouette 2	1575	instructional, frame only	8-96
❏ 7661		Alouette 2	1718	instructional	4-98
❏ 7693		Alouette 2	1836	instructional	4-98
❏ 7718	37	Alouette 2	1872	dumped, near tower	6-97
❏ 8035		H-34G	58-1100	preserved, at gate	4-98
❏ 9057		FWP149D	074	instructional, D-EIOV ntu	4-98

GERMANY - 166

❑ 9163	FWP149D	185	instructional		4-98
❑ 9835	G91R/3	0054	instructional, ex 3001		4-98
❑ 9920	OV-10B	338-5	instructional, ex D-9549		4-98

BERLIN

Town: The Museum fur Verkehr und Technik has been renamed as the Deutsches Technikmuseum Berlin. Outside is the former Tempelhof C-47. A new building is being built next to the current location. In here, the aircraft which are currently stored at Gatow (which see) will be displayed.

❑ N951CA	C-47B	16954/34214	preserved, marked as 5951, ex 45-0951		11-97

The new Alliierten Museum of Berlin is located along the Clayalee in the Zehlendorf part of Berlin. The museum is scheduled to open in 1998.

❑ ..	L-19E	..	preserved, ex French Army		9-97
❑ WG466	Chipmunk T.10	C1/0516	preserved, ex RAF		9-97
❑ TG503	Hastings T.5	..	preserved, ex Gatow, ex RAF		9-97
❑ ..	UH-1H		preserved, ex US Army		9-97

During 1992 an art festival was held behind the Reichstag, near the hospital. Under art project Mutonia two former NVA MiG-21s were taken from the military range at Waldsieverdorf near Strausberg and painted in bright colours were MiG-21SPS 756 (4307) and 763 (4313). Former Soviet Mi-2 (34 yellow, c/n 511618090) was also part of the project. They were all scrapped when the festival ended late 1992.

Airfield - Gatow: On 7th July 1994 the RAF returned the airfield of Gatow back to the Germans. The barracks were renamed General Steinhoff kaserne. In early 1996 the Luftwaffe Museum from Uetersen openend its gates here and a large collection can now be seen. Some 56 aircraft were transported from Uetersen by air underneath a CH-53. Currently only one hangar is open to the public, but more are expected to be opened in the near future. The museum is closed on Mondays, but open between 09:00 and 15:00 on the other days. **Inside** are:

❑ 2037	F-104G	2044	preserved, ex Uetersen	5-98
❑ 2219	MiG-21SPS-K	94A6804	preserved, nose only, ex 989/NVA	5-98
❑ 2801	TF-104G	5931	preserved, cockpit only, crashed 26-4-89	3-98
❑ 4150	Alpha Jet	0150	preserved	5-98
❑ 9620	L-18C	18-3443	preserved, ex Schwenningen, ex AS-525	5-98
❑ BB-250	CL-13 Sabre 5	895	preserved, marked as 'BB-150', ex Uetersen	5-98
❑ 163	MiG-15UTI	922257	preserved, ex Uetersen	5-98
❑ 950	MiG-21PFM	761402	preserved, ex Fassberg	5-98
❑ 225	Yak-11	72232	preserved, ex Uetersen, ex Kamenz, ex Bautzen	5-98
❑ 191904	Me163B-1a	191904	preserved, ex Oldenburg	3-98
❑ D-EGUF	Bü181B-1	17	preserved, marked as 'NF-IR', ex Uetersen	3-98
❑ D-FABU	Harvard 4	CFF4-465	preserved, ex AA-615, ex 52-8544	5-98
❑ D-1979	Grunau Baby IIb	3	preserved, glider, marked as 'WL-VII-21'	5-98
❑ G-AZMH	MS500	637	preserved, marked as '7A-WN', ex Duxford	5-98
❑ ..	Avro 504K	..	preserved, replica, marked as 'E3349'	5-98
❑ ..	DFS230A	..	preserved, replica, glider, marked as 'KA-1-52'	5-98
❑ ..	Fokker E.III	..	preserved, replica, marked as '603/15'	5-98
❑ ..	Fokker Dr.I	..	preserved, replica, marked as '152/17'	5-98
❑ ..	Fokker D.VII	..	preserved, replica, marked as '7775/18'	3-98
❑ ..	Grunau SG-38	..	preserved, replica, glider, marked as '4'	5-98
❑ ..	HA1112M1L	..	preserved, marked as '10575/4'	3-98
❑ ..	Junker J.9	..	preserved, replica	5-98
❑ ..	Rumpler Taube	..	preserved, replica	5-98

Outside:

❑ 72	10-SA	Super Mystère B2	72	preserved, ex Uetersen, ex Montpellier, ex FAF	5-98
❑ XV278		Harrier GR.1	..	preserved, ex Uetersen, ex Gütersloh, ex RAF	3-98

GERMANY - 167

☐ XG152	L	Hunter F.6A	..	preserved, ex Uetersen, ex Gütersloh, ex RAF	5-98
☐ XN730	J	Lightning F.2A	95107	preserved, ex Uetersen, ex RAF	5-98
☐ WV865		Sea Hawk FGA.6	6110	preserved, ex Uetersen, ex Royal Navy	5-98
☐ 0101		CL-13B Sabre 6	1591	preserved, ex Manching	5-98
☐ 1606		HFB320	1048	preserved, ex Uetersen	5-98
☐ 1626		HFB320	1063	preserved, ex Uetersen	5-98
☐ 2002		F-104G	2002	preserved, ex DB-127, ex Erding, ex Karlsruhe	5-98
☐ 2002		MiG-23MF	0390213299	preserved, ex Laage, ex 577/NVA	5-98
☐ 2013		MiG-23ML	0390324624	preserved, ex Laage, ex 333/NVA	5-98
☐ 2051		MiG-23BN	0393214225	preserved, ex Drewitz, ex 710/NVA	5-98
☐ 2063		MiG-23UB	A1037902	preserved, ex Laage, ex 105/NVA	5-98
☐ 2209		MiG-21SPS-K	94A7009	preserved, ex Fassberg, ex Neuhardenburg	3-98
☐ 2377		MiG-21UM	02695156	preserved, ex Neubrandenburg, ex 256/NVA	5-98
☐ 2453		MiG-21bis	N75035841	preserved, ex Drewitz, ex 990/NVA	5-98
☐ 2454		F-104G	8202	preserved, ex Uetersen	5-98
☐ 2511		Su-22M-4	25018	preserved, ex Uetersen, ex 613/NVA	5-98
☐ 2544		Su-22M-4	31406	preserved, ex Laage, ex 798/NVA	5-98
☐ 2552		Su-22UM-3	17532367001	preserved, ex Laage, ex 112/NVA	5-98
☐ 2651		F-104G	7311	preserved, ex Uetersen	5-98
☐ 2790		TF-104G	5920	preserved, ex Fassberg	5-98
☐ 2848		L-39V	630705	preserved, ex Rothenburg, ex 170/NVA	5-98
☐ 3215		G91R/3	483	preserved, ex Uetersen, ex Schwenningen	5-98
☐ 5209		An-26SM	11402	preserved, ex 369/NVA, ex DDR-SBL	5-98
☐ 5407		Pembroke C.54	P66/102	preserved, marked as 'XA-109', ex Uetersen	5-98
☐ 9103		FWP149D	124	preserved, ex Uetersen, ex Fürstenfeldbruck	5-98
☐ 9301		Mi-8T	031233	preserved, ex Briest, ex 398/NVA	5-98
☐ 9314		Mi-8T	10543	stored, ex Briest, ex 927/NVA	5-98
☐ 9351		Mi-8S	105104	preserved, ex 914/NVA	5-98
☐ 9392		Mi-9	340006	preserved, ex Basepohl, ex 411/NVA	5-98
☐ 9456		T-33A	9120	preserved, marked as '9455', ex Manching	5-98
☐ 9469		T-33A	9257	preserved, marked as 'EB-399', ex Uetersen	5-98
☐ 9643		Mi-24P	340333	preserved, ex Basepohl, ex 387/NVA	5-98
☐ 9912		G91R/3	554	preserved, ex Uetersen, ex 3284	5-98
☐ 9935		Canberra B.2	R3/EA3/6652	preserved, ex D-9567, ex 0003	3-98
☐ 9940		G91T/3	621	preserved, ex Uetersen, ex Schwenningen	5-98
☐ 9941		G91T/3	0027	preserved, ex Uetersen, ex 3425	5-98
☐ BF-106		F-84F	..	preserved, ex Uetersen, ex 52-6804	5-98
☐ BR-239		G91R/4	0113	preserved, marked as '3541', ex Uetersen	5-98
☐ DD-339		F-84F	..	preserved, as 'DD-313/52-6674', ex 52-6774	5-98
☐ EB-244		RF-84F	..	preserved, ex 'EB-344', Uetersen, ex 52-7346	5-98
☐ 55-4881		F-86K	232-121	preserved, marked as 'JD-249', ex Uetersen	5-98
☐ 812		An-2	18120	stored, marked as '8120', ex Uetersen	5-98
☐ 822		An-2	117012	preserved, unmarked, ex Kamenz	5-98
☐ 995		An-14A	601005	preserved, ex Uetersen, ex Bautzen	5-98
☐ 208		Il-28B	55006448	preserved, ex Bautzen	5-98
☐ 311		L-29	692053	preserved, ex Uetersen, ex Bautzen	5-98
☐ 338		L-29	591525	preserved, ex Uetersen, ex Bautzen	5-98
☐ 569		Mi-4A	13146	preserved, ex Uetersen, ex Bautzen	5-98
☐ 521		Mi-24D	110171	preserved, marked as '5211', ex Bad Düben	5-98
☐ 07		MiG-17F	..	stored, ex Stade, ex Hoyerswerda	5-98
☐ 615		Lim-5P	1D-0208	preserved, marked as '091', ex Uetersen	5-98
☐ 905		Lim-5	1C-0829	preserved, ex Uetersen	5-98
☐ 335		MiG-19PM	650929	preserved, ex Uetersen, ex Bautzen	5-98
☐ 596		MiG-21M	960708	preserved, ex Kamenz	3-98
☐ 645		MiG-21F-13	741924	preserved, ex Uetersen, ex Bautzen	5-98

GERMANY - 168

❏ 779	MiG-21SPS	94A4502	stored, ex Bad Düben		3-98
❏ D-9539	CL-13B Sabre 6	1603	preserved, ex Uetersen		5-98

A large number of other aircraft are currently held in storage.

❏ 3905	MiG-15bis	623905	stored, ex Uetersen, ex Czech AF	97
❏ 52	CM170	52	stored, ex Montelimar, ex French AF	97
❏ 57	NC701 Martinet	57	stored, French built Siebel 204, ex French	97
❏ 81	N1101	81	stored, ex French	97
❏ B.2I-117 'GI-AD'	CASA 2111D	..	stored, ex Spanish AF	97
❏ XM556	Skeeter AOP.12	S5/5110	stored, ex British Army	97
❏ 2806	L-39ZO	731006	stored, ex Uetersen, ex Rothenburg, ex 144/NVA	97
❏ 2906	F-104F	5055	stored, ex Uetersen, ex 59-5002	97
❏ 3272	G91R/3	542	stored, ex Uetersen	97
❏ 3562	RF-4E	4144	expected, still at Kaufbeuren, ex 69-7509	—
❏ 3413	G91T/3	0015	stored, forward fuselage only	97
❏ 3452	G91T/3	612	stored, forward fuselage only	97
❏ 4026	Alpha Jet	0026	stored, ex Fürstenfeldbruck, wfu 18-1-94	3-98
❏ 5738	Do27A-4	467	stored, ex Uetersen, ex YA-904, D-EGAQ	97
❏ 5920	Do28D-2	4195	stored, ex Uetersen	97
❏ 7583	Alouette 2	1496	stored, ex Uetersen	97
❏ 7804	Sycamore Mk.52	13442	stored, ex Uetersen	97
❏ 8034	H-34G	58-1099	stored, ex Uetersen	97
❏ 8308	H-21C	WG08	stored, ex Uetersen	3-98
❏ 9444	T-33A	8960	stored, ex Fürstenfeldbruck, ex EB-399	97
❏ 9463	Mi-2	563148103	stored, ex Briest, ex 393/NVA	97
❏ 9820	Bo105M	S-90	expected, still at Diepholz	—
❏ 9925	OV-10B	338-10	stored, ex Laatzen, ex D-9554	97
❏ 9933	OV-10B	338-18	stored, ex Uetersen, ex D-9562	97
❏ 529	Mi-1MU	05017	stored	97
❏ 346	Lim-5	1C-0723	stored, ex Drewitz, ex Rothenburg	97
❏ 472	MiG-21SPS-K	94A7005	stored, cockpit only	97
❏ 686	MiG-21MF	966301	stored, ex Kamenz	97
❏ 574	MiG-23MF	0390213294	stored, ex Finow, ex Kamenz, crashed 7-7-80	97
❏ 25	Yak-18A	616	stored, ex Uetersen	97
❏ AA-014	CM170	229	stored, ex Uetersen	97
❏ UA-110	Gannet AS.4	F9391	stored, marked as 'UA-106', ex Uetersen	97
❏ YD-101	Do29	001	stored, marked as 'YA-101', ex Uetersen	97
❏ D-EBUC	Sportavia C1	V1	stored	97
❏ D-EFLG	FWP149D	097	preserved, ex 9078 and marked as such	97
❏ D-EJUX	Do27	464	stored, ex Koblenz, ex 5735	97
❏ D-3547	Doppelraab IV	..	stored, glider	97
❏ D-9534	Dornier DS10	2	stored	97

Additional non museum aircraft can be found at the airfield. The stored aircraft are for Deutsches Technik Museum Berlin (see page 166). Starfighter 2206 is mounted on a pole in the former RAF barracks at the airfield.

❏ 323	Lim-2	..	stored, ex Polish AF	97
❏ T.2B-108	CASA 352L	4145	stored, marked as 'D-2201', ex Spanish AF	97
❏ WF382	Varsity T.1	541	stored, ex dump, ex RAF	97
❏ ZD215	C-47B	15686/27127	stored, ex RAF, ex A65-69/RAAF, ex 43-49866	97
❏ 2206	F-104G	7076	preserved, ex Erding, ex Jever, ex Leck	11-97
❏ 2649	F-104G	7309	stored, ex Manching	97
❏ DDR-SAM	VEB-14P	14803045	stored, ex Eilendorf, ex 422/NVA	97

Airfield – Schönefeld: Schönefeld was the main base of East Germany's Interflug and just before the reunion of East and West Germany became the storage base of the Mi-2s of the Grenztruppen (Borderguards) of HS-16 from Nordhausen. Some of these Mi-2s were later used by the Polizie; 302 (563402034) to D-HZPF, 304 (563404044) to D-HZPG, 307 (563821114) to D-HZPH, 311 (568837104) to D-HZPI, 312 (568838104) to D-HZPJ, 314 (569341085) to DDR-VGF (later to D-HZPK), 322 (569342085) to DDR-VGH (later to D-HZPL), 401 (563150103) to D-HZPM (to Borkum), 420 (562817043) to D-HZPN, 500 (543048083) to D-HZPO, 504 (562945063) to D-HZPQ (to Hermeskeil) and 557 (543625074) to D-HZPR. 501 went to the nearby museum of Diepensee. 555 is preserved at Hermeskeil and 556 went to Bückeburg. Stored D-HZPC went to Aschersleben. Instructional Mi-2 DDR-VP (511029039, incomplete markings and ex Soviet) was last noted in December 1990 and may have been scrapped. Mi-8S 9342 (10532, ex Cottbus) was noted here as stored in late 1994 and became ES-PMA in 1995. Stored Interflug Il-18D DDR-STD (180002302) went to Harbke, while instructional Il-62 DDR-SEC went to Grossmachnow.

☐ 9363	Mi-8TB	10555	stored, ex Cottbus, ex 126/NVA	7-95
☐ 9386	Mi-8TB	10581	preserved, ex Basepohl, ex 764/NVA	9-97
☐ D-HOZC	Mi-8T	10513	stored, ex 9419, ex 627/NVA	5-98
☐ D-HOZH	Mi-8T	10534	stored, ex 9424, ex 985/NVA	2-98

Airfield – Tegel: The French military left Base Aérienne 165 (on the north side of Tegel airport) in 1995, leaving only a Noratlas behind. The other two French aircraft have not been seen for some time. Preserved Beaver 58-2020 went to Tempelhof for restoration, while CM170 545 moved to Luneville.

☐ 25		N2501F	25	stored	6-92
☐ 126	63-VM	N2501F	126	stored, ex preserved, ex French AF	11-97
☐ 013		MH1521M	013	preserved, ex French AF	6-92
☐ D-ENYH		Do27B	191	preserved, on roof of the terminal, ex 5558	9-96
☐ N130KR		B707-458	18071	preserved, marked as 'D-ABOC', ex N32824	5-98

Airfield - Tempelhof: The memorial to the Berlin Airlift for many years consisted of a C-47B Skytrain and a C-54G Skymaster. By 1995 the situation changed, the C-54 was parked in a hangar and the C-47 relocated to the Deutsches Technikmuseum Berlin.

☐ 45-0557	C-54G	36010	stored, ex preserved, ex USAF	11-97
☐ 58-2020	U-6A	1350	under restoration, ex Tegel, ex US Army	11-97
☐ F-PCDA	Klemm L-25	138	preserved, in terminal, ex EZ-AAB	6-97
☐ N106TA	N262A-21	34	dumped, blue c/s	6-97

BERNSDORF
It has been reported that a Tu-134 arrived here by road in the early 1990s. Unfortunately it is not known in which 'Bernsdorf' the Tu-134 can be found. There are several Bernsdorfs in the former East Germany.

| ☐ DDR-SCZ | Tu-134 | 9350913 | preserved, ex Dresden, ex DM-SCZ, ex 177/NVA | — |

BIBERACH
Biberach is located some 40km south of Ulm. An maschinenbau firm here has Tu-134 DDR-SCL in use as an eye-catcher.

| ☐ DDR-SCL | Tu-134A | 1351305 | preserved, ex Lahr, ex DM-SCL, ex 182/NVA | 9-92 |

BIRKENFELD
The Heinrich Herz kaserne still has its F-104 Starfighter at the gate. Birkenfeld is located some 30km north west of Ramstein.

| ☐ 2427 | F-104G | 8169 | preserved, at gate, marked as '2374' | 9-97 |

BITBURG
The USAFE airfield of Bitburg was closed in April 1994. Nowadays it is normally in use as an industrial park, but in 1997 some of Spangdahlem's aircraft were based here. All aircraft, with the exception of the unknown

Sabre, have now gone from here. Mystère 4As 54, 66, 187, 194/8-MM and 202 were the first ones to go in the early 1990s. F-4C 63-7421 moved to Hermeskeil, while the fate of 63-7576 (625, l/n May 1993) is unknown. Instructional F-15A 73-0095 moved to Spangdahlem, 74-0098 went to Ramstein, while 74-0109 is now at the museum at Speyer. F-100D 54-2239 (ex Lahr) went to Toulouse. Surprisingly, a visit in April 1996 found two Phantoms remaining, RF-4Cs 68-0577 (3517) and 69-0360 (3777), stored in a sorry state in the western shelter area. These were scrapped during the summer of 1996 by Kiemele from Seifertshofen. The Sabre has been relocated from the base to the American housing area near the Mötsch part of Bitburg town.

| ❏ .. | F-86F | .. | preserved, marked as '51-13036/FU-036' | 5-98 |

BLOMBERG
Since December 1988 a Dutch air defence unit at the Willem Versteegh barracks had Starfighter D-8282 on their site. It had left by the early 1990s for Budel. Its place was taken during the night of 1st September 1993 by former NVA MiG-21SPS 2240 (ex Rothenburg). When the unit was relocated to De Peel they took their MiG with them.

BOCHUM
Along highway 43, at the Bochum Wattenscheid exit, a private company put a Sabre on display. It is a former Air Classik aircraft and is painted in pink colours. The aircraft is mounted on a pole with on each wing three life-size female mannequins each holding flags *(reductio ad absurdum!)*.

| ❏ BB-131 | CL-13 Sabre 5 | 931 | preserved, unmarked, ex 'JA-102', ex Marl | 9-97 |

The Ikarus Flugmuseum at Marl was closed in the late 1980s. Some of its inmates moved to a location in southern Bochum. All were short-lived here and moved on. CASA 352L T.2B-257 (marked as 'D-ADAM') went to Sinsheim and Javelin FAW.9 XH768/E and C-47A N8041B (marked as 'N569R') moved to Seifertshofen. The fate of Thunderstreak DD-248 (ex 52-6783, marked as '5') is unknown.

BONN - HANGELAR
The gate of the Bundesgrenzschutz part of this airfield is guarded by an Alouette 2.

| ❏ D-HBJI | Alouette 2 | 2113 | preserved, at gate, ex 7759 | 1-93 |

The former collection of Butzweilerhof has been spilt. A part is now displayed at Merseburg, while other aircraft are now at this airfield. The museum's owners are hoping to set up a new museum at the town of St Augustin and the aircraft here are all currently stored. More aircraft are expected to arrive from Butzweilerhof and even some from Merseburg may return.

❏ 636	Mi-8T	10519	stored, ex Butzwielerhof, ex Bad Düben	5-98
❏ 764	MiG-21SPS	94A4314	stored, nose only, ex Butzwielerhof, ex Dresden	5-98
❏ 829	MiG-21SPS	94A4705	stored, ex Butzw, ex Uetersen, ex Bautzen	5-98
❏ 564	MiG-23MF	0390213089	stored, ex Butzweilerhof, ex Kamenz	5-98
❏ EB-319	RF-84F	..	stored, as 'ED-119', ex Butzw., ex Osnabruck	5-98
❏ D-BAKE	F27-100	10257	stored, ex Butzwielerhof, ex OE-HLA	5-98

BORKHEIDE
Compare the size of this former civil grass airfield and the size of the Il-18, and it is hard to believe that the airliner has landed here. It is preserved at the Luftfahrtmuseum Hans Grade. Grade was one of the early German aviation pioneers. Borkheide is located south west of Potsdam.

❏ D-HOAJ	Ka-26	7404804	preserved, ex DDR-SPJ, ex DM-SPJ	4-96
❏ DM-SQI	Zlin 37	09-05	preserved, never entered DDR- or D-register	4-96
❏ DDR-SQL	Zlin 37	09-08	stored	4-96
❏ DDR-SQM	Zlin 37	10-16	stored	4-96
❏ DDR-STE	Il-18D	182005101	preserved, arr 24-11-89, ex DM-STE, ex NVA	4-96

GERMANY - 171

BORKUM
Preserved at a playground in the town is a former Polizei Mi-2.
❏ D-HZPM   Mi-2   563150103   preserved, ex Schönefeld, ex 401/NVA   6-95

BRANDIS
As with all other ex Soviet airfields, Brandis also is nearly empty. The Russians left only the gate-guarding Il-2 behind. Stored with the aero club should also be two Ka-26s.
❏ 02 yel   Il-2   ..   preserved, ex Soviet AF   5-98

BRAUNSCHWEIG
D-BABM should still be stored here as a spares source for the DFVLR's flying test-bed D-ADAM. It was not seen here during 1995, but may well be kept in one of the hangars. Stored HFB320 D-CARA (1021) has been scrapped. A Do27 is preserved at the gate of the LBA, the German Federal Aviation Authority.
❏ D-BABM   VFW614   G13   stored, ex F-GATH   88
❏ D-EITE   Do27   286   preserved, marked as 'LBA', ex 5618   8-95

BREMEN
Town: The Hochschule fur Technik (Technical High School) still had an F-84F as instructional airframe in 1985. It is not known if the aircraft, DF-316 (ex 53-7058), is still there.

Airfield - Lemwerder: Used as a training aid by the airport firemen till at least the late 1980s was ex Air Services N2501D D-ANAS (014). By the early 1990s the aircraft had expired. In use as an instructional airframe at the former Vereinigte Flugtechnische Werke (VFW) plant at the airfield was Gannet UA-112. This aircraft was donated to the Koblenz museum in 1985, but never made it there. In the mid 1990s an unmarked Gannet arrived at Speyer. As there are only a few Gannets in Germany, the aircraft at Speyer is thought to come from here. Recently a Junkers W33 was obtained from the USA (on loan) and will be placed on display in the near future.
❏ D-1167   Junkers W33   2504   stored   97

BREMERHAVEN
Inside one of the town discos is a MiG-15 painted in fantasy colours. It hangs from the ceiling and is an ex Czech aircraft.
❏ 1170   MiG-15SB   141170   preserved, ex Druztová, ex Czech AF   97

BREMGARTEN
This RF-4E airfield closed in January 1993. Known departures included G91R/3 3058 to Seifertshofen, 3178 to Butzweilerhof, 3212 to Neubiberg and 3308 (ex Fürstenfeldbruck) to München. Also to München went Thunderflash EA-301 (marked as 'EA-101'). Instructional P149D 9186 left in the 1980s for Norvenich. Thunderflash EA-341 (marked as 'EA-241') is now at Hermeskeil and Starfighter 2004 (marked as '2404') went to Siegburg.
❏ 3237   G91R/3   506   dumped, ex Husum   9-92

BRIEST
Brandenburg Briest used to house a large number of Mi-2s and Mi-8s of the former NVA units HTG-34 and HAG-35. Only a handful of helicopters were kept operational after October 1990 and on 30th June 1994 the last Mi-8s were retired. By early 1996 all the helicopters had left the airfield as follows.
  ..   Mi-8T   ..   dumped, l/n 6-90, ex Soviet   scrapped
  9301   Mi-8T   031233   stored, ex 398/NVA, l/n 5-94   to Berlin

9302	Mi-8T	..1333	stored, ex 399/NVA, l/n 9-92	gone
9305	Mi-8T	10515	stored, ex 631/NVA, l/n 5-94	to Rothenburg
9306	Mi-8T	10518	stored, ex 634/NVA, l/n 5-94	to Friedrichsthal
9308	Mi-8T	10538	stored, ex 922/NVA, l/n 5-94	to Bad Ischl
9311	Mi-8T	10542	stored, ex 925/NVA, l/n 5-94	to Diepensee
9312	Mi-8T	10541	stored, ex 926/NVA, l/n 5-94	to Friedrichsthal
9314	Mi-8T	10543	stored, ex 927/NVA, l/n 5-94	to Berlin
9315	Mi-8T	10544	stored, ex 928/NVA, l/n 5-94	to Friedrichsthal
9316	Mi-8T	10545	stored, ex 930/NVA, l/n 5-94	to Rothenburg
9318	Mi-8T	10547	stored, ex 932/NVA, l/n 5-94	to Friedrichsthal
9319	Mi-8PS	10552	stored, ex 970/NVA, l/n 4-93	to Bad Oeynhausen
9320	Mi-8PS	10522	stored, ex 973/NVA, l/n 5-94	to Pergau
9330	Mi-8T	10510	stored, ex 389/NVA, l/n 5-94	to Friedrichsthal
9332	Mi-8T	10516	stored, ex 632/NVA, l/n 5-94	boom to Diepensee
9334	Mi-8T	10530	stored, ex 902/NVA, l/n 5-94	to Stade
9335	Mi-8T	10531	stored, ex 903/NVA, l/n 5-94	to Friedrichsthal
9338	Mi-8PS	10550	stored, ex 962/NVA	to Bad Ischl
9343	Mi-8PS	10533	stored, ex 977/NVA, l/n 4-93	to Rothenburg
9344	Mi-8PS	10585	stored, ex 990/NVA, l/n 5-94	to Bad Ischl
9406	Mi-8TB	10566	stored, ex 810/NVA, ex Parow, l/n 9-94	gone
9415	Mi-8PS	0826	stored, ex 397/NVA, l/n 5-94	to Rothenburg
9416	Mi-8PS	10520	stored, ex 971/NVA, l/n 5-94	gone
9417	Mi-8T	105101	stored, ex 391/NVA, l/n 10-92	gone
9418	Mi-8T	0323	stored, ex 394/NVA	to Sinsheim
9420	Mi-8T	10525	stored, ex 909/NVA, l/n 8-91	to Hermeskeil
9421	Mi-8T	10526	stored, ex 910/NVA, l/n 5-94	to Seifertshofen
9422	Mi-8T	10528	stored, ex 912/NVA, D-HOZF ntu, l/n 10-93	gone
9424	Mi-8T	10534	stored, ex 985/NVA, l/n 10-92	to D-HOZH
9450	Mi-2S	563401044	stored, ex 301/NVA, l/n 5-94	to Seifertshofen
9451	Mi-2S	563403034	stored, ex 303/NVA, l/n 5-94	to Archen
9452	Mi-2S	563405044	stored, ex 305/NVA, l/n 10-93	to Well
9453	Mi-2S	563820114	stored, ex 306/NVA, l/n 4-93	to D-HVAC
9454	Mi-2S	563822114	stored, ex 308/NVA, l/n 5-94	to Drewitz
9455	Mi-2S	563823114	stored, ex 309/NVA, l/n 10-93	to Rothenburg
9456	Mi-2S	563824114	stored, ex 310/NVA, l/n 5-94	tailboom to Schaijk
9457	Mi-2S	564411105	stored, ex 348/NVA, l/n 5-94	to Peenemünde
9458	Mi-2S	564413105	stored, ex 352/NVA, l/n 4-93	to Rothenburg
9459	Mi-2S	562632112	stored, ex 382/NVA, l/n 10-93	to Rothenburg
9460	Mi-2S	562633112	stored, ex 383/NVA, l/n 10-93	to Rothenburg
9461	Mi-2S	562635112	stored, ex 385/NVA, l/n 10-93	to Rothenburg
9462	Mi-2S	563147103	stored, ex 392/NVA, l/n 5-94	to Drewitz
9463	Mi-2S	563148103	stored, ex 393/NVA, l/n 5-94	to Berlin
9464	Mi-2F	562818043	stored, ex 423/NVA, l/n 7-91	to D-HVAA
9465	Mi-2F	562944063	stored, ex 503/NVA, l/n 7-91	to D-HVAB
9466	Mi-2F	562946063	stored, ex 506/NVA, l/n 8-91	to D-HVAE
9470	Mi-2F	514416125	stored, ex 335/NVA, l/n 4-93	to Rothenburg
9471	Mi-2S	564410105	stored, ex 347/NVA, l/n 10-93	to Rothenburg
9472	Mi-2S	562248032	stored, ex 379/NVA, l/n 4-93	to Rothenburg
9473	Mi-2See	552701122	stored, ex 388/NVA, l/n 10-93	to Kaltwasser
9480	Mi-2F	514415125	stored, ex 328/NVA, l/n 10-92	to Rothenburg
9481	Mi-2S	562247032	stored, ex 377/NVA, l/n 4-92	to D-HVAD
9482	Mi-2S	562250032	stored, ex 381/NVA, l/n 10-93	to Schaijk
9483	Mi-2See	552649122	stored, ex 386/NVA, l/n 10-93	to Rothenburg
396	Mi-8T	0726	instructional, l/n 4-91	to Diepensee
..	MiG-17	2095 ?	dump, l/n 10-92	scrapped

GERMANY - 173

BRÜGGEN
Many changes since the first edition of this book. Most of the older instructional airframes with 431 MU left, as have all the Lightning decoys. Harrier GR.3 XW922 is now at Laarbruch, while XV782, XV789 and XV793 have all been scrapped. Hunters XE608/XX (41H-679966) and XL566/86 were scrapped. Lightning T.4s XM970/T (95055), XM973/V (95073), F.2As XN783/G (95136), XN789/J (95142), XN792/M (95145) and F.6 XS901/BK (95247) have also been scrapped. Wessex HU.5 XT467/F went to Laarbruch, while the older Whirlwind HAR.10 XP403 (WA391) disappeared. Harrier T.4 XW927 arrived here on 12th April 1992 and went to Hermeskiel in early 1997. The Jaguar is a composite, based upon the nose and forward fuselage section of XW563. It is painted in the colours of a former based Jaguar, XX822.

☐ XZ998		V	Harrier GR.3	..	dumped, ex Swanton Morley	4-97
☐ XW563			Jaguar S	07	preserved, marked as 'XX822/AA', see note	10-97
☐ XV425		CD	Phantom FGR.2	3093	preserved	10-97
☐ XV475		H	Phantom FGR.2	3314	preserved, arr 11-7-91	10-97
☐ XV481		H	Phantom FGR.2	3355	instructional, arr 16-2-92	11-96
☐ XV485		M	Phantom FGR.2	3377	preserved, arr 11-7-91	10-97
☐ XV569		BQ	Phantom FG.1	2970	instructional, ex Wildenrath	4-97
☐ XZ630			Tornado	P12	instructional, weapons-loader	10-94
☐ ZE357		N	F-4J(UK)	2529	instructional	10-97

BÜCHEL
Airfield: Büchel is home of the Tornados of JBG33. The base is along Bundesstrasse 259, some 60km south west of Koblenz. Some of the instructional aircraft are in use as decoys, while others are at the Lehrwerkstatt (technical training school). Instructional F-104G 2388 (8095) was scrapped in October 1992. F-84F DD-354 (as 'DC-101', ex 52-6707) was only temporarily here and returned to Norvenich.

☐ 2058		F-104G	2067	instructional, marked as '3033', will be scrapped	9-96
☐ 2167		F-104G	7036	preserved, at main gate, marked as '4033'	8-97
☐ 2411		F-104G	8151	instructional, ex Fassberg	4-96
☐ 2438		F-104G	8181	instructional, ex Fassberg	4-96
☐ 2586		F-104G	9061	instructional	8-97
☐ 2626		F-104G	9178	instructional	9-96
☐ 2665		F-104G	7411	instructional, arr 6-4-92, will be scrapped	9-96
☐ ..		F-104G	..	instructional, marked as '2444', nose only	9-96
☐ 3162		G91R/3	430	instructional, will be scrapped	9-96
☐ 3455		G91T/3	0615	instructional	9-96
☐ 9165		FWP149D	188	instructional	9-96
☐ 9433		T-33A	8197	dumped, ex instructional, ex 52-9966	3-90
☐ 9434		T-33A	8198	preserved, marked as 'DC-382', ex 52-9967	9-96
☐ 9903		G91R/3	328	instructional, ex 3067, arr 11-12-91	5-92
☐ 9910		G91R/3	482	instructional, ex 3214, arr 7-10-91	5-92
☐ 9943		G91R/3	453	instructional, ex 3185	9-96
☐ JD-395	3	T-33A	1657	preserved, ex instructional, ex 58-0688	9-96
☐ ..		MiG-21	..	instructional, nose only	9-96

Barracks: JBG33's barracks is in Cochem-Brauheck, just off the road between the base and Cochem.

☐ DC-319		F-84F	..	preserved, ex 53-7045, wfu 2-5-65	8-97
☐ 2503		F-104G	8261	preserved, marked as '2533'	5-92

BÜCKEBURG
Town: In the town of Bückeburg, the Hubschrauber Museum (Helicopter Museum) is located in Sableplatz, near the town centre. The museum is open daily between 09:00 and 17:00. VFW H3A D-9544 (E2) has gone.

☐ 58-5348		OH-13H	2361	preserved, ex US Army	5-98
☐ 55-4109		OH-23C	814	preserved, ex US Army	5-98

GERMANY - 174

☐ 62-4547	HH-43F	174	preserved, ex USAF		5-98
☐ 67-16955	TH-55A	1062	preserved, ex US Army		5-98
☐ XN348	Skeeter AOP.12	S2/7154	preserved, ex British Army		5-98
☐ 7505	Alouette 2	1193	preserved		5-98
☐ 7717	Alouette 2	1871	preserved, unmarked, crashed 13-3-69		5-98
☐ 7820	Sycamore Mk.52	13478	preserved, ex G-18-151		5-98
☐ 8109	H-34G	58-1679	preserved, ex 150807		5-98
☐ 8306	H-21C	WG6	preserved, demo rig		4-97
☐ 8307	H-21C	WG7	preserved		5-98
☐ 9823	Do34 Kiebitz	P01	preserved		5-98
☐ 556	Mi-2	543624074	preserved, ex Schönefeld, DDR-VGO ntu		5-98
☐ 100406	Fa330A-1	100406	preserved, ex Cranfield		5-98
☐ CCCP-05712	SM-1	..	preserved		5-98
☐ D-9505	Bo103	V1	preserved, ex D-HECA		5-98
☐ D-9515	Bo46	V2	preserved, marked as 'D-9514 (V1)'		5-98
☐ D-9543	VFW H3E	E1	preserved		5-98
☐ D-HAJU	Havertz HZ-5	5	preserved		4-97
☐ D-HAJY	Bo105B	V3	preserved, arr 4-5-70		4-97
☐ D-HBEC	Bo108A	VT002	preserved, unmarked		4-97
☐ D-HBKA	BK117A-1	P2	preserved, unmarked		4-97
☐ D-HIBY	VFW H2	V1	preserved		9-96
☐ D-HIDI	WGM-21	V1	preserved		4-97
☐ D-HOAL	Ka-26	7505201	preserved, ex DDR-SPL, ex DM-SPL		7-97
☐ D-HOBB	Air&Space 18A	18-26	preserved, ex N6120S		4-97
☐ D-HOCY	TRS-1 Hummel	..	preserved		9-96
☐ D-HOHS	HSX-2	..	preserved, marked as 'D-HOKY'		4-97
☐ F-BNAY	SO1221S Djinn	FR8/1109	preserved, marked as '7'		5-98
☐ ..	Bo102B	4502	preserved, unmarked		4-97
☐ ..	Bensen B-8M	..	preserved		4-97
☐ ..	Flying Jeep	..	preserved, Bölkow built		9-96
☐ ..	Fw61	..	preserved, replica, as 'D-EBVU', ex 'D-EKRA'		5-98
☐ ..	Georges G1 Papillon	..	preserved		4-97
☐ ..	Georges G.2	V1	preserved		4-97
☐ ..	Merckle SM.67	V3	preserved, as 'D-9506', parts of D-9506 (V2)		5-98
☐ ..	Nagler Rolz NR54	V2	preserved		4-97
☐ ..	Wagner Rotocar 3	..	preserved, marked as 'TT-046'		4-97

Airfield: The German Army airfield of Bückeburg is the home of the main HEER training unit, the HFWS. Also a number of smaller units are based here, including a technical school. Dumped Bo105P 8682 (6082) in a maintenance hangar and the civil Piaggio D-EBSN (169) have not been seen for some time and are presumably scrapped. Former gate guard Alouette 2 'PQ-131' has been replaced by 7609. This 'PQ-131' is now at Dessau (together with the stored 7528). Instructional Alouette 2 7623 (1610) departed by 1984. See also the Stop Press listing on page 458.

☐ 7555	Alouette 2	1406	instructional, pod only, inside cinema	11-95
☐ 7566	Alouette 2	1448	stored	1-95
☐ 7609	Alouette 2	1563	preserved, at gate	5-98
☐ 7692	Alouette 2	1834	instructional	9-97
☐ 9162	FWP149D	184	instructional	5-98
☐ 9380	Mi-8PS	10597	preserved, ex stored, ex Basepohl, ex 732/NVA	5-98
☐ 9649	Mi-24P	340339	preserved, ex Basepohl, ex 464/NVA	5-98
☐ 9917	OV-10B	338-2	instructional, ex D-9546	5-98
☐ 9929	OV-10B	338-14	instructional, ex D-9558	5-98
☐ 9931	OV-10B	338-16	instructional, ex D-9560	9-96
☐ 9942	G91R/3	0059	instructional, ex 3006	4-98
☐ 9945	G91R/3	499	instructional, ex Fürstenfeldbruck, ex 3230	5-98

BURGAU
Preserved outside on a pole at the Auto Motorrad Museum in the Bleichstrasse in Burgau (some 45km west of Augsburg) is a Fiat. It was not noted in April 1991, but may have moved inside.

| ☐ 3193 | G91R/3 | 461 | preserved, ex Leipheim, with tail from 3047 (304) | 85 |

BUTZWEILERHOF
In the northern part of Köln, named Ossendorf, the Luftfahrtmuseum Butzweilerhof has been established. The former Belgian Army airfield and barracks are now completely owned by the German Army. The Belgiams had Alouette 2 A-08 and Islander B-06 stored here. These aircraft moved to Brussels. The site of the museum is within the German barracks and was leased. The lease ended early January 1997 and the museum was closed. A large number of aircraft have moved on to the new museum at Merseburg (which see), while other have moved for storage to Bonn-Hangelar (which see, these are all for the new museum at St Augustin).

☐ 7589	Alouette 2	1514	preserved	4-96
☐ D-HOAR	Ka-26	7404619	preserved, ex DDR-SPR, ex DM-SPR	4-96
☐ D-6011	Fauvel AV36CR	..	preserved, glider	4-96
☐ D-9568	X-113AM	V1	preserved, Ground Effect Vehicle	4-96
☐ ..	X-114	..	preserved, Ground Effect Vehicle	4-96

CÄMMERSWALDE
Cämmerswalde is a small village, south west of Dresden, near the border with the Czech Republic. The village should still have the German-built Il-14.

| ☐ DM-SAB | VEB-14P | 14803008 | preserved, wfu 10-70 | 5-98 |

CELLE
During 1996 only the Alouette 2 was noted. All hangars were checked and there was no sign of the previous instructional airframes; the Do27A-1 5616 (284), Bo105C 9827 (S-315) and CL-13B Sabre D-9523 (1784).

| ☐ 7668 | 26 | Alouette 2 | 1734 | instructional | 8-96 |

COLEMAN BARRACKS
We will probably never know which T-33 was used here on the fire dump. It expired in the late 1980s. It was succeeded by the F-105 which used to be preserved by the CH-47 unit. When this unit returned to the USA, the F-105 was placed on the dump. In 1997 it was removed from the dump and currently awaits restoration. During 1995, 1996 and 1997 a number of UH-1Hs were scrapped here, complete details are unknown.

| ☐ 63-8362 | F-105F | .. | stored, ex preserved, ex Ramstein | 5-98 |
| ☐ 73-22092 | UH-1H | 13575 | preserved | 5-98 |

COTTBUS
In the first days after the famous October 1990 reunion local aviation enthusiasts set up an aviation museum on the airfield here. It expanded during the years and opened its gates by 1994 to the general public. It can be found on the opposite side of the German Army barracks at the airfield. In the 'early' days the museum borrowed helicopters from the airfield. The Polish Lim-5 is not part of the museum, but is stored here on behalf of a private owner.

☐ 1103	C	Lim-5	1C-1103	stored, ex Polish AF	11-97
☐ 32 yellow		Mi-2	511019039	preserved, ex Soviet AF	11-97
☐ 2062		MiG-23UB	A1037901	preserved, ex 104/NVA, ex Laage	9-97
☐ 2399		MiG-21US	01685134	preserved, ex 215/NVA, ex Rothenburg	11-97
☐ 2421		MiG-21bis	N75051407	preserved, ex 848/NVA, ex Drewitz	11-97
☐ 2504		Su-22M-4	25511	preserved, ex Laage, ex 365/NVA	11-97
☐ 7622		Alouette 2	1609	preserved	11-97

GERMANY - 176

❏ 9368	Mi-8TB	10560	preserved, ex 132/NVA and marked as such		11-97
❏ 9397	Mi-9	340004	stored, ex 407/NVA and marked as such		11-97
❏ 9398	Mi-9	340001	preserved, ex 482/NVA and marked as such		11-97
❏ 380	Mi-2	562249032	preserved		11-97
❏ 996	An-14A	600904	preserved, ex Dresden		11-97
❏ 538	Mi-4A	07142	stored, wfu 1-12-79		11-97
❏ 154	MiG-15UTI	1615393	stored		11-97
❏ 437	Lim-5P	1C-0212	stored, marked as '2001', ex '850', ex Dresden		11-97
❏ 537	Lim-5	1C-0917	stored, ex Drewitz		11-97
❏ 653	MiG-21MF	965311	preserved, ex Kamenz		11-97
❏ 981	MiG-21SPS-K	94A6704	preserved		11-97
❏ 986	MiG-21SPS-K	94A6715	stored, ex gate guard		11-97
❏ 98	Yak-11	68210	stored, ex Cottbus town		9-97
❏ D-EOON	Zlin 37A	06-18	preserved, ex DDR-OON		9-97
❏ D-ESLQ	Zlin 37A	14-20	preserved, ex DDR-SLQ		9-97
❏ DDR-TCA	PZL 106A	07810140	preserved, ex Baschutz, D-FOCA ntu		9-97

By mid 1996 only one helicopter was still to be found on the airfield. It is preserved near the main gate.

❏ 9370	Mi-8TB	10577	preserved, ex 751/NVA		5-97

All the rest of the stored helicopters have gone. The fates of some are still unknown, presumably these have also moved to Friedrichsthal.

9333	Mi-8T	10517	stored, ex 633/NVA, l/n 7-93	to Friedrichsthal
9337	Mi-8PS	10549	stored, ex 961/NVA, l/n 3-92	to Rothenburg
9342	Mi-8PS	10532	stored, ex 976/NVA, l/n 6-93	to Schönefeld
9360	Mi-8PS	10599	stored, ex 739/NVA, l/n 6-93	to Fichtelberg
9361	Mi-8TB	10553	stored, ex 124/NVA, l/n 4-95	gone
9362	Mi-8TB	10554	stored, ex 125/NVA, l/n 2-94	gone
9363	Mi-8TB	10555	stored, ex 126/NVA, l/n 4-95	to Schönefeld
9364	Mi-8TB	10556	stored, ex 128/NVA, l/n 2-94	gone
9365	Mi-8TB	10557	stored, ex 129/NVA, l/n 2-94	to Friedrichsthal
9366	Mi-8TB	10558	stored, ex 130/NVA, l/n 2-94	gone
9367	Mi-8TB	10559	stored, ex 131/NVA, l/n 7-93	gone
9369	Mi-8TB	10562	stored, ex 134/NVA. l/n 8-93	gone
9372	Mi-8TB	10587	stored, ex 933/NVA, l/n 2-95	to Friedrichsthal
9373	Mi-8TB	10589	stored, ex 935/NVA, l/n 2-94	to Friedrichsthal
9374	Mi-8TB	10590	stored, ex 936/NVA, l/n 2-95	to Friedrichsthal
9375	Mi-8TB	10592	stored, ex 938/NVA, l/n 4-95	gone
9376	Mi-8TB	10594	stored, ex 940/NVA, l/n 2-94	gone
9395	Mi-9	340002	stored, ex 402/NVA, l/n 4-95	to Hermeskeil
9396	Mi-9	340003	stored, ex 405/NVA, l/n 4-95	to Cerbaiola
9601	Mi-24D	110156	stored, ex 390/NVA, l/n 4-95	to Poland
9602	Mi-24D	110157	stored, ex 396/NVA, l/n 8-93	to Hungary
9603	Mi-24D	B4069	stored, ex 414/NVA, l/n 4-95	to Poland
9604	Mi-24D	110160	stored, ex 424/NVA, l/n 2-94	to Hungary
9605	Mi-24D	110161	stored, ex 433/NVA, l/n 4-95	to Hungary
9606	Mi-24D	110163	stored, ex 446/NVA, l/n 2-94	to Poland
9607	Mi-24D	110165	stored, ex 487/NVA, l/n 2-94	to Hungary
9608	Mi-24D	110170	stored, ex 498/NVA, l/n 4-95	to Poland
9609	Mi-24D	110169	stored, ex 520/NVA, l/n 4-95	to Poland
9610	Mi-24D	110172	stored, ex 522/NVA, l/n 2-94	to Hungary
9611	Mi-24D	110173	stored, ex 523/NVA, l/n 2-94	to Hungary
9612	Mi-24D	340269	stored, ex 524/NVA, l/n 4-95	to Poland
9613	Mi-24D	340227	stored, ex 525/NVA, l/n 4-95	to Poland
9614	Mi-24D	730209	stored, ex 534/NVA, l/n 4-95	to Poland

GERMANY - 177

9615	Mi-24D	730210	stored, ex 536/NVA, l/n 4-95		to Poland
9616	Mi-24D	730208	stored, ex 538/NVA, l/n 4-95		to Poland
9617	Mi-24D	730211	stored, ex 539/NVA, l/n 4-95		to Poland
9618	Mi-24D	730212	stored, ex 540/NVA, l/n 2-94		to Hungary
9619	Mi-24D	730213	stored, ex 543/NVA, l/n 2-94		to Poland

DARMSTADT
Just north of the former US Army airfield at Darmstadt, the Technische Hochschule Windkanal (Windtunnel) can be found. The Fouga in front of the buildings is mounted on a pole.

❏ AA-162	CM170	062	preserved		1-98

DERMSDORF
Dermsdorf is a small town north west of Sömmerda in the former DDR. At the local airfield five former NVA fighters are nicely lined up.

❏ 2229	MiG-21SPS	94A4510	preserved, ex Rothenburg, ex 771/NVA	8-97
❏ 2388	MiG-21U-400	661016	preserved, ex Rothenburg, ex 251/NVA	8-97
❏ 2395	MiG-21U-600	664620	preserved, ex Rothenburg, ex 289/NVA	8-97
❏ 2517	Su-22M-4	26204	preserved, ex Laage, ex 682/NVA	8-97
❏ 560	MiG-21M	960410	preserved, ex Kamenz	8-97

DESSAU
The Förderverein Technikmusuem Hugo Junkers was founded on 12th August 1992. The museum emphasises the impact Junkers had on the local economy with its aircraft, locomotive and other factories. The aircraft here have become of secondary importance (except for the Ju52) to the museum. The former Bückeburg gate guard 'PQ-131' has been checked here and underneath the paint the serial 7700 can be read, This cannot be its former serial as this Alouette 2 went to Portugal and later became F-GLPI. It has been reported that this 'PQ-131' is 7526, but this still has te be confirmed. The Ju52 arrived from Norway, where it was recovered from a frozen lake. Going the other way was MiG-21SPS 2237 which is now at <u>Gardermoen</u>. Ka-26 D-HZPT and PZL 104 DDR-WBY went to <u>Aschersleben</u>. The ex Soviet Mi-24V 01 yellow moved to <u>Bad Soolen Allendorf</u>. The museum is open van Monday to Thursday from 10:00 to 16:00.

❏ 16 red		MiG-15UTI	461810	preserved, ex Wunsdorf, ex Soviet AF	8-97
❏ 28 red		Yak-27R	0708	preserved, ex Wunsdorf, ex Soviet AF	8-97
❏ 2005		MiG-23MF	0390213100	preserved, ex Laage, ex 585/NVA	8-97
❏ 2359		MiG-21UM	05695168	preserved, ex Rothenburg, ex 212/NVA	8-97
❏ 2389		MiG-21U-400	661118	preserved, ex Rothenburg, ex 258/NVA	8-97
❏ 2410		MiG-21US	05685139	preserved, ex Rothenburg, ex 246/NVA	8-97
❏ 2430		MiG-21bis	N75033211	preserved, ex Drewitz, ex 875/NVA	8-97
❏ 2509		Su-22M-4	25916	preserved, ex Laage, ex 600/NVA	8-97
❏ 2549		Su-22UM-3	17532367003	preserved, ex Laage, ex 127/NVA	8-97
❏ 2820		L-39ZO	731009	preserved, ex Rothenburg, ex 161/NVA	8-97
❏ 7526 ?		Alouette 2	1300	preserved, marked 'PQ-131', see note	8-97
❏ 7528		Alouette 2	1310	preserved, ex Bückeburg	8-97
❏ 390		Mi-8T	0223	preserved, ex Bad Düben	8-97
❏ 670		MiG-21MF	966206	preserved, ex Bad Düben	8-97
❏ 673		MiG-21MF	966207	preserved, ex Bad Düben	8-97
❏ 725		MiG-21SPS	94A4212	preserved, ex Bad Düben	8-97
❏ 6134	1Z-BY	Ju52/3mg4e	6134	stored, arr 1-95	8-97
❏ D-ESMX		Zlin 37	05-12	preserved, ex DDR-SMX, ex DM-SMX	8-97
❏ D-HZPE		Mi-2	539811066	preserved, ex DDR-VPJ	8-97
❏ DDR-SUC		Zlin 37A	18-21	preserved, ex Leipzig, D-ESUC ntu	8-97
❏ DDR-TEJ		PZL 106BR	10880223	preserved, ex Leipzig, D-FOEJ ntu	8-97

DETMOLD

The British Army vacated the airfield in late 1994, taking two of their preserved helicopters, Sioux AH.1 XT550/D (WA439) and Skeeter AOP.12 XL739 (S2/5071) with them to their new home at Wattisham, UK. Wessex HU.5 XT764 moved to Gütersloh. The remaining helicopters were presumably sold and scrapped.

❏ XP900		Scout AH.1	F9501	instructional	6-87
❏ XP903		Scout AH.1	F9504	instructional	9-91
❏ XR637		Scout AH.1	F9537	instructional	9-91
❏ XW615		Scout AH.1	F9743	instructional	5-89
❏ XS571	614	Wasp HAS.1	F9582	instructional	5-89
❏ XT436	506	Wasp HAS.1	F9606	instructional, ex Soest	—
❏ XV627	321	Wasp HAS.1	F9722	instructional	5-89
❏ XT474		Wessex HU.5	WA296	instructional	10-91

DIEPENSEE

On the south side of Berlin Schönefeld airport, the Aero Park Brandenburg is located. It is open daily from April to September from 09:00 to 19:00. There is also Mi-8 tailboom (9332) in the museum. Ka-26 D-HOAY (7001309) was exchanged with Weston-super-Mare, UK, for the Whirlwind.

❏ 35 red	Yak-27R	0214	preserved, ex Werneuchen	5-98
❏ 2381	MiG-21UM	10695162	preserved, ex Rothenburg, ex 207/NVA	5-98
❏ 9311	Mi-8T	10542	preserved, ex Briest, ex 925/NVA	5-98
❏ 001	SM-1Sz	..	stored, ex Rothenburg, ex Kamenz, ex Bautzen	—
❏ 501	Mi-2	562819043	preserved, ex Schönefeld, D-HZPP ntu	5-98
❏ 396	Mi-8T	0726	preserved, ex Briest	5-98
❏ 698	MiG-21SPS	94A4503	preserved, ex Kamenz	5-98
❏ D-ESMU	Zlin 37A	08-06	preserved	5-98
❏ D-ESUU	Zlin 37A	19-18	preserved	5-98
❏ D-FONF	An-2T	117419	preserved, ex DDR-SKF, ex 811/NVA	5-98
❏ D-HOAO	Ka-26	7404617	preserved, ex DDR-SPO, ex DM-SPO	5-96
❏ D-HOAQ	Ka-26	7404618	preserved, ex DDR-SPQ, ex DM-SPQ	5-98
❏ D-HOAS	Ka-26	7504620	preserved, ex DDR-SPS, ex DM-SPS	2-95
❏ D-HOAX	Ka-26	7303806	preserved, ex DDR-SPX, ex DM-SPX	5-98
❏ D-HOXB	Mi-8T	0623	preserved, ex DDR-SPB, ex DM-SPB	5-98
❏ DM-SCM	Tu-134A	3351904	preserved, forward fuselage, crashed 22-11-77	3-98
❏ DDR-STB	Il-18D	180002001	preserved, ex DM-STB	3-98
❏ G-AYNP	Whirlwind 3	WA71	preserved, ex Weston-super-Mare, ex Redhill	5-98

DIEPHOLZ

For several years, the Luftwaffe has based a UH-1D unit here. One of their helicopters, 7038, is stored at the airfield and will be converted into a bar. 7321 was in May 1998 seen with Unsere Luftwaffe. The stored F-104G 2625 moved to Ankum.

❏ 7038	UH-1D	4369	stored, ex 64-13662	11-97
❏ 7321	UH-1D	8441	preserved, marked as '7030'	11-97
❏ 9820	Bo105M	S-90	under restoration, to go to Berlin	11-97

DRESDEN

Town: The Millitarhistorisches Museum at the Olbrichtplatz traces the history of the (East) German military services. MiG-21SPS 764 (nose only) moved to Butzweilerhof and An-14A 996 and Lim-5P 437 (marked as '850') went to Cottbus in 1997. The museum is open daily between 09:00 and 17:00, Sundays only between 09:00 and 12:00.

❏ 7524	Alouette 2	1298	preserved, inside, upstrairs	2-98
❏ 9901	G91R/3	0060	preserved, ex 3007	2-98

❑ 9921	OV-10B	338-6	preserved, arr 9-3-92, ex D-9550	2-98
❑ JB-110	CL-13B Sabre 6	1734	preserved, ex Oldenburg	2-98
❑ 826	An-2	17906	stored, wfu 20-4-90	4-91
❑ 313	L-29	692054	preserved	2-98
❑ 792	Mi-4A	0251	preserved, marked as '785', ex DM-SPB	2-98
❑ 300	Lim-5	1C-0630	preserved	2-98
❑ 671	MiG-21F-13	741620	preserved, inside, marked as '268'	2-98
❑ 868	MiG-21SPS	94A1103	preserved	2-98
❑ 525	Grunau SG-38	83	preserved, glider	97
❑ ..	Yak-18	..	preserved	97
❑ DM-WAA	CCS-13	420-32	preserved, inside, marked as '250', ex SP-AFN	2-98

A transport museum in the town has acquired the Baaden 152 fuselage from the storage at Rothenburg. It is somewhere in Dresden under restoration and will join the museum in 1999. Currently the museum has only one aircraft on display.

❑ ..	Aero 45	..	preserved, marked as 'DM-VMD'	8-96
❑ ..	Baaden 152	011	under restoration, ex Rothenburg	96

Airfield – Klotzsche: During 1992 and 1993 a total of 141 MiG-21s were scrapped at this airfield, all came from Drewitz (which see, 128 known) and Rothenburg (which see, 3 known). Of approximately 25 scrapped airframes noted in April 1993, only three could be identified. All the aircraft from the NVA days have gone, with the exception of the VEB-14, which is preserved near the control tower. It was joined in early 1996 by an ex Laage MiG-23 and some stored ex Rothenburg MiG-21s (including three unmarked cockpits). Noted here in the early 1990s were a number of ex NVA MiG-21s which were not delivered to Iraq. Their identities are still not confirmed, but should include MiG-21U-600 278 (663316), MiG-21PFM 828 (760609), 858 (761010) and 870 (760914). MiG-21bis 2429 (N75033205) was on overhaul during the reunion and never made it back to his homebase of Drewitz. From here it moved to Geel. Other aircraft not surviving after 1991 comprise some dumped and stored aircraft with the overhaul plant here. Gone are L-29 336 (591524), MiG-21F-13 629 (741917), MiG-21PFM 851 (761008) and 862 (760811), MiG-21SPS-K 979 (94A6505) and 474 (94A7007) and MiG-23MF 596 (0390213354, marked as '1596'). All have been scrapped. An-2 469 (113901) became D-FGGG, while stored MiG-15UTI 127 and dumped MiG-21UM 285 went to Cerbaiola. Four Mi-8s, 9350 (10733), 9352 (105104), 9354 (105106) and an unknown one, were noted here in storage in late 1996. Of these, 9350 moved to Merseburg, one has gone to Friedrichshafen and one to the back garden of one of the former Mi-8 pilots (where to ?). Also the fourth Mi-8 has gone.

❑ 1101	Tu-154M	799	stored, ex 114/NVA, ex DDR-SFA	2-98
❑ 2035	MiG-23ML	0390324050	preserved, ex Laage, ex 601/NVA	2-98
❑ 2230	MiG-21SPS	94A4310	stored, ex Rothenburg, ex 780/NVA	2-98
❑ 2286	MiG-21M	960513	stored, ex Rothenburg, ex 588/NVA	8-96
❑ 2402	MiG-21US	08685145	preserved, ex Rothenburg, ex 218/NVA	2-98
❑ 2403	MiG-21US	09685145	stored, ex Rothenburg, ex 219/NVA	8-96
❑ ..	MiG-21F-13	..	stored, marked as '705', ex preserved	8-96
❑ DDR-SAL	VEB-14P	14803026	preserved, marked as 'DDR-ZZB', ex DM-SAL	1-98

DREWITZ
Currently the only *Wrecks & Relics* aircraft here is a former instructional MiG-21SPS from Kamenz. It is parked on the old flightline near the aero club.

❑ 449	MiG-21SPS	94A6712	preserved, ex Kamenz	9-97

Since late 1990 more than 150 MiG-21s and MiG-23s were stored at Drewitz. The MiG-21s fell under the CFE treaty and large numbers were scrapped. Most were not scrapped here, but at Dresden. Only a handful of these MiG-21s were identified at Dresden (marked with *). The art of scrapping was first tried out on 2294 which was later placed on the dump.

2038	MiG-23BN	0393211085	stored, ex 689/NVA, l/n 4-93	to Rothenburg
2039	MiG-23BN	0393211087	stored, ex 690/NVA, l/n 4-93	to Speyer

GERMANY - 180

2040	MiG-23BN	0393211088	dumped, ex 691/NVA, l/n 6-92	scrapped
2041	MiG-23BN	0393214101	stored, ex 692/NVA, l/n 6-92	gone
2042	MiG-23BN	0393214210	stored, ex 694/NVA, l/n 4-93	to Seifertshofen
2043	MiG-23BN	0393214211	stored, ex 695/NVA, l/n 4-93	to Rothenburg
2044	MiG-23BN	0393214212	stored, ex 696/NVA, l/n 4-93	to Fichtelberg
2046	MiG-23BN	0393214214	stored, ex 698/NVA, l/n 4-93	to Hermeskeil
2047	MiG-23BN	0393214217	stored, ex 701/NVA, l/n 4-93	to Oberschleissheim
2049	MiG-23BN	2963222830	stored, ex 705/NVA, l/n 4-93	to Speyer
2050	MiG-23BN	0393214220	stored, ex 707/NVA, l/n 4-93	to Seifertshofen
2051	MiG-23BN	0393214225	stored, ex 710/NVA, l/n 4-93	to Berlin
2053	MiG-23BN	0393215721	stored, ex 715/NVA, l/n 6-92	gone
2054	MiG-23BN	0393215729	stored, ex 718/NVA, l/n 4-93	to Rothenburg
2055	MiG-23BN	0393215732	stored, ex 720/NVA, l/n 4-93	to Finow
2201	MiG-21SPS-K	94A6703	stored, ex 429/NVA, l/n 6-92	scrapped at Dresden
2202	MiG-21SPS-K	94A6709	stored, ex 441/NVA, l/n 6-92	scrapped at Dresden
2203 *	MiG-21SPS-K	94A7003	stored, ex 466/NVA, l/n 6-92	scrapped at Dresden
2204	MiG-21SPS-K	94A7210	stored, ex 484/NVA, l/n 6-92	scrapped at Dresden
2205	MiG-21SPS-K	94A7213	stored, ex 489/NVA, l/n 6-92	scrapped at Dresden
2206 *	MiG-21SPS-K	94A7303	stored, ex 560/NVA, l/n 6-92	scrapped at Dresden
2207	MiG-21SPS-K	94A7305	stored, ex 572/NVA, l/n 6-92	scrapped at Dresden
2246	MiG-21M	962104	stored, ex 410/NVA, l/n 6-92	scrapped at Dresden
2247	MiG-21M	962111	stored, ex 419/NVA, l/n 4-93	scrapped at Dresden
2248	MiG-21M	963204	stored, ex 428/NVA, l/n 4-93	scrapped at Dresden
2249 *	MiG-21M	963205	stored, ex 431/NVA, l/n 6-92	scrapped at Dresden
2250	MiG-21M	963206	stored, ex 432/NVA, l/n 4-93	scrapped at Dresden
2251	MiG-21M	963207	stored, ex 435/NVA	to Fassberg
2254	MiG-21M	963212	stored, ex 461/NVA, l/n 4-93	scrapped at Dresden
2255	MiG-21M	963211	stored, ex 465/NVA, l/n 6-92	scrapped at Dresden
2256	MiG-21M	963214	stored, ex 468/NVA, l/n 4-93	scrapped at Dresden
2258	MiG-21M	963301	stored, ex 491/NVA, l/n 4-93	scrapped at Dresden
2259	MiG-21M	963302	stored, ex 493/NVA, l/n 6-92	scrapped at Dresden
2260	MiG-21M	963303	stored, ex 497/NVA, l/n 4-93	scrapped at Dresden
2263	MiG-21M	960307	stored, ex 508/NVA	to Fassberg
2264	MiG-21M	960308	stored, ex 515/NVA	to Fassberg
2265	MiG-21M	960401	stored, ex 526/NVA, l/n 4-93	scrapped at Dresden
2266	MiG-21M	963310	stored, ex 527/NVA, l/n 6-92	scrapped at Dresden
2267	MiG-21M	960404	stored, ex 541/NVA, l/n 4-93	scrapped at Dresden
2268	MiG-21M	963311	stored, ex 542/NVA, l/n 6-92	scrapped at Dresden
2269	MiG-21M	960405	stored, ex 549/NVA, l/n 4-93	scrapped at Dresden
2271	MiG-21M	963313	stored, ex 566/NVA, l/n 4-93	scrapped at Dresden
2272	MiG-21M	960503	stored, ex 570/NVA, l/n 4-93	scrapped at Dresden
2273	MiG-21M	963314	stored, ex 573/NVA, l/n 6-92	scrapped at Dresden
2274	MiG-21M	960504	dumped, ex 575/NVA, 2274 ntu, l/n 2-92	scrapped
2275	MiG-21M	960506	stored, ex 578/NVA, l/n 4-93	scrapped at Dresden
2276	MiG-21M	963315	stored, ex 580/NVA, l/n 4-93	scrapped at Dresden
2290	MiG-21M	962106	stored, ex 413/NVA, l/n 6-92	scrapped at Dresden
2291	MiG-21M	960402	stored, ex 531/NVA, l/n 6-92	scrapped at Dresden
2292	MiG-21M	960510	stored, ex 583/NVA, l/n 6-92	scrapped at Dresden
2293	MiG-21M	960512	stored, ex 587/NVA, l/n 6-92	scrapped at Dresden
2294	MiG-21M	960602	dumped, ex 594/NVA, l/n 6-92	scrapped
2295	MiG-21M	960801	stored, ex 595/NVA, l/n 6-92	scrapped at Dresden
2296	MiG-21M	960705	stored, ex 597/NVA, l/n 6-92	scrapped at Dresden
2297	MiG-21M	960707	stored, ex 603/NVA, l/n 6-92	scrapped at Dresden
2298	MiG-21M	960711	stored, ex 609/NVA, l/n 6-92	scrapped at Dresden
2299	MiG-21M	960712	stored, ex 614/NVA, l/n 6-92	scrapped at Dresden

GERMANY – 181

2301	MiG-21M	960713	stored, ex 616/NVA, l/n 6-92	scrapped at Dresden
2302	MiG-21M	960715	stored, ex 621/NVA, l/n 6-92	scrapped at Dresden
2303	MiG-21MF	967603	stored, ex 427/NVA, l/n 6-92	scrapped at Dresden
2304	MiG-21MF	967604	stored, ex 430/NVA, l/n 6-92	scrapped at Dresden
2305	MiG-21MF	967605	stored, ex 437/NVA, l/n 6-92	scrapped at Dresden
2306	MiG-21MF	967607	stored, ex 448/NVA, l/n 6-92	scrapped at Dresden
2307	MiG-21MF	967608	stored, ex 460/NVA, l/n 6-92	scrapped at Dresden
2308	MiG-21MF	967610	stored, ex 467/NVA, l/n 6-92	nose to Speyer
2309	MiG-21MF	967609	stored, ex 470/NVA, l/n 6-92	scrapped at Dresden
2310	MiG-21MF	967612	stored, ex 473/NVA, l/n 6-92	scrapped at Dresden
2311	MiG-21MF	967613	stored, ex 477/NVA, l/n 6-92	scrapped at Dresden
2312	MiG-21MF	967614	stored, ex 478/NVA, l/n 6-92	scrapped at Dresden
2313	MiG-21MF	967615	stored, ex 490/NVA, l/n 6-92	scrapped at Dresden
2314	MiG-21MF	968609	stored, ex 509/NVA, l/n 6-92	scrapped at Dresden
2315	MiG-21MF	968615	stored, ex 510/NVA, l/n 6-92	scrapped at Dresden
2316	MiG-21MF	968611	stored, ex 511/NVA, l/n 6-92	scrapped at Dresden
2318	MiG-21MF	968614	stored, ex 514/NVA, l/n 6-92	scrapped at Dresden
2319	MiG-21MF	968610	stored, ex 516/NVA, l/n 6-92	to Rothenburg
2320	MiG-21MF	968608	stored, ex 518/NVA, l/n 6-92	scrapped at Dresden
2321	MiG-21MF	969009	stored, ex 535/NVA, l/n 6-92	scrapped at Dresden
2322	MiG-21MF	968612	stored, ex 548/NVA, l/n 6-92	scrapped at Dresden
2323	MiG-21MF	965306	stored, ex 649/NVA, l/n 6-92	scrapped at Dresden
2324	MiG-21MF	965307	stored, ex 650/NVA, l/n 6-92	scrapped at Dresden
2325	MiG-21MF	965308	stored, ex 651/NVA, l/n 6-92	scrapped at Dresden
2326	MiG-21MF	965310	stored, ex 652/NVA, l/n 6-92	scrapped at Dresden
2327 *	MiG-21MF	965313	stored, ex 657/NVA, l/n 6-92	scrapped at Dresden
2328	MiG-21MF	965315	stored, ex 659/NVA, l/n 6-92	scrapped at Dresden
2329	MiG-21MF	965401	stored, ex 660/NVA, l/n 6-92	scrapped at Dresden
2330	MiG-21MF	965402	stored, ex 662/NVA, l/n 6-92	scrapped at Dresden
2331	MiG-21MF	965403	stored, ex 664/NVA, l/n 6-92	scrapped at Dresden
2332	MiG-21MF	965404	stored, ex 665/NVA, l/n 6-92	scrapped at Dresden
2333	MiG-21MF	966205	stored, ex 667/NVA, l/n 6-92	scrapped at Dresden
2334	MiG-21MF	966209	stored, ex 675/NVA, l/n 6-92	scrapped at Dresden
2335	MiG-21MF	966210	stored, ex 680/NVA, l/n 6-92	scrapped at Dresden
2336	MiG-21MF	966213	stored, ex 683/NVA, l/n 6-92	scrapped at Dresden
2337	MiG-21MF	966214	stored, ex 683/NVA, l/n 4-93	to Rothenburg
2350	MiG-21UM	02695174	stored, ex 167/NVA, l/n 6-92	scrapped at Dresden
2351	MiG-21UM	05695174	stored, ex 168/NVA, l/n 6-92	scrapped at Dresden
2352	MiG-21UM	03695174	stored, ex 169/NVA, l/n 6-92	scrapped at Dresden
2353	MiG-21UM	03695165	stored, ex 210/NVA, l/n 6-92	scrapped at Dresden
2354	MiG-21UM	516915001	stored, ex 232/NVA, l/n 6-92	scrapped at Dresden
2355	MiG-21UM	03695156	stored, ex 257/NVA, l/n 6-92	scrapped at Dresden
2356	MiG-21UM	04695156	stored, ex 259/NVA, l/n 6-92	scrapped at Dresden
2357	MiG-21UM	05695156	stored, ex 262/NVA, l/n 6-92	scrapped at Dresden
2358	MiG-21UM	06695156	stored, ex 264/NVA, l/n 6-92	scrapped at Dresden
2362	MiG-21UM	516921051	stored, ex 228/NVA, l/n 6-92	scrapped at Dresden
2363	MiG-21UM	516995031	stored, ex 243/NVA, l/n 6-92	scrapped at Dresden
2364	MiG-21UM	516995036	stored, ex 245/NVA, l/n 6-92	scrapped at Dresden
2365	MiG-21UM	516995041	stored, ex 249/NVA, l/n 6-92	scrapped at Dresden
2366	MiG-21UM	516931001	stored, ex 267/NVA, l/n 4-93	scrapped at Dresden
2367	MiG-21UM	516931041	stored, ex 268/NVA, l/n 6-92	scrapped at Dresden
2368	MiG-21UM	516931046	stored, ex 269/NVA, l/n 6-92	scrapped at Dresden
2369	MiG-21UM	516931051	stored, ex 270/NVA, l/n 4-93	scrapped at Dresden
2370	MiG-21UM	06695165	stored, ex 211/NVA, l/n 4-93	scrapped at Dresden
2371	MiG-21UM	516915071	stored, ex 226/NVA, l/n 6-92	scrapped at Dresden

GERMANY - 182

2372 *	MiG-21UM	516915076	stored, ex 227/NVA, l/n 6-92	scrapped at Dresden
2373	MiG-21UM	516915021	stored, ex 234/NVA, l/n 6-92	scrapped at Dresden
2374	MiG-21UM	09695155	stored, ex 252/NVA, l/n 6-92	scrapped at Dresden
2375	MiG-21UM	10695155	stored, ex 253/NVA, l/n 4-93	scrapped at Dresden
2376	MiG-21UM	01695156	stored, ex 254/NVA, l/n 6-92	scrapped at Dresden
2384	MiG-21UM	01695163	stored, ex 203/NVA, l/n 6-92	scrapped at Dresden
2385 *	MiG-21UM	07695168	stored, ex 213/NVA, l/n 6-92	scrapped at Dresden
2413	MiG-21bis	N75051035	stored, ex 792/NVA, l/n 6-92	scrapped at Dresden
2414	MiG-21bis	N75051108	stored, ex 794/NVA, l/n 6-92	scrapped at Dresden
2415	MiG-21bis	N75051121	stored, ex 797/NVA, l/n 6-92	scrapped at Dresden
2416	MiG-21bis	N75051345	stored, ex 835/NVA, l/n 4-93	scrapped at Dresden
2417	MiG-21bis	N75051347	stored, ex 837/NVA, j/n 6-92	scrapped at Dresden
2418	MiG-21bis	N75051378	stored, ex 838/NVA, l/n 6-92	to Rothenburg
2419	MiG-21bis	N75051384	stored, ex 840/NVA, l/n 6-92	scrapped at Dresden
2421	MiG-21bis	N75051407	stored, ex 848/NVA, l/n 4-93	to Cottbus
2422	MiG-21bis	N75051426	stored, ex 849/NVA, l/n 6-92	scrapped at Dresden
2423 *	MiG-21bis	N75058003	stored, ex 850/NVA, l/n 6-92	scrapped at Dresden
2424	MiG-21bis	N75058015	stored, ex 853/NVA, l/n 4-93	to Hermeskeil
2425	MiG-21bis	N75058087	stored, ex 856/NVA, l/n 4-93	to Rothenburg
2426	MiG-21bis	N75033148	stored, ex 871/NVA, l/n 6-92	scrapped at Dresden
2427	MiG-21bis	N75033151	stored, ex 872/NVA, l/n 6-92	scrapped at Dresden
2428	MiG-21bis	N75033169	stored, ex 873/NVA, l/n 4-93	to Tannheim
2430	MiG-21bis	N75033211	stored, ex 875/NVA, l/n 4-93	to Dessau
2431	MiG-21bis	N75033213	stored, ex 876/NVA, l/n 6-92	scrapped at Dresden
2432	MiG-21bis	N75033219	stored, ex 879/NVA, l/n 6-92	scrapped at Dresden
2433	MiG-21bis	N75033305	stored, ex 881/NVA, l/n 6-92	scrapped at Dresden
2434	MiG-21bis	N75033397	stored, ex 882/NVA, l/n 6-92	scrapped at Dresden
2435	MiG-21bis	N75033419	stored, ex 886/NVA, l/n 6-92	scrapped at Dresden
2436	MiG-21bis	N75033190	stored, ex 887/NVA, l/n 6-92	scrapped at Dresden
2437	MiG-21bis	N75033445	stored, ex 892/NVA, l/n 6-92	scrapped at Dresden
2438	MiG-21bis	N75033507	stored, ex 893/NVA, l/n 6-92	scrapped at Dresden
2439	MiG-21bis	N75033515	stored, ex 895/NVA, l/n 4-93	scrapped at Dresden
2440	MiG-21bis	N75033522	stored, ex 899/NVA, l/n 6-92	scrapped at Dresden
2441	MiG-21bis	N75035201	stored, ex 900/NVA, l/n 6-92	scrapped at Dresden
2442	MiG-21bis	N75035213	stored, ex 904/NVA, l/n 6-92	scrapped at Dresden
2443	MiG-21bis	N75035284	stored, ex 916/NVA, l/n 6-92	scrapped at Dresden
2444	MiG-21bis	N75035289	stored, ex 917/NVA, l/n 6-92	scrapped at Dresden
2445	MiG-21bis	N75035291	stored, ex 920/NVA, l/n 6-92	scrapped at Dresden
2446	MiG-21bis	N75035304	stored, ex 933/NVA, l/n 4-93	scrapped at Dresden
2447	MiG-21bis	N75035374	stored, ex 936/NVA, l/n 6-92	scrapped at Dresden
2448	MiG-21bis	N75035399	stored, ex 946/NVA, l/n 6-92	scrapped at Dresden
2449	MiG-21bis	N75035407	stored, ex 951/NVA, l/n 6-92	scrapped at Dresden
2450	MiG-21bis	N75035422	stored, ex 954/NVA, l/n 4-93	scrapped at Dresden
2451	MiG-21bis	N75035445	stored, ex 956/NVA, l/n 6-92	scrapped at Dresden
2452	MiG-21bis	N75035502	stored, ex 987/NVA, l/n 6-92	scrapped at Dresden
2453	MiG-21bis	N75035841	stored, ex 990/NVA, l/n 4-93	to Berlin

After the disposal of the MiG-21s and MiG-23s, the first of the (then) still operational Mi-2s and Mi-8s arrived here in 1994 for storage. In 1996 a dozen of these helicopters were still present. By 1997 all the helicopters had gone, presumably scrapped. Of the other aircraft at Drewitz from the past, the preserved Lim-5 537 went to Cottbus, while 346 (1C-0723, ex Rothenburg) went to Berlin. Lim-5s 656 (1C-0851) and 777 (1C-0825) have both scrapped. Dumped MiG-21PFM 918 (94A1115) may also have been scrapped.

9304	Mi-8T	10512	stored, ex 626/NVA, l/n 1-95	gone
9307	Mi-8T	10537	stored, ex 921/NVA, l/n 1-95	gone
9309	Mi-8T	10539	stored, ex 923/NVA, l/n 4-96	to OK-FXA

9310	Mi-8T	10540	stored, ex 924/NVA, l/n 1-95	gone
9317	Mi-8T	10546	stored, ex 931/NVA, l/n 4-96	to OK-FXE
9331	Mi-8T	10514	stored, ex 630/NVA, l/n 1-95	gone
9336	Mi-8PS	10548	stored, ex 960/NVA, l/n 1-95	gone
9340	Mi-8PS	10523	stored, ex 974/NVA, l/n 4-96	gone
9345	Mi-8PS	10586	stored, ex 993/NVA, l/n 1-95	gone
9346	Mi-8PS	10584	stored, ex 998/NVA, l/n 4-96	gone
9371	Mi-8TB	10579	stored, ex 755/NVA, l/n 1-95	to Berlin, to ES-PMC
9388	Mi-8TB	10588	stored, ex 934/NVA, l/n 1-95	to Berlin, to ES-PMB
9411	Mi-8TB	10573	stored, ex 830/NVA, l/n 3-95	gone
9423	Mi-8T	10529	stored, ex 913/NVA, D-HOZG ntu, l/n 1-95	gone
9454	Mi-2S	563822114	stored, ex Briest, ex 308/NVA, l/n 5-94	gone
9462	Mi-2S	563147103	stored, ex Briest, ex 392/NVA, l/n 4-96	gone

DÜREN
About 20km east of Aachen, an F-104F is displayed inside Luftwaffe Munitions Depot 61 barracks in the Gürzenich part of Düren.

☐ 2914	F-104F	5066	preserved, ex 59-5011	2-98

DÜSSELDORF
Of all the Air Classik aircraft mentioned in the first edition, only CASA352 T.2B-165 (marked as 'D-CIAK') survived into 1990s. It has now joined the 'warbirds' circuit as HB-HOY.

EGGEBEK
Eggebek (some 25km from the Danish border) is home to the Marine Tornados of MFG2. Although the instructional Tornado '3809' looks quite authentic, it is not a real. Dumped F-104 2125 moved to Flensburg.

☐ RB-363	Sea Hawk Mk.101	6707	preserved, at main gate	5-98
☐ 2292	F-104G	7175	instructional, ex Fassberg	6-97
☐ 2309	F-104G	7192	stored, ex preserved	5-98
☐ 2446	F-104G	8190	instructional	5-98
☐ 2672	F-104G	7418	preserved, at main gate, in Viking c/s	5-98
☐ EB-322	RF-84F	..	fire dump, ex 53-7661	8-96
☐ ..	Tornado	..	instructional, mock-up as '3809', ex Jagel	5-98

EILENBURG
In the town of Eilenburg, north west of Leipzig, preserved Dresden-built Il-14, DDR-SAM, was damaged by local hooligans during 1992/1993. The aircraft moved to Berlin for restoration.

ERDING
Serial number 2473 was allocated to the Starfighter at the gate, but was never carried as the F-104 went out of service before it could be applied. At Erding restoration work is carried out on F-104s; 2002 (ex Karlsruhe) was restored here and went to Berlin, together with 2206 (ex Jever, ex Leck). Other Starfighters leaving here are 2100 (ex Lechfeld) which went to Karlsruhe, 2507 (cockpit only) went to Fürstenfeldbruck as '7500' and the dumped F-104F 2921 (marked as '4711') went to Söllingen. TF-104Gs 2813 and 2822 went to Seifertshofen and F-104G 2376 (marked as 'BB-371', ex Jever) and 2392 (also ex Jever) both went to Baarlo. The two T-33s also left, 9429 went to Hermeskeil and 9447 to Oberschleissheim.

☐ 2258	F-104G	7139	instructional, on left side only 2 of serial shown	9-96
☐ 2291	F-104G	7174	instructional, silver c/s	9-96
☐ 2468	F-104G	8217	instructional	9-96

GERMANY - 184

❏ 2473	F-104G	8222	preserved, at main gate, ex '8058', ex EB-121	9-96
❏ 2652	F-104G	7312	instructional	9-96
❏ 2911	F-104F	5061	under restoration, marked as 'BB-374', ex Köln	11-97
❏ 4312	Tornado	4012	instructional	4-97
❏ 5615	Do27A-1	283	instructional, unmarked	9-96

ERFURT
A former Interflug Il-18 is still preserved on this airfield. Il-62 DDR-SEH, which was also here, can now be found at town of Alach.

❏ DDR-STG	Il-18V	182004402	preserved, ex DM-STG	8-96

ESCHERHAUSEN
Near the local airfield, some 30km south west of Hildesheim, a Viscount is in use as a restaurant.

❏ D-ANAB	Viscount 814	369	preserved, wfu 4-12-69	2-98

ESSEN – MÜLHEIM
The unmarked ex Polish Air Force Mi-2 is still stored at this airfield, painted red. The Friendships are all in storage with WDL.

❏ 5824	Mi-2R	565824128	preserved, unmarked, ex Polish AF	3-98
❏ D-AELL	F27-200	10414	stored, ex PH-FON	3-98
❏ D-BAKA	F27-100	10198	stored, ex PH-LIP	2-98
❏ D-BAKE	F27-200	10263	stored, fuselage only, ex PH-FGE	11-96
❏ D-BAKH	F27-200	10233	stored, ex PH-FEY	2-98
❏ D-BAKJ	F27-200	10321	stored, ex ST-AWB	3-98
❏ D-BAKK	F27-200	10229	stored, ex G-BHMW	3-98
❏ D-BAKU	F27-200	10137	stored, fuselage only, ex TN-ACR	7-96

FALKENBERG
With the departure of the Soviets and their MiG-29s, the dump should also have been cleared.

❏ 03 red	Il-14	..	dumped, ex Soviet AF	5-93
❏ .1 red	MiG-15UTI	..	dumped, ex preserved, ex Soviet AF	5-93
❏ 60 red	MiG-23UB	3901320	dumped, ex Soviet AF	5-93
❏ ..	MiG-23UB	4706	dumped, front fuselage, ex Soviet AF	4-93

FASSBERG
The technical training unit here is the Technische Schule der Luftwaffe 3 (TSLw3), which trains both Luftwaffe and Heer personnel, also uses airframes obtained on loan from active units. A large number of helicopters have been noted in Hangar 8. Of these UH-1D 7084 (8144, ex Ahlhorn) was last noted in 1989. Bo105s 8004, 8010, 8011, 8012, 8013, 8605, 8606, 8613, 8620, 8623, 8677, 8688, 8689 and 8712 were here for a short time and have returned to active service.

❏ 7047	UH-1D	8107	instructional, old c/s	9-96
❏ 7090	UH-1D	8150	instructional, old c/s	9-96
❏ 7099	UH-1D	8159	instructional, lizard c/s	4-94
❏ 7168	UH-1D	8228	instructional, old c/s	9-96
❏ 7384	UH-1D	8504	instructional	9-96
❏ ..	UH-1D	'8001'	instructional, composite a/c	4-94
❏ 7602	Alouette 2	1547	stored	10-94
❏ 8002	Bo105M	5002	instructional	9-96
❏ 8006	Bo105M	5006	instructional	9-96

GERMANY - 185

☐ 8028	Bo105M	5028	instructional	4-94
☐ 8046	Bo105M	5046	instructional	9-96
☐ 8404	CH-53G	V65-002	instructional	9-96
☐ 8608	Bo105P	6008	instructional	9-96
☐ 8615	Bo105P	6015	instructional	9-96
☐ 9908	G91R/3	467	instructional, ex 3199	9-96
☐ ..	Bo105C	S.8	instructional	4-94

In hangar 7 are a number of aircraft for battle damage training and the training of fire fighting crews.

☐ 2037	MiG-23ML	0390324249	instructional, ex Laage	9-96
☐ 2160	F-104G	7029	instructional, BDR flight	1-98
☐ 2171	F-104G	7040	instructional	1-98
☐ 2245	F-104G	7123	instructional, BDR flight	1-98
☐ 2327	F-104G	8001	instructional, BDR flight	1-98
☐ 2489	F-104G	8239	instructional	1-98
☐ 8413	CH-53G	V65-011	instructional	9-96
☐ 8628	Bo-105P	6028	instructional	9-96
☐ 9382	Mi-8TB	10563	instructional, BDR flight, ex Basepohl	9-96
☐ 9806	Tornado	P07	instructional, BDR flight, ex München	9-96
☐ 9926	OV-10B	338-11	instructional, BDR flight, ex D-9555	9-96
☐ 9932	OV-10B	338-17	instructional, BDR flight, ex D-9561	9-96

The apprentice school occupy Hangar 6. Sabre D-9523 moved to Celle, while D-9593 went to Uetersen.

☐ 2708	TF-104G	5709	instructional, ex 61-3038, crashed 12-5-66	1-98
☐ 5503	Do27	106	instructional, marked as '1104'	9-96
☐ 5556	Do27B-1	189	instructional, D-EDNW ntu?	9-96
☐ 9190	P149D	275	instructional	9-96
☐ D-9542	CL-13B Sabre 6	1740	instructional	9-96
☐ ..	Alpha Jet	..	instructional, rig marked as 'SO-2A'	9-96
☐ ..	Alpha Jet	..	instructional, rig marked as 'SO-3A'	9-96

More destructive fire fighting is done outside, noted in their area were:

☐ 2007	F-104G	2007	instructional	1-98
☐ 2251	MiG-21M	963207	instructional, ex 435/NVA and marked as such	9-96
☐ 2257	MiG-21M	963215	instructional, ex 483/NVA, ex Neubrandenburg	9-96
☐ 2263	MiG-21M	960307	instructional, ex 508/NVA and marked as such	9-96
☐ 2264	MiG-21M	960308	instructional, ex 515/NVA and marked as such	9-96
☐ 2265	F-104G	7146	instructional	1-98
☐ 2529	F-104G	8306	instructional	1-98
☐ 2540	F-104G	8327	instructional	1-98
☐ 2617	F-104G	9159	instructional	1-98
☐ 3098	G91R/3	364	instructional, fuselage only, ex Husum	9-96
☐ 9905	G91R/3	378	instructional, ex 3112	9-96

Over the years more aircraft have been noted at Fassberg, but their exact location on the airfield is unknown. Of these MiG-21s 950 and 2209 moved recently to Gatow and Starfighter 2126 went to Ahaus.

☐ 75-0635	F-4E	4960	instructional	2-98
☐ 75-0636	F-4E	4962	instructional	2-98
☐ 2262	MiG-21M	963306	dumped, ex 505/NVA, ex Neubrandenburg	4-95
☐ 2477	F-104G	8226	dumped	6-93
☐ 2581	F-104G	9049	dumped	6-93
☐ 2583	F-104G	9054	dumped	6-93
☐ 2647	F-104G	7307	dumped	1-98
☐ 3232	G91R/3	501	preserved, marked as '3105'	6-97
☐ 4078	Alpha Jet	0078	instructional	8-93

GERMANY - 186

☐ 4175	Alpha Jet	0175	instructional		6-93
☐ 5039	C-160D	D61	instructional, rear fuselage only		7-96
☐ 5901	Do28D-2	4176	instructional, ex Uetersen		9-96
☐ 7640	Alouette 2	1656	instructional, BDR flight		4-94
☐ D-EAXT	FWP149D	171	instructional, BDR flight, ex 9149		9-96
☐ D-ICDY	Do28D-2	4164	instructional, ex Uetersen, ex 5889		..

With a fire training school its not surprising that they have used up some airframes! Aircraft known to have been used here and which have perished include; F-104Gs 2383 (8086), 2494 (8247, marked as 'DA-06'), TF-104G 2795 (5925), F-104Fs 2904 (5050), 2910 (5060), 2913 (5066), G91R/3s 3012 (0067), 3053 (310), 3090 (354), 3229 (498), RF-84F EB-313 (ex 51-1923) and CL-13B Sabres JB-223 (1761) and JB-235 (1745). Other instructional airframes have moved on to various locations; F-104G 2290 to Neuburg, while 2443 and 2512 are now at Jagel and 2292 went to Eggebeck. F-104G 2394 moved to Hopsten and Büchel received 2411 and 2438 whilst 2498 went to Wittmundhaven. TF-104G 2790 is in the museum at Berlin. Alpha Jets 4075 and 4076 had returned to operational service by 1991, while 4061 moved to Hermeskeil.

FICHTELBERG

The Automobil Museum Fichtelberg has expanded its collection of cars and motors with a number of ex NVA hardware. All the aircraft are parked outside. Fichtelberg is located some 25km north east of Bayreuth.

☐ 2044	MiG-23BN	0393214212	preserved, ex Drewitz, ex 696/NVA	5-96
☐ 2060	MiG-23UB	A1038280	preserved, ex Laage, ex 109/NVA	5-96
☐ 2404	MiG-21US	07685147	preserved, ex Rothenburg, ex 221/NVA	5-96
☐ 2551	Su-22UM-3	17532366510	preserved, ex Laage, ex 111/NVA	5-96
☐ 2824	L-39ZO	831116	preserved, ex Rothenburg, ex 166/NVA	5-96
☐ 2839	L-39ZO	831129	preserved, ex Rothenburg, ex 197/NVA	5-96
☐ 7624	Alouette 2	1611	preserved	5-96
☐ 9360	Mi-8PS	10599	preserved, ex Cottbus, ex 739/NVA	5-96
☐ D-ESUH	Zlin 37A	18-26	preserved, ex DDR-SUH, ex DM-SUH	4-96
☐ D-HOAT	Ka-26	733404	preserved, ex DDR-SPT, ex DM-SPT	4-96

FINOW

Directly after the departure of the Soviet MiG-29s from this base in 1993, the Luftfahrthistorische Sammlung Finow was opened. The Soviet MiG-23S received its 08 red code when repianted at the NVA werft at Dresden (before going to Bad Düben). Its last original VVS code was in blue, the number unknown. MiG-23MF 574 was borrowed from the Luftwaffe museum and is now in Gatow. The museum hopes to receive An-22 RA-09315 (00340302) in 1998, while the collection will also be expanded with that from Grossrohrsdorf (which see). Noted in 1998 was L-39ZO 2818, a repainted 2837 or a new aircraft (ex Seiferyshofen) The museum is in the western shelter area and is open from Thursday to Sunday between 10:00 and 17:00 hrs.

☐ 08 red	MiG-23S	220001013	preserved, ex Bad Düben, ex Soviet AF	5-98
☐ 2055	MiG-23BN	0390215732	preserved, ex 720/NVA, ex Drewitz	5-98
☐ 2057	MiG-23UB	A1038506	preserved, ex Laage, ex 103/NVA	5-98
☐ 2238	MiG-21SPS	94A5509	preserved, ex Rothenburg, ex 897/NVA	3-98
☐ 2287	MiG-21M	960514	preserved, ex Rothenburg, ex 589/NVA	5-98
☐ 2361	MiG-21UM	516915006	preserved, ex Rothenburg, ex 233/NVA	5-98
☐ 2837	L-39ZO	831127	preserved, ex Rothenburg, ex 195/NVA	10-97
☐ 7680	Alouette 2	1765	preserved, in shelter	3-98
☐ 482	VEB-14P	14803035	preserved, ex Bautzen	3-98
☐ 395	Mi-8T	0423	preserved, unmarked, no boom	5-98
☐ 135	MiG-15UTI	2212100	preserved	5-98
☐ 08	MiG-17F	1C-0630	preserved, ex Uetersen, ex Bautzen	3-98
☐ 708	MiG-21F-13	741611	preserved, unmarked, ex Holzdorf	5-98
☐ D-CARE	HFB320	1022	preserved, ex Uetersen	5-98

☐ D-ESLZ	Zlin 37	15-14	preserved, ex DDR-SLZ, ex DM-SLZ	5-98
☐ D-ESVT	Zlin 37A	20-23	stored, ex DDR-SVT, ex DM-SVT	10-97
☐ D-FOAA	PZL 106A	48039	preserved, ex DDR-TAA	5-98
☐ D-FOLO	PZL M18A	1Z019-23	stored, ex DDR-TLO, ex DM-TLO	10-97
☐ D-HOAU	Ka-26	7303405	preserved, ex DDR-SPU, ex DM-SPU	5-98
☐ D-HOXA	Mi-8T	0211	preserved, unmarked, ex DDR-SPA	5-98
☐ DDR-SCH	Tu-134	9350906	preserved, ex DM-SCH	5-98
☐ DDR-WLA	L-200A	170712	preserved, rebuild, parts from L-200D 171214	5-98

FINSTERWALDE
This former VVS airfield had three MiG-27Ds on its dump during 1993. Only the construction numbers of these aircraft are known, 61912520025, 61912525043 and 61912525052. All were scrapped.

FLENSBURG
The Marineschule on the east side of this town had two aircraft preserved in its grounds. Starfighter 2574 (marked as '2681') went to the new museum at Nordholz and Sea Hawk VB-136 (marked as 'MS-001') is now in the museum at Schwenningen. Still at the scrapyard North Schrott near the airfield, along the railroad track, is at least one Starfighter. Their other F-104, 2075 (2088, marked as '2476', ex Leck) was scrapped in late 1996.

☐ 2125	F-104G	6686	stored, ex Eggebeck	7-97

FRANKFURT
Town: Somewhere by a supermarket in Frankfurt should be a MiG-21 which was bought in December 1991 in St Petersburg (Russia) for some 40,000 German Marks. It has not yet been seen.

Airfield – Rhein Main: A large part of the Air Classik collection could be found at the Frankfurt terminal (see the first edition). Due to renovation work at the terminals, the aircraft on display had to move. Some aircraft were auctioned while others disappeared. Of the known departures, the museum at Butzweilerhof received Do27 D-EFHO, An-2 HA-MHL and replicas of a Fokker D.VI, Fokker D.VII, Fokker Dr.1 and Me163B-1a. CASA 2111 BR.2I-14 and G91R 3243 went to Hermeskeil, CASA 352 D-CIAS went to Lahr and L-1049 F-BHML ended up at München, together with DC-3 N65371 (marked as 'N569R'). The replicas of a Ju87 and Me262A went to Schwenningen and the replica Ryan NYP is now at Laatzen.

☐ 5A-DGK	DC-8-21	45300	preserved, nose only, ex TL-AHI, ex N8033U	96
☐ D-HAUD	S-58C	58-388	preserved, on viewing area, ex B-12/Belgian	12-97

At the other side of the airport near the military gate are a C-47 (ex 43-49081 and marked as such) and C-54 (ex 44-09063 and marked as such) which are part of the Berlin Luftbrucken Denkmal, a memorial for the Berlin airlift. USAF VC-140B Jetstar 61-2491 (5027) which used to be here for battle damage training has not been seen since the late 1980s.

☐ N88887	C-54E	27389	preserved, ex EL-AJP, ex N88887, ex ZS-LMH	4-96
☐ N1350M	C-47B	14897/26342	preserved, ex T.3-64/Spanish AF, ex EC-ASF	4-96

A further aircraft is on the airfield. The Viscount is still in use for cabin training and can be found near the freight terminals. Devon G-DEVN was stored near the commuter ramp, but moved to Butzweilerhof.

☐ D-ANAF	Viscount 814	447	instructional, unmarked, for cabin crew	5-96

FRIEDRICHSHAFEN
During the Aero show at Friedrichshafen in 1995 the cockpit section of a Su-22 was noted. The aircraft belongs to an unknown local collector. Also in the area, maybe with the same collector, should be a Mi-8S.

☐ 2543	Su-22M-4	26818	stored, cockpit only, ex Laage, ex 769/NVA	4-95
☐ ..	Mi-8S	..	stored, ex Dresden	—

FRIEDRICHSTHAL
A large number of Mi-8s were brought here during the mid 1990s for scrapping. One of them was saved and put on display at the scrappers yard. Friedrichsthal is located some 15km north of Schwedt an der Oder.

❏ 9318	Mi-8T	10547	preserved, ex Briest, ex 932/NVA	10-97

Up to 30 Mi-8s and some ex NVA fighters were reported to have been scrapped at this yard, however only twelve Mi-8s are defenitely known to have ended their lives here.

9306	Mi-8T	10518	ex Briest, ex 634/NVA	scrapped
9312	Mi-8T	10541	ex Briest, ex 926/NVA	scrapped
9315	Mi-8T	10544	ex Briest, ex 928/NVA	scrapped
9330	Mi-8T	10510	ex Briest, ex 389/NVA	scrapped
9333	Mi-8T	10517	ex Cottbus, ex 633/NVA	scrapped
9335	Mi-8T	10531	ex Briest, ex 903/NVA	scrapped
9365	Mi-8TB	10557	ex Cottbus, ex 129/NVA	scrapped
9372	Mi-8TB	10587	ex Cottbus, ex 933/NVA	scrapped
9373	Mi-8TB	10589	ex Cottbus, ex 935/NVA	scrapped
9374	Mi-8TB	10590	ex Cottbus, ex 936/NVA	scrapped
9385	Mi-8TB	10580	ex Basepohl, ex 763/NVA	scrapped
9387	Mi-8TB	10582	ex Basepohl, ex 768/NVA	scrapped

FRITZLAR
This German HEER base has only one *Wrecks & Relics* aircraft, an Alouette which is stored in a hangars.

❏ 7536	Alouette 2	1337	stored	7-94

FÜRSTENFELDBRUCK
The last airfield left with operational Alpha Jets was Fürstenfeldbruck. In the summer of 1997 these stopped flying from here and the base was put on a care and maintenance status. The base had a relatively high number of *Wrecks & Relics* aircraft, including a Lehrwerkstatt (Apprentice School) which was housed in Hangar 6. Over the years a number of aircraft have left the base. F-104F 2918 (5058) was scrapped at München, together with G91R/3 3037 (335). G91R 3308 went to Bremgarten, FWP149D 9103 moved to Uetersen and P149D 9213 went to Koblenz. G91R 9945 has gone to Bückeburg and 9947 went to Oberpfaffenhofen. Instructional Alpha Jet 4026 is now at Gatow, as is T-33A 9444. The F-104G Starfighter cockpit of 2507 (ex Erding) moved on to Neubiberg. Decoy G91R/3 9902 moved on to Ankum, while G91T/3 3450 (610, I/n 1980) left for Ota in the early 1980s.

❏ 2262	F-104G	7143	preserved, marked as '2236'	10-96
❏ 3198	G91R/3	466	preserved	10-96
❏ 3252	G91R/3	521	preserved, on pole	10-96
❏ 3402	G91T/3	0002	preserved, in blue/white Bavaria c/s	8-97
❏ 5619	Do27A-1	287	stored, unmarked, ex instructional	6-95
❏ 9094	FWP149D	114	preserved, D-EOGE ntu	10-96
❏ 9205	P149D	300	stored, with aero club, ex instructional	6-95
❏ 9422	T-33A	7981	preserved, marked as 'AB-773', ex 52-9930	10-96
❏ DD-344	F-84F	..	preserved, as 'DD-244/51-1665', ex 52-6737	10-96
❏ EB-343	RF-84F	..	preserved, at gate, as 'BD-119', ex 51-17041	10-96

From the mid-1990s a large number of Alpha Jets were put in storage at Fürstenfeldbruck. The final four WDT-61 aircraft are expected also to be stored here.

❏ 4001	Alpha Jet	0001	expected, still and Manching	—
❏ 4002	Alpha Jet	0002	expected	—
❏ 4003	Alpha Jet	0003	stored, shelter	12-97
❏ 4005	Alpha Jet	0005	stored, shelter	12-97
❏ 4007	Alpha Jet	0007	stored, shelter	12-97

GERMANY - 189

☐ 4009	Alpha Jet	0009	stored, shelter	12-97
☐ 4011	Alpha Jet	0011	stored, shelter	12-97
☐ 4012	Alpha Jet	0012	stored, shelter	12-97
☐ 4013	Alpha Jet	0013	stored, hangar 4	12-97
☐ 4016	Alpha Jet	0016	stored, shelter	12-97
☐ 4018	Alpha Jet	0018	stored, shelter	12-97
☐ 4020	Alpha Jet	0020	stored, hangar 4	12-97
☐ 4022	Alpha Jet	0022	stored, shelter	12-97
☐ 4023	Alpha Jet	0023	stored, hangar 4	12-97
☐ 4024	Alpha Jet	0024	stored, hangar 4	12-97
☐ 4027	Alpha Jet	0027	stored, shelter	12-97
☐ 4029	Alpha Jet	0029	stored, hangar 4	12-97
☐ 4031	Alpha Jet	0031	stored, hangar 4	12-97
☐ 4034	Alpha Jet	0034	stored, shelter	12-97
☐ 4035	Alpha Jet	0035	stored, hangar 4	12-97
☐ 4036	Alpha Jet	0036	stored, hangar 4	12-97
☐ 4038	Alpha Jet	0038	stored, hangar 4	12-97
☐ 4040	Alpha Jet	0040	stored, shelter	12-97
☐ 4042	Alpha Jet	0042	stored, hangar 4	12-97
☐ 4043	Alpha Jet	0043	stored, hangar 4	12-97
☐ 4044	Alpha Jet	0044	stored, shelter	12-97
☐ 4046	Alpha Jet	0046	stored, hangar 4	12-97
☐ 4049	Alpha Jet	0049	stored, hangar 4	12-97
☐ 4050	Alpha Jet	0050	stored, hangar 4	12-97
☐ 4051	Alpha Jet	0051	stored, hangar 4	12-97
☐ 4052	Alpha Jet	0052	stored, hangar 4	12-97
☐ 4056	Alpha Jet	0056	expected	—
☐ 4057	Alpha Jet	0057	stored, shelter	12-97
☐ 4058	Alpha Jet	0058	stored, shelter	12-97
☐ 4063	Alpha Jet	0063	stored, shelter	12-97
☐ 4065	Alpha Jet	0065	stored, shelter	12-97
☐ 4068	Alpha Jet	0068	stored, shelter	12-97
☐ 4069	Alpha Jet	0069	stored, hangar 4	12-97
☐ 4070	Alpha Jet	0070	stored, shelter	12-97
☐ 4072	Alpha Jet	0072	stored, shelter	12-97
☐ 4074	Alpha Jet	0074	stored, hangar 4	12-97
☐ 4076	Alpha Jet	0076	stored, shelter	12-97
☐ 4077	Alpha Jet	0077	stored, shelter	12-97
☐ 4079	Alpha Jet	0079	stored, hangar 4	12-97
☐ 4081	Alpha Jet	0081	stored, shelter	12-97
☐ 4082	Alpha Jet	0082	stored, hangar 4	12-97
☐ 4085	Alpha Jet	0085	stored, shelter	12-97
☐ 4088	Alpha Jet	0088	stored, hangar 4	12-97
☐ 4089	Alpha Jet	0089	stored, shelter	12-97
☐ 4090	Alpha Jet	0090	stored, hangar 4	12-97
☐ 4093	Alpha Jet	0093	stored, shelter	12-97
☐ 4094	Alpha Jet	0094	stored, shelter	12-97
☐ 4095	Alpha Jet	0095	stored, hangar 4	12-97
☐ 4098	Alpha Jet	0098	stored, hangar 4	12-97
☐ 4102	Alpha Jet	0102	stored, shelter	12-97
☐ 4104	Alpha Jet	0104	stored, hangar 4	12-97
☐ 4105	Alpha Jet	0105	stored, hangar 4	12-97
☐ 4106	Alpha Jet	0106	stored, hangar 4	12-97
☐ 4107	Alpha Jet	0107	stored, hangar 4	12-97
☐ 4109	Alpha Jet	0109	stored, shelter	12-97

GERMANY - 190

❏ 4111	Alpha Jet	0111	stored, hangar 4		12-97
❏ 4112	Alpha Jet	0112	stored, hangar 4		12-97
❏ 4114	Alpha Jet	0114	stored, shelter		12-97
❏ 4115	Alpha Jet	0115	stored, hangar 4		12-97
❏ 4116	Alpha Jet	0116	stored, shelter		12-97
❏ 4119	Alpha Jet	0119	stored, hangar 4		12-97
❏ 4120	Alpha Jet	0120	stored, shelter		12-97
❏ 4121	Alpha Jet	0121	stored, hangar 4		12-97
❏ 4124	Alpha Jet	0124	stored, shelter		12-97
❏ 4125	Alpha Jet	0125	stored, shelter		12-97
❏ 4126	Alpha Jet	0126	stored, shelter		12-97
❏ 4128	Alpha Jet	0128	stored, hangar 4		12-97
❏ 4129	Alpha Jet	0129	stored, shelter		12-97
❏ 4132	Alpha Jet	0132	stored, shelter		12-97
❏ 4134	Alpha Jet	0134	stored, shelter		12-97
❏ 4135	Alpha Jet	0135	stored, shelter		12-97
❏ 4136	Alpha Jet	0136	stored, shelter		12-97
❏ 4137	Alpha Jet	0137	stored, shelter		12-97
❏ 4138	Alpha Jet	0138	stored, shelter		12-97
❏ 4139	Alpha Jet	0139	expected, still at Manching		—
❏ 4140	Alpha Jet	0140	stored, shelter		12-97
❏ 4141	Alpha Jet	0141	stored, hangar 4		12-97
❏ 4142	Alpha Jet	0142	stored, shelter		12-97
❏ 4143	Alpha Jet	0143	stored, shelter		12-97
❏ 4145	Alpha Jet	0145	stored, shelter		12-97
❏ 4147	Alpha Jet	0147	stored, shelter		12-97
❏ 4148	Alpha Jet	0148	stored, hangar 4		12-97
❏ 4149	Alpha Jet	0149	stored, shelter		12-97
❏ 4151	Alpha Jet	0151	stored, shelter		12-97
❏ 4153	Alpha Jet	0153	stored, shelter		12-97
❏ 4154	Alpha Jet	0154	stored, hangar 4		12-97
❏ 4155	Alpha Jet	0155	stored, shelter		12-97
❏ 4156	Alpha Jet	0156	stored, hangar 1		12-97
❏ 4157	Alpha Jet	0157	stored, hangar 1		12-97
❏ 4159	Alpha Jet	0159	stored, shelter		12-97
❏ 4161	Alpha Jet	0161	stored, shelter		12-97
❏ 4162	Alpha Jet	0162	stored, shelter		12-97
❏ 4163	Alpha Jet	0163	stored, shelter		12-97
❏ 4164	Alpha Jet	0164	stored, shelter		12-97
❏ 4165	Alpha Jet	0165	stored, hangar 4		12-97
❏ 4166	Alpha Jet	0166	stored, shelter		12-97
❏ 4167	Alpha Jet	0167	stored, shelter		12-97
❏ 4168	Alpha Jet	0168	stored, shelter		12-97
❏ 4171	Alpha Jet	0171	stored, shelter		12-97
❏ 4172	Alpha Jet	0172	stored, shelter		12-97
❏ 4173	Alpha Jet	0173	stored, hangar 1		12-97
❏ 4174	Alpha Jet	0174	stored, shelter		12-97
❏ 4175	Alpha Jet	0175	stored, shelter		12-97

GANDERKESEE

The blue and unmarked G91R/3 which was preserved here with the local aero club was not 3120 (386) as mentioned in the first edition of this book. The contruction plate was misread and the G91R/3 is 3074 (336). The aircraft moved on from here to Augsburg and later to Tannheim, Lahr and finally ended up at the new museum at Rothenburg.

GEISELWIND
Preserved on a pole in the Freizeitland park here, some 20km east of Wurzburg, is an unmarked G91R. It has recently changed its Frecce Tricolori colours for those of a giraffe. The aircraft carries no construction plate beneath the cockpit (therefore ex Italian ?) and also has no markings inside the cockpit.

| ❏ .. | G91R | .. | preserved, unmarked | 5-97 |

GERMERSHEIM
Germersheim can be found along Bundesstrasse 9, some 20km north of Karlsruhe. F-84F DD-308 (marked as 'BA-102') moved to Sinsheim by 1987 and was replaced at the gate by a Starfighter. The F-104G is at the Hans Graf von Sponeck kaserne.

| ❏ 2466 | F-104G | 8215 | preserved, at gate | 8-97 |

GILCHING
A unmarked ex Rothenburg MiG-21UM is mounted on the roof of a garage in this town. Gilching is close to the airfield of Oberpfaffenhoffen.

| ❏ 2360 | MiG-21UM | 516921056 | stored, ex Rothenburg, ex 229/NVA | 8-97 |

GOSLAR
CL-13B Sabre JB-107 (1732) is still at the barracks of Luftwaffenausbildungsregiment 5 in the northern part of Goslar (about 75km south east of Hannover) on the Marienburger Strasse.

| ❏ JB-107 | CL-13B Sabre 6 | 1732 | preserved, marked as 'GS-338' | 2-98 |

GROSS DOLLN
This former Soviet base, some 15km south of Templin, was also abandoned during 1994. The Su-7 at the gate should have gone by now.

| ❏ 07 red | Su-7U | .. | preserved, at gate, ex Soviet AF | 3-94 |

GROSSENBRODE
The Fehmarnsund kaserne is in Grossenbrode which is about 60km east of Kiel, on the German Ostsee coast.

| ❏ 9465 | T-33A | 9156 | preserved, at gate, ex 54-1539 | 6-97 |

GROSSENHAIN
One of the few former Soviet bases where an original preserved Soviet aircraft can still be found is at the now private airfield of Grossenhain. Grossenhain is situated some 25km north west of Dresden. Su-7 01 red, which also used to be preserved here, has gone to Beelitz.

| ❏ 47 red | MiG-17F | .. | preserved, ex Soviet AF | 8-96 |

GROSSER WESERBOGEN
Preserved at this airfield south of Porta Westfalica (near Bad Oeynhausen) is a former Rothenburg Albatros.

| ❏ 2816 | L-39ZO | 731020 | preserved, ex Rothenburg, ex 157/NVA | 5-98 |

GROSSMACHNOW
The remains of Il-62 DDR-SEB (00704) were noted at Grossmanchnow (south of Berlin). The ex Rangsdorf Il-62 was burned out here on 2nd August 1990. In its place nearby came a 'new' Il-62.

| ❏ DDR-SEC | Il-62 | 10903 | preserved, ex Schönefeld, ex DM-SEC | 3-98 |

GROSSROHRSDORF
The private collector Helmut Hubner had, as well as a large collection of aviation memorabilia, three aircraft in his back garden. These could be found along the main road through Grossrohrsdorf, which is located west of Dresden. In 1997 Mister Hubner died and his collection may move to Finow.

❑ 340	L-29		591526	preserved	10-92
❑ 836	MiG-21SPS		94A4706	preserved, forward fuselage only	10-92
❑ DM-WKX	Zlin 526A		1041	preserved	10-92

GÜTERSLOH
With the departure of the RAF Harriers from here, all of the *Wrecks & Relics* aircraft moved out. Only two are current with the new residents, the British Army. Harrier GR.1 XV278 went to Uetersen, while GR.3 XW917 moved to Laarbruch. Uetersen also received Hunter F.6A XG152. Scrapped here were Canberra B(I).8 XM244, Harrier GR.3 XZ989, and Whirlwind HAR.10s XP347 (WA363) and XP358 (WA374).

❑ XT548	D	Sioux AH.1	WA437	preserved, ex Hildesheim, ex Middle Wallop	6-97
❑ XT764		Wessex HU.5	WA486	stored, ex Detmold	6-97

GUT KUMMERSDORF
This small village is located just south of the former Soviet transport base of Sperenberg, 50km south of Berlin.

❑ 17 white	Mi-2	511415020	preserved, ex Soviet	3-94

HAHN
Germany seems to be full of former military bases. Not only the Soviets, but also the Americans left a number of bases behind. Hahn is now a civil airfield. Instructional F-105 63-8357 went to Hermeskeil. Gone are: Mystère 4A 40, F-4C 64-0879 (1276), F-16A 78-0016 (61-22), F-101F 58-0318 and T-33A 54-1576 (9265).

HALLE
Preserved Il-14P DM-SAF has moved to Pulsforde in late 1991.

HAMBURG
Town: In the Osdorf area of the city is the Reichspresident Ebert kaserne. All their aircraft have left. Sabre BB-250 went to Uetersen, while Sabre 0107 (1668) and F-84F DC-233 (ex 53-6977) are unaccounted for. F-104G 2643 at the Lettow Vorbeck kaserne in the eastern part of Hamburg left for Hamburg Heide.

Another barracks (the Heide/Wulf Isebrand kaserne) is in the northern part of Hamburg along road 203 towards Rendsburg. Non of the airframes are visible from outsite, but a friendly duty officer in 1996 arranged an escort into the barracks. According to the Luftwaffe museum, their T-33 9402 should also be here.

❑ 2643		F-104G	7303	preserved	9-96
❑ 9402		T-33A	7374	preserved, marked as 'ND-204', ex Uetersen	97
❑ BR-362		G91R/4	0124	preserved, marked as '3541', ex Husum	9-96
❑ JA-111		CL-13B Sabre 6	..	preserved	9-96

Airfield - Finkenwerder: One Noratlas was noted in February 1998. It should be ex French Air Force N2501F, which may now be in German colours. German N2501D 5348 (178) was scrapped around 1986.

❑ 157	62-KS	N2501F	157	preserved, ex French AF, see note	4-89
❑ F-SDSG	3	Boeing 377SG	003	preserved, Super Guppy	2-98
❑ ..		HFB320	..	preserved	2-98

Airfield – Fuhlsbüttel: At the international airport, Lufthansa uses a Boeing 707 for cabin crew training. Since early 1997 two German Air Force 707s are stored here.

GERMANY - 193

☐ 1001	B707-307C	19997	stored	2-97
☐ 1004	B707-307C	20000	stored	2-97
☐ D-ABOD	B707-430	17720	instructional, wfu 12-75	1-97

HAMMELBURG
On the northern edge of a military exercise area, just south of Hammelburg (some 25km north west of Schweinfurt), the small civil airfield has a Thunderstreak in false marking.

☐ DE-107	F-84F	..	preserved, marked as 'DE-175', ex 51-1724	5-92

HANAU
In a small museum near the main gate of this US Army airfield (on the north side of town) a Royal Netherlands Air Force Piper Super Cub was painted in American colours. The museum was not to be found in 1997.

☐ R-160	L-21B	18-3850	preserved, marked as '64415', PH-KNM ntu	4-92
☐ 79-23200	AH-1F	22245	stored	3-97
☐ 70-15622	OH-58C	41173	preserved	12-97

HARBKE
Just outside Harbke, on the road to Helmstedt (at the old east-west border), an Il-18 is in use as a restaurant. Displayed outside is an unmarked MiG-21and some aero engines.

☐ 2235	MiG-21SPS	94A5206	preserved, ex Rothenburg, ex 878/NVA	4-98
☐ DDR-STD	Il-18D	180002302	preserved, ex Schönefeld	4-98

HERMESKEIL
The famous museum of the Junior family keeps on expanding and is one of Germany's main aviation museums. Since the first edition the collection has more than doubled and consist now about 100 aircraft and helicopters. Aircraft having left over the years include MiG-21MF 2343 (ex Rothenburg), G91R 3243 (ex Frankfurt), Sabre JA-339, F-105F 63-8357 (ex Hahn), all to Savigny lès Beaune in exchange for the Mirages and Super Mystère. Bü181B D-EBAM (76) returned to its private owners. The museum at Savigny lès Beaune delivered composite Mirage 5BA BA-35 (fuselage) with a tail from a Mirage 5F, wings of a Mirage 3E and the nose of a Mirage 3C. The museum is open between 1st April and 1st November, from 09:00 to 18:00 hours.

Inside:

☐ 031	SM-1	401031	preserved, ex Hungarian AF	5-98
☐ 301	MiG-15UTI	3501	preserved, ex Polish AF	5-98
☐ C-541	C3605	321	preserved, ex Swiss AF	5-98
☐ J-1797	Vampire FB.54	967	preserved, ex Swiss AF	5-98
☐ J-4098	Hunter F.58	..	preserved, ex Swiss AF	5-98
☐ XP352	Whirlwind HAR.10	WA368	preserved, with boom of XD186, ex Abingdon	5-98
☐ 56-4014	F-100F	243-290	preserved, marked as '56-3944', ex FAF	5-98
☐ 2046	MiG-23BN	0390324214	preserved, ex Drewitz, ex 698NVA	5-98
☐ 2379	MiG-21UM	03695163	preserved, cockpit, ex Rothenburg, ex 205/NVA	—
☐ 2491	F-104G	8241	preserved, bare metal, crashed 4-4-78	5-98
☐ 2830	L-39ZO	232303	preserved, ex Rothenburg, ex 180/NVA	5-98
☐ 3086	G91R/3	350	preserved, nose only	5-98
☐ 4061	Alpha Jet	0061	preserved, ex Fassberg	5-98
☐ 7813	Sycamore Mk.52	13466	preserved, incomplete cockpit only	5-98
☐ 7833	Sycamore Mk.52	13493	preserved, marked as 'D-HFUM'	5-98
☐ 8311	H-21C	WG11	preserved, cockpit only, ex Niedermendig	5-98
☐ 9439	T-33A	8901	preserved, ex Erding, ex 53-5562	10-97
☐ 9916	OV-10B	338-1	preserved, ex D-9545	5-98

GERMANY - 194

	Reg	Code	Type	Serial	Notes	Date
☐	D-CIAD		CASA 352L	128	preserved, ex T.2B-127/Spanish AF	5-98
☐	D-EDEW		J-3C-65	10506	preserved, ex D-EDET, ex 43-29215	5-98
☐	D-EFSV		Do27A-1	339	preserved, ex 5653	8-97
☐	D-EHCI		Blériot XI	01	preserved, replica, ex Schwenningen	5-98
☐	D-ENHO		N1002	91	preserved, marked as 'KG-EM', ex F-BDUP	5-98
☐	D-FDEM		AT-6F	121-42500	preserved, ex D-IDEM	5-98
☐	D-HOBC		Brantly B2	093	preserved	5-98
☐	D-HZPS		Ka-26	7404609	preserved, ex DDR-VPK	5-98
☐	D-HZPQ		Mi-2	562945063	preserved, ex Schönefeld, ex 504/NVA	5-98
☐	D-YLAS		Landmann La11	9	preserved	5-98
☐	D-7160		Grunau Baby IIB	030195	preserved, glider	5-98
☐	D-8045		Hütter H17B	..	preserved, glider	5-98
☐	D-8518		Bergfalke II	02	preserved, glider	5-98
☐	D-9511		S-64	64-003	preserved, cockpit only	5-98
☐	G-55-2		Vi-22	UMA-01	preserved, ex Koblenz, ex G-BKLZ	5-98
☐	HA-ANA		An-2P	1G162-10	preserved, ex 10/Hung AF	5-98
☐	N56786		PT-18 Kaydet	75-0521	preserved, ex 40-1964	5-98
☐	OO-ZOR		Ka4	211	preserved, glider, ex D-5499	5-98
☐	..		Fokker Dr.I	..	preserved, replica, marked as '157/17'	5-98

Outside:

	Reg	Code	Type	Serial	Notes	Date
☐	BR.21-14	'G1-FL'	CASA 2111	535	preserved, ex Spanish AF	5-98
☐	BA-35	'13-PL'	Mirage 5BA	35	preserved, ex Savigny lés Beaune, ex Belg AF	5-98
☐	FX-60		F-104G	9103	preserved, ex Belgian AF	5-98
☐	2139		Mi-4	02139	preserved, bare metal, ex Czech AF	5-98
☐	310		Mirage 3R	310	preserved, as '304/33-TN', ex Savigny, ex FAF	5-98
☐	173	12-YP	Super Mystère B2	173	preserved, ex Savigny lés Beaune, ex FAF	5-98
☐	61125		F-102A	..	stored, ex Ramstein, ex Greek AF	5-98
☐	413	C	Lim-5	1C-0413	preserved, ex Tannheim, ex Polish AF	10-97
☐	1217		MiG-21F-13	741217	preserved, bare metal, ex Polish AF	5-98
☐	3076		VEB-14P	14803076	preserved, ex Polish AF	5-98
☐	09		Su-7BM	5309	preserved, ex Oleśnica, ex Polish AF	5-98
☐	J-1635		Venom FB.50	845	stored, ex Swiss AF	4-97
☐	XM264		Canberra B(I).8	EEP71624	preserved, ex Laarbruch, ex RAF	5-98
☐	XL450	R-042	Gannet AS.4	F9433	stored, ex Mönchengladbach	5-98
☐	XW927		Harrier T.4	212015	stored, ex Brüggen, ex RAF	10-97
☐	XF418		Hunter F.6A	..	preserved, ex Wildenrath, ex RAF	5-98
☐	XN782	H	Lightning F.2A	95135	preserved, ex RAF	5-98
☐	XD186		Whirlwind HAR.10	WA29	stored, ex Chivenor, ex RAF	10-97
☐	XP339		Whirlwind HAR.10	WA335	preserved, ex Manching, ex RAF	5-98
☐	63-7421	SA	F-4C	358	preserved, ex Bitburg, ex USAF	5-98
☐	63-7583		F-4C	635	preserved, ex Zweibrücken, ex USAF	5-98
☐	68-0587	SW	RF-4C	3566	preserved, ex Zweibrücken, ex USAF	5-98
☐	62-4417		F-105F	..	preserved, ex Soesterberg, ex USAF	5-98
☐	2001		MiG-23MF	0390213095	preserved, ex Laage, ex 568/NVA	5-98
☐	2006		MiG-23MF	0390213096	preserved, ex Laage, ex 586/NVA	5-98
☐	2019		MiG-23ML	0390324617	preserved, ex Laage, ex 343/NVA	5-98
☐	2043		F-104G	2050	preserved, ex Schwenningen, ex Unsere Luftw.	10-97
☐	2236		MiG-21SPS	94A5209	preserved, ex Rothenburg, ex 889/NVA	5-98
☐	2344		MiG-21MF	96002003	preserved, ex Rothenburg, ex 775/NVA	5-98
☐	2408		MiG-21US	02685139	preserved, ex Rothenburg, ex 238/NVA	5-98
☐	2424		MiG-21bis	N75058015	preserved, ex Drewitz, ex 853/NVA	5-98
☐	2506		Su-22M-4	25713	stored, ex Laage, ex 370/NVA	5-98
☐	2516		Su-22M-4	26103	preserved, ex Laage, ex 678/NVA	5-98
☐	2661		F-104G	7407	preserved, ex Memmingen	10-97

❏ 2825	L-39ZO	831118	preserved, ex Rothenburg, ex 173/NVA	5-98
❏ 3170	G91R/3	438	preserved, marked as '5-257'	5-98
❏ 5208	An-26T	10706	preserved, ex 368/NVA, ex DDR-SBB	5-98
❏ 5421	Pembroke C.54	1013	preserved	5-98
❏ 5424	Pembroke C.54	1016	preserved, unmarked, in RAF c/s	5-98
❏ 8321	H-21C	WG21	preserved, unmarked, at entrance	5-98
❏ 9303	CM170	080	stored	10-97
❏ 9395	Mi-9	340002	stored, ex Cottbus, ex 402/NVA	5-98
❏ 9420	Mi-8T	10525	preserved, ex Briest, ex 909/NVA	10-97
❏ 9502	Mi-14PL	B4002	preserved, ex Parow, ex 618/NVA	5-98
❏ 9517	T-33A	1730	preserved, marked as '393', ex 58-0681	5-98
❏ 9650	Mi-24P	340340	preserved, ex Basepohl, ex 480/NVA	5-98
❏ 555	Mi-2	543620074	preserved, ex Schönefeld	5-98
❏ BD-248	G91R/4	0122	stored, marked as '7500', ex Leipheim	8-97
❏ BF-105	F-84F	..	preserved, ex 52-6778	5-98
❏ EA-341	RF-84F	..	preserved, marked as 'EA-241', ex 52-7377.	5-98
❏ JC-101	CL-13B Sabre 6	1696	preserved, ex Rottenburg	5-98
❏ D-ACUT	N2501D	065	preserved, ex 5256	5-98
❏ D-ALIN	L-1049G	4604	preserved	5-98
❏ D-ANAM	Viscount 814D	368	preserved, marked as '814'	5-98
❏ DDR-STH	Il-18V	184007305	preserved, ex Augsburg, ex DM-STH	5-98
❏ DDR-SCK	Tu-134A	1351304	stored, ex Augsburg, ex DM-STK	5-98
❏ G-ARVF	VC10-1101	808	preserved	5-98
❏ G-BDIW	Comet 4	6470	preserved, ex XR398/RAF	5-98
❏ G-NAVY	Sea Heron C.20	04406	stored, ex Staverton, ex XJ348, ex G-AMXX	5-98
❏ N62443	C-47A	19460	preserved, ex 111/RJordAF and marked as such	5-98
❏ RA-21133	Mi-6A	5309	preserved	5-98
❏ ..	Concorde	..	preserved, replica, marked as 'G-SST/F-WTSA'	5-98

HILDESHEIM
The British Army's 1 Regiment moved during the summer of 1993 from here to Gütersloh, taking their Sioux XT548 with them. The fate of the two instructional airframes, Scout AH.1 XP852 (F9478) and Wasp HAS.1 XT438/465 (F9608), is unknown.

HOHN
Beside the based Transalls, only two Skyservants can be found on the airfield. Formerly stored G91T 9940 moved on to Schwenningen.

❏ 5834	Do28D-2	4109	preserved	5-98
❏ 5923	Do28D-2	4198	stored, fuselage only	5-98

Just to the north of the airfield is the Hugo Junkers kaserne. The Noratlas is preserved at the gate, while the French-built Ju52 is preserved inside the barracks.

❏ 6320	AAC.1	053	preserved, marked as '1Z-IK', ex Port AF	8-96
❏ 5355	N2501D	185	preserved, marked as '5255'	12-97

HOLZDORF
Holzdorf, some 25km north of Torgau, was the home of Jagtfliegergeschwader 1 (JG-1) of the NVA. After October 1990 the unit's MiG-21MFs were relocated to storage at Drewitz. Since 1997 LTG62 has based some of its UH-1Ds at this airfield. The stored Mi-24 will become the gate guard. The MiG-21M has been reported as 2283, but this aircraft should still be Wolls Petersdorf. Of the decoy MiG-21F-13s, 708 moved on to Finow. What has happened to the remaining decoy MiG-21Fs is unknown. Only one may still be here as a second MiG

GERMANY - 196

was noted in November 1997 at the fire dump.

❏ 9622	Mi-24D	110158	stored, ex Basepohl, ex 408/NVA	11-97
❏ ..	MiG-21M	..	stored, see note	11-97
❏ 622	MiG-21F-13	740902	stored	9-90
❏ 678	MiG-21F-13	..	stored	9-90
❏ 741	MiG-21F-13	741612	stored	9-90
❏ 877	MiG-21PFM	761012	preserved, at gate, marked as '1982'	9-90

HOPSTEN
Starfighter 2239 was dumped at the airfield and the remains went to Baarlo in the Netherlands. Instructional F-104G 2530 went to the nearby General Wever kaserne. The current status of 2606 and the Fiat are unknown.

❏ 75-0629	F-4E	4947	stored	2-98
❏ 2330	F-104G	8005	stored	8-97
❏ 2394	F-104G	8113	preserved, arr 3-3-94, ex Fassberg	5-95
❏ 2606	F-104G	9134	stored, ex Norvenich	90
❏ 3125	G91R/3	392	preserved, at gate	9-86

The airfield's barracks, the General Wever kaserne, are in the nearby town of Rheine. A small museum has been set up here with the following three aircraft on display.

❏ DF-126	F-84F	..	preserved, marked as 'DF-240', ex 52-7614	5-97
❏ 2259	F-104G	7140	preserved, marked as 'DF-101' ('7010')	5-97
❏ 2530	F-104G	8307	preserved, forward fuselage, ex instructional	10-96

HOYERSWERDA
Hoyerswerda's MiG-17 07 became the first former NVA aircraft to arrive in West Germany. It came from what was still at that time East Germany in late August 1990. It moved to Stade.

HUSUM
Another airfield which closed in the 1990s was the northern base of Husum. The local unit, JBG41 with its Alpha Jets, was disbanded and the base officially closed on 31st January 1993. Husum's Fliegerhorst kaserne was some 5km from the base, along road 200. F-104G 2505 went to a private collector in Niederalteich. Of the G91R/3s, 3017 (0073) and 3019 (0075) were scrapped, while 3098 went to Fassberg and 3237 moved to Bremgarten. G91R/4 BR-239 (marked as '3541') went to Uetersen and BR-362 (also marked as '3541') is now at Hamburg. Thunderstreak DE-121 (ex 52-6752) was last noted here in December 1990 and its current whereabouts are unknown.

INGOLSTADT
The base of Manching - Ingolstadt is located on the east side of highway 9. On the western side, in the southern part of Ingolstadt, is Oberstimm. Here, the Max Immelmann kaserne serves the Manching - Ingolstadt base. The Thunderflash is still at the gate.

❏ EA-236	RF-84F	..	preserved, ex EB-336, ex 52-7375	9-97

ISERLOHN
On display with the Truppendienstlichen Fachschule der Luftwaffe at the B Hulsman kaserne could be G91R/3 3139 (407). It has not been seen in a long time. Iserlohn is some 20km south east of Dortmund.

ITZEHOE
UH-1s of the German Army's are based here. Former equipment in the shape of two Alouettes can still be seen.

☐ 7525	Alouette 2	1299	dumped, special c/s		9-97
☐ 7603	Alouette 2	1548	preserved		9-97

JAGEL
The T-33 preserved at one of the smaller gates is marked on one side as an AG52 aircraft (EB-396) and on the other side as AG51 (EB-297). Is this the same aircraft as ED-733? The two stored Starfighters are parked in the middle of this Tornado base. Tornado mock-up, marked '3809', moved on to Eggebeck, while the real Sea Hawk VA-234 went to Nordholz. The airfield's barracks are at Kropp.

☐ 2081	F-104G	2094	preserved, at gate, marked as '2381' (6381)	5-98
☐ 2277	F-104G	7159	preserved, marked as '2477', at barracks, ex Leck	8-96
☐ 2443	F-104G	8186	stored, ex Fassberg	5-98
☐ 2512	F-104G	8275	stored, ex Fassberg	9-97
☐ 9459	T-33A	9140	preserved, at small gate, as 'EB-396/EB-397'	8-96
☐ EA-251	RF-84F	..	preserved, at gate	8-96
☐ EB-250	RF-84F	..	preserved, at gate, ex 52-7355, ex Leck	5-98
☐ ED-733	T-33A	..	stored, on field, see note	6-94

JAHNSDORF
The civil airfield of Jahnsdorf is located some 20km south of Chemnitz. The Delfin is painted completely blue and can be identified by is construction number on the plate in the left wheelbay.

☐ 323	L-29	691499	preserved, unmarked	9-97

JEVER
Besides the Tornados of JBG38, the base also houses LwW62 (Luftwaffen Werft 62/Luftwaffe Works 62) which is responsible for the maintenance of Tornados and F-4Fs. F-104s 2206, 2376 (marked as 'BB-371') and 2392 moved to Erding for restoration.

☐ 2155	F-104G	7024	instructional, in hangar	8-96
☐ 2667	F-104G	7413	decoy, ex instructional	6-97
☐ 2909	F-104F	5059	preserved, at gate, ex 59-5006	6-97
☐ 3258	G91R/3	528	decoy	6-97
☐ 3270	G91R/3	540	stored, ex instructional	6-97
☐ 9938	G91R/3	437	decoy, ex 3169	6-97
☐ JB-114	CL-13B Sabre 6	1730	preserved, on pole, marked as 'BB-103'	8-96

KALKAR
Some 15km east of Kleve, near the Dutch border, in the village of Kalkar is the Von Seydlitz kaserne which is occupied by Luftwaffendivision 3. 4259 was the former postal code for Kalkar.

☐ 2905	F-104F	5054	preserved, ex '4259', ex 59-5001	11-97
☐ 3310	G91R/3	581	preserved	11-97

KALTWASSER
A former Briest Mi-2 surfaced in late 1994 in the small village of Kaltwasser (cold water). This village is located some 12km north of Görlitz near the Polish border, not far away from the airfield of Rothenburg. The chopper is in a children's park in blue/green/white colours.

☐ 9473	Mi-2See	552701122	stored, unmarked, ex Briest, ex 388/NVA	9-97

KAMENZ
In the NVA-era Kamenz was the home base of TAS-45 with An-2s, L-410s and Zlin 43s and the Offiziershoch-

schule Franz Mehring with a number of instructional airframes. After October 1990 the An-2s were placed in stored. Nearly all of them have since entered the civil register: 440 (117411) became D-FKMA, 451 (17308) became D-FKMB, 452 (17309) became F-AZHB, 454 (17612) became D-FKMC, 455 (17710) became D-FUKM (D-FKMF ntu), 456 (17812) became D-FKMD, 457 (17805) became D-FKME, 804 (1G160-01) became D-FBAW, 805 (1G160-02) became D-FKMH (later to YV-558C), 845 (17207) became D-FKMI, 855 (19319) became D-FKMF (later to YV-889C) and 857 (17912) became D-FKMK (later to YV-860C). 812 went to Uetersen and 822 is now preserved at Berlin. The based Z-43s became civil within a few months and the Turbolets were relocated to Marxwalde. None of the instructional airframes have been noted since 1993. A number of them became part of collections, while the others were presumably scrapped. MiG-21U 297 went to Rothenburg, MiG-21M 560 to Dermsdorf and 596 went to Berlin. MiG-21MF 686 is also at Berlin, while 653 has moved to Cottbus. Of the instructional MiG-23MFs, 564 went to Butzweilerhof and 574 to Finow. Stored L-29 366 (591534) disappeared after June 1992, as did MiG-21U-400 260 (660819), MiG-21U-600 294 (664816) and MiG-21F-13s 605 (741009), 635 (741524) and 685 (741005). The MiG-21SPSs were better off; 499 moved to Drewitz, 698 is now at Diepensee, 791 is at the museum at Nordholz and 959 went to Uetersen. Two former Bautzen gate guards were under restoration here. Of these Yak-11 225 went to Uetersen and the SM-1 001 moved to Rothenburg. Instructional Mi-8T 393 (10527, marked as '911') went to Peenemünde. The stored Yak-18 42 has gone, as have the unknown MiG-15bis (736047) on the dump.

❏ D-FKMG   An-2T   12801   stored, unmarked, ex 801/NVA   3-95

KARLSRUHE
At barracks in the north of the city, Luftwaffenunterstutzungsgruppe Sud had F-86K Sabre 55-4881 (one of a batch supplied to the WGAF and not put into service as they had more aircraft than pilots at the time). It has moved on to Uetersen. The oldest F-104G in Germany is now preserved at Berlin, 2002 (2002) moved from here to Erding for restoration. 2100 was exchanged for the 2002.

❏ 2100   F-104G   6621   preserved, marked as 'DB-101', ex Erding   7-95

KASSEL
During the 1980s G91 3147 was preserved near the control tower of the civil airfield at Kassel. Nowadays it can be found at the Speyer museum. At the Eurocopter factory gate a former HEER Alouette 2 is preserved.

❏ 7698   Alouette 2   1843   preserved   97

KAUFBEUREN
Kaufbeuren is the home of the WGAF's TSLw1. The unit uses a selection of dedicated ground instructional airframes and more modern types borrowed from active units in order to teach the skills required to service and maintain modern aircraft. The WGAF is unusual in borrowing active aircraft for instructional use, most other air arms preferring to use only retired airframes. Two instructional Starfighters have left Kaufbeuren, 2136 went to Lechfeld and 2586 is now at Büchel. The fate of the RF-84F BD-701 (ex 51-17021) is unknown, it was last seen here in the late 1980s. Tornados 4373, 4395, 4465, 4490 and 4576 have all returned to operational use.

❏ 2006	F-104G	2006	instructional	8-96
❏ 2398	F-104G	8124	preserved, at gate	5-97
❏ 3562	RF-4E	4144	instructional, allocated for Berlin Gatow	5-97
❏ 3704	F-4F	4346	instructional	5-97
❏ 3714	F-4F	4381	instructional	5-97
❏ 4334	Tornado	4034	instructional	5-97
❏ 4361	Tornado	4061	instructional	5-97
❏ 4413	Tornado	4113	instructional	5-97
❏ 4564	Tornado	4264	instructional	5-97

KERPEN
The Boelcke kaserne, which is JBG31's barracks, is a few miles north of the Norvenich air base.

GERMANY - 199

☐ 2919	F-104F	5073	preserved, ex 'DA-101', ex 59-5020	2-98
☐ BF-108	F-84F	..	preserved, marked as 'DA-127', ex 53-7102	83

KIEL - HOLTENAU
Still preserved here at a secondary gate of the German Navy base of MFG5 is an H-34G.

☐ 5914	Do28D	4189	preserved	5-98
☐ 8059	H-34G	58-1515	preserved, at gate, ex 150743	5-98

KOBLENZ
Along the Rhein in the Mayener Strasse 85 in Koblenz, the Wehrtechnische Studiensammlung has been established. In this museum full of military hardware, mostly from the two world wars, a section with preserved aircraft can also be found. The Do34 is prototype of a battlefield observation system. It looks like a flying barrel with a military serial number. The ex Oberpfaffenhofen gate guard Do-29 YD-101 (marked as 'YA-101') was briefly preserved here. It has moved on to <u>Uetersen</u>. The second Vi-22, G-55-2, has gone to <u>Hermeskeil</u>, while Do27A D-EJUX went to <u>Gatow</u>. The Alpha Jet has construction number 0001 painted on its tail, but the construction plate uncovered its real number, 03. The museum is open daily between 09:30 and 16:30 hours (but closed between 24th December and 1st January).

☐ NF11-14	BV	Meteor NF.11	..	preserved, ex Manching, ex French AF	8-97
☐ 10	10-RD	Mirage 3C	10	preserved, ex French AF	8-97
☐ 199	MF-070	N2501F	199	preserved, ex Manching, ex French AF	8-97
☐ 2048	7	MiG-23BN	0393214218	preserved, ex Manching, ex 702/NVA	8-97
☐ 2420	2	MiG-21bis	N75051402	preserved, ex Manching, ex 846/NVA	8-97
☐ 3206		G91R/3	474	preserved	8-97
☐ 7552		Alouette 2	1402	preserved	8-97
☐ 9213		P149D	309	preserved, ex Fürstenfeldbruck	8-97
☐ 9824		Do34 Kiebitz	P02	preserved	8-97
☐ 9833		Mi-24P	340330	preserved, ex 9640, ex Manching, ex Basepohl	8-97
☐ 9836		F-104G/CCV	8100	preserved, ex 2391	8-97
☐ 9855		Alpha Jet	03	preserved, ex Leipheim	8-97
☐ 9856		Firebird M1	1065108	stored	5-90
☐ 9930		OV-10B	338-15	preserved, ex D-9559	8-97
☐ D-9564		VAK-191B	V2	preserved	8-97
☐ G-55-1		Vi-22	UMA-02	preserved	8-97

KOLBERMOOR
In this town near Rosenheim in southern Germany a former Zruč An-2 is preserved.

☐ OK-RYA	An-2	117010	preserved, ex Zruč, ex Letňany	97

KÖLN - BONN
Of the *Wrecks & Relics* aircraft at the military side of this civil airport not much is known as the area is not really accessible. In the early 1980s three Sabres were seen on the airfield, BB-237 (1111), JC-361 (1805) and an unknown one. The dumped Sabre is one of these, the other two were scrapped. The preserved Starfighter 2911 is currently at <u>Erding</u> for restoration.

☐ 3129	G91R/3	396	preserved, at gate	12-82
☐ ..	Sabre	..	dumped, burned wreck	5-95

LAAGE
Laage now houses the MiG-29s and F-4Fs of JG73. Two former Preschen (where the MiG-29s were based) and two former Pferdsfeld (where the F-4Fs were based) aircraft are now preserved here.

GERMANY - 200

☐ 2011	MiG-23ML	0390324619	preserved, ex 330/NVA and marked as such		10-97
☐ 2502	F-104G	8255	preserved, marked as 'BB-105' ex Pferdsfeld		10-97
☐ 2538	Su-22M-4	31205	stored, ex 734/NVA and marked as such		10-97
☐ 2915	MiG-29	2960526300	stored		9-97
☐ 619	MiG-21F-13	741004	preserved, marked as '335', ex Preschen		10-97
☐ ..	CL-13B Sabre 6	..	preserved, marked as 'JC-102', ex Pferdsfeld		10-97
☐ ..	MiG-19PM	..	preserved, marked as '391', ex Preschen		10-97

During the NVA days, JBG77 and MFG28 with their Su-22s, flew from here. After October 1990 the Su-22s, together with the MiG-23s from Peenemünde, were stored here. Nearly all have found new homes.

2001	MiG-23MF	0390213095	stored, ex 568/NVA, l/n 6-92	to Hermeskeil
2002	MiG-23MF	0390213299	stored, ex 577/NVA, l/n 5-94	to Berlin
2004	MiG-23MF	0390213098	stored, ex 584/NVA, l/n 12-90	to Peenemünde
2005	MiG-23MF	0390213100	stored, ex 585/NVA, l/n 5-94	to Dessau
2006	MiG-23MF	0390213096	stored, ex 586/NVA, l/n 4-94	to Hermeskeil
2007	MiG-23MF	0390213300	stored, ex 591/NVA, l/n 5-94	to Rothenburg
2008	MiG-23MF	0390213351	stored, ex 592/NVA, l/n 5-94	to Seifertshofen
2009	MiG-23MF	0390213352	stored, ex 593/NVA, l/n 5-94	to Rothenburg
2010	MiG-23ML	0390324623	stored, ex 329/NVA, l/n 5-94	to Peenemünde
2012	MiG-23ML	0390324621	stored, ex 331/NVA, l/n 4-93	to Cuatro Vientos
2013	MiG-23ML	0390324624	stored, ex 333/NVA, l/n 8-93	to Berlin
2014	MiG-23ML	0390324627	stored, ex 336/NVA, l/n 5-94	to Bad Ischl
2015	MiG-23ML	0390324630	stored, ex 338/NVA, l/n 3-91	to United States
2016	MiG-23ML	0390324635	stored, ex 339/NVA, l/n 3-91	to United States
2017	MiG-23ML	0390324636	stored, ex 340/NVA, l/n 5-94	to Rothenburg
2018	MiG-23ML	0390324637	stored, ex 341/NVA, l/n 12-90	gone
2019	MiG-23ML	0390324617	stored, ex 343/NVA, l/n 5-94	to Hermeskeil
2020	MiG-23ML	0390324618	stored, ex 345/NVA, l/n 12-90	gone
2021	MiG-23ML	0390324638	stored, ex 349/NVA, l/n 12-90	to Brasschaat
2022	MiG-23ML	0390324639	stored, ex 350/NVA, l/n 12-90	gone
2023	MiG-23ML	0390324640	stored, ex 353/NVA, l/n 6-92	gone
2024	MiG-23ML	0390324250	stored, ex 471/NVA, l/n 5-94	to Tyndall, FL
2025	MiG-23ML	0390324254	stored, ex 475/NVA, l/n 4-93	gone
2026	MiG-23ML	0390324255	stored, ex 488/NVA, l/n 5-94	gone
2027	MiG-23ML	0390324018	stored, ex 550/NVA, l/n 6-92	to Sinsheim
2028	MiG-23ML	0390324019	stored, ex 551/NVA, l/n 3-91	to United States
2029	MiG-23ML	0390324027	stored, ex 554/NVA, l/n 12-90	to Neubrandenburg
2031	MiG-23ML	0390324031	stored, ex 563/NVA, l/n 5-94	gone
2032	MiG-23ML	0390324033	stored, ex 567/NVA, l/n 3-91	to United States
2033	MiG-23ML	0390324038	stored, ex 569/NVA, l/n 5-94	to Ankum
2034	MiG-23ML	0390324040	stored, ex 576/NVA, l/n 6-92	gone
2035	MiG-23ML	0390324050	stored, ex 601/NVA, l/n 5-94	to Dresden
2036	MiG-23ML	0390324051	stored, ex 606/NVA, l/n 3-91	to United States
2037	MiG-23ML	0390324249	stored, ex 610/NVA, l/n 4-93	to Fassberg
2056	MiG-23UB	A1038504	stored, ex 100/NVA, l/n 5-94	to Seifertshofen
2057	MiG-23UB	A1038506	stored, ex 103/NVA, l/n 5-94	to Finow
2058	MiG-23UB	A1038034	stored, ex 106/NVA, l/n 4-94	to Bad Oeynhausen
2059	MiG-23UB	A1038221	stored, ex 107/NVA, l/n 5-94	to Seifertshofen
2060	MiG-23UB	A1038280	stored, ex 109/NVA, l/n 5-94	to Fichtelberg
2061	MiG-23UB	A1037826	stored, ex 102/NVA, l/n 4-94	to Stade
2062	MiG-23UB	A1037901	stored, ex 104/NVA, l/n 5-93	to Cottbus
2063	MiG-23UB	A1037902	stored, ex 105/NVA, l/n 5-94	to Berlin
2501	Su-22M-4	25307	stored, ex 360/NVA, l/n 5-94	to Rothenburg
2502	Su-22M-4	25509	stored, ex 362/NVA, l/n 5-94	to Peenemünde
2503	Su-22M-4	25510	stored, ex 363/NVA, l/n 5-94	gone

GERMANY - 201

2504	Su-22M-4	25511	stored, ex 365/NVA, l/n 8-93	to Cottbus
2505	Su-22M-4	25512	stored, ex 366/NVA, l/n 5-94	to Nordholz
2506	Su-22M-4	25713	stored, ex 370/NVA, l/n 3-95	to Hermeskeil
2507	Su-22M-4	25714	stored, ex 546/NVA, l/n 5-94	gone
2508	Su-22M-4	25715	stored, ex 574/NVA, l/n 5-94	gone
2509	Su-22M-4	25916	stored, ex 600/NVA, l/n 5-94	to Dessau
2510	Su-22M-4	25017	stored, ex 605/NVA, l/n 5-94	gone
2511	Su-22M-4	25018	stored, ex 613/NVA	to Uetersen
2512	Su-22M-4	25019	stored, ex 636/NVA, l/n 5-94	gone
2513	Su-22M-4	25020	stored, ex 641/NVA, l/n 5-94	to Seifertshofen
2514	Su-22M-4	26001	stored, ex 644/NVA, l/n 4-93	to Speyer
2515	Su-22M-4	26102	stored, ex 645/NVA, l/n 5-94	gone
2516	Su-22M-4	26103	stored, ex 678/NVA, l/n 8-93	to Hermeskeil
2517	Su-22M-4	26204	stored, ex 682/NVA, l/n 5-94	to Dermsdorf
2518	Su-22M-4	26205	stored, ex 686/NVA, l/n 5-94	to Cuatro Vientos
2519	Su-22M-4	26206	stored, ex 700/NVA, l/n 5-94	gone
2520	Su-22M-4	26307	stored, ex 704/NVA, l/n 4-93	to Sinsheim
2521	Su-22M-4	26408	stored, ex 711/NVA, l/n 5-94	gone
2522	Su-22M-4	26409	stored, ex 721/NVA, l/n 7-92	gone
2523	Su-22M-4	26510	stored, ex 723/NVA, l/n 5-94	to Bad Ischl
2524	Su-22M-4	30913	stored, ex 378/NVA, l/n 5-94	gone
2525	Su-22M-4	30914	stored, ex 380/NVA, l/n 3-91	to United States
2526	Su-22M-4	30915	stored, ex 537/NVA, l/n 12-90	to 9815
2528	Su-22M-4	30917	stored, ex 598/NVA, l/n 8-93	to Stade
2529	Su-22M-4	30918	stored, ex 629/NVA, l/n 7-91	to 9814
2530	Su-22M-4	30920	stored, ex 674/NVA, l/n 5-94	gone
2531	Su-22M-4	31001	stored, ex 706/NVA, l/n 12-90	to 9817
2532	Su-22M-4	31002	stored, ex 716/NVA, l/n 5-94	gone
2533	Su-22M-4	31203	stored, ex 724/NVA, l/n 3-91	to United States
2534	Su-22M-4	26511	stored, ex 725/NVA, l/n 5-94	gone
2535	Su-22M-4	26512	stored, ex 727/NVA, l/n 8-94	to Rothenburg
2536	Su-22M-4	31204	stored, ex 728/NVA, l/n 7-92	gone
2537	Su-22M-4	26613	stored, ex 730/NVA, l/n 5-94	gone
2539	Su-22M-4	26614	stored, ex 737/NVA, l/n 5-94	gone
2540	Su-22M-4	26715	stored, ex 741/NVA	to Uetersen
2541	Su-22M-4	26716	stored, ex 743/NVA, l/n 5-94	to Seifertshofen
2542	Su-22M-4	26817	stored, ex 757/NVA, l/n 5-94	to Rothenburg
2543	Su-22M-4	26818	stored, ex 769/NVA, l/n 5-94	to Friedrichshafen ?
2544	Su-22M-4	31406	stored, ex 798/NVA, l/n 5-94	to Berlin
2545	Su-22M-4	31407	stored, ex 798/NVA, l/n 12-90	to 9810
2546	Su-22M-4	31508	stored, ex 824/NVA, l/n 3-94	gone
2547	Su-22UM-3	17532369809	stored, ex 113/NVA, l/n 5-94	to Bad Ischl
2548	Su-22UM-3	17532367002	stored, ex 119/NVA, l/n 5-94	to Rechlin
2549	Su-22UM-3	17532367003	stored, ex 127/NVA, l/n 5-94	to Dessau
2550	Su-22UM-3	17532371002	stored, ex 146/NVA, l/n 5-94	to Bad Oeynhausen
2551	Su-22UM-3	17532366510	stored, ex 111/NVA, l/n 5-94	to Fichtelberg
2552	Su-22UM-3	17532367001	stored, ex 112/NVA, l/n 5-94	to Berlin
2553	Su-22UM-3	17532370810	stored, ex 137/NVA	to 9816
2554	Su-22UM-3	17532371001	stored, ex 138/NVA, l/n 7-91	to 9811

LAARBRUCH
The RAF Harriers will be short lived as the aircraft will be relocated to Cottesmore and Wittering in the UK late in 1999. Until then the base will hold a number of *Wrecks & Relics* aircraft. Some of the aircraft have gone. Buccaneer S.1 XN956 was scrapped, along with Lightning F.2 XN732 (95109, marked as '17'), XN788 (95141)

and Lightning F.6 XR758 (95223). Canberra B(I).8 XM264 is now at Hermeskeil, while Wessex HU.5 XT467 (WA289, ex Brüggen) returned to the UK and is now at Odiham. Harrier XZ993/H returned to St Athan, UK. Instructional F-4J(UK)s ZE350/T (1692), ZE352/G (1870) and ZE363/W (3338) went to Pendine in the UK.

❑ XW917		Harrier GR.3	..	preserved, at gate, ex Gütersloh	12-97
❑ XW922		Harrier GR.3	..	dumped, ex instructional, ex Brüggen	2-96
❑ XW924		Harrier GR.3	..	preserved, 3sq area, ex instructional	12-97
❑ XZ999	H	Harrier GR.3	..	instructional	5-94
❑ XE606		Hunter F.6A	41H679964	preserved, 4sq area, marked as 'XJ637/A'	12-97
❑ XV412	P	Phantom FGR.2	2981	instructional, 4sq area, arr 10-7-91	12-97

LAATZEN

Luftfahrt Museum Laatzen (south of Hannover) is unusual as it does not have the almost standard row of ex NVA aircraft on display. Instead they have an Finnish Air Force Fw-44 and a Hungarian Yak-18. The MiG-15 has been painted in Soviet marks. The Bf109 is an original aircraft recovered from a lake. OV-10B 9925 went to Berlin. The museum is at the Ulmerstrasse 2 and open Tuesday to Sunday between 10:00 and 17:00.

❑ SZ-30	Fw44J	..	preserved, ex Finnish AF	3-98
❑ 5418	Yak-18U	5418	preserved, ex Hungarian AF	3-98
❑ 712	MiG-15bis	31530712	preserved, marked as '022', ex Hungarian AF	3-98
❑ 2045	F-104G	2053	preserved, ex Uetersen	3-98
❑ 5868	Do28D-2	4143	preserved, outside	3-98
❑ 7502	Alouette 2	1180	preserved	3-98
❑ 9068	FWP149D	086	preserved	3-98
❑ 688	MiG-21F-13	741006	preserved, outside, ex Preschen	3-98
❑ TC-FNS	HFB320	1026	preserved, ex D-CARY and marked as such	4-97
❑ D-EJOF	HM-504A2	..	preserved, replica	4-97
❑ DM-2188	Meise	0188	preserved, glider	4-97
❑ DM-2559	Lom 57/I Libelle	059	preserved, glider	4-97
❑ G-FXIV	Spitfire FR.XIVc	..	preserved, ex MV370 and marked as such	3-98
❑ HA-MHM	An-2R	1G123-03	preserved	4-97
❑ OE-2129	Grunau Baby IIb	..	preserved	4-97
❑ ..	Bf109G-2	14753	preserved, marked as '-3'	3-98
❑ ..	Bf109G		preserved, replica, fuselage	3-98
❑ ..	Fokker Dr.I	..	preserved, replica, marked as '425/17'	3-98
❑ ..	Fokker E.III	..	preserved, replica	3-98
❑ ..	Horvath III	..	preserved, replica	4-97
❑ ..	Junkers F13A	..	preserved, replica, marked as 'D-1'	4-97
❑ ..	Klemm L-25D	..	preserved, replica	4-97
❑ ..	Nieuport 17	..	preserved, replica	3-98
❑ ..	Ryan NYP	..	preserved, replica, marked as 'NX211', ex Frankf.	4-97
❑ .	Sopwith F.1 Camel	..	preserved, replica, marked as 'B7220'	3-98
❑ ..	SV-4C	..	preserved, marked as '863/GI'	3-98

LAHR

Nothing is known as to what happened to the two preserved aircraft when the Canadians left the base here; Starfighter '104784' and CL-13B Sabre JC-373 (1638, marked as '23444'). They were not here when the base was checked in 1995. The stored F-100D 54-2239 went to Bitburg in the late 1980s. After the departure of the Canadians the base became an industrial estate and civil airfield. Aircraft of the former Tannheim collection were noted here in 1995. The hangars were said to contain some of the collection's flyable aircraft. These left for Sandown in the UK. Two of the stored aircraft moved on, G91R 3074 to Rothenburg and CASA 352L D-CIAS (ex Frankfurt) to München. The stored Tu-134A DDR-SCL did not belong to the Tannheim collection, it arrived much earlier and should now be in Biberach.

❑ 140	MiG-15UTI	..	stored, marked as '44', ex Tannheim	5-98

LANDSBERG

Landsberg houses the Transall and UH-1D equipped LTG61. Two of the preserved aircraft reflect an earlier era when Landsberg was the base for the FFS-A flying school.

☐ 128	62-KY	N2501F	128	preserved, marked ex 'GA-125', ex F-MC/FAF	9-97
☐ 7156		UH-1D	8216	stored, wreck, crashed 5-12-94	1-95
☐ 7837		Sycamore Mk.52	13503	preserved, marked as 'LB-105', ex Ahlhorn	9-97
☐ AA-152		CM170	052	preserved	9-97
☐ AA-666		Harvard 4	CCF4-458	preserved, marked as '68-4623', ex 52-8537	9-97

LANDSHUT

Landshut's Ellersmuhle airfield is a civilian and had a MiG-21U stored. The aircraft was sold by the NVA to the werft at Dresden, who had sold it to Iraq. Due to the Gulf War, the deal fell through and the aircraft was never delivered. MiG-21U 291 (4718) was last noted here in August 1995 and is now at Mojave (USA) as N121TJ.

LANGENBERNSDORF

Langenbernsdorf is a village east of Zwickau. South of here, near Hotel Waldperie, is an Il-14.

☐ DM-SAZ	VEB-14P	14803001	preserved, marked as 'DDR-SAZ', wfu 2-67	5-98

LAUDA

In Lauda, which is about 45km south west of Wurzburg, the gate of the Tauberfranken kaserne (home of Fernmelderegiment 32) is guarded by a Sabre.

☐ JB-371	CL-13B Sabre 6	1611	preserved, marked as '371'	8-97

LAUPHEIM

The Belgian S-58 is painted to represent one of the former aircraft of the based German Army unit HFR25.

☐ 7521	Alouette 2	1287	preserved, ex stored	3-98
☐ 9076	FWP149D	094	stored, in hangar, D-EJAI ntu	7-94
☐ D-HAUC	S-58C	58-350	preserved, marked as '8025', ex B-14/BelgAF	3-98

LECHFELD

JBG32's former HFB320ECMs were stored here from late 1993. All were sold by late 1995 and left for the States: 1621 (1058) became N322AF, 1623 (1060) N321AF, 1624 (1061) N320AF, 1625 (1062) N323AF, 1627 (1064) N324AF and 1628 (1065) N325AF. Preserved F-104G 2047 moved on to the museum of Schwenningen. F-104G 2100 went to Erding.

☐ 2138	F-104G	7006	instructional	9-92
☐ 2186	F-104G	7055	instructional	10-97
☐ 2525	F-104G	8301	instructional	10-93
☐ 2528	F-104G	8305	instructional	9-92
☐ 2620	F-104G	9168	instructional	9-96
☐ 2920	F-104F	5075	instructional, marked as '4711', ex Sollingen	8-96
☐ 5855	Do28D-2	4130	instructional	10-93
☐ 9116	FWP149D	137	instructional, D-EKAD ntu	10-93
☐ 9119	FWP149D	140	instructional	10-92
☐ 9904	G91R/3	343	instructional, ex 3081, arr 28-1-92	10-97
☐ 9944	G91R/3	486	decoy, ex 3218	1-92

Located at the southern side of the base is the Schwabstadlkaserne.

☐ 2136	F-104G	7004	preserved, marked as '2002', ex Kaufbeuren	9-97
☐ 9482	T-33A	1650	preserved, as 'DB-396', ex '9403', ex 57-0681	7-97

| ☐ BF-104 | F-84F | .. | preserved, marked as 'DB-132', ex 52-6764 | 7-97 |
| ☐ DD-367 | F-84F | .. | preserved, as 'DB-232', ex 'DB-032', ex 51-1645 | 7-97 |

Near the northern side of the base is the Ulrich kaserne, home of Fernmelde Lehr und Versuchs Regiment 61. They may still have a Pembroke, which is not visible from the road.

| ☐ 5426 | Pembroke C.54 | 1018 | preserved | 88 |

LECK

The RF-4Es of AG52 have been withdrawn from service with the Luftwaffe and the northern base of Leck is now closed. All *Wrecks & Relics* aircraft have gone. To Jagel went Starfighter 2277, T-33A 9459 and Thunderflash EB-250. Jever received Starfighter 2206 and G91R 3099 (marked as '3599') went to Schwenningen. Starfighter 2075 (marked as '2476') moved to Flensburg. The fate of G91R/3 3025 (0082) is unknown, it is definitely no longer at Leck.

LEIPHEIM

The base of Leipheim was closed during the mid 1990s. Before that there were a number of G91Rs at the base. Of these 3103 went to Uetersen and BD-248 (marked as '7500') to Hermeskeil. Last noted in 1990 were 3037 (0097), 3102 (368), 3154 (422), 3188 (456) and 3261 (531). All these have gone. Only 3138 may still linger. The preserved TF-104G 2726 has moved on to Memmingen. The instructional USAF F-4C 63-7467 went to the museum at Seifertshofen, Alpha Jet 9855 went to Koblenz.

| ☐ 3138 | G91R/3 | 406 | preserved, at gate | 9-97 |
| ☐ .. | F-84F | .. | preserved | 8-93 |

By 1992 the Germans had withdrawn their Do28s. More than 60 of these aircraft were stored here. Nearly all were auctioned and all left by late 1994.

5808	Do28D-2	4083	stored, l/n 12-93	to E3-IAAB
5809	Do28D-2	4084	stored, l/n 6-94	to D-IDNB, to HA-ACP
5814	Do28D-2	4089	stored, l/n 6-94	to D-IECE, to Panama
5815	Do28D-2	4090	stored, l/n 5-93	to D-IEDI, to Nassau
5818	Do28D-2	4093	stored, l/n 8-94	gone
5820	Do28D-2	4095	stored, l/n 7-94	to D-IBDA, to Venezuela
5823	Do28D-2	4098	stored, l/n 6-94	to D-IECB, to CC-PHI
5828	Do28D-2	4103	stored, l/n 6-94	to D-IECD, to USA
5829	Do28D-2	4104	stored, l/n 6-94	to D-IECF, to USA
5830	Do28D-2	4105	stored, l/n 6-94	to D-IDNH, to HK-4053X
5832	Do28D-2	4109	stored, l/n 9-93	to D-IMOC for Lufthansa GIA
5836	Do28D-2	4111	stored, l/n 12-93	to D-IEDD, to YV-764CP
5837	Do28D-2	4112	stored, l/n 6-94	to D-IDNG
5838	Do28D-2	4113	stored, l/n 12-93	to D-IDAH
5839	Do28D-2	4114	stored, l/n 6-94	to D-IEDE, to N952
5846	Do28D-2	4121	stored, l/n 12-93	to D-IDRS, to HA-ACU
5849	Do28D-2	4124	stored, l/n 12-93	to D-IDRN, to Venezuela
5850	Do28D-2	4125	stored, l/n 8-93	to D-IDRC, to HA-ACL
5852	Do28D-2	4127	stored, l/n 12-93	to D-IDRH, to RP-C1179
5853	Do28D-2	4128	stored, l/n 6-94	to D-IFDC, to CC-PWR
5854	Do28D-2	4129	stored, l/n 6-94	to D-IDRR, to HP-1269
5859	Do28D-2	4134	stored, l/n 8-93	to D-IDRD, to HA-ACM
5860	Do28D-2	4135	stored, l/n 6-94	to D-IDRL, to Venezuela
5861	Do28D-2	4136	stored, l/n 6-94	to D-IEDF, to Nassau
5862	Do28D-2	4137	stored, l/n 6-94	to D-IBDB, to RP-C678
5865	Do28D-2	4140	stored, l/n 8-93	gone
5866	Do28D-2	4141	stored, l/n 6-94	to D-IDOI, to N876

5867	Do28D-2	4142	stored, l/n 6-94	to D-IBDC
5869	Do28D-2	4144	stored, l/n 6-94	to D-IEDG, to Venezuela
5871	Do28D-2	4146	stored, l/n 6-94	to D-IECC, to 5Y-MIG
5872	Do28D-2	4147	stored, l/n 5-93	to D-IDRE, to N8132K
5873	Do28D-2	4148	stored, l/n 6-94	to D-IDNF, to HK-3991
5874	Do28D-2	4149	stored, l/n 9-93	to Lufthansa as GIA
5878	Do28D-2	4153	stored, l/n 8-93	to D-IDRF
5879	Do28D-2	4154	stored, l/n 12-93	to D-IDBC, to Colombia
5880	Do28D-2	4155	stored, l/n 6-94	to D-IFDA, to RP-C682
5882	Do28D-2	4157	stored, l/n 6-94	to D-IDOC, to HA-ACR
5884	Do28D-2	4159	stored, l/n 12-93	to D-IDRT, to Venezuela
5885	Do28D-2	4160	stored, l/n 6-94	to D-IFDB, to Philippines
5886	Do28D-2	4161	stored, l/n 6-94	to D-IDNE, to HK-3992
5887	Do28D-2	4162	stored, l/n 6-94	to D-IDND, to HK-3981
5892	Do28D-2	4167	stored, l/n 12-93	to D-IDBH, to Wunsdorf
5894	Do28D-2	4169	stored, l/n 6-94	to D-IDNC, to HK-3982
5898	Do28D-2	4173	stored, l/n 6-94	to D-IBDD, to N53U
5899	Do28D-2	4174	stored, l/n 6-94	to D-IDBE, to RP-C673 ?
5900	Do28D-2	4175	stored, l/n 12-93	to E3-IAAS
5903	Do28D-2	4178	stored, l/n 5-93	to D-IDRP, to RP-C673
5904	Do28D-2	4179	stored, l/n 12-93	to E3-IAMA
5905	Do28D-2	4180	stored, l/n 12-93	to E3-IANA
5907	Do28D-2	4182	stored, l/n 6-94	to D-IEDH, to Colombia
5908	Do28D-2	4183	stored, l/n 6-94	to D-IFDD, to CC-PWS
5909	Do28D-2	4184	stored, l/n 12-93	to D-IDRG, to HR-ANV
5910	Do28D-2	4185	stored, i/n 6-94	to D-IECA, to Panama
5912	Do28D-2	4187	stored, l/n 9-93	to D-IDES, to Colombia
5915	Do28D-2	4190	stored, l/n 12-93	to D-IBDF
5916	Do28D-2	4191	stored, l/n 12-93	to E3-IAME
5917	Do28D-2	4192	stored, l/n 12-93	to E3-IADE
5918	Do28D-2	4193	stored, l/n 12-93	to D-IDRV, to USA
5921	Do28D-2	4196	stored, l/n 9-93	to YS-402P, to HK3953P
5924	Do28D-2	4199	stored, l/n 3-91	to F-GHYI?

LEIPZIG - HALLE

The Aero Park at Schkeuditz airport (some 15km west of Leipzig) was set up during the late 1980s. The aircraft were clearly visible as they were parked on a grass area between the highway and terminal. All aircraft have to move because of the extension of the airfield. The two smaller aircraft, Zlin 37 DDR-SUC and PZL 106 DDR-TEJ, left for Dessau, while the Il-62 moved to a far corner of the airfield. The new museum at nearby Merseburg has shown interest in the Il-18 and Tu-134 and, if funds will become available, the aircraft are likely to move to there.

❏ DDR-SCF	Tu-134	9350905	preserved, ex DM-SCF, canx 5-6-86	2-98
❏ DDR-SEF	Il-62	31402	stored, ex preserved, ex DM-SEF, wfu 29-3-90	2-98
❏ DDR-STA	Il-18D	180001905	preserved, ex DM-STA, canx 26-9-88	2-98

LEMWERDER

The sale of the A340 fell through in 1995 and it has been stored here ever since. The TriStar was a trade-in for an Airbus and is reported that it will be used as a restaurant somewhere in the area. Airbus and DASA left the field in 1995 and it is now privately owned. Maintenance of Airbuses is still carried out here. The stored VFW614 OY-TOR went to Speyer and F-GATI (G-15) returned to flying status.

❏ F-WWDA	A340-311	003	stored	5-96
❏ VR-HOI	L1101-1	1039	stored, ex N318EA	5-96

MAHLWINKEL
An unmarked An-12 (c/n 9900902) was stored here during days of the Soviets. When they left, the An-12 was scrapped, it was last noted here in 1993.

MANCHING - INGOLSTADT
A large number of the *Wrecks & Relics* aircraft from this airfield, have been relocated. The museum at Koblenz received Meteor NF11-14, N2501F 199, MiG-23BN 2048, MiG21bis 2420 and Mi-24P 9833. To Berlin went Sabre 0101, F-104G 2649 and T-33A 9456 (marked as '9455'), while Whirlwind HAR.10 XP339 went to Hermeskeil. The French at Montelimar received MiG-23MF 2003/6, T-33A 9454 and G91T 9858. Starfighter 2156 has gone to Butzweilerhof. Starfighter 2917 (marked as '2121') went to Norvenich. The wings of the RAF Canberra were fitted to the Luftwaffe's own 9934.

❏ WK123		CY	Canberra TT.18	..	stored, fuselage only	3-93
❏ 1608			HFB320	1025	instructional, ex D-9537, ex D-CARU	9-95
❏ 2008			F-104G	2008	preserved, at gate, ex decoy	8-97
❏ 2030		3	MiG-23ML	0390324028	stored, ex 558/NVA	8-97
❏ 2052		4	MiG-23BN	0393215600	decoy, ex 712/NVA	9-95
❏ 2049			F-104G	2057	instructional, unmarked	9-94
❏ 2104			F-104G	6625	preserved, marked as 'MBB', at MBB gate	8-96
❏ 2251			F-104G	7131	instructional, with DASA	5-97
❏ 2317			MiG-21MF	968613	stored, ex 513/NVA	9-92
❏ 2373			F-104G	8072	instructional, with DASA	8-97
❏ 2521			F-104G	8295	preserved, ex '2104', at DASA gate	8-97
❏ 3121			G91R/3	388	dumped, ex preserved, ex instructional	8-97
❏ 4001			Alpha Jet	0001	stored, to go to Fürstenfeldbruck	8-97
❏ 4139			Alpha Jet	0139	stored, to go to Fürstenfeldbruck	8-97
❏ 5638			Do27A-1	318	preserved	9-95
❏ 5877			Do28D-2	4152	stored, ex instructional	8-97
❏ 9066			FWP149D	084	stored	8-97
❏ 9802			Tornado	P13	instructional	7-94
❏ 9804		11	F-104G	7406	instructional, in special c/s, ex 2660	9-95
❏ 9811		2	Su-22UM-3	17532371001	stored, ex 2554, ex 138/NVA, ex Laage	8-97
❏ 9815		1	Su-22M-4	30915	stored, ex 2526, ex 537/NVA, ex Laage	8-97
❏ 9816		7	Su-22UM-3	17532370810	stored, ex 2553, ex 137NVA, ex Laage	8-97
❏ 9831			Mi-24D	340278	stored, ex 9639, ex Basepohl	6-97
❏ 9834			Mi-24P	340337	stored, ex 9647, ex Basepohl	8-97
❏ 9861		5	Su-20	72412	decoy, unmarked, ex Egypt AF	8-97
❏ DA-129			F-104G	9059	decoy, crashed 17-5-66	9-95
❏ YA-207			CM170	071	instructional	7-87
❏ D-9565			VAK191B	V3	preserved, at WTD61 gate	3-93

MANNHEIM
Town: The Bundesakademie fur Wehrtechnik had two airframes. CM191 D-9532 is now on display at Speyer, while the G91T may be in storage for Speyer.
❏ 9857   G91T/3   0007   preserved, ex 3407   3-89

Airfield: Stored at the local airfield was a former Swiss Air Force C3605 C-535 (315). It has moved on to Speyer airfield and went flying as D-FOXY.

MARL
The Ikarusflug Museum was set up here in 1985 using mostly ex Air Classik exhibits. Within a few years the museum was closed and all the airframes moved on. CASA 352 T.2B-257, Javelin XH768, Sabre BB-131, F-84

DD-248 and C-47 N8041B all went to Bochum. DH.89A D-IGUN has gone to Augsburg and the replica Me163 went to Frankfurt, while the fates of the other aircraft mentioned in the first edition are unknown.

MARXZELL
The Fahrzeug Museum Marxzell has an ex Soviet MiG-15 on their parking lot. In 1995 the Sycamore was in a part of the museum which was not yet open to the public. It is on public view now.

☐ 75	MiG-15UTI	..	preserved, ex Soviet AF		9-95
☐ 7818	Sycamore Mk.52	13476	stored, marked as 'D-HELM'		9-95

MEMMINGEN
Canberra B.2 9934 arrived here for storage on 27th October 1993, it moved later to the museum at Schwenningen. Starfighter 2661 went to Hermeskeil. F-104s 2550 (8345) and 2726 (5727), both last noted in September 1997, were scrapped by late 1997. Starfighter 2460 (8208) was donated in September 1990 to the TestFuchs company somewhere in Austria.

☐ 2169	F-104G	7038	instructional	10-97
☐ 2255	F-104G	7135	instructional, special c/s	10-97
☐ 2261	F-104G	7142	instructional	8-96
☐ 2417	F-104G	8158	preserved, marked as '2334'	8-96
☐ 2419	F-104G	8161	stored, ex instructional, in 25 years c/s	10-97
☐ 9189	P149D	274	instructional	10-97
☐ DD-320	F-84F	..	preserved, marked as 'DD-113', ex 52-6669	8-97

MENDIG
The two choppers are preserved, while the dumped Do27A 5678 (385) has not been seen for a long time and may have been scrapped.

☐ 7676	Alouette 2	1760	preserved	3-98
☐ 8332	H-21C	WG32	preserved, at gate	3-98

MENGEN
At some 70km south west of Ulm the Luftwaffen Ausbildungs Regiment 4 has its base on the south side of the civil airfield. Its been reported that the Thunderstreak is not visible from outside due to lot of trees.

☐ DD-380	F-84F	..	preserved, ex 51-1702, see note	—

MERSEBURG
A new museum is currently being set up at this former Soviet airfield. The museum is not yet officially open, but visitors are welcome at weekends (between 10:00 and 16:00). A number of former Butzweilerhof aircraft arrived here on long term loan. Already one aircraft has left, Ka-26 D-HOAV (7303406) has gone to Hungary.

☐ A-16	Alouette 2	1595	preserved, ex Butzweilerhof , ex Belgian Army	2-98
☐ 06	SM-1	W05006	preserved, ex Vecsés, ex Hungarian AF	2-98
☐ 2156	F-104G	7025	preserved, ex Butzweilerhof, ex Manching	2-98
☐ 3178	G91R/3	446	preserved, ex Butzweilerhof, ex Bremgarten	2-98
☐ 3211	G91R/3	479	preserved, nose only, ex Butzw., ex Uetersen	2-98
☐ 7586	Alouette 2	1511	preserved, ex Butzweilerhof	2-98
☐ 7710	Alouette 2	1862	preserved	2-98
☐ 9350	Mi-8S	10733	stored, ex Dresden, ex 911/NVA	2-98
☐ D-EFHO	Do27B-1	177	preserved, ex Butzw., ex Frankfurt, ex 5547	2-98
☐ D-ESRF	Zlin 37	16-03	preserved, ex Butzweilerhof, ex DDR-SRF	2-98
☐ D-HOAI	Ka-26	7605406	stored, ex DDR-SPI, ex DM-SPI	2-98
☐ D-HZPR	Mi-2	543625074	preserved, ex 557/NVA	2-98

GERMANY - 208

☐ D-KNPF	Knechtel KN-1	V-1		stored, ex Butzweilerhof	2-98
☐ D-5526	..	..		stored, glider	2-98
☐ G-DEVN	Devon C.2	04269		preserved, ex Butzw., ex Frankfurt, ex WB533	2-98
☐ HA-MHL	An-2R	1G123-02		stored, ex Butzweilerhof, ex Frankfurt	2-98
☐ HA-OMD	Aero 145	..		preserved, ex Cespel	2-98
☐ ..	Bell 47G	..		stored, frame only	2-98
☐ ..	Bo105	..		stored, test frame/mock up	2-98
☐ ..	Fokker D.VI	..		preserved, replica, marked as '17', ex Butzw.	2-98
☐ ..	Fokker D.VII	..		preserved, replica, as '5290/18', ex Butzw.	2-98
☐ ..	Fokker Dr.I	..		preserved, replica, ex Butzw., ex Frankfurt	2-98
☐ ..	Me163B	..		preserved, replica, marked as 'VD-ER, ex '54'	2-98

MESSTETTEN

Fernmelderegiment 31 should still be based at the Zollernal kaserne in Messtetten, which is some 80km south of Stuttgart.

☐ 2599	F-104G	9118	preserved, marked as '2176'	90
☐ DD-306	F-84F	..	preserved, marked as 'DD-113', ex 52-6639	88

MINDEN

Town: The Potts Park, an amusement park in the Dutzen part of Minden, still has a former Hamburg Noratlas in their grounds. The aircraft is in fairly good condition.

☐ D-ACUG	N2501D	043	preserved, ex 5241	4-97

Airfield: The British Army base here was closed in 1993 and the fate of the relics, Scout AH.1 XP898 (F9499) and Gazelle AH.1 XX376 (1254) is not known.

MÖNCHENGLADBACH

One of the few former Air Classik aircraft which is still current (see Düsseldorf and Frankfurt) is the Lightning at this airfield. The stored Pembroke 5408 has moved on to Nordholz and the preserved Gannet XL450 moved to Hermeskeil.

☐ XN784	Lightning F.2A	95137	preserved, marked as 'XN781'	8-97
☐ 5414	Pembroke C.54	1006	stored, in pink c/s	8-97

MORGENRÖTHE RAUTENKRANZ

In honour of the first (East) German cosmonaut Sigmund Jahn, who was born in this town, a museum was established at the Bahnhofstrasse. Preserved outside is an ex NVA MiG-21.

☐ 737	MiG-21F-13	741608	preserved	5-98

MOSBACH

A Starfighter has been reported as preserved in this village.

☐ 2347	F-104G	8026	preserved	9-94

MÜNCHEN

Town: On an island in the Isar river in the town centre is the Deutsches Museum (open daily between 09:00 and 17:00). The site and size of the location were not suitable for any expansion. Therefore a new dedicated aviation museum was set up at Oberschleissheim, just north of München. A part of the Deutsche Museum collection found a new home there, including: Draken 35086, Sabre 0105, C-47 1401, F-104G 2903, Alouette 2 7584, H-34 8073, RF-84F EB-231, D-9531 Do31, VAK-191 D-9563 and CASA 2111 D-CAGI. Tornado 9806 moved

moved on to Fassberg. The fate of Do335A-2 240102 (marked as 'VG-PH') is unknown. The museum also has a large number of aircraft in storage. One of these is SBLim-2 770 (712270).

☐ 363		AAC.1	363	preserved, ex French AF	9-97
☐ E.3B-555		CASA 1131E	2169	preserved, ex Spanish AF	9-97
☐ AM210		Me163B-1a	120370	preserved, ex RAF, ex 120370	9-97
☐ 53-4458		UH-19B	..	preserved, ex USAF	9-97
☐ 2090		F-104G	6607	preserved, nose only	8-96
☐ 2153		F-104G	7022	preserved	9-97
☐ 5666		Do27B-1	360	preserved, ex 'D-EHAV'	9-97
☐ AS-058		AB47G-2	258	preserved	9-97
☐ 1258		Arado 66D	1258	preserved, fuselage	9-97
☐ 500071	3-I	Me262A-1a	500071	preserved	9-97
☐ D-CLOU		HFB320	1002	preserved, ex 'D-CASEK'	8-96
☐ D-EAWI		Fokker Dr.I	..	preserved, replica, marked as '425/17'	8-96
☐ D-EEWQ		Quickie	1	preserved	8-96
☐ D-EMDU		Klemm L-25e	980	preserved, ex SE-ANF, ex D-EMDU	8-96
☐ D-ENAY		Fw44J	45	preserved, ex D-EGAM, ex SE-BWH	8-96
☐ D-ENEK		Thubulent D	103	preserved	8-96
☐ D-HAPE		Bo105P	V4	preserved, ex 9832	8-96
☐ D-HOPA		Do32E	..	preserved, composite of D-HOPA/HOPF/HOPS	8-96
☐ D-779		Me M17	25	preserved, glider	8-96
☐ D-2054		Ju A50ci	3575	preserved, glider, ex HB-UXI, ex CH-538	8-96
☐ D-6426		HKS3	..	preserved, glider	8-96
☐ D-9093		Fs24 Phönix	V1	preserved, glider, ex D-8258	8-96
☐ D-9099	15	Ka6BR	378	preserved, glider	8-96
☐ D-9518		VJ101C	X2	preserved	9-97
☐ HB-ARU		Fi156C-3	4299	preserved, ex A-96/Swiss AF	9-97
☐ OY-AIJ		N1002	77	preserved, marked as 'D-IBFW', ex F-BEAI	9-97
☐ N7515A		B707-123B	17642	preserved, cockpit only	8-96
☐ VQ-FAB		Do A Libelle II	117	preserved	8-96
☐ ..		Akaflieg Hannover	..	preserved, glider	97
☐ ..		Ba349		preserved	9-95
☐ ..		Bf109E-3	790	preserved, marked as '2804', ex 6-106/SpaAF	9-97
☐ ..		Blériot XI	..	preserved	9-95
☐ ..		DOWA 81		preserved	9-85
☐ ..		Fa330A-1	42 ?	preserved	9-97
☐ ..		Finsterwalder Bergfex	..	preserved, glider	97
☐ ..		Fokker D.VII	..	preserved, replica?, as '4404/18', ex Spain	9-97
☐ ..		Grade A	..	preserved	9-95
☐ ..		Grunau SG-38	..	preserved, glider	97
☐ ..		Ju F13fe	..	preserved, marked as 'D-366'	8-96
☐ ..		Rumpler C.IV	310	preserved	9-97
☐ ..		Rumpler Taube	19	preserved	9-95
☐ ..		Wright Type A	..	preserved, replica	9-95

Somewhere near München, a private owner has a Starfighter.

☐ 2490		F-104G	8240	preserved	1-97

Town - Langwied: At a scrapyard, a number of dismantled aircraft were noted in early 1995. All had gone by late 1995. The report of G91R 9947 here seems incorrect as this aircraft is still at Oberpfaffenhoffen. Known to have been scrapped here are F-104F 2918 (5071, ex Fürstenfeldbruck), RF-84F EA-301 (ex Bremgarten, ex 53-6719, marked as 'EA-101') and G91R/3s 3073 (335, ex Fürstenfeldbruck) and 3308 (579, ex Bremgarten).

Airfield – Franz Jozef Strauss: The new international airport of München, replacing the now closed München Riem, received three aircraft from the former Air Classik collection at Frankfurt. The DC-3 is currently painted

GERMANY - 210

in Swiss Air Lines colours and came from Frankfurt

❏ D-CIAS	CASA 352L	54	preserved, ex Lahr, ex Frankfurt, ex N88927	2-98
❏ D-IMOC	Do28D-2	4109	instructional, ex Leipheim, ex 5832	93
❏ F-BHML	L-1049G	4671	preserved, marked as 'D-ALAP', ex 'D-ADAM'	2-98
❏ N65371	DC-3	4828	preserved, marked as 'HB-IRN', ex 'N569R'	2-98

MÜNSTER
The Wehrwissenschaftliche Dienststelle der Bunderswehr could still have some Starfighters. 2463 arrived in the early 1980s and may have gone by now.

❏ 2224	F-104G	7099	instructional, ex Westerland	—
❏ 2463	F-104G	8212	instructional, marked as 'DA-04', ex Norvenich	—
❏ 2568	F-104G	9015	instructional, marked as '6951' ('1234')	—

NEUBIBERG
The instructional G91R/3 3212 (ex Bremgarten) made a surprising move to a supermarket in Roma. Nothing is known of the remaining airframes here. Alouette 2 PP-144 (1763), F-104G 2059 (2068) and the unfinished Do27 marked '5663' may have perished a long time ago. The cockpit of Starfighter 2507 arrived here from Fürstenfeldbruck in the early 1990s. It may still be here.

❏ 2507	F-104G	8265	stored, cockpit only, ex Fürstenfeldbruck	—

NEUBRANDENBURG
The former NVA base is also known as Trollenhagen. The southern part of the airfield is still used by the military, while a civil terminal has been built on the northern side. From here the stored MiG-23ML is visible. The Strafighter is preserved near the main gate. Known departures include: MiG-21SPS-Ks 2209 to Fassberg and 2219 to Berlin; MiG-21Ms 2257 and 2262 went to Fassberg, while the museum at Berlin received MiG-21UM 2377.

❏ 2029	MiG-23ML	0390324027	stored, ex Laage, ex 554/NVA	8-96
❏ 2112	F-104G	6661	preserved, ex Wilhelmshafen	5-96
❏ 2208	MiG-21SPS-K	94A6702	stored, ex 404/NVA and marked as such	5-94
❏ 2210	MiG-21SPS-K	94A7010	stored, ex 481/NVA and marked as such	5-94
❏ 2211	MiG-21SPS-K	94A7211	stored, ex 486/NVA and marked as such	5-94
❏ 2212	MiG-21SPS-K	94A7214	stored, ex 492/NVA and marked as such	5-94
❏ 2213	MiG-21SPS-K	94A7304	stored, ex 565/NVA and marked as such	6-94
❏ 2214	MiG-21SPS-K	94A6414	stored, ex 967/NVA and marked as such	5-94
❏ 2215	MiG-21SPS-K	94A6506	stored, ex 980/NVA and marked as such	1-93
❏ 2216	MiG-21SPS-K	94A6705	stored, ex 982/NVA and marked as such	5-94
❏ 2217	MiG-21SPS-K	94A6713	stored, ex 983/NVA and marked as such	5-94
❏ 2218	MiG-21SPS-K	94A6803	stored, ex 988/NVA and marked as such	5-94
❏ 2252	MiG-21M	963209	stored, ex 438/NVA	5-94
❏ 2253	MiG-21M	963210	stored, ex 445/NVA and marked as such	5-94
❏ 2261	MiG-21M	963305	stored, ex 499/NVA	6-94
❏ 2270	MiG-21M	960501	stored, ex 561/NVA	5-94
❏ 823	MiG-21PFM	760605	dumped	1-93
❏ 825	MiG-21PFM	760606	preserved	5-94
❏ 843	MiG-21PFM	761005	stored	5-94
❏ 885	MiG-21PFM	761106	stored	5-94
❏ 890	MiG-21PFM	761107	stored	5-94
❏ 896	MiG-21PFM	761110	stored	5-94
❏ 908	MiG-21PFM	761113	stored	5-94
❏ 947	MiG-21PFM	761211	stored	5-94
❏ ..	MiG-17F	..	preserved, marked as '003'	5-96

NEUBURG
Neuburg is home of the Phantoms of JG74. Arriving here in 1997 was a former 1st GAFTS/49th FW F-4E.

☐ 75-0633	F-4E	4956	stored	9-97
☐ 2290	F-104G	7173	instructional, ex Fassberg	9-97
☐ 2331	F-104G	8006	instructional	7-94
☐ 2362	F-104G	8044	decoy, marked as 'JA-240'	5-94
☐ 2653	F-104G	7313	instructional	9-97
☐ 9164	FWP149D	178	stored, near control tower	5-94
☐ 55-4928	F-86K	232-168	preserved, at gate, marked as 'JD-119'	9-97

Preserved at the nearby Wilhelm Frank kaserne, the staff buildings for JG74, is a Sabre which was delivered but never flown by the German Air Force. Its correct USAF serial is still not known.

☐ 2357	F-104G	8037	preserved, at gate	8-96
☐ 55-4932 ?	F-86K	232-172	preserved, at gate, marked as 'JG-74', see note	8-96

NEUHARDENBERG
All the time that East Germany was still intact, the airbase was known as Marxwalde and was home of a transport and MiG-21bis unit. After the reunification of East and West the base was used for a few years by the military. Of the stored L-410UVP-Ts, 5301 (810726) became ES-EPA, 5302 (810727) became ES-EPI, 5303 (820737) went to the Air Force of Lithuania as 45, 5304 (810738) as 01 and 5305 (910739) as 02. 5307 (831136) was sold as YS-406. Stored MiG-21F-13 679 (marked as '696') went to Uetersen. The two MiG-21s with false serials should be 661 (742001) and 745 (741921). No *Wrecks & Relics* aircraft were seen in 1996.

☐ 135	MiG-15UTI	..	presrved	9-90
☐ 642	MiG-21F-13	741923	stored	4-91
☐ 716	MiG-21F-13	740911	stored	4-91
☐ ..	MiG-21F-13	..	stored, marked as '049'	4-91
☐ ..	MiG-21F-13	..	stored, marked as '555'	4-91

NEUHAUSEN OB ECK
The Alouette 2 which was kept in a sorry state in an 'outhouse' has probably gone as this airfield is closed.

☐ 7645	Alouette 2	1672	stored, ex Roth, crashed 5-6-75	4-91

NEU ULM
Located close to the local civil airfield is the Konigs Schutzenhaus. Here a G91 is mounted on a pole. The location is about 5km south of Neu Ulm on the road to Reutti.

☐ 3256	G91R/3	526	preserved, with tail from 3186 (454)	9-97

NIERDERALTEICH
A private collector in this village has a complete F-104G, the cockpit of another one and some additional parts.

☐ 2449	F-104G	8193	preserved, cockpit only, ex Pferdfeld	98
☐ 2505	F-104G	8263	preserved, ex Husum	98

NIEDERSTETTEN
The army units at the Hermann Kohl kaserne and airfield have an Alouette 2 mounted at the main gate. Another Alouette 2, a UH-1 and unknown Piaggio are stored at the base.

☐ 7301	UH-1D	8421	stored, wreck, crashed 1-5-93	7-94
☐ 7711	Alouette 2	1863	preserved, at gate	3-98
☐ 7715	Alouette 2	1869	stored, ex instructional	11-94
☐ ..	P149D	..	stored, dismantled	9-95

GERMANY - 212

NORDHAUSEN
In late 1990 the Mi-2s and Mi-8s of HS-16 were stored here. They moved to Schönefeld soon afterwards.

NORDHOLZ
The Marineflieger has their Atlantic, Do228 and Sea Lynx unit, MFG3, based here.

❏ 3439		G91T/3	0043	preserved, with aero club	8-96
❏ 5911		Do28D-2	4186	stored	8-95
❏ 5925		Do28D-2/OU	4200	stored, unmarked	8-97
❏ 6101		Br1150	2	stored	96

In the mid-1990s the first aircraft arrived for the newly established Deutsches Luftschiff und Marine Fliegermuseum Aeronauticum. The museum opened its gates on 7th March 1997 and for 1998 is open from 1st March to 30th June and from 1st September to 31st October on weekdays between 13:00 and 17:00 and on Sundays (and holidays) the gate open between 10:00 and 18:00. Between 1st July and 31st August it is open daily between 10:00 and 18:00.

❏ 388		CM170	388	preserved, marked as 'SC-101/JC-601', ex FAF	5-98
❏ IN238	W	Sea Hawk Mk.100	6684	preserved, marked as 'VA-134', ex Indian Navy	5-98
❏ 2505		Su-22M-4	25512	preserved, ex 366/NVA, ex Laage	5-98
❏ 2574		F-104G	9035	preserved, as '2681' ('8160'), ex Flensburg	5-98
❏ 5408		Pembroke C.54	P66/105	preserved, ex Monchengladbach	5-98
❏ 5919		Do28D-2/OU	4194	preserved, wfu 24-11-95	5-98
❏ 5922		Do28D-2	4197	preserved	5-98
❏ 9077		FWP149D	096	preserved	5-98
❏ 9401		Mi-8S	105100	preserved, ex Parow, ex 733/NVA	5-98
❏ 9410		Mi-8TB	10572	stored, ex Parow, ex 827/NVA	5-98
❏ 9414		Mi-8TB	10575	preserved, ex Parow, ex 834/NVA	5-98
❏ 791		MiG-21SPS	4613	stored, ex Kamenz	8-95
❏ UA-113		Gannet AS.4	F9395	preserved, ex gate, ex XG853	5-98
❏ VA-234		Sea Hawk Mk.100	6667	stored, marked as 'VA-229', ex Jagel	5-98

NORVENICH
After the USAF gave up the FOL (forward operating location) here in the late 1980s, the instructional F-4C 63-7536 moved to Seifertshofen. Not seen for some time and all scrapped are Starfighters 2042 (2049, l/n 8-87), 2387 (8094, scrapped 1992),), and 2481 (marked as (DA-05'). F-104G 2388 went to Büchel, together with F-84F DD-354 (marked as 'DA-101/DA-322'). 2455 (8203, marked as 'DA-03') went to Leewarden as 'LETS-1' and 2463 (8212, marked as 'DA-04') went to Münster. The F-84F returned here as 'DC-101'. The P149 was noted inside one of the former USAF shelters. Starfighter 2452 has the tail from 2481 (8231).

❏ 2452	F-104G	8199	preserved, marked as 'DA-235', ex 'N104RB'	12-97
❏ 2630	F-104G	9182	instructional	12-97
❏ 2917	F-104F	5070	preserved, ex '2121', Manching, 59-5017	3-98
❏ 5841	Do28D-2	4116	instructional, ex Wunstorf, arr 28-4-93	9-93
❏ 9186	P149D	270	instructional, ex Bremgarten	5-97
❏ 9911	G91R/3	518	preserved, ex 3249	9-93
❏ DD-354	F-84F	..	preserved, marked as 'DC-101', ex 52-6707	3-98

OBERPFAFFENHOFEN
Nearly all the Dorniers from the first edition have left their place at the gate. Former guard Do29 YD-101 (marked as 'YA-101') went to Koblenz, while Do28 D-IKAS (4030) was scrapped in 1996.

❏ HD.5-3	Do24T-3	5344	preserved, N99222 ntu, ex Spanish AF	5-98
❏ 3135	G91R/3	403	stored, ex preserved	5-98
❏ 4014	Alpha Jet	0014	instructional	5-98

GERMANY - 213

☐ 4017	Alpha Jet	0017	stored		8-97
☐ 9946	G91R/3	511	instructional, ex 3242		8-97
☐ 9947	G91R/3	..	instructional, ex Fürstenfeldbruck		12-95
☐ D-9530	Do31	E1	preserved		8-97
☐ D-CATI	Do328	3002	stored		12-95
☐ D-ICOG	Do228-100	7001	stored, ex LN-HPG		12-95
☐ D-IFNT	Do28TNT	4043	stored, ex preserved, ex AF206/Zambian AF		12-95

OBERSCHLEISSHEIM

This Schleissheim airfield was founded in 1912. Between the wars it housed a number of training units and after the Second World War it was a US Army airfield. Currently it is still in use by the Bundesgrenzshutz and the local aero club. As the Deutsch Museum location at München became too small, a new museum was built here to house a second aviation section. The HA1112 is currently with DASA at Augsburg for restoration. The museum is open daily between 09:00 and 17:00 (except last admission 16:30).

1st hall:

☐ A-208		Bf108B-2	2064	preserved, as 'D-1010', ex Swiss AF	9-97
☐ 1401		C-47D	15544/26989	preserved	9-97
☐ D-CAGI		CASA 2111B	025	preserved, unmarked, ex G-AWHA, ex B.2I-77	9-97
☐ D-1469		Meise	001	preserved, glider	9-97
☐ 8103/18		Halberstadt CL.IV	..	preserved	9-97
☐ ..		HA1112K-1L	..	preserved, unmarked, see note	9-97

2nd hall:

☐ 51-100 ?		HA300	V1	preserved, ex Egypt AF	9-97
☐ D-1256		HF24 Marut Mk.1	..	preserved, ex Indian AF	9-97
☐ 003		SBLim-2	1A-08003	preserved, ex Polish AF	9-97
☐ 35086	48	J-35A	35086	preserved, ex München, ex RSweAF	9-97
☐ XK824		Grasshopper TX.1	1043	preserved, glider, ex RAF	9-97
☐ 67-0260	SP	F-4E	2972	preserved, ex Spandahlem, ex USAF	9-97
☐ 03 red		An-2	1G59-29	preserved, ex Soviet AF	9-97
☐ 0105		CL-13B Sabre 6	1659	preserved, ex KE-105 and marked as such	9-97
☐ 2047		MiG-23BN	0393214217	preserved, ex Drewitz, ex 701/NVA	9-97
☐ 2340		MiG-21MF	966215	preserved, ex Rothenburg, ex 687/NVA	9-97
☐ 2903		F-104F	5049	preserved, ex München, ex 59-4996	9-97
☐ 3401		G91T/3	0001	preserved, nose only	9-97
☐ 7584		Alouette 2	1497	preserved	9-97
☐ 8073		H-34G	58-1557	preserved, ex 150808	9-97
☐ 9447		T-33A	8967	preserved, ex Erding, ex 53-5628	9-97
☐ 9907		G91R/3	460	preserved, ex 3192	9-97
☐ EB-231		RF-84F	..	preserved, ex München, ex 52-7783	9-97
☐ D-CATD		Do24ATT	5345	preserved, rebuild of Do24T-3 HD.5-4/Spa AF	9-97
☐ D-ECUX		Fw44J	91	preserved, ex SE-BXN, ex 639/RSweAF	9-97
☐ D-EBCQ		LF-1 Zaunkönig	V2	preserved	97
☐ D-EFWC		MBB223PFM	151	preserved, ex EC-51A	9-97
☐ D-EHOG		Pützer Motorraab	03	preserved	9-97
☐ D-EOMA		J-3C-65	12622	preserved, ex F-BFBI	9-97
☐ D-ESOZ		Zlin 37A	08-18	preserved, ex DDR-SOZ	9-97
☐ D-HOAZ		Ka-26	7605615	preserved, ex DDR-SPZ	9-97
☐ D-1001		Mü-10 Milan	5	preserved, glider, ex D-14-126	9-97
☐ D-1065		Grunau Baby IIb	..	preserved, glider	9-97
☐ D-1085		Mü-13E Bergfalke I	2	preserved, glider	9-97
☐ D-1220		Doppelraab IV	..	preserved, glider	9-97
☐ D-6007		Gö IV Goevier 3	406	preserved, glider	9-97

	GERMANY - 214				
☐ D-6171	Kranich II	533	preserved, glider		97
☐ D-8129	Hütter H17A	..	preserved, glider		9-97
☐ D-8273	Fauvel AV36CR	213	preserved, glider		9-97
☐ D-8362	Kaiser Ka1	2	preserved, glider		9-97
☐ D-8802	Dittmar Condor IV	24/53	preserved, glider		9-97
☐ D-9531	Do31E	E3	under restoration, ex München		9-97
☐ D-9563	VAK-191B	V1	preserved		9-97
☐ DDR-WQV	Yak-50	781206	preserved, ex DM-WQV		9-97
☐ N3480V	Ce195	7180	preserved		97
☐ ..	Do27	..	preserved, nose only		9-97
☐ ..	Grunau SG-38	..	preserved, glider		97
☐ ..	Vollmoller	..	preserved, damaged		97

OLDENBURG
It is believed that after the departure of the Alpha Jets, all the non-operational aircraft left. Unfortunately, only a few ex Oldenburg aircraft have surfaced elsewhere. Sabre JB-110 went to Dresden and Me163B 191904 is now at Berlin. G91R 3100 (l/n 5-93) will go to Ankum. G91R/3s 3060 (320), 3087 (351, marked as '3259') and 3101 (367) and Sabre JB-371 (1813) have all not been reported since the early 1990s. As the base is still in use (by Luftlandebrigade 31) some aircraft may still be here.

OSCHERSLEBEN
Preserved in this former East German town is a former Interflug Tu-134.
☐ DDR-SCB	Tu-134	8350503	preserved, ex DM-SCB	5-98

OSNABRUCK
The former gate guard of Fernmelderegimen 71, RF-84F EB-319, moved to the museum at Butzweilerhof. Its not known if its place has been taken by another aircraft.

PADERBORN - LIPPSTADT
Stored away from the main buildings on this regional airliner airfield is an Aero Commander.
☐ D-IIWE	RC500B	1301-116	stored	12-95

PAROW
Within a year of East and West Germany becoming one country, most of the helicopters of the former East German Navy unit were withdrawn from service. Only a handful continued flying until 30th September 1994 when the last Mi-8 landed here. No stored helicopters remain on the airfield today.

9401	Mi-8PS	105100	stored, ex 773/NVA, l/n 9-91	to Nordholz
9402	Mi-8T	10535	stored, ex 906/NVA, l/n 4-91	to D-HOWA
9403	Mi-8T	10536	stored, ex 907/NVA, l/n 4-91	to D-HOWB
9404	Mi-8TB	10564	stored, ex 807/NVA, l/n 9-91	gone
9405	Mi-8TB	10565	stored, ex 808/NVA, l/n 9-94	gone
9406	Mi-8TB	10566	stored, ex 810/NVA, l/n 9-92	to Briest
9407	Mi-8TB	10567	stored, ex 812/NVA, l/n 9-94	gone
9408	Mi-8TB	10568	stored, ex 814/NVA, l/n 9-94	to Dax
9409	Mi-8TB	10569	stored, ex 818/NVA, l/n 9-91	gone
9410	Mi-8TB	10572	stored, ex 827/NVA, l/n 7-92	to Nordholz
9412	Mi-8TB	10574	stored, ex 831/NVA, l/n 9-94	gone
9414	Mi-8TB	10575	stored, ex 834/NVA, l/n 7-92	to Nordholz
9502	Mi-14PL	B4002	stored, ex 618/NVA, l/n 7-91	to Hermeskeil

GERMANY - 215

9503	Mi-14PL	B4003	stored, ex 619/NVA, l/n 7-92	to Rothenburg
9504	Mi-14PL	B4004	stored, ex 620/NVA, l/n 9-91	to Rothenburg
9505	Mi-14PL	B4005	stored, ex 625/NVA, l/n 9-91	to Rothenburg
9506	Mi-14PL	B4006	stored, ex 637/NVA, l/n 7-91	to Speyer
9507	Mi-14PL	B4008	stored, ex 640/NVA, l/n 7-91	to 13790/US Army
9508	Mi-14PL	B4009	stored, ex 643/NVA, l/n 7-91	to US Army
9509	Mi-14BT	B4010	stored, ex 646/NVA, l/n 9-91	to Rothenburg
9510	Mi-14BT	B4011	stored, ex 647/NVA, l/n 9-91	to Rothenburg
9511	Mi-14BT	B4012	stored, ex 648/NVA, l/n 7-91	to Rothenburg
9512	Mi-14BT	B4013	stored, ex 653/NVA, l/n 9-91	to Rothenburg
9514	Mi-14BT	B4014	stored, ex 654/NVA, l/n 9-91	to Rothenburg
9515	Mi-14BT	B4015	stored, ex 655/NVA, l/n 9-91	to Rothenburg

PÄTZ

Just south of Königs Wusterhausen (south west of Berlin) is the small town of Pätz. An unknown Mi-8 is preserved at a local transport company along road number 179.

❏ ..	Mi-8	4038	preserved, marked as 'D-PÄTZ'	2-98

PEENEMÜNDE

Town: Historisch-Technisches Informationszentrum Geburtsort der Raumfahrt was opened here at the site of the former secret rocket research factories. Displayed around the original buildings are model rockets and former NVA aircraft. The museum is open from Tuesday to Sunday between 09:00 and 16:00.

❏ 2010	MiG-23ML	0390324623	preserved, ex Laage, ex 329/NVA	10-97
❏ 2502	Su-22M-4	25509	preserved, ex Laage, ex 362/NVA	10-97
❏ 2801	L-39ZO	731001	preserved, ex Rothenburg, ex 139/NVA	10-97
❏ 9457	Mi-2S	564411105	preserved, ex Briest, ex 348/NVA	10-97
❏ 393	Mi-8T	10527	preserved, marked as '911', ex Kamenz	10-97
❏ 009	MiG-17F	..	preserved	10-97
❏ 992	MiG-21SPS-K	94A6806	preserved	10-97
❏ 332	MiG-23ML	0390324625	preserved, crashed 4-1-88	10-97
❏ D-ESUJ	Zlin 37A	18-28	preserved, ex DDR-SUJ	10-97
❏ D-FONB	An-2TD	1G180-41	preserved, ex DDR-SKB, ex 799/NVA	10-97
❏ D-HOAW	Ka-26	7001404	preserved, ex DDR-SPW, ex 404/HungarianAF	10-97

Airfield: The large airfield of Peenemünde also had a range. A number of aircraft were in use as range targets here. An unidentified Il-28 could be seen from the former flightline in 1990. The airfield was closed in the early 1990s. The airfield was since then used as military storage site of former East German Army equipment. Four aircraft are in storage for the museum.

❏ 2004	MiG-23MF	0390213098	stored, ex Laage, ex 584/NVA	—
❏ 677	MiG-21F-13	741619	stored	10-90
❏ 693	MiG-21F-13	740815	stored	10-90
❏ 934	MiG-21PFM	761205	stored	—

PERGAU

Preserved in this town is a former Briest Mi-8.

❏ 9320	Mi-8PS	10552	preserved, ex Briest	11-97

PFERDSFELD

The former Phantom base (the F-4Fs are now at Laage) of Pferdsfeld still has a number of non-operational fighters, of which the F-4 is the latest arrival. Starfighter 2502 (ex Westerland) moved on to <u>Laage</u>. A few miles down the road, outside the airfield, is JBG35's kaserne at Sobernheim. The kaserne's CL-13B Sabre 6 'JC-102' also found a new home at Laage. The dumped Starfighter 2351 went to Baarlo, while the nose of Starfighter 2449 went to Niederalteich.

☐ 3033	G91R/3	0093	instructional	9-96
☐ 3303	G91R/3	574	dumped, ex preserved	7-97
☐ 3746	F-4F	4461	dumped, ex instructional, ex 72-1156	7-97
☐ 9909	G91R/3	468	instructional, ex 3200	10-94

PINNEBERG
T-33 9464 and Sabre BB-239 (marked as 'JB-111') in this Hamburg suburb have moved on to Schwenningen.

PRESCHEN
The former NVA MiG-29 base was closed in the mid-1990s after the 'Fulcrums' had moved on to the former NVA Su-22 base of Laage. They took both their gate guards with them, MiG-19PM '391' and MiG-21F-13 619 (marked as '335') are now at Laage. The stored MiG-21F-13 688 went to Laatzen.

☐ 713	MiG-21F-13	740914	stored	7-94
☐ 726	MiG-21F-13	741606	stored	7-94
☐ 736	MiG-21F-13	741607	stored	7-94

PREUSSISCH OLDENDORF
Outside the Mobel Holstein store is a DC-6 furnished as restaurant. The town is some 30km east of Osnabruck.

☐ D-ABAH	DC-6	42855	preserved, ex Frankfurt, ex N90702, ex HP-361	4-97

PULSFORDE
Just east of Zerbst (itself a former Soviet fighter base) is the village of Pulsforde. Here an Il-14 is in use as a restaurant. The aircraft arrived in 1991 and came from Halle

☐ DM-SAF	VEB-14P	14803016	preserved, ex Halle	5-95

RAMSTEIN
One of the last remaining USAF airfields in Germany is Ramstein. Just as the operational fleet of aircraft has been reduced, so has the non operational fleet. T-33A FT-38 (9585) has not been noted since 1987, as with F-4E 74-1641 (4884). F-105F 63-8362 has gone to Coleman Barracks. More recent departures are the two ex Greek aircraft, F-84F 26789 went to Toulouse and F-102A 61125 is now at Hermeskeil.

☐ 64-0917	SA	F-4C	1385	instructional	7-94
☐ 68-0554	ZR	RF-4C	3369	instructional, arr 15-3-91	11-94
☐ 74-0098		F-15A	72/A59	instructional, ex Bitburg	10-95
☐ 57-0386		F-101B	564	instructional	7-97

RANGSDORF
As well as a helicopter maintenance unit, Rangsdorf also had a famous scrapyard. Over the years a large number of Soviet aircraft, or more accurately, major components of aircraft, have been noted here (comprising of crashed aircraft, former decoys, etc). All were processed here quite rapidly. The only long term inmate was gate guarding Yak-27R 88 red (0616), which was finaly scrapped at the end of the Soviet occupation of Germany.

RECHLIN
At the local Rechlin-Lärz airfield (some 30km north east of Wittstock) a local museum is being set up. Two air-

craft are already on outside display.

☐ 2548	Su-22UM-3	17532367002	preserved, ex Laage, ex 119/NVA	97
☐ D-ESSD	Zlin 37A	17-22	preserved, ex DDR-SSD	97

REINSDORF
The An-2 which is stored at this (East) German airfield is owned by an ex NVA MiG-21 pilot. Stored Mi-24V 01 yellow (ex Soviet, l/n 5-93) went to <u>Dessau</u>.

☐ RA-05825	An-2	1G63-32	stored, ex 05 red, ex D-FMGM	8-95

RIBNITZ DAMGARTEN
This well known Soviet MiG-29 airfield had a number of MiG-23s and Il-28s which were dumped on the airfield. All have presumably been scrapped when the based MiG-29s returned to the 'Rodina'.

☐ 11 red	Il-28	56542	dumped	7-92
☐ 02 red	Il-28U	69601	dumped	7-92
☐ 22 blue	Il-28R	44040176	dumped	7-92
☐ ..	Il-28	54005917	dumped	7-92
☐ 01 blue	Il-28U	63006305	dumped	7-92
☐ 11 yellow	MiG-23	5104	dumped	7-92
☐ 20 yellow	MiG-23M	03671	dumped	7-92

ROTENBURG - WÜMME
Processed for spares at this HEER field in the early 1990s were a number of Alouette 2, including 7510 (1216), 7543 (1352), 7556 (1416), 7577 (1479), 7606 (1560), 7626 (1613), 7640 (1656, to <u>Fassberg</u>), 7647 (1674) and 7658 (1707). The T-33 is confirmed as being 9520 (see EWR-1) and has moved to <u>Ankum</u>.

ROTH
Preserved at the air force barracks within this army airfield (the Otto Lilienthal kaserne) are a number of fighters. Preserved by the army itself is only an Alouette 2, while the instructional example, 7645, went to <u>Neuhausen ob Eck</u>.

☐ 2336	F-104G	8011	preserved, marked as '2399'	8-97
☐ 7067	UH-1D	8127	instructional, crashed 4-5-89	6-97
☐ 7598	Alouette 2	1543	preserved	9-97
☐ 8044	Bo105M	5044	instructional	7-97
☐ 8754	Bo105P	6154	instructional	7-97
☐ BB-130	CL-13 Sabre 5	838	preserved, marked as 'JA-130'	9-97
☐ DD-379	F-84F	..	preserved, marked as 'DA-379', ex 51-1796	8-97
☐ EA-315	RF-84F	..	preserved, marked as 'EA-105', ex 53-7693	8-97

ROTHENBURG
During the NVA years the base was used by a training wing with MiG-21s and during the last years of the NVA the L-39s from Bautzen were also based here. After reuniting East and West Germany Rothenburg became one of the storage airfields. The civil Aerotec company on the airfield was responsible for the de-militarisation of these aircraft. This company also de-militarised some aircraft from Laage and Drewitz before transporting them to museums and private collectors. During 1996 a museum was set up with a dozen ex NVA aircraft. A G91 is also preserved here, being one the few ex West German aircraft to be found in the former East Germany.

☐ 2017	MiG-23ML	0390324636	preserved, ex Laage, ex 340/NVA	9-97
☐ 2038	MiG-23BN	0393211085	preserved, ex Drewitz, ex 689/NVA	9-97
☐ 2043	MiG-23BN	0393214211	preserved, ex Drewitz, ex 695/NVA	9-97
☐ 2220	MiG-21SPS-K	94A6808	preserved, ex 463/NVA and marked as such	9-97

GERMANY - 218

❑ 2242	MiG-21SPS	94A5604	preserved, ex 940/NVA and marked as such	9-97
❑ 2285	MiG-21M	960508	preserved, ex 581/NVA and marked as such	9-97
❑ 2349	MiG-21MF	96002170	preserved, ex 784/NVA and marked as such	9-97
❑ 2387	MiG-21U-400	661017	preserved, ex 244/NVA and marked as such	9-97
❑ 2398	MiG-21U-600	664818	preserved, ex 296/NVA and marked as such	9-97
❑ 2407	MiG-21US	05685134	preserved, ex 236/NVA and marked as such	9-97
❑ 2418	MiG-21bis	N75051378	preserved, ex 838/NVA and marked as such	9-97
❑ 2805	L-39ZO	731005	preserved, ex 143/NVA	9-97
❑ 3074	G91R/3	336	preserved, ex Lahr, ex Tannheim, ex Augsburg	9-97
❑ 9460	Mi-2S	562633112	preserved, ex Briest, ex 383/NVA	9-97
❑ 9504	Mi-14PL	B4004	preserved, ex Parow, ex 620/NVA	9-97

Most of the stored MiGs, Suhkois and L-39s have gone from the airfield, only some 20 are left. Not part of the NVA storage here are the Mil helicopters, these were acquired by the Aerotec company. A number of Mi-14s were converted to waterbombers, while the others were used a spares source. The Mi-2s and Mi-8s were bought for onward sale.

❑ 2007	MiG-23MF	0390213300	stored, ex Laage, ex 591/NVA	9-97
❑ 2221	MiG-21SPS-K	94A7215	stored, ex 545/NVA and marked	9-97
❑ 2222	MiG-21SPS	94A4006	stored, ex 703/NVA, Hermeskeil owned	9-97
❑ 2227	MiG-21SPS	94A4303	stored, ex 742/NVA	9-97
❑ 2231	MiG-21SPS	94A4504	stored, ex 783/NVA	8-96
❑ 2241	MiG-21SPS	94A5602	stored, ex 937/NVA	9-97
❑ 2288	MiG-21M	960706	stored, ex 602/NVA	9-97
❑ 2346	MiG-21MF	96002045	stored, ex 779/NVA	8-96
❑ 2347	MiG-21MF	96002037	stored, ex 781/NVA	9-97
❑ 2392	MiG-21U-600	663219	stored, ex 275/NVA	9-97
❑ 2393	MiG-21U-600	663220	stored, ex 276/NVA	5-97
❑ 2396	MiG-21U-600	664719	stored, ex 292/NVA	9-97
❑ 2406	MiG-21US	04685134	stored, ex 230/NVA, Hermeskeil owned	9-97
❑ 2425	MiG-21bis	N75058087	stored, ex Drewitz, ex 856/NVA	9-97
❑ 2535	Su-22M-4	26512	stored, ex Laage, ex 727/NVA	9-97
❑ 2542	Su-22M-4	26817	stored, ex Laage, ex 757/NVA	9-97
❑ 2829	L-39ZO	232302	stored, ex 178/NVA	8-96
❑ 2849	L-39V	630715	stored, ex 171/NVA	9-97
❑ 9305	Mi-8T	10515	overhaul, ex Briest, ex 631/NVA	8-96
❑ 9316	Mi-8T	10545	overhaul, ex Briest, ex 930/NVA	8-96
❑ 9343	Mi-8S	10533	stored, ex Briest, ex 977/NVA	9-97
❑ 9415	Mi-8S	0826	stored, ex Briest, ex 397/NVA, D-HOZI ntu	9-97
❑ 9455	Mi-2S	563823114	stored, ex Briest, ex 309/NVA	5-97
❑ 9459	Mi-2S	562632112	stored, ex Briest, ex 382/NVA	5-97
❑ 9461	Mi-2S	562634112	stored, ex Briest, ex 385/NVA	9-97
❑ 9470	Mi-2F	514416125	stored, ex Briest, ex 355/NVA	8-96
❑ 9471	Mi-2S	564410105	stored, ex Briest, ex 347/NVA	5-97
❑ 9472	Mi-2S	562248032	stored, ex Briest, ex 379/NVA	9-97
❑ 9480	Mi-2F	514415125	stored, ex Briest, ex 328/NVA	5-97
❑ 9483	Mi-2See	552649126	stored, ex Briest, ex 386/NVA	4-96
❑ 9503	Mi-14PL	B4003	stored, ex Parow, ex 619/NVA	9-97
❑ 9505	Mi-14PL	B4005	stored, ex Parow, ex 625/NVA	9-97
❑ 9514	Mi-14BT	B4014	stored, ex Parow, ex 654/NVA	9-97
❑ 9515	Mi-14BT	B4015	stored, ex Parow, ex 655/NVA	9-97
❑ 339	L-29	692061	stored	9-97
❑ 297	MiG-21U-600	664819	stored, ex Kamenz	9-97
❑ 623	MiG-21F-13	741916	stored	9-97
❑ D-HZPO	Mi-2	543048083	stored, ex Schönefeld, ex 500/NVA	8-96
❑ S9-TAH	Mi-8PS	10549	stored, ex Cottbus, ex 9337, ex 961/NVA	9-97

GERMANY - 219

More than 70 MiG-21s and 55 L-39s have departed from here, but not all their new homes are yet known. The Baade 152 is a 50-seat airliner which was built at Dresden but after the crash of one of the prototypes the project was abandoned. The fuselage of one of the unfinished aircraft which was stored here moved to Dresden.

2009	MiG-23MF	0390213352	stored, ex 593/NVA, ex Laage	to Wolls Petersdorf
2054	MiG-23BN	0393215729	stored, ex 718/NVA, ex Drewitz	to Cerbaiola
2223	MiG-21SPS	94A4209	stored, ex 717/NVA, l/n 2-94	to Ankum
2223	MiG-21SPS	94A4209	stored, ex 717/NVA, l/n 2-94	to Ankum
2224	MiG-21SPS	94A4213	stored, ex 729/NVA, l/n 10-94	to New Mexico, USA
2225	MiG-21SPS	94A4301	stored, ex 738/NVA, l/n 4-93	to Speyer
2226	MiG-21SPS	94A4302	stored, ex 740/NVA, l/n 2-94	to Cuatro Vientos
2228	MiG-21SPS	94A4309	stored, ex 760/NVA, l/n 10-94	gone
2229	MiG-21SPS	94A4510	stored, ex 771/NVA, l/n 8-94	to Dermsdorf
2230	MiG-21SPS	94A4310	stored, ex 780/NVA, l/n 4-93	to Dresden
2232	MiG-21SPS	94A4506	stored, ex 833/NVA, l/n 10-94	to Bad Ischl
2233	MiG-21SPS	94A5202	stored, ex 861/NVA, l/n 10-94	to Sinsheim
2234	MiG-21SPS	94A5204	stored, ex 869/NVA, l/n 8-94	to Seifertshofen
2235	MiG-21SPS	94A5206	stored, ex 878/NVA, l/n 4-93	to Harbke
2236	MiG-21SPS	94A5209	stored, ex 889/NVA, l/n 4-93	to Hermeskeil
2237	MiG-21SPS	94A5210	stored, ex 891/NVA, l/n 2-94	to Dessau
2238	MiG-21SPS	94A5509	stored, ex 897/NVA, l/n 2-94	to Finow
2239	MiG-21SPS	94A5510	stored, ex 898/NVA, l/n 2-94	to Uithuizen
2240	MiG-21SPS	94A5511	stored, ex 919/NVA, l/n 4-93	to Blomberg
2243	MiG-21SPS	94A5606	stored, ex 948/NVA, l/n 4-93	to Baschutz
2244	MiG-21SPS	94A6408	stored, ex 953/NVA, l/n 5-93	to Bad Oeynhausen
2245	MiG-21SPS	94A6410	stored, ex 963/NVA, l/n 8-93	to Seifertshoven
2277	MiG-21M	962308	stored, ex 425/NVA, l/n 4-93	scrapped at Dresden
2278	MiG-21M	960309	stored, ex 517/NVA, l/n 4-93	scrapped at Dresden
2279	MiG-21M	960406	stored, ex 552/NVA, l/n 4-93	scrapped at Dresden
2280	MiG-21M	960407	stored, ex 553/NVA, l/n 2-93	to Speyer
2281	MiG-21M	960409	stored, ex 559/NVA, l/n 4-93	scrapped
2282	MiG-21M	960502	stored, ex 562/NVA, l/n 4-95	scrapped
2283	MiG-21M	960505	stored, ex 571/NVA, l/n 10-94	to Wolls Petersdorf
2284	MiG-21M	960507	stored, ex 579/NVA, l/n 2-94	to Volkmarsen
2286	MiG-21M	960513	stored, ex 588/NVA, l/n 2-94	to Dresden
2287	MiG-21M	960514	stored, ex 588/NVA, l/n 2-94	to Finow
2289	MiG-21M	961111	stored, ex 611/NVA, l/n 4-93	scrapped
2319	MiG-21MF	968610	stored, ex 516/NVA, ex Drewitz, l/n 2-94	scrapped
2337	MiG-21MF	966214	stored, ex 685/NVA, ex Drewitz, l/n 2-94	scrapped
2338	MiG-21MF	965314	stored, ex 658/NVA, l/n 10-94	to Bad Wirchshofen?
2339	MiG-21MF	966211	stored, ex 681/NVA, l/n 3-95	scrapped
2340	MiG-21MF	966215	stored, ex 687/NVA, l/n 8-93	to Oberschleissheim
2341	MiG-21MF	96001012	stored, ex 767/NVA, l/n 10-94	to Seifertshofen
2342	MiG-21MF	96001039	stored, ex 772/NVA, l/n 2-94	to Cerbaiola
2343	MiG-21MF	96001091	stored, ex 774/NVA, l/n 2-94	to Hermeskeil
2344	MiG-21MF	96002003	stored, ex 775/NVA, l/n 8-93	to Hermeskeil
2345	MiG-21MF	96002009	stored, ex 776/NVA, l/n 2-94	to Trenton, Canada
2348	MiG-21MF	96002112	stored, ex 782/NVA, l/n 5-94	to Bad Oeynhausen
2359	MiG-21UM	05695168	stored, ex 212/NVA, l/n 10-94	to Dessau
2360	MiG-21UM	516921056	stored, ex 229/NVA, l/n 2-94	to Gilching
2361	MiG-21UM	516915006	stored, ex 233/NVA, l/n 4-93	to Finow
2378	MiG-21UM	02695163	stored, ex 204/NVA, l/n 3-95	to Bad Ischl
2379	MiG-21UM	03695163	stored, ex 205/NVA, l/n 5-95	to Hermeskeil
2380	MiG-21UM	04695163	stored, ex 206/NVA, l/n 2-94	to Baschutz
2381	MiG-21UM	10695162	stored, ex 207/NVA, l/n 10-94	to Diepensee
2382	MiG-21UM	516915011	stored, ex 231/NVA, l/n 4-93	to Cerbaiola

GERMANY - 220

2383	MiG-21UM	07695156	stored, ex 266/NVA, l/n 2-94	to Bad Oeynhausen
2386	MiG-21U-400	661119	stored, ex 237/NVA, l/n 5-95	to Bad Ischl
2388	MiG-21U-400	661016	stored, ex 251/NVA, l/n 2-94	to Dermsdorf
2389	MiG-21U-400	661118	stored, ex 258/NVA, l/n 8-93	to Dessau
2390	MiG-21U-600	662617	stored, ex 265/NVA, l/n 2-94	to Ankum
2391	MiG-21U-600	662619	stored, ex 272/NVA, l/n 8-93	to Bad Oeynhausen
2394	MiG-21U-600	663820	stored, ex 281/NVA, l/n 2-94	to Montelimar
2395	MiG-21U-600	664620	stored, ex 289/NVA, l/n 10-94	to Dermsdorf
2397	MiG-21U-600	664817	stored, ex 295/NVA, l/n 2-94	to Soltau
2399	MiG-21US	01685134	stored, ex 215/NVA, l/n 4-93	to Cottbus
2401	MiG-21US	07685145	stored, ex 217/NVA, l/n 2-94	to Ankum
2402	MiG-21US	08685145	stored, ex 218/NVA, l/n 2-93	to Dresden
2403	MiG-21US	09685145	stored, ex 219/NVA, l/n 8-93	to Dresden
2404	MiG-21US	07685147	stored, ex 221/NVA, l/n 2-94	to Fichtelberg
2405	MiG-21US	02685134	stored, ex 225/NVA, l/n 2-93	to Bayreuth
2408	MiG-21US	02685139	stored, ex 238/NVA, l/n 7-94	to Hermeskeil
2409	MiG-21US	04685139	stored, ex 242/NVA, l/n 4-93	to Ankum
2410	MiG-21US	05685139	stored, ex 246/NVA, l/n 10-94	to Dessau
2411	MiG-21US	06685139	stored, ex 248/NVA, l/n 2-94	to Seifertshofen
2412	MiG-21US	01685148	stored, ex 250/NVA, l/n 2-94	to Seifertshofen
2501	Su-22M-4	25307	stored, ex 360/NVA, ex Laage	to Cerbaiola
2801	L-39ZO	731001	stored, ex 139/NVA, l/n 6-92	to Peenemünde
2802	L-39ZO	731002	stored, ex 140/NVA, l/n 4-95	to G-BWTS
2803	L-39ZO	731003	stored, ex 141/NVA, l/n 6-92	to F-GOJS
2804	L-39ZO	731004	stored, ex 142/NVA, l/n 6-92	to Baschutz
2806	L-39ZO	731006	stored, ex 144/NVA, l/n 7-91	to Uetersen
2807	L-39ZO	731008	stored, ex 145/NVA, l/n 6-92	to Hungary as spare
2808	L-39ZO	731010	stored, ex 147/NVA, l/n 6-92	to Speyer
2809	L-39ZO	731012	stored, ex 149/NVA, l/n 4-95	to France
2810	L-39ZO	731013	stored, ex 150/NVA, l/n 4-95	to G-BWTT
2811	L-39ZO	731014	stored, ex 151/NVA, l/n 4-95	to France
2812	L-39ZO	731015	stored, ex 152/NVA, l/n 6-92	to Bad Oeynhausen
2813	L-39ZO	731016	stored, ex 153/NVA, l/n 6-92	to Hungary as spare
2814	L-39ZO	731017	stored, ex 154/NVA, l/n 3-95	to Bad Ischl
2815	L-39ZO	731018	stored, ex 155/NVA, l/n 6-92	to Hungary as spare
2816	L-39ZO	731020	stored, ex 157/NVA	to Grosser Weserbogen
2817	L-39ZO	731021	stored, ex 158/NVA, l/n 7-94	to Seifertshofen
2818	L-39ZO	731022	stored, ex 159/NVA, l/n 6-92	to Seifertshofen
2819	L-39ZO	731011	stored, ex 160/NVA, l/n 6-92	to Hungary as spare
2820	L-39ZO	731009	stored, ex 161/NVA, l/n 6-92	to Dessau
2821	L-39ZO	831125	stored, ex 162/NVA, l/n 6-92	to Hungary as 125
2822	L-39ZO	831114	stored, ex 164/NVA, l/n 6-92	to Hungary as 114
2823	L-39ZO	831115	stored, ex 165/NVA, l/n 6-92	to Hungary as 115
2824	L-39ZO	831116	stored, ex 166/NVA, l/n 2-94	to Fichtelberg
2825	L-39ZO	831118	stored, ex 173/NVA, l/n 6-92	to Hermeskeil
2826	L-39ZO	831119	stored, ex 174/NVA, l/n 6-92	to Hungary as 119
2827	L-39ZO	232301	stored, ex 175/NVA, l/n 6-92	to Sinsheim
2828	L-39ZO	831120	stored, ex 177/NVA, l/n 6-92	to Hungary as 120
2830	L-39ZO	232303	stored, ex 180/NVA, l/n 6-92	to Hermeskeil
2831	L-39ZO	232304	stored, ex 182/NVA, l/n 6-92	to Ankum
2832	L-39ZO	831122	stored, ex 188/NVA, l/n 6-92	to Hungary as 122
2833	L-39ZO	831123	stored, ex 189/NVA, l/n 6-92	to Bad Oeynhausen
2834	L-39ZO	831124	stored, ex 191/NVA, l/n 6-92	to Hungary as 124
2835	L-39ZO	831132	stored, ex 192/NVA, l/n 6-92	to Hungary as 132
2836	L-39ZO	831126	stored, ex 194/NVA, l/n 6-92	to Hungary as 126

GERMANY - 221

2837	L-39ZO	831127	stored, ex 195/NVA, l/n 6-92	to Finow
2838	L-39ZO	831128	stored, ex 196/NVA, l/n 6-92	to Hungary as 128
2839	L-39ZO	831129	stored, ex 197/NVA, l/n 6-92	to Fichtelberg
2840	L-39ZO	831130	stored, ex 198/NVA, l/n 6-92	to Hungary as 130
2841	L-39ZO	831131	stored, ex 199/NVA, l/n 6-92	to Hungary as 131
2842	L-39ZO	831134	stored, ex 214/NVA, l/n 6-92	to Hungary as 134
2843	L-39ZO	831135	stored, ex 216/NVA, l/n 6-92	to Hungary as 135
2844	L-39ZO	831137	stored, ex 271/NVA, l/n 6-92	to Hungary as 137
2845	L-39ZO	831138	stored, ex 273/NVA, l/n 6-92	to Hungary as 138
2846	L-39ZO	831139	stored, ex 277/NVA, l/n 6-92	to Hungary as 139
2847	L-39ZO	831140	stored, ex 279/NVA, l/n 6-92	to Hungary as 140
2848	L-39V	630705	stored, ex 170/NVA, l/n 6-92	to Berlin
2850	L-39ZO	831121	stored, ex 187/NVA, l/n 7-94	to Seifertshofen
2851	L-39ZO	831133	stored, ex 200/NVA, l/n 6-92	to Hungary as 133
2852	L-39ZO	831136	stored, ex 222/NVA, l/n 6-92	to Hungary as 136
9458	Mi-2S	564413105	stored, ex 352/NVA, ex Briest, l/n 10-94	to S9-TAL
9509	Mi-14BT	B4010	stored, ex 646/NVA, ex Parow, l/n 4-94	to S9-TAJ
9510	Mi-14BT	B4011	stored, ex 647/NVA, ex Parow, l/n 4-94	to S9-TAF
9511	Mi-14BT	B4012	stored, ex 648/NVA, ex Parow, l/n 4-94	to S9-TAG
9512	Mi-14BT	B4013	stored, ex 653/NVA, ex Parow, l/n 10-94	to S9-TAI
345	L-29	591528	dumped, reamins only, ex Bautzen, l/n 6-92	scrapped
359	L-29	591531	dumped, remains only, ex Bautzen, l/n 2-93	scrapped
361	L-29	390734	dumped, remains only, l/n 2-93	scrapped
001	Mi-1	..	stored, ex Kamenz, l/n 5-95	to Diepensee
31	MiG-15bis	..	dumped, l/n 7-91	gone
..	MiG-17	54311819?	dumped, l/n 7-91	gone
346	Lim-5	1C-0723	stored, l/n 9-91	to Drewitz
235	MiG-21U-400	..	dumped, fuselage only, l/n 2-93	scrapped
239	MiG-21U-400	660917	dumped, fuselage only, l/n 2-93	scrapped
247	MiG-21U-400	661020	dumped, fuselage only, l/n 2-93	scrapped
261	MiG-21U-600	662420	dumped, fuselage only, l/n 2-93	scrapped
274	MiG-21U-600	663218	dumped, fuselage only, l/n 2-93	scrapped
282	MiG-21U-600	663916	dumped, fuselage only, l/n 2-93	scrapped
284	MiG-21U-600	664519	dumped, fuselage only, l/n 2-93	scrapped
293	MiG-21U-600	664720	dumped, fuselage only, l/n 2-93	scrapped
701	MiG-21F-13	741706	dumped, ex gate, marked as '7011'	scrapped
812	MiG-21PFM	760514	stored, l/n 2-94	scrapped
935	MiG-21SPS	94A5601	dumped, l/n 7-91	scrapped
D-HZPF	Mi-2	563402034	stored, ex 302/NVA, l/n 4-96	returned to D-HZPF
..	Baade 152	011	stored, fuselage only, l/n 4-95,	to Dresden

ROTTENBURG AN DER LAABER
Displayed at the Generaloberst Weise kaserne was Sabre JC-101, it moved to Hermeskeil.

SCHÖNHAGEN
Preserved at the entrance of the former military and now civil airfield is an ex NVA L-29 Delfin.

❏ 370	L-29	591535	preserved, unmarked	9-97

SCHWELM
Directly at an Autobahn exit is a Noratlas in use as a restaurant.

❏ 5237	N2501D	39	preserved	91

SCHWENNINGEN

The International Luftfahrt Museum (at the airfield, some 70km east of Freiburg) is open daily between 09:00 and 19:00. Seen in the past were G91 3215 (ex Oldenburg, to Uetersen), T-33A 9402 (ex Westerland, to Uetersen), L-18C 9620 (ex Westerland, to Berlin), G91T 9940 (ex Hohn, to Uetersen) and Harvard D-FABU (to Berlin). Others which have gone include: Blériot D-EHCI and F-104G 2043 went to Hermeskeil and P149 D-EHMG, AT-6 D-FOBY and L-60 OE-BVL all went to Stuttgart. Klemm D-ENAE moved to Cuatro Vientos.

❏ 2613	An-2	1G26-13	preserved, ex Polish AF		9-96
❏ 1019	Lim-2	1B-01019	preserved, in false Soviet markings, ex Polish AF		9-96
❏ 1225	MiG-21F-13	741225	preserved, marked as '1981', ex Hungarian AF		9-96
❏ E.3B-526	CASA 1131E	..	preserved, marked as 'D-EBZE', ex Spanish AF		—
❏ ..	Skeeter AOP.12	..	preserved, ex British Army (not XL762)		9-96
❏ J-1068	Vampire FB.6	979	preserved, ex Luzern, ex Swiss AF		9-96
❏ J-4062	Hunter F.58	..	preserved, ex Swiss AF		—
❏ 2047	F-104G	2055	preserved, ex Lechfeld		9-96
❏ 2728	TF-104G	5730	preserved, fuselage only		12-95
❏ 3099	G91R/3	365	preserved, marked as '3599', with tail from 3299		10-93
❏ 3195	G91R/3	463	preserved, tail from 3256 (526), ex Sonthofen		9-96
❏ 7546	Alouette 2	1363	preserved		1-97
❏ 9464	T-33A	9152	preserved, ex Pinneberg, ex 54-1435		9-96
❏ 9918	OV-10B	338-3	preserved, ex D-9547		1-97
❏ 9934	Canberra B.2	R3/EA3/6651	preserved, ex Memmingen		1-97
❏ BB-239	CL-13 Sabre 5	840	preserved, marked as 'JB-111', ex Pinneberg		9-96
❏ VB-136	Sea Hawk Mk.100	6686	preserved, marked as 'MS-001', ex Flensburg		9-96
❏ D-EADL	Wiegel Harz Fink	01	preserved		12-95
❏ D-EBCG	LF-1 Zaunkönig	V4	preserved, ex D-ECER		12-95
❏ D-EDEV	RW3	004	store		12-95
❏ D-EDIB	Bü181B-1	25039	preserved, ex 25039/RSweAF		9-96
❏ D-EEHG	FWP149D	065	preserved, ex 9050		9-96
❏ D-EEVP	Evans VP-1	1	preserved		12-95
❏ D-EMKA	Do27B-1	152	preserved, ex 5532		9-96
❏ D-EHOX	Do27B-1	269	preserved, ex 5602		9-96
❏ D-EKLB	Grade Monoplane	7	preserved, replica		12-95
❏ D-ERLA	SV-4C	622	preserved, ex F-BDFH		9-96
❏ D-ERSV	Zlin 37A	16-20	preserved, ex DDR-SRV, ex DM-SRV		9-96
❏ D-HMIA	Mitka Mi-1	..	preserved		12-95
❏ D-KFFS	Fs26 Moseppl	V1	preserved		12-95
❏ D-7033	Grunau SG-38	..	preserved, glider		12-95
❏ D-8290	Fauvel AV36	..	preserved, glider		12-95
❏ D-9204	Grunau Baby IIb	03	preserved, glider		12-95
❏ F-BSEI	Zlin 526	1051	preserved, ex OO-PTZ, ex OK-XRG		12-95
❏ HB-DEG	Mooney M.20E	345	preserved		9-96
❏ HB-RAY	P2-05	35	preserved, ex U-115/Swiss AF		9-96
❏ HB-RBV	P3-05	468-17	preserved, ex A-830/Swiss AF		9-96
❏ ..	Bo-105	..	preserved, with tailboom from D-HMMM		9-96
❏ ..	Bf109E	..	preserved, replica, marked as '6'		9-96
❏ ..	Europa Jet	..	preserved, mock-up, marked as '9107', in bag		4-93
❏ ..	Fokker E.III	..	preserved, replica, marked as '36/15'		9-96
❏ ..	Fw190A-3	..	preserved, replica, marked as '<-11'		4-91
❏ ..	Ju87	..	preserved, replica, marked as 'T6-KL', ex Frankf		12-95
❏ ..	Me262A-1a	..	preserved, replica, marked as '5', ex Frankfurt		12-95

SECKENHAUSEN

A unknown red MiG-21 is parked at the Ferrari car dealer in the Haupstrasse. Seckenhausen is south of Bremen.
❏ ..   MiG-21   ..   preserved, unmarked   3-98

SEE
The MiG-17 here has windows in the fuselage and instead of an engine has some seats for the children to sit on.
☐ 22　　　　　　　MiG-17　　　　　　　..　　　　　preserved　　　　　　　　　　　　　　　8-96

SEIFERTSHOFEN
Only a few aircraft are really preserved in Herr Kiemeles Schwabisches Bauern und Technik Museum as most of the other aircraft are stored and awaiting sale. Except for the museum's aircraft, only the MiG-15s are stored inside. The aircraft shown as 'in compound' are stored in a yard across the road. French C-45 44562 was sold off, as was G91R 3058 (318, ex Bremgarten). F-4Cs 63-7467 (453, ex Leipheim) and 63-7536 (562, ex Norvenich) were both scrapped in 1995. The BMW dealer from Baarlo did some shopping here and bought F-4 64-0745 (ex Spangdahlem), MiG-23BN 2050 (ex Laage), MiG-23UB 2059 (ex Laage), MiG-21SPS 2245 (ex Rothenburg) and MiG-21US 2411 (ex Rothenburg). MiG-23UB 2056 (ex Laage) went to Nieuw Loosdrecht, while the last Phantom here, F-4C 64-0757, went to Cerbaiola. Also F-104G 2628 and Javelin XH768/E have gone to Cerbaiola. Starfighters 2813 and 2822 (both ex Erding) also passed through here in 1997 on their way to Baarlo. Scrapped were Pembroke 5417 (1008) and Be95-A55 D-IKUC (TC-194). Ce310B D-IDIX (35527) was sold in 1996.

☐ FT-09		T-33A	5995	stored, in compound, ex Belgian AF	5-97
☐ FT-26		T-33A	9064	stored, in compound, ex Belgian AF	5-97
☐ FU-160		F-84F	..	preserved, inside, ex Belgian AF, ex 53-6899	5-97
☐ 54-2185	11-MO	F-100D	223-65	preserved, marked as '54-2136', ex French AF	5-97
☐ 618		N1300	618	preserved, ex French AF	5-97
☐ 144685		SP-2H	726-7136	stored, fuselage in sections, ex French Navy	9-96
☐ 51-17487	33-XG	T-33A	7381	stored, in compound, ex French AF	5-97
☐ 307		MiG-15	..	stored, marked as '1973', ex Hungarian AF	5-92
☐ 915		MiG-21F-13	741915	stored, marked as '1981', ex Hungarian AF	5-97
☐ 143	D	SH-34J	58-1153	preserved, inside, ex RNethNavy	5-97
☐ 01		MiG-15	..	stored, Soviet c/s, ex ?	1-92
☐ 57-0348		F-101B	..	stored, marked as '85-0701', ex Spangdahlem	5-97
☐ 2008		MiG-23MF	0390213351	stored, ex Laage, ex 592/NVA	5-97
☐ 2042		MiG-23BN	0393214210	stored, ex Drewitz, ex 694/NVA	5-97
☐ 2234		MiG-21SPS	94A5204	stored, ex Rothenburg, ex 869/NVA	5-97
☐ 2341		MiG-21MF	96001012	stored, ex Rothenburg, ex 767/NVA	5-97
☐ 2412		MiG-21US	01685148	stored, ex Rothenburg, ex 250/NVA	5-97
☐ 2513		Su-22M-4	25020	stored, ex Laage, ex 641/NVA	5-97
☐ 2541		Su-22M-4	26716	stored, ex Laage, ex 743/NVA	5-97
☐ 2817		L-39ZO	731021	stored, ex Rothenburg, ex 158/NVA	1-97
☐ 2818		L-39ZO	731022	stored, ex Rothenburg, ex 159/NVA	1-97
☐ 2850		L-39ZO	831121	stored, ex Rothenburg, ex 187/NVA	1-97
☐ 5205		An-26T	10509	stored, ex 376/NVA, ex DDR-SBD	5-97
☐ 5210		An-26M	14208	stored, in compound, ex 369/NVA	5-97
☐ 5343		N2501D	173	preserved	5-97
☐ 8318		H-21C	WG-18	preserved, with tail from 8310, parts from 8315	5-97
☐ 9421		Mi-8T	10526	stored, ex Briest, ex 910/NVA, D-HOZE ntu	5-97
☐ 9450		Mi-2S	563401044	stored, ex Briest, ex 301/NVA	5-97
☐ AS-930		Do27B-1	278	stored, ex preserved, marked as 'HF-201'	9-96
☐ D-5383		Bergfalke II	..	preserved	5-97
☐ N8041B		C-47A	10100	stored, marked as 'N569R', ex Bochum	5-97
☐ SP-WUS		PZL 106A	37016	stored	5-97

SEMBACH
The times of A-10s and OV-10s here are long gone. From the early 1990s no more operational aircraft were based here and by 1996 the USAFE closed all its business here. All the non operational aircraft have not been

GERMANY - 224

noted since 1990, these include the F-101B 58-0267/02 (639), T-33A 54-1581 (9270) and CT-39A 62-4471 (276-24). The instructional F-4C 64-0922 moved to Mangam.

| ☐ 52-5372 | F-86F | 193-101 | preserved, ex C.5-163/Spanish AF | 4-97 |

SIEGBURG
Preserved in this town, west of Bonn, at the Brückberg kaserne is a former Bremgarten F-104G.

| ☐ 2004 | F-104G | 2004 | preserved, marked as '2404', ex Bremgarten | 2-98 |

SINSHEIM
A number of aircraft from the Auto und Technik Museum can be seen from the Highway 6. Displayed is a varied collection of aircraft of which a relatively high number are painted with false markings. The museum is daily open between 09:00 and 18:00. Several aircraft have departed to the sister museum at Speyer.

Hall 1:

☐ 099	AT-16ND	14-555	preserved, marked as 'FT454', ex RNethNavy	11-97
☐ B.2I-82	CASA 2111B	045	preserved, marked as '5J-GN', ex Spanish AF	11-97
☐ T.2B-209	CASA 352L	..	preserved, marked as 'D-AQUI', ex Spanish AF	11-97
☐ 1006	Lim-2	1B-01006	preserved, marked as '12', ex Polish AF	11-97
☐ C-501	C3605	281	preserved, ex Swiss AF	11-97
☐ J-1628	Venom FB.50	838	preserved, ex Swiss AF	11-97
☐ J-1798	Venom FB.54	968	preserved, ex Swiss AF	11-97
☐ 2249	F-104G	7129	preserved	11-97
☐ EB-302	RF-84F	04530	preserved, forward fuselage only, ex 51-1862	11-97
☐ 1379	Ju88A-4	1379	preserved, marked as '4V-UH'	11-97
☐ 1301643	Ju87B	1301643	preserved, unmarked, wreck, crashed 16-8-44	11-97
☐ D-EDON	DH.82A	82043	preserved, marked as 'DE623', ex N6779	11-97
☐ D-HBAU	Ka-26	7303204	preserved, as 'CCCP-26001', ex CCCP-24054	11-97
☐ D-IKER	Dove 7	04530	preserved, unmarked, ex Speyer, ex G-ARUE	11-97
☐ D-1619	Bergfalke II/55	374	preserved, glider	11-97
☐ D-1876	Ka4	3036	preserved, glider	11-97
☐ G-AWHS	Ha1112M1L	228	preserved, marked as '-4', ex C.4K-170/Spa AF	11-97
☐ HA-ANB	An-2T	1G168-05	preserved, marked as '03 red'	11-97
☐ HB-USE	Zlin 381	325	preserved, local built Bü181D, ex OK-DRB	11-97
☐ N3951A	B707-123B	17647	preserved, cockpit only	11-97
☐ N9012P	CASA 352L	148	preserved, marked as 'RJ-NP', ex T.2B-140	11-97
☐ ..	Fokker E.III	..	preserved, replica, unmarked	11-97
☐ ..	Fw190A-3	..	preserved, replica, marked as '10-'	11-97
☐ ..	HM-14	..	preserved, marked as 'D-EMIL'	11-97
☐ ..	Kurir L	..	preserved, as 'PT-TP', Czech built Fi156	11-97
☐ ..	MS500	..	preserved, marked as 'D-EMWF'	11-97

Hall 2:

☐ J-1603	Venom FB.50	813	preserved, bare metal c/s, ex Swiss AF	11-97
☐ 3264	G91R/3	534	preserved, with tail from 3295 (566)	11-97
☐ D-EAPT	Frebel F5 Aeolus	1	preserved	11-97
☐ D-HAUF	S-58C	58-356	preserved, ex B-11/Belgian AF, ex OO-SHI	11-97
☐ D-8055	Grunau Baby IIB	..	preserved, glider	11-97
☐ D-8117	Kaiser Ka1	..	preserved, glider	11-97
☐ D-8182	Gruanu SG-38	..	preserved, glider	11-97
☐ D-8771	L-13 Blanik	..	preserved, glider	11-97
☐ D-9187	L-Spatz 55	565	preserved, glider	11-97
☐ ..	BK117A	..	preserved, mock-up, marked as 'DRF-4'	11-97

GERMANY - 225

Outside:

☐ 0833	Il-14P	14600833	preserved, marked as 'CCCP', ex Polish AF	11-97
☐ 2027	MiG-23ML	0390324018	preserved, ex Laage, ex 550/NVA	11-97
☐ 2233	MiG-21SPS	94A5202	preserved, ex Rothenburg, ex 861/NVA	11-97
☐ 2520	Su-22M-4	26307	preserved, marked as '798', ex 704/NVA	11-97
☐ 2827	L-39ZO	232301	preserved, in Czech markings, ex 175/NVA	11-97
☐ 8317	H-21C	WG17	preserved, unmarked	11-97
☐ 9418	Mi-8T	0323	preserved, ex Briest, ex 394/NVA	11-97
☐ 9936	Canberra B.2	R3/EA3/6644	preserved, ex D-9566, ex 0002, ex YA-152	11-97
☐ JD-397 7	T-33A	1758	preserved, ex 58-0709/USAF	11-97
☐ D-CAIL	CASA 352L	148	preserved, ex 'D-2527', ex Marl, ex 'D-ADAM'	11-97
☐ D-CAKE	Pembroke C.54	P66/93	preserved, ex 5402	11-97
☐ F-BGNU	Viscount 708	38	preserved	11-97
☐ HA-LBH	Tu-134A	0350925	preserved, ex HA-925	11-97
☐ N8041A	DC-3A	14005/25450	preserved, marked as 'D-CADE', ex T.3-62	11-97
☐ OK-PAI	Il-18E	181003105	preserved, ex OK-BYP, ex CCCP-75754	11-97

SOEST
During 1993 the British Army's 3 Regiment was relocated to Wattisham, taking the Sioux XT190 (WA349) with them. Wasp HAS.1 XT436 went to <u>Detmold</u>, but the fate of Scout AH.1 XP897 (F9498) is unknown.

SÖLLINGEN
Town: On 7th November 1995 the former Söllingen gate guard was erected as a monument in the village.

☐ 104785	CF-104	1085	preserved, ex airfield	5-98

Airfield: This former Canadian base has be reopened as a replacement of the airfield of Baden Baden Oos and Karlsruhe Forchheim and is named Baden Airpark. All of the former gate guards, CF-100 Mk.5 18784, CF-104 104785 (1085, to town), T-33 21417 (417, to Elvington, UK) and CL-13B Sabre 0103 (1605, as '23605'), have gone. A large number of instructional Starfighters were noted in the 1980s. Of these 104648, 104750 and 104799 went to <u>Savigny lès Beaune</u>, while the former gate guard 104785 moved to the village. 104790 is now at Hamilton, Canada as '104756'. The fate of the rest, Canadian Starfighters 104634, 104639/856B, 104653/822C, 104706, 104790, 104805, 104822/848C, 104835/849C, 104843/851C, 104880/850C and Luftwaffe F-104F 2921 (5076, l/n October 1992), is unknown. They are no longer here and may have all been scrapped.

SOLTAU
Parked inside on concrete blocks at the local Heidepark is a MiG-21U 'Mongol'. The aircraft is painted in a non standard colours.

☐ 2397	MiG-21U-600	664817	preserved, ex Rothenburg, ex 295/NVA	97

SONTHEIM
A car garage in the Gundefingerstrasse at the east side of town has a G91 to attract passers-by.

☐ 3010	G91R/3	0063	preserved	7-95

SONTHOFEN
Several airframes have been reported at the Jagerkaserne. The current status of UH-1D 7355 (8475) seen here in the 1980s is required. The preserved G91R 3195 went to <u>Schwenningen</u>.

☐ 2098	F-104G	6619	preserved, marked as '2203'	97
☐ 2902	F-104F	5048	instructional, ex 59-4995	2-93
☐ YA-005	CL-13 Sabre 5	819	preserved, marked as 'GAF-001'	9-92

GERMANY - 226

SPANGDAHLEM
The last remaining USAFE fighter base in Germany is Spangdahlem and it still houses a fair amount of *Wrecks and Relics*. The dumped T-33A 51-9147 (6931) was scrapped in the mid-1980s. F-101B 57-0348 (marked as '86-0701') went to Seifertshofen, together with F-4Cs 64-0745 and 64-0757. Instructional F-4E 67-0260 is now at the museum at Oberschleissheim.

☐ 76-0550		A-10A	97	instructional	8-97
☐ 76-0553	SP	A-10A	100	instructional	4-96
☐ 77-0264	SP	GA-10A	189	instructional	8-97
☐ 66-0308	SP	F-4E	2505	preserved, at gate, arr 12-2-89	8-97
☐ 69-7236	SP	F-4G	3904	instructional	7-93
☐ 73-0095	SP	F-15A	36/A29	instructional, ex Bitburg, arr 9-93	8-97
☐ 74-0085	SP	F-15A	57/A46	instructional	6-97
☐ 78-0057	SP	F-16A	61-53	preserved, at gate, marked as '85-1552/SP'	8-97
☐ 62-4446	SP	F-105G	..	preserved, at gate, ex FK088/MASDC	8-97

SPERENBERG
The former headquarters of the Soviet troops in (East) Germany was located at Sperenberg. Based here also were a number of transport aircraft. All (including the *Wrecks & Relics* aircraft?) have left the base.

☐ 50 red	MiG-21MF	..	preserved, at gate, ex Soviet AF	5-95
☐ 05 red	Il-14G	147001823	stored, ex Soviet AF	5-95
☐ 64	An-12BP	0901209	dumped, in woods, ex Soviet AF	5-95

SPEYER
The Auto und Technik Museum from Sinsheim has set up a second museum at Speyer. The openings times are daily between 09:00 and 18:00 hrs, the same as those at Sinsheim. Dove D-IKER is now at Sinsheim. Note that the L-39 is preserved at the hotel entrance, just east of the museum (in the same area).
Inside:

☐ 0805		L-29	290805	preserved, marked as 'OK-02', ex Czech AF	3-98
☐ 387		CM170	387	preserved, ex French AF	3-98
☐ 355		Mirage 3R	355	preserved, marked as '33', ex 33-TD, ex FAF	3-98
☐ 432	13-QJ	Mirage 3E	432	preserved, arr 1993, ex French AF	3-98
☐ 45 yellow		Mi-2	511622100	preserved, ex Soviet AF	3-98
☐ 18 red		MiG-15UTI	0415320	preserved, ex Soviet AF	3-98
☐ A-808		P3-03	326-8	preserved, ex Swiss AF	3-98
☐ J-1081		Vampire FB.6	992	preserved, ex Swiss AF	3-98
☐ 0104		CL-13B Sabre 6	1613	preserved, marked as '23042', ex Sinsheim	3-98
☐ 2827		TF-104G	5957	preserved, ex 63-8453	3-98
☐ 7535		Alouette 2	1327	preserved	3-98
☐ 9928		OV-10B	338-13	preserved, ex D-9557	3-98
☐ 6821		Ju52/3mg4e	6821	preserved, marked as 'CA-JY', ex 'VB-UB'	3-98
☐ D-ECYV		Bü181B-1	331381	preserved, ex SL-AAS	4-97
☐ D-EJHD		Jurca MJ-5	D-29	preserved	8-97
☐ D-EJZO		M-1D Sokol	..	preserved	8-97
☐ D-KONY		Raab Krahe 2	222	preserved, ex Sinsheim, ex OE-9008	8-97
☐ D-1948		Meise	9	preserved, glider	8-97
☐ D-8594		Ka4	762	preserved, glider	8-97
☐ D-9532		CM191B	V2	preserved, ex Mannheim	3-98
☐ N211EL		Beech E50	EH-56	preserved, marked as 'D-ITMS', ex A-711/SwiAF	3-98
☐ SP-ZBH		PZL 106BR	07810145	preserved, ex Sinsheim	8-97
☐ ..		Alpha Jet	..	preserved, marked as '33'	3-98
☐ ..		Fi156C-3	5440	preserved, marked as '127/H3-BF', ex Sinshiem	3-98
☐ ..		Fokker Dr.I	..	preserved, replica, marked as '152/17'	8-97

GERMANY - 227

☐ ..	HM-14	..	preserved, marked as 'D-EMIL'		6-96
☐ ..	SV-4C	..	preserved, replica, marked as '1010/IF'		8-97

Outside:

☐ 33 red	Mi-8T	3135	preserved, marked as 'CCCP-06181', ex Soviet	8-97
☐ 154	N2501F	154	preserved, ex Saintes, ex French AF	3-98
☐ 37446	AJ37	37446	preserved, ex Swedish AF	5-98
☐ J-4072	Hunter F.58	..	preserved, ex Swiss AF	3-98
☐ 63-7423	F-4C	364	preserved, ex Zweibrücken, ex USAF	3-98
☐ 63-7446	F-4C	413	preserved, ex Ahlhorn, ex USAF	3-98
☐ 74-0109	F-15A	83/A70	preserved, ex Bitburg, ex USAF	3-98
☐ 58-0265	F-101B	637	preserved, ex Ahlhorn, ex USAF	3-98
☐ 2039	MiG-23BN	0393211087	preserved, marked as '2202', ex 690/NVA	3-98
☐ 2049	MiG-23BN	2963222830	preserved, ex Drewitz, ex 705/NVA	3-98
☐ 2201	F-104G	7070	preserved	3-98
☐ 2225	MiG-21SPS	94A4301	preserved, ex Rothenburg, ex 738/NVA	3-98
☐ 2280	MiG-21M	960407	preserved, forward fuselage only, ex 553/NVA	3-98
☐ 2308	MiG-21MF	967610	preserved, nose only, ex Drewitz	7-97
☐ 2514	Su-22M-4	26001	preserved, ex Laage, ex 644/NVA	3-98
☐ 2808	L-39ZO	731010	preserved, ex Rothenburg, ex 147/NVA	3-98
☐ 3147	G91R/3	415	preserved, ex Kassel	3-98
☐ 5204	An-26S	10409	preserved, ex 375/NVA, ex DDR-SBN	3-98
☐ 9401	T-33A	7365	preserved, marked as '63659/TR-659/43'	3-98
☐ 9506	Mi-14PL	B4006	preserved, ex Parow, ex 637/NVA	3-98
☐ DD-308	F-84F	..	preserved, marked as 'BA-102', ex 52-6816	3-98
☐ UA-112	Gannet AS.4	F9394	preserved, unmarked, ex Bremen	3-98
☐ ..	Do24T-3	..	preserved, fuselage only, from Müritz Lake	3-98
☐ D-ILUX	RC680F	1087-65	preserved, ex N6245X	8-97
☐ F-BFGX	DC-3A	11722	preserved, ex SE-BAW, ex 42-68795	3-98
☐ F-BTTB	Mercure 100	2	preserved, arr 12-6-95	8-97
☐ OY-TOR	VFW614	G-004	preserved, ex Lemwerder, ex D-BABD	8-97
☐ RA-41343	An-2TP	1G65-18	preserved	8-97

ST AUGUSTIN

In March 1998 it was announced that the remainder of the Butzweilerhof collection will set up a new museum in this small town south west of Bonn, not very far from the airfield of Hangelar. A number of former Butzweilerhof aircraft are already on temperary storage at Bonn-Hangelar (which see).

STADE

The Technik und Verkehrs Museum is located in the Freiburgerstrasse in Stade and is open Thursday to Friday (10:00-16:00) and weekends between 10:00 and 18:00. The MiG-17 is owned by a collector from nearby Drochtersen and was stored at the back of the museum. In late 1997 the MiG-17 moved to the Luftwaffe museum at Berlin Gatow.

☐ 2061	MiG-23UB	A1037826	preserved, ex Laage, ex 102/NVA	9-96
☐ 2528	Su-22M-4	30917	preserved, ex Laage, ex 598/NVA	9-96
☐ 9334	Mi-8T	10530	preserved, ex Briest, ex 902/NVA	9-96

STÖLLN

An aerial photo from October 1995 still showed a complete Il-62 at Stölln, which is just east of Rhinow. It is part of the Otto Lilienthal Gedenkstätte.

☐ DDR-SEG	Il-62	31403	preserved, arr 23-10-89, ex DM-SEG	10-95

STOLZENAU
A Dutch air defence unit was based at Stolzenau. Due to defence cuts the unit was relocated to Holland, taking their gate guarding Starfighter with them. F-104G, D-8051, is now at Gilze Rijen.

STRAUSBERG
After October 1990 two L-410UVP-T Turbolets were stored at this former NVA headquarters airfield, east of Berlin. Both aircraft were sold within a year; 5306 (831135) went to the air force of Letvia as 46, while 5308 (831137) went the other way and is now YS-407. The airfield was completely empty when checked in 1996.

STUTTGART
Town: The Daimler-Benz Museum in the Unterturkheim part of town may still have a Halberstadt and Klemm.

Airfield: The airfield has set up a collection of aircraft (the Albatros Flugmuseum) on their spectator terrace. The ex Malev Tu-154 is used by the airfield's fire service as a trainer.

❑ D-EDHS	Do27A-4	371		preserved	7-96
❑ D-EHMG	P149D	320		preserved, ex Schwenningen, ex Düsseldorf	7-96
❑ D-FOBY	AT-6A	77-4176		preserved, ex Schwenningen, ex Düsseldorf	7-96
❑ G-AVHE	Viscount 812	363		preserved, forward fuselage, ex N251R	7-96
❑ HA-LCB	Tu-154B-2	046		instructional, fire trainer, arr 20-1-95	8-97
❑ OE-BVL	L-60 Brigadyr	150401		preserved, ex Schwennigen, ex OK-LGL	7-96
❑ ..	Albatros L.13	..		preserved, replica, marked as 'D-961'	7-96
❑ ..	Junkers F13	..		preserved, replica, marked as 'D-1'	7-96

TANNHEIM
Some of the museum collction from Augsburg arrived here in 1993. This was short lived here as most aircraft had left by 1995. Josef Koch set up a new museum at Sandown on the Isle of Wight, UK. All of the Sandown aircraft should have been here, but not all of them were seen here. Maybe some went direct to Lahr (where some aircraft of the collection were noted in 1996, including MiG-15UTI 140 (marked as '44') and G91R 3074 which is now at Rothenburg) and from there to the UK. MS505 73 is now flying in the UK as G-BWRF, while MS733 read as 71/YT was noted at the Freimann railway maintenance building in München on 23rd October 1997. Former Augsburg museum Lim-5 413 was noted here in July 1994 and is now at Hermeskeil. The fate of the two fighters is not known, they did not move to the UK.

❑ 2046	F-104G	2054	preserved, ex Augsburg	8-93
❑ 2428	MiG-21bis	N75033169	preserved, ex Drewitz, ex 873/NVA	7-94

TEMPLIN
At this former Soviet airbase a Su-7U was preserved. It may have gone by now.

❑ 07 red	Su-7U	..	preserved	3-94

ÜBERLINGEN
At the ABIG Werke, a factory in Überlingen which is some 20km north west of Friedrichshafen, a former NVA Mi-9 is preserved.

❑ 9394	Mi-9	340008	preserved, ex Basepohl, ex 426/NVA	8-96

UETERSEN
On 17th May 1995 the first aircraft (a Delfin and Starfighter) were moved by HEER CH-53s to the new location of the Luftwaffe museum at Berlin Gatow. By late 1996 all aircraft intended for Berlin had left, with the exception of the Mi-14 and N2501. These will go in the near future. The Su-22 and Sabre will stay at Uetersen.

Some aircraft have gone from here to other places such as MiG-21SPS 959 (6503) which went to Duxford, UK, together with Sea Fury D-CACY (ES-3617). F-104G 2045 is now at <u>Laatzen</u>, Do28Ds 5901 and D-ICDY are now at <u>Fassberg</u>. The <u>Finow</u> museum acquired MiG-17F 08 (ex Bautzen) and HFB320 Hansajet D-CARE. MiG-21SPS 829 (ex Bautzen) and G91R 3211 (nose only) have gone to <u>Butzweilerhof</u>. T-33A 9402 went to <u>Hamburg</u>, while <u>Ankum</u> received MiG-21F-13 679 and MiG-21SPS 969. Fate of the G91R is unknown.

❑ 2540	Su-22M-4	26715	preserved, ex Laage, ex 741/NVA	97
❑ 3103	G91R/3	369	stored, wreckage, ex Leipheim	—
❑ 9501	Mi-14PL	B4001	stored, arr 24-4-91, ex 617/NVA	97
❑ 9914	N2501D	152	stored, ex D-9580, ex 5330	97
❑ JB-110	CL-13B Sabre 6	1643	preserved	97

ULM
Displayed on a pole at the entrance to Boelcke kaserne (in Romerstrasse) is a F-84F. The barracks is on the south western outskirts of Ulm.

❑ DD-373	F-84F	..	preserved, ex 51-1816	4-91

UMMENDORF
A Starfighter has been reported with LwWerft32 at Ummendorf near Biberach.

❑ 2243	F-104G	7121	preserved	3-91

UNSERE LUFTWAFFE
Not a location, but the title of the Luftwaffe's equivalent to the RAF Exhibition Flight. They use airframes as travelling exhibits. Redundant airframes are F-104G 2043 (to <u>Schwenningen</u>) and 2462 (8211, as '3105', gone).

❑ 2064	F-104G	2075	preserved, forward fuselage only	7-96
❑ 4059	Alpha Jet	0059	preserved	5-98
❑ 7321	UH-1D	8441	preserved, marked as '7030'	5-98
❑ 8693	Bo105P	60-93	preserved	5-98
❑ 9801	Tornado	P11	preserved, marked as '4400', ex '4800'	5-98
❑ 9804	Tornado	P01	preserved, marked as '4448', ex '4300'	5-90

VOLKMARSEN
Some 30km west of Kassel is the village of Volkmarsen. The ex Rothenburg MiG-21M is displayed on top of some containers at a second hand shop. The aircraft carries false Soviet stars.

❑ 2284	MiG-21M	960507	preserved, ex Rothenburg, ex 579/NVA	1-97

WERNEUCHEN
After the departure of the based Soviet fighters, the preserved Yak-27R 35 red moved on (to <u>Diepensee</u>). The stored Yak-28BI 22 red (l/n 3-93, 6941607) was scrapped locally.

WESTERLAND
Westerland air base on the island of Sylt houses the Marineflieger's technical training school. No reports have come from the school since 1985 and an update of the situation here would be appreciated. The Atlantic here was the first prototype and has now been painted in spurious markings as '6100' to precede the first WGN production aircraft 6101. F-104G 2502 went <u>Pferdsfeld</u> and the museum at <u>Schwenningen</u> received T-33A 9402 and L-18C 9620. Turkey took delivery of 2214 (7085), while 2224 and 2568 (as 6951') went to <u>Münster</u>.

❑ 2184	F-104G	7053	instructional, with parts of 2568	82
❑ 2298	F-104G	7181	instructional	7-87
❑ 2324	F-104G	7208	instructional	90

❏ 8961	Sea King Mk.41	WA765	instructional, crashed 16-1-74		85
❏ 9181	P149D	264	instructional		85
❏ JA-301	CL-13B Sabre 6	1647	instructional, fire dump		83
❏ 01	Br1150	01	instructional, marked as '6100' (03), ex 'UC-301'		85

WILDENRATH
The RAF moved out of this base many years ago. None of the *Wrecks & Relics* aircraft remain. Hermeskeil received Hunter XF418. To Brüggen went Phantom XV569. Lightning T.4s XM955/T (95104), F.2A XN778/A (95131) and F.6 XR727/BH (95210) were scrapped, as was Pembroke C.1 WV701 (PAC/66/4).

WILHELMSHAFEN
F-104G 2112, which was preserved at the Marine arsenal has, moved on to the gate at Neubrandenburg.

WITTMUND
Town: Still preserved at the gate of JG71's barracks in the town of Wittmund are a Sabre and Starfighter. Inside should be the Richthofen museum, to whom the other aircraft should belong.

❏ C.4K-134	HA1112-M1L	194	preserved, marked as '027083/12'	8-96
❏ 2086	F-104G	6602	preserved	6-97
❏ JA-112	CL-13B Sabre 6	1724	preserved	6-97
❏ JB-112	CL-13B Sabre 6	1775	preserved, marked as 'JA-110', ex Ahlhorn	—
❏ ..	Fokker Dr.I		preserved, replica, marked as '102/17'	—

Airfield: The cockpit of F-104 2631 should still be displayed in the JG71 Traditions Room. The rest of the aircraft should be scrapped by now.

❏ 0106	CL-13B Sabre 6	1664	preserved, marked as 'JA-106'	8-96
❏ 2246	F-104G	7124	preserved, cockpit only, in traditionsraum	91
❏ 2301	F-104G	7184	instructional	4-95
❏ 2485	F-104G	8235	preserved, marked as 'JG71'	8-96
❏ 2498	F-104G	8251	instructional, ex Fassberg	4-95
❏ 2631	F-104G	9183	stored, see notes	1-92
❏ ..	F-104G	..	dumped, no tail	8-96
❏ 3268	G91R/3	538	decoy	4-89

WOLGAST
Both the An-2 and Zlin 37 are painted in the bright white or yellow Opel car colours and are marked 'Nur Opel fahren ist schöner' (Only driving an Opel is nicer). Wolgast is in north west Germany on the road towards to the Peenemünde museum.

❏ D-FONI	An-2T	17807	preserved, completely white, ex DDR-SKI	10-97
❏ ..	Zlin 37	..	preserved, completely yellow	10-97

WOLLS PETERSDORF
A second-hand car dealer has bought two MiGs from Rothenburg in 1995. The garage Lorenz is east of Halle, along the main road number 100 towards Bitterfeld.

❏ 2009	MiG-23MF	0390213352	preserved, red c/s, ex Rothenburg, ex 593/NVA	2-98
❏ 2283	MiG-21M	960505	preserved, ex Rothenburg, ex 571/NVA	2-98

WÜNSDORF - ZOSSEN
All the aircraft of the Soviet's 16 Vozdushnaya Armiya Museum were sold in 1994 to other museums. Only the

MiG-15UTI 16 red and Yak-27R 28 red have surfaced so far, both are now at museum at Dessau.

❏ 01 red	Mi-2	..	preserved, ex Soviet	2-93
❏ 50 red	MiG-21MF	..	preserved, on pole, ex Soviet	2-93
❏ 24 red	MiG-23UB	..	preserved, ex Soviet	2-93
❏ 07 red	Su-7BM	..	preserved, ex Soviet	2-93
❏ --	Il-2	..	preserved, replica, marked as '50 red', ex Soviet	2-93

WUNSTORF

The Transall base of Wunstorf still has a number of instructional airframes from the BAW (Berufs Ausbildungs Werkstatt). Scrapped is the instructional G91R/3 3300 (571, marked as '3309') and the Do27 of which the complete serial will never be known (5.35, second digit is 5 or 6). Do27A 5616 went to Celle and Do28D 5841 has gone to Norvenich.

❏ 3036	G91R/3	0096	instructional	6-94
❏ 3137	G91R/3	405	instructional, ex decoy	10-94
❏ 3245	G91R/3	514	instructional	6-94
❏ 3447	G91T/3	607	instructional	6-94
❏ 5870	Do28D-2	4145	instructional	8-94
❏ 5890	Do28D-2	4165	instructional	6-94
❏ 5892	Do28D-2	4167	preserved, used D-IDBH for ferry	6-94
❏ 9035	FWP149D	049	preserved, at gate marked 'AC-404', ex '3051'	5-98
❏ 9044	FWP149D	058	instructional, unmarked, ex D-EEHG?	6-94
❏ 9056	FWP149D	073	instructional, marked as 'BAW'	6-94

Next to the gate is a museum set up in honour of the Ju52. The K-65 is only on temporary loan.

❏ 66	N2501F	66	preserved, marked as 'GR-248', outside, ex FAF	5-98
❏ 5895	Do28D-2	4170	preserved, ex instructional	5-98
❏ 7068	UH-1D	8128	preserved	5-98
❏ 9303	Mi-8T	10511	preserved, outside, ex 400/NVA	5-98
❏ 6693	Ju52/3mg4e	6693	preserved, marked as 'DB-RD'	9-97
❏ D-EMAV	K-65 Cap	741	preserved, Czech built Fi156C-3	5-97

WUPPERTAL

Preserved in an industrial estate near one of the autobahn exits of Wuppertal (at the south side of town?) is an unmarked Harvard.

❏ D-FABY	AT-6C	88-9811	preserved, ex F-BJBL, ex H-30/BelgAF	9-93

WÜRZBURG

A new museum is currently being set up in this town. It will have the remains of about a dozen crashed aircraft, from mostly the Second World War period, on display. They were all excavated in Germany. The only complete aircraft will be the former Baschutz MiG-21 dual.

❏ 2380	MiG-21UM	04695165	stored, ex Baschutz, ex Rothenburg, ex 206/NVA	97

ZEPFENHAHN

Behind hangars at this airfield east of Rottweil (some 20km east of Schwenningen) is an Italian-built P149.

❏ 9214	P149D	310	stored	7-92

ZWEIBRÜCKEN

The USAF had left the airfield in the late 1980s and their instructional airframes went to Speyer (F-4C 63-7423), Hermeskeil (F-4C 63-7583 and RF-4C 68-0587) and Kleine Brogel (F-101B 58-0322).

GREECE

Reports from the Greek military airfields are becoming more and more frequent, although it must be said that these are mostly from official pre-arranged base visits. Greece is still a country which does not understand the interest of enthusiasts in military aviation. Photographing aircraft and making notes can still lead to much trouble with the local authorities. Extra care has to be taken when visiting this country.

ACHARNAE
Near Tatoi/Dekélia is an unknwon B720 preserved as 'Club B720'.
☐ ..   B720-051B   ..   preserved, marked as 'Club B720', ex Hellinikón11-97

AGRINION
The airfield of Agrinion is located between Aráxos and Patras and acts as a relief landing field for Aráxos. Since the early 1990s it has also housed the Hellenic Air Force F-104 store. Only one departure is known, F-104G 32730 went to Messolongi.

☐ 51-6158		F-86D	173-302	stored	4-97
☐ 51-6234		F-86D	173-378	stored	4-97
☐ 51-8330		F-86D	173-463	stored	4-97
☐ 5719	TF-719	TF-104G	5719	stored, ex 2718/WGAF	4-97
☐ 5733	TF-733	TF-104G	5733	stored, ex Aráxos, ex 2731/WGAF	4-97
☐ 5906	TF-906	TF-104G	5906	stored, ex 2777/WGAF	4-97
☐ 5910	TF-910	TF-104G	5910	stored, ex 2781/WG Navy	4-97
☐ 5916	TF-5916	TF-104G	5916	stored, ex 2786/WGAF	4-97
☐ 5917	TF-917	TF-104G	5917	stored, ex 2787/WGAF	4-97
☐ 5928	TF-5928	TF-104G	5928	stored, ex 2798/WGAF	4-97
☐ 5954	TF-954	TF-104G	5954	stored, ex 2824/WGAF	4-97
☐ 5959	TF-959	TF-104G	5959	stored, ex Athíne, ex 2829/WGAF	4-97
☐ 6629	RF-6629	RF-104G	6629	stored, ex 2107/WG Navy	4-97
☐ 6672	RF-672	RF-104G	6672	stored, ex 2117/WG Navy	4-97
☐ 6674	RF-6674	RF-104G	6674	stored, ex 2119/WG Navy	4-97
☐ 6676		RF-104G	6676	stored, ex 2121/WG Navy	11-95
☐ 6690	FG-6690	F-104G	6690	stored, ex 2129/WG Navy	4-97
☐ 6693	FG-6693	F-104G	6693	stored, ex 2132/WG Navy	11-95
☐ 7087	FG-087	F-104G	7087	stored, ex 2216/WG Navy	4-97
☐ 7088	FG-7088	F-104G	7088	stored, ex 2217/WG Navy	4-97
☐ 7090	FG-7090	F-104G	7090	stored, ex 2219/WG Navy	4-97
☐ 7152	FG-152	F-104G	7152	stored, ex Aráxos, 2271/WG Navy	4-97
☐ 7163	FG-7163	F-104G	7163	stored, ex Athíne, ex 2281/WG Navy	4-97
☐ 7168	FG-168	F-104G	7168	stored, ex 2286/WG Navy	4-97
☐ 7172	FG-172	F-104G	7172	stored, ex 2289/WG Navy	4-97
☐ 7176	FG-176	F-104G	7176	stored, ex 2293/WG Navy	4-97
☐ 7180	FG-180	F-104G	7180	stored, ex 2297/WG Navy	4-97
☐ 7183	FG-183	F-104G	7183	stored, ex 2300/WG Navy	4-96
☐ 7195	FG-195	F-104G	7195	stored, ex Aráxos, ex 2312/WG Navy	4-97
☐ 7201	RF-7201	RF-104G	7201	stored, ex 2317/WG Navy	4-97
☐ 7203	FG-203	F-104G	7203	stored, ex 2319/WG Navy	4-97
☐ 7206	RF-206	RF-104G	7206	stored, ex 2322/WG Navy	4-97
☐ 7409	FG-7409	F-104G	7409	stored, ex Athíne, ex 2663/WG Navy	4-96
☐ 7421	FG-7421	F-104G	7421	stored, ex 2675/WG Navy	4-97
☐ 7422	FG-422	F-104G	7422	stored, ex 2676/WG Navy	4-97
☐ 7424	FG-424	F-104G	7424	stored, ex 2678/WG Navy	4-97

GREECE - 233

☐ 7425	FG-425	F-104G	7425	stored, ex 2679/WG Navy	4-97
☐ 7426	FG-7426	F-104G	7426	stored, ex 2680/WG Navy, with tail from 7428	4-97
☐ 7427	FG-427	F-104G	7427	stored, ex 2681/WG Navy	4-97
☐ 7428	FG-428	F-104G	7428	stored, ex 2682/WG Navy, with tail from 7426	4-97
☐ 7429	FG-429	F-104G	7429	stored, ex 2683/WG Navy	4-96
☐ 7433	FG-433	F-104G	7433	preserved, ex 2687/WG Navy	4-96
☐ 7434	FG-434	F-104G	7434	stored, ex 2688/WG Navy	4-97
☐ 7435	FG-7435	F-104G	7435	stored, ex Athíne, ex 2689/WG Navy	4-97
☐ 8176	RF-176	RF-104G	8176	stored, ex 2433/WG Navy	4-97
☐ 13044	TF-044	TF-104G	5715	stored, ex 2714/WGAF, ex 61-3044	4-97
☐ 22317	FG-317	F-104G	6016	stored, ex 62-12317	4-97
☐ 32706	FG-706	F-104G	6058	stored, ex 63-12706	4-96
☐ 32717	FG-717	F-104G	6069	stored, ex Aráxos, with tail from 32712	4-97
☐ 32724	FG-724	F-104G	6076	stored, ex Aráxos, ex 63-12724	4-97
☐ 32734	FG-734	F-104G	6086	stored, ex C.8-16/SpaAF, ex 63-12734	4-97
☐ 33643	FG-643	F-104G	6092	stored, ex C-8-12/SpaAF, ex 63-13643	4-97
☐ 47781	FG-781	F-104G	6126	stored, ex 64-17781	4-97

ANDRAVIDA

Andravida is the home base of 338 and 339MPK with their Phantoms. It also houses the Greek Top Gun School named KEAT. F-84 26858 is sectioned, the aft fuselage has been scrapped and its tail is preserved (seen April 1996), while the forward fuselage was in use as a rescue trainer in November 1993.

☐ 69-7471		RF-4E	4060	stored, ex 3524/WGAF	4-96
☐ 69-7534		RF-4E	4197	stored, ex 3587/WGAF	4-96
☐ 26858		F-84F	..	sectioned, see note, ex 52-6858	11-93
☐ 37145	145	F-84F	..	preserved, near main gate, ex 53-7145	4-96
☐ 19243	243	CL-13 Sabre 2	143	preserved, at gate	4-96
☐ 35736	TR-736	T-33A	9075	dumped, ex 53-5736	11-95

ARÁXOS

Based here are former US Navy Corsairs, which replaced the Starfighters in the 1990s. A large number of the latter found their way to the nearby airfield of Agrinion, although more than 30 of these fighters are still here. Starfighters stored here before moving on included 5908, 7151, 32720 and 7415 which all went to Tatoi. 5953 has gone to Limnos, while 5955, 5965 and 6692 went to Lárissa. Stored here before going to Agrinion were 5733, 7152, 32717 and 7195.

☐ 51-6216	216	F-86D	173-360	dumped	8-90
☐ 5717	TF-717	TF-104G	5717	stored, ex 2716/WGAF	4-97
☐ 5912	TF-5912	TF-104G	5912	stored, ex 2783/WGAF	4-97
☐ 6639	FG-6639	F-104G	6639	stored, ex 2109/WG Navy	4-93
☐ 6642	FG-6642	F-104G	6642	stored, ex 2111/WG Navy	4-93
☐ 6664	FG-6664	F-104G	6664	stored, wreck, ex 2115/WG Navy	8-95
☐ 6665	FG-6665	F-104G	6665	stored, ex 2116/WG Navy	4-97
☐ 6697	FG-697	F-104G	6697	stored, ex D-6697/RNethAF	4-97
☐ 7082	FG-082	F-104G	7082	stored, ex 2212/WG Navy, tail only	4-97
☐ 7094	FG-7094	F-104G	7094	stored, ex 2221/WG Navy, tail only	4-97
☐ 7097		F-104G	7097	stored, ex Kato Achia, ex 2222/WG Navy	4-97
☐ 7106	FG-106	F-104G	7106	stored, ex 2229/WG Navy, tail only	4-97
☐ 7153	FG-153	F-104G	7153	stored, ex 2272/WG Navy	4-97
☐ 7155	FG-155	F-104G	7155	stored, ex 2274/WG Navy	4-97
☐ 7205	FG-205	F-104G	7205	stored, ex 2321/WG Navy	4-97
☐ 7207		RF-104G	7207	stored, ex Athíne, ex 2323/WG Navy	8-92
☐ 7420		F-104G	7420	stored, ex 2674/WG Navy, tail only	4-97

☐ 22278	TF-278	TF-104G	5523	stored, ex CE.8-1/SpAF, ex 62-12278	4-97
☐ 22302	FG-302	F-104G	6001	dumped, ex 62-12302	4-97
☐ 22303	FG-303	F-104G	6002	stored, ex 62-12303	4-97
☐ 22304	FG-304	F-104G	6003	stored, ex 62-12304	4-97
☐ 22306	FG-306	F-104G	6005	stored, ex 62-12306	4-97
☐ 22307	FG-307	F-104G	6006	preserved, marked as '12307', ex 62-12307	4-97
☐ 22310	FG-310	F-104G	6009	stored, ex 62-12310, tail from 32710	4-97
☐ 22311	FG-311	F-104G	6010	stored, ex 62-12311	4-97
☐ 22314	FG-314	F-104G	6013	stored, ex 62-12314	4-97
☐ 22315	FG-315	F-104G	6014	dumped, ex 62-12315	4-97
☐ 32708	FG-708	F-104G	6060	stored, ex 63-12708	4-97
☐ 32710	FG-710	F-104G	6062	stored, ex 63-12710	4-96
☐ 32712	FG-712	F-104G	6064	stored, ex 63-12712, tail to 32717	4-97
☐ 32713	FG-713	F-104G	6065	stored, ex 63-12713	4-97
☐ 32722	FG-722	F-104G	6074	dumped, ex 63-12722	4-97
☐ 32725	FG-725	F-104G	6077	dumped, ex 63-12725	8-95
☐ 33638	FG-638	F-104G	6087	dumped, ex C.8-17/SpAF, ex 62-13638	4-97
☐ 33639	FG-639	F-104G	6088	dumped, ex C.8-18/SpAF, ex 62-13639	4-97
☐ 47787	FG-787	F-104G	6132	stored, ex 64-17787, no tail	4-97
☐ 47788	FG-788	F-104G	6133	stored, ex 64-17788, no tail	4-97

ATHÍNE
Town: Polemico Moussio in the town centre has lost two of its aircraft to the museum at Tatoi, SB2C-5 83321 and Spitfire IXc MJ755 both have departed.

☐ 689071		F-5A	N6442	preserved, ex Néa Ankhíalos, ex RJordAF	11-97
☐ 53-7216	216	F-84F	..	preserved, ex Tatoi, ex DB-344/WGAF	11-97
☐ 51-6171		F-86D	173-315	preserved, marked as '51671'	11-97
☐ 6695	FG-695	F-104G	6695	preserved, ex D-6695/RNethAF	11-97
☐ 49-3500		T-6G	168-644	preserved, marked as '32803'	11-97
☐ G-776		DH.82A	..	preserved, ex T6776 ?	4-96
☐ ..		Farman MF-7	..	preserved, replica	4-95

At Katheki Street a new museum, the Parko Ellinikou Stratou, has been founded, dedicated to the Greek Army. Allocated for this museum in September 1993 were F-104G 6681, F-84F 37182 and T-33A 58594 (all still at Hellinikón ?).

Preserved with the Greek CAA near the airport on the road towards Glyfada is one of their former C-47s.
☐ SX-ECF		C-47C	33206	preserved, ex N73	11-97

Town - Pefki: At the corner of Kazatzaki and Appolonus is a park with a children's playground. In there, a vandalised T-33 is te be found.
☐ 61592		T-33A	9942	stored, ex 56-1592	2-97

Town – Kifissia: In the north east side of Athíne an unidentified B720 is preserved in a fairground.
☐ ..		B720-051B	..	preserved, unmarked, ex Hellinikón	11-97

Airfield – Tatoi (Dekélia): The Helliniki Aeroporia Moussio (Hellenic Air Force Museum) is located at this base in the north western suburbs of Athíne and has expanded over the years. RF-84F 37682 is now at Sedes, while the museum should also have T-6G 51-15057.

☐ L9044		Blenheim IV	..	preserved, wreck	11-97
☐ 301		Lim-2	1B-00301	preserved, marked as '925', ex Polish AF	11-97
☐ KJ960		C-47B	14807/26252	preserved, ex white c/s, ex Sedes, ex 43-48991	11-97
☐ 10541		F-5A	N6202	preserved, ex RJordAF, ex IIAF, ex 65-10541	11-97
☐ 110822	998	F-84G	..	preserved, dark blue c/s, ex 51-10822	11-97

GREECE - 235

☐ 26361	FU-361	F-84F	..	preserved, ex instructional, ex 52-6361	4-97
☐ 26595	595	F-84F	..	preserved, ex 52-6595	11-97
☐ 37216		F-84F	..	preserved, ex Athíne, ex DB-334, ex 53-7216	4-97
☐ 17011		RF-84F	..	preserved, ex Lárissa, ex 51-17011	11-97
☐ 19169		CL-13 Sabre 2	69	preserved, marked as '12910' in USAF c/s	11-97
☐ 19202	202	CL-13 Sabre 2	102	preserved	11-97
☐ 52-10067		F-86D	190-892	preserved	11-97
☐ ..998	998	F-86D	..	preserved	11-97
☐ 5908	TF-5908	TF-104G	5908	preserved, ex Aráxos, ex 2779/WG Navy	11-97
☐ 7151		F-104G	7151	preserved, Olympus c/s, ex 2270/WG Navy	11-97
☐ 7415	FG-415	F-104G	7415	preserved, ex Aráxos, ex 2669/WG Navy	11-97
☐ 32720		F-104G	6072	preserved, tiger c/s, ex Aráxos, ex C.8-5/SpaAF	11-97
☐ ..		F-104G	..	preserved, cockpit only, marked as '2691'	11-97
☐ 120		G159	120	preserved, ex P9 (unmarked)	11-97
☐ 13952		UH-19B	55-460	preserved, ex 51-3952	11-97
☐ 5322		OH-13H	2335	preserved, ex 58-5322/US Army	11-97
☐ 5385		OH-13H	2398	preserved, ex 58-5385/US Army	11-97
☐ 83321		SB2C-5	..	preserved, ex Athíne, ex 83321/US Navy	11-97
☐ MJ755		Spitfire LF.IXc	CBAF.IX1285	preserved, ex Athíne	11-97
☐ 493424		T-6G	168-548	preserved, ex store, ex 49-3424	11-97
☐ 517190	190	HU-16B	250	preserved, ex 51-7190	11-97
☐ 517204	204	HU-16B	274	preserved, ex store, ex RNoAF, ex 51-7204	11-97
☐ SX-AGU		PA-18-135	18-5066	preserved, ex Greek AF, ex SX-ADS	4-97

The museum has also a storage hangar in which a Bell 47 can be found. The aircraft is marked 066, but has the boom of 2064 and the cockpit construction plate reads 2060. Only one of the T-6 was seen is November 1997.

☐ 19347		CL-13 Sabre 2	247	stored, ex Athíne	11-97
☐ 246		L-21B	..	stored, frame only, ex Army	11-97
☐ 751		T-6D	..	stored, ex 49-2751	4-97
☐ 830		T-6G	..	stored	4-97
☐ 2060		AB47J-2	2060	stored, composite, marked as '066'	5-96

The 123 Pterix Technical School also has a place at the airfield.

☐ KK169		C-47B	15415/26860	preserved, ex instructional, ex KK169/RAF	11-97
☐ 49111		C-47B	14927/26372	instructional, ex Sedes, ex 43-49111/USAAF	11-97
☐ 26743	743	F-84F	..	instructional, ex DD-326/WGAF, ex 52-6743	11-97
☐ ..752	752	F-84F	..	instructional	11-92
☐ 54035		TF-102A	..	instructional, ex Elefsís, ex 55-4035	11-95
☐ 6699	FG-699	F-104G	6699	instructional, ex D-6699/RNethAF	11-97
☐ 53-258		N2501D	001A	instructional, ex 5358/WGAF	11-97
☐ 16714	TR-714	T-33A	6046	instructional, ex 14693/RCAF, ex 51-6714	11-97
☐ 16717	TR-717	T-33A	6049	instructional, ex 51-6717	4-96

Separate from the above school is the Ikarus School, which has its own aircraft.

☐ 97166		RF-5A	RF1046	instructional, ex 69-7166	4-97
☐ ..		F-84F	..	instructional	4-97
☐ 37665		RF-84F	..	instructional, ex EB-328/WGAF, ex 53-7665	4-97
☐ O-61233		F-102A	..	instructional, ex 56-1233	4-97
☐ 5958	TF-958	TF-104G	5958	instructional, ex Athíne, ex 2828/WGAF	4-97
☐ ..		T-6	..	instructional	4-97
☐ 36129	TR-129	T-33A	9750	instructional, ex 53-6129	4-97

There are still more aircraft at the base. The two OH-13s are with the Ikarus or with the 123 Pterix school. T-6G 93014 and three unknown T-6s were seen in November 1997.

☐ 4458		AB205A	4458	stored, wreck	4-96

GREECE - 236

❏ KK171		C-47B	15417/26862	dumped, crashed 27-12-91, ex Sedes, ex KK171	4-96
❏ 26837	837	F-84F	..	preserved, at gate, ex 52-6837	11-97
❏ 4944		OH-13H	2527	instructional, ex 59-4944	4-96
❏ 4964		OH-13H	2547	instructional, ex 59-4964	4-96
❏ 93014		T-6G	168-116	dumped, burnt fuselage at end of runway	11-97
❏ 93416		T-6G	168-540	dumped, in 360 Moira area, ex 49-3416	11-90
❏ 93434		T-6G	168-558	stored, with aero club, ex 49-3434	4-96
❏ 93514		T-6G	168-658	dumped, with aero club, ex 49-3514	4-96
❏ ..		T-6	..	stored, on rebuild with aero club	4-96
❏ 1534		G164 Ag-Cat	1534	dumped, ex N8806	8-91
❏ 122		PZL M18	1Z011-22	stored, no engin es	11-97

Airfield - Hellinikón: After some time in storage during the 1980s C-47Bs 92622, KK156 (15295/26740), KN575 (16451/33199) and KP255 (16809/33557) have all returned to operational status. RF-84Fs 17011 and 37683 have moved to Lárissa and F-84F 26425/425 is now at Néa Ankhíalos. The dumped F-84F 37216 went to Tatoi, together with TF-104G 5958/TF-958. Other F-104s moving on were 5959, 7163, 7409 and 7435, all are now at Agrinion. At the civil side a number of stored B720-051Bs were scrapped in 1992, these were SX-DBG (18352), SX-DBH (18353), SX-DBK (18356), SX-DBL (18420) and SX-DBM (18687). Of these five B707s one went to Athína - Kifissia and one to Acharne.

❏ 92621		C-47A	26110 ?	dumped, ex 49-2621	3-98
❏ 92625		C-47	4499	dumped, ex 49-2625, ex 41-18371	3-98
❏ 49-2630		C-47A	12452	dumped, ex 49-2630	3-98
❏ 92637		C-47A	11983	dumped, ex Sedes, ex NC62386, ex 49-2637	3-98
❏ 4086		Do28D-2	4086	stored, ex 5811/WGAF	11-97
❏ 26439	439	F-84F	..	dumped, no tail, ex 52-6439	4-96
❏ 26467	467	F-84F	..	dumped, ex 52-6467	11-97
❏ 26659	659	F-84F	..	preserved, ex 52-6659	4-97
❏ 26681	681	F-84F	..	dumped, ex 52-6681	4-96
❏ 26701	701	F-84F	..	dumped, ex 52-6701	7-90
❏ 26703	703	F-84F	..	dumped, ex 52-6703	11-97
❏ 26727	727	F-84F	..	dumped, ex 52-6727	11-97
❏ 26761	761	F-84F	..	dumped, no tail, ex 52-6761	4-93
❏ 26773	773	F-84F	..	dumped, no tail, ex 52-6773	4-96
❏ 26797	797	F-84F	..	dumped, ex 52-6797	11-97
❏ 26811	811	F-84F	..	dumped, ex 52-6811	3-90
❏ 27086	086	F-84F	..	dumped, ex 52-7086	4-96
❏ 27091	091	F-84F	..	dumped, ex 52-7091	4-96
❏ 27162		F-84F	..	dumped, ex 52-7162	7-90
❏ 28730		RF-84F	..	dumped, ex 52-8730	3-98
❏ 28740	740	RF-84F	..	dumped, ex 52-8740	3-98
❏ 37182	182	F-84F	..	dumped, allocated for Athíne, ex 53-7182	2-92
❏ 37209	209	F-84F	..	dumped, ex 53-7209	11-97
❏ 2188		F-104G	7057	stored, ex 2188/WGAF	4-96
❏ 2210		F-104G	7080	stored, ex 2210/WGAF	4-96
❏ 6666	FG-666	F-104G	6666	dumped, ex D-6666/RNethAF, fuselage only	4-96
❏ 6670	FG-6670	F-104G	6670	dumped, ex D-6670/RNethAF	1-98
❏ 6679	FG-679	F-104G	6679	dumped, fuselage only, ex 2124/WG Navy	2-92
❏ 6680	FG-680	F-104G	6680	dumped, ex D-6680/RNethAF	4-96
❏ 6681	FG-6681	F-104G	6681	stored, no tail, ex D-6681/RNethAF	3-98
❏ 6700	FG-700	F-104G	6700	dumped, ex D-6700/RNethAF	4-96
❏ 7167	FG-167	F-104G	7167	dumped, ex 2285/WG Navy	4-97
❏ 22274	TF-274	TF-104G	5519	dumped, ex 62-12274	3-98
❏ 32709	FG-709	F-104G	6061	dumped, ex 63-12709	4-97
❏ 32727	FG-727	F-104G	6079	dumped, ex C.8-6/SpaAF, ex 63-12727	11-93
❏ ..		UH-1	..	dumped, on a hill	11-90

GREECE - 237

☐ 16771	TR-771	T-33A	6103	dumped, ex preserved, ex 51-6771	11-97
☐ 17556	TR-556	T-33A	7701	dumped, ex M-17/RNethAF, ex 51-17556	4-97
☐ 29962	TR-962	T-33A	8193	dumped, ex 9432/WGAF, ex 52-9962	11-97
☐ 35126	TR-126	T-33A	8465	dumped, ex 53-5126	3-98
☐ 35328	TR-328	T-33A	8667	dumped, ex 53-5328	4-96
☐ 35780	TR-780	T-33A	9119	stored, ex 9455/WGAF, ex 53-5780	3-98
☐ 58594	TR-594	T-33A	1643	stored, ex 9495/WGAF, allocated for Athíne	3-98
☐ 58645	TR-645	T-33A	1694	stored, ex 58-0645	3-98
☐ 61755	TR-755	T-33A	1105	stored, ex 56-1755	3-98
☐ 517498	TR-498	T-33A	7478	dumped, ex M-14/RNethAF, ex 51-17498	4-95
☐ 2144		YS-11A	2144	stored, ex SX-BBK	3-98
☐ SX-BBN		SC7 Skyvan 3	SH1869	stored, ex G-AXLB, ex G-14-41	6-88
☐ SX-BBO		SC7 Skyvan 3	SH1870	stored, ex G-AXLS, ex G-14-42	6-88
☐ SX-BGC		SD330-100	3065	stored	4-95
☐ SX-BGD		SD330-100	3066	stored	4-95
☐ SX-CBA		B727-284	20003	stored	11-97
☐ SX-CBB		B727-284	20004	stored	11-97
☐ SX-CBE		B727-284	20201	stored	11-97
☐ SX-CBF		B727-284	19536	stored	11-97
☐ SX-CBG		B727-230	20918	stored	11-97
☐ SX-ECD		C-47B	14787/26232	stored, ex KJ950, ex 43-48971	11-97
☐ N1130J		BAC111-215AU	96	stored, SX-BAR ntu	11-97

ELEFSÍS

All the N2501 Noratlasses from EWR-1 have gone except those listed. Furthermore, four F-102 Delta Daggers have left Elefsís: 54035 went to <u>Tatoi</u>, 61025 has gone to <u>Tripolis</u>. 61052 went all the way to <u>Woensdrecht</u>, while 61232 is now at <u>Lárissa</u>. Of the five stored/dumped N2501s only four were left in November 1997. The serials of the HU-16s were not read of in November 1997

☐ 92626	C-47		preserved, at gate, ex 49-2626	11-97
☐ 13012	C-47A	13012	stored, ex SX-TAE (wings/tail only by 11-93)	4-93
☐ 948	C-130B	3624	stored, ex 61-0948, wfu ?	4-96
☐ 0296	C-130B	3597	stored, ex 60-0296, wfu ?	4-96
☐ 4082	Do28D	4082	stored, ex 5807/WGAF	4-96
☐ 4087	Do28D	4087	stored, ex 5812/WGAF	4-96
☐ 4097	Do28D	4097	stored, ex 5822/WGAF	4-96
☐ 4102	Do28D	4102	stored, ex 5827/WGAF	4-96
☐ 4108	Do28D	4108	stored, ex 5833/WGAF	4-96
☐ 4110	Do28D	4110	stored, ex 5835/WGAF	4-96
☐ 4117	Do28D	4117	stored, ex 5842/WGAF	11-95
☐ 4120	Do28D	4120	stored, ex 5845/WGAF	4-96
☐ 4131	Do28D	4131	stored, ex 5856/WGAF	4-96
☐ 4138	Do28D	4138	stored, ex 5863/WGAF	4-96
☐ 60981	F-102A	..	decoy, ex 56-0981, to go to Messolongi	11-97
☐ 61007	F-102A	..	decoy, ex 56-1007	11-97
☐ 61059	F-102A	..	decoy, ex 56-1059	11-97
☐ 62326	TF-102A	..	decoy, ex 56-2326	11-97
☐ 62327	TF-102A	..	decoy, ex 56-2327	11-97
☐ 62335	TF-102A	..	decoy, ex 56-2335	11-97
☐ 52-128	N2501D	030	preserved, white c/s, ex 5228/WGAF	11-97
☐ 52-188	N2501D	106	dumped, ex 5288/WGAF	4-97
☐ 52-189	N2501D	107	dumped, wreck, camo c/s, ex 5289/WGAF	4-97
☐ 52-228	N2501D	148	dumped, ex 5328/WGAF	4-97
☐ 53-234	N2501D	157	stored, ex 5334/WGAF	4-97
☐ 53-239	N2501D	162	stored, wreck, camo c/s, ex 5339/WGAF	4-97

☐ 510044	HU-16B	122	stored, ex RNoAF		11-97
☐ 510068	HU-16B	147	stored, ex RNoAF		11-97
☐ 510070	HU-16B	149	stored, ex RNoAF		11-97
☐ 517177	HU-16B	227	stored, ex RNoAF		11-97
☐ 517201	HU-16B	268	stored, ex RNoAF		11-97
☐ 517207	HU-16B	278	stored, ex RNoAF		11-97
☐ 152183	P-3A	5153	instructional, ex Tanágra, ex 152183/USN		11-97
☐ 2136	YS-11A	2136	stored, ex SX-BBG		4-97
☐ 2145	YS-11A	2145	stored, ex SX-BBL		4-96
☐ 2153	YS-11A	2153	stored, ex SX-BBP		4-97

HERÁKLION
No recent reports from this airfield, however a F-86 and two F-84s were noted in 1985.

IOÁNNINA
The F-84 is in a park near a sport complex (Ioánina Tennis Club) on the road out of the town to the west.

☐ 36741	F-84F	..	preserved, ex 53-6741		4-96

IRAKLION
The Thunderstreak ending with 745 can not be 36745 as this aircraft was seen at Soúda in the same month.

☐ 27468		RF-84F	..	preserved, on base, ex Soúda, ex 52-7468	4-97
☐ ..745	745	F-84F	..	preserved	7-95
☐ 13371		F-5A	N7004	preserved, ex Kastelli, ex 371/RNoAF	4-97

KALAMÁTA
Still preserved at the gate of this training base is a T-33. Four aircraft have been noted on dumps.

☐ 27114		F-84F	..	dumped, ex preserved, in sections near tower	4-96
☐ 35786	TR-786	T-33A	9125	preserved, at gate, ex 53-5786	4-97
☐ ..		F-84F	..	dumped	7-95
☐ ..		F-84F	..	dumped	7-95
☐ ..		F-104G	..	dumped	7-95

KASTÉLLION
Preserved with the 133 Sminarkia Mahis was F-5A 13371 which was last noted in 1995. It is now at Iraklion.

KATAHAS
On the roof of the 'Cafe DC-3', some 200 meters from the road from Lárissa to Thessaloníki is a wingless C-47.

☐ 92641	C-47	..	preserved, ex 49-2641	4-96

KATO ACHIA
Preserved on the town square of Kato Achia, located 10 kilometres east of Aráxos, was F-104G 7097. The Starfighter was last noted in November 1993 and has since gone to Aráxos.

KOTRONI
On top of the Mount Pentelli (near Marathon) is the home base of the Greek Navy. On this airfield, from which

nothing can be seen from outside, are the hulks of two crashed choppers.

☐ ΠΝ-02	Alouette 3	2220	dumped, wreck near hangars, crashed 1994	3-98
☐ ΠΝ-24	AB212ASW	..	dumped, wreck near hangars, crashed 1980	3-98

LÁRISSA

Town: At barracks in town two fighters are preserved. The F-5 is the original 063 (see Stilisa).

☐ 89063		F-5A	N6434	preserved, at barracks in town, ex RJordAF	3-98
☐ 6668	FG-668	F-104G	6668	preserved, at barracks in town, ex D-6668	3-98

Airfield: This large airfield is home of some Phantom and F-16 units. The four stored RF-4Es were also seen in November 1997, but not read of. The base can be split into several parts, noted in the shelter areas were:

☐ 3507	RF-4E	3927	stored, ex 3507/WGAF, ex 69-7454	4-96
☐ 3513	RF-4E	4007	stored, ex 3513/WGAF, ex 69-7460	11-95
☐ 3546	RF-4E	4114	stored, ex 3546/WGAF, ex 69-7493	11-95
☐ 3584	RF-4E	4191	stored, ex 3584/WGAF, ex 69-7531	11-95
☐ 13376	F-5A	N7009	stored, ex 376/RNoAF, ex 64-13376	4-96
☐ 10569	F-5A	N7017	stored, in 349Mira area, ex 569/RNoAF	4-96
☐ 69132	F-5A	N6236	stored, in 349Mira area, ex Néa Ankhíalos	4-96
☐ 89066	F-5A	N6437	stored, in 349Mira area, ex RJordAF	4-96
☐ 97164	RF-5A	RF1044	stored, ex 69-7164	4-96
☐ 97165	RF-5A	RF1045	stored, ex 69-7165	4-96

Parked in a line at the far side of the airfield were a number of aircraft. Of these NF-5A 3073 was delivered in crates to Greece and has never flown in HAF service. RF-84F 37575 moved to Messolongi.

☐ 3073		NF-5A	3073	dumped, ex K-3073/RNethAF	3-98
☐ 01399		F-5A	N6561	stored, ex 70-1399	4-96
☐ 01617		RF-5A	RF1057	stored, ex 70-1617	4-96
☐ 10566		F-5A	N7014	stored, ex 566/RNoAF, ex 65-10566	4-96
☐ 10574		F-5A	N7022	stored, ex 574/RNoAF, ex 65-10574	4-96
☐ 50476		F-5A	N6137	stored, ex 65-10476	4-96
☐ 89064		F-5A	N6435	stored, ex RJordAF, ex 68-9064	4-96
☐ 11294		RF-84F	..	stored, ex 51-11294	11-97
☐ 26679		F-84F	..	stored, ex 52-6679	4-96
☐ 26680		F-84F	..	stored, ex 52-6680	11-92
☐ 12601	FG-601	F-104G	4001	stored, ex 61-2601	11-97
☐ 29900	TR-900	T-33A	7796	stored, ex M-20/RNerthAF, ex 52-9900	4-96
☐ 36082	TR-082	T-33A	9703	dumped, ex 63-6082	3-98
☐ 58602	TR-602	T-33A	1651	stored, ex 9502/WGAF, ex 58-0602	4-96
☐ 58639	TR-639	T-33A	1688	stored, ex 58-0639	4-96

More aircraft can be found elsewhere. Stored RF-84F 17011 went to Tatoi in the early 1990s, while 37575 went to Messolongi. Two T-33s were still noted on the dump in November 1997, these may be 16867 and 29561.

☐ 01618	RF-5A	RF1058	stored, ex 70-1618	3-98
☐ 01619	RF-5A	RF1059	stored, ex 70-1619	11-95
☐ 38405	F-5A	N6042	stored, ex 63-8405	3-98
☐ 38430	F-5A	N6067	stored, ex 63-8430	3-98
☐ 69232	F-5B	N8040	preserved, ex 66-9232	4-94
☐ 689081	F-5A	N6452	stored, ex RJordAF, ex 68-9081	3-98
☐ 697170	RF-5A	RF1050	stored, special colours, ex 69-7170	3-98
☐ 26688	F-84F	..	dumped, ex 52-6688	3-98
☐ 27470	RF-84F	..	stored, marked as '52470', ex 52-7470	11-97
☐ 28736	RF-84F	..	stored, ex 52-8736	11-97
☐ 37588	RF-84F	..	preserved, gate guard, ex 53-7588	3-98
☐ 37683	RF-84F	..	decoy, ex Athíne, ex EB-324/WGAF	11-97

GREECE - 240

☐ 5708	TF-708	TF-104G	5708	stored, ex 2707/WGAF, ex 61-3037	11-97
☐ 5955	TF-5955	TF-104G	5955	stored, ex Aráxos, ex 2825/WG Navy	11-97
☐ 5965	TF-5965	TF-104G	5965	stored, ex Aráxos, ex 2835/WGAF	11-97
☐ 16867	TR-867	T-33A	6199	dumped, ex 51-6867	11-95
☐ 29561	TR-561	T-33A	7721	dumped, ex 52-9561	3-98
☐ 35494	TR-494	T-33A	8833	preserved, at gate, ex 53-5494	3-98
☐ 35845	TR-845	T-33A	9246	dumped, ex 53-5845	6-88
☐ ..633	TR-633	T-33A	..	dumped	11-93
☐ SX-CGV		Aero Commander	..	stored	8-90

There is a small museum locally known as the memorial display, near the VIP ramp at the airfield.

☐ 69209		F-5A	N7032	preserved, ex 209/RNoAF, ex 66-9209	3-98
☐ 28728		RF-84F	..	preserved, silver c/s, ex 52-8728	3-98
☐ ..		F-84F	..	preserved, marked as '37050/050', ex stored	3-98
☐ 19752	FS-752B	F-84G	..	preserved, ex 51-9752	3-98
☐ 19409	409	CL-13 Sabre 2	309	preserved, ex 19409/RCAF	3-98
☐ 0-61232		F-102A	..	preserved, ex Elefsís, ex 56-1232	3-98
☐ 6692	FG-6692	F-104G	6692	preserved, ex Aráxos, ex 2131	3-98
☐ 29913	TR-913	T-33A	7884	preserved, ex M-29/RNethAF, ex 52-9913	3-98

LIMNOS
Based at the Limnos Város airport is the 130 Sminakia Mahis. Preserved with them is a two-seat Starfighter.

☐ 26690	690	F-84F	..	dumped, ex 52-6690	8-95
☐ 26890		F-84F	..	dumped, ex Préveza, ex 52-6890	8-95
☐ 37222	222	F-84F	..	dumped, ex Préveza, ex 53-7222	8-95
☐ 5953	TF-953	TF-104G	5953	preserved, at gate, ex Aráxos, ex 2823/WGAF	4-97

MÉGARA
Mégara is where one of the Greek Army's regiments is based. From here they fly with all types of Greek Army helicopters, Be200s and U-17s. An U-17 was used as a gate guard, but was noted on the dump in late 1993.

☐ ΕΣ-268		U-17A	..	dumped, ex gate guard	11-93

MESSOLONGI
A museum was set up here on the site of a famous battle near the harbour.

☐ 70376		F-5A	N6498	preserved, ex RJordAF, ex IIAF, ex 70-1376	11-97
☐ 26914	914	F-84F	..	preserved, ex Préveza, ex 52-6914	11-97
☐ 37575		RF-84F	..	preserved, ex Lárissa, ex EB-255/WGAF	11-97
☐ 32730	FG-730	F-104G	6082	preserved, ex Agrinion, ex C.8-7/SpaAF	11-97
☐ 35265	TR-265	T-33A	8604	preserved, ex 53-5265	11-97

NÉA ANKHÍALOS
Four of the stored F-5As have moved on, 063 went to Lárissa, 69132 to Lárissa and 69228 is now at Vólos. 89071 is at the museum in Athíne.

☐ 080	F-5A	N6451	stored, desert c/s, ex RJordAF, ex 68-9080	5-96
☐ 10542	F-5A	N6203	stored, blue c/s, ex RJordAF, ex 65-10542	5-96
☐ 10577	F-5A	N7025	stored, ex 577/RNoAF, ex 65-10577	5-96
☐ 22550	F-5A	N6408	stored, ex RJordAF, ex IIAF, ex 67-22550	5-96
☐ 89070	F-5A	N6441	stored, blue c/s, ex RJordAF, ex 68-9070	5-96
☐ 689072	F-5A	N6443	stored, blue c/s, ex RJordAF, ex 68-9072	5-96
☐ 689073	F-5A	N6444	stored, ex RJordAF, ex IIAF, ex 68-9073	5-96

☐ 689077		F-5A	N6448	preserved, on pole at HQ buildings, ex RJordAF	5-96
☐ 89079		F-5A	N6450	stored, ex RJordAF, ex 68-9079	5-96
☐ 89088		F-5B	N8066	stored, camo c/s, ex 68-9088	5-96
☐ ..		F-5A	..	preserved, at gate	11-97
☐ 19448		CL-13 Sabre 2	348	preserved, on pole near HQ buildings	5-96
☐ 26425	425	F-84F	..	dumped, in 330 Mira area, no tail, ex Athíne	4-96

PRÉVEZA
Since the mid 1980s a large numbers of Thunderstreaks have been stored here. Most of them had gone by 1997 including 26890 and 37222 which went to <u>Limnos</u> and 26914 went to <u>Messolongi</u>. Préveza is also used by a NATO E-3 detachment.

☐ LX-N90457		E-3A	22852	stored, wreck, crashed 14-7-96	4-97
☐ 26550		F-84F	..	stored, ex 52-6550	11-93
☐ 26565		F-84F	..	dumped, ex 52-6565	4-96
☐ 26585		F-84F	..	stored, ex 52-6585	11-93
☐ 26623	623	F-84F	..	dumped, ex 52-6623	4-96
☐ 26676	676	F-84F	..	stored, ex 52-6676	8-90
☐ 26809		F-84F	..	dumped, ex 52-6809	4-96
☐ 26824		F-84F	..	stored, ex 52-6824	4-97
☐ 26828		F-84F	..	dumped, ex 52-6828	4-96
☐ 26831		F-84F	..	dumped, ex 52-6831	4-97
☐ 26840	840	F-84F	..	stored, ex 52-6840	4-96
☐ 26857		F-84F	..	dumped, ex 52-6857	4-96
☐ 26883		F-84F	..	stored, ex 52-6883	11-93
☐ 26891	891	F-84F	..	stored, ex 52-6891	11-93
☐ 26904	904	F-84F	..	dumped, ex 52-6904	4-96
☐ 26926		F-84F	..	dumped, burned, ex 52-6926	4-96
☐ 26939		F-84F	..	dumped, ex 52-6939	4-96
☐ 26941	941	F-84F	..	dumped, ex 52-6941	4-96
☐ 26959		F-84F	..	dumped, ex 52-6959	4-96
☐ 26960		F-84F	..	dumped, ex 52-6960	4-96
☐ 27089		F-84F	..	dumped, ex 52-7089	4-96
☐ 28951		F-84F	..	dumped, ex 52-8951	4-96
☐ 36545		F-84F	..	dumped, ex 53-6545	4-96
☐ 36611		F-84F	..	stored, ex 53-6611	4-96
☐ 36663		F-84F	..	dumped, ex 53-6663	4-96
☐ 36676		F-84F	..	stored, ex 53-6676	11-93
☐ 36689	689	F-84F	..	preserved, at main gate, ex 53-6689	4-96
☐ 36796		F-84F	..	dumped, ex 53-6796	4-96
☐ 36798		F-84F	..	dumped, ex 53-6798	4-96
☐ 37175		F-84F	..	stored, ex 53-7175	4-97
☐ 37201		F-84F	..	stored, ex 53-7201	4-96
☐ 37230		F-84F	..	stored, ex 53-7230	4-97
☐ 51-6149		F-86D	173-293	stored, wreck	4-97
☐ 51-6206		F-86D	173-350	stored	4-97
☐ 51-8297		F-86D	173-430	stored, wreck	4-97
☐ ..-.329		F-86	..	stored	4-96

SOÚDA
This airfield is often referred to as Soúda Bay, but this only the name for the naval base. The Thunderflash was noted at Iraklion during the same month.

☐ 27468		RF-84F	..	preserved, on base, ex 52-7468, see Iraklion	4-97

GREECE - 242

❏ 36744	744	F-84F	..	dumped, wreck, ex 53-6744	5-98
❏ 36745	745	F-84F	..	preserved, at gate, ex 53-6745	4-97
❏ 36797		F-84F	..	preserved, marked as '554', ex 53-6797	4-97
❏ 37218		F-84F	..	preserved, marked as '815', ex 53-7218	4-97
❏ 17521	TR-521	T-33A	7581	dumped, ex 51-17521	6-88
❏ 17550	TR-550	T-33A	7695	stored, ex 9412/WGAF, with tail from 58597	9-93
❏ 29925	TR-925	T-33A	7896	stored, ex 9420/WGAF, ex 52-9925	5-98
❏ 35029	TF-029	T-33A	8368	stored, ex 53-5029	6-96
❏ 35490	TR-490	T-33A	8829	stored, ex 53-5490	5-98
❏ 41575	TR-575	T-33A	9264	stored, ex 41575/FAF, ex 54-1575	5-98
❏ 58520	TR-520	T-33A	1569	stored, ex 9489/WGAF, ex 58-0520	5-98
❏ 58597	TR-597	T-33A	1646	dumped	5-98
❏ 58599	TR-599	T-33A	1648	stored, ex 9499/WGAF, ex 58-0599	5-98
❏ 58691	TR-691	T-33A	1740	stored, ex 9525/WGAF, ex 58-0691	5-98
❏ 59039	TR-039	T-33A	6823	stored, ex 51-9039	5-98

STEFANAVÍKION
Stefanavíkion houses a Greek Army regiment and the flying training school. Some *Wrecks and Relics* aircraft are to be found at this airfield, although Aero Commander EΣ-316 was reported back in service in 1997.

❏ EΣ-241		L-21B	..	stored	4-93
❏ EΣ-253		L-21B	..	preserved, near HQ buildings	7-95
❏ EΣ-288		RC680FL	..	stored	7-95
❏ EΣ-315		RC680FL	1837-154	stored, in hangar	4-97
❏ EΣ-316		RC680FL	1851-155	stored, see notes	4-96
❏ EΣ-701		OH-13S	3994	preserved, near HQ buildings as 'EΣ-101'	7-95

STILISA
In a park next to the E75 main road through the town (on the waterfront) a Freedom Fighter is preserved.

❏ ..		F-5A	..	preserved, marked as '06Σ'	11-97

TANÁGRA
This airfield is the home base of a wing operating two squadrons of Mirage 2000s and one of Mirage F1s. The preserved Mirage 2000CG was replaced after its crash by a new aircraft with the same serial number.

❏ 3551		RF-4E	4122	instructional, ex 3551/WGAF, ex 69-7498	4-96
❏ 19235	FU-235	CL-13 Sabre 2	135	preserved, at 114 Pterix, ex '52235'	3-98
❏ 0-61001		F-102A	..	decoy, at 332 Mira, ex 56-1001	3-98
❏ 0-61024		F-102A	..	dumped, ex decoy, ex 56-1024	3-98
❏ 0-61034		F-102A	..	decoy, at 331 Mira, ex 56-1034	11-97
❏ 0-61039		F-102A	..	decoy, at 331 Mira, ex 56-1039	3-98
❏ 0-61040		F-102A	..	decoy, at 332 Mira, ex 56-1040	3-98
❏ 0-61079		F-102A	..	decoy, at 331 Mira, ex 56-1079	3-98
❏ 0-61106		F-102A	..	preserved, at 114 Pterix, ex stored, ex 56-1106	3-98
❏ 5901	TF-901	TF-104G	5901	decoy, at 332 Mira, ex 2772/WGAF	3-98
❏ 5961	TF-5961	TF-104G	5961	decoy, at 332 Mira, ex 2831/WGAF	3-98
❏ 6662	FG-662	RF-104G	6662	preserved, at 114 Pterix, ex 2113/WG Navy	3-98
❏ 6691	RF-691	RF-104G	6691	preserved, gate guard, ex 2130/WG Navy	3-98
❏ 215 (1)		Mirage 2000EG	..	preserved, cockpit and wings only, at 331 Mira	4-96
❏ 35118	TR-118	T-33A	8457	decoy, ex 53-5118	3-98
❏ 35777	TR-777	T-33A	9116	decoy, ex 9453/WGAF, ex 53-5777	3-98
❏ 35801	TR-801	T-33A	9158	decoy, ex 53-5801	3-98
❏ 35893	TR-893	T-33A	9369	stored, ex 53-5893	3-98

☐ ..		T-33A	..	decoy	11-97
☐ 517183		HU-16B	235	instructional, at 342 Mira, ex RNoAF	3-98

At the Hellenic Aerospace Industries area a number of P-3s and Corsairs are in storage for spares. Of these P-3A 152183 moved to Elefsís. The HAI instructional school was closed in 1993 and their aircraft are now stored or dumped. All Corsairs were noted in November 1997, but none of them were actually read off.

☐ 154379		TA-7C	C-019	stored, for spares?, ex AMARC	4-96
☐ 154424		TA-7C	C-064	stored, for spares?, ex AMARC	4-97
☐ 156767		TA-7C	C-034	stored, for spares?, ex AMARC	4-96
☐ 156800		TA-7C	C-067	stored, for spares?, ex AMARC	4-96
☐ 156805		A-7E	E-005	stored, for spares?, ex AMARC	4-97
☐ 156827		A-7E	E-027	stored, for spares?, ex AMARC	4-97
☐ 156833		A-7E	E-033	stored, for spares?, ex AMARC	4-97
☐ 156851		A-7E	E-051	stored, for spares?, ex AMARC	4-96
☐ 157480		A-7E	E-136	stored, for spares?, ex AMARC	4-97
☐ 157486		A-7E	E-142	stored, for spares?, ex AMARC	4-97
☐ 157496		A-7E	E-152	stored, for spares?, ex AMARC	4-96
☐ 157502		A-7E	E-158	stored, for spares?, ex AMARC	7-95
☐ 158021		A-7E	E-270	stored, for spares?, ex AMARC	4-96
☐ 158824		A-7E	E-313	stored, for spares?, ex AMARC	7-95
☐ 158825		A-7E	E-314	stored, for spares?, ex AMARC	4-96
☐ 158829		A-7E	E-318	stored, for spares?, ex AMARC	4-96
☐ 159639		A-7E	E-381	stored, for spares?, ex AMARC	4-96
☐ 159997		A-7E	E-446	stored, for spares?, ex AMARC	4-96
☐ 160864		A-7E	E-525	stored, for spares?, ex AMARC	4-96
☐ 160868		A-7E	E-529	stored, for spares?, ex AMARC	4-96
☐ ..		F-4	..	stored, camo c/s, front and other bits	4-96
☐ 26722	722	F-84F	..	stored, ex instructional, ex 52-6722	11-97
☐ 0-61056		F-102A	..	stored, ex instructional, ex 56-1056	11-97
☐ EΣ-738		OH-13H	1904	stored, ex instructional, ex 56-2192	4-97
☐ 219		L-21B	..	stored	11-90
☐ 244		L-21B	..	stored	11-90
☐ 151366		P-3A	5079	stored, for spares, ex 151366/US Navy	11-97
☐ 151389		P-3A	5102	stored, for spares, ex 151389/US Navy	11-97
☐ 152181		P-3A	5151	stored, for spares, ex 152181/US Navy	11-97
☐ 21269	TR-269	T-33AN	269	stored, ex instructional, ex 21269/RCAF	11-97

THESSALONÍKI
Airfield – Sedes: The museum at Sedes has already lost two aircraft, CL-13B Sabres 19168 and 19347 both went to Tatoi.

☐ KN542		C-47B	16398/33146	preserved, ex KN542/RAF	11-96
☐ 3031		NF-5A	3031	preserved, ex K-3031/RNethAF	11-96
☐ 89065		F-5A	N6436	preserved, ex Micrá, ex RJordAF	11-96
☐ 37682		RF-84F	..	preserved, ex Tatoi, ex EB-334/WGAF	4-96
☐ 0-61031		F-102A	..	preserved, ex 56-1031	11-96
☐ 21367	TR-367	T-33AN	367	preserved, ex 21367/RCAF	4-96
☐ 35629	TR-629	T-33A	8968	preserved, ex 9448/WGAF, ex 53-5629	11-96
☐ 516713	TR-713	T-33A	6045	preserved, ex 51-6713	11-96
☐ 53-241		N2501D	164	preserved, ex 5341/WGAF	11-96

Of the other airframes at the airfield, C-47 92613 has gone to Micrá, 92637 is now at Athíne and Tatoi received C-47Bs 49111, KJ960 and KK171. Former preserved F-84F 26866 has gone to Micrá.

☐ 12351		C-47A	12351	dumped, ex SX-BAE, ex KG344, ex 42-92539	11-96
☐ 16348		C-47B	20814	dumped, ex 43-16348	11-96

☐ 92615		C-47	..	dumped, ex 49-2615	11-96
☐ 92619		C-47	..	dumped, ex 49-2619	11-96
☐ 92620		C-47	..	dumped, ex 49-2620, SX-PAC ntu	11-96
☐ 92623		C-47	..	dumped, ex 49-2623	11-96
☐ 92627		C-47	..	dumped, ex 49-2627	11-96
☐ 92628		C-47	..	stored, ex 49-2628	11-93
☐ 92629		C-47	..	dumped, ex stored, ex 49-2629	11-96
☐ 92634		C-47A	13837	dumped, ex stored, ex 49-2634, ex NC54339	11-96
☐ 92638		C-47	..	dumped, ex 49-2638	11-96
☐ KK181		C-47B	15427/26872	dumped, ex KK181/RAF, ex 43-49611	4-95
☐ KN475		C-47B	16204/32952	dumped, ex preserved, ex KN475	11-96
☐ KN516		C-47B	16308/33056	dumped, ex KN516/RAF, ex 44-76724	4-93
☐ KN616		C-47B	16515/33263	dumped, ex KN616/RAF, ex 44-76931	11-96
☐ 58642	TR-642	T-33A	1691	dumped, ex 9509/WGAF, ex 58-0642	9-93
☐ ..		F-5A	..	dumped, burnt fuselage	9-93

Airfield – Makedonia: This airfield was renamed from Micrá. Only one of the stored F-5As is known to have moved on, 89065 is now at Sedes.

☐ 3033		NF-5A	3033	stored, ex K-3033/RNethAF	8-97
☐ 3047		NF-5A	3047	stored, ex K-3047/RNethAF	8-97
☐ 3062		NF-5A	3062	stored, ex K-3062/RNethAF	8-97
☐ 01612		F-5B	N8082	stored, ex RJordAF, ex IIAF, ex 70-1612	4-97
☐ 21218		F-5A	N6405	stored, ex RJordAF, ex IIAF, ex 67-21218	4-97
☐ 38381		F-5A	N6018	stored, ex RJordAF, ex IIAF, ex 63-8381	4-97
☐ 38409		F-5A	N6046	stored, ex 63-8409	4-97
☐ 38410		F-5A	N6047	stored, ex 63-8410	4-97
☐ 38419		F-5A	N6056	stored, ex 63-8419	11-96
☐ 69135		F-5A	N6239	stored, ex 66-9135	11-96
☐ 69230		F-5B	N8038	stored, on its belly, w/o 14-7-92, ex 66-9230	4-96
☐ 89056		F-5A	N6427	stored, ex 68-9056	4-96
☐ 89068		F-5A	N6439	stored, ex RJordAF, ex IIAF, ex 68-9068	4-97
☐ 89078		F-5A	N6449	stored, ex RJordAF, ex IIAF, ex 68-9078	8-97
☐ 89084		F-5A	N6455	stored, ex RJordAF, ex IIAF, ex 68-9084	8-97
☐ 26866		F-84F	..	preserved, at main gate, ex 52-6866	11-97
☐ ..		F-86	..	dumped, burnt out wreck	4-96
☐ 17555	TR-555	T-33A	7700	dumped, ex 9413/WGAF, ex 51-17555	4-96
☐ 35784	TR-784	T-33A	9123	dumped, ex 53-5784	4-96

The Macedonian Flying Club near the civil terminal has two preserved aircraft. The preserved Sabre is a mark 2, while the serial '19494' is that of a Sabre 4.

☐ 92613		C-47	..	preserved, ex Sedes, ex 49-2613	11-97
☐ 19294		CL-13 Sabre 2	..	preserved, marked as '19494'	3-98
☐ EL-AIW		SE210-6N	106	stored, unmarked, ex I-DABS	11-97

TRIPOLOS

The 124 Pterix has a collection of aircraft. F-104G 33638 was allocated to this collection but is still at Aráxos.

☐ 92632		C-47	..	preserved, marked as '925432', ex 49-2632	11-97
☐ 26900	900	F-84F	..	preserved, ex 52-6900	11-97
☐ 37660		RF-84F	..	preserved, ex 53-7660	11-97
☐ 0-61025		F-102A	..	preserved, ex Elefsís, ex 56-1025	11-97
☐ 32715	FG-715	F-104G	6067	preserved, gate, ex C.8-1/SpaAF, ex 63-12715	11-97
☐ 36131	TR-131	T-33A	9752	preserved, ex 53-6131	11-97
☐ 53-240		N2501D	163	preserved, ex 5340/WGAF	11-97

VÓLOS
Next to the main road through the town (not in the harbour area) is a Freedom Fighter.
❑ 69228   F-5A   N7051   stored, ex Néa Ankhíalos, ex 228/RNoAF   11-97

VONITSA
A small air force camp here just outside the town of Vonitsa, 15km east of Préveza, has a Thunderstreak. Best thing is to follow the signs to the 133 HAF unit.
❑ 26540   F-84F   ..   preserved, ex 52-6540   4-97

HUNGARY

Since the ending of the Warsaw Pact, Hungary has willingly cut most links with that past. Display of the Soviet red star in public is no longer allowed and all the tanks, aircraft and other military items which used to be displayed on village greens and in school yards have had to be removed. Although a large number of these were transported to Vecsés, many others have been scrapped and only a few of these aircraft are still current. The Hungarian Air Force has the annoying habit of applying year numbers on their aircraft which are used for display. This makes the identification of some of these aircraft quite difficult.

AJKA
Owned by a Malev pilot was a small open air museum with some MiGs and an An-2. The aircraft were preserved in the Haditechnikai Park in Ajka, which is some 35km north of Veszprém (along road No.8). By May 1997 the location was empty. All aircraft, MiG-15bis 061, MiG-17PF 315, MiG-21U-400 1418 and An-2 HA-MHF, have since been noted at Kecel.

ALSÓNÉMEDI
The owner of the aircraft collection at Vecsés is building a new house here. Behind his house just outside of Alsónémedi (on the south east side) is a small private airstrip with a number of former Vecsés aircraft preserved. Its strange to see four MiG-15UTIs lined up which all carry the marking '1975'.

❏ 20 red	Yak-28	7961004	stored, ex Soviet AF, ex Vecsés	3-98
❏ 011	SBLim-2	1A-10011	stored, marked as '1975', ex Vecsés	3-98
❏ 027	SBLim-2	1A-11027	stored, marked as '1975', ex Vecsés	3-98
❏ 060	MiG-15bis	3060	stored, marked as '1977', ex Vecsés	3-98
❏ 201	MiG-15UTI	3201	stored, marked as '1975', ex Vecsés	3-98
❏ 754	MiG-15UTI	612754	stored, marked as '1975', ex Vecsés	3-98
❏ ..	MiG-15bis	2666	stored, marked as '14', ex Vecsés	3-98
❏ 404	MiG-17PF	0404	stored, marked as '1975' and '17', ex Vecsés	3-98
❏ 32 ?	MiG-19PM	651032	stored, marked as '1975', ex Vecsés	3-98
❏ 308	MiG-21F-13	741308	stored, marked as '1981' and '15', ex Vecsés	3-98
❏ HA-PZK	PZL 101	21012	stored, frame only, ex Vecsés	3-98

BÉKÉSCSABA
The preserved MiG-15bis on the airfield has also been reported as 825.
❏ 807	MiG-15bis	..	preserved	2-95

BÓCSA
Along the main road in this village (40km south west of Kecskemét) is a former Malev Li-2 preserved.
❏ HA-LIV	Li-2P	18439310	preserved, ex Dunaújváros, ex MN310	3-98

BÖRGÖND
During 1993 the based Mi-2s all moved to Szolnok. Since then the airfield is in use by civilians, but the MiG-15 is still preserved on a pole.
❏ 071	MiG-15bis	3153071	preserved, on pole	3-98

BUDAÖRS
The Air Force have relocated their operational aircraft from here, the based Zlin 43s moved to Tököl. Left

HUNGARY - 247

behind is a MiG-21F preserved on pole and a stored SM-1. The two Tu-A were used by Malev in South Pole expeditions. Stored MiG-15 677 and PZL M15 CCCP-15187 moved on to Cespel, while the Li-2 HA-LIQ went to Budapest Ferihegy. Both aircraft were last noted in July 1992.

❏ 37	SM-1	W04037	stored	9-97
❏ 305	MiG-21F-13	741305	preserved	9-97
❏ 2202	Tu-A3	..	stored	9-97
❏ 2302	Tu-A3	..	stored	9-97

BUDAPEST
Town: The Közlekedési Múzeum (Transport Museum) has, besides a large number of smaller civil aircraft, one complete military SM-1. Also here is the cockpit of the MiG-21PF which came from Kecskemét.

❏ 30	SM-1	401030	preserved	8-97
❏ 14	MiG-21PF	761514	preserved, cockpit only, ex Kecskemét	8-97

Town – Cespel: Cespel is a south eastern suburb of Budapest. Located here on along a large north-south road is the Kossuth Lajos Aviation Technical School. Most aircraft are parked outside, except for the Zlins and one SM-1 which are parked inside. In early 1998 Aero 45 HA-OMD moved to Merseburg. After restoration the WSK M15 CCCP-11587 (ex Budaörs) moved on to Szolnok.

❏ 408	Ka-26	7001408	instructional	3-98
❏ 376	L-29	591376	instructional	3-98
❏ 31	SM-1	W04031	instructional	3-98
❏ 033	SM-1	W04033	instructional	9-97
❏ 677	MiG-15bis	..	instructional, ex Budaörs	3-98
❏ 4407	MiG-21MF	964407	instructional	3-98
❏ CCCP-25625	Mi-8	9775212	instructional	3-98
❏ HA-ANF	An-2M	500403	instructional	9-97
❏ HA-BRB	L-60T	150910	instructional	3-98
❏ HA-PXA	PZL 101	..	instructional	7-97
❏ HA-PXB	PZL 101	..	instructional	3-98
❏ HA-MHG	An-2M	601220	instructional	9-97
❏ HA-TRL	Zlin 226MS	370	stored, frame only	7-97
❏ HA-TRN	Zlin 526F	1243	under restoration	7-97
❏ HA-TRR	Zlin 226	..	stored, frame only	7-97
❏ HA-TRT	Zlin 326	833	stored, frame only	7-97
❏ N14111	PA31-310	31-442	instructional	3-98

Airfield - Ferihegy: The airport is split in two parts. On the older part (Ferihegy 1) three MiGs were stored since the late 1980s. These three aircraft, MiG-15 061, MiG-17 315 and MiG-21U 1418, all left for Várpalota. The Tu-134 is in use by Malev for cabin training.

❏ HA-LBG	Tu-134	0350924	instructional	9-97
❏ HA-MOG	Il-18B	184007103	stored	7-97

On Ferihegy 2, a small museum is setting up. Its officially not yet open but all aircraft can be seen from outside.

❏ 04 red	Il-14G	147001821	preserved, ex Soviet AF	3-98
❏ HA-MHI	An-2M	701647	preserved	3-98
❏ HA-MOA	Il-18B	180001903	preserved	3-98
❏ HA-LBE	Tu-134	9350802	preserved	3-98
❏ HA-LCG	Tu-154B-2	75A-127	preserved	3-98
❏ HA-LIQ	Li-2P	23441206	preserved, ex Budaörs, ex MN206	3-98

DUNAÚJVÁROS
Preserved on a pole at the civil airfield is a MiG-15bis in false markings. Underneath the '1952' and '1977' marks

other false markings can be read in the form of '1973'.
| ❏ .. | MiG-15bis | .. | preserved, marked as '1952/1977', ex '1973' | 7-97 |

ESZTERGOM
Stored on this airfield, along the Slovak border, is an An-2.
| ❏ CCCP-82805 | An-2 | 1G166-23 | stored | 7-97 |

GYÖR - ÖTTEVÉNY
On the west side of Györ, along the main road from Bratislava to Budapest, is a large restaurant next to a tourist park called 'Elvis Park'. The restaurant is built around the Il-18.
| ❏ HA-MOI | Il-18V | 187010002 | preserved | 7-97 |

JAKABSZÁLLÁS
At this small civil airfield (south west of Kecskemét) a MiG-15 is preserved. Its complete serial is unknown.
| ❏ ..14 | MiG-15 | .. | preserved, marked as '1978' | 9-97 |

KAPOSVÁR
A large number of An-2s can always been found on this airfield as maintenance on this type is carried out here. The MiG was not noted in 1995 and may have gone.
❏ ..	MiG-19PM	..	preserved, marked as '1978'	7-92
❏ HA-MEH	An-2	1G190-17	stored, wreck	7-92
❏ HA-MHS	An-2	1G145-44	stored, wreck	6-95

KECEL
In a field at this location a new museum was noted in 1997. Hopefully this will be the final location as some of the aircraft have travelled through Hungary in the past few years.
❏ ..	Mi-1	..	preserved	3-98
❏ 061	MiG-15bis	3061	preserved, ex Ajka, ex Várpolata, ex Ferihegy	3-98
❏ ...	MiG-15	..	preserved, marked as '1963'	3-98
❏ 315	MiG-17PF	0315	preserved, ex Ajka, ex Várpolata, ex '1974'	3-98
❏ 847	MiG-17PF	0847	preserved	3-98
❏ ...	MiG-19	..	preserved, marked as '1974'	3-98
❏ 1418	MiG-21U-400	661418	preserved, ex Ajka, ex Várpolata, ex Ferihegy	3-98
❏ 8202	MiG-21MF	968202	preserved, ex Kecskemét	3-98
❏ ..	MiG-21F-13	..	preserved, marked as '1977'	3-98
❏ HA-MHF	An-2M	601219	preserved, ex Ajka, ex Vecsés	97

KECSKEMÉT
During 1993 the first MiG-29s arrived at Kecskemét, to replace the MiG-21MFs and MiG-21UMs. A number of the operational MiG-21s were transferred to Pápa, while the remainder were stored and will be sold or broken up. The first to leave was MiG-21UM 0565 which went to St Isidorushoeve. MiG-21MF 4406 is now preserved at the museum at Graz and 9309 went to Szolnok. Some of the older MiG-21 types have not been seen in recent years and may all be scrapped. Of these MiG-21PF 14 was broken up in late 1994 and the cockpit moved to Budapest. MiG-21PF 409 went to Taszár and MiG-21F-13 915 went to Seifertshoven. MiG-21MF 8202 is now at Kecel with both 9507 and 9603 at Pápa. MiG-21UM 3036 moved to 's Graveland. Czechoslovakian Air Force MiG-21F-13 0220 was used for many years as instructional airframe before moving in 1992 to Vecsés. Twelve aircraft were cut-up for the CFE treaty and are currently stored. It has been reported that all remaining MiG-21PFs and MiG-21bis 3955 and 4112 were scrapped in November 1994.

HUNGARY - 249

☐ S1	L-39ZO	731018	stored, for spares, ex Rothenburg, ex 2815/GAF	5-97
☐ S2	L-39ZO	731011	stored, for spares, ex Rothenburg, ex 2819/GAF	5-97
☐ S3	L-30ZO	731008	stored, for spares, ex Rothenburg, ex 2807/GAF	5-97
☐ 47	Il-28	..	dumped	9-97
☐ 684	MiG-15bis	2684	dumped, marked as '1977', ex preserved	9-97
☐ 725	MiG-15bis	31530725	preserved, marked as '1976'	5-97
☐ ..	SBLim-2	1A-06002	dumped, marked as '1951', ex preserved	3-98
☐ ..	MiG-17PF	..	stored, marked as '1976', ex preserved	9-97
☐ ..	MiG-19PM	..	preserved, marked as '1976'	5-97
☐ 10	MiG-21PF	760510	stored	8-94
☐ 25	MiG-21F-13	..	preserved	9-97
☐ 504	MiG-21PF	760504	stored	7-93
☐ 508	MiG-21PF	760508	stored, marked as '905'	8-91
☐ 509	MiG-21PF	760509	stored	8-94
☐ 602	MiG-21PF	761602	stored	8-94
☐ 805	MiG-21F-13	741815	dump, in MiG-29 area	9-97
☐ 909	MiG-21F-13	741909	preserved, marked as '1977/1978', ex stored	3-98
☐ 0158	MiG-21UM	01695158	stored, dismantled	9-97
☐ 1601	MiG-21PF	761601	stored	8-94
☐ 1605	MiG-21PF	761605	stored	8-94
☐ 3955	MiG-21bis	N75033955	stored, CFE a/c, with tail from 9178	9-95
☐ 4112	MiG-21bis	N75034112	stored, CFE a/c	9-95
☐ 4603	MiG-21MF	964603	stored, old markings, CFE a/c	9-95
☐ 4607	MiG-21MF	964607	stored, old markings, CFE a/c	9-95
☐ 4608	MiG-21MF	964608	stored, old markings, CFE a/c	9-95
☐ 8108	MiG-21MF	968108	stored, old markings, CFE a/c	9-95
☐ 8109	MiG-21MF	968109	stored, old markings, CFE a/c	9-95
☐ 8110	MiG-21MF	968110	stored	3-98
☐ 8111	MiG-21MF	968111	stored, old markings, CFE a/c	9-95
☐ 8113	MiG-21MF	968113	stored, old markings, CFE a/c	9-95
☐ 8115	MiG-21MF	968115	stored, old markings, CFE a/c	9-95
☐ 8203	MiG-21MF	968203	stored, old markings, CFE a/c	9-95
☐ 9310	MiG-21MF	969310	stored, old markings, CFE a/c	9-95
☐ 9311	MiG-21MF	969311	stored, dismantled	9-97
☐ 9312	MiG-21MF	969312	stored	9-97
☐ 9506	MiG-21MF	969506	stored	6-95
☐ 9509	MiG-21MF	969509	stored	3-98
☐ 9511	MiG-21MF	969511	stored, dismantled	9-97
☐ 9513	MiG-21MF	969513	stored	9-97
☐ 9514	MiG-21MF	969514	stored, dismantled	3-98
☐ 9602	MiG-21MF	969602	stored	3-98

OROSZLÁNY
Stored on this mainly glider field is a An-2 in Aeroflot colours.

☐ HA-ABH	An-2	..	stored	7-97

OZIGETVAR
A small museum here had a MiG-15 (reported as 981) which was transported to Vecsés in the early 1990s.

PÁPA
The gate guards here are long term inmates, while the MiG-21s, MiG-23s and Su-22's are 'new' Relics. Some

of stored MiG-21s may return to flying status and be swopped with currently operational aircraft.

❏ 779	MiG-15UTI	..	preserved, at gate, marked as '1950', ex '1974'		3-98
❏ ..	MiG-19PM	..	preserved, at gate, marked as '1975'		3-98
❏ 47	MiG-21bis	..	stored		5-97
❏ 806	MiG-21F-13	741806	preserved, at gate, marked as '1976'		3-98
❏ 913	MiG-21F-13	741913	dump		9-97
❏ 1844	MiG-21bis	N75061844	stored		7-97
❏ 1867	MiG-21bis	N75061867	stored		7-97
❏ 1874	MiG-21bis	N75061874	stored		7-97
❏ 1900	MiG-21bis	N75061900	stored		7-97
❏ 1904	MiG-21bis	N75061904	stored		7-97
❏ 1953	MiG-21bis	N75061953	stored		7-97
❏ 1968	MiG-21bis	N75061968	stored		7-97
❏ 2098	MiG-21bis	N75062098	stored		7-97
❏ 2105	MiG-21bis	N75062105	stored		7-97
❏ 3964	MiG-21bis	N75033964	stored		7-97
❏ 4025	MiG-21bis	N75034025	stored		7-97
❏ 6145	MiG-21bis	N75046145	stored		7-97
❏ 6253	MiG-21bis	N75046253	stored		7-97
❏ 6305	MiG-21bis	N75046305	stored		7-97
❏ 6327	MiG-21bis	N75046327	stored		7-97
❏ 6384	MiG-21bis	N75046384	stored		7-97
❏ 4409	MiG-21MF	964409	stored		7-97
❏ 8114	MiG-21MF	968114	stored		7-97
❏ 8201	MiG-21MF	968201	stored		7-97
❏ 8204	MiG-21MF	968204	stored		7-97
❏ 9099	MiG-21bis	N75049099	stored		7-97
❏ 9125	MiG-21bis	N75049125	stored		7-97
❏ 9307	MiG-21MF	969307	stored		7-97
❏ 9315	MiG-21MF	969315	stored		7-97
❏ 9507	MiG-21MF	969507	stored, ex Kecskemét		7-97
❏ 9510	MiG-21MF	969510	stored		7-97
❏ 9515	MiG-21MF	969515	stored		7-97
❏ 9603	MiG-21MF	969603	stored, ex Kecskemét		7-97
❏ 9604	MiG-21MF	969604	stored		7-97
❏ 01	MiG-23MF	..	stored		7-97
❏ 02	MiG-23MF	..	stored		7-97
❏ 03	MiG-23MF	..	stored		7-97
❏ 06	MiG-23MF	0390217165	stored		7-97
❏ 07	MiG-23MF	0390217166	stored, wfu 2-5-96		7-97
❏ 08	MiG-23MF	..	stored		7-97
❏ 10	MiG-23MF	..	stored		9-97
❏ 12	MiG-23MF	..	stored		7-97
❏ 20	MiG-23UB	19015091	stored		7-97
❏ 02	Su-22M-3	52102	stored		7-97
❏ 03	Su-22M-3	52303	stored		7-97
❏ 04	Su-22M-3	52304	stored		7-97
❏ 05	Su-22M-3	52305	stored		7-97
❏ 08	Su-22UM-3K	17532390304	stored		7-97
❏ 09	Su-22UM-3K	17532390305	stored		7-97
❏ 10	Su-22M-3	52610	stored		7-97
❏ 11	Su-22M-3	52611	stored		7-97
❏ 12	Su-22M-3	52612	stored		7-97
❏ 14	Su-22M-3	52814	stored		7-97
❏ 15	Su-22M-3	52815	stored		7-97

PÉCS
The Li-2 is preserved in the Vidampark. This park is on a hillside just outside the town of Pécs. The ex Soviet Mi-2 is stored at the local airfield. The exact location of the MiG-17 is unknown, it was reported at the airfield, but is currently not there. There is also doubt if the serial is correct.

❏ 31 red	Mi-2	543722064	stored, rotorless, ex Soviet	9-97
❏ 522 ?	MiG-17	..	preserved	—
❏ HA-LIS	Li-2P	23441301	preserved, ex MN301	4-96

SOLT
A petrol station on the ringroad around Solt has two MiG-15UTIs. Both have been here for a long time. 202 was repainted in the mid 1990s and placed on a pole. It received a false serial and had fallen off its pole by September 1997.

❏ 028	SBLim-2	1A-06028	stored	6-96
❏ 202	MiG-15UTI	3202	preserved, marked as '501'	9-97

SZEGED
At an unknown location in this town is a Li-2 preserved.

❏ MN306	Li-2P	18439306	preserved, ex HA-LIU	2-92

SZENTKIRÁLYSZABADJA
Some people have some difficulties in pronouncing the name of this airfield and threrefore call it Veszprém, which is not correct. All the helicopters are stored rotorless and the ex NVA Mi-24s have recently moved into some hangars for further storage. Some of the stored Hungarian helicopters may return to operational service once their long awaited overhaul has been carried out.

❏ 9602	Mi-24D	110157	stored, ex Cottbus, ex 396/NVA	2-97
❏ 9604	Mi-24D	110160	stored, ex Cottbus, ex 424/NVA	—
❏ 9605	Mi-24D	110161	stored, ex Cottbus, ex 433/NVA	5-97
❏ 9607	Mi-24D	110165	stored, ex Cottbus, ex 487/NVA	5-97
❏ 9610	Mi-24D	110172	stored, ex Cottbus, ex 522/NVA	4-96
❏ 9611	Mi-24D	110173	stored, ex Cottbus, ex 523/NVA	4-96
❏ 9618	Mi-24D	730212	stored, ex Cottbus, ex 540/NVA	5-97
❏ 9620	Mi-24D	B4001	stored, ex Basepohl, ex 403/NVA	4-96
❏ 9624	Mi-24D	B4071	stored, ex Basepohl, ex 417/NVA	4-96
❏ 9625	Mi-24D	B4072	stored, ex Basepohl, ex 418/NVA	4-96
❏ 9627	Mi-24D	110162	stored, ex Basepohl, ex 434/NVA	4-96
❏ 9628	Mi-24D	110164	stored, ex Basepohl, ex 447/NVA	4-96
❏ 9632	Mi-24D	110168	stored, ex Basepohl, ex 496/NVA	4-96
❏ 9636	Mi-24D	340275	stored, ex Basepohl, ex 532/NVA	4-96
❏ 9641	Mi-24P	340331	stored, ex Basepohl, ex 358/NVA	4-96
❏ 9642	Mi-24P	340332	stored, ex Basepohl, ex 361/NVA	4-96
❏ 9644	Mi-24P	340334	stored, ex Basepohl, ex 415/NVA	4-96
❏ 9645	Mi-24P	340335	stored, ex Basepohl, ex 422/NVA	4-96
❏ 9646	Mi-24P	340336	stored, ex Basepohl, ex 439/NVA	4-96
❏ 9648	Mi-24P	340338	stored, ex Basepohl, ex 444/NVA	4-96
❏ 130	Mi-8TB	0130	stored	3-98
❏ 230	Mi-8TB	0230	stored	7-92
❏ 428	Mi-8TB	0428	stored	7-92
❏ 736	Mi-8TB	0736	stored	7-92
❏ 6220	Mi-8TB	226220	stored	3-98
❏ 6223	Mi-8TB	226223	stored	3-98
❏ 10418	Mi-8TB	10418	stored	3-98

HUNGARY – 252

❏ 10419	Mi-8TB	10419	stored		3-98
❏ 10429	Mi-8TB	10429	stored		3-98
❏ 703	Mi-17	104M03	stored		3-98
❏ 706	Mi-17TPB	104M06	stored		3-98
❏ 005	Mi-24D	K4005	stored		5-97
❏ 007	Mi-24D	K4007	stored		5-97
❏ 104	Mi-24D	K20104	stored		5-97
❏ 105	Mi-24D	K20105	stored		5-97
❏ 107	Mi-24D	K20107	stored		3-98
❏ 108	Mi-24D	K20108	stored		5-97
❏ 109	Mi-24D	K20109	stored		5-97
❏ 114	Mi-24D	K20114	stored		5-97
❏ 115	Mi-24D	K20115	stored		3-98
❏ 116	Mi-24D	K20116	stored		3-98
❏ 711	Mi-24V1	220711	stored		3-98
❏ 714	Mi-24V1	220714	stored		3-98

SZOLNOK
Airfield: On the south western side of the town is the military base of Szolnok. Besides a number of operational aircraft the base houses the Magyar Repüléstörténeti Múzeum Alapítvany. After some four years as a closed storage site on the airfield, the museum moved during 1993 to a location near the main gate and opened its gates during early 1994. Currently on display are a number of the 'hardware' items of the museum. At this location the museum hopes to build a large museum hall in which also the more vulnerable aircraft, which are currently still held is storage, will be displayed. The two ex NVA aircraft have never flown with the Hungarian Forces and were only obtained to fill the gap in the museum's collection. Li-2P 209 (18433209, ex HA-LIX) moved to Budapest Ferihegy to become airworthy again. The museum is open (during the period from 1st March to 30th November) from Monday to Thursday from 09:00 to 14:30 and on Fridays between 09:00 and 13:00s.

❏ 534	Mi-4	04146	preserved, marked as '13', ex NVA, arr 1-12-79	3-98
❏ 426	VEB-14P	14803022	preserved, ex NVA, arrived 1-12-79	3-98
❏ 23	Aero 45	..	preserved	3-98
❏ 907	An-24V	77303907	preserved	3-98
❏ 55	Il-28	56455	preserved	3-98
❏ 505	Ka-26	7001505	preserved	3-98
❏ 379	L-29	591379	preserved	3-98
❏ 10422	Mi-8TB	10422	preserved, wfu after landing accident	3-98
❏ 512	MiG-15bis	4512	preserved, ex '1974'	3-98
❏ 203	MiG-15UTI	3203	preserved	3-98
❏ 405	MiG-17PF	0405	preserved	3-98
❏ 28	MiG-19PM	651028	preserved	3-98
❏ 813	MiG-21F-13	741813	preserved	3-98
❏ 814	MiG-21F-13	741814	preserved	3-98
❏ 911	MiG-21F-13	741911	preserved	3-98
❏ 4419	MiG-21U-600	664419	preserved	3-98
❏ R-06	PZL 104	59062	preserved	3-98
❏ CCCP-15187	WSK M15	IS019-10	preserved, ex Cespel, ex Budaörs	3-98
❏ HA-LBF	Tu-134	0350923	preserved, wfu 12-88	3-98
❏ HA-MOE	Il-18B	182005505	preserved	3-98

As mentioned above the museum still has storage facilities on the airfield from which sightings are very rare.

❏ J-4022	Hunter F.58	41H-697389	stored, ex Swiss AF	7-97
❏ 09	SM-1	W05009	stored, serial R-09 ntu	7-92
❏ R-10	Mi-2	535431127	stored, fuselage only, crashed 2-4-87	4-94
❏ ..	L-60	..	stored	7-92
❏ ..	Yak-11	..	stored, marked as '1949'	7-92

☐ ..	Yak-12R	..	stored	8-91
☐ ..	Yak-18	..	stored	7-92
☐ HA-PAO	CSS-13	0448	stored, Polish built Polikarpov Po-2, ex 48	9-96
☐ HA-TRI	Zlin 326	..	stored	8-91

Based at Szolnok is the Hungary military school. This school has a number of instructional airframes in a hangar and a storage compound. The L-39 may have moved to Kecskemét.

☐ 2813	L-39ZO	731016	instructional, ex Rothenburg, ex 153/NVA	7-97
☐ 19	SM-1	W02019	stored, in compound	6-96
☐ 36	SM-1	W04036	stored, in compound	7-97
☐ 032	SM-1	401032	stored, dumped	7-92
☐ 036	Mi-8T	201036	instructional, in hangar	2-98
☐ 328	Mi-8T	0328	instructional, in hangar	7-97
☐ 10420	Mi-8T	10420	instructional, in hangar	7-97
☐ 406	MiG-21PF	760406	stored, in compound	7-97
☐ 808	MiG-21F-13	741808	stored, in compound	7-97
☐ 816	MiG-21F-13	741816	instructional, in hangar	9-96
☐ 819	MiG-21F-13	741819	stored, in compound	2-98
☐ 822	MiG-21F-13	741822	stored, in compound	2-98
☐ 824	MiG-21F-13	741824	stored, in compound	7-97
☐ 1511	MiG-21PF	761511	instructional	6-91
☐ 1512	MiG-21PF	761512	instructional, in hangar	7-97
☐ 2221	MiG-21F-13	742221	stored, in compound	2-98
☐ 2311	MiG-21F-13	742311	stored, in compound	7-97
☐ 2316	MiG-21F-13	742316	stored, in compound	9-96
☐ 3537	MiG-21bis	N75033537	stored, in compound	2-98
☐ 3732	MiG-21bis	N75033732	instructional, in hangar	7-97
☐ 3745	MiG-21bis	N75033745	stored, in compound	2-98
☐ 3945	MiG-21bis	N75033945	instructional, in hangar	7-97
☐ 4605	MiG-21MF	964605	stored, in compound	2-98
☐ 4606	MiG-21MF	964606	instructional, in hangar	7-97
☐ 9306	MiG-21MF	969306	instructional	6-95
☐ 9309	MiG-21MF	969309	instructional, in hangar	9-96
☐ 9512	MiG-21MF	969512	instructional, in hangar	7-97

Furthermore a number of aircraft are stored at the airfield, most of which have run out of airframe hours. The MiG-15 is dumped on the northern side of the airbase.

☐ 202	An-26	02202	stored	3-98
☐ 203	An-26	02203	stored, wfu since summer 1995	3-98
☐ 204	An-26	02204	stored, wfu since summer 1995	3-98
☐ 208	An-26	02208	stored	3-98
☐ 209	An-26	02209	stored	3-98
☐ 7831	Mi-2	517831092	stored	7-97
☐ 7834	Mi-2	517834092	stored	7-97
☐ 7835	Mi-2	517835092	stored	7-97
☐ 8344	Mi-2	518344093	stored	7-97
☐ 8911	Mi-2	519811104	stored	5-97
☐ 8912	Mi-2	519812104	stored	7-97
☐ 8913	Mi-2	519813104	stored	7-97
☐ 8914	Mi-2	519814104	stored	7-97
☐ 8915	Mi-2	519815104	stored, derelict	2-98
☐ 8916	Mi-2	519816104	stored, derelict	2-98
☐ 9408	Mi-2	519408095	stored	3-98
☐ 9409	Mi-2	519409095	stored, derelict	7-97
☐ 9410	Mi-2	519410095	stored	3-98

HUNGARY - 254

❏ 9411	Mi-2	519411095	stored, derelict	7-97
❏ 9412	Mi-2	519412095	stored	3-98
❏ 10030	Mi-2	5110030116	stored, derelict	3-98
❏ 330	Mi-8T	0330	stored	2-98
❏ 416	Mi-8S	0416	stored	3-98
❏ 730	Mi-8S	0730	stored	2-98
❏ 936	Mi-8T	300936	stored, derelict	2-98
❏ 2639	Mi-8S	22639	stored	7-97
❏ 10417	Mi-8TB	10417	preserved, near tower	3-98
❏ 10424	Mi-8TB	10424	stored	3-98
❏ 10427	Mi-8TB	10427	stored	2-98
❏ 10439	Mi-8TB	10439	stored	9-96
❏ 724	MiG-15bis	31530724	preserved, near gate	3-98
❏ 338	MiG-15	..	stored	9-97
❏ ..	MiG-15	..	preserved, marked as '1951/1975', near tower	3-98
❏ ..	MiG-15	..	dumped	9-97
❏ 224	MiG-21F-13	741224	preserved, op pole	4-96

Airfield – **Szandaszölös**: Close to the main airfield of Szolnok, along the ring road, is a large exercise field which is often used by the helicopters from Szolnok. Preserved here next to the control tower is a Li-2 which used to fly with Malev between 1957 and May 1964, its civil marks are still visible.

❏ HA-LIP	Li-2P	18439504	preserved, ex MN504	5-97

TASZÁR

Taszár became public news when the Americans arrived here in 1996 for their part in IFOR/SFOR. The operational flying by the Hungarian Air Force was transferred to Pápa. Serial of the MiG-19PM 35 seems doubtfull as this aircraft has crashed on 24th October 1964.

❏ ..	MiG-15UTI	..	preserved, marked as '1978'	7-96
❏ 817	MiG-15bis	31530817?	preserved, marked as '7104'	7-96
❏ 35	MiG-19PM	651035	preserved, marked as '1973', see note	6-95
❏ 37	MiG-19PM	651037	preserved, at barracks	7-96
❏ 409	MiG-21PF	760409	preserved, marked as '1989', ex Kecskemét	6-95
❏ 505	MiG-21PF	760505	preserved, marked as '1989'	8-94
❏ 801	MiG-21F-13	741801	preserved, marked as '1975'	6-95
❏ 5721	MiG-21bis	75035721	preserved	9-97
❏ 5809	MiG-21bis	75035809	dump, wreck, crashed 29-6-90	6-91
❏ 01	Su-22UM-3	52101	preserved	9-97
❏ 07	Su-22UM-3	17532390303	dumped, wreck, crashed 25-5-95	7-95

TÖKÖL

Stored at former Soviet airfield of Tököl are a dozen Mi-8s which were not accepted by the Hungarian Air Force. Three of these have gone to the Peruvian Navy as HT-450 (10455), HT-451 (10456) and HT-452 (10457). Anonther four (10451, 10452, 10453 and 10454) were sold in late 1997 to Ethiopian Air Force and left on 22nd November 1997. Of the reaiming four, three have been reported to have sold to Russia.

❏ 908	An-24V	77303908	stored	3-98
❏ 046	MiG-15UTI	..	instructional	7-97
❏ 313	MiG-21F-13	741313	instructional	7-97
❏ 1319	MiG-21U-400	661319	instructional	7-92
❏ ..	Mi-8P	10448	stored, ntu	7-97
❏ ..	Mi-8T	10449	stored, ntu	12-96
❏ ..	Mi-8T	10450	stored, ntu	12-96
❏ ..	Mi-8T	10458	stored, ntu	12-96

VARPALOTA
In this village near Lake Balaton, three MiGs were preserved for a short time. All three, MiG-15 061, MiG-17 315 and MiG-21 1418, moved to Ajka.

VECSÉS
Close to the airfield of Ferihegy is the famous yard at Vecsés. After the Soviets left Hungary, most of the aircraft which were preserved at various towns were collected here. A large number of these aircraft carry year numbers as serials and are not in too good a shape, making identification very difficult. In 1997 the owner of the yard moved a number of aircraft to his new home at Alsónémedi. SM-1 13 moved to Merseburg, while An-2M HA-MHF moved to Ajka. During a visit in March 1998 three MiG-15s (all marked as '1974') remained unidentified. The following is believed to be the correct version:

☐ 0220	MiG-21F-13	560220	stored, ex Kecskemét, ex Czechoslovak AF	3-98
☐ 13	SM-1	W02013	stored	3-98
☐ 062	MiG-15bis	3062	stored, marked as '1974'	7-97
☐ 065	MiG-15bis	3065	stored, marked as '1974'	7-97
☐ 067	MiG-15bis	3067	stored, marked as '1978'	3-98
☐ 069	MiG-15bis	3069	stored, marked as '1974'	10-94
☐ 115	MiG-15bis	..	stored, marked as '1977'	3-98
☐ 613	MiG-15bis	..	stored, marked as '1974'	10-94
☐ 708	MiG-15bis	31530708	stored, marked as '1977', also reported as 708	3-98
☐ 713	MiG-15bis	3153713	stored, marked as '1983'	3-98
☐ 771	MiG-15UTI	..	stored	3-98
☐ ..	MiG-15bis	3056 ?	stored, marked as '981', ex '1981', ex Ozigetvar	3-98
☐ ..	MiG-15bis	26..	stored, marked as '1974'	4-95
☐ ..	MiG-15	..	stored, marked as '1978'	10-94
☐ ..	MiG-15	..	stored, marked as '1978'	4-95
☐ 402	MiG-17PF	0402	stored, marked as '1975'	3-98
☐ ..	MiG-17PF	..	stored, marked as '1975'	3-98
☐ 36	MiG-19PM	651036	stored, marked as '1978'	3-98
☐ 1218	MiG-21F-13	741218	stored	3-98
☐ 1320	MiG-21U-400	661320	stored	3-98
☐ 2319	MiG-21F-13	742319	stored, marked as '1981'	3-98
☐ HA-MHB	An-2M	601215	stored	9-97
☐ HA-MHJ	An-2M	601302	stored	3-98
☐ ..	PZL 101	..	stored	9-97

ICELAND

Iceland is one of those countries seldom visited by aviation enthusiasts from other parts of Europe. Sightings from here are rare. There are also not that many *Wrecks & Relics* on the islands to report on.

HNJOTUR
The Egil Òlaffson Folk museum on the extreme west side of the island has an An-2 on display.
❑ RA-50502	An-2	..	preserved		94

KEFLAVÍK
The formerly preserved and later dumped F-102A 56-1378 has last been noted in May 1990 and will have been scrapped by now. Also the dumped C-118B 533247 (44618) has gone. The Phantom is painted in the colours of a former 57th FIS aircraft which used to be based at Keflavík.

❑ 72-1407	F-4E	4398	preserved, marked as '66-300', ex USAF	7-97
❑ 17191	C-117D	43379	preserved, at gate, ex USNavy	7-97
❑ 151367	UP-3A	5080	preserved, ex USNavy	7-97
❑ N904WA	BN-2A-21	904	stored, wreck, crashed 17-3-96	7-97

REYKJAVÍK
Reykjavík airport is the main airfield on Iceland and has beside a number of light civil aircraft, one larger *Wrecks & Relics* inmate. It is a former Icelandair Dakota.

❑ TF-ISB	DC-3	9860	stored, ex G-AKSM	7-97

ITALY

Since the early 1990s, a large number of military aircraft have been disposed of, by Italy, like other countries. Types like the F-104G and all versions of the G91 have gone from the list of operational aircraft. Luckily these aircraft were not scrapped straight away and many lingered on for many years. Large numbers of these aircraft can still be found on airfields like Améndola, Cérvia and Villafranca. New collections have been set up in Italy of which those at Cerbaiola and Castel Volturno are the most notable. The museum of Vigna di Valle closed in November 1996 for total refurbishment. A new fourth hall will be added, while the other three will be restored. The museum is hoping to open its gates again in the fall of 1998.

ACQUI TERME
Preserved at the airfield (about 25km south west of Alessandria) and cared for by the preservation group Associazione Arma Aeronautica is a T-6.

❑ MM53766	RM-10	T-6G/D	..	preserved	9-95

ALBANO TERME (or Monte Venda)
South west of Padova, alongside a railway, is a military compound. Mounted on a pole and clearly visible from outside is a freshly painted MB326. It lost its MM number in the process. Preserved nearby and maybe not visible from outside is an Agusta Bell.

❑ MM80221	53-96	AB47J	1136	preserved	2-96
❑ MM.....	51-69	MB326	..	preserved, camo c/s	2-96

ALESSANDRIA
The Instituto Tecnico a Volta Di Alessandria has a Fiat-built Sabre preserved in front of the school building. Some smaller aircraft may be inside.

❑ MM53-8274	5-57	F-86K	221-2	preserved	9-95

ALGHERO
Five Harvards formerly with the Scuola Volo Basico Iniziale Elica based at Alghero-Fertilia have been noted here in storage.

❑ MM53674		T-6G	..	stored, in hangar	96
❑ MM53838	SL-3	T-6H-4M	..	stored	97
❑ MM53861	SL-45	T-6C	..	stored, in hangar	96
❑ MM53881	SL-46	T-6H-4M	..	stored	97
❑ MM54117	SL-33	T-6G	..	stored, in hangar	96

AMÉNDOLA
The AMI's last G91T flight was from here on 30th September 1995. But even before that date a large number of G91Ts were in storage here. Most of these aircraft came under the CFE treaty. The scrapping of these aircraft is carried out in three different batches. The aircraft were cut into three parts and dumped in an area near the flightline. Two early inmates of the scrap dump, G91T/1s MM6321/SA-21 (51) and MM54413/SA-113 (140) were scrapped in the early 1990s. Of the stored G91Ts 27 (of which 24 are known, all CFE aircraft) were sold to a scrapyard in Roma (which see). Three of the other CFE aircraft should also have left for Roma. An other one te leave is MM54403/SA-103 which went to the museum at Cerbaiola near San Marino. It has been reported that all stored G91Ys have been sold and that more have already left the base. The sole G91Y carries two different codes, one on each side, while the T-33A carries a false serial. MM54-2951 is quoted as its original serial. The TF-104s on the dump remain a mystery. The example coded '20-21' carries a contruction number on number on

its plate from an operational aircraft, while the other Starfighter is completely unmarked. Both may have moved to Grosseto by now. Améndola is the largest Italian airbase and not much can be noted from outside The preserved T-33 can be seen in a distance.

❑ MM6316	32-16	G91T/1	46	stored	10-97
❑ MM6317	60-20	G91T/1	47	dumped, in scrap compound	9-96
❑ MM6323	60-23	G91T/1	53	stored, CFE a/c, code 12, first batch	10-97
❑ MM6325	32-25	G91T/1	55	stored	10-97
❑ MM6326	60-26	G91T/1	56	stored	10-97
❑ MM6327	60-27	G91T/1	57	stored, CFE a/c, code 9, first batch	10-97
❑ MM6329	32-29	G91T/1	59	stored	10-97
❑ MM6330	32-30	G91T/1	60	stored	10-97
❑ MM6332	32-32	G91T/1	62	stored	9-96
❑ MM6333	32-33	G91T/1	63	stored	10-97
❑ MM6337	32-37	G91T/1	67	stored	10-97
❑ MM6338	SA-38	G91T/1	68	dumped, in scrap compound	9-96
❑ MM6341	32-41	G91T/1	71	stored	4-95
❑ MM6343	32-43	G91T/1	73	stored	10-97
❑ MM6344	32-44	G91T/1	74	stored	9-96
❑ MM6348	SA-48	G91T/1	78	preserved	10-97
❑ MM6349	32-49	G91T/1	79	stored	10-97
❑ MM6350	32-50	G91T/1	80	stored	10-97
❑ MM6360	32-60	G91T/1	90	stored	10-97
❑ MM6361	32-61	G91T/1	91	stored	10-97
❑ MM6362	32-62	G91T/1	92	stored, ex Prática di Mare	10-97
❑ MM6363		G91T/1	93	preserved, unmarked, special c/s	10-97
❑ MM6364	32-64	G91T/1	94	stored	10-97
❑ MM6368	32-68	G91T/1	98	stored	10-97
❑ MM6425	32-75	G91T/1	106	stored	10-97
❑ MM6426	60-76	G91T/1	107	stored	10-97
❑ MM6427	60-77	G91T/1	108	stored	10-97
❑ MM6432	60-82	G91T/1	113	stored	10-97
❑ MM6436	60-86	G91T/1	117	stored	10-97
❑ MM6439	32-89	G91T/1	101	stored	10-97
❑ MM6952	32-22/8-62	G91Y	2059	preserved, ex Bríndisi	10-97
❑ MM54392	60-92	G91T/1	119	stored	10-97
❑ MM54395	32-05	G91T/1	122	stored	9-96
❑ MM54396	32-06	G91T/1	123	stored	10-97
❑ MM54397	60-97	G91T/1	124	stored	10-97
❑ MM54398	60-98	G91T/1	125	stored, CFE a/c, code 11, marked as 'MM54410'	10-97
❑ MM54399	32-07	G91T/1	126	stored	10-97
❑ MM54400	32-10	G91T/1	127	stored	10-97
❑ MM54401	32-11	G91T/1	128	stored	10-97
❑ MM54404	60-104	G91T/1	131	stored	10-97
❑ MM54412	60-112	G91T/1	139	stored	10-97
❑ MM54414	60-114	G91T/1	141	stored, CFE a/c, code 8, third batch	9-96
❑ MM54415	60-115	G91T/1	142	stored, CFE a/c, code 4, third batch	10-97
❑ MM54-2951		T-33A	9451	preserved, marked as 'MM54-1951/SA-951'	10-97
❑ MM.....	20-21	TF-104G	5209	dumped, in scrap compound, see note	9-96
❑ MM.....		TF-104G	..	dumped, in scrap compound, see note	9-96

ATINA - CASTELLIRI
Alongside the SS509 road, which goes to the north from Atina, is a furniture shop named Valentino Corsi. You cannot miss the Harvard which is preserved outside. Much more difficult to see is the AB47 which is right next door at the Hamilton disco. It is parked outside on the dance floor at the back of the premises.

☐ MM53835		T-6H-4M	CCF4-400	preserved, ex 51-17218, ex Castrette	9-96
☐ MM80240	SE-42	AB47J	1147	preserved	9-96

AUGUSTA
Located in an old fortress close to the harbour is the Museo della Piazzaforte. They have one helicopter.

☐ MM5001N	5-01	A106	..	preserved, ex Catania	..

AVIANO
Due to the crisis in the former Yugoslavia, this base has become world famous. Of the *Wrecks & Relics* aircraft, the Hunter and F-4C are owned by a private collector and may have moved on. It is reported that they will be part of the new La Comina museum near Udine. During May 1991 an AB204B (MM80470/3-70) was reported as being preserved at the Zapala barracks. These barracks are the Italian area on the east side of the base. In 1994 the helicopter was not to be found and in 1995 the area became part of the American base. The Sabre was preserved in the housing area in the town of Aviano. The serial 53-8189 is often quoted for it, but that is not from a Sabre. Former instructional airframe RF-4Cs 68-0567/ZR (3469, dep 12-91) and 68-0580/ZR (3532, dep 10-91), which both arrived on 23-3-91, have been relocated to Davis Monthan, USA. F-16A 79-0340 isd still on the AMARC inventory list at stored at Davis Monthan. The F-100D is a former Danish Air Force aircraft, which never made it to Turkey. It went unreparable at Sigonalla and was preserved for some time at the Vigna di Valle museum. The dumped T-33As 51-17483 (7377) and 55-3104 (9674) have been scrapped, while F-105G 62-4442 (marked as '67-040/AV') has gone.

☐ J-4068		Hunter F.58	..	stored, arr 12-12-94, see notes	5-98
☐ 63-7512		F-4C	525	instructional, marked as '63-040/AV', see notes	1-91
☐ 79-0340		F-16A	61-125	instructional, ex AMARC FG012	11-95
☐ 54-2290	SS	F-100D	223-170	preseved, at gate, ex Vigna di Valle	5-98
☐ ..		F-86K	..	preserved, marked as '55-847/AV'	7-97

BACHERO DI CÍNGOLIA
Listed in the previous edition under Cíngoli, Bachero di Cíngolia is the more correct location and is some 8km north of Cíngolia. The AB47J-3 helicopter is parked outside the disco park and has been rebuilt with parts of AB47G MM80166, including the tailpiece with its serial. The construction number confirms it to be MM80296 (ex Roma Quatro Miglio). The rest of MM80166 (248) was scrapped locally. The DC-6 is within the compound.

☐ MM61965		DC-6A	44251	preserved, marked as 'I-LOVE', ex I-DIMA	9-95
☐ MM80296	33	AB47J-3	2047	preserved, with parts of MM80166	9-95

BAREGGIO
The Silvani di San Martino company has a number of facilities (all close to each other) in this town. Somewhere inside was a RF-84F Thunderflash. It has been sold to a private collector in 1995 and the aircraft is still somewhere in the area.

☐ MM52-7471	3-14	RF-84F	..	preserved, see not	98

BARI - PALESE
During 1995 and 1996 only three Harvards were noted, these three, MM53143/RB-6, MM54110/RB-8 and MM54111/RB-7, all went to <u>Roma Guidonia</u> There was no trace of the others. MM53866 is at the Facoltà di Ingegneria dell'Università di Bari as an instructional airframe. T-6 MM54101/RB-11 went to <u>Loreto</u>.

☐ MM53694	RB-3	T-6G	..	stored	8-94
☐ MM53806	RB-3	T-6H-4M	..	stored	4-89
☐ MM53866	RB-5	T-6C	..	instructional	6-97
☐ MM54135	RB-1	T-6H-2M	..	stored	8-94
☐ MM54136	RB-2	T-6H-2M	..	stored	8-94

BASILIANO
This town and its G91T are some 8km west of Udine along the SS-13 road.
❑ MM6428   60-78   G91T/1   109   preserved, ex Roma, ex Améndola, CFE code 6   5-98

BERGAMO
Town: The Museo del Risorgimento di Bergamo (in the old town) has an Ansaldo A-1 Balilla on display.
❑ 16553   Ansaldo A-1   ..   preserved   —

Airfield – Orio el Serio: Preserved inside the military part of Bergamo airfield is an Agusta Bell. Also noted on the dump in 1996 were two aircraft. No serials were noted and they looked like a crashed A-129 and an SM1019 or Piper Cub.
❑ MM80485   CC-16 AB47G-3B-1   1630   preserved, ex Salerno   2-96

BIBANO
A private collector has an ex Treviso G91. Bibano is west of Conegliano, roughly halfway between Aviano and Treviso.
❑ MM6290   2-26   G91R/1A   154   preserved, with tail from MM6418, ex Treviso   5-97

BIELLA - CERRIONE
The preserved HU-16 MM50-174 has been exchanged. The Albatross went to Cámeri. Preserved now on this civil airfield is a Fiat in Frecce Tricolori colours. It was last noted at Treviso San Angelo with the code 2-26 and has the tail from MM6290.
❑ MM6418   8   G91R/1B   222   preserved, with tail from MM6290 (89)   9-95

BOLOGNA
Town: Preserved in the backyard of a garden centre somewhere in Bologna is an ex AMI P166. More information about its exact location would be appreciated.
❑ MM61914   VV-14 P166M-APM   426   preserved   10-94

Airfield – Borgo Panigale: The military part of the airfield has processed a lot of aircraft in the past. Of these only the two AB204s (MM80300 and MM80381, at the military gate) and the dumped fire brigade AB204 (MM80521) were seen over the last few years. Known departures include SM1019 MM57195 to San Possidonio and AB204B MM80303 to Fossano. The stored O-1E Bird Dog MM61-2987/EI-20 turned civil as I-EIAI.

❑ MM57201	EI-408	SM1019	1-008	dumped	3-98
❑ MM57226	EI-433	SM1019	1-034	dumped	7-92
❑ MM57250	EI-457	SM1019	1-058	dumped	7-92
❑ MM57264	EI-471	SM1019	1-074	dumped, crashed 17-8-84	7-92
❑ MM80288	EI-205	AB204B	3045	dumped	9-90
❑ MM80300	EI-206	AB204B	3047	preserved, at gate	5-98
❑ MM80316	EI-216	AB204B	3074	stored	9-91
❑ MM80324	EI-224	AB204B	3095	stored	6-88
❑ MM80326		AB204B	3078	dumped	9-90
❑ MM80333	SA-3	AB204B	3108	dumped	9-90
❑ MM80334	SE-54	AB204B	3111	stored	9-90
❑ MM80381	EI-227	AB204B	3100	preserved, at gate	5-98
❑ MM80388	EI-234	AB204B	3116	dumped	7-90
❑ MM80472	1-72	AB204B	3207	dumped	9-90
❑ MM80521	I-VFMZ	AB204AS	3233	stored, in hangar near fire department	2-96
❑ MM.....	72-51	AB204B	..	dumped	7-90
❑ MM.....	72-65	AB204B	..	dumped	7-90

BORGO PIAVE
Close to the town of Latina is the village of Borgo Piave. Preserved here is a former Grosseto Starfighter.
| ☐ MM6589 | 4-49 | F-104G | 6589 | preserved, ex Grosseto | 5-98 |

BRACCIANO
The ALE base of Bracciano has been used over the years to store ex ALE aircraft and helicopters. In 1996, besides the SM1019 on the gate, only three AB47/OH-13s were noted in store. The rest have moved on or are stored inside. Former stored AB204B MM80317 went to Salerno and MM80385 to Forli. OH-13H MM80811 has gone to Roma, MM80812 went to Caserta and MM80816 is now at Comignago. O-1E MM61-2985/EI-29 returned to life as I-EIAK.

☐ MM57207		SM1019E	1-014	preserved, at gate, as 'ESERCITO', ex EI-414	5-97
☐ MM57260	EI-467	SM1019E	1-070	stored	9-91
☐ MM80098	29	AB47G-2	215	stored	9-92
☐ MM80308	EI-208	AB204B	3051	stored	10-89
☐ MM80309	EI-209	AB204B	3052	stored	9-92
☐ MM80311	EI-211	AB204B	3059	stored	10-89
☐ MM80313	EI-213	AB204B	3065	stored	10-89
☐ MM80321	72-52	AB204B	3087	stored	9-93
☐ MM80386	EI-232	AB204B	3114	stored	9-92
☐ MM80387	EI-233	AB204B	3115	stored	9-92
☐ MM80399	EI-245	AB204B	3153	dumped	1-87
☐ MM80400	EI-246	AB204B	3158	dumped	1-87
☐ MM80428	57	AB47G-3B-1	1628	stored	9-96
☐ MM80810	72	OH-13H	..	dumped	1-87
☐ MM80817	79	OH-13H	..	stored	9-96
☐ MM80821	83	OH-13H	..	stored	9-96
☐ MM61-2980	EI-31	O-1E	305M-0026	stored	9-92
☐ MM.....	28	AB47/H13	..	stored	9-91
☐ MM.....	SE-60	AB204B	..	stored	9-92
☐ MM.....	72-54	AB204B	..	stored	9-92
☐ MM.....	72-55	AB204B	..	stored	9-92
☐ MM.....	72-57	AB204B	..	stored	9-92
☐ MM.....	72-58	AB204B	..	stored	9-92

BRENO
In the Camonica Valley, Caravelle I-DAXU was used as a restaurant. The aircraft moved to Cergnago by 1997.

BRÍNDISI
Because of its close proximity to the town of Bríndisi, the 32 stormo moved to the middle of nowhere at Améndola. They took only one G91Y (MM6952/32-22) with them, the rest and the two MB326s were left behind. Thirteen of these were cut in half for the CFE treaty, four others are stored without wings and tail. One G91Y and one G91PAN are preserved inside the base. The T-33 MM51-6660/36-64 (5992), S-2s MM133078/41-40 (49) and MM133103/41-42 (74) and two HU-16s MM51-7157/15-10 (207) and MM51-7252/15-12 (342) have all gone. Only one HU-16 is still here.

☐ MM6242	32-13	G91PAN	8	preserved	9-96
☐ MM6442	32-14	G91Y	2004	stored, CFE a/c, code 6	9-96
☐ MM6443	32-11	G91Y	2005	stored, CFE a/c, code 7	9-96
☐ MM6445	32-03	G91Y	2007	stored, CFE a/c, code 8	9-96
☐ MM6453	32-10	G91Y	2015	stored, CFE a/c, code 9	9-96
☐ MM6454	32-14	G91Y	2016	stored, CFE a/c, code 2	9-96
☐ MM6455	32-01	G91Y	2017	preserved	9-96

ITALY - 262

☐ MM6460	32-05	G91Y	2022	stored, CFE a/c, code 10	9-96
☐ MM6461	32-05	G91Y	2023	stored, no wings and tail	9-96
☐ MM6462	32-05	G91Y	2024	stored, CFE a/c, code 5	9-96
☐ MM6463	32-06	G91Y	2025	stored, CFE a/c	9-96
☐ MM6471	32-07	G91Y	2033	stored, CFE a/c	9-96
☐ MM6475	32-22	G91Y	2037	stored, CFE a/c	9-96
☐ MM6478	32-01	G91Y	2040	stored, CFE a/c	9-96
☐ MM6489	32-12	G91Y	2051	stored, no wings and tail	9-96
☐ MM6490	32-06	G91Y	2052	stored, CFE a/c, code 4	9-96
☐ MM6493	32-21	G91Y	2055	stored, CFE a/c, code 3	9-96
☐ MM6494	32-13	G91Y	2056	stored, no wings and tail	9-96
☐ MM6955	32-04	G91Y	2062	stored, no wings and tail	9-96
☐ MM54186	32-51	MB326	6191	stored	9-96
☐ MM54211	32-54	MB326	..	stored	9-96
☐ MM51-7175	15-11	HU-16A	G-225	dumped	9-96

CADIBONA
This village is located along the A6 strada statele near Savona. A scrapyard has the wreck of an MB326.

☐ MM54176	00	MB326	..	stored	—

CAGLIARI - ELMAS
During a visit in 1996 only one SM1019 was seen. The second SM1019 may still be present, but was not seen from outside. The third SM1019 MM57227/EI-434 (1-035) was returned to operational service in the early 1990s. The stored BN-2A-21 I-KIMO (726) was last noted in October 1993 and was flown out as N352SP.

☐ MM53799	SL-10	T-6H-4M	CCF4-296	stored, in hangar, ex 51-17114	4-98
☐ MM57247	EI-454	SM1019E	1-055	stored, wfu 30-4-93	4-93
☐ MM57262	EI-469	SM1019E	1-072	stored	2-96
☐ MM80502	CC-31	AB47J-3	2119	preserved, with Carabinieri	12-96
☐ MM144716	86-6	S-2A	677	preserved	4-98
☐ MM53-5668		RT-33A	9007	preserved, with aero club, ex Decimomannu	12-96

CÁMERI
The base of Cámeri houses a large number of preserved aircraft, mostly parked near the operations building. The instructional Starfighter MM6634 is earmarked to go to Piacenza as a gate guard. Other instructional airframes have left, F-104G MM6533 went to <u>Ghedi</u> and MM6599 is now at <u>Vigna di Valle</u>. Stored P166M MM61928/53-34 (438) had returned to operational service by September 1992.

☐ MM587		Tornado	P09	preserved, marked as 'MM-P009', travelling a/c	2-96
☐ MM6504	53-21	F-104G	6504	preserved, on pole in Tiger c/s	3-98
☐ MM6559	3-41	F-104G	6559	instructional	9-92
☐ MM6634	3-33	RF-104G	6634	instructional	8-97
☐ MM6716	53-21	F-104S	1016	stored, tiger c/s, may be preserved in future	3-98
☐ MM19668		CL-13 Sabre 4	568	preserved, unmarked in Lanceri Neri c/s	8-97
☐ MM54106		T-6G	182-183	preserved, as 'MM53802/53-22', ex 51-14492	3-98
☐ MM54209		MB326	..	stored	2-96
☐ MM54218		MB326	..	stored	10-96
☐ MM54274	53-31	MB326	6439	preserved, ex stored	3-98
☐ MM61886	53-76	P166ML1	394	preserved, unmarked	3-98
☐ MM80257	53-95	AB47J-3	2038	preserved	3-98
☐ MM50-174	15-2	HU-16A	G-65	preserved, ex Biella	3-98
☐ MM51-17534	534	T-33A	7594	preserved	8-97
☐ MM53-8316	51-21	F-86K	221-44	preserved	8-97

CAPUA

The Scuola Specialisti (part of the AMIs Scuola Sottufficiali AM) is currently rebuilding its site here. This forced instructional MB326s (MM54188 and MM54220) and P166 MM61884 to move to Caserta, while the remaining instructional airframes were stored outside. The exception is the DC-6 which is preserved. When the worked is completed here the location at Caserta will be closed and all the school's activity will be at Capua.

☐ MM6282	2-34	G91R/1	48	stored	3-98
☐ MM19664	4-50	CL-13 Sabre 4	..	stored	3-98
☐ MM53679	SL-36	T-6G	..	stored	3-98
☐ MM53733	27	P148	181	stored	3-98
☐ MM53741	21	P148	189	stored	3-98
☐ MM61875	303-38	P166ML1	383	stored	3-98
☐ MM61922	31-26	DC-6	43216	preserved, ex I-LADY	3-98
☐ MM133069	AS-2	S-2A	40	stored	3-98
☐ MM51-6576	SS-2	T-33A	5908	stored	3-98
☐ MM51-9033	SS-1	T-33A	6817	stored	3-98
☐ MM52-7395	3-12	RF-84F	..	stored	3-98
☐ MM53-6733	50-28	F-84F	..	stored	3-98
☐ MM53-8297	51-60	F-86K	221-25	stored	3-98
☐ MM.....		T-6	..	stored, near aero club	10-97

CARPI

The aero club still have their Fiat. The aircraft is an original Frecce Tricolori aircraft and is still in these colours.

☐ MM6264	6	G91PAN	30	preserved	2-96

CARRARA SAN PELAGIO

The Museo dell'Aria is located in the Castello San Pelagio (south of Padova), from where in August 1918 a famous air raid was launched against Vienna. The AB204 is fitted with the tail boom of a crashed example (thought to be MM80467). The RF-84 is preserved in front of the castle at the former airfield. The museum is open between 09:30-12:30 and 14:30-19:00 from Tuesday to Sunday.

☐ MM53823	SC-70	T-6H-4M	..	preserved	9-96
☐ MM80227	SE-40	AB47J	1122	preserved	9-96
☐ MM80281		AB204B	3049	preserved, see note	9-96
☐ MM51-7253	15-14	HU-16A	G-343	preserved	9-96
☐ MM52-7339	69-5	RF-84F	..	preserved, marked as 'MM56-9745'	9-96
☐ 24525		Ansaldo SVA5	..	preserved	9-96
☐ I-AIAD		Avia FL3	A-10	preserved	9-96
☐ I-ATAA		MB308	5851/78	preserved, ex MM52931	9-96
☐ I-AVMM		EC38-56C	008	preserved	9-96
☐ I-CARD		Mantelli AM10	001	preserved	9-96
☐ I-DAWN		Fauvel AV36	42	preserved	9-96
☐ I-FRIF		UC-61K	1036	preserved, ex KK418 and marked as such	9-96
☐ I-LOLE		P66B	..	preserved	9-96
☐ I-TRIW		CVV6	0003	preserved, glider, ex MM100005, ex I-AECE	9-96

CASALE

The Thunderstreak here was not found in 1995. Casale is halfway between Vercelli and Alessandria.

☐ MM53-6970	36-40	F-84F	..	preserved, with tail of MM53-6579	9-88

CASERTA

The main base of the Scuola Sottufficiali is at Caserta. When, in the near future, the site at Capua is finished, the school (with its aircraft) will move to the new location. Till that time the school is divided into three parts, all close together and surrounded by a high wall. With a little effort, some of the aircraft can be seen, including the MB326s at a secondary gate. The fate of the CL-13 Sabre 4 MM19596/4-46 (496) and AB204B MM80339 (3137) is unknown, both were last noted in 1989. The OH-13H MM80791/SE-7 moved to Latina to become a gate guard there.

❏ MM586		Tornado	P05	preserved	5-97
❏ MM6507	51-01	F-104G	6507	instructional, ex Grazzanise	5-97
❏ MM6552	9-10	F-104G	6552	preserved, ex Grazzanise	5-97
❏ MM54188	8-74	MB326	6193	preserved, ex Capua	5-97
❏ MM54220	32-53	MB326	..	preserved, ex Capua	5-97
❏ MM61884		P166ML1	392	stored, ex Capua	5-97
❏ MM61921	36-66	P166ML1	433	stored	5-97
❏ MM80049		AB47G	..	stored, ex instructional	5-97
❏ MM80213		AB47J	1123	preserved, ex instructional	5-97
❏ MM80335	43	AB204B	3112	stored, ex instructional	5-97

CASSINO
Cassino (not Cascino as mentioned in the first edition) can be found some 50km east of Frosinone. A search for the helicopter in 1996 did not find it, but there are a number of military barracks here and it may well be inside one of them.

❏ MM80217	SE-44	AB47J	1127	preserved, ex Frosinone	8-91

CASTEL DEL RIO
A local war museum (Museo della Guerra, some 25km south west of Imola) acquired T-6H-4M Texan MM54146 from Guidonia by 1987. The aircraft carries USAAF markings, acquired during the filming of La Pelle at Guidonia.

❏ MM54149		T-6H-2M	..	preserved, marked as '284491/145'	5-98

CASTELLO DI ANNONE
Preserved here at the gate of the 3 Magazzino Sussidiario is a clearly visible T-6G Texan. Not much more can been seen of these barracks and it is not known if the Piaggio P166s are still here. Castello di Annone is some 30km west of Alessandria.

❏ MM53670	RM-16	T-6G	..	preserved, at gate	9-95
❏ MM61888	RM-80	P166M	396	stored	3-86
❏ MM61907	53-75	P166M	419	stored	3-86

CASTEL VOLTURNO
The official address is Mondragone Sud, but the large Ditellandia park can be found near Castel Volturno, some 13km south east of Mondragone. The first aircraft in this swimming pool/museum/zoo/etc were noted in 1990. T-6C MM53863/RR-64 was scrapped and P148 MM53558/RR-71 (134) went to somewhere on Sicily. MM53562 (138) was last noted in May 1995 and moved to the new museum at Pontedera. The park is open every day in July and August (between 09:00 and 20:00) and only on Saturday and Sunday during the rest of the year. Noted inside the park are:

❏ MM53549	72	P148	125	preserved, ex Latina	5-97
❏ MM53580	14	P148	148	stored, ex Latina, ex RR-74	5-97
❏ MM53589	RR-73	P148	165	preserved, ex Latina	5-97
❏ MM53593	RR-72	P148	169	preserved, ex Latina, ex RR-91	5-97
❏ MM53724	15	P148	172	preserved, ex Latina	5-97
❏ MM61644	RR-15	C-45F	..	preserved, ex Guidonia	5-97
❏ MM61675	RM-50	C-45F	..	preserved, ex Guidonia	5-97

ITALY - 265

☐ MM61878	SP-35	P166ML1	386	preserved, ex Latina	7-97
☐ MM61881	SP-34	P166ML1	389	preserved, ex Guidonia	7-97
☐ MM61906	53-72	P166ML1	418	preserved, ex Guidonia	7-97
☐ MM80170	21	AB47J	1087	preserved	5-97
☐ MM51-11260	17	RF-84F	..	preserved, ex Castrette	5-97
☐ MM55-4858	51-51	F-86K	221-98	preserved, ex Castrette, tail from MM55-4869	5-97
☐ I-SATT		Macchi AL60B-2	6161/16	preserved	5-97

Outside the park, on and above (some 15 aircraft are mounted on poles, see photo on front cover) the parking area are a number of aircraft. The MB326 is a purpose built instructional airframe, has never flown and has no MM number allocated. The unidentified OH-13Hs were only seen in January 1997. They have all moved on to an unknown location.

☐ MM6309	2-56	G91R/1B	173	preserved, ex Treviso, with tail from MM6406	5-98
☐ MM6376	2-01	G91R/1B	180	preserved, ex Treviso, with tail from MM6414	5-98
☐ MM6414	2-52	G91R/1B	218	preserved, ex Treviso, with tail from MM6395	5-98
☐ MM6420	2-43	G91R/1B	224	preserved, ex Treviso, with tail from MM6424	5-98
☐ MM53547	19	P148	123	preserved, ex Latina	5-98
☐ MM53561	18	P148	137	preserved, ex Latina	5-98
☐ MM53573	6	P148	149	preserved, ex Latina	5-98
☐ MM53729	20	P148	177	preserved, ex Latina	5-98
☐ MM53734	71	P148	182	preserved, ex Latina	5-98
☐ MM53736	3	P148	184	preserved, ex Latina, ex RR-50	5-98
☐ MM53739	23	P148	187	preserved, ex Latina	5-98
☐ MM53743	10	P148	191	preserved, ex Latina	5-98
☐ MM61646	RR-23	C-45F	..	preserved, ex Guidonia	5-98
☐ MM61677	RR-22	C-45H	..	preserved	5-98
☐ MM61717	RR-26	C-45F	..	preserved, ex Guidonia	5-98
☐ MM61722	RR-27	C-45H	..	preserved, ex Guidonia	1-97
☐ MM61754	RR-28	C-45	..	preserved, ex Guidonia	7-97
☐ MM61800	CR-51	C-47	4260	preserved, ex Prática di Mare, ex I-ELFO	5-98
☐ MM61815	14-47	C-47B	15355/26800	preserved, ex Guidonia, ex I-TRAS	5-98
☐ MM61876	303-32	P166ML1	384	preserved, ex Pisa	5-98
☐ MM61900	31-6	DC-6B	43152	preserved, ex Furbara, ex I-DIMC	5-98
☐ MM61902	303-36	P166ML1	404	preserved, ex Pisa	5-98
☐ MM50-182	15-7	HU-16A	..	preserved, ex Guidonia	5-98
☐ MM53-3219	46-96	C-119G	11235	preserved	5-98
☐ MM.....	2	OH-13H	..	preserved	1-97
☐ MM.....	4	OH-13H	..	preserved	1-97
☐ MM.....	70	OH-13H	..	preserved	1-97
☐ MM.....	71	OH-13H	..	preserved	1-97
☐ MM.....	75	OH-13H	..	preserved	1-97
☐ I-GRAC		A109A	7125	preserved, wreck	5-98
☐ ..		MB326	..	preserved, instructional cockpit only	5-97

Directly attached to the park is a scrapyard (where it all started), which still holds some surplus aircraft. Five P148s can be seen from the road side (MM53550, MM53551, MM53566, MM53582 and MM53725), the others maybe only from inside the yard.

☐ MM53550	9	P148	126	stored, ex Latina	5-98
☐ MM53551	4	P148	127	stored, ex Latina	5-98
☐ MM53566	74	P148	142	stored, ex Latina	5-98
☐ MM53582	7	P148	158	stored, ex Latina	5-98
☐ MM53583	RR-74	P148	159	stored, ex Latina	5-98
☐ MM53725	RR-78	P148	173	stored, ex Latina	5-98
☐ MM53737	28	P148	185	stored, ex Latina	5-98
☐ MM61708	RB-16	C-45F	..	stored, ex Guidonia	5-98

| ITALY - 266 |

If you take the first road to the west of the park towards Cancello, after a few kilometers you will find the Ditellandia ultra light airfield, marked as Campo di Volo. Dumped there is yet another P148.
❑ MM53568   8   P148   144   stored, ex Latina   5-98

CATANIA – FONTANA ROSSA
None of the Trackers which were noted on the dump survived into the 1990s, they comprised MM133085/41-41 (56), MM133097/41-44 (68), MM133106/41-23 (77), MM133113/41-25 (84), MM136558/41-28 (467), MM136561/41-45 (470), MM136727/41-30 (636), MM136728/41-29 (637), MM136734/AS-10 (643), MM144696/41-32 (657), MM144702/41-31 (663), MM144717/41-57 (678), MM148294/41-43 (733), MM148299/41-46 (738), MM148301/41-36 (740) and MM148303/41-38 (742). Of the several AB204ASWs which were noted in storage here in April 1989 only four MM80363 (at Táranto), MM80366 (to gate), MM80371 (at Luni) and MM80519 (still here), were seen in 1990s. The rest (MM80301/3-01, MM80372/3-11, MM80375/3-14 and MM80376/3-15) have not been noted since April 1986. The preserved gate guard A106 MM5001N moved on to Augusta.
❑ MM5002N   5-02   A106   ..   stored, under restoration   88
❑ MM80366   3-05   AB204ASW   3106   preserved, at navy gate   9-96
❑ MM80518   3-30   AB204ASW   3228   stored   8-94

CAVALCASELLE
At the most southern part of the Lago di Garda is the village of Cavalcaselle. Here along the main road 11 an unmarked frame of a M416 (Fokker S11) is preserved.
❑ MM.....   M416   ..   preserved   6-95

CECCHINA
In Cecchina, 20km south east of Roma, an AB47 should be preserved. Its exact whereabouts are not known.
❑ MM80417 GdiF-43 AB47G-2   294   preserved, ex Roma   1-94

CEOLINI
Ceolini is a village near Pordenone, some 10km south of Aviano. Stored here at a sports field is a Caravelle.
❑ I-DABM   SE210-6N   143   preserved, ex F-BJSO, ex F-WJSO   4-97

CERBAIOLA
The name of this location is not on the road maps, but it is easy to find. It is located close to the village of Cerasolo along road 72 from Rimini to San Marino. Some of the aircraft are displayed on the hill site and give a good indication of where the museum is located. The Museo dell'Aviazione opened its gates on 2nd May 1995. Most aircraft of this still expanding museum are in superb condition. The Harvard came from Loretto (which see) and it is not known (not even to the museum owners) which one it is. The opening times are between 09:00 to 19:00 (daily between 1st April to 31st October and the rest of the year only on Saturday and Sunday).
❑ 0316   MiG-19S   150316   preserved, ex Presov, ex Czech AF   5-98
❑ 5308   Su-7BM   ..   preserved, ex Vodochody, ex Czech AF   5-98
❑ 2054   MiG-23BN   0393215729 preserved, ex 718/NVA and marked as such 5-98
❑ 2342   MiG-21MF   96001039   preserved, ex 772/NVA and marked as such 5-98
❑ 2382   MiG-21UM   516915011 preserved, ex Rothenburg, ex 231/NVA   5-98
❑ 2501   Su-22M-4   25307   preserved, marked as '27 yellow', ex 360/NVA 5-98
❑ 2628   F-104G   9180   preserved, ex Seifertshofen, ex WGAF   5-98
❑ 9396   Mi-9   340003   preserved, ex Cottbus, ex 405/NVA   5-98
❑ 127   MiG-15UTI   022532   preserved, ex Dresden, ex NVA   5-98
❑ 285   MiG-21UM   516999436 preserved, cockpit only, ex Dresden, ex NVA 5-98
❑ 870   MiG-21PFM   0914   preserved, marked as '3-0914', ex NVA   5-98

ITALY - 267

☐ 212		Lim-6bis	1F-0212	preserved, ex Rovereto, ex Polish AF	5-98
☐ 7		Il-28	..	preserved, ex Oleśnica, ex Polish AF	5-98
☐ ..		TS-8	..	preserved, marked as '0309', ex Łódź, ex PolAF	5-98
☐ 158830	AC-403	A-7E	E-319	preserved, ex Sigonella, ex US Navy	5-98
☐ 64-0757		F-4C	1047	preserved, ex Seifertshofen, ex USAF	5-98
☐ XH768	E	Javelin FAW.9	..	preserved, ex Seifertshofen, ex Bochum, ex RAF	5-98
☐ MM6249	2	G91PAN	15	preserved	5-98
☐ MM6389	2-04	G91R/1B	193	preserved, ex Treviso, with tail from MM6302	5-98
☐ MM53727	9-22	P148	175	preserved, marked as 'MM43740/19', ex Latina	5-98
☐ MM54216	51-78	MB326	6185	preserved, ex Istrana	5-98
☐ MM54403	60-103	G91T/1	130	preserved, ex Améndola, ex CFE a/c, code3	5-98
☐ MM61913	303-37	P166ML1	425	preserved, ex Guidonia	5-98
☐ MM80479	9	AB47G-2	302	preserved, ex Frosinone	5-98
☐ MM52-7459	3-09	RF-84F	..	preserved	5-98
☐ MM53-6646		F-84F	..	preserved, marked as 'MM53-6591', ex Vedelago	5-98
☐ MM54-1602	SST-2	T-33A	9338	preserved, ex Prática di Mare	5-98
☐ MM.....		T-6	..	preserved, as 'MM53042/AA-31', ex Loretto	5-98
☐ I-AEKE		G46-3A	69	preserved, ex Fiabilandia (Rimini), ex MM52805	5-98
☐ I-COFR		R4D-5	12679	preserved, ex Loretto, ex N711TD	5-98
☐ I-FALK		SAIMAN S202M	3	preserved, ex MM52161 and marked as such	5-98
☐ OK-FDC		L-410AS	750408	preserved, in camo markings, ex Vodochody	5-98
☐ SP-SAE		Mi-2	535437127	preserved, marked as '155'	5-98
☐ SP-TCD		An-2R	1G173-03	preserved	5-98
☐ SP-TCG		An-2R	1G188-60	preserved, ex SP-TTC, in Croation colours	5-98
☐ SP-TSB		Mi-2	543042123	preserved, ex SP-SDN	5-98
☐ ..		Fokker Dr.I	..	preserved, replica, ex Bologna	5-98

CERGNAGO
In use as a restaurant here is the former Breno Caravelle I-DAXU.

☐ I-DAXU		SE210-6N	79	preserved, ex Breno	97

CÉRVIA - SAN GIORGIO
The former G91 base still has a large number of 'Ginas' on their premises. In a compound between the main gate and the 23 Gruppo area a number of G91Ys were stored since 1994. In May 1995 an additional five G91s were noted here, but they had moved to the opposite side of the runway by October 1995. The compound was taken over by an air defence unit in 1996 and only the two MB326s and MM6451 were still in the compound. The G91Ys are by then in use as decoys and parked on several locations on the airfield These aircraft move around the airfield frequently. The two Starfighters are stored near the operational Starfighter unit (23 Gruppo). G91Y MM6444 moved to Lugo di Romagna and MM6491 went to Latina. The stored MB326 MM54272 went to Ravenna by early 1998.

☐ MM6524	3-46	F-104G	6524	stored, ex Rimini	10-97
☐ MM6551	3-51	F-104G	6551	stored, ex Rimini	10-97
☐ MM6446	8-05	G91Y	2008	stored	2-96
☐ MM6447	8-07	G91Y	2009	stored	10-95
☐ MM6448	8-06	G91Y	2010	stored, ex preserved	5-98
☐ MM6450	8-10	G91Y	2012	stored, as decoy	10-96
☐ MM6451	8-11	G91Y	2013	stored	8-97
☐ MM6459		G91Y	2021	preserved, near fire station	10-97
☐ MM6465	8-24	G91Y	2027	stored, as decoy	10-96
☐ MM6470	8-30	G91Y	2032	stored	10-97
☐ MM6486	8-46	G91Y	2048	stored, as decoy	10-97
☐ MM6487	8-47	G91Y	2049	stored	5-98

ITALY - 268

☐ MM6956	8-61	G91Y	2063	preserved, at gate	5-98
☐ MM6957	8-64	G91Y	2064	stored, ex decoy	10-97
☐ MM6959	8-66	G91Y	2066	stored, ex decoy	7-97
☐ MM54224	8-72	MB326	6307	stored	7-97
☐ MM53-10524	8-34	F-84F	..	preserved, at gate, with tail from 53-6619	5-98

On the opposite side of the runway 21 G91Ys are stored. These aircraft are part of the CFE treaty and their storage area is surrounded by barbed wire. Although the term fuselage only is correct for this storage area, the wings and tails from these aircraft could be found just behind the main gate (on the left side). These parts had moved in 1995 to the opposite side of the runway as well. Only 18 of the 21 stored aircraft were confirmed in October 1995. Of the three unconfirmed airframes (marked with 'see notes'), two are in the storage area, while the third is the burned out wreck on the north side of the airfield. Its not known which is which. This wreck was no longer noted in 1996.

☐ MM6449	8-14	G91Y	2011	stored, fuselage only	10-95
☐ MM6452	8-55	G91Y	2014	stored, fuselage only	10-95
☐ MM6457	8-44	G91Y	2019	stored, see notes	10-95
☐ MM6466	8-26	G91Y	2028	stored, complete aircraft	10-95
☐ MM6467	8-23	G91Y	2029	stored, fuselage only	10-95
☐ MM6468	8-25	G91Y	2030	stored, fuselage only	10-95
☐ MM6469		G91Y	2031	stored, complete aircraft	10-95
☐ MM6472	8-32	G91Y	2034	stored, complete aircraft	10-95
☐ MM6476	32-31	G91Y	2038	stored, fuselage only	10-95
☐ MM6477	32-16	G91Y	2039	stored, fuselage only	10-95
☐ MM6479	8-27	G91Y	2041	stored, fuselage only	10-95
☐ MM6480	8-40	G91Y	2042	stored, fuselage only	10-95
☐ MM6481	32-10	G91Y	2043	stored, fuselage only	10-95
☐ MM6482	8-42	G91Y	2044	stored, see notes	10-95
☐ MM6488		G91Y	2050	stored, complete aircraft, bare metal	10-95
☐ MM6492	8-55	G91Y	2054	stored, fuselage only, 50000 hours marks	10-95
☐ MM6495	8-57	G91Y	2057	stored, see notes	10-95
☐ MM6951	8-61	G91Y	2058	stored, fuselage only	10-95
☐ MM6953	8-63	G91Y	2060	stored, fuselage only	10-95
☐ MM6954	8-64	G91Y	2061	stored, fuselage only	10-95
☐ MM6958	8-65	G91Y	2065	stored, complete aircraft	10-95

CESENA
Along the main road from Cérvia airfield towards the town centre, is a Fiat preserved in a schoolyard on the right hand side of the road.

☐ MM6375	G91R/1B	179	preserved	5-98

CISTERNA DI LATINA
Preserved in this town north west of Latina are two P148s. The first one is along the main road (SS7) through the town at a car dealer. The second one is with a scrap dealer along the road to Cori, just outside Cisterna. This aircraft is missing its tail.

☐ MM53544	77	P148	120	preserved	5-97
☐ MM53546	16	P148	122	preserved, no tail	5-97

COAZZE
T-6H-4M MM53820 arrived in the mid 1980s at a restaurant in Celotte di Forno near Coazze. It was handed over to GAVS (an aircraft restoration group) who will restore it. This will presumably take place in Leini (Revelli Metallik) near Torino.

ITALY - 269

CODRÓIPO
Preserved in a school/playing field just south of the village centre is a G91R painted in the colours of the Frecce Tricolori. Frecce's base, Rivolto, is only a few kilometers away.

❏ MM6398	1	G91R/1B	202	preserved	5-95

COMIGNAGO
The L'Aeroplano museum was opened in 1993 and has more than a thousand aircraft models on show, Beside these there are some real aircraft and replicas. These are all in good condition, with the exception of the AS61. This helicopter was never completed on the production line at Vergiate. The museum is some 20km west of Gallarate and is open from Tuesday to Sunday, between 10:00 and 16:00.

❏ J-1742		Venom FB.54	912	preserved	9-97
❏ MM6306	2-23	G91R/1A	170	preserved, with tail from MM6283 (49)	9-97
❏ MM61872	SP-81	P166ML1	380	preserved	9-97
❏ MM80728	PS-28	AB47J-3B-1	2123	preserved, ex Prática di Mare	9-97
❏ MM80319	EI-219	AB204B	3084	preserved	9-97
❏ MM80816	78	OH-13H	2402	preserved, ex Bracciano, ex 58-5389	9-97
❏ OO-HDA		FWP149D	151	preserved, as 'AS-441', ex D-EHET, ex 9129	9-97
❏ I-SABA		Be65	LC-91	preserved, ex HB-GAZ	10-96
❏ ..		A129	..	preserved, wooden mock-up, marked as 'EI-901'	10-96
❏ ..		A129	..	preserved, wooden mock-up	10-96
❏ ..		AS61	..	preserved, uncompleted airframe	9-97
❏ ..		SIAI S211	..	preserved, replica	9-97

CONEGLIANO D'OTRANTO (or Corigliano d'Otranto)
Preserved in 1987 at some gardens in this village (near Galatina) was an SNJ-5 Texan. This, the sole Italian SNJ-5, was by that time in very poor condition.

❏ MM54292	SC-79	SNJ-5	..	preserved, bad condition	12-87

DECIMOMANNU
Decimomannu is a well visited base by all NATO partners as well some other countries like Switzerland, to make use of the ranges at Sardine. The British and Germans have preserved two aircraft on the base. G91T/1 MM6435/60-85 (116) was scrapped in late 1993 and some pieces of the aircraft were still noted in November 1994. RT-33A MM53-5668 went to Elmas. Starfighter MM6819 was damaged here on 13th July 1994. It was still stored here in June 1995, but since moved to Grosseto.

❏ 2154		F-104G	8278	preserved, in GAF area	12-96
❏ XV758	R	Harrier GR.3	712021	stored, ex preserved, inside RAF hangar	12-96
❏ MM....	37-25	F-104S	..	stored, no nose and no tail	3-92
❏ MM54193	SST-1	MB326	6198	stored, near tower	12-96
❏ MM54278	SST-4	MB326	6443	stored	12-96
❏ MM54372	SST-3	MB326	..	stored	11-94
❏ MM54376	SST-5	MB326	..	stored	12-96
❏ MM54377	SST-3	MB326	..	stored	9-91
❏ MM80256	AWTI	AB47J-3	2035	preserved, at gate	12-96
❏ MM51-17455	SST-1	T-33A	7148	preserved, at gate, ex Napoli	12-96

FAGIOLI DI SAN ILARIO D'ENZA
As explained in EWR-1, the exact location of ex German Army H-34s is not known. Faggioli is a hill some 20km south east of Imola, but the village of San Ilario d'Enza is not mentioned on the maps. Does anybody have any clues? The H-34Gs 8072 (58-1553), 8093 (58-1605) 8203 (58-1619), 8204 (58-1733) and 8205 (58-1737) were last seen in 1986.

FALCONARA
Former Altair SE210 Caravelle I-GISE was in use as a pizzeria at the airport by 1988. The aircraft moved to Loreto in the early 1990s.

FIORANO MODENESE
Preserved near the Ferrari test circuit (Pista di Fiorano, south of Modena) is an F-104 in red colours.
❏ MM6546   4-27   F-104G   6546   preserved   11-95

FIRENZE - PERETOLA
Preserved outside (airport side) the local AVES unit is an SM1019.
❏ MM57255   EI-462   SM1019E   1-063   preserved   3-98

FONTE
In the Onè part of Fonte (named Onè di Fonte) a former Istrana AB47 is stored. Fonte is some 25km west north west of Treviso.
❏ MM80162   53-94   AB47J   1064   stored, ex Istrana   6-95

FORLÌ
The G-46 which was preserved with the AAAV (Italian state ATC service) at the airfield here moved into a hangar during the 1990s were also a former San Possidonio G-46 under restoration is. A third unknown G46 was dumped outside. The AB204 comes from Bracciano.
❏ MM80385   EI-231   AB204B   3110   instructional, forward fuselage only   6-96
❏ I-AEHL   G46-1B   23   stored, ex MM52790 and marked as such   2-96
❏ I-LEOR   G46-4B   180   under restoration, ex MM53312   6-96
❏ ..   G46   ..   dumped, unmarked   96

FOSSANO
Preserved at an AMI barracks in this town (north east of Cuneo) should be an Agusta Bell. It was not noted during a search in late 1995.
❏ MM80303   AB204B   3056   preserved, ex Bologna   1-95

FROSINONE
A large number of stored AB47/OH-13s were noted here in the 1990s. The report that the unmarked OH-13H at the gate was MM80177 is incorrect. This machine is preserved at the nearby village of Isola del Liri. The construction plate of the gate guard has been checked and confirmed the helicopter as being MM80790. Formerly stored AB47s which have gone include MM80217 to Cassino, MM80240 to Atina Castelliri and MM80479 which is now at Cerbaiola. OH-13H MM80796 has moved to San Possidonio by early 1998, together with some five other (see Stop Press).
❏ MM80046   30   AB47G-2   296   stored   10-94
❏ MM80048   AB47G   ..   stored   10-94
❏ MM80052   25   AB47G   287   preserved   3-98
❏ MM80106   18   AB47G-2   163   stored   10-94
❏ MM80107   AB47G-2   172   stored   10-94
❏ MM80112   32   AB47G-2   195   stored   10-94
❏ MM80208   SE-40   AB47J   1113 stored, in hangar   10-97
❏ MM80282   SE-63   AB204B   3053 preserved, at 208 gruppo area   3-98
❏ MM80474   SE-41   AB47G-2   297   stored, in hangar, ex code 31   3-98
❏ MM80784   20   OH-13H   ..   stored   10-94

| ITALY - 271 |

☐ MM80787		OH-13H	..	stored	2-96
☐ MM80790	37	OH-13H	2262	preserved, at main gate, ex 58-1498	5-98
☐ MM80793	12	OH-13H	..	stored	10-94
☐ MM80796		OH-13H	2139	stored, ex 57-6210	10-94

GALLARATE

The 2 Deposito Centrale AM still has a large number of aircraft on display at their barracks in Gallarate (some 30km north west of Milano).

☐ MM6239	2	G91PAN	5	preserved	12-95
☐ MM54143	RB-3	T-6H-2M	..	preserved	12-95
☐ MM61882		P166M	390	preserved	12-95
☐ MM80224	RM-93	AB47J	1139	preserved	12-95
☐ MM80278	SE-56	AB204B	3030	preserved	12-95
☐ MM80477	21	AB47G-2	300	preserved	12-95
☐ MM53-8299	51-6	F-86K	221-27	preserved, ex Bergamo	12-95
☐ MM.....	24	MB326	..	preserved	12-95

GAMBÉTTOLA

The scrapyard at Gambéttola (some kilometers east of Cesena) has not been visited recently. All aircraft may have expired. In EWR-1 one the name Gambéttola was mis-spelled as Gambetta.

☐ MM6399	2-21	G91R/1B	203	stored	84
☐ MM6707	51-05	F-104S	1007	stored, crashed 13-5-74	7-82
☐ MM52777	ZP-24	G46-1B	10	stored	84
☐ MM52798	ZP-15	G46-1B	31	stored	84
☐ MM80188	SC-2	AB47J	1101	stored	7-82

GHEDI

The base of Ghedi, which is largely known because of its based 6 Stormo Tornados and *Deny Flight* F-16s from Turkey. The Tornado is a wooden mock-up painted in full AMI colours and is missing its tail and canopy. It may be in use as weapons loader. Starfighter MM6533 (ex Cámeri) has been moved to Villafranca, but returned here in 1997.

☐ MM6522	5-01	F-104G	6522	preserved, ex Rimini	3-98
☐ MM6533	3-35	F-104G	6533	dumped, ex Villafranca, ex Ghedi, ex Cámeri	3-98
☐ MM6601	6-26	F-104G	6601	preserved, at gate	5-98
☐ MM54165	6-62	MB326	6141	stored	3-98
☐ MM54194	6-74	MB326	6199	stored	3-98
☐ MM54200	55	MB326	6205	dumped, fuselage only	2-96
☐ MM54284	6-75	MB326	6649	instructional	3-98
☐ MM53-6740		F-84F	..	preserved, special c/s	11-95
☐ MM53-6855	5-855	F-84F	..	preserved, ex Rimini	5-98
☐ ..		Tornado	..	stored, wooden mock-up, no tail	7-97

GIOIA DEL COLLE

Gioia del Colle is one of the AMIs main Tornado bases. Besides these operational aircraft, the base has a fair number of *Wrecks & Relics*. RT-33A MM53-5396 was under restoration for display in a local village and may have left the base by now. Also preserved here is the fin of Sabre MM19600.

☐ MM6545	3-32	F-104G	6545	instructional	7-94
☐ MM6835	36-12	F-104S	1135	preserved, ex stored	1-98
☐ MM6928		F-104S	1228	dumped	7-94
☐ MM7045		Tornado	5054	dumped, crashed 26-7-84	7-94

ITALY - 272

☐ MM54281	36-73	MB326	6446	preserved, ex stored	5-96
☐ MM61871	36-74	P166ML1	379	stored, in scrap compound	9-95
☐ MM61883	36-76	P166ML1	391	stored, in scrap compound	9-95
☐ MM61903	36-73	P166ML1	405	stored, in scrap compound	9-95
☐ MM.....	36-75	P166ML1	..	preserved, at main gate	11-96
☐ MM51-9253	36-65	T-33A	7037	stored	6-92
☐ MM53-5396	36-69	RT-33A	8735	stored, under restoration	10-94
☐ MM53-5430	36-68	RT-33A	8768	preserved, at main gate	7-97
☐ MM53-6653	36-06	F-84F	..	preserved, unmarked, in 156 gruppo area	11-96
☐ MM53-6858	36-39	F-84F	..	preserved	11-96
☐ MM53-8308	36-06	F-86K	221-36	preserved, at main gate, unmarked	11-96
☐ MM54-1603	36-65	RT-33A	9339	instructional	4-94

GORIZIA
Close to the border crossing to Slovenia, just south of the town of Gorizia, is a former AMI airfield. Preserved here with the local aero club is an Albatross.

☐ MM51-037	15-9	HU-16A	..	preserved	3-96

GRAZZANISE
A few hundred meters north of the airbase are the Grazzanise barracks. Preserved here is the Sabre, all the other aircraft are/were at the airbase. MB326 MM54157 moved on to Istrana. Three T-Birds were offered for sale in September 1986 and MM53-5587 (8926) and MM55-3077/9-30 (9618) were sold to the USA. The fate of the third, MM51-17470/9-31 (7364) is unknown. Two Starfighters, MM6507/3-50 and MM6552/4-50, moved on to the nearby school at Caserta, where they received new codes. T-33 MM55-2980 is now at Latina.

☐ MM6542	4-53	F-104G	6542	dumped	3-98
☐ MM6565	6-05	F-104G	6565	dumped, fuselage only	5-96
☐ MM6765		F-104S	1065	stored, fuselage & tail only, crashed 10-8-82	8-93
☐ MM19523	13-30	CL-13 Sabre 4	423	preserved, at barracks, ex XB620/RAF	5-98
☐ MM54098	9-09	T-6G	..	preserved	5-98
☐ MM54240	9-04	MB326	..	stored	5-97
☐ MM54378		MB326	..	stored	9-95
☐ MM55-3033	4-333	T-33A	9574	stored	9-95

GRONTARDO
Preserved in Grontardo should be an ex Treviso G91R. The village is located 10km north east of Cremona.

☐ MM6314	2-57	G91R/1A	178	stored, ex Treviso	5-89

GROSSETO
Grosseto is home of the last operational TF-104Gs. F-104G MM6505 is used as a recruiting aid and travels through the country. It is marked on its aft fuselage as an F-104G. F-104G MM6589 is now at Borgo Piave. F-104G MM6527 and F-104S MM6916/53-02 went to Prática di Mare. The sale of the T-33s at Grazzanise also included MM51-8829 (6613) from here. This T-33 also went to the USA. Some of these stored F-104S's have left, MM6722/5-45, MM6767/51-05 and MM6772/9-52 were all last noted in juli 1997, while MM6802/4-1, MM6872/4-20 and MM6914/9-42 all left on 9th September 1997.

☐ MM6505	4-1	F-104G	6505	preserved, travelling exhibition aircraft	3-98
☐ MM6528	4-48	F-104G	6528	instructional, ex preserved	9-97
☐ MM6576	3-45	F-104G	6576	dumped	5-97
☐ MM6631	3-30	RF-104G	6631	stored, no tail	10-97
☐ MM6732	4-5	F-104S	1032	stored	10-97
☐ MM6764	51-03	F-104S	1063	stored	10-97

ITALY - 273

☐ MM6778	4-4	F-104S	1078	stored	10-97
☐ MM6782	37-15	F-104S	1082	stored	10-97
☐ MM6787	4-7	F-104S	1087	stored	7-97
☐ MM6791	36-14	F-104S	1091	stored	3-98
☐ MM6814	53-16	F-104S	1114	stored	10-97
☐ MM6819	9-31	F-104S	1119	stored, ex Decimomannu	3-98
☐ MM6826	4-51	F-104S	1126	stored, in hangar	5-97
☐ MM6931		F-104S	1231	dumped	8-94
☐ MM19841	4-44	CL-13 Sabre 4	741	preserved, at main gate, ex XB954/RAF	5-98
☐ MM54190		MB326E	6195	preserved, on pole, ex stored	5-98
☐ MM54283	4-83	MB326	..	stored, ex code 9-03	10-97
☐ MM54230	4-27	TF-104G	5776	stored, ex 63-12689	10-97
☐ MM54231	20	TF-104G	5778	preserved, on pole in special c/s, ex 63-12690	3-98
☐ MM54233	4-30	TF-104G	5781	stored, ex 63-12692	7-97
☐ MM54235	4-31	TF-104G	5783	stored, ex 63-12694	10-97
☐ MM54257	4-39	TF-104G	5208	stored, wfu after belly landing 4-97	5-97
☐ MM54261	4-42	TF-104G	5212	stored	5-97
☐ MM54556	4-47	TF-104G	5743	stored, ex 2741/WGAF	5-97

GROTTÁGLIE
Noted stored in a hangar in 1994 after an accident was an AB212ASW. It is strange to note that this main Italian Navy base has no preserved aircraft, while the smaller base of Luni has several airframes.

☐ MM80941	7-09	AB212ASW	..	stored, in hangar, crashed 20-7-94	8-94

IMOLA
Preserved at a war memorial should have been T-6 MM54149 (not MM54146, see first edition and Castel de Rio). Nothing was here in 1998. Mix-up with Castel del Rio?

ISOLA DEL LIRI
Placed on top of a Q8 filling station is an Agusta Bell. The garage is on the main road to Frosinone.

☐ MM80177	SE-43	AB47J	1074	preserved	9-96

ISOLA SACRE
Nicely restored and placed under a glass housing is a former Ciampino Convair. Isola Sacre is south east of the Roma Fiumicino airport and this sole surviving AMI Convair 440 is in use as a very good restaurant and cabaret. Please call for reservations; Italy 06-6522201.

☐ MM61833	31-1	CV440	442	preserved	5-98

ISTRANA
Dumped at the back of the AMX hangar is an F-104G with no tail. It should be MM6633 (which was noted here some years earlier with the tail of MM6561). By 1996 T-33 MM51-9090 was completely burned and replaced on the fire dump by another T-33 and an MB326 from Grazzanise. One of the Istrana MB326s is said to have moved to Rivolto, this may be MM54180. F-104 MM6518 has moved to Villafranca to be scrapped. Two former dumped aircraft have been scrapped, T-6G MM54096/RR-58 and T-33A MM51-8936/51-86 (6720). MB326 MM54216 moved on to Cerbaiola, while MM54389 went to Vigna di Valle. AB47J MM80162 is now at Fonte and AB47J-3 MM80246 moved just down the road to Pease.

☐ MM6417		G91R/1B	221	preserved, unmarked	3-98
☐ MM6529		F-104G	6529	stored, special c/s	10-97
☐ MM6561	3-35	F-104G	6561	stored, with tail from MM6633	10-94

ITALY - 274

☐ MM6633	3-26	F-104G	6633	dumped, no tail	3-98
☐ MM6739	51-01	F-104S	1039	stored	10-96
☐ MM6870	51-04	F-104S	1170	stored	10-96
☐ MM8307		MC200	..	stored, marked as 'MM7707/359-8', flyable	2-96
☐ MM54157		MB326	6162	dumped, fire dump, ex Grazzinise	9-96
☐ MM54180	53-30	MB326E	6184	stored	8-94
☐ MM54195	51-76	MB326	6200	stored	10-97
☐ MM54217	51-75	MB326	..	preserved, at gate	5-98
☐ MM54243	6-76	MB326	6329	stored	5-98
☐ MM54374	51-77	MB326	6487	stored	5-98
☐ MM51-9030	51-85	T-33A	6814	dumped, burned, ex stored	8-94
☐ MM51-9037	51-86	T-33A	6821	stored	8-94
☐ MM52-9898	51-86	T-33A	7794	preserved	5-98
☐ MM53-5238	51-78	RT-33A	8577	dumped, fire dump, ex stored, orange c/s	10-97
☐ MM53-5322	51-76	RT-33A	8661	preserved, orange c/s	3-98
☐ MM53-6704	51-26	F-84F	..	preserved, near gate	3-98
☐ MM53-8278	51-01	F-86K	221-6	preserved, with parts from MM55-4863	3-98

LAMPEDUSA
Preserved at the airfield on the island of Lampedusa (in the southern Mediterranean) is HU-16 MM50-177 which arrived on 2nd November 1979, making it the last AMI Albatross flight.

☐ MM50-177	15-4	HU-16A	..	preserved	7-96

LANCENIGO
Planned here was an explanation of how to get to the Harvard in Signor Bettiols market garden. Although close to the Treviso Nord autostrada exit, it was not to easy to find. This will not be necessary now, as T-6H-4M MM53849 moved to Villorba by September 1996.

L'AQUILA
The aero club here has a G91R, which has the tail from MM6303. To make the puzzle complete, its own tail is fitted to MM6303 which is at Sassuolo.

☐ MM6312	2-63	G91R/1A	176	preserved, with tail from MM6303	2-98

LATINA
Town: On 28th March 1994 a P148 arrived from the airfield for gate guard duties at the Istituto Professionale di Stato per l'Agricoltura e l'Ambiente San Benedetto. If this name is too difficult to use, just follow the signs to Istituto Statale Agrario (which is the same). The MB326 is at a roundabout in town along the main road.

☐ MM53590	70-07	P148	166	preserved, at gate, ex airfield store	5-98
☐ ..		MB326	..	preserved, unmarked	5-98

Airfield: In 1995 only one P148 was noted dumped on the airfield, but by 1996 some wings and other P148 parts were dumped next to it. The G91T was first noted in September 1996. The P166 MM61873 is still mounted on the gate. The bulk of the stored P148s here went to Castel Volturno; MM53547, MM53549, MM53550, MM53551, MM53558, MM53561, MM53562, MM53566, MM53568, MM53573, MM53580, MM53582, MM53583, MM53589, MM53593, MM53724, MM53725, MM53729, MM53734, MM53736, MM53737, MM53739 and MM53743. MM53590 went to Latina town and MM53727 to Cerbaiola. The museum at Comignago received P166ML1 MM61872, while MM61878 went to Castel Volturno.

☐ MM6339	32-39	G91T/1	69	preserved	3-98
☐ MM6491	8-27	G91Y	2053	preserved, near gate, ex Cervia	5-98
☐ MM53560	75	P148	136	stored	4-86

☐ MM53570		P148	146	stored	4-86
☐ MM53572	25	P148	148	stored	4-86
☐ MM53584	29	P148	160	stored	97
☐ MM53587	11	P148	163	stored	4-86
☐ MM53740	77	P148	188	stored	4-86
☐ MM.....	33	P148	..	dumped	5-97
☐ MM61873	SP-30	P166ML1	381	preserved, on the gate	5-98
☐ MM61877	SP-37	P166ML1	385	dumped, gone by 3-98?	10-97
☐ MM61904		P166ML1	407	stored	8-94
☐ MM80791	SE-7	OH-13H	2405	preserved, near gate, ex Caserta, ex 58-5392	3-98
☐ MM55-2980	9-33	T-33A	9477	preserved, ex Grazzanise	3-98

LARIANO
Some 40km south east of Roma is the village of Lariano. Preserved here should be an AB47.

☐ MM80414	GdiF-40	AB47J	2102	preserved	7-93

LECCE
Town: A local company with the name Sorlini have taken delivery of a former Guidonia Harvard.

☐ MM53659	RB-9	T-6G	..	stored, ex Guidonia	4-95

Airfield – Galatina: The airfield houses the Italian jet training wing which flies with the MB339. Two aircraft are preserved at the gate, while the restoration projects have moved on. Spitfire MM4084 and P-47D MM4653 are both now at <u>Vigna di Valle</u> and G59-4A went to <u>Leinì</u>.

☐ MM53831	SL-25	T-6H-4M	..	preserved, at gate	3-98
☐ MM54154	85	MB326	6159	preserved, at gate	3-98
☐ MM.....		MB326	..	stored, in hangar	9-96

LEINÌ
The Torino branch of the Gruppo Amici Velivoli Storici (GAVS) have their workshop at Leinì, which is some kilometers north of Torino. The workshop is not open to the public.

☐ MM6265		G91R	31	preserved, silver colours, ex Treviso	5-98
☐ MM53265		G59-4A	74	under restoration, maybe converted to G55	98
☐ MM53820		T-6H-4M	..	under restoration, ex Coazze, ex Castrette	5-98
☐ MM61714	RR-18	C-45B	6979	under restoration, ex Guidonia, ex KJ493/RAF	5-98
☐ I-AEEL		L-5 Sentinel	1538	under restoration, ex MM52882, ex 42-99119	5-98
☐ I-BIOU		Avia FL3	..	stored, ex MM56233	5-98
☐ I-CNVU		CVT M200	..	stored, glider	5-98
☐ I-IBAE		SIAI S205/20R	06-005	stored	5-98
☐ I-MCPV		CVV8	011	stored	5-98
☐ I-SART		Bensen B-8	..	stored, ex I-STAR	5-98

LEVALDIGI
Preserved on the airfield of Levaldigi (near Fossano) in fairly good condition is Albatross MM50-180. In the first edition this HU-16 could be found under the heading of Cuneo, which is some 20km south of Levaldigi.

☐ MM50-180	15-6	HU-16A	..	preserved	9-95

LIDO DI IÉSOLO
The resort of Lido di Iésolo (west of Venezia) has a go-kart track where a Thunderflash is preserved.

☐ MM52-7399	3-4	RF-84F	..	preserved, with tail from MM52-7456	3-98

LIVORNO - CAMP DARBY
As a follow on from the first EWR-1, more fates are known for the aircraft which were on sale here. L-18C MM52-2409/EI-82 (18-2009) became N917BS and MM52-2413/EI-83 (18-2013) I-BDUE. L-21B MM53-7735/EI-116 (18-3335) became N145C, MM54-2382/EI-193 (18-3582) has become N4036S, MM54-2396/EI-205 (18-3596) became C-GFRF, MM54-2400/EI-209 (18-3600) became N1394V and MM54-2581/EI-259 (18-3981) became N3930L, O-1E MM61-2961/EI-22 (305M-0007) became N3234K, while MM61-2969/EI-16 went to Viterbo.

LONATE POZZOLO
This G59 is preserved at the local war memorial, near the cemetery, and is only a fuselage hulk with one wing.

❑ MM53136	S3-158	G59-2B	45	preserved, wreck	9-96

LORETO
Preserved at the Air Park near Porto Recanati were two T-6s. One was in USAAF colours, the other one in German Second World War colours. It is not known which MM serial belongs to which T-6. In 1996 one T-6 had moved to Cerbaiola. The other T-6 was still here in September 1996. C-47 I-COFR also went to Cerbaiola.

❑ MM53145	T-6C/D	..	preserved, unmarked, see note	5-95
❑ MM54101	T-6G	..	preserved, unmarked, ex Bari, see note	5-95
❑ MM51-17488	T-33A	7468	preserved, Thunderbirds c/s	9-96
❑ OK-CFE	Tu-134A	2351602	preserved, used ferry regi OK-9522, wfu 9-11-90	9-96
❑ I-GISE	SE210-3	208	preserved, ex Falcona, ex F-BNKB	9-96
❑ I-VFEI	AB47J-3	2041	preserved	9-96
❑ I-ADRI	P64B	23	preserved	9-96

Preserved since the late 1980s are two former AMI jets at the AMI Scuola Perfezionamento Sottuficiali.

❑ MM6301	2-25	G91R/1A	165	preserved	5-89
❑ MM6632	3-24	F-104G	6632	preserved	5-89

LUGO DI ROMAGNA
Preserved at the local airfield is an ex Cérvia G91Y. The aircraft is painted in special colours.

❑ MM6444	8-03	G91Y	2006	preserved, ex Cérvia	3-97

LUNI
Of the preserved helicopters here two are parked at the gate, while MM80360 is also visible from outside. Gone are the stored SH-34Js MM149082/4-04 and MM154618/4-10 and dumped AB212ASW MM81084/7-39.

❑ MM5020N	6-19	SH-3D	6077	stored	3-98
❑ MM80087	1-01	AB47G	..	preserved, marked as 'MM83087'	3-98
❑ MM80360	2-07	AB47J-3	2073	preserved, at gate, marked as 'MM80306'	5-98
❑ MM80371	3-10	AB204ASW	3140	preserved, at gate, ex Catania	3-98
❑ MM80941	7-09	AB212ASW	..	stored, crashed 20-7-94	3-98
❑ MM80949	7-18	AB212ASW	..	stored, dismantled	12-95
❑ MM80951	7-20	AB212ASW	..	stored	3-98
❑ MM81093	7-48	AB212ASW	..	stored, dismantled	12-95
❑ MM153622	4-14	SH-34J	..	preserved, at gate	5-98

MACERATA
Preserved at the Scuola Addestramento Reclute AM should still be an MB326 and an anonymous T-6G.

❑ MM54153	84	MB326	6158	preserved	12-85
❑ MM.....	303-56	T-6G	..	preserved	12-85

MANERBIO
Preserved at the Bocciodromo near the village centre is another ex Treviso Fiat.
❑ MM6308   2-21   G91R/1A   172   preserved, tail from MM6403 (207), ex Treviso   5-97

MANIAGO
North east of Aviano is a range which is mainly in use by the Italian Army. The range (Poligono Tiri) has some small barracks (not the large barracks in Maniago town), where in 1991 a G91 was noted. Only the code is known. Suggestions that this may be MM6376 seem incorrect as it was still at Treviso at the same time.
❑ MM....   2-01   G91R   ..   preserved   91

MANTAGRAVA
Reported here is a preserved G91. The name Mantagrava was not in the standard road maps and may therefore not be correct. Any ideas on the correct location would be welcome.
❑ MM6291   2-24   G91R/1   155   stored, ex Treviso   7-90

MANTOVA
Preserved at the Galleria Storica del Corpo Nazionale dei Vigili del Fuoco (national fire brigade museum) is a red AB47. Mantova is some 30km south of Verona.
❑ I-VFEN   AB47G-3B-1   1597   preserved   94

MASERA
Preserved at the local airfield is a G91Y. It is parked right along the E62 road from Switzerland and is a bit difficult to see (you have to look down from the road!).
❑ MM6474   8-30   G91Y   2036   preserved   5-98

MEGLIADINO SAN FIDÉNZIO
The Sabre MM54-4815 can be found along road 10 from Legnago to Monsélice. It is preserved here in a private garden just north of Megliadino San Fidénzio.
❑ MM55-4815 51-50   F-86K   221-55   preserved, with tail from MM54-1256   9-95

MILANO
Town: The Museo Nazionale della Scienza e della Tecnica has also an aviation section. Of this collection the Caproni TM2 and the aircraft marked with an * will move in the near future to a new department of the museum at Milano Malpensa airport. This new museum will be opened in the year 2000. The museum can be found at the Via San Vittore, which is near the San Ambrogio subway station. Opening times are Tuesday to Friday between 09:30 and 17:00 (weekends to 18:30). Vampire FB.50 MM6153 has not been seen since 1977.
❑ MM253   Magni PM 3/4   ..   preserved, unmarked   9-96
❑ MM511   Caproni TM2   2   preserved   9-96
❑ MM6112   76   Vampire FB.52A   ..   preserved   9-96
❑ MM6382   2-20   G91R/1B   186   preserved, outside, tail from MM6420 (224)   9-96
❑ MM54114 RM-18 T-6G   ..   preserved   9-96
❑ MM92166   MC205V   ..   preserved, as 'MM9327/81-5', ex 1243/EgyptAF   9-96
❑ MM53-6805 50-30   F-84F   ..   preserved, outside   9-96
❑ MM55-4812 51-3   F-86K   221-52   preserved, outside   9-96
❑ I-BORA   Macchi Ni10   15179   preserved, marked as 'Ni1467'   9-96
❑ I-CIER   Cierva C.30A   753   preserved, ex MM30030, ex I-CIER   9-96
❑ I-CUPI   SAIMAN S202/M   ..   preserved, ex MM52162   9-96
❑ I-FOGL   *   DH.80A   2114   preserved   9-96

☐ I-HAWK	* Pasotti F9	D2	preserved		9-96
☐ I-GION	* SM102	17	preserved, ex MM61810		9-96
☐ I-PAIN	Ambrosini S7	1/01	preserved		9-96
☐ I-TITI	* Muegyetemi M.24	..	preserved		9-96
☐ ..	Blériot XI	..	preserved, replica		9-96
☐ ..	Breda Ba15	..	preserved		9-96
☐ ..	Caproni Campini CC2	..	preserved, fuselage only		9-96
☐ ..	Farman HF-4	..	preserved, replica, marked as 'I-FARM'		9-96
☐ ..	* FN333 Riviera	..	preserved		9-96
☐ ..	Junkers J4	..	preserved, parts only		9-96
☐ ..	Ricci 6	..	preserved, replica		9-96

Airfield – Bresso: At the military Army airfield of Bresso there is also a large civil area. Here a Thunderflash is preserved in good condition.

☐ MM52-7390	RF-84F		..	preserved, with aero club	5-98

Airfield – Linate: At the military side of the international airport Linate, an Agusta Bell is preserved well inside the barracks.

☐ MM80160 RM-92	AB47J		1060	preserved	5-98

Airfield - Malpensa: Malpensa airport has seen major reconstruction over the past years. Stored Hercules MM62000/46-14 (4497) disappearded in the late 1980s. Agusta (at Cascina Costa part of the airfield) here uses two fuselages for crash test duties. One was a A129 fitted with a boom marked as EI-905. As the fuselage was in a different colour to the boom it my not be the original boom.

☐ MM..... EI-905	A129		..	instructional, see note	3-98
☐ F-BHRS	SE210-3		54	dumped, crashed 2-7-83	7-95

MODENA
Preserved near the town centre at the site of the former airfield is an unmarked Fiat G46.

☐ MM.....	G46-5	..	preserved, unmarked	5-95

MONTEGALDELLA
The firm Sorlini Motori Avio is restoring Harvard I-TSIX (ex MM53875) to flying condition here. They acquired MM53815 from Roma for use as spares source.

☐ MM53815	T-6H-4M	..	stored, ex Roma Urbe	97

MONTE SAN SAVINO
Bird Dog MM61-2988/EI-123 (305M-0036) which was preserved here at a factory near the railway station has been restored to flying condition as I-EIAT. Unfortunately it crashed on 10th June 1995.

NAPOLI - CAPODICHINO
The situation here is not very clear, mainly because the international airfield is largely surrounded by a large wall. From outside only the preserved T-33 is clearly visible. In July 1995, from within the airfield, three MB326s (one remains unknown) and one Tracker were noted. The report of three additional S-2As (MM136588/41-28, MM144696/41-32 and MM148299/41-46) seems to be an error, as these three aircraft were noted at Catania at the same time. The dumped T-33A MM51-17455 moved on to Decimomannu.

☐ MM54170	170	MB326	6174	preserved	3-90
☐ MM54275	4-08	MB326	6440	stored	7-95
☐ MM.....	5-54 ?	MB326	..	dumped	7-95
☐ MM.....	32-52	MB326	..	stored	6-92

☐ MM.....	50	MB326	..	dumped	3-94
☐ MM144710	41-33	S-2A	671	stored, near aero club	7-95
☐ MM148295	41-35	S-2A	734	dumped, near Aeritalia factory	7-92
☐ MM51-9031	3-931	T-33A	6815	dumped	8-87
☐ MM51-17536	4-16	T-33A	7596	preserved, at gate	7-97
☐ MM..-.....	36-67	T-33A	..	dumped	86

NOVI LIGURE
South east of Alessandria is the village of Novi Ligure. Stored at some barracks at the airfield is an AB204.

☐ MM80395	EI-241	AB204B	3126	preserved	9-95

ORTE
Near the village of Orte are some military barracks with at least two preserved aircarft at the gate..

☐ MM6391	2-54	G91R/1B	195	preserved, ex Treviso	5-98
☐ MM....	3-36	F-104G	..	preserved	5-98

ORZINUOVI
Preserved at the Plaza Aeronautica in the southern part of Orzinuovi is a Thunderflash. Orzinuovi is south west of Brescia.

☐ MM52-7397	3-13	RF-84F	..	preserved	5-97

PALERMO - BOCCADIFALEE
The airfield of Palermo is used by a Carabinieri unit, who should have an AB47 preserved at their barracks.

☐ MM80501	CC-30	AB47J	2118	preserved	—

PARMA
The hangars of this civil airfield were checked in September 1995, but the two T-6s were not found. The G91 was still preserved at the attached military barracks.

☐ MM6287	2-31	G91R/1	53	preserved	5-98
☐ MM54102	RR-51	T-6G	..	stored, ex Guidonia	2-94
☐ MM54109	RR-56	T-6G	..	stored, ex Guidonia	2-94

PEASE
Parked alongside the main road (SS53) from Treviso to Istrana is an Agusta Bell with local car dealer.

☐ MM80246	AB47J-3	2017	preserved, unmarked, ex Istrana	5-98

PERDASDEFOGU
This T-33 (airlifted from Decimonannu) may still be current as gate guard at the range near Perdasdefogu, which itself is some 75km north east of Decimomannu.

☐ MM51-8832	51-87	T-33A	6616	preserved, at gate	1-88

PESCARA
The G91PAN which was stored here was searched for in 1995 and not found. It may have been that the aircraft was temporarily inside for restoration.

☐ MM6253	1	G91PAN	19	preserved	7-94

ITALY - 280

PIACENZA
Town – San Polo: Halfway between the airbase and Piacenza town is the little village of San Polo. On the west side of San Polo are the administration buildings of the airbase. Preserved here as gate guard is a Thunderstreak.
❏ MM53-6810   50-21   F-84F   ..   preserved   5-98

Airfield - San Damiano: No gate guards, decoy and instructional airframes at this AMI Tornado base, except one MB326 which is stored in one of the hangars. F-104 MM6634 is earmarked to become a gate guard here, but is currently still stored at Cámeri.
❏ MM54185   5-50   MB326   6190   stored   5-98

PISA
Town: Since at least 1964 the local technical college (ITI Da Vinci) had a CL-13 Sabre for instructional purposes. It was still current in 1997.
❏ MM19782   4-83   CL-13 Sabre 4   682   preserved, ex XB894/RAF   3-98

Airfield – San Giusto: All the C-119s, except the preserved MM53-3200, have gone. The scrapped aircraft included C-119Gs MM52-6002/46-31 (10827), MM52-6024/46-91 (10995), MM53-7828/46-97 (11245) and MM53-7845/46-94 (11266). The C-119Js were 51-8046 (10924), MM51-8113/46-69 (116), MM51-8125/46-51 (128), MM51-8128/46-52 (131), MM51-8130/46-53 (133), MM51-8152/46-53 (155), MM51-8156/46-58 (159), MM52-5866/46-58 (11025), MM52-5897/46-65 (11064), MM52-5947/46-66 (11134), MM53-8098/46-67 (201) and MM53-8103/46-68 (206). 51-8046 was never flown operationally by the AMI and only used for spares. C-119G MM52-6030 went to <u>Vergiate</u> and MM53-3219 to <u>Castel Volturno</u>. Also going to <u>Castel Volturno</u> were P166s MM61876 and MM61902. Dumped AB205 MM80776/CC-35 (4172, crashed 10-9-88) was last noted in December 1988. Note that all maintenance on the G222s is carried out at Pisa, which will result in sightings at Pisa of G222s without vital parts. Those aircraft noted as such in the past nearly all returned to service.
❏ MM60216   AM-96   G222L   4041   stored, ex Somali AF   5-98
❏ MM61707   RR-04   C-45   ..   dumped   12-93
❏ MM53-3200 46-38 C-119G   11213   preserved   5-98

PONTEDERA
A new museum will be set up here in honour of Rinaldo Piaggio. Two of the aircraft are coming from Castel Volturno, a P148 and a P166.
❏ MM53562   26   P148   138   stored, ex Castel Volturno   97
❏ MM......   P166   ..   expected, ex Castel Volturno   —

PORDENONE
Giovanni Follador is a private collector from Pordenone. He has acquired a Hunter which arrived at Aviano in December 1994. He also owns the former Aviano instructional Phantom 63-7512. It is not known where this F-4C currently is. It may have moved to Pordenone or it may be at a new museum near/in Udine (see Aviano).
❏ J-4068   Hunter F.58   ..   expected, still at Aviano   —
❏ 63-7512   F-4C   525   expected, marked as '63-040/AV', still at Aviano ? —

POZZUOLI
On the road from Napoli to Pozzouli, near a place called Angano (which could not be found on a map), a G91 is preserved on a pole. Its identity is not yet known.
❏ MM....   G91   ..   preserved, on pole   7-95

PRÁTICA DI MARE
Prática di Mare is the largest military airbase in Italy. Besides the Italian Air Force there are also large quantities

of Carabinieri, Polizia and Guardia di Finanze helicopters here. All these units have their own preserved aircraft. The dump is located on the opposite side of the runway. Noted inside the compound in September 1996 was a PD808GE1, an HH-3F and the crashed remains of a G222. Outside was a preserved T-33, a stored Skytrain and a dumped PD808VIP. Also preserved here is the tailboom of MM80185/31-32, an AB47J. PS-001 is used by helicopters for hoist practice. Starfighter RS-03 was noted as MM6508 in February 1996, while in September it was marked MM6527. Airframes known to have left include G91T/1 MM6362 to Améndola and F-104G MM6567 to Villafranca by 1995. C-45 MM61758 has gone to Tor Sapienza (Roma) for scrapping together with AB47Js MM80198, MM80218, MM80248 and MM80254. Skytrain MM61800 went to Castel Volturno and T-33A MM54-1602 has gone to Cerbaiola. AB47G-2 MM80042/EL-4 (285) was scrapped locally, while AB47J-3B-1 MM80728 moved on to Comignago. All the other airframes noted in the first issue have not been seen in the 1990s and all have gone.

❏ MM577	RS-48	PD808TA	501	stored	3-98
❏ MM578	RS-49	PD808TA	502	dumped	3-98
❏ MM579	RS-11	G91Y	2001	preserved, at training school	5-98
❏ MM582	RS-06	G222	4001	dumped, ex I-MARD	6-92
❏ MM6248	7	G91PAN	14	preserved, at gate, Frecce Tricolori c/s	5-98
❏ MM6288	RS-01	G91T	1	preserved, at officers mess	5-98
❏ MM6527	RS-03	F-104G	6527	preserved, behind officers mess, ex Grosseto	5-98
❏ MM6660	RS-02	F-104S	6660	preserved, at RSV buildings	5-98
❏ MM6916	4-03	F-104S	1216	preserved, ex Grosseto	3-98
❏ MM53882	RS-22	G80-3B	2	under restoration, ex Vigna di Valle	5-98
❏ MM54201		MB326	..	preserved, at officers mess	5-98
❏ MM54222	6-09	MB326	..	stored	5-98
❏ MM54385		MB326E	..	preserved, at fire department buildings	5-98
❏ MM54391	34	MB326K	6478	stored	5-98
❏ MM61643		C-45	..	stored	6-92
❏ MM61775	14-40	C-47A	19016	stored, near flightline, ex 42-100553	5-98
❏ MM61825	14-50	ECM-47	4221	stored, outside dump compound, ex 41-7742	5-98
❏ MM61893	14-46	C-47	4236	stored, ex 41-7749	5-98
❏ MM61948		PD808VIP	506	stored, fuselage only	5-98
❏ MM61949		PD808VIP	507	dumped, dump compound	3-98
❏ MM61951		PD808VIP	509	stored	3-98
❏ MM61956		PD808TP	514	stored	3-98
❏ MM61957		PD808TP	515	stored	3-98
❏ MM61958		PD808GE1	505	preserved	5-98
❏ MM61959		PD808GE1	516	stored	3-98
❏ MM61963		PD808GE1	520	dumped, dump compound	3-98
❏ MM62015		PD808RM	522	stored	3-98
❏ MM62016		PD808RM	523	stored	5-98
❏ MM62017		PD808RM	524	stored, in hangar, special c/s	3-98
❏ MM62129	RS-44	G222TCM	4036	dumped, dump compound, crashed 5-6-91	5-98
❏ MM80043	26	AB47G-2	288	preserved	3-98
❏ MM80157	31-48	AB47J	1054	preserved	3-98
❏ MM80166	GdiF-1	AB47G-2	248	preserved, marked 'MM80084'	5-98
❏ MM80185	31-34	AB47J	1097	stored, l/n 12-86, by 9-96 tailboom only	9-96
❏ MM80294	CC-1	AB47J-3	2039	preserved, at Carabinieri platform	5-98
❏ MM80476	32	AB47G-2	299	preserved, at Carabinieri buildings	5-98
❏ MM80482	CC-13	AB47G-3B-1	1615	preserved, at Carabinieri platform	9-96
❏ MM80781	CC-37	AB205A-1	4237	stored, in Carabinieri hangar	2-96
❏ MM80804	3	OH-13H	2270	preserved	5-98
❏ MM80856	GdiF-57	NH500MC	..	dumped, wreck at Guardia de Finanza hangars	3-98
❏ MM80987	15-16	HH-3F	6214	dumped, dump compound, crashed 19-8-85	9-96
❏ MM81170	GdiF-122	A109A-II	7305	dumped, wreck at Guardia de Finanza hangars	2-96
❏ MM.....	GdiF-135	A109A-II	..	dumped, wreck at Guardia de Finanza hangars	2-96
❏ MM81173	CC-98	A109A-II	7297	stored, wreck	3-98

ITALY - 282

☐ MM81314	CC-10	AB412	..	stored, wreck, crashed 10-6-89	3-98
☐ MM51-4418	418	T-33A	5713	stored	9-96
☐ MM51-4514	14-22	T-33A	5809	preserved, near dump compound	5-98
☐ MM51-9140	6-31	T-33A	6924	stored	5-98
☐ MM53-8146	46-35	EC-119G	249	stored	5-98
☐ MM.....	PS-001	AB204B	..	instructional, used for hoist training	9-96
☐ MM.....	PS-B01	AB204B	..	preserved, on pole	3-98
☐ PS-A91		P64B-200	05	stored, inside Polizia hangar, ex MM57191	3-98
☐ PS-A92		P64B-200	06	stored, inside Polizia hangar, ex MM57192	3-98

PRATO
Alongside, but not visible from, the A11 motorway should still be the hulk of a Thunderstreak. The F-84 is at Casale, a hamlet near Prato, which is some 10km north west of Firenze.

☐ MM53-6579	36-40	F-84F	..	preserved	9-88

RAVENNA
In the Russi di Ravenna part of town an ex Cérvia MB326 is preserved.

☐ MM54272	8-73	MB326	6437	preserved, ex Cérvia	98

REGGIO NELL'EMILIA
Noted at the aero club should still be a former Guidonia Texan.

☐ MM54148	RR-54	T-6H-2M	..	preserved, with aero club, ex Guidonia	5-98

REZZATO
Preserved at a monument since at least 1983 is a former Ghedi T-33. Rezzato is some kilometers east of Brescia.

☐ MM51-17484	6-30	T-33A	7378	preserved	5-95

RIETI
Preserved on the military side of the aerodrome here (some 70km north east of Roma) by 1985 was a G91R.

☐ MM6277	2-70	G91R/1	43	preserved	7-97

RIMINI
Town: On the western side of the ring road around Rimini is a furniture shop with a preserved Thunderstreak.

☐ MM53-6637		F-84F	..	preserved, as 'MM56-9474/69-4', tail of 53-6695	7-97

Airfield – Miramare: During the mid 1990s the operational Starfighter unit was relocated to Cérvia, taking the F-104Gs MM6524 and MM6551 with them, while MM6522 (ex Villafranca) moved to <u>Ghedi</u> and MM6550 to <u>Villafranca</u> and F-84 MM53-6855 went to <u>Ghedi</u>. The two F-86Ks noted here may be the same aircraft.

☐ MM6820	51-07	F-104S	1120	stored, arr 19-1-94 for spares	1-94
☐ MM54387		MB326E	..	stored	8-92
☐ MM51-17477	477	T-33A	7371	dumped, crashed 10-6-78	9-96
☐ MM53-8291	5-52	F-86K	221-19	preserved, on pylon	9-96
☐ MM53-8301	5-52	F-86K	221-29	preserved, near gate	10-95

RIVOLTO
Rivolto is the home-base of the 313 Gruppo which flies the MB339 as Frecce Tricolori and has a large number

of preserved aircraft. Some near the gate are painted in colours of former display teams. The Sabre in Lanceri Neri colours is also reported as MM19595, but this seems unlikely as this aircraft was last noted in 1963 as B of Frecce Tricolori. Also the serial MM19680 seems unlikely as that aircraft is still current at the technical school at Udine. The serial for the second Sabre is not correct as MM19664 is confirmed at Capua. The completely burned out wreck of the MB326 is said to come from Istrana and may be MM54180.

☐ J-1170	4-41	Vampire FB.6	679	preserved, near gate, ex Swiss AF	3-98
☐ MM6241	3	G91PAN	7	preserved	5-98
☐ MM6381	2-41	G91R/1B	185	preserved, at 14 Gruppo area	7-97
☐ MM6416	2-14	G91R/1B	220	preserved, on pole at 14 Gruppo area	3-98
☐ MM19664 ?		CL-13 Sabre 4	..	preserved, as 'MM19685/4-20', see note	3-98
☐ MM19680 ?		CL-13 Sabre 4	..	preserved, near gate, Lanceri Neri, see note	3-98
☐ MM19724	1	CL-13 Sabre 4	624	preserved, near gate, Frecce Tricolori	3-98
☐ MM53822	SC-55	T-6H-4M	..	preserved, near tower	3-98
☐ MM52-6020	46-84	C-119G	10950	preserved, near tower	3-98
☐ MM52-7474	3-18	RF-84F	..	preserved, parts used from 52-7463	3-98
☐ MM53-5795	36-68	RT-33A	9134	preserved	3-98
☐ MM53-6845		F-84F	..	preserved, near gate, Diavoli Rossi	3-98
☐ MM54-1292		F-86K	213-62	preserved, marked as 'MM55-4818/51-36'	3-98
☐ MM55-3080	51-85	T-33A	9621	preserved	3-98
☐ MM.....		MB326	..	dumped, burned out, ex Istrana ?	9-96

ROMA
Town: The Instituto Tecnico Galileo Galilei, a Technical College in Roma, may still have a G46. Their CL-13 Sabre MM19666/4-11 went to Duxford, UK, in the 1980s.

☐ MM.....	AA-77	G46-1B	..	instructional, may be MM52795 (28)	—

Another college is the IAS F. de Pinedo. The Viscount is still current, but their G46-1B I-DEPI (11) has moved on some years ago.

☐ I-LIRG		Viscount 798	284	instructional, ex N6594C	98

In 1998 a G91T was noted some 15km from the airfield of Fiumicino at the Roma motor ringway exit 31. The aircraft is mounted on a rotating pole inside some kind of cement works.

☐ MM6323	60-23	G91T/1	53	preserved, CFE a/c, code 12, ex Améndola	5-98

At the Via Prenestina 699/703 (ringway exit 16) is the Bentivoglio Elio scrapyard. In October 1997 the yard bought 30 G91Ts from Améndola and three G91Rs for onward sale. All are CFE aircraft and only 24 are known. Two are known to have moved on. These are the above mentioned MM6323/60-23 and MM6428/60-78 which went to <u>Basiliano</u>. G91Rs have been fitted with different tails. The owner has bought seven G91Ys from Brindisi and these will start to arrive in in June 1998.

☐ MM6315	60-15	G91T/1	45	stored, CFE a/c, code 2, third batch	5-98
☐ MM6322	32-22	G91T/1	52	stored, CFE a/c, code 5, third batch	5-98
☐ MM6340	60-40	G91T/1	70	stored, CFE a/c, code 4, first batch	5-98
☐ MM6342	60-42	G91T/1	72	stored, CFE a/c, code 8, first batch	5-98
☐ MM6346	60-46	G91T/1	76	stored, CFE a/c, code 4, second batch	5-98
☐ MM6355	60-55	G91T/1	85	stored, CFE a/c, code 1, third batch, tail only	5-98
☐ MM6356	60-56	G91T/1	86	stored, CFE a/c, code 1, second batch	5-98
☐ MM6357	60-57	G91T/1	87	stored, CFE a/c, code 3, second batch	5-98
☐ MM6359	60-59	G91T/1	89	stored, CFE a/c, code 10, first batch	5-98
☐ MM6369	60-69	G91T/1	99	stored, CFE a/c, code 5, first batch	3-98
☐ MM6373	60-73	G91T/1	104	stored, CFE a/c, code 5, second batch	5-98
☐ MM6408	2-07	G91R/1B	..	stored, ex Treviso, with tail from MM6389	5-98
☐ MM6388	2-42	G91R/1B	..	stored, ex Treviso, with tail from MM6396	3-98
☐ MM6429	60-79	G91T/1	110	stored, CFE a/c, code 7, second batch	5-98
☐ MM6430	60-80	G91T/1	111	stored, CFE a/c, code 3, first batch	5-98

☐ MM6431	60-81	G91T/1	112	stored, CFE a/c, code 2, second batch	3-98
☐ MM6434	60-84	G91T/1	115	stored, CFE a/c, code 1, first batch	3-98
☐ MM6437	60-87	G91T/1	118	stored, CFE a/c, code 8, second batch	3-98
☐ MM54394	60-94	G91T/1	121	stored, CFE a/c, code 2, first batch	9-96
☐ MM54405		G91T/1	132	stored, bare metal, CFE a/c, code 7, first batch	3-98
☐ MM54407		G91T/1	134	stored, bare metal, CFE a/c, code 6, first batch	3-98
☐ MM54408	60-108	G91T/1	135	stored, CFE a/c, code 10, third batch	3-98
☐ MM54409	60-109	G91T/1	136	stored, CFE a/c, code 6, third batch	3-98
☐ MM54410	60-110	G91T/1	137	stored, CFE a/c, code 9, third batch	3-98
☐ MM54416	60-116	G91T/1	143	stored, CFE a/c, code 7, as SA-116 on LH-side	3-98

Town – Centocelle: In the eastern part of Roma the Comando 2 Regio Aerea of the Italian Air Force have their headquarters. Mounted at the gate is a Starfighter.

☐ MM6611	3-52	RF-104G	6611	preserved	10-93

Town – Citta Militare: A large complex of military barracks is situated on the south side of Roma. It is located inside the Roma highway orbital (exit 25), near the Laurentina exit. Within these barracks is the Museo Storico della Motorizzazione Militare, a museum of the Italian Army. Although it is inside the barracks it can be visited on weekends between 09:00-12:00.

☐ MM80263	EI-24	AB47J-3	2026	preserved	9-95
☐ MM80390	EI-236	AB204B	3120	preserved, ex Salerno	9-95
☐ MM80811	73	OH-13H	..	preserved, ex Bracciano	9-95
☐ MM54-2397	EI-206	L-21B	18-3597	preserved, ex I-EIJA	9-95
☐ MM..-....		L-18C	..	preserved, marked as 'EI-00'	9-95

Town – Tor Sapienza: In 1994 a scrap dealer in Tor Sapienza took delivery of some former Prática di Mare inmates. A number of these have been identified, although more may have passed through here (see Prática di Mare). Some parts of the crashed G222 MM62129/RS-44 were noted here, the rest is still at Prática di Mare.

☐ MM61758	CR-30	C-45	..	dumped, ex Prática di Mare	7-94
☐ MM80198	SE-51	AB47J	1092	dumped, ex Prática di Mare	4-94
☐ MM80218	512-44	AB47J	1128	dumped, ex Prática di Mare	4-94
☐ MM80248	15-34	AB47J-3	2019	dumped, ex Prática di Mare	4-94
☐ MM80254	15-35	AB47J-3	2030	dumped, ex Prática di Mare	4-94

Town – Quatro Miglio: An auction was held here in 1987 and all helicopters left within a year. AB47 MM80417 went to Cecchina and MM80414 to Lariano. AB47G-2 MM80128/GdiF-10 (231), AB47J-3 MM80304/GdiF-36 and NH500MC MM81016/GdiF-87 all have disappeared.

Town – Viale Manzoni: A supermarket with the name Babilonia in the Viale Manzoni part of Roma (along the SS59) bought a G91 from Germany. This was easier than getting one from the AMI at Treviso.

☐ 3212		G91R/3	480	preserved, ex Neubiberg, ex WGAF	12-96

Airfield – Ciampino: The international airfield on the south side of Roma has an HU-16 as gate guard. DC-6B MM61964/SM-23 (44253) was scrapped and the T-33A MM51-9249 moved on to Vigna di Valle.

☐ MM51-035	15-8	HU-16A	..	preserved	5-98
☐ I-VFED		AB47G-2	176	preserved	5-98

Airfield – Fiumicino: Close to the airport are some military barracks with some aircraft inside. Caravelle I-DABU was stored at the international airport, but went to San Angelo in Villa in the 1980s

☐ MM53801	4-01	T-6H-4M	..	preserved	98
☐ MM....	8-06	G91Y	..	preserved, at gate	5-98
☐ MM....		P148	..	preserved	5-98

Airfield – Furbara: Douglas DC-6B MM61900 which was abandoned on this Roma airfield by 1982, moved to Castle Volturno in the mid 1990s.

ITALY - 285

Airfield – Guidonia: The long lines of stored aircraft here have gone. Only a handful now remain. The aircraft marked 'scrapyard' can be found at a yard on the edge of the airfield on the eastern side. This is near a large public swimming pool. The sighting of MM61885 may not be correct as this aircraft should have been scrapped in the 1980s. The departed aircraft include T-6Gs MM53665 and MM53669 to San Possidonio, MM54102 and MM54109 to Parma, MM53659 went to Lecce, Castel Volturno received MM53863 and MM54108, while MM54105/RR-54 is just gone. The T-6H-2M MM54141 was last noted in August 1990 and has gone to Sulmona. MM54148 went to Reggio Nell'Emilia. T-6H-4Ms MM53786/RB-13 and MM53811/RB-13 were sold of in the mid 1980s and have left. Castel Volturno also received a large number of C-45s, including MM61644, MM61646, MM61675, MM61677, MM61708, MM61754, MM61717 and MM61722. MM61714 went to Torino. Other C-45s having left are MM61672/CR-36, MM61710/CR-35, MM61716 and MM61743/RM-57 (to Alisarda Airlines as 'I-SARE'). C47s MM61897/CR-53, MM61894/CR-50 (4261), C-53s MM61764/14-48, MM61765/14-49 (11681), MM61766/RR-04 and MM61818/RR-02 (7397) have been scrapped. Saved was C-47 MM61815 which is now at Castel Volturno. Guidonia was also the final station for some P166s. Of these, MM61904 went to Latina, MM61881 and MM61906 went to Castel Volturno, Forcelle received MM61911 and Cerbaiola has MM61913. The MM61889/SP-31 (397), MM61908/53-73 (420) and MM61916/303-36 (428) have all gone and have maybe been scrapped. HU-16A MM50-182 went to Castel Volturno in 1995.

☐ MM8071		Re2001	3	under restoration, ex Vigna di Valle	—
☐ MM53143	RB-6	T-6D	..	stored, ex Bari	3-98
☐ MM53825	303-50	T-6H-4M	..	preserved, main gate	3-98
☐ MM53839	RR-1	T-6H-4M	..	stored	7-92
☐ MM54110	RB-8	T-6G	..	stored, ex Bari	3-98
☐ MM54111	RB-7	T-6G	..	stored, ex Bari	3-98
☐ MM54155	40	MB326	6160	stored, near hangars	8-94
☐ MM61693	RR-16	C-45	..	stored	12-93
☐ MM61727	RM-51	C-45	..	stored	12-93
☐ MM61762	RR-11	C-45	..	preserved	5-98
☐ MM61776	14-45	C-47A	19194	under restoration, ex stored, ex PH-BTH	5-97
☐ MM61777	14-43	C-47A	9910	stored, ex PH-TBK, ex 42-24048	5-98
☐ MM61799	14-41	C-47B	15262/26707	stored, ex 43-49446	5-98
☐ MM61826	14-44	C-47	4380	stored, ex I-RIBE, ex I-LIRA, ex 41-18342	5-98
☐ MM61874		P166ML1	382	stored, near railway station	11-96
☐ MM61879	36-75	P166ML1	387	stored	10-89
☐ MM61885	303-33	P166ML1	393	stored, should be scrapped, correct sighting ?	93
☐ MM61895	14-42	C-47	6011	stored, ex I-LICE, ex I-VARO, ex 41-18650	5-98
☐ MM61905	SP-33	P166ML1	408	dumped	9-96
☐ MM61909	303-28	P166ML1	421	stored, at scrapyard	8-94
☐ MM61910	SP-41	P166ML1	422	stored, ex Viterbo	9-92
☐ MM61912	VV-12	P166ML1	424	stored	7-93
☐ MM61915	303-21	P166ML1	427	stored	6-92
☐ MM61918	303-33	P166ML1	430	stored, at scrapyard	7-97
☐ MM61920	53-35	P166ML1	432	stored, at scrapyard	7-97
☐ MM61923	31-8	DC-6B	43217	stored, ex I-DIMT, ex I-LOVE	5-98
☐ MM61924	303-10	P166ML1	434	stored, near railway station	11-96
☐ MM61925	36-65	P166ML1	435	stored, near railway station	11-96
☐ MM61926	VV-26	P166ML1	436	stored, near railway station	11-96
☐ MM61927	15-43	P166ML1	437	stored	—
☐ MM61928	303-39	P166ML1	438	stored, at scrapyard, ex Cámeri	7-97
☐ MM61930	53-35	P166ML1	440	stored, near railway station	11-96
☐ MM61932	36-66	P166ML1	442	stored, near railway station	3-98
☐ MM100006	VV-03	CVV6	..	dumped, glider	10-89
☐ MM100013	VV-04	CVV6	..	dumped. glider	10-89
☐ MM.....		P148	..	stored, fuselage only	3-98

Airfield – Urbe: A smaller civil airfield of Roma is Urbe in the northern part of town. T-6 MM53815 was preserved at the Air Force military barracks at this airfield. It moved to Montegaldella in 1996.

ROVERETO
The Museo Storico Italiano della Guerra Rovereto (located in the Castello di Rovereto, itself about 50km up the A22 motorway from Verona) displays Macchi-Nieuport Ni10-18mg 13469.

❏ 13469	Macchi Ni10-18mg	..	preserved	—

SABÁUDIA
About 15km south of Latina, a F-84 is preserved with the Scuola di Artiglieria Contra-Aerea. The aircraft is not visible from outside of the military army school. However it was confirmed in 1996 by the guards that the aircraft was still preserved.

❏ MM53-6988	50-19 F-84F	..	preserved	96

SALERNO - PONTECAGNANO
Stored here, at this combined Army/Carabinieri/Vigili del Fuoco base, for a number of years already are some AB204s. The AB47 is preserved at the Carabinieri gate which is just north of the Army gate. Other airframes noted here were AB204Bs MM80390 which went to <u>Roma - Citta Militare</u> and MM80398 is now at <u>Solbiate Olbano</u>. AB47G-3B-1 MM80485 moved on to <u>Bergamo</u>.

❏ MM80284	EI-201	AB204B	3027	stored	7-97
❏ MM80317	EI-217	AB204B	3079	stored, ex Bracciano	7-97
❏ MM80318	EI-218	AB204B	3083	stored	7-97
❏ MM80323	EI-223	AB204B	3092	stored	7-97
❏ MM80397	EI-243	AB204B	3130	stored	7-97
❏ MM80491	CC-24	AB47G-3B-1	1637	preserved, at Carabinieri gate	7-97

SALTO DI QUIRRA
On the island of Sardinia, 672 Squadriligia were reported to possess two AB47Js coded PI-02 and PI-04 in July 1985. These were formerly in service with the Poligono Interforce.

❏ MM80244	PI-02	AB47J-3	2015	stored	7-85
❏ MM.....	PI-04	AB47J	..	stored	7-85

SAN ANGELO IN VILLA
Mounted on high poles is a Caravelle at the 4R Rufa furniture shop. It is clearly visible from the motorway between Sora and Frosinone. The aircraft is not at Giglio di Veroli as mentioned in the first edition, although the two villages are close together.

❏ I-DABU	SE210-6N	77	preserved, unmarked, ex Roma Fiumicinio	9-96

SAN POSSIDONIO
The amusement park Deltaland has its aviation department here with the name Parco Velivolo Storici. It is an open air museum at the local airstrip. Also noted here in February 1996 was the boom of AB206 MM80592. G46-4 MM53312/ZC-10 moved on to <u>Forli</u>.

❏ MM6305	2-57	G91R/1A	169	preserved	5-98
❏ MM6520	3-44	F-104G	6520	stored, ex Villafranca	2-96
❏ MM6647	3-22	F-104G	6647	stored, ex Villafranca	5-98
❏ MM53665	RR-66	T-6G	..	preserved, ex Guidonia	2-96
❏ MM53669	RR-70	T-6G	..	stored, ex Guidonia	2-96
❏ MM54168	6-43	MB326E	6172	preserved, ex Ghedi	5-98
❏ MM57195	EI-402	SM1019E	1-002	preserved, ex Bologna	5-98
❏ MM61911	SP-36	P166ML1	423	preserved, ex Guidonia, ex I-PIAA	5-98
❏ MM80382	EI-228	AB204B	3104	preserved	5-98
❏ MM80812		OH-13H	2614	preserved, ex Caserta, ex Bracciano, ex 58-1548	2-96

| ☐ I-ADRO | | G46-3B | 46 | stored, ex Fidenza, ex MM52803 | 2-96 |
| ☐ I-VFET | VF-06 | AB205A-1 | 4503 | stored, wreck, behind shed at far side of airfield | 5-98 |

SASSUOLO
Just north of the town of Sassuolo (some 15km south of Modena) is a small civil airfield. Preserved here is a G91R with the tail from the Fiat at l'Aquila.

| ☐ MM6303 | 2-61 | G91R | 167 | preserved, with tail from MM6312 | 9-95 |

SAVONA
Signor Flavio Cabib at Via Nostra Signora del Monte 60 has a composite F-84F Thunderstreak acquired from Castrette in 1977. The front fuselage is from MM53-6695/36-35 with parts from 53-6623 and 53-6972.

| ☐ MM53-6695 | 36-35 | F-84F | .. | preserved, marked as '69623' | 5-98 |

SIGONELLA
On the US Navy base of Sigonella a former Villafrance dump and CFE Starfighter arrived on 28th June 1996 for use by the local fire brigade. Stored US Navy A-7E 158830 went to Cerbaiola. The two stored S-2 Trackers, MM133212/41-27 (183) and MM148300/41-37 (739), have not been seen since April 1986.

☐	AA-...	F/A-18	..	dumped, ex US Navy	7-93
☐ 159361		CT-39A	306-65	stored, ex US Navy	9-96
☐ MM6535	3-40	F-104G	6535	instructional, ex Villafranca	9-96

SOLBIATE OLONA
Solbiate Olona can be found along the Milano to Gallarate E62 highway. Along this highway, on the northern side are some army barracks with a preserved AB204.

| ☐ MM80398 | EI-244 | AB204B | 3152 | stored, ex Salerno | 10-95 |

SORA
Sora is located north east of Frosinone where in 1991 two AB47s were noted with a local scrapdealer.

| ☐ MM80136 | SE-48 | AB47J | 1035 | stored | 12-91 |
| ☐ MM80230 | SE-45 | AB47J | 1131 | stored | 12-91 |

SULMONA
Preserved on a square in this town is a former Guidonia Texan.

| ☐ MM54141 | RR-59 | T-6H-2M | .. | preserved, Guidonia | 5-92 |

TÁRANTO
When entering the city over the big bridge on road number 172, the large navy barracks can be found at the end of the bridge. Although the base belongs to the navy, the gate is guarded by an MB326. The navy AB204AS has not been seen for some time, but may be well preserved inside the barracks.

☐ MM6275	2-30	G91R/1	41	preserved, ex Treviso	9-96
☐ MM54152	89	MB326	6157	preserved	9-96
☐ MM80363	3-02	AB204AS	3082	preserved, ex Catania	10-88

TORINO
Town: It has been reported that the Politecnico Institute at Torino acquired a former NVA MiG-21 in mid 1996.

The serial was reported as 779. This raises some questions as there are two candidates with this serial, of which the MiG-21SPS is still at Berlin and the MiG-21MF (as 2346) was still at Rothenburg in August 1996.

❏ 779		MiG-21	..	preserved, ex NVA, see note	96

The Fiat Engineering facility in town should have a preserved G91R.

❏ MM....	2-13	G91R	..	preserved	—

The technical institute Carlo Grassi at the Via Reiss-Romoli 305 has in the courtyard an MB326.

❏ MM54177	53	MB326	..	instructional	8-96

Airfield – Caselle: Preserved with the AMI facility is Sabre MM19534. Near the Aeritalia (or Alenia as it is now called) facility on the south side of the airfield was the dumped fuselage of a Fiat G91Y marked only 'Y-12'. It has been suggested that this airframe was damaged on the production line and that a replacement aircraft was built for the AMI. The wreck was last noted in 1989. The preserved Flying Boxcar MM51-8121 was last noted in 1988 and moved on to Castrette. The registration of the C-47 is incorrect as it belongs to a Curtiss from 1929.

❏ MM6311	3	G91R/1A	175	preserved, with Alenia, Frecco Tricolori c/s	92
❏ MM19534	4-74	CL-13 Sabre 4	..	preserved	3-94
❏ MM61714	RR-18	C-45F	6979	under restoration, ex Guidonia	11-91
❏ N8333 ?		C-47	..	stored, in covers	3-94
❏ ..		G91Y	..	stored, marked as 'Y-12', at Aeritalia hangar	7-89

TRAPANI - BIRGI

Trapani on Sicily is a seldom visited base. A dozen *Wrecks & Relics* aircraft can still be found at the base.

❏ MM6532	4-51	F-104G	6532	dumped	12-97
❏ MM6540	4-51	F-104G	6540	stored	12-97
❏ MM6544	37-00	F-104G	6544	preserved, ex stored	9-97
❏ MM6578	37-00	F-104G	6578	stored	12-97
❏ MM6649	4-49	RF-104G	6649	dumped	12-97
❏ MM6794	37-20	F-104S	1094	stored, wfu	12-97
❏ MM6798	37-01	F-104S	1098	stored, wfu	12-97
❏ MM6910	37-12	F-104S	1210	stored, wfu	12-97
❏ MM54244	37-32	MB326	..	stored	9-95
❏ MM54245	37-31	MB326	..	stored	9-95
❏ MM80259	SE-39	AB47J-3	2042	dumped	5-90

TRENTO

Due to reconstruction work at Milano Malpensa the Museo Caproni moved in early 1990 from Vizzola Ticino to Trento. A purpose built building was erected at the local airfield and the museum was opened in October 1992. A number of the aircraft came from the former site at Vizzola Ticino. The museum is open 09:00-13:00 and 14:00-18:00 and is only closed on Mondays. The Agusta A101 is also listed here, although the helicopter is not part of the Museo Caproni. This stored A101 belongs to Agusta, which hopes to open their own Agusta museum at Milano Malpensa airport in the near future.

❏ L-113		SM79	..	preserved, ex Lebanese AF, ex LR-AMC	5-97
❏ MM194		Fokker D.VIII	2916	preserved, frame only	5-97
❏ MM6609	3-01	F-104G	6609	preserved, outside on a pole	5-97
❏ MM70019		Breda Ba19	..	preserved, marked as 'I-ABCT'	5-97
❏ MM80358		A101G	..	stored, on airfield	5-97
❏ 11777		Ansaldo SVA5	..	preserved	5-97
❏ I-AABO	3	Macchi M20	..	preserved, ex Vizzola Ticino, ex I-BERG	5-97

Continued at page 321

GREECE

T-33A 35777 'TR-777'
decoy at Tanágra.
Chris Schmidt

C-47 Skytrain 92638
stored at Thessaloniki-Sedes.
Stephan de Bruijn

L-21B Super Cub EΣ253
preserved at Stafanavikion.
Alan Warnes/Airforces Monthly

HUNGARY

MiG-19UTI '1975' (one of *four* so marked!) at Alsónémedi.
Berry Vissers

MiG-21F-13 224 preserved at Szolnok.
Tieme Festner

MiG-21MF 8202 outside the museum at Kecel.
Berry Vissers

HUNGARY

L-29 376 at the technical school in Cespel, Budapest.
Berry Vissers

Aero 45 23 on display at Szolnok.
Otger van der Kooij

Mi-8TB 10417 preserved in front of Szolnok's control tower.
Berry Vissers

ITALY

F-84F Thunderstreak MM53-6619 (with the tail of MM52-10524 and the rudder from yet another example!) at Villorba.
Otger van der Kooij

RF-84F Thunderflash MM52-7339, marked as 'MM56-9745', at Carrara San Pelagio.
Berry Vissers

F-86K Sabre MM54-1292 '51-36' (with the tail of 55-4818) at Rivolto. *Otger van der Kooij*

ITALY

F-104G Starfighter MM6529 at Istrana. *Paul Gross*

G91PAN MM6241 '3' on display at Rivolto. *Otger van der Kooij*

G91T MM6359 '60-59' along with many others, at Améndola. *Otger van der Kooij*

ITALY

MB326 MM54374 '51-77' heading a line-up at Istrana.
Paul Gross

PD808GE1 MM61963 on the dump at Pràtica di Mare.
Otger van der Kooij

T-33A MM54-2951 'SA-951', marked as 'MM54-1951' at Améndola.
Otger van der Kooij

ITALY

C-47 Skytrain MM61800 'CR-51' and C-47B Skytrain MM61815 '14-47' at Castel Volturno.
Otger van der Kooij

P166ML1s MM61874 and MM61930 '53-35' stored at Roma-Guidonia.
Berry Vissers

HU-16A Albatross MM51-7253 '15-14' in the garden at Carrara San Pelagio.
Berry Vissers

ITALY

SM1019E MM57195 'EI-402'
at Deltaland in San Possidonio.
Berry Vissers

AB47J-3 MM80296 '33'
outside the disco
at Bachero di Cingolia.
Otger van der Kooij

AB204B MM80282 'SE-63'
at Frosinone.
Berry Vissers

C-45H Expediter MM61734
'RR-25' displayed at
Vigna di Valle.
Otger van der Kooij

ITALY

A-7E Corsair 158830 'AC-403' of VA-72 at the Cerbaiola museum near San Marino.
Otger van der Kooij

G222L MM60216 'AM-96', in Somali markings, stored at Pisa.
Paul Gross

Caravelle 6N I-DABU on its high perch at San Angelo in Villa.
Otger van der Kooij

MALTA

O-1E Bird Dog 9H-ACB stored at Luqa.
Berry Vissers

DHC-4A Caribou 5H-AAB stored at Hal Far. It later moved to Luqa and was scrapped.
Berry Vissers

C-7A Caribou N888NC at Luqa; it later flew on to the USA.
Berry Vissers

NETHERLANDS

F-16B J-260 displayed at Woensdrecht.
Erik Leijdens

NF-5A Freedom Fighter K-3020 at the technical school at Hoofddorp.
Otger van der Kooij

F-84F Thunderstreak P-224 used for camouflage training at Reek.
Paul van den Hurk

NETHERLANDS

Hunter F.4 N-138 on a pole at Leeuwarden. *Gerard Post*

US-2N Tracker 159 'V' at Schiphol's Aviodome. *Anton Homma*

Meteor T.7 I-320 on the gate at Leeuwarden. *Johan Mulder*

NETHERLANDS

F-15A Eagle 74-0083, painted as '77-0132' of the 32nd TFS at Soesterberg.
Gerard Post

Su-20 9862 in Soviet markings at Leeuwarden.
Johan Mulder

MiG-21SPS 2239 preserved at Uithuizen.
Roger Seroo

F-104G Starfighter D-8063 instructional airframe at Volkel.
Berry Vissers

NORWAY

RF-5A Freedom Fighter 113 stripped out at Ørland.
Berry Vissers

F-86F Sabre 25202 'MU-F' preserved at Ørland.
Tieme Festner

F-86K 41290 'ZK-Z' preserved by the transport unit at Gardermoen.
Stephan de Bruijn

NORWAY

CF-104s 870 and 890 at Bodo.
Tieme Festner

O-1A Bird Dog 4953 at the former Gardermoen museum.
Berry Vissers

UH-1B Iroquois 591 pole-mounted at Bardufoss.
Berry Vissers

POLAND

MiG-15UTI 197 preserved at Deblin.
Otger van der Kooij

Lim-6MR 523 preserved in the domestic area at Siemirowice.
Otger van der Kooij

MiG-21PF 2412 stored at Bydgoszcz.
Otger van der Kooij

MiG-21F-13 2307 and MiG-21U-400 1318 at Drzonów.
Lubuskie Muzeum Wojskowe Drzonów

MiG-23MF 120 at the Kraków museum.
Otger van der Kooij

Su-7BM 02 was stored at Mierzecice before being scrapped.
Chris Schmidt

Il-28U S3 at Kraków.
Otger van der Kooij

POLAND

Mi-4ME 617, with instructional code 'C' at Kraków.
Otger van der Kooij

Yak-23A 06 at the museum in Drzonów. *Lubuskie Muzeum Wojskowe Drzonów*

TS-11-100 Iskra 209 heading a line-up at Bydgoszcz.
Otger van der Kooij

POLAND

Mirage 5BA BA-21 at Deblin, the first NATO fighter preserved in Poland.
Otger van der Kooij

PA-35 Pocono N3535C, the only one built, at Widelka.
Dimitri Schmidt

An-2 5155 preserved at the aero club at Piotrków Trybunalski.
Otger van der Kooij

PORTUGAL

G91R/4 (possibly 5431) at Alverca.
Rui Ferreira

Stack of unidentified G91s in the scrapyard at Arranhó.
Rui Ferreira

T-37C 02411 displayed at Alverca.
Rui Ferreira

PORTUGAL

T-33A 1905, one of many stored at Beja.
Tony Seeley

Anonymous Alpha Jet in store at Beja.
Rui Ferreira

T-38A Talon 2602 preserved near the main gate at Beja.
Jack Bosma

PORTUGAL

Tiger Moth '102'
of the Museo do Ar
on an outing from Sintra.
Rui Ferreira

UH-19D Chickasaw
MM57-5979 at Alverca
in Portuguese colours as '9101'.
Tony Seeley

Do27A-4 3487 fuselage
stored at Ota.
Paul Gross

F-86F Sabre 5347 stored at Beja.
Tony Seeley

SLOVAK REPUBLIC

MiG-15bis 3717 preserved at Vyhne. *Otger van der Kooij*

MiG-21PF 0306 at Trencin carrying the Slovak cross. *Otger van der Kooij*

MiG-21U 2419 at Sliac with Czech roundel removed. *Otger van der Kooij*

MiG-21U 4820 heading two others at Kosice. *Otger van der Kooij*

SLOVAK REPUBLIC

Su-7BM 5316 stored at Trencin, awaiting the new museum.
Otger van der Kooij

Li-2 (possibly 2107) on show at the Dukelse Muzeum in Svidnik in Soviet markings.
Otger van der Kooij

Avia 14T 3156 at Podlipniky.
Otger van der Kooij

SPAIN

F-4C Phantom C.12-37 '12-29' at Cuatro Vientos.
Otger van der Kooij

SF-5A Freedom Fighter hulk A.9-037 at Talavera la Real.
Chris Schmidt

F-86F Sabre C.5-2 marked as 'C.5-199' '732-1', at Talavera la Real. *Dimitri Schmidt*

SPAIN

HA200 Saeta A.10B-89 'EVA.9-1' on display outside the Santa Ann Hospital, Motril.
Mike Bursell

Do27B-1 U.9-76 '407-72' stored at Cuatro Vientos.
Chris Schmidt

CASA 207 Azor T.7-16 at the El Avion cafe, San Juan de Aznalfarache.
Mike Bursell

SPAIN

Aztec 250E E.19-2 '42-51' stored at Cuatro Vientos. *Dimitri Schmidt*

AB47G-2 HE.7-11 '78-02' displayed on the military side of Cuatro Vientos. *Chris Schmidt*

AB205A HE.10A-1 '78-60' displayed in Granada. *Otger van der Kooij*

CV-990 Coronado EC-BQQ at Palma de Mallorca. *Simon Ellwood*

SWEDEN

A32A Lansen 32070 'H' in F15 markings at Söderhamn. *Berry Vissers*

J35F Draken 35475 '11' in all-red colours on the dump at Angelholm. *Paul Gross*

SAAB 35 35-0, unflown prototype, dumped at Linköping-Vårdsberg. *Otger van der Kooij*

J35FJ Draken 35541 '43' at the Linköping museum. *Alan Warnes/Airforces Monthly*

SWEDEN

AJ37 Viggen prototype 37-1 '51' at the Linköping museum.
Chris Schmidt

Hkp1 (Vertol 44) 01009 '09' at Kåremo. It is now at Visby.
Otger van der Kooij

Hkp2 Alouette II 02034 '34' preserved on the *Småland* in Göteborg harbour.
Berry Vissers

SWEDEN

FFVS J22-1 22185 'K'
with the collection at Angelholm.
Berry Vissers

SAAB 91B Safir 50068 in the
shopping centre at Linköping.
Otger van der Kooij

MiG-21SMT '11 yellow'
at Arboga.
Otger van der Kooij

SWITZERLAND

Vampire FB.6 J-1190 preserved at Sion.
Erik Jan Engelen

Twin Pioneer 3 HB-HOX awaiting restoration at Sion.
Richard Nels

Venom FB.50 J-1559 stored at Hesirau.
Otger van der Kooij

TURKEY

F-84F Thunderstreak 37040 '6-040' preserved near the gate at Bandirma.
Stephan de Bruijn

F-104G Starfighter 12344 '3-344' on a pole at Istanbul-Yesilköy.
Alan Warnes/Airforce Monthly

S-2E Tracker 152368 '2368' among others stored at Topel.
Alan Warnes/Airforces Monthly

☐ I-ACSN	MB308	5885/112	preserved, ex Vizzola Ticino, ex MM53065	5-97
☐ I-AIAE	Avia FL3	A-16	preserved, ex Vizzola Ticino	5-97
☐ I-AXAQ	G51bis	805	preserved	5-97
☐ I-BIOL	SAIMAN S202M	5	preserved, ex MM52163, ex Vizzola	5-97
☐ I-CERM	Bü131B	57	preserved, ex Vizzola Ticino, ex HB-UTN	5-97
☐ I-DISC	Caproni Ca100	3752	preserved, ex Vizzola Ticino, ex MM56237	5-97
☐ I-ELIO	SM80bis	30003	preserved, ex Vizzola Ticino, ex I-TATI	5-97
☐ I-FACT	Caproni Trento F5	..	preserved, ex I-RAIA, ex MM553	5-97
☐ I-POLO	Caproni Ca193	5736	preserved, ex Vizzola Ticino, ex MM56701	5-97
☐ I-RUMM	Jurca MJ-5	67	preserved	5-97
☐ I-WEST	Caproni Ca163	..	preserved, ex Vizzola Ticino	5-97
☐ ..	Bristol Coanda Monoplane	..	preserved, ex Vizzola Ticino	5-97
☐ ..	Caproni Ca6	..	preserved, ex Vizzola Ticino	5-97
☐ ..	Caproni Ca9	..	preserved, ex Vizzola Ticino	5-97
☐ ..	Caproni Vizzola C22J	..	preserved, replica	5-97

A second building is planned for the museum in which the remainder of the aircraft will be displayed. Currently they are all in storage in the town of Trento.

☐ MM8287	Re2000 Falco 1	..	stored, ex Vizzola Ticino	97
☐ MM54142	T-6H-2M	..	stored, ex Vizzola Ticino	97
☐ MM53-8300	F-86K	221-28	stored, ex Vizzola Ticino	97
☐ 16552	Ansaldo A-1	..	stored, ex Vizzola Ticino	97
☐ I-AEVO	SM102	..	stored, ex Vizzola Ticino, ex MM61829	97
☐ I-ANIC	Dove 6	04495	stored, ex Vizzola Ticino	97
☐ I-DIAN	Viberti Musca 1	12	stored, ex Vizzola Ticino	97
☐ I-DASI	GP2 Asiago	..	stored, ex Vizzola Ticino, ex MM30098	97
☐ I-GENI	F24C8C	2262	stored, ex Vizzola Ticino	97
☐ I-LANC	Ambrosini S2S	..	stored, ex Vizzola Ticino	97
☐ I-MANN	Libellula I	1	stored, ex Vizzola Ticino	97
☐ I-MARY	Caproni Ca113	3473	stored, ex Vizzola Ticino	97
☐ I-RENI	Vizzola II	003	stored, ex Vizzola Ticino	97
☐ I-SELI	P53 Aeroscooter	002	stored, ex Vizzola Ticino	97
☐ I-SIBI	RC-3 Seabee	331	stored, ex Vizzola Ticino, ex HB-SEA	97
☐ I-VERG	GP2 Asiago	02/GS/04	stored, ex Vizzola Ticino	97
☐ I-ZUME	GP2 Asiago	..	stored, ex Vizzola Ticino, ex MM30096	97
☐ ..	Caproni Ca53	..	stored, ex Vizzola Ticino	97

TREVISO - SAN ANGELO

Due to the close proximity of the town of Treviso and after the withdrawal of the G91R from operational service, 2 Stormo moved to Rivolto. The base was left with some MB326s and a large number of G91Rs.

☐ MM6269	2-67	G91R/1	35	stored	7-92
☐ MM6285	2-30/2-60	G91R/1	51	preserved, at officers mess	4-92
☐ MM6299	2-60	G91R/1A	163	stored	4-92
☐ MM6302	2-64	G91R/1A	166	stored, tail to Cerbaiola	9-92
☐ MM6377	2-12	G91R/1B	181	stored, no tail	7-97
☐ MM6390	2-06	G91R/1B	194	stored	8-94
☐ MM6393	2-50	G91R/1B	196	stored	8-94
☐ MM6396	2-03	G91R/1B	200	stored, tail to Roma	8-94
☐ MM6401	2-46	G91R/1B	205	stored, tail to MM6410	8-94
☐ MM6395	2-16	G91R/1B	199	stored, tail to Castel Volturno	3-92
☐ MM6406	2-43	G91R/1B	210	stored, tail to Castel Volturno	8-94
☐ MM6409	32-01	G91R/1B	213	preserved, ex stored	8-94
☐ MM6410	2-53	G91R/1B	214	preserved, with tail from MM6401	5-98
☐ MM6411	2-12	G91R/1B	215	stored	8-94

☐ MM6412	2-22	G91R/1B	216	stored	8-94
☐ MM6414	2-52	G91R/1B	218	stored, tail to Castel Volturno	8-94
☐ MM6424	2-20	G91R/1B	228	stored, tail to Castel Volturno	10-93
☐ MM54202	2-83	MB326	..	stored	8-94
☐ MM54206		MB326	..	stored	8-94
☐ MM54238	2-82	MB326	..	stored	10-93

A large number of the stored G91Rs have found new homes.

MM6265	2-34	G91R	31	stored, l/n 4-92	to Leini
MM6280	2-33	G91R/1	46	stored, dep 15-6-92	to Vigna di Valle
MM6275	2-30	G91R/1	41	stored, l/n 11-90	to Táranto
MM6276	2-75	G91R/1	42	stored, l/n 3-88	to Villorba
MM6283	2-70	G91R/1	49	stored, l/n 6-89	to Comignago
MM6287	2-31	G91R/1	53	stored, l/n 11-90	to Parma
MM6290	2-26	G91R/1A	154	stored, l/n 5-90	to Bibano
MM6291	2-24	G91R/1A	155	stored, l/n 1-90	to Mantagrava
MM6298	2-26	G91R/1A	162	stored, l/n 7-92	to Bevilacqua airstrip (Verona) ?
MM6303	2-63	G91R/1A	167	stored, l/n 3-89	to Sassoulo
MM6306	2-23	G91R/1A	170	stored, l/n 11-90	to Comignago
MM6308	2-61	G91R/1A	172	stored, l/n 10-93	to Manerbio
MM6309	2-56	G91R/1A	173	stored, l/n 8-94	to Castel Volturno
MM6311	2-62	G91R/1A	175	stored, l/n 7-92	to Torino
MM6312	2-61	G91R/1A	176	stored, l/n 11-90	to L'Aquila
MM6375	2-41	G91R/1B	179	stored, l/n 11-90	to Cesena
MM6376	2-01	G91R/1B	180	stored, l/n 4-92	to Castel Volturno
MM6381	2-41	G91R/1B	185	stored, l/n 2-92	to Rivolto
MM6382	2-20	G91R/1B	186	stored, l/n 7-92	to Milano
MM6387	2-11	G91R/1B	191	stored, l/n 9-97	to Ta'Qali
MM6388	2-42	G91R/1B	192	stored, l/n 8-94	to Roma
MM6389	2-04	G91R/1B	193	stored, l/n 8-94 (tail to Roma)	to Cerbaiola
MM6408	2-07	G91R/1B	212	stored, l/n 11-90	to Roma
MM6416	2-14	G91R/1B	220	stored, l/n 8-94	to Rivolto
MM6417	2-10	G91R/1B	221	stored, l/n 1-90	to Istrana
MM6418	2-26	G91R/1B	222	stored, l/n 11-90	to Biella Cerrione
MM6420	2-43	G91R/1B	224	stored, l/n 7-92 (tail to Milano)	to Castel Volturno

TRINO - VERCELLESE
The colours of Frecce Tricolori are highly favoured by the Italians as this 2 Stormo G91R has also been repainted in these colours. It is preserved at the Plazza dell'Aeronautica, along the Via Guiseppe Saragat.

☐ MM6274	6	G91R/1	40	preserved, Frecce Tricolori colours	9-95

UDINE
Town: Inside the large school complex of the Instituto Tecnico a Malignani is an aeronautics and engineers department. They have a number of instructional airframes which are in perfect shape and are fully functional. The real Sabre MM19680 is here (see Rivolto).

☐ MM6272	2-31	G91R/1	38	instructional, arr 19-9-85	9-96
☐ MM19680	2-37	CL-13 Sabre 4	580	instructional, ex XB826/RAF	9-96
☐ MM52792		G46	25	instructional, arr 4-8-59	9-96
☐ MM54266	12	MB326	..	instructional, arr 17-3-80	9-96
☐ MM61755	303-12	C-45	..	instructional, arr 28-5-82, ex Roma	9-96
☐ MM80798		OH-13H	1892	instructional, unmarked, ex 56-2180	9-96
☐ I-AEAM		AM-3C	6530	instructional, arr 28-7-79	9-96

Airfield – Campoformido: Preserved at the aero club of this ALE-field is a Flying Boxcar.
☐ MM52-6029 46-93 C-119G   11030   preserved   5-98

VEDELAGO
The shop in this small town just to the west of Istrana has sold its F-84F Thunderstreak to Cerbaiola. The MM53-6646 was painted as 'MM53-6634/51-30'.

VENEGONO
By late 1988 the Aermacchi factory here had acquired an MB326 for display purposes. It is mounted on a pole outside the factory.
☐ MM.....   MB326   ..   preserved, on pole   10-95

Inside the factory is a small museum, with some of the locally built aircraft.
☐ MM572   MB326E   6153   preserved, marked as 'MM571', ex Capua   5-98
☐ MM92166   MC205   ..   preserved   10-95
☐ I-BIOI   MB308   122   preserved, ex MM53075   10-95
☐ I-FABR   MB308   001   preserved   10-95
☐ I-MABD   Macchi AL60   ..   preserved   10-95
☐ ..   Bazzochi EB-4   ..   preserved   10-95

VENEZIA - TESSERA
The engineless stored United Arab Emirates G222 301 (4006) was scrapped in the early 1990s.

VERGIATE
The stored and derelict uncompleted SF-260 airframes and the C-119s mentioned in the first edition, comprising C-119Js MM51-8144/46-55 (147), MM51-8158/46-62 (161), MM52-5849/46-59 (11008), MM52-5851/46-60 (11010), EC-119Js MM52-5884/46-63 (11051), MM52- 5896/46-64 (11063) and C-119G MM52-6030/46-95 (10996), have all gone. Stored C-45 MM61702/RB-20 has also been scrapped. These have been replaced by modern hardware in the form of five Iraqi AB212s which were never delivered. These AB212s, 1 (5183), 2, 3 (5206), 4 (5207) and 5 (5208), were seen in 1995 and bought by the Italian Army in late 1997 for whom they will be made airworthy again. Noted at the same time were two Zambia SF-260s in long term storage.
☐ AF536   SF-260TP   ..   stored, Zambia AF   11-95
☐ AF538   SF-260TP   ..   stored, Zambia AF   11-95

VERONA
Town: Preserved somewhere in town could still be an P148.
☐ MM53545   9-75 P148   121   preserved   12-88

Airfield – Boscomantico: The ALE airfield of Verona has a number of stored AB204s
☐ MM80285   EI-202 AB204B   3031   stored, unmarked   3-96
☐ MM80286   EI-203 AB204B   3036   stored, unmarked   3-96
☐ MM80307   EI-207 AB204B   3050   stored, unmarked   3-96
☐ MM80312   EI-212 AB204B   3063   stored, unmarked   3-96
☐ MM80315   EI-215 AB204B   3076   stored, unmarked   3-96
☐ MM80383   EI-229 AB204B   3105   stored, unmarked   3-96
☐ MM80389   EI-235 AB204B   3119   stored, unmarked   3-96
☐ MM80392   EI-238 AB204B   3122   stored, unmarked   3-96
☐ MM80396   EI-242 AB204B   3129   stored, unmarked   3-96
☐ MM.....   55   AB204B   ..   stored, unmarked   3-96

VESTONE
Along the road from Vestone centre to Lavenone is an MB326 preserved on a pole.
- ❏ MM54219   6-45 MB326   ..   preserved   4-95

VICENZA
The headquarters of the southern NATO forces is located at Vicenza airfield. It has been reported that the preserved aircraft are with the Museo Storico dell'Aeroporto Tommasa dal Molin and not visible from outside. Also the tail of MB326E MM54384 is preserved.
- ❏ MM6292   2-22   G91R/1   156   preserved   3-98
- ❏ MM6577 3-01/3-31 F-104G   6577   preserved, unmarked   3-98
- ❏ MM80357   31-10   AB204B   3085   preserved   3-98

VIGNA DI VALLE
The AMI museum (Museo Storico Aeronautica Militare) is located at the historic seaplane base at the Lago di Bracciano, some 25km north of Roma. The museum was closed on November 20th 1996 for major rework. They hope to open their gates again in the fall of 1998. The Super Sabre 54-2290 was on delivery to Turkey but did not make it further then Sigonella where it stopped due to technical problems. In went to <u>Aviano</u> in 1996 and became a gate guard there. Re2001 MM8071 has gone to <u>Guidonia</u>, whilst Sabre MM19666/4-11 (566) is now in the RAF museum at Hendon as XB812. G80-3B MM53882 went to <u>Pratica di Mare</u>, while airworthy MC200 MM8307 (marked as 'MM7707/359-8') is now at <u>Istrana</u>. Below is the situation as of late 1996. This will all have changed when the museum re-opens in 1998.

Outside:
- ❏ 29543   0   J-29F   29543   preserved, ex Swedish AF   9-96
- ❏ MM580   RS-11   G91Y   2002   preserved   12-96
- ❏ MM583   RS-07   G222   4002   preserved   12-96
- ❏ MM6405   2-05   G91R/1B   209   preserved, arr summer 1992   9-96
- ❏ MM6413   G91R/1B   217   preserved, special c/s   9-96
- ❏ MM6501   3-11   F-104G   9998   preserved   12-96
- ❏ MM6599   53-21   F-104G   6599   preserved, ex Cámeri   12-96
- ❏ MM53886   19   G82   3   preserved   9-96
- ❏ MM61890   P166ML1   398   preserved   9-96
- ❏ MM50-179   15-5   HU-16A   G-70   preserved   12-96
- ❏ MM51-9249 51-88   T-33A   7033   preserved, ex Roma Ciampino   12-96
- ❏ N7486C   PV-2   15-1248   preserved, ex BuNo37282   9-96

Hall 1:
- ❏ L-127   Lohner L-1   ..   preserved   9-96
- ❏ 19309   Hanriot HD.1   515   preserved,   9-96
- ❏ 23174   Caproni Ca33   ..   preserved, marked as '4166'   9-96
- ❏ 11721   1   Ansaldo SVA5   ..   preserved   9-96
- ❏ ..   Blériot-SIT XI-2 ..   preserved, marked as 'BL246', is maybe BL160 9-96

Hall 2:
- ❏ C.1-328   HA132L   328   preserved, marked as 'MM4666/92', ex Span AF 9-96
- ❏ MM76   M39-II   5   preserved   9-96
- ❏ MM105   10   M67   3   preserved   9-96
- ❏ MM130bis   Fiat C29   2   preserved   9-96
- ❏ MM181   MC72   5   preserved   9-96
- ❏ MM487   1   Caproni Campini CC1 4850 preserved, unmarked   9-96
- ❏ MM1208   94-6   Ansaldo AC2   ..   preserved   9-96
- ❏ MM6250   9   G91PAN   16   preserved, Frecce Tricolori c/s   9-96

	Reg	Code	Type	Serial	Status	Date
☐	MM27050	ORB-23	IMAM Ro43	..	preserved	9-96
☐	MM45425		Z506S	..	preserved, marked as 'MM45442/84-4'	9-96
☐	I-AEEU		L-5 Sentinel	..	preserved, ex MM52848	9-96
☐	I-DASM		FN305	766	preserved, ex MM52757 and marked as such	9-96

Hall 3:

	Reg	Code	Type	Serial	Status	Date
☐	L-112		SM79		preserved, as 'MM24327/278-2', ex Lebanese AF	9-96
☐	MM558		Ambrosini S7	2	preserved, unmarked	9-96
☐	MM4084		Spitfire IX	CBAF-IX-1780	preserved, ex MK805 and marked as such	9-96
☐	MM4323		P-51D	..	preserved, unmarked	9-96
☐	MM4653	51-19	P-47D	..	preserved, ex Lecce, ex 44-89746	9-96
☐	MM9667		MC202T-AS	..	preserved, marked as 'MM7844/91-3'	9-96
☐	MM53276		G59-4B	61	preserved	9-96
☐	MM54097	RR-67	T-6G	..	preserved	9-96
☐	MM61187		SM82PW	..	preserved, marked as 'MM61850/14'	9-96
☐	MM61734	RR-25	C-45H	..	preserved	9-96
☐	MM61804	142-5	G212CA	19	preserved	9-96
☐	MM80113	12	AB47G-2	196	preserved	9-96
☐	MM80187	SE-38	AB47J	1100	preserved	9-96
☐	MM100042	2	CVV8	001	preserved, glider	9-96
☐	MM136556	41-6	S-2A	465	preserved	9-96
☐	I-AELM		G46-4A	192	preserved, ex MM53286 and marked as such	9-96

Hall 4:

	Reg	Code	Type	Serial	Status	Date
☐	J-1107		Vampire FB.6	616	preserved, ex Swiss AF	9-96
☐	MM554	RS-10	MB323	1	preserved, upstairs, ex I-RAIE	9-96
☐	MM555		P150	1/170	preserved, upstairs	9-96
☐	MM561		Aerfer Sagittario II	2	preserved, upstairs	9-96
☐	MM569		Aerfer Ariete	2	preserved, upstairs	9-96
☐	MM6085		Vampire FB.52A	..	preserved, unmarked, composite	9-96
☐	MM6152		Vampire NF.54	12094	preserved	9-96
☐	MM6280	2-33	G91R/1	46	preserved, arr summer 1992	9-96
☐	MM7001	RS-01	Tornado	PS14	preserved, ex MM588	9-96
☐	MM9546	97-2	MC205-V	..	preserved, ex 'MM9345'	9-96
☐	MM19792	13-1	CL-13 Sabre 4	692	preserved, ex XB915/RAF	9-96
☐	MM53526	RR-80	G59-4A	131	stored, behind Hall 4	9-96
☐	MM53778	RM-41	G59-4B	185	preserved	9-96
☐	MM53888	RS-21	G82	5	preserved, upstairs	9-96
☐	MM54389		MB326E	..	preserved, ex Istrana	9-96
☐	MM54390	RS-33	MB326K	6477	preserved, ex I-IVOA, ex I-AMKK	9-96
☐	MM61933	53-34	P166ML1	443	preserved, arr 4-92	9-96
☐	MM91001		H359B Courier	529	preserved, ex N4178R	9-96
☐	MM100028		CVV6	2-131	preserved, glider	9-96
☐	MM51-11049	51-18	F-84G	..	preserved	9-96
☐	MM52-7458	3-05	RF-84F	..	preserved	9-96
☐	MM53-5594	9-35	RT-33A	8933	preserved	9-96
☐	MM53-6892	36-38	F-84F	..	preserved	9-96
☐	MM55-4868	51-62	F-86K	221-108	preserved	9-96
☐	I-ADOD		Avia FL3	A34-28-28	preserved	9-96
☐	I-AELY		M416	1059	preserved, ex MM53762	9-96
☐	I-AEDA		SM56	5611	preserved	9-96
☐	I-BFFI		Fiat G5bis	2	preserved, ex MM290	9-96
☐	I-BIGI	37	SIAI 3-VI	001	preserved	9-96
☐	I-FIAT		G49-2	2	preserved, upstairs, ex MM566	9-96
☐	I-GIAB		M416	1042	stored, behind Hall 4, ex MM53244	9-96

ITALY - 326

☐ I-GORI	MB308	5878/105	preserved, ex MM53058		9-96
☐ I-GTAB	Caproni Ca100	1	preserved, ex Vizzola		9-96
☐ ..	Blériot XI	..	preserved, replica		9-96
☐ ..	Curtiss A1	..	preserved, replica		9-96
☐ ..	SPAD VII	..	preserved, replica		9-96

The museum should have a number of additional aircraft, which have not been seen recently.

☐ MM6415		G91R/1B	219	stored	7-92
☐ MM53283		G46-4A	189	stored	—
☐ MM53292		G46-4A	198	stored	—
☐ MM53885		G82	2	stored	—
☐ MM54290	RS-20	MB326K	6403	stored	—
☐ MM80005	140-6	P136F	110	stored	9-92
☐ MM80078	140-15	P136L-1	204	stored	9-92
☐ MM80083		P136L-1	213	stored, wreck	—
☐ MM51-8128		C-119J	131	stored	—
☐ G-FIST		Fi156C-3	5802	stored, ex D-EDEC, ex I-FAGG, ex MM12822	—
☐ I-AECG		CCV6	..	stored, glider, ex MM100007	—
☐ I-AEEL		L-5 Sentinel	..	stored, wreck, ex MM52882	—
☐ I-AEPF		M416	6005/61	stored, wreck, ex MM53457	—
☐ I-AELI		M416	1058	stored, ex MM53761	—
☐ I-AELS		M416	6013/69	stored, wreck, ex MM53444	—
☐ I-ECIN		A102	004	stored, ex MM80201	7-92
☐ I-DONT		MB308	2	stored	—
☐ I-FAMA		UC-61K	1055	stored, ex MM56698, ex KK437, ex 44-83094	—
☐ I-FOLK		MS30L	..	stored, glider	—
☐ I-MIBO		Milita MB3	..	stored	—
☐ I-SARD		SAIMAN S202M	508	stored, ex MM51497	—
☐ I-REDI		P53 Aeroscooter	1	stored	—

VIGNALE
A priest here (in the Asti area) has a T-6D Texan in his garden.

☐ MM53043	RR-57	T-6D	..	preserved, ex EX361/RAF, ex 41-33334	1-85

VILLAFRANCA DI VERONA
Town: The administration barracks of Villafranca airbase are in the nearby village of Villafranca di Verona. Preserved here at the gate is an F-104G on a pole.

☐ MM6547	3-01	F-104G	6547	preserved	10-95

Airfield: Some 17 CFE Starfighters were stored here for some years in a special compound. By late 1995 some of them had moved on. F-104G MM6520 (CFE code 4) moved to San Possidonio, together with RF-104G MM6647 (CFE code 7). F-104G MM6535 (CFE code 10) went to Sigonella. F-104G MM6550/3-21 (6550) arrived 2nd October 1995 from Rimini and was scrapped soon afterwards. Also MM6529 (6529, ex Istrana) was reported to have come here for scrapping (in late 1996), but this was incorrect. The aircraft is still at Istrana. Dumped ex Ghedi F-104 MM6533 has returned to Ghedi.

☐ MM6502		F-104G	6502	stored, CFE a/c, code 5	5-98
☐ MM6513	3-23	F-104G	6513	preserved, with 132 Gruppo	5-98
☐ MM6514	3-54	RF-104G	6514	stored, CFE a/c, code 3	7-97
☐ MM6518	3-37	F-104G	6519	stored, ex Istrana	3-98
☐ MM6525	3-01/3-30	F-104G	6525	preserved	9-96
☐ MM6558	3-53	F-104G	6558	stored, CFE a/c, code 1	5-98
☐ MM6563	3-43	F-104G	6563	stored, CFE a/c, code 16	3-98

☐ MM6567	3-47	F-104G	6567	stored, ex Prática di Mare	3-98
☐ MM6568		F-104G	6568	stored, fuselage only, CFE a/c, code 6	3-98
☐ MM6577		F-104G	6557	dumped, burned	10-95
☐ MM6579	28	F-104G	6579	preserved, with 28 Gruppo, special c/s	3-98
☐ MM6588	3-07	F-104G	6588	stored, CFE a/c, code 11	3-98
☐ MM6593	3-05	F-104G	6593	stored, CFE a/c, code 14	5-98
☐ MM6595	3-36	F-104G	6595	stored, CFE a/c, code 9	3-98
☐ MM6598	3-12	F-104G	6598	stored, CFE a/c, code 2	5-98
☐ MM6603	3-13	F-104G	6603	stored, CFE a/c, code 17	3-98
☐ MM6635	3-30	RF-104G	6635	stored, CFE a/c, code 13	5-98
☐ MM6637	3-23	RF-104G	6637	stored, CFE a/c, code 8	3-98
☐ MM6643	3-31	RF-104G	6643	stored, CFE a/c, code 15	5-98
☐ MM6651	3-47	RF-104G	6651	stored, CFE a/c, code 12	3-98
☐ MM54239	39	MB326	6362	stored	5-98
☐ MM54246	46	MB326	..	stored	5-98
☐ MM51-9141	3-141	T-33A	6925	stored	3-98
☐ MM51-9145	3-145	T-33A	6929	preserved	3-98
☐ MM52-7403	3-28	RF-84F	..	preserved	3-98
☐ MM52-7466	3-06	RF-84F	..	stored	5-98
☐ MM55-3030	3-330	T-33A	9571	stored	3-98

VILLORBA CASTRETTE
The large number of aircraft mentioned in the first edition has been reduced dramatically. Only a handful have survived of which the Thunderflash, Thunderstreak and Sabre are preserved as eye-catchers along road number 13. Of these the F-84F is not fitted with its original tail and even this tail does not have its original rudder. As part of the serial is on this rudder, this will lead to complications. The tail is from 52-10524, therefore on the right side it reads 30524 (the 3 is on the rudder) and on the left side 20523 (the 3 is on the rudder). More complicated than this it will not get! The Harvard has recently arrived from nearby Lancenigo. The bare and unmarked frame of the M416 (Fokker S11) was preserved next to the fighters, but it collapsed and was dumped by September 1996.

☐ MM53849	SC-78	T-6H-4M	..	stored, ex Lancenigo	9-96
☐ MM51-8121	46-50	C-119J	124	stored, ex Torino	9-96
☐ MM53-7585	3-07	RF-84F	..	preserved, tail from 52-6972	9-96
☐ MM53-6619	50-19	F-84F	..	preserved, tail from 52-10524	9-96
☐ MM55-4869	51-56	F-86K	221-109	preserved, tail from 54-1288	9-96
☐ ..		M416	..	dumped, ex preserved, unmarked	9-96

The fate of most of the other aircraft here is not known. Seen over the years were:

MM6276	2-75	G91R/1	42	stored, l/n 10-93	gone
MM53657	SC-61	T-6G	..	stored, l/n 10-93	gone
MM53792	SC-65	T-6H-4M	..	stored, l/n 1984	gone
MM53797	SC-67	T-6H-4M	..	stored, l/n 10-81	gone
MM53820		T-6H-4M	..	stored, l/n 1985	to Coazze
MM53828	SL-69	T-6H-4M	CCF4-386	stored, l/n 1984	returned to service as MM604
MM53832	SC-72	T-6H-4M	..	stored, l/n 1-84	gone
MM53841	SL-9	T-6H-4M	..	stored, l/n 1984	gone
MM53844	SL-30	T-6H-4M	CCF4-407	stored, ex 51-17225, l/n 1-87	gone
MM53847	SC-77	T-6H-4M	CCF4-410	stored, ex 51-17212, l/n 10-93	gone
MM54144	SL-51	T-6H-2M	..	stored, l/n 1984	gone
MM80812		OH-13H	2614	stored, ex Bracciano, l/n 7-92	to San Possidonio
MM80331	15-26	AB204B	3094	stored, l/n 7-88	gone
MM52-7399	3-16	RF-84F	..	stored, tail from other a/c, l/n 1981	gone
LR-LWD		SE210-6R	114	stored, l/n 5-90, ex OY-SBZ, ex N1020U	gone

VITERBO

As can be expected of the main Italian Army base, there is a large number of non-operational airframes here. Of these, the P166 and two T-6s were unmarked except for their strange codes. Note that the base is split up in two parts, both parts being used as airfields. The preserved P166 MM61910/SP-41 moved on to Guidonia.

☐ MM53818	303-31	T-6H-4M	..	preserved	9-92
☐ MM54139		T-6H-2M	..	preserved, marked as '93141/TA-141'	9-92
☐ MM.....	3-2	T-6H	..	preserved, at barracks on north side	2-96
☐ MM.....	3-4	T-6H	..	preserved, at barracks on north side	2-96
☐ MM.....	4-4	P166M	..	preserved, at barracks on north side	2-96
☐ MM57231	EI-438	SM1019E	1-039	stored	12-96
☐ MM57237	EI-444	SM1019E	1-045	stored	2-96
☐ MM57254	EI-461	SM1019E	1-062	stored	12-96
☐ MM80273	33	AB47G-3B	1527	preserved, at south gate, marked as 'MM80573'	5-97
☐ MM80320	EI-220	AB204B	3086	preserved, on south side	2-96
☐ MM80322	EI-222	AB204B	3091	preserved, at north gate	2-96
☐ MM80345		AB47G-3B-1	1567	preserved, on south side	7-97
☐ MM80394	EI-240	AB204B	3125	preserved	8-93
☐ MM80492		AB47G-3B-1	1638	preserved, at gate	2-98
☐ MM80586	EI-525	AB206C-1	9011	instructional	5-85
☐ MM80611	EI-550	AB206C-1	9043	stored, wreck in hangar, crashed 18-11-95	2-96
☐ MM52-2378		L-18C	18-1978	preserved, at south gate, marked as 'I-EIMU'	2-98
☐ MM52-2415	EI-85	L-18C	18-2015	stored, in hangar, ex I-EIRO	2-96
☐ MM52-2425	EI-94	L-18C	18-2025	stored, in hangar, ex I-EITE	7-97
☐ MM53-7736	EI-117	L-21B	18-3336	stored, in hangar, ex I-EIHG	2-96
☐ MM54-2552	EI-252	L-21B	18-3952	stored, in hangar	2-96
☐ MM61-2956	EI-21	O-1E	305M-0002	preserved, at gate	8-94
☐ MM61-2958	EI-19	O-1E	305M-0004	stored	7-94
☐ MM61-2964	EI-27	O-1E	305M-0010	stored	12-96
☐ MM61-2969	EI-16	O-1E	305M-0015	stored, ex Livorno	8-89
☐ MM61-2977	EI-28	O-1E	305M-0023	stored	7-94
☐ MM61-2989	EI-43	O-1E	305M-0037	stored	7-94
☐ MM61-3000	EI-14	O-1E	305M-0050	stored	12-96

VIZZOLA TICINO

The Caproni Museum here was situated on the western side of Milano Malpensa airport in a hangar within the Caproni factory. By 1992 a new site was opened at Trento and nearly all of the aircraft have moved there (see Trento for the ex Vizzola Ticino aircraft). An exception is Caproni I-GTAB which went to Vigna di Valle.

MALTA

The *Wrecks & Relics* situation in Malta has changed over the years. The demise of Newcal Caribou's overhaul at Luqa means that the large number of stored propliners has gone. Only a few remain there. Luckily this has been compensated with the opening of the Malta Aviation Museum at Ta' Qali. This museum was founded in 1994 and opened its gates to the public in April 1996.

BIRZEBUGGIA
The scrapyard of Sylvian is on top of a hill (along the Tal – Brolli Road), at the north east side of Birzebuggia. An unknown DC-3, marked '140', was last noted in May 1994 and has been scrapped.
❏ N491F　　　　　　Aero Commander　　..　　　　　stored, poor state　　　　　　　　　　　　　2-98

HAL FAR
Based here was the Fire & Safety Training School. This school has closed down and most of the aircraft have gone. Beech C-45H N945F and C-47A C-FITH moved to Ta' Qali, while C-47A N535M (20558), C-47B N565 (14955/26400) and DHC-4s N3262W (244, ex Luqa) and N3262X (11, ex Luqa) have been scrapped.
❏ N84897　　　　　　DHC-4　　　　　283　　　　stored, fuselage only, ex Luqa　　　　　　10-97

KIRKOP
In this village close to the airfield of Luqa a L-1049, which arrived in Malta on 16th February 1968, was used as restaurant since 1973. On 30th January 1997 the aircraft was badly damaged by arson. The remains (tail, wings and engines) of CS-TLC (4618, marked as '5T-TAF') were transported to the museum at Ta' Qali.

LUQA
The Armed Forces of Malta have their main base here. A number of their aircraft are in temporary storage here, mostly awaiting spares or heavy maintenance. Spares may be taken out on a rotational basis of operational aircraft. Only one Bird Dog is really out of service.
❏ 9H-ACB　　　　　O-1E　　　　　　305M-0029　stored, for spares, wfu 5-5-93, ex MM61-2983　　98

Newcal Aviation (NCA) it based on the former Safi airfield which is adjacent to Luqa airfield. NCA specialised the overhaul of DHC-4/C-7 Caribous. This work moved to the United States. Stored DHC4/C-7s N51NC (261, ex T.9-04), N54NC (49, ex T.9-18), N57NC (72, ex T.9-22), N96NC (238, ex 5B-CIN), N300NC (303, ex 5Y-BET), N555NC (15, ex T.9-13), N666NC (27, ex T.9-26) and N888NC (58, ex T.9-28) all left for the United States. C-47A N9050T (12472) moved to Fleet in the UK during 1992, while DHC-4s N3262W (ex 5H-AAB), N3262X (ex 5H-AAC) and N84897 moved to Hal Far. The stored DHC-3's, N18NC (132, ex EA51/Ethiopian AF), N24NC (350, ex EA53), N28NC (349, ex EA52) and N44NC (347, ex EA50), were all shipped to New Jersey, USA. NCA still does maintenance on types like CASA 212s, DHC-6s and smaller aircraft. B720 9H-AAM was burnt by vandals. The aircraft has been donated to the Ta' Qali museum, but only its tail section can be salvaged. This will be restored in Air Malta colours and displayed at Ta' Qali.
❏ T9-ABA　　　　　CASA 212-200　　302　　　　stored, ex F-GHOX, arr 11-6-96　　　　　　2-98
❏ T9-ABC　　　　　C-47B　　　　　　16187/32925　stored, ex 3C-JJN　　　　　　　　　　2-98
❏ UR-67060　　　　L-410UVP　　　　810636　　stored, unmarked, arr 27-11-97　　　　　　2-98
❏ 5A-DMJ　　　　　CASA 212-100　　133　　　　stored, ex 9H-AAT　　　　　　　　　　　2-98
❏ 9H-AAM　　　　　B720-040B　　　　18378　　dumped　　　　　　　　　　　　　　　2-98

TA'QALI
Raymond Polidano and a group of volunteers were rebuilding Spitfire EN199 here. The aircraft was damaged in

a gale on 23rd December 1946 and was SOC on 30th January 1947. It had joined the Malta Aviation Museum, which was founded on 1st January 1994, by 1997. The museum can be found in Hut 161, Crafts Village and is open daily (except Good Friday and Christmas Day) between 09:00 and 17:00. Most of the aircraft are in the restoration process, but are all in the public viewing area (except the Cessna).

❏ MM6387	2-11	G91R/1B	191	preserved, marked as 'MM6377', ex Treviso	2-98
❏ Z3055		Hurricane IIa	..	preserved	2-98
❏ Z3571		Hurricane IIa	..	preserved, frame and tail only	2-98
❏ EN199	R-B	Spitfire LF.IXc	3677	preserved	2-98
❏ WZ550	N-F	Vampire T.11	15109	preserved, ex Booker, arr 25-10-96	2-98
❏ C-FITH		C-47A	20228	preserved, ex Hal Far, ex Luqa	2-98
❏ N495F		C-45H	AF-888	preserved, ex Hal Far, ex N114G	2-98
❏ N3548D		Ce170		stored	2-98
❏ ..		HM-14	..	stored	2-98

VALETTA

The National War Museum at St Elmo's fort still have their Sea Gladiator on display.

❏ N5520	Sea Gladiator	..	preserved, fuselage only	2-98
❏ BR108	Spitfire Mk.5b	..	preserved, cockpit and engine only	2-98

NETHERLANDS

The Dutch Air Force currently operates only one fighter type, the days of Starfighters and Freedom Fighters have gone. A handful of these are still to be found in the Netherlands, although their numbers are decreasing. Sadly a number of aircraft and helicopters were scrapped at Den Dolder. For the Air Force it was not cost effective to maintain these aircraft. Included in this useless scrapping was an unique Fokker S-14 Machtrainer. On the bright side, more and more people in the Netherlands take an interest in aviation and want to have their own aircraft. The collection at Baarlo is a good example of this.

AALSMEERDERBRUG
Inside an aviation shop at the Molenweg 249, the cockpit of a C-47 is preserved. It is in use as a demonstrator for (home) flight simulators.
- G-BPMP   C-47A   10073   preserved, cockpit only, ex N54607   11-97

ABBENES
A group of volunteers is restoring the Aviodome's Fokker S-11 locally. The aircraft was in bad condition after its long spell outside the Aviodome and later at Hoofddorp.
- PH-UET   S-11   6198   under restoration, marked as 'E-42'   9-95

ARCEN
Along the road from Well into Arcen is the Peeters' nursery. Parked outside here was former NVA Mi-2 9451. The aircraft was sold early 1997 and moved on to Hoornsterzwaag.

ASSEN
The aircraft at the car museum all returned to Gilze Rijen. The former exhibits were S-11 E-24, F-84G K-6, Hunter F.4 N-122, F-84F P-248, L-18C R-87 and U-6A S-9.

BAAK
Three former Czech Air Force MiG-21s arrived in The Netherlands in 1996. One of them is parked along the main road (N314) between Zutphen and Doetinchem in the village of Baak.
- 0812   MiG-21PF   760812   preserved, ex Sobeslav, ex Czech AF   1-97

BAARLO
The local BMW dealer in the Napoleonsbaan acquired an F-104G from Pferdsfeld for display. He noted that the German Starfighter was built by Fokker and decided to paint it in Royal Netherlands Air Force colours. The state of the aircraft was far from ideal and a second F-104 was bought from Belgium. This aircraft was painted as D-8030. The other German Starfighters were used to rebuild the Belgian aircraft. Ex NVA MiG-23BN 2050 moved on to Hoek van Holland and MiG-21SPS 2245 to Zwannenburg.

FX-45	F-104G	9088	preserved, marked as 'D-8030', ex BelgAF	4-98
64-0745	GF-4C	1028	stored, ex Seifertshofen, ex USAF	11-97
2059	MiG-23UB	A1038221	stored, ex Seifertshofen, ex Laage, ex 107/NVA	4-98
2239	F-104G	7117	stored, arrived 13-10-95, ex Hopsten, ex WGAF	4-98
2351	F-104G	8030	stored, ex Pferdsfeld, ex WGAF	4-98
2376	F-104G	8075	stored, marked as 'BB-371', ex Erding, ex Jever	4-98
2392	F-104G	8102	stored, ex Erding, ex Jever	4-98
2411	MiG-21US	06685139	stored, ex Seifertshofen, ex 248/NVA	4-98

NETHERLANDS - 332

❑ 2813	TF-104G	5943	expected, ex Seifertshofen, ex Erding, ex WGAF	—
❑ 2822	TF-104G	5952	stored, ex Seifertshofen, ex Erding, ex WGAF	4-98

BADHOEVEDORP
The Aviodome (Schiphol) has a storage site in this town. Some of the museum's smaller aircraft are held here.

BEST
The Bevrijdende Vleugels museum (Liberating Wings) was relocated to Best from Veghel in 1997. The L-4 Cub and Spitfire are currently stored, but will be put on display in late 1998. The DC-3 is still airworthy and flying around in the Netherlands, but it will retire in the near future and will be displayed in the museum. Also expected is a Fieseler Storch.

❑ PH-NLA	L-4J	12472	stored, marked as '32-S', ex Veghel, ex OO-AVL	98
❑ N32MS	DC-3C	4978	expected, still airworthy, marked as '290321'	—
❑ ..	Spitfire LF.IXc	..	stored, replica, marked as 'P7981', ex Veghel	98
❑ ..	Fi156	..	expected	—

BREDA
The Koninklijke Militaire Academie had lost its gate guarding F-104G in 1993, D-8245 moved to <u>Soesterberg</u>.

BUDEL
The Sabre is long term inmate at the German barracks between Weert and Eindhoven. It was joined by a Starfighter from Blomberg in early 1991.

❑ JC-240	CL-13B Sabre 6	1704	preserved, on pole, ex WGAF	12-96
❑ D-8282	F-104G	8282	preserved, ex Blomberg, ex Gilze Rijen	7-97

CAPELLE AAN DE IJSSEL
Preserved in a shop-window at the Dump 2000 store in the Alexandrium shopping centre is a former Polish Air Force Mi-2.

❑ SP-FSO	Mi-2URN	562640112	preserved, ex 2640/Polish AF	11-97

CRAILO
The NBC training Hunter F.4 N-129 was last noted in June 1996. It moved to <u>Vijfeiken</u> shortly afterwards. The ex Woensdrecht NF-5A is still present inside the barracks.

❑ K-3044	NF-5A	3044	instructional, crashed 4-5-90, arr 5-91	11-96

DEELEN
Preserved on the gate at the airfield side of Deelen airbase is a Starfighter which used to be disguised as a 'Bolfighter' (a crossbreed of a Bölkow 105 and a Starfighter). Preserved Hunter F.4 N-138 moved to <u>Leeuwarden</u> in 1991, while decoy F-84F P-243 was scrapped in the early 1990s.

❑ D-8098	F-104G	8098	preserved, at base, ex 'Bolfighter'	6-96
❑ D-8244	F-104G	8244	stored, ex Schaarsbergen, 'Gorbi Jet' markings	3-98

DE KOOY
Preserved inside the traditiekamer (traditions room) at this Lynx base are two former Royal Navy airframes in Dutch colours.

NETHERLANDS - 333

❏ WV828	Sea Hawk FGA.6	7474	preserved, marked as '118/D', ex Valkenburg	9-97
❏ XT795	Wasp HAS.1	F9677	preserved, marked as '235/K', ex Fleetlands	9-97

DELFT
The Technical University of Delft has a large collection of airframes and better parts of airframes. Beside the aircraft mentioned below there are many smaller pieces of aircraft here, ranging from gliders to parts of F28s.

❏ CC-054	Do27A	231	instructional, as 'OC-054', fuselage, ex G Army	6-92
❏ D-8048	F-104G	8048	instructional, ex Ypenburg, arr 1989	6-92
❏ D-8114	F-104G	8114	instructional, ex Ypenburg, arr 18-6-84	6-92
❏ 207	SP-2H	7257	instructional, nose only	6-92
❏ 5N-ACQ	Do28A-1	3006	instructional, ex PH-ACU, ex D-IHYL	6-92
❏ OO-SID	RC680	357-46	instructional, ex Schiphol, ex NC6846S	3-92
❏ PH-LIZ	G164A Ag-Cat	680	instructional, ex N67072, ex 5Y-API	6-92
❏ PH-NIV	F27-100	10101	instructional, cockpit section only	—
❏ PH-STA	Ce150M	1257	instructional, forward fuselage only	—
❏ PH-HGV	Ce171L	0838	instructional, forward fuselage only	—
❏ PH-UEA	SAAB 91A	91125	instructional, ex SE-BFU	6-92
❏ PH-UEU	S-11	6194	instructional, forward fuselage only	—
❏ PH-VRS	Beagle B-121 Pup	B-119	instructional, ex G-AXPC	—

DEN DOLDER
At the northern side of Soesterberg Air Base, just outside the perimeter fence is the Hoffland scrapyard. During recent years the yard has processed a lot of aircraft which should have been preserved.

63-7453	F-4C	426	ex Soesterberg, ex USAF, l/n 1-95	scrapped
B-46	Bo105C	S.246	burnt out 3-2-92 at Deelen, l/n 12-92	scrapped
D-8257	F-104G	8257	ex Soesterberg, l/n 4-95	scrapped
D-8326	F-104G	8326	ex Volkel, l/n 12-92	scrapped
D-8341	F-104G	8341	ex Volkel, l/n 12-92	scrapped
J-007	F-16A	6D-163	remains only, crashed 10-1-92, l/n 12-92	scrapped
J-054	F-16A	6D-137	remains only, crashed 11-2-92, l/n 1-93	scrapped
J-056	F-16A	6D-139	remains only, crashed 19-4-89, l/n 1-93	scrapped
J-880	F-16A	6D-97	remains only, crashed 21-5-91, l/n 12-92	scrapped
K-3007	NF-5A	3007	ex Woensdrecht, l/n 12-92	scrapped
K-3008	NF-5A	3008	ex Woensdrecht, l/n 12-92	scrapped
K-3028	NF-5A	3028	ex Woensdrecht, l/n 12-92	scrapped
K-3030	NF-5A	3030	ex Woensdrecht, ntu Turkey AF, l/n 4-98	scrapped
K-3040	NF-5A	3040	ex Woensdrecht, l/n 12-92	scrapped
L-18	S-14	7363	ex Soesterberg, ex Ypenburg, l/n 4-94	scrapped
P-276	F-84F	..	ex Oss, ex 53-7000, l/n 1992	scrapped
51-8756	T-33A	6540	ex Deelen, l/n 1-93	scrapped

DEN HAAG
Most of the airframes of the famous Anthony Fokker School moved to Hoofddorp to form the Nederlands Luchtvaart College. To Hoofddorp went Alouette 3 A-293 (this helicopter arrived in Den Haag April 1994 and was exchanged for the UH-34J), AT-16ND B-165, Starfighters D-5810 and D-8259, NF-5A K-3020 (arrived January 1992, ex Gilze Rijen), T-33A M-50, F-84Fs P-134 and P-254, PA27 5N-AEZ, Robin PH-DIS, Ce150 PH-ALB and SAAB 91 PH-RLE. UH-34J 134 went to Soesterberg and U-6A Beaver S-6 moved to Gilze Rijen. Two Ypenburg Starfighters D-8084 (8084, with tail from D-8312) and D-8258 (8258) moved in 1993 to Switzerland for testing by the Oerlikon company. The damaged airframes returned to Ypenburg in 1994 (via Woensdrecht). Both aircraft were scrapped in June 1994 by the Pamatex company (on the same street as the former Anthony Fokker School). This company had also scrapped some SP-2H Neptunes in the early 1980s.

NETHERLANDS - 334

DEN HAM
Preserved as an eye-catcher in this village is a Beech which has been reported as being in a neglected state.
❑ PH-UBY   Beech D18S   A-101   preserved   —

DEN HELDER
During 1994 and 1995 the 'Van Normandie tot Nieuwendiep' exhibition in a museum here displayed L-18C R-87. The exhibition closed on 8th May 1995 and the Super Cub was moved to Soesterberg.

DE PEEL
De Peel is a reserve airfield which houses some air defence units. Preserved with one of these squadrons is an ex NVA MiG-21SPS.
❑ 2240   MiG-21SPS   94A5511   preserved, ex Blomberg, ex 919/NVA   12-96

DOETINCHEM
The former Slusovice MiG-21PF 0404, after its arrival in The Netherlands in March 1996, was on display at the Boudewijn Smeitink antique shop for only a few months. The aircraft moved on to Ulft in July 1996.

DORDRECHT
An ex Polish Air Force MiG-15 was bought by the G den Otter company in 1992. This Lim-5 1220 (1C-1220, with wings of SBLim-2 712273) moved on to Twello. The yard was completely empty by late 1995.

DRIEBRUGGEN
Another ex Polish MiG was bought by Bruin Metaal in Driebruggen. This aircraft, Lim-6R 418, has also been relocated. It went to Hoornsterzwaag.

DUIVEN
The first Polish MiGs in the Netherlands arrived in 1991at Duiven, making national newspaper headlines as the aircraft were considered dangerous weapons and should not have been be imported. Lim-5 003 moved to Texel, while the SBLim-1 012 (1A-09012, l/n March 1992) was sold in the USA as N9012.

EINDHOVEN
Troopship C-10 was the first of the KLu F-27s to be withdrawn from use. It was stored here before it moved to Soesterberg on 15th October 1993. Starfighter D-8091 which arrived here in May 1990 moved on 26th March 1991 to Vlissingen. NF-5 K-3068 was placed at the gate on 5th February 1992, its younger brother K-3003 arrived here on 11th April 1991 for instructional purposes. It moved on to Gilze Rijen to become the gate guard there. K-3073 (3073) was stored here for a while, before being collected on 12th July 1991 by a Greek C-130 and flown to Greece. Decoy F-84Fs P-164 and P-200 have both been scrapped, while P-277 moved on to Deelen. The real Spitfire H-15 (MK959) was swopped for three plastic ones and a plastic Hurricane. H-15 went to the UK for restoration to flying condition. The ex Veghel Mitchell nose is under restoration in the Duke of Brabant Air Force hangar. The cockpit of Jetstream G-CTRX is at the Flash Aviation Shop at the Zeelsterstraat in Eindhoven.

❑ C-8	F27-300M	10158	preserved	4-98
❑ C-12	F27-300M	10162	stored, ex Soesterberg	4-98
❑ K-85	F-84G	..	preserved, at gate, ex 51-10178	4-98
❑ K-3068	NF-5A	3068	preserved, at gate, ex instructional	4-98
❑ P-231	F-84F	..	preserved, at gate, ex 53-6742	4-98

NETHERLANDS - 335

☐ B-9	B-25D	100-20754	under restoration, nose only, ex Veghel, ex M-9	12-96
☐ G-CTRX	HP137-1	246	stored, fuselage only, see note	9-97
☐ ..	Spitfire LF.IXc	..	preserved, replica, marked as 'H-15'	4-98

GAANDEREN
Some 5km from Doetinchem is the town of Gaanderen. Preserved at an antiques wholesaler is a Czech MiG-21.

☐ 0404	MiG-21PF	760404	preserved, ex Ulft, ex Doetinchem, ex Slusovice	6-98

GELDROP
The car scrapyard De Kinderen at the Spaarpot 110 (a streetname) at Geldrop acquired a Nomad from Zestienhoven. This Nomad was used as a spares source for the now defunct Holland Aero Lines.

☐ N1022K	N24A	71	stored, ex Zestienhoven, ex VH-MXI	1-97

GILZE RIJEN
Former Prinsenbosch gate guard F-104G D-8084 arrived in August 1992. In 1993 it went to Ochsenboden (Switzerland). NF-5s K-3007, K-3008, K-3028, K-3056, K-3063 and K-3065 were robbed of all spares which were sent to Turkey. The hulks went to <u>Woensdrecht</u> by early 1992. Stored K-3020 went to the school in <u>Den Haag</u>. The instructional K-3071 (marked as 'NAC-7') also moved to <u>Woensdrecht</u>. NF-5B K-4003 left inside crates for the USA to become N3206Y, while stored K-4010 and K-4018 went to <u>Woensdrecht</u> for a major overhaul before going to Venezuela. Stored K-4021 went to Turkey by air. The Gilze Rijen Historical Flight (formally Stichting Vliegsport Gilze Rijen) made the former Den Haag AT-016 B-181 airworthy again as PH-IBI. L-21 R-122 also become civil and is now PH-PPW. The fate of stored R-116 and R-206 are unknown. Their Harvard B-176 went to <u>Lelystad</u>.

☐ B-184	AT-16ND	14A-1100	stored, ex FS960, ex 43-12801	5-98
☐ K-3003	NF-5A	3003	preserved, gate, marked as 'K-3066'	4-98
☐ K-3034	NF-5A	3034	instructional	11-93
☐ P-191	F-84F	..	preserved, at gate, ex 53-6916	4-98

The DVM store has been cleared, the aircraft having moved to locations all over the Netherlands. Harvard B-177 is now preserved at <u>Woensdrecht</u>. B-182 was made airworthy again as PH-TBR. TF-104G D-5803 went to <u>Twenthe</u>, D-5804 went to <u>Soesterberg</u>, D-5805 moved to <u>Woensdrecht</u> and D-5806 passed through here in March 1992, it arrived from Ypenburg and went on to <u>Schiphol</u>. F-104G D-8051 arrived from Stolzenau before going to <u>Leeuwarden</u>. Fokker S-11 E-14 arrived here after its display at Assen and is now stored at <u>Soesterberg</u>. Meteor F.8 I-189/7E-5 is also at <u>Soesterberg</u>, together with F-84E K-6, Hunter F.4s N-122 (ex Assen) and N-144, T-33A M-5 (ex Soesterberg), F-84F P-248 (ex Assen), L-21B R-120, U-6A S-6 (ex Den Haag) and ex Belgium RF-84F FR-31 The stored remains of crashed F-16As J-007 (w/o 10th January 1992) and J-056 (w/o 19th April 1989) were scrapped at <u>Den Dolder</u>. Fokker S-14 L-17 moved to <u>Rosmalen</u>. Hunter N-305 is now on display at <u>Soesterberg</u>. U-6A S-9 (965) was stored here before it went flying again as G-BUVF.

's GRAVELAND
Temporarily stored here at the Cannenburgerweg 59 is a Hungarian MiG-21UM. The aircraft will move to a new museum in Oud Loosdrecht as soon as there is space available.

☐ 3036	MiG-21UM	516903036	preserved, ex Kecskemét, ex Hungarian AF	3-98

HAARLEM
De Hogere Technische School (technical college) exchanged their SAAB 91 PH-RLR (91382) for an ex RAF Jet Provost. The NHI H-3 is an all Dutch built helicopter.

☐ XW406	Jet Provost T.5	EEP/3P/1028	instructional, arr 21-3-95, ex RAF	10-97
☐ A-366	Alouette 3	1366	instructional, arr 22-2-95, white c/s	10-97

NETHERLANDS - 336

☐ K-3045	NF-5A	3045	instructional, arr 1993, ex Woensdrecht	10-97
☐ PH-NFT	NHI H-3	2001	instructional	10-97

HEILOO
Parked in some 'volkstuintjes' (allotment gardens) along the railroad track is the former A Fokker school Aztec.

☐ 5N-AEZ	PA27-250	27-2053	preserved, ex Hoofddorp, ex Den Haag	4-98

HEINO
In use as a garden shed in the area is the former Zwolle An-2 SP-WWI.

☐ SP-WWI	An-2	1G173-60	stored, ex Zwolle	2-98

HOEK VAN HOLLAND
At the Nieuwe Oranjekanaal is the greenhouse builder Knijnenburg. Displayed here on a pole is a MiG-23BN painted in its former NVA colours.

☐ 2050	MiG-23BN	0393214220	preserved, ex Baarlo, ex 707/NVA	2-98

HOEVEN
The leisure centre Bosbad Hoeven near Seppe airfield should still have their two aircraft on display.

☐ N-226	Hunter F.6	8858	preserved	—
☐ 190	CS-2A	DH14	preserved, marked as '151', ex 1515/RCanNavy	—

HOOFDDORP
The new Nederlands Luchtvaart College (Dutch Aviation College) was officially opened on 10th October 1995. The first aircraft arrived on 11th May 1995 from Den Haag. The college is a co-operation between all major Dutch airlines, the former Fokker company and the A Fokker School from Den Haag. The KLM Neptune and Tracker, which are still at Schiphol, are also expected here as soon as the extension of the hangar is finished. The college can be found at the Opaallaan 25 in Hoofddorp. Aztec 5N-AEZ moved to Heiloo and CS-2A 184 went to Valkenburg and Starfighter D-8245 (8245) left in February 1998 for an unknown location.

☐ A-293	Alouette 3	1293	instructional, ex Den Haag	2-98
☐ B-165	AT-16ND	14-764	instructional, marked as 'A-50', ex Den Haag	2-98
☐ D-5810	TF-104G	5810	instructional, ex Den Haag	2-98
☐ K-3020	NF-5A	3020	instructional, ex Den Haag	2-98
☐ M-50	T-33A	5679	instructional, ex Den Haag, ex 51-4384	2-98
☐ P-134	F-84F	..	instructional, ex Den Haag, ex 52-7185	2-98
☐ P-254	F-84F	..	instructional, ex Den Haag, ex 53-6600	2-98
☐ PH-ALB	Ce150E	60797	instructional, marked as 'PH-SKS', ex 'PH-AFS'	2-98
☐ PH-DIS	Robin HR200/100	47	instructional, ex Den Haag, ex D-EEAX	2-98
☐ PH-LEG	CeF150M	1403	instructional	10-95
☐ PH-RLE	SAAB 91D	91372	instructional, marked as 'PH-AFS', ex Den Haag	10-95
☐ TF-ODI	HP137-1	210	instructional, cockpit only	2-98
☐ TF-ODM	HP137-1	202	instructional	2-98
☐ TF-ODN	HP137-1	214	instructional	2-98

Outside a home for the elderly on the Bornholm part of Hoofddorp, the Aviodome's Fokker S-11 PH-UET (marked as 'E-42') was preserved in 1992. It moved to Abbenes by the mid 1990s.

HOORNSTERZWAAG
JAWA importer D de Haan acquired Lim-6 418 (1C-0418) from Driebruggen in early 1996. He sold the aircraft

in the autumn of 1996 to a collector in Peutie. The MiG was replaced by a Mi-2 in early 1997.
☐ 9451   Mi-2   563403034   preserved, ex Arcen, ex Briest, ex 303/NVA   4-97

LEEUWARDEN
Town: F-104G D-8268 was under restoration here locally. The F-104 came from Twenthe and went Zwolle.

Airfield: One of the unique aircraft in The Netherlands is the ex Luftwaffe Su-20 which is preserved near the flightline. The Su-20 is ex Egyptian Air Force and painted in Soviet colours. The German instructional Starfighter 2455 (marked as 'LETS-1', with tail from 2475) is now at Woensdrecht, while the preserved F-104G D-8257 went to Soesterberg in 1994. F-16B J-260 was stored here after an accident on 21st December 1992. In early 1996 the aircraft moved to Woensdrecht. F-84F P-194 on the dump was scrapped in the early 1990s.

☐ 9862	Su-20	72410	preserved, as '56 red', ex WGAF, ex Egypt AF	3-98
☐ D-8051	F-104G	8051	preserved, ex Gilze Rijen, arr 28-7-93	3-98
☐ I-320	Meteor T.7	..	preserved, marked as 'I-323', ex VW417	3-98
☐ J-219	F-16A	6D-8	instructional, wfu 19-4-95	2-97
☐ J-245	F-16A	6D-34	instructional, wfu 27-4-95	2-97
☐ N-138	Hunter F.4	8638	preserved, ex Schaarsbergen, arr 4-91	3-98
☐ ..	Spitfire	..	preserved, replica, marked as 'MJ964/3W-V'	4-96

LELYSTAD
The Vroege Vogels, Pioneer and Wings over Holland associations have a lot of vintage aircraft flying or in temporary storage.
☐ 130082   HUP-2   259   stored, ex Zwolle, ex Schiphol, ex French Navy   3-98
☐ B-176   AT-16ND   14-719   stored, for sale 6-96, ex Gilze Rijen   6-96

MAASBRACHT
Still preserved on a pole near the river Maas is a Harvard. It is in the grounds of Dumphandel Valkenburg and is located close to the highway exit of Maasbracht.
☐ B-178   AT-16ND   14-739   preserved, ex FH105, ex 42-12492   9-97

MIDDEN ZEELAND
Stored here is the bare frame of a former KLu Piper Cub. The aircraft used to be stored at Hilversum.
☐ R-131   L-21B   18-3821   stored, frame only, ex Hilversum, ex 54-2421   6-95

NIEUW LOOSDRECHT
In January 1997 an ex NVA MiG-23UB 2056 was first noted at Nieuw Loosdrecht. The aircraft was temporarily stored here before moving to Hilversum for an auction in early June 1997. It was sold to a new owner in Bree.

NIEUW MILLINGEN
The major Dutch military radar unit (Dutch Mil) received their gate guard in May 1996.
☐ D-8053   F-104G   8053   preserved, at gate, ex Woensdrecht   9-96

NIJMEGEN
F-104G D-8053 at an Air Force training school went to Woensdrecht in 1995. At the industrial estate Westkanaaldijk (along the Hooglandseweg) is the Kleijngeld scrapyard which still has a Beech.
☐ PH-UBX   Beech D18S   A-105   stored   1-97

OIRSCHOT
Stored at Oirschot during the first part of the 1990s were two airframes, T-33A 51-4114 (KLu 'reserve' airframe) and Hunter F.4 N-129. These were used for several accident rescue exercises all over the country. The T-33 was noted in Amsterdam harbour in May 1992 and Deventer later that year. Both aircraft moved on to Vijfeiken, the Hunter doing this via Crailo.

OSS
The preserved F-84F P-276 at the Mini Delta Airport Model Club was transported to Den Dolder in 1992 where it was scrapped.

OUDE TONGE
Stored on the industrial complex, inside a building, is the cockpit of a B707 which was scrapped at Schiphol in the summer of 1993.

❑ OD-AFY	B707-327C	19108	stored, cockpit only, ex N7099	11-93

OUD LOOSDRECHT
A new museum will be set up here in the near future. Also the MiG-21 at 's Graveland belongs to this collection.

❑ 1220	Lim-5	1C-1220	stored, ex Twello, ex Dordrecht	10-97

OVERLOON
The Oorlogsmuseum still has its three aircraft. The ex India Air Force Spitfire is painted in RAF colours.

❑ HS649	Spitfire LF.XIVc	6S-672268	preserved, ex IndianAF, ex NH649	11-97
❑ B-199	AT-16ND	14-610	preserved, marked as '12964', ex FE876	11-97
❑ B-6	B-25D	..	preserved, ex FR193 and marked as such	11-97

PRINSENBOSCH
The preserved Starfighter at these barracks (D-8084) moved to Gilze Rijen by the early 1990s.

PURMEREND
In November 1985 a DC-8 arrived here from Luxemburg. The aircraft is still in use as a restaurant.

❑ TU-TCB	DC-8-53	45671	preserved, ex F-BJCB	11-96

REEK
The two Thunderstreaks are still in use at Reek for instructional purposes. The aircraft can be found at the army training area of the Opleidingsrichting Mineurs (ORM), just south of Reek.

❑ P-224	F-84F	..	instructional, ex 53-6604	11-97
❑ P-229	F-84F	..	instructional, marked as 'P-312', ex 53-6582	11-97

RIJSSEN
At the former Lambourghini dealer in the Nijverheidsstraat, a Rotorway Exec helicopter (N650PG) was preserved in 1996. Also noted was the cockpit of an unknown jet fighter and a plastic fuselage of an F-5.

ROSMALEN
The national Dutch car museum Autotron has a number of aircraft on display. The Anson, S-14 and F-84 are on

NETHERLANDS - 339

on loan from the MLM. Rosmalen is just east of 's Hertogenbosch.
☐ VM352	Anson C.19	..	preserved, marked as 'D-26'	5-95
☐ L-17	S-14	7356	preserved, outside	4-98
☐ P-263	F-84F	..	preserved, outside, ex 53-6780	4-98
☐ PH-HOE	S-11	6195	preserved, ex E-6	5-95
☐ ..	Nieuport 11	..	preserved, replica, marked as 'N220'	5-95
☐ ..	SV-4	..	preserved, unmarked, ex La Ferté Alais	5-95

ROTTE
On the edge of the Rottemeren recreation park (north of Rotte, which itself north is of Rotterdam) is the Hofman car garage. Displayed outside is a Lim-6bis, while stored Lim-5 1213 went to Tilburg in 1993.
☐ 417	Lim-6bis	1J-0417	preserved, ex Polish AF	4-98

ROTTERDAM
Town: At the Charloise Lagedijk 638 in Rotterdam south, outside a garage, is still a red/white/blue Retriever.
☐ 130076	HUP-2	253	preserved, unmarked, ex French Navy	3-98

Airfield – Zestienhoven: F27-100 F-WKPX arrived on 29th April 1993 for spares recovery and its cockpit went to Schiedam. The rest was scrapped. A second spare source F27 is still on the dump.
☐ LX-OOO	F27-100	10269	dumped, marked as 'G-HOST', ex LX-LGB	4-98

SCHAARSBERGEN
The Luchtmacht Electronische en Technische School (LETS, Air Force electronic and technical school) at Schaarsbergen (very close to Deelen) is still using a number of ground instructional airframes. Of these Starfighters D-8061 and D-8318 moved to Soesterberg in 1995, while D-8244 went to Deelen. NF-5A K-3003 was here only between 22nd January 1987 and 11th April 1991. It went to Eindhoven.
☐ A-319	Alouette 3	1319	instructional, ex Soesterberg, ex Woensdrecht	12-96
☐ A-343	Alouette 3	1343	instructional, ex Woensdrecht	4-97
☐ D-8256	F-104G	8256	preserved, at gate LETS, ex instructional	1-98
☐ J-015	F-16A	6D-171	instructional, arr 15-1-98	1-98
☐ J-238	F-16A	6D-27	instructional, arr 17-2-94, ex Woensdrecht	1-98
☐ J-263	F-16B	6E-5	instructional, arr 6-9-95, ex Volkel	1-98
☐ G-AZTI	Bo105D	S.34	instructional, ex Soesterberg	12-96

The Brandweerschool (fire school) has only a few airframes left. The F-104G D-8133 (8133) was completely burned by the late 1980s and was scrapped. The KLu 'reserve' T-33A 51-8756 was moved to Den Dolder.
☐ A-274	Alouette 3	1274	instructional, with tail from A-350	1-98
☐ D-8281	F-104G	8281	instructional, fire training	1-98
☐ N-150	Hunter F.4	8652	instructional, fire training	1-98
☐ P-277	F-84F	..	preserved, ex instructional, ex Eindhoven	1-98

SCHAIJK
Along road N321 from Heesh to Grave is the small village of Schaijk (some 13km from Volkel). Preserved on a platform at a small animal farm here is a white ex NVA Mi-2. The aircraft was painted in a garage in Oss.
☐ 9482	Mi-2	562250032	preserved, with tailboom from 9456, ex Briest	11-97

SCHIEDAM
The cockpit of a Fokker Friendship is preserved at a shop in the Noordvest 133.
☐ F-WKPX	F27-100	10121	preserved, cockpit only, ex Rotterdam	8-96

SCHIPHOL

On 23rd July 1993 the undercarriage of B707-327C OD-AFY failed during taxiing. The aircraft was scrapped at Schiphol and the cockpit went to Oude Tonge. B707-336C SU-DAC (19843) of Zarkani was stored at Schiphol from July 1993. It finally left in February 1996 as SU-PBA. The Dutch Dakota Association has a DC-2 stored in their hangar at Schiphol Oost.

❑ PH-AJU	DC-2-112	1288	stored, ex Badhoevendorp, ex VH-CRH	1-97

The Fokker company went into receivership in March 1996. Fokker S-14 PH-XIV, which is owned by the Aviodome and was at Fokker for restoration, moved to hangar 9 at Schiphol in November 1997 for further storage. The Fokker Spin PH-WEY (an original airframe which was found in Poland) has been bought by the Fokker Heritage Foundation and will be preserved. It is currently on display at the Aviodome. F-84F P-172 was in use as an instructional airframe, together with the tail from P-230. Four prototypes are stored at Schiphol.

❑ P-172	F-84F	..	instructional, ex Deelen, ex 53-6678	5-96
❑ PH-MKC	Fokker 70	11243	stored	1-98
❑ PH-MKH	Fokker 100	11242	stored	1-98
❑ PH-OSI	Fokker 50	10688	stored	1-98
❑ PH-OSO	Fokker 50	10685	stored	1-98

The instructional school of KLM became part of the NLC in Hoofddorp in 1995. Their two airframes in hangar 9 at Schiphol are expected to move to Hoofddorp in the near future.

❑ 151	US-2N	712	instructional, KLM c/s, ex 147641	2-98
❑ 210	SP-2H	7263	instructional, KLM c/s	2-98
❑ PH-XIV	S-14-2	6289	stored, ex Fokker, ex Aviodome, ex 'K-1'	2-98

The Aviodome collection is still expanding. One of their major changes was the swopping of two DC-3s, both of which were painted in the same markings. DC-3C PH-PBA (19434, marked as 'PH-TCB') was removed from the museum in February 1996 and left for Coventry to become airworthy again. It was replaced by DC-3 G-BVOL which is also painted as 'PH-TCB'. Sea Hawk WV828 was transported to Valkenburg for restoration, before moving to De Kooij. HUP-2 130082 went to Zwolle, Sea Fury 6-43 departed to Soesterberg. Of the civil aircraft, Auster J/5 PH-NFH (1845) went to storage at Badhoevendorp and SAAB 91 PH-RLN is now in use as an instructional airframe in Zwolle. Fokker S-11 PH-EUT went to Hoofddorp. The prototype Fokker S-14 went to Fokker at Schiphol Oost for restoration. RC680 OO-SID went to Delft.

❑ E-410	69	Hunter F.51	41H680269	preserved, with tail from WV395, ex Danish AF	2-98
❑ XJ350		Sea Devon C.20	04453	preserved, marked as 'PH-MAD', ex Royal Navy	2-98
❑ XN600		Jet Provost T.3A	..	preserved, cockpit only, ex RAF	2-98
❑ 634		Fokker C.VD	..	preserved, ex '618'	2-98
❑ A-38		DH.82A	83101	preserved, ex R5242	2-98
❑ B-73		AT-16ND	14A-1268	preserved, ex Badhoevendorp, ex Veghel	11-94
❑ D-5806		TF-104G	5806	preserved, ex Gilze Rijen, ex Ypenburg	2-98
❑ E-9		S-11	6200	preserved, ex Badhoevedorp	2-98
❑ H-8		Spitfire LF.IXc	CBAFIX970	preserved, marked as 'H-53/MH424'	2-98
❑ 076	V	UH-19F	..	preserved, ex 8-2, ex 133777	2-98
❑ 159	V	US-2N	720	preserved, outside, ex 148281	2-98
❑ CF-AUV		Fokker B.IVA	906	preserved	11-96
❑ F-BCDB		DH.89A	6897	preserved, ex G-AKDW and marked as such	—
❑ G-BVOL		DC-3	9836	preserved, marked as 'PH-TCB', ex Coventry	11-96
❑ N7904C		DHC-2	1288	preserved, marked as 'JZ-PAD', cabin only	11-96
❑ OO-SCD		Dove 6	04117	preserved, nose only, ex VP-KDE, ex G-AMFU	11-96
❑ OY-ASE		Fokker F.VIIa	5054	preserved, marked as 'H-NACT', ex SE-ASE	11-96
❑ PH-FCX		F27-100	10183	preserved, outside, arr 5-88, ex Ypenburg	11-96
❑ PH-NHI		NHI H-3	3001	preserved	
❑ PH-TRO		SE210-3	33	preserved, cockpit only, ex HB-ICW	11-96
❑ PH-VRV		Beagle B-121 Pup	B-120	preserved, ex D-EBYZ, ex G-35-120	—
❑ PH-WEY		Fokker Spin 3	..	preserved, ex Krakow	97
❑ PJ-AIS		Fokker F.VIII	5312	under restoration, forward fuselage only	—

NETHERLANDS - 341

☐ SE-AFI	Cierva C.30A	735	preserved, ex 99/RNoAF, ex LN-BAD	2-98
☐ ..	Evans VP-1	V.2567	preserved, marked as 'PH-VPI', PH-GOO ntu	—
☐ BAPC.22	HM-14	WM.1	preserved, marked as 'G-AEOF'	—
☐ ..	Fokker Dr.I	..	preserved, replica, marked as '152/17'	11-96
☐ ..	Fokker F.II	..	preserved, replica, marked as 'H-NABC'	11-97
☐ ..	Fokker Spin 3	..	preserved, replica	11-96
☐ ..	Wright Flyer	..	preserved, replica	11-96

SOESTERBERG

Town – Kamp Zeist: The Militaire Luchtvaart Museum is located at a military complex along the highway from Utrecht to Amersfoort. The museum is open from April to December from Tuesday to Friday between 10:00 and 16:30 and on Sunday between 12:00 and 16:30. Catalina 16-212 and Dragonfly WG752 moved to Valkenburg for restoration. The four (USAF) fighters outside are painted in colours to represent aircraft of the former USAFE 32nd TFS, based at Soesterberg. The museum used the Finnish Air Force Fokker C.X FK-113 for a few years (l/n October 1996) as specimen for their own replica. The aircraft returned to Tikkakoski.

Hall A:

☐ H-1	Spitfire LF.IXc	CBAFIX907	preserved, ex MJ143, ex 3W-1	11-97
☐ M-464	B-25J	108-37333	preserved, ex N5-264, ex 44-31258	11-97
☐ 44-12125	P-51D	111-30258	preserved, marked as 'H-307'	11-97
☐ N4729V	Fokker D.VII	2523	preserved, marked as '266', ex NX3764	11-97
☐ PH-OTA	Dominie Mk.3	6740	preserved, ex NF869, ex PH-TGC, ex V-3	11-97
☐ ..	Farman HF-20	..	preserved, replica, marked as 'LA-2', ex 'LA-16'	11-97
☐ ..	Fokker C.X	..	preserved, replica under restoration	9-96
☐ ..	Fokker D.XXI	..	preserved, replica, marked as '221'	11-97
☐ ..	Fokker G.1A	..	preserved, replica, marked as '330'	11-97

Hall B:

☐ A29		Alouette 2	1753	preserved, marked as 'H-4', ex Belgium Army	11-97
☐ 11253		RF-84F	134	preserved, ex GreekAF, ex P-5	11-97
☐ MM53-8305		F-86K	207-33	preserved, marked as 'Q-305', ex ItAF	11-97
☐ HD.5-1		Do24T-3	5342	preserved, marked as 'X-24', ex Spanish AF	11-97
☐ XE489		Sea Hawk FGA.6	6021	preserved, marked as '131/D', tail from WM983	11-97
☐ B-103		AT-16ND	14A-1459	preserved, marked as '099/K', ex FT419	11-97
☐ D-8022		F-104G	8022	preserved, arr 15-3-84	11-97
☐ E-22		S-11	6213	preserved	11-97
☐ H-20		Alouette 3	1320	preserved, arr 13-12-93	11-97
☐ I-69		Meteor F.4	..	preserved, ex VZ409	11-97
☐ K-171	DU-24	F-84G	..	preserved, ex 'K-40/DU-24', ex 51-10806	11-97
☐ K-4011		NF-5B	4011	preserved, ex Gilze Rijen	11-97
☐ N-305		Hunter T.7	41H-693457	preserved	11-97
☐ O-36		OH-23C	937	preserved, ex 57-6521	11-97
☐ 134	V	UH-34J	58-1597	preserved, ex Den Haag, ex 150729	11-97
☐ 160	V	US-2N	721	preserved, ex 148282	11-97
☐ 218		SP-2H	7157	preserved, forward fuselage only, ex 146438	11-97
☐ 6-43		Sea Fury FB.51	6310	preserved, ex Schiphol	11-97
☐ PH-UFC		DH.82A	86587	preserved, ex A-10, ex PG690	11-97

Outside:

☐ K-688	C-47A	20118	preserved, marked as 'X-5', ex RDanAF	11-97
☐ 61052	F-102A	..	preserved, marked as 'FC-032', ex Greek AF	11-97
☐ 5307	F-86F	191-876	preserved, as '25385/FU-385', ex PortAF	11-97
☐ 47 red	MiG-21PFM	940MS13	preserved, ex Twenthe, ex Soviet AF	11-97
☐ 74-0083	F-15A	55/A44	preserved, marked as '77-0132/CR'	11-97

NETHERLANDS - 342

❏ 54-2265		F-100D	223-145	preserved, marked as '41871', ex FAF	11-97
❏ C-10		F27-300M	10160	preserved, ex Eindhoven	11-97
❏ R-87		L-18C	18-3185	stored, in outbuilding, ex Den Helder, ex Assen	12-96
❏ 201	V	SP-2H	7241	preserved	11-97
❏ OY-AVO		Lockheed L12A	1306	preserved, marked as 'L2-100', ex LN-BFS	11-97
❏ ..		Spitfire LF.IXc	..	preserved, replica, marked as 'MJ881/3W-B'	11-97

Airfield: Stored Alouette 3 A-319 (ex Woensdrecht) moved to Deelen for instructional purposes, while A-293 went to Den Haag. Troopship C-12 was declared wfu here on 18th August 1994. It was flown out in September 1996 to Eindhoven, while the stored C-10 went to Zeist in January 1997. After the 32nd FS had left Soesterberg, their *Wrecks & Relics* aircraft were disposed of. F-4C 63-7453 was allocated to the museum at Kamp Zeist, but was found in a too bad a state and was scrapped at Den Dolder. Instructional F-15A 73-0091 (31/A25) moved to Lakenheath, while F-15A 74-0083 is now at the museum. F-105G 62-4417 went to Hermeskeil. The instructional Bo105 G-AZTI went to Deelen. The Hunter used to be preserved near the gate, but has been removed. Its current location is unknown.

❏ D-8245		F-104G	8245	stored, in shelter, ex Breda	95
❏ N-258		Hunter F.6	8934	preserved, at gate, MLM owned, to ..	7-96

The museum has some storage facilities at Soesterberg air base (mainly in hangar 2). The former Polish Air Force Lim-2 301 (1B-00301) was swopped in 1995 for an ex Greek F-102A and went to Tatoi. TF-104 D-5804 was last noted in November 1995 and has since resurfaced in Bad Oeyenhausen in Germany. F-104G D-8331 went to Oklahoma, USA. Former Zeist museum F-84E K-6 was swopped for the ex Belgian Army Alouette 2 and the F-84 moved to Florennes. The former Ypenburg gate guard Fokker S-14 L-18 arrived here on 1st June 1992 for storage. It was scrapped at Den Dolder in 1994. The AB204B is a former Royal Netherlands Navy example which went to Sweden as instructional airframe many years ago.

❏ FR-31		RF-84F	..	stored, ex Gilze Rijen, ex Saffraanberg	2-97
❏ J-4073		Hunter F.58	..	stored, arr 16-6-95, ex Swiss AF	12-96
❏ A-391		Alouette 3	1391	stored, wfu 18-7-93	2-97
❏ A-465		Alouette 3	1465	stored, arr 8-8-95, ex Woensdrecht	2-97
❏ B-175		AT-16ND	14-765	stored, no fuselage skin, ex FH131	2-97
❏ D-8061		F-104G	8061	stored, ex Deelen	2-97
❏ D-8257		F-104G	8257	stored, nose only, ex Leeuwarden	11-95
❏ D-8318		F-104G	8318	stored, ex Deelen	2-97
❏ E-24		S-11	6215	stored	2-97
❏ I-187		Meteor F.8	6464	stored, outside, marked as 'I-147'	2-97
❏ I-189	7E-5	Meteor F.8	6468	stored, ex Woensdrecht	2-97
❏ K-3011		NF-5A	3011	stored, ex Woensdrecht	2-97
❏ L-11		S-14	7356	stored	2-97
❏ N-122		Hunter F.4	8622	stored, ex Gilze Rijen, ex Assen	2-97
❏ N-144		Hunter F.4	8644	stored	2-97
❏ M-5		T-33A	6812	stored, ex 51-9028	2-97
❏ P-226		F-84F	..	stored, ex 53-6612	1-97
❏ P-230		F-84F	..	stored, cockpit only (tail at Schiphol)	95
❏ P-248		F-84F	..	stored, ex Gilze Rijen, ex Assen, ex 53-6584	1-97
❏ R-11		Auster AOP.3	350	stored, ex MZ236, ex 'R-5', ex 'R-55'	2-97
❏ S-6		U-6A	959	stored, ex Den Haag	2-97
❏ 225		AB204B	3023	expected, ex Linköping, ex Skavsta, ex RNN	—
❏ PH-UDT		Beech T7	A-472	store, in shelter, marked as 'G-10'	2-97
❏ ..		Spitfire	..	stored, replica, marked as 'MK.../OU-U'	96

St ISIDORUSHOEVE

The Nissan car dealer named Auto Actief at this little village near Haaksbergen received a former Kecskemet MiG-21UM. The aircraft arrived in the summer of 1996.

❏ 0565		MiG-21UM	05695165	preserved, ex Kecskemét	11-97

TEXEL
The museum at the airfield on this island sold their Lim-5 003 (1C-0003, ex Duiven) to the USA in January 1995. The Dijkhaster is a homebuilt which has made only one (unofficial) flight.

☐ D-8266	F-104G	8266	preserved, ex Ypenburg, arr 3-5-89	5-96
☐ PH-COA	Cessna 140	..	preserved	6-91
☐ PH-COR	Dijkhaster FB25	..	preserved, not an official registration	5-96
☐ PH-NEI	S-13	6288	preserved, cockpit only	8-96

TIENDEVEEN
Skyvan OE-FDI made a forced landing in a field in the Netherlands, after running out of fuel. The damaged aircraft was brought to Tiendeveen were it acts as a clubhouse for the local model aircraft club.

☐ OE-FDI	SC7-3-400	1869	preserved	12-96

TILBURG
In the mid 1990s Lim-5 1213 arrived here at the Broekhovenseweg from Schiedam. The aircraft was sold in 1995 and has moved on to Peutie. The NF-5A is at the local technical school (MTS) at Stappegoorweg 181.

☐ K-4028	NF-5B	4028	instructional, arr 15-5-90, ex Woensdrecht	12-92

TWELLO
Near the airfield of Teuge, the MDT Materiaal Dienst Twello (Molenstraat 85A), had an ex Polish Lim-5. The aircraft (1220) had left by December 1996 for Oud Loosdrecht.

TWENTHE
Former Soviet Air Force MiG-21PFM 47 red, which arrived here for display from Stargard in Poland, was donated to the museum at Soesterberg. Its place at Twenthe was taken by an ex NVA MiG-21. The instructional F-104 D-8268 went to Leeuwarden in 13th November 1989. Both F-84s, P-166 and P-209, have been scrapped.

☐ 473	MiG-21SPS-K	7006	preserved, ex Bad Düben, ex NVA	9-96
☐ D-8338	F-104G	8338	instructional, marked as 'BEEZ'	1-97
☐ J-221	F-16A	6D-10	instructional, wfu 27-4-95	2-97
☐ J-234	F-16A	6D-23	stored, wfu 2-96	2-97
☐ J-247	F-16A	6D-36	instructional, wfu 28-4-95	2-97
☐ J-262	F-16B	6E-4	stored, wfu 18-3-98	3-98
☐ K-3029	NF-5A	3029	preserved, at gate, ex 'TPG-29'	1-97
☐ P-171	F-84F	..	preserved, ex tiger c/s, ex 53-6687	6-97
☐ Q-283	F-86K	213-53	preserved, ex 54-1283	9-96

Museum Zuid Kamp is located on the Overmaatweg, close to the airbase of Twenthe. Although the museum is on military grounds it is freely accessible for the public. The museum also has some parts of Typhoon RB396.

☐ D-5803	TF-104G	5803	preserved, ex Gilze Rijen	5-96
☐ I-19	Meteor T.7	..	preserved, MLM owned, ex WH223	5-96
☐ ..	Spitfire	..	preserved, replica, marked as 'H-27/PT987'	5-96

UITHUIZEN
A small museum from the Stichting '39-'45 received an ex NVA MiG-21SPS on 28th February 1995.

☐ 2239	MiG-21SPS	94A5410	preserved, ex Rothenburg, ex 898/NVA	7-96

Just north of Uithuizen on an estate lives Marthe Röling, a painter who has done some work for the Netherlands Air Force. In return the Air Force gave her a Starfighter.

☐ D-8300	F-104G	8300	preserved, ex Weesp, ex Ypenburg	8-96

ULFT
Ex Czech MiG-21PF 0404 arrived here on 15th July 1996 and moved in April 1997 to Gaanderen.

VALKENBURG
A number of aircraft have been restored at Valkenburg in the past years. They include Dragonfly WG752, Sea Hawk WV828 (now at De Kooij) and the Catalina from Soesterberg.

❏ WG752		Dragonfly HR.3	WA232	stored, marked as '8-1', ex RAF	8-97
❏ 184	H	CS-2A	DH5	stored, arr 5-2-98, ex Hoofddorp, ex Den Haag	2-98
❏ 216	V	SP-2H	7143	preserved, at gate, ex 144692	2-98
❏ 16-212		PBY-5A	1679	under restoration, ex Soesterberg, ex Hoeven	11-97

VEGHEL
The Bevrijdende Vleugels museum (Liberating Wings) moved to Best in 1997, taking their L-4J and Spitfire with them. Before this move the museum returned Auster AOP.3 MZ236 to the MLM and the aircraft is now stored at Soesterberg. They also returned Harvard B-73 to the Aviadome at Schiphol. The nose of the B-25D went to Eindhoven.

VIJFEIKEN
Late in 1996 two instructional airframes were noted at the barracks at Vijfeiken. The village is located near Gilze-Rijen and Dongen. The aircraft are used for all kinds of exercises and may travel throughout the country.

❏ 51-4114	T-33A	5408	stored, ex Oischot, ex Deelen, ex Soesterberg	10-97
❏ N-129	Hunter F.4	8629	stored, ex Crailo, ex Oirschot	10-97

VLISSINGEN
A fire school is located near the Olau terminal in the port of Vlissingen. In July 1995 one NF-5 was still present. Which one has left is not known.

❏ D-8091	F-104G	8091	instructional, ex Eindhoven, arr 26-3-91	8-97
❏ K-3043	NF-5A	3043	instructional, ex Woensdrecht, see note	1-93
❏ K-3063	NF-5A	3063	instructional, ex Woensdrecht, see note	1-93

VOLKEL
Two Starfighters have left Volkel. Instructional airframes D-8326 and D-8341 were both scrapped at Den Dolder in 1992. D-8312 is preserved with the 312sq. Stored F-16B J-263 went to Deelen on August 15th 1995. F-16A J-257 (6D-46) was used as an instructional airframe during 1994 and 1995, but has returned to operational service. Not longer operational were the stored J-212 and J-214 which both moved to Woensdrecht in late 1997.

❏ D-8063	F-104G	8063	instructional	10-97
❏ D-8279	F-104G	8279	preserved, at gate	10-97
❏ D-8312	F-104G	8312	preserved, tail from D-8084, ex Woensdrecht	4-97
❏ J-213	F-16A	6D-2	stored, wfu 13-3-98	3-98
❏ J-222	F-16A	6D-11	instructional, wfu 11-5-95	10-97
❏ J-223	F-16A	6D-12	stored	2-97
❏ J-229	F-16A	6D-18	instructional, weapons loader, wfu 28-4-95	2-97
❏ J-242	F-16A	6D-31	instructional, weapons loader, wfu 27-4-95	10-97

WEESP
The artist Marthe Röling used to live here with her Starfighter D-8300. She and the Starfighter moved to Uithuizen.

WELL
Near the border with Germany (close to Laarbruch is the village of Well. Preserved here was Mi-2 9452. The chopper had moved by December 1997 to Zwijndrecht.

WESTERSCHOUWEN
Harvard B-193 is preserved at Camping Duinrand at the Hogeweg 96, Westerschouwen.
❏ B-193   AT-16ND   14-770   preserved, ex FH136, ex 42-12523   5-97

WOENSDRECHT
The Depot Vliegtuig Materieel (DVM) arrived here in 1994 from Gilze Rijen. The DVM carries out storage, maintenance and crash investigation tasks. F-104s D-8084 and D-8258 arrived here from Switzerland on 13th September 1993 and moved on to Ypenburg. D-8053 (ex Nijmegen) went to Nieuw Millingen in 1996. D-8312 departed late 1996 and is now at Volkel. The 'LETS-1' has the tail from 2475 (8224). Stored Meteor I-189 (ex Soesterberg) moved back to Soesterberg on 13th November 1996. F-16A J-238 crashed on 18th December 1992 and was rebuilt here for instructional purposes with the tail from the crashed J-645, moving to Deelen on 17th February 1994. J-645 was last noted on the dump in 1994 and was scrapped soon afterwards. NF-5A K-3030 went to Den Dolder in April 1998, where also two other NF-5s were noted (presumably K-3060 and K-3071). F27-300M C-6 (10156) was sold in August 1997 and became C-GWXD. Early 1998 C-1 (10152) went as ZS-OEH and C-5 (10155) became ZS-OEI. The DVM restored ex Greek F-102A 61052, it moved on to Soesterberg in late 1997. Over the years a very large quantity of aircraft have been (temporarily) stored with Fokker here. Nearly all have moved on (details can be obtained from the compiler).

❏ 2455	F-104G	8203	stored, marked as 'LETS-1', ex Leeuwarden	3-98
❏ B-177	AT-16ND	14-733	preserved, marked as 'B-77/FS839', ex FE999	3-98
❏ C-2	F27-100	10149	stored, wfu 26-9-96	3-98
❏ C-3	F27-100	10150	stored, wfu 15-5-96	12-97
❏ C-4	F27-300M	10154	stored, wfu 4-9-96	12-97
❏ C-7	F27-300M	10157	stored, wfu 13-6-96	4-98
❏ C-9	F27-300M	10159	stored, wfu 27-10-95	12-97
❏ C-11	F27-300M	10161	stored, wfu 6-96	4-98
❏ D-5805	TF-104G	5805	stored, travelling exhibit, ex Gilze Rijen	12-97
❏ J-212	F-16A	6D-1	stored, ex Volkel, arr 27-11-97	4-98
❏ J-213	F-16A	6D-2	stored	4-98
❏ J-214	F-16A	6D-3	stored, ex Volkel	4-98
❏ J-228	F-16A	6D-17	stored, awaiting fate	12-97
❏ J-260	F-16B	6E-2	preserved, ex Leeuwarden, crashed 21-12-93	4-98
❏ K-3060	NF-5A	3060	stored, ntu Turkey	2-98
❏ K-3071	NF-5A	3071	stored, marked as 'NAC-7', ex Gilze Rijen	2-98
❏ M-54	T-33A	7150	preserved, marked as 'M-51/51-6528', ex 'M-52'	4-98
❏ P-170	F-84F	..	stored, ex 53-6673	11-95
❏ PH-RLD	SAAB 91D	91370	instructional, with Fokker	2-96

A large number of Alouettes have passed through the DVM at Woensdrecht. All of them have since departed.

A-177	Alouette 3	1177	stored, dep 20-3-97	to SE-JEF, to N7049Y
A-208	Alouette 3	1208	stored, arr 11-3-96, dep 31-10-96	to SE-JCR
A-217	Alouette 3	1217	stored, dep 24-7-96, dep 23-7-96	to Eurocopter
A-218	Alouette 3	1218	stored, arr 11-3-96, dep 31-10-96	to SE-JCS
A-226	Alouette 3	1226	stored, dep 24-4-95	to Marseille
A-227	Alouette 3	1227	stored, arr 18-3-96, dep 20-6-96	to N227RA
A-235	Alouette 3	1235	stored, arr 18-3-96, dep 22-1-97	to SE-JED
A-266	Alouette 3	1265	stored, dep 9-8-94	to 1266/Pakistan Navy
A-267	Alouette 3	1267	stored, dep 1-5-95	to Marseille
A-281	Alouette 3	1281	stored, dep 1-5-95	to Marseille

A-307	Alouette 3	1307	stored, dep 23-7-96	to Eurocopter, to I-BXWA
A-319	Alouette 3	1319	stored, l/n 7-93	to Soesterberg
A-324	Alouette 3	1324	stored, dep 10-7-96	to Eurocopter, to HB-XQD
A-336	Alouette 3	1336	stored, dep 1-5-95	to Marseille
A-343	Alouette 3	1343	stored, dep 8-10-96	to Deelen
A-350	Alouette 3	1350	stored, to go to Sweden	12-97
A-366	Alouette 3	1366	stored, dep 22-2-95	to Haarlem
A-374	Alouette 3	1374	stored, dep 6-8-96	to Eurocopter
A-383	Alouette 3	1383	stored, dep 1-5-95	to Marseille
A-390	Alouette 3	1390	stored, dep 16-7-96	to Eurocopter
A-398	Alouette 3	1398	stored, arr 25-3-96, dep 31-10-96	to SE-JVT
A-407	Alouette 3	1407	stored, dep 6-8-96	to Eurocopter
A-414	Alouette 3	1414	stored, dep 1-5-95	to Marseille
A-451	Alouette 3	1451	stored, to go to Sweden	12-97
A-452	Alouette 3	1452	stored, dep 23-7-96	to F-GIYE
A-453	Alouette 3	1453	stored, dep 16-7-96	to Eurocopter
A-464	Alouette 3	1464	stored, dep 16-7-96	to F-GNVC
A-465	Alouette 3	1465	stored, dep 8-8-95	to Soesterberg
A-471	Alouette 3	1471	stored, dep 9-97	to SE-JEI
A-482	Alouette 3	1482	stored, arr 25-3-96, dep 29-7-96	to SE-JCM
A-483	Alouette 3	1483	stored, arr 25-3-96, dep 22-1-97	to SE-JEE
A-488	Alouette 3	1488	stored, arr 11-3-96, dep 22-1-97	to SE-JCU
A-489	Alouette 3	1489	stored, arr 11-3-96, dep 9-7-96	to SE-JEJ
A-494	Alouette 3	1494	stored, dep 9-97	to SE-JEK
A-495	Alouette 3	1495	stored, dep 12-3-96	to Sweden
A-499	Alouette 3	1499	stored, arr 18-3-96, dep 27-6-96	to N499RA
A-500	Alouette 3	1500	stored, dep 9-97	to SE-JEL
A-514	Alouette 3	1514	stored, arr 18-3-96, dep 31-10-96	to Sweden
A-515	Alouette 3	1515	stored, arr 19-11-96, dep 20-3-97	to SE-JEG
A-521	Alouette 3	1521	stored, to go to Sweden	12-97
A-528	Alouette 3	1528	stored, dep 29-7-96	to SE-JCN, to N528RA
A-529	Alouette 3	1529	stored, dep 29-7-96	to SE-JCO, to N529RA
A-535	Alouette 3	1535	stored, dep 7-10-96	to F-GMTA
A-536	Alouette 3	1536	stored, dep 10-7-96	to F-GPGH
A-542	Alouette 3	1471	stored, dep 20-3-97	to Sweden
A-549	Alouette 3	1549	stored, dep 5-7-96	to Alverca
A-550	Alouette 3	1550	stored, dep 5-7-96	to Alverca
H-67	Alouette 3	1367	stored, arr 21-3-94, dep 10-8-94	to Pakistan Navy
H-75	Alouette 3	1375	stored, arr 21-3-94, dep 10-8-94	to Pakistan Navy
H-81	Alouette 3	1381	stored, arr 21-3-94, dep 9-8-94	to Pakistan Navy

Before the influx of Alouettes and even before the arrival of the DVM, Woensdrecht held a large number of NF-5s in storage (most of which arrived in the late 1980s).

K-3003	NF-5A	3003	stored, l/n 1-87	to Deelen
K-3007	NF-5A	3007	stored, ex Gilze Rijen	to Den Dolder
K-3008	NF-5A	3008	stored, ex Gilze Rijen	to Den Dolder
K-3011	NF-5A	3011	stored, Turkey ntu, l/n 10-90	to Soesterberg
K-3013	NF-5A	3013	stored, dep 10-7-91	to Turkey (via Eindhoven)
K-3015	NF-5A	3015	stored, dep 27-11-92	to Turkey
K-3018	NF-5A	3018	stored, dep 3-12-92	to Turkey
K-3022	NF-5A	3022	stored, dep 27-4-92	to Turkey
K-3027	NF-5A	3027	stored, dep 10-7-91	to Turkey (via Eindhoven)
K-3028	NF-5A	3028	stored, ex Gilze Rijen	to Den Dolder
K-3032	NF-5A	3032	stored, dep 27-4-92	to Turkey
K-3035	NF-5A	3035	stored, dep 20-12-91	to Turkey

NETHERLANDS - 347

K-3040	NF-5A	3040	stored, Turkey ntu, l/n 1992	to Den Dolder
K-3043	NF-5A	3043	stored, Turkey ntu, l/n 1992	to Vlissingen
K-3044	NF-5A	3044	stored, l/n 5-91	to Crailo
K-3045	NF-5A	3045	stored, Turkey ntu, l/n 1992	to Haarlem
K-3048	NF-5A	3048	stored, dep 8-10-92	to Turkey
K-3056	NF-5A	3056	stored, ex Gilze Rijen	scrapped
K-3057	NF-5A	3057	stored, dep 24-6-92	to Venezuela as 6324
K-3063	NF-5A	3063	stored, ex Gilze Rijen	to Vlissingen
K-3065	NF-5A	3065	stored, ex Gilze Rijen	scrapped
K-3070	NF-5A	3070	stored, dep 8-10-92	to Turkey
K-4002	NF-5B	4002	stored, dep 16-9-92	to Venezuela as 1721
K-4006	NF-5B	4006	stored, dep 23-3-93	to Venezuela as 6340
K-4008	NF-5B	4008	stored, dep 22-12-92	to Venezuela as 2111
K-4010	NF-5B	4010	stored, dep 27-1-93	to Venezuela as 6292
K-4012	NF-5B	4012	stored	to Zwolle
K-4018	NF-5B	4018	stored, dep 8-5-92	to Venezuela as 1711
K-4020	NF-5B	4020	stored	to Eindhoven and to Turkey 4-4-91
K-4028	NF-5B	4028	stored	to Tilburg
K-4030	NF-5B	4030	stored, dep 2-3-93	to Venezuela as 6372

YPENBURG
Ypenburg is where it all started for the compiler and therefore it is sad to see that where once was the runway and tower, modern houses are being built. The only current *Wrecks & Relics* aircraft is a Starfighter of the Van Weerden Poelman scouting group. The gate guard Fokker S-14 L-18 has moved to Soesterberg.
❑ D-8062　　　F-104G　　　8062　　　preserved　　　　　　　　　　　　3-98

All the stored Starfighters had left by 1993. In 1994 two aircraft (D-8084 and D-8258) returned here from Ochsenboden in Switzerland (via Woensdrecht) where they were used by the Oerlikon company as targets. Both aircraft were transported to Den Haag for scrapping.

D-5803	TF-104G	5803	stored, arr 26-11-84, dep 1989	to Gilze Rijen
D-5804	TF-104G	5804	stored, arr 23-11-84, dep 8-5-92	to Gilze Rijen
D-5806	TF-104G	5806	stored, arr 18-10-84, dep 4-3-92	to Gilze Rijen
D-5810	TF-104G	5810	stored, arr 23-11-84, dep 11-6-88	to Den Haag
D-8048	F-104G	8048	stored, arr 12-10-84, dep 1989	to Delft
D-8051	F-104G	8051	stored, arr 8-11-84, dep 3-90	to Stolzenau
D-8053	F-104G	8053	stored, arr 11-10-84, dep 1988	to Nijmegen
D-8084	F-104G	8084	stored, arr 18-10-84, dep 12-5-89	to Prinsenbosch
D-8091	F-104G	8091	stored, arr 25-10-84, dep 5-90	to Eindhoven
D-8258	F-104G	8258	stored, arr 26-11-84, dep 12-92	to Ochsenboden
D-8259	F-104G	8259	stored, arr 11-10-84, dep 9-9-85	to Den Haag
D-8266	F-104G	8266	stored, arr 9-11-84, dep 3-5-89	to Texel
D-8268	F-104G	8268	stored, arr 25-10-84, dep 13-11-89	to Twenthe
D-8281	F-104G	8281	stored, arr 1-11-84, dep 4-12-89	to Schaarsbergen
D-8300	F-104G	8300	stored, arr 1-11-84, dep 18-8-89	to Weesp
D-8312	F-104G	8312	stored, arr 18-10-84, dep 1993	to Gilze Rijen
D-8326	F-104G	8326	stored, arr 2-11-84, dep 1990	to Volkel
D-8331	F-104G	8331	stored, arr 8-11-84	to Soesterberg

ZORGVLIED
A recreation and camping site named Groot Bartje in this small village has a MiG-21PF at their gate. The aircraft, which arrived by February 1996, has the wings of 0404 (see Ulft).
❑ 0403　　　MiG-21PF　　　..　　　preserved, ex Slusovice, ex Neubuz　　　5-97

ZWANNENBURG
At the lasergames Silverstone on the Werenweg 21-23 (near Halfweg) is a former Baarlo MiG-21.
❑ 2245   MiG-21SPS   6410   preserved, ex Baarlo, ex Seifert., ex 963/NVA   1-98

ZWIJNDRECHT
The former Well Mi-2 9452 should be somewhere in this town.
❑ 9452   Mi-2   563405044   preserved, ex Well, ex Briest, ex 305/NVA   98

ZWOLLE
The College voor Beroepsonderwijs Zwolle (CBZ) near the Zwolle centrum exit of the highway has a number of aircraft as instructional airframes. F-104G D-8268 was placed on a pole in January 1996 in a non standard grey colour scheme. HUP-2 82 was late noted here in 1995 and went to Lelystad.
❑ D-8268   F-104G   8268   preserved, on pole, ex Leeuwarden, ex Twenthe   10-97
❑ K-4012   NF-5B   4012   instructional, ex Woensdrecht   10-97
❑ PH-NDC   S-12   6287   instructional, ex Lelystad, ex Badhoevendorp   10-97
❑ PH-RLN   SAAB 91D   91-379   instructional, ex Schiphol   10-97

The Harley Davidson Indian American Motorcycle Museum at the Oude Almeloseweg 2 had the fuselage of a second An-2 SP-WWI. By late 1996 the aircraft moved to Heino. Three other aircraft are still current.
❑ D-EADK   FWP149D   023   stored, ex 9013/WGAF   10-97
❑ SP-FAX   An-2   1G26-20   preserved, ex SP-UXD, ex 2620/Polish AF   10-97
❑ SP-TSC   Mi-2   524738046   preserved, ex Twenthe   10-97

NORWAY

In the early 1990s it was decided that a new Norwegian Air Force museum was to be built at Bodø, to replace the collection at Gardermoen. The airfield of Gardermoen was rebuilt to become the new Oslo international airport. Due to the fact that Bodø is somewhat in the middle of nowhere, the number of visitors to the museum was disappointing. In 1997 it has been decided that there would again be an Air Force museum at Gardermoen! This new museum will display aircraft which are currently held in store (at Bodø and Gardermoen) as well as new acquisitions.

ANDØYA
This northern P-3 Orion base should still have a Starfighter as an instructional airframe.

| ❑ 755 | | CF-104 | 1055 | instructional, ex 12755/RCAF | 5-89 |

BANAK LAKSELV
The Thunderflash became part of the RNorAF museum in 1985 and may still be preserved at the airfield.

| ❑ 27332 | AZ-S | RF-84F | .. | preserved, ex 52-7332 | 85 |

BARDUFOSS
Bardufoss is the home base of the military Bell 412 and Lynx helicopters. Also here is a training school with a number of instructional airframes.

❑ 584		UH-1B	230	instructional, ex 60-3584	12-95
❑ 586		UH-1B	232	instructional, ex 60-3586	8-94
❑ 591		UH-1B	237	preserved, at gate, ex 60-3591	5-95
❑ 132		F-5A	N7076	instructional, wfu 23-2-95, ex 69-7132	2-95
❑ 41313	SO-I	F-86K	213-83	preserved, ex 54-1313, with tail from 54-1334	8-94
❑ 836		CF-104	1136	instructional, ex 12836/RCAF	8-94
❑ 324		SAAB 91B-2	91324	instructional, ex 6324/AE-U	12-95
❑ N9023W		WW1123A	10	instructional, unmarked	12-95

BERGEN
Preserved Sabre 31082/AH-A moved to the museum at Sola. The Fw190F-8 931862 which was under restoration here was sold to the USA and became N91FW.

| ❑ 801 | CF-104 | 1101 | stored, ex Gardermoen, 12801/RCAF | — |

BODØ
Town: The Nowegian Aviation Musuem (formely known as the Norsk Luftfartssenter) is in the southern parts of Bodø town. The museum is open (Mid June to Mid August) between 10:00 and 20:00. On Saturdays the museum closes at 17:00. The rest of the year it is open on Wednesday between 10:00 and 19:00, Saterdays between 11:00 and 17:00, Sundays between 11:00 and 18:00 and during the rest of the week between 10:00 and 16:00. Safir LN-FMU moved on to Værnes. The PT-26 is the former 261 (ex FZ708) and has parts from LN-BIS (T40-269, ex 119). Some of the other aircraft may move to Gardermoen.

❑ 881478 '4D-AM'	Ju88A-4	881478	preserved, wreck, ex BH-QQ/Luftwaffe	98
■ AA-622	T-6J	CCF4-491	preserved, marked as '8570/BS-M', ex WGAF	98
❑ 6306	Ju52/3mg3e	5664	preserved, marked as 'LN-DAF', ex Port AF	98
■ N6972	DH.82A	82210	preserved, marked as '141', ex RAF	98
❑ TW117	Mosquito T.3	..	preserved, marked as 'TD753/KK-T'	98
❑ 56-6953	U-2CT	393	preserved, arr 28-3-94, ex USAF	98
❑ 103	Avro 504K	..	preserved	98

NORWAY - 350

☐ N5641	HE-G	Gladiator II	..	preserved, ex Gardermoen	98
☒ 349		Fokker C.VD	133	preserved, ex SE-ALS, ex Gardermoen	98
☐ 33		FF9 Kaie	77	preserved, ex Gardermoen	98
☐ MH350	FN-T	Spitfire LF.IXe	CBAF.IX490	preserved	98
☐ V0184	ZK-U	Vampire FB.52	V0184	preserved, ex Oslo	98
☐ 337		SAAB 91B-2	91337	preserved	98
☐ 102		RF-5A	RFG1003	preserved, ex Kjevik, ex 68-9102	98
☐ 28465	MU-P	F-84G	..	preserved, forward fuselage, ex Gardermoen	98
☐ 117047	AZ-A	RF-84F	..	preserved, nose only, ex Værnes, 51-17047	98
☒ 31206	FN-D	F-86F	202-135	preserved, ex Gardermoen, 53-1206	98
☐ 4637		CF-104D	5307	preserved, ex 12637/RCAF	98
☐ 079		UH-1B	1203	preserved, white UN colours, ex 64-14079	98
☒ 835		L-18C	18-3235	preserved, ex 53-4835	98
☒ 712		O-1A	21386	preserved, ex 50-1712	98
☒ 261		PT-26B	..	preserved, marked as '163/DM-L', ex Garderm.	98
☐ C-FIKT		DHC-3	81	preserved, marked as 'AM-O' ex Gardermoen	98
☐ C-FIZO		PBY-6A	2009	preserved, marked as 'FP535/X', ex N10013	98
☐ LN-BND		PA22	22-8947	preserved	98
☐ LN-DBW		C.5 Polar	1/1948	preserved, ex LN-11	98
☐ LN-GAH		Grunau ESG-9	..	preserved, glider	98
☐ LN-GBM ?		Grunau SG-38	..	preserved, glider	98
☐ LN-GGS		L-13 Blanik	..	preserved, glider	98
☐ LN-GGV		FES-530	0245	preserved, glider	98
☐ LN-HAK		PA28-140B	28-20733	preserved, ex SE-EOF	98
☐ LN-LMN		DHC-6-300	127	preserved	98
☐ LN-ORW		Bell 47G	632	preserved	98
☐ LN-TVY		Ce337D	1084	preserved, ex N12500	98
☐ ..		F-104	..	preserved, cockpit only	7-97
☐ ..		Hurricane IIb	..	preserved, replica, marked as 'BD374/FN-D'	98
☐ ..		Spitfire F.Vc	..	preserved, replica, marked as 'BR81C/AH-V'	98

Airfield: Preserved outside the civil terminal is a former Soviet MiG-15. F-104G Starfighter 800, previously stored here, should still be current at the Asphaugen Videregænde School, located somewhere between the airbase and the town.

☐ 34 blue		MiG-15UTI	771 ?	preserved, ex 34 red, ex Soviet AF	98
☐ 2564		F-104G	9010	preserved, at gate, marked as '104', ex WGAF	98
☐ 800		CF-104	1100	instructional, ex 12800/RCAF, see note	5-89
☐ 597		UH-1B	243	instructional	5-95

Furthermore there are a number of aircraft stored or under restauration at the airfield, owned by the museum. The Mirage 3E carries the code from the EdC00.070 unit of Châteaudun, the unit who ferried the aircraft from Nancy to Bodø.

☐ AR-120		RF-35	351120	stored, arr 21-12-93, ex RDanAF	98
☐ 2237		MiG-21SPS	5210	stored, ex Gardermoen, ex Dessau, ex 891/NVA	98
☐ 588	MV	Mirage 3E	588	stored, ex French AF, see note	98
☐ 850	8J-NL	Arado 66C	850	stored, ex Luftwaffe	98
☐ ..		Arado 66C	..	stored, fuselage only, ex Luftwaffe	98
☐ ..		Arado 66C	..	stored, fuselage only, ex Luftwaffe	98
☐ ..		Arado 66C	..	stored, fuselage only, ex Luftwaffe	98
☐ 13470		Bf109G-2	13470	stored, ex Luftwaffe	98
☐ 6304		Ju52/3mg3e	5661	under restaration, ex Port AF	98
☐ L2910	8K	Blackburn Skua	..	stored	98
☐ WD955		Canberra T.17A	..	stored, arr 3-95, ex RAF	98
☐ 33		Pe-2FT	2/225	under restauration, ex Soviet	98
☐ ..		Pe-2FT	16/141	stored, ex Soviet	98

			NORWAY - 351		
☐ J-4027		Hunter F.58	..	stored, ex Swiss AF	98
☐ 67-0333	GA	F-4E	3168	stored, ex Davis Monthan, ex USAF	98
☐ 57-2247		T-37B	40180	stored, ex Davis Monthan, ex USAF	98
☐ 41245	RI-Z	F-86K	213-15	stored, ex Oslo, 54-1245	98
☐ ..		VL Sääski II	..	stored, replica, marked as '40'	98
☐ 870		CF-104	1170	stored, ex 12870/RCAF	98
☐ 890		CF-104	1190	stored, ex 12890/RCAF	98

GARDERMOEN
Displayed at the transport squadron's HQ buildings is Sabre 41290. In 1998 Gardermoen became the new international airport of Oslo. The based 335skv will move to a corner of the airfield and probably take their Sabre with them (this aircraft was preserved near the terminal). Stored in one of the hangars was former Rygge F-5, 207 (N7030), which was last noted in June 1995. This is now in Hawthorne, USA

☐ 41290	ZK-Z	F-86K	213-60	preserved, ex 54-1290	6-95

In the early 1990s the decision was made to move the museum of Gardermoen to Bodø. A large number of aircraft have moved to this new location (which see). Other departures from here include Starfighters 801 (to Bergen) and 882 (to Kongsberg). By 1997 the decision had been made to build another museum at Gardermoen. Displayed here will be some of the stored aircraft, a number of new airframes and some from Bodø. The aircraft marked * will most likely be included in the new museum, as well as a CF-104, F-5A, J-4A, Tiger Moth, SAAB 91 and an Il-2M3 (ex Kirkenes). All the aircraft listed below are from the times when the Forsvarsmuseet Flysamlingen at Gardermoen was still open.

☐ ..		*	Arado 66C	..	stored, frame only, ex Luftwaffe	83
■ 1526	5J-CN	*	He111P-1	1526	stored, ex Luftwaffe	6-94
☐ 6134	CO-EI		Ju52/3mg4e	6134	stored, ex Luftwaffe	91
☐ 6657	CA-JY	*	Ju52/3mg4e	6657	under restoration, ex Luftwaffe	91
☐ 6791	VB-UB		Ju52/3mg4e	6791	stored, ex Luftwaffe	89
☐ 55657	1Z-BY		Ju52/3mg4e	55657	under restoration, ex Luftwaffe	88
☐ 881033	4D-FH	*	Ju88C	881033	under restoration, ex Luftwaffe	—
■ 117546	DP-K	*	T-33A	7691	stored, ex French AF, ex 51-17546	6-94
☐ 28456	93	*	Sk28C-3	..	stored, marked as 'PX-A', ex RSweAF	—
■ 93797	BW-K	*	C-47A	13749	stored, ex 42-93797	6-94
■ 111209	MU-S	*	F-84G	..	stored, nose only, ex 51-11209	6-94
☐ 117051	T3-E	*	RF-84F	..	stored, no rear fuselage, ex 51-17051	6-89
■ 117053	AZ-G	*	RF-84F	..	stored, ex 51-17053	6-94
■ 41274	RI-T	*	F-86K	213-44	stored, ex 54-1274	6-94
☐ 25069	AH-D	*	F-86F	191-765	stored, ex 52-5069	89
■ 469		*	TF-104G	5779	stored, ex 2771/WGAF, ex 63-8469	6-94
■ 966		*	UH-1B	1090	stored, ex 64-13966	8-94
☐ 64279	HA-B	*	UH-19D	..	stored, ex 56-4279	89
☐ 845	AI-F	*	L-18C	18-3245	stored, ex 53-4845	6-94
■ 320	U	*	Northrop N-3PB	320	stored, crashed 1-4-43	6-94
☐ 4953		*	O-1A	21838	stored, ex 51-4953	8-94
☐ 103		*	PT-19	T40-208	stored, ex 208, ex 42-76477	6-94
■ 105		*	Avro 504K	..	stored, maybe ex B5405	89
■ 131		*	RAF BE.2e	..	stored, ex 59, ex A1380	6-94
■ 25		*	Farman F.46-80R	25	stored	6-94
☐ 329			SAAB 91B-2	91329	stored, ex 57-329/AB-U	6-94
☐ PL979	ZB-A	*	Spitfire PR.XI	6S/583719	stored, wfu 18-5-54	6-94
■ P42408	PX-E	*	Vampire F.3	EEP42408	stored, ex VT833	6-94
☐ G-BMEW		*	C-60A	18-2444	under restoration, ex OH-SIR, ex OH-MAP	88
■ LN-BDP			Norseman IV	64	stored, ex 2491/RCAF and marked as such	6-94
☐ LN-DAV		*	S.1A Cadet	203/1041	stored, ex 505 and marked as such, ex Jarlsberg	—
☐ LN-LFK			SAAB 91B-2	91341	stored, ex 341 and marked as such	6-94

NORWAY - 352

❏ LN-ORM	*	Bell 47D-1	642	stored, composite, marked as 'BE-D'	89
❏ N90FW	*	Fw190A-3	731283	stored	—
❏ --	*	Kjølseth PK X-1	1	stored	—
❏ --		Larsen Special	..	stored, uncompleted airframe	88

GRIMSTAD

A private collector here has acquired, besides components and memorabilia, a Sabre and the forward fuselage of a Freedom Fighter and Thunderflash.

❏ 101		F-5A	RFG1002	preserved, ex Raufoss, ex 68-9101	—
❏ 41338		F-86K	213-108	preserved, forward fuselage only, ex 54-1338	—
❏ ..		RF-84F	..	preserved, forward fuselage only	—

HVALSMOEN

Preserved here (exact location unknown) should still be an ex 334 Skv F-86K Sabre. Hvalsmoen is some 40km north west of Oslo

❏ 41241	RI-Z	F-86K	213-11	preserved, ex 54-1241	—

KIRKENES

The Grenseland museum at this extreme northerly town has an Il-2. This aircraft will become part of the new Gardermoen museum in due course.

❏ 2	Il-2M3	303560	preserved, ex Soviet AF	—

KJELLER

Still guarding the gate at the Air Force depot is a Starfighter.

❏ 766	CF-104	1066	preserved, marked as '104', ex 12766/RCAF	5-95

KJEVIK

The airfield of Kjevik has a large instructional school. Of their aircraft, RF-5A 102 went to Bodø in 1993. Stored UH-1B 995 (515) was sold to the USA as N96NW. The Norseman IV is under restoration for the museum at Sola.

❏ F-500		F-86D	173-633	stored, ex Karup, RdanAF, 51-8500	2-98
❏ F-977		F-86D	173-121	stored, ex Karup, RdanAF, 51-5977	2-98
❏ F-984		F-86D	173-128	stored, ex Karup, RdanAF, 51-5984	2-98
❏ 81-0683	LF	F-16A	61-364	instructional, ex USAF	2-98
❏ 130		Bell 412SP	33130	instructional, crashed 4-1-88	2-98
❏ 105		RF-5A	RFG1006	instructional, arr 5-85, ex 68-9105	2-98
❏ 563		F-5A	N7011	instructional, arr 2-83, ex 65-10563	2-98
❏ 572		F-5A	N7020	instructional, arr 1974, ex 65-10573	2-98
❏ 595		F-5B	N9004	instructional, arr 7-95, ex 65-10595	2-98
❏ 900		F-5A	N7063	dumped, ex instructional, ex 67-14900	2-98
❏ 117055	AZ-H	RF-84F	..	preserved, ex 51-17055	2-98
❏ 4818		CF-104	1118	stored, ex instructional, ex 12818/RCAF	2-98
❏ 588		UH-1B	810	instructional, ex 63-8588	6-89
❏ 235		Lynx Mk.86	235	instructional, with parts of 350	2-98
❏ 840		MFI-15	15840	instructional, crashed 22-11-85	2-98
❏ 641		O-1A	..	instructional, wingless, ex 51-12641	6-89
❏ 333		SAAB 91B-2	91333	stored, ex instructional, ex LN-SAP	4-95
❏ 068		Sea King Mk.43	WA749	instructional, crashed 10-11-86	2-98
❏ LN-BDR		Norseman IV	92	under restoration, for Sola museum	2-98

NORWAY - 353

KONGSBERG
An arms factory (Våpenfabrik) has taken delivery of a former Gardermoen Starfighter.

❏ 882		CF-104	1182	stored, ex Gardermoen, ex 12882/RCAF	—

ØRLAND
Ørland is home of one of the Air Force's F-16 squadrons and has a number of *Wrecks & Relics* aircraft spread around the base. The F-16 fuselage is of an unknown crashed example.

❏ 113		RF-5A	RFG1014	instructional, arr 5-5-87, ex 68-9113	10-97
❏ 210		F-5A	N7033	instructional, ex Rygge, 69-9210	9-97
❏ 215		F-5A	N7038	preserved, wfu 1994, ex 66-9215	10-97
❏ ..		F-16	..	instructional, fuselage only	9-97
❏ 22912	MU-J	F-84G	..	preserved, ex 52-2912	9-97
❏ 25202	MU-F	F-86F	191-898	preserved, ex 52-5202	9-97

OSLO
Town: The Norsk Teknisk Museum is located in the northern part of town and is open daily (except Mondays) between 10:00 and 19:00 (from June to August). In 'winter' the closing time may vary between 15:00 and 21:00. Preserved Gemini LN-TAH went to Sola

❏ 759		CF-104	1059	preserved, ex 12759/RCAF	5-95
❏ P42459	PX-H	Vampire F.3	EEP42459	preserved, nose only	5-95
❏ LN-BAH		Loening C2	308	preserved, ex NC10239	5-95
❏ LN-BWT		Aero C-104	189	preserved, ex OK-BFJ	5-95
❏ LN-HAL		G44A Widgeon	1332	preserved, ex SE-ARZ	5-95
❏ LN-KLH		SE210-3	3	preserved	5-95
❏ LN-ORD		Bell 47J	1562	preserved	5-95
❏ ..		Blériot XI	794	preserved	—
❏ ..		FF7 Hauk	..	preserved, nose only	5-95

The Forsvarsmuseet is located near the town centre at the harbour side. The museum is located inside a former fortress. Of the collection the Sabre 41245, Spitfire MH350 and Vampire V1084 have moved to Bodø. Their place should have been taken by a replica Spitfire (note that there is also a replica Spitfire at Bodø marked AH-V, same aircraft) and a PT-26.

❏ 151		DH.82A	161	preserved, ex SE-ANL	89
❏ LN-OAU		PT-26B	..	preserved, ex 205/Cl-L and marked as such	—
❏ ..		Spitfire F.Vc	..	preserved, replica, marked as 'AH-V'	—

Airfield – Fornebu: This airfield will be replaced by Gardermoen in 1998 as the international airport. What will happen to the Starfighter and UH-1B here is unknown.

❏ 889	CF-104	1189	instructional, ex 12889/RCAF	8-94
❏ 937	UH-1B	457	instructional, ex 62-1937	8-94
❏ LN-KLG	Convair 440	506	instructional, fire dump	8-96

RAUFOSS
An ammunition factory here received two ex Sola RF-5As in September 1987. 101 and 110 (RFG1011) were both said to have been scrapped at Gjovik in 1989, but the complete 101 is now privately owned in Gridstad.

RYGGE
This F-5 base has dispatched one of its former stored Freedom Fighters to Ørland, where 210 can now be found.

❏ 225	F-5A	N7048	stored, for spares, wfu 1-3-89, ex 66-9225	—
❏ 594	F-5B	N9003	stored, wfu 15-5-94, ex 65-10594	—

NORWAY - 354

❏ 895		F-5A	N7058	preserved, on pole, ex 67-14895	6-95
❏ 886		CF-104	1186	stored, in shelter, ex 12886/RCAF	5-93
❏ 025		UH-1B	545	preserved, in 720 Skv area, ex 62-2025	5-93
❏ 961		UH-1B	1085	stored, ex 64-13961	5-93

SKEDSMO
A school at this village has a complete F-5 on their grounds.

❏ 898	F-5A	N7061	instructional, ex 67-14898	6-94

SOLA
The Flyhistorik Museum Sola is open daily between 22th June and 9th August between 12:00 and 16:00. During the winter (29th November to 26th April) it is closed, while during the rest of the year it is only open on Sundays between 12:00 and 16:00.

❏ 2605	IV	Harvard II	66-2338	stored, ex RCAF	97
❏ AR-114		S-35XD	351114	preserved, arr 21-2-94, ex RDanAF	97
❏ DT-571		T-33A	5903	preserved, marked as '16571/DP-X', ex RDanAF	97
❏ L-857		PBY-5A	928	preserved, marked as '382/KK-A', ex RDanAF	97
❏ ..	63-?L	Arado 66		under restauration, ex Luftwaffe	97
❏ 4246	PI-OT	Arado 96B-1	4246	preserved, ex Luftwaffe	97
❏ ..	6W-?N	Arado 196A-2	..	preserved, frame only, ex Luftwaffe	97
❏ 14141	DG-UF	Bf109G-1	14141	preserved, ex Luftwaffe	97
❏ J-4110		Hunter F.58A	..	preserved, arr 16-6-95, ex Swiss AF, ex XF318	97
❏ U-1217		Vampire T.55	977	preserved, ex Swiss AF	97
❏ 580		UH-1B	226	preserved, ex 60-3580	97
❏ 688		UH-1B	268	stored, ex 61-0688	97
❏ 220		F-5A	N7043	preserved, marked as '14896/DP-C', ex 66-9220	97
❏ 117045	AZ-N	RF-84F	..	preserved, ex 51-17045	97
❏ 28470	WH-Z	F-84G	..	preserved, with tail from 110161, ex 52-8470	97
❏ 31082	AH-A	F-86F	202-11	preserved, ex Bergen, ex 53-1082	97
❏ 41266	ZK-L	F-86K	213-36	preserved, ex 54-1266	97
❏ 730		CF-104	1030	preserved, ex 12730/RCAF	97
❏ 336		SAAB 91B-2	91336	preserved, ex Gardermoen	97
❏ F-BDHZ		MS502	540	preserved, composite, with parts from a MS500	97
❏ HB-LOU		CeFT337HP	0023	stored, ex OO-ADI, ex F-GCVP	97
❏ LN-BFY		Heron 1B	14015	preserved, marked as 'LN-PSG', ex G-AOXL	97
❏ LN-BNR		FWP149D	088	preserved, ex D-EBDA, ex 9070/WGAF	97
❏ LN-FAD		Piel CP301	L-355/67	preseved	97
❏ LN-FAG		Taylorcraft A	416	stored	97
❏ LN-GBH		Bergflake II/55	202	preserved, glider, ex D-1271	97
❏ LN-GBI		Elliott AP8	018	preserved, glider	97
❏ LN-MAF		BN-2A-21	441	stored, ex G-BCZS	97
❏ LN-KLK		Convair 440	357	preserved, ex SE-BSR	97
❏ LN-LMD		RC680FL	1401-56	preserved, ex N8484A, ex CF-SHC	97
❏ LN-NAU		L-4H	11673	preserved, ex 43-30382	97
❏ LN-SCA		Meise	527	preserved, glider, ex LN-GAR	97
❏ LN-SUF		F27-100	10298	preserved, ex PH-SAN	97
❏ LN-TAH		M65 Gemini 1A	6528	stored, ex Oslo, ex G-AKKA	97
❏ SE-HME		Bell 47G-3B-1	WA583	preserved, marked as 'LN-ORB', ex G-BBZL	97

Adjacent to the museum, in a seperate part of the same building, is a small military instructional school which has three airframes.

❏ 717	CF-104	1027	instructional, ex 12717/RCAF	97

		NORWAY - 355			

| ☐ 6326 | AG-I | SAAB 91B-2 | 91326 | instructional, crashed 26-1-62 | 97 |
| ☐ N95B | | Jet Commander | .. | instructional | 97 |

TORP
The RF-84F here is still part of the Forsvarsmuseet Flysamlingen of Gardermoen and may still be current here.

| ☐ 28723 | AZ-X | RF-84F | .. | instructional, ex 52-8723 | 85 |

VÆRNES
The nose of RF-84F 117047 moved to <u>Bodø</u> by the early 1990s, whilst the remainder of it was scrapped here.

| ☐ LN-FMU | | SAAB 91B | 91267 | stored, ex Bodø, ex Gardermoen, 50058 | 6-96 |

POLAND

The Polish *Wrecks & Relics* scene is largely dominated by the military base of Mierzęciece. Nearly 300 redundant aircraft were noted there over the last decade. Large quantities of the Polish Lims from Mierzęcice can be found all over the world. An-2s, MiG-21s, Su-7s and TS-11s were also stored at this base for disposal. As with most other former Eastern Bloc states, aircraft can be found in the villages all over the country in use as monuments or as 'play-things'. Poland has several museum collections, of which Kraków is the largest.

ANDRYCHÓW
Preserved in the town centre, along road number 69, should still be one of Poland's many Lims.
❑ 1527   Lim-2   1B-01527   preserved   10-91

BABIE DOŁY
Just north of Gdynia is the navy airfield of Babie Doły which houses 1DLMW with its An-2s, An-28s, Mi-2s and MiG-21s. The gate is guarded by a MiG-21PF on a pole. The correct identity is not yet confirmed but it could be 0901 which is known to be preserved here, but has never been seen as such.
❑ ..   Lim-1   ..   preserved, at base housing area, marked as '001'   3-96
❑ 926   MiG-17PF   0926   preserved, ex Gdynia   —
❑ 0901   MiG-21PF   760901   preserved, at gate, marked as '0716', see note   4-98

Close by, in a small village a TS-8 should be preserved.
❑ ..   TS-8   ..   preserved, marked as '001'   98

BABIMOST
The Lim at the civil side is painted in the same colours as the surrounding buildings, a non standard green/grey/brown camouflage. The stored VEB-14 3069 moved on to <u>Drzonów</u>.
❑ 605   Lim-6R   1J-0605   preserved, at barracks, ex '1982'   8-95
❑ 627   Lim-6R   1J-0627   preserved, at terminal   8-97

Near the main gate a scrapyard was discovered in 1997. The dumped Mi-2s all came from Łódź.
❑ 1246   Mi-2P(VIP)   531246109   dumped, cabin and tail, ex Łódź   8-97
❑ 2124   Mi-2URP   562124121   dumped, cabin, ex Łódź   8-97
❑ 2125   Mi-2T   562125121   dumped, complete, ex Łódź   8-97
❑ 2127   Mi-2URN   562127121   dumped, complete, ex Łódź   8-97
❑ 3104   Mi-2URN   563104093   dumped, ex Łódź, soc 30-12-95   8-97
❑ 4602   Mi-2TSz   544602016   dumped, cabin and tail, ex Łódź   8-97
❑ 4720   Mi-2T   514720036   dumped, cabin, ex Łódź   8-97
❑ 6001   Mi-2Ch   516001039   dumped, cabin and tail, ex Łódź   8-97

BALICE
During the early 1990s a number of fighters were stored at this military transport base on behalf of the museum at Kraków. All these aircraft are now at <u>Kraków</u> including MiG-15UTI 2004, Su-7BM 06, Su-7UM 116 and Su-7BKL 806. Preserved VEB-14P 3073 (14803073) was sold to serve as a restaurant, this may be the unidentified aircraft noted at <u>Sanok</u>.
❑ 5705   An-2TD   1G157-05   stored, to go to Kraków, wfu 15-5-96   —
❑ 3037   VEB-14P   14803037   stored   4-98
❑ 3065   VEB-14P   14803065   stored   4-98
❑ 3067   VEB-14P   14803067   stored   4-98
❑ 6604   MiG-21MF   966604   stored, ex Mierzęcice   8-96

BIAŁA BŁOTA
Preserved in this small village, some 5km south of Bydgoszcz, is an L-410 in Pepsi colours.
☐ SP-FTN	L-410MU	781102	preserved, ex CCCP-67291	9-96

BIAŁA PODLASKA
The gate at this TS-11 training base is guarded by an unknown Polish Lim-5.
☐ ..	Lim-5P	..	preserved, at gate, marked as '5053'	4-96

BYDGOSZCZ
This airfield is in use as a maintenance base. Types like the Mi-2, TS-11 and Su-22 are overhauled here. A line up of aircraft are displayed between the hangars.

☐ 7448	An-2T	1G74-48	preserved	4-98
☐ 4	CCS-13	..	preserved, local built Polikarpov Po-2	8-97
☐ 4710	Mi-2P	534710036	preserved	4-98
☐ 1302	Lim-5	1C-1302	preserved	4-98
☐ 2001	MiG-21M	962001	preserved	4-98
☐ 139	MiG-23MF	0390217139	preserved	4-98
☐ 03	Su-7BM	5303	preserved, marked as '3117'	4-98
☐ 7125	Su-20	74105	preserved	4-98
☐ 209	TS-11-100	1H-0209	preserved	4-98

Over the years a large number of aircraft have been noted at the base. Of these a few have found new homes. Lim-6R 611 (1J-0611) moved to Topeka in the USA, while Lim-5 605 (1C-0605) went to Schenectady, USA and Lim-6bis 325 (1F-0352) went to Quonset. MiG-21PF 1811 went to Mierzęcice, 2406 (762406) also moved to Schenectady and 2410 (762410) now flies as N21MF. The barracks opposite the base are no longer in military hands, therefore the fate of MiG-21PF 0608 is unknown. Stored Su-7BKL 816 (7816, l/n 2-93) has been scrapped.

☐ 0614	Mi-2RL	510614018	stored	8-97
☐ 1242	Mi-2P(VIP)	531242119	stored	8-97
☐ 2950	Mi-2RM	552950063	stored, ex navy	8-97
☐ 4509	Mi-2RL	554509125	stored	8-97
☐ 4708	Mi-2P	534708036	stored	8-97
☐ 7837	Mi-2Ch	517837102	stored	8-97
☐ 207	Lim-5	1C-0207	stored	8-91
☐ 414	Lim-5	1C-0414	stored	8-91
☐ 1230	Lim-5	1C-1230	stored	8-91
☐ 1418	Lim-5R	1C-1418	stored	8-91
☐ 1427	Lim-5R	1C-1427	stored	8-91
☐ 1623	Lim-5R	1C-1623	stored	8-91
☐ 1626	Lim-5	1C-1626	stored	8-91
☐ 1712	Lim-5R	1C-1712	stored, ex navy	8-91
☐ 1714	Lim-5R	1C-1714	stored	8-91
☐ 1718	Lim-5R	1C-1718	stored, ex navy	8-91
☐ 1723	Lim-5	1C-1723	stored	8-91
☐ 215	Lim-6bis	1F-0215	stored	8-91
☐ 413	Lim-6R	1J-0413	stored	8-91
☐ 419	Lim-6R	1J-0419	stored	8-91
☐ 0608	MiG-21PF	760608	preserved, at barracks opposite the base	5-92
☐ 0907	MiG-21PF	760907	stored	8-97
☐ 2408	MiG-21PF	762408	stored	8-97
☐ 2411	MiG-21PF	762411	stored	3-95
☐ 2412	MiG-21PF	762412	stored	8-97

POLAND - 358

❏ 516	TS-11-100 bisB	1H-0516	stored		8-96
❏ 519	TS-11-100 bisB	1H-0519	stored		8-96
❏ 808	TS-11-100 bisB	1H-0808	stored		8-96
❏ 814	TS-11-100 bisB	1H-0814	stored		8-96
❏ 905	TS-11-200Art bisC	2H-0905	stored		3-95

On 28th February 1997 the last operational Su-20s were ferried here for storage. 6265 has moved to Dęblin.

❏ 6135	Su-20	74725	stored, arr 28-2-97	97
❏ 6136	Su-20	74726	stored, arr 28-2-97	97
❏ 6138	Su-20	74828	stored, arr 28-2-97	97
❏ 6250	Su-20	74210	stored, arr 28-2-97	97
❏ 6252	Su-20	76302	stored, arr 28-2-97	97
❏ 6256	Su-20	74416	stored, arr 28-2-97	97
❏ 6259	Su-20	74209	stored, arr 28-2-97	97
❏ 6262	Su-20	74312	stored, arr 28-2-97	97
❏ 6264	Su-20	74314	stored, arr 28-2-97	97

CHORZÓW
Somewhere in this town, just north-west of Katowice, a MiG-21 should be preserved.

❏ 1913	MiG-21PF	761913	preserved	—

CZAPLINEK
Along road number 163, from Wałcz to Kołobrzeg, the small town of Czaplinek can be found. Along the main road through the town a MiG-17 is preserved.

❏ 102	Lim-6M	1D-0102	preserved	10-97

DARŁÓWKO
Preserved in the village of Darłówko (near the airfield of Darłowo), is an unidentified Il-28.

❏ ..	Il-28	..	preserved	..

DĘBLIN
The Wyższa Szkoła Oficerska Sił Powietrznych is based at Dęblin. This training base has a number of airframes on display outside the main gate.

❏ 86	LWD Junak 3	..	preserved, at gate, marked as '12'	4-98
❏ 1929	Lim-2	1B-01929	preserved, at gate, marked as '1980'	4-98
❏ 948	MiG-17PF	0948	preserved, at gate	4-98
❏ 405	SM-2	S204005	preserved, at gate	4-98
❏ 910	TS-8	1E-0910	preserved, at gate, ex 0910	4-98
❏ 801	TS-11-100 bisB	1H-0801	preserved, at gate	4-98
❏ 25	Yak-18	..	preserved, at gate, marked as '26'	4-98
❏ 087	Yak-23	..	preserved, at gate	4-98

A second line of preserved aircraft is in front of the main hangars.

❏ BA-21		Mirage 5BA	21	preserved, ex Belgian AF, ex Weelde	4-98
❏ 65		Il-28E	2212	preserved, see Wrocław	4-98
❏ 197		MiG-15UTI	142697	preserved	4-98
❏ 9113		MiG-21MF	969113	preserved, ex instructional	4-98
❏ 021		MiG-23MF	0390224121	preserved	4-98
❏ 210	7	TS-11-100	1H-0210	preserved, demo team c/s	4-98
❏ 6265		Su-20	74415	preserved, ex Bydgoszcz	4-98

The school has a small hangar where some instructional airframes are parked. The MiG-21MF is parked outside and carries no serial.

❏ 2702	Mi-2RM	552702122	instructional, ex navy		8-97
❏ 6602	MiG-21MF	966602	instructional, marked as 'Szkolny'		8-97
❏ 007	PZL 130TM	01880007	instructional		8-97
❏ 614	TS-11-100 bisB	1H-0614	instructional		8-97
❏ 901	TS-11-200Art bisC	2H-0901	instructional		97
❏ 1105	TS-11-200SB bisD	1H-1105	instructional		8-97

Over the years a number of other aircraft have been noted at the base. Some of them may still be around, but were not seen in August 1997.

❏ 0001	Lim-5	1C-0001	instructional, marked as '001' on portside	8-95
❏ 305	Lim-5	1C-0305	stored, fuselage only, ex Mierzęciece	2-95
❏ 307	MiG-15/17	..	dumped, see Drzonów and Łódź	5-93
❏ 197	MiG-15UTI	142697	stored	8-95
❏ 309	MiG-17PF	0309	preserved, front fuselage only	2-95
❏ 8010	MiG-21MF	968010	stored	9-94
❏ 8011	MiG-21MF	968011	instructional	8-95
❏ 8014	MiG-21MF	968014	instructional	8-95
❏ 8705	MiG-21MF	968705	instructional	8-95
❏ 0528	TS-11-100 bisB	1H-0528	instructional	8-95
❏ 2002	TS-11-200SB bisD	3H-2002	stored, wreck in hangar, crashed 5-7-95	8-95
❏ ..	TS-11	..	preserved, marked as '1979'	2-95
❏ SP-PWC	I-22 Iryda	1ANP01-04	instructional	8-95

On the south side, just outside of the base, a school has a pole-mounted TS-8 in their front yard.

❏ ..	TS-8	..	preserved, marked as '1977'	8-97

DRZONÓW

The Lubuskie Muzeum Wojskowe can be found it the small village of Drzonów. Drzonów is some kilometers west of Zielona Góra, on the south west side of the country. The museum is open from Wednesday till Friday between 09:00 and 15:30. On Saturday and Sunday it is open between 10:00 and 16:00.

❏ 11	Avia B33	5339	preserved, Czech built Il-10	4-96
❏ 50	Il-28	56583	preserved	8-97
❏ 1	LET C-11	64233	preserved, Czech built Yak-11	8-97
❏ 104	Mi-2M2	ZD0104054	preserved	8-97
❏ 1624	Mi-2Sz	541624100	preserved	8-97
❏ 314	Mi-4A	03141	preserved	8-97
❏ 112	SBLim-2M	1A-10012	preserved, ex navy	8-97
❏ 1809	Lim-2	1B-01809	preserved	8-97
❏ 8020	SBLim-2Art	1A-08020	preserved, ex navy	8-97
❏ 307	MiG-17PF	0307	preserved	8-97
❏ 635	Lim-6R	1J-0635	preserved	8-97
❏ 1721	Lim-5R	1C-1721	preserved, ex Mierzęcice	8-97
❏ 1318	MiG-21U-400	661318	preserved	8-97
❏ 2307	MiG-21F-13	742307	preserved	8-97
❏ 7815	MiG-21PFM	94N7815	preserved, ex Mierzęcice	8-97
❏ 1105	SM-1	S1-01105	preserved	8-97
❏ 1005	SM-2	S2-01005	preserved	8-97
❏ 905	Su-7UM	2905	preserved	8-97
❏ 0310	TS-8	1E-0310	preserved, inside	8-97
❏ 0506	TS-11-100 bisB	1H-0506	preserved	8-97
❏ 0530	TS-11-200Art bisC	2H-0530	preserved	8-97

POLAND - 360

☐ 0823	TS-11-200Bis R	4H-0823	preserved, ex navy		8-97
☐ 3069	VEB-14P	14803069	preserved, ex Babimost		8-97
☐ 06	Yak-23A	807	preserved		8-97
☐ SP-BRN	Yak-18	6131	preserved, inside, marked as '08', ex DM-WDG		8-97
☐ SP-CXH	Yak-12A	12640	preserved, unmarked, ex SP-BKK		8-97
☐ SP-LAS	Li-2P	18423203	preserved		8-97

ELBLĄG
Stored at this dual military/civil airfield along the road to Kaliningrad is an engineless An-2.

☐ 9859	An-2T	1G98-59	stored	4-96

GDYNIA
On the south side of the southern pier, where a warship is on display, the Muzeum Marynarki Wojennej is located. The aircraft overlook the sandy beach. The preserved MiG-17PF 926 should have moved to Babie Doły, it is no longer here. The museum is open all week (except Mondays) between 10:00 and 17:00.

☐ 69	Il-28R	41302	preserved, ex navy	4-98
☐ 1717	Mi-4ME	02177	preserved, ex navy	4-98
☐ 316	Lim-6bis	1F-0316	preserved, ex navy	4-98
☐ 8905	MiG-21bis	75078905	preserved, crashed 13-10-94, ex navy	4-98
☐ 4017	SM-1W	S1-04017	preserved, ex navy	4-98
☐ 414	TS-11-100 bisA	1H-0414	preserved, ex navy	4-98
☐ 2	Yak-9P	..	preserved, ex navy	4-98

GLIWICE
The airfield and its MiG-21 are just south of Gliwice, near road number 91. Gliwice itself is west of Katowice.

☐ 5615	MiG-21PFM	94A5615	preserved, ex Mierzęcice	10-97

GOLENIÓW
North of Szczecin, near the German border, is the town of Goleniów. In the town, and not on the airfield, a MiG-15 is preserved.

☐ ..	Lim-2	..	preserved, marked as '1967'	10-97

GOSZCZYŃ
Along road 7 from Radom to Warszawa is the village of Goszczyń. Preserved here, and clearly visible from road number 7, with the Arm Ex Trading company is an unmarked TS-11.

☐ ..	TS-11	..	preserved, unmarked	8-97

GRODZISKO
There are serveral villages named Grodzisko. In one of them (which one?) an An-24 is dumped along the main road though the village.

☐ SP-LTC	An-24B	67302208	dumped	4-96

INOWROCŁAW
In this town a Yak-23 marked as '1988' was noted on a pole in May 1992. By 1994 this aircraft had gone from its pole and was noted in Bydgoszcz for an accident exercise where the Yak-23 had collided with a tram! The final fate of the aircraft is unknown.

JÓZEFÓW
Along one of the smaller roads from Warszawa to Dęblin, a MiG-21 is preserved. This road, number 801, runs on the east side of the Wisła river.

☐ 0613	MiG-21PF	760613	preserved	8-97

KALISZ
Bt taking road number 450 from the centre of Kalisz to Grabów, near the edge of the town is a school with a MiG-17 on a pole. Its exact version is not known, either a Lim-5, Lim-6M or Lim-6R.

☐ 517	MiG-17	..	preserved	8-97

KOŁOBRZEG
The Muzeum Oreza Polskiego can be found in the centre of the coastal town of Kołobrzeg. The museum is near the river Parsęta and the large cathedral.

☐ 64	Il-28	2113	preserved, marked as '52'	8-97
☐ 1705	Lim-2	1B-01705	preserved, marked as '1984', ex '1980'	8-97
☐ 728	MiG-19PM	620728	preserved, marked as '723'	8-97
☐ 3010	SM-2	S2-03010	preserved, marked as '417'	8-97
☐ 0710	TS-8	1E-0710	preserved, ex '315', ex 710	8-97
☐ SP-BFA	CCS-13	8-0511	preserved, inside, local build Polikarpov Po-2	8-97
☐ ..	Fw190F-8	..	preserved, inside, wreck	8-97

KOZIENICE
North east of Radom is the town of Kozienice. A Lim-5 can be found mounted on a pole on one of the smaller roads in the west side of this town.

☐ 1212	Lim-5	1C-1212	preserved	8-97

KRAKÓW
In the Bielany part of Kraków (towards Balice) a Yak-23 is preserved. The aircraft is marked with the number of the regiment which restored it.

☐ ..	Yak-23	..	preserved, marked as '1616'	4-98

Outside the District Military Museum are some heavy guns, tanks and a Lim-5. The museum is in the Ul. Rakowicka street, which is close to the museum of Kraków.

☐ 506	Lim-5R	1C-0506	preserved, ex museum	8-97

The Muzeum Lotnictwa Polskiego (till 1988 known as Muzeum Lotnictwa i Astronautyki) is located at the former airfield of Rakowici. The airfield as such is no longer recognisable. The museum has a large collection of aircraft and is open on Tuesday to Friday between 09:00 and 16:00, Saturday between 10:00 and 15:00 and on Sunday between 10:00 and 16:00. Admission is only 4 zloty. On Monday it is free of charge, but then only the outside display can be viewed. The museum is closed in the winter (between 1st December and 30th April).

Outside:

■ 87916		A-37B	43063	preserved, ex Vietnam, ex 68-7916	4-98
■ 00852		F-5E	R1033	preserved, ex Vietnam, ex 73-0852	4-98
■ 3078		VEB-14P	14803078	preserved, ex 001	4-98
■ 72		Il-28R	41909	preserved	4-98
☐ S3		Il-28U	69216	preserved	4-98
☐ 027		Li-2T	18439102	preserved, unmarked	4-98
■ 511		Mi-4A	15114	preserved	4-98
■ 617	C	Mi-4ME	06175	preserved	4-98

POLAND - 362

☐ 014		SBLim-2	1A-06014	preserved	4-98
☒ 018		SBLim-2M	1A-06018	preserved, ex Mierzęcice, ex navy	4-98
☐ 035		SBLim-2	1A-06035	preserved	4-98
☐ 304		MiG-15UTI	3404	preserved	4-98
☒ 712		Lim-1	1A-07012	preserved	4-98
☒ 1230		Lim-2	1B-01230	preserved	4-98
☒ 2004		MiG-15UTI	27004	preserved, ex Balice	4-98
☐ 105		Lim-6bis	1F-0105	preserved	4-98
☐ 415		Lim-5	1C-0415	preserved	4-98
☒ 606		Lim-6M	1D-0606	preserved	4-98
☒ 618		Lim-6MR	1D-0618	preserved	4-98
☐ 1023		Lim-5	1C-1023	preserved	4-98
☐ 1414		Lim-5R	1C-1414	preserved, ex Mierzęcice	4-98
☐ 1508		Lim-5	1C-1508	preserved	4-98
☐ 1910		Lim-5R	1C-1910	preserved, ex Mierzęcice	4-98
☒ 905		MiG-19PM	650905	preserved	4-98
☒ 01		MiG-21PFM	94ML01	preserved	4-98
☒ 809	C	MiG-21F-13	740809	preserved	4-98
☒ 1125		MiG-21R	94R021125	preserved	4-98
☒ 1217		MiG-21U-400	661217	preserved	4-98
☒ 1901		MiG-21PF	761901	preserved	4-98
☐ 2004		MiG-21PF	762004	preserved, ex Mierzęcice	4-98
☐ 2009		MiG-21PF	762009	preserved, ex Mierzęcice	4-98
☒ 4401		MiG-21US	01685144	preserved	4-98
☒ 6504		MiG-21MF	966504	preserved	4-98
☒ 6513		MiG-21PFM	94A6513	preserved, ex Mierzęcice	4-98
☐ 6614		MiG-21PFM	94A6614	preserved, ex Mierzęcice	4-98
☒ 120		MiG-23MF	0390217120	preserved	4-98
☒ 01		Su-7BM	5301	preserved, ex Mierzęcice	4-98
☒ 06		Su-7BM	5306	preserved, ex Balice	4-98
☒ 116		Su-7 M	2UU6	preserved, ex Balice	4-98
☒ 806		Su-7BKL	7806	preserved, ex Balice	4-98
☐ 807		Su-7BKL	7807	preserved, ex Mierzęcice	4-98
☒ 4242		Su-20	6602	preserved	4-98
☒ 1007		TS-11-100 bisB	1H-1007	preserved	4-98
☒ ..		WSK M15	1C006-01	preserved	4-98
☒ SP-PBL		MD12F	0004	preserved	4-98

Inside:

☒ SM411	AU-Y	Spitfire LF.XVIe	..	preserved	4-98
☒ Y61501		T-6G	168-87	preserved, ex Tunesian AF, ex 49-2983/TA-983 and marked as such, ex AT-6C 41-32501	4-98
☒ 4		Avia B33	3061	preserved, Czech built Il-10	4-98
☒ 4		Polikarpov Po-2LNB	641-646	preserved, ex 02 , ex Soviet AF	4-98
☒ 8.63		PZL P11c	562	preserved, marked as 2	4-98
☒ 4316		Mi-2URP	564316105	preserved	4-98
☒ 0309		TS-8	1E-0309	preserved	4-98
☒ 36		Yak-11	64236	preserved	4-98
☒ 16		Yak-23	1216	preserved	4-98
☒ B.350/17		LVG B.IIa	..	preserved	4-98
☒ C.197/15		Albatros C.I	..	preserved	4-98
☒ C.12250/17		Aviatik C.III	1996	preserved	4-98
☒ CL.15459/17		Halberstadt CL.II	1046	preserved	4-98
☒ D.2225/18		LFG Roland D.VIb	..	preserved	4-98
☒ ..		Tu-2S	..	preserved	4-98

POLAND - 363

	Registration	Type	Serial	Status	Date
■	B7280	Camel F.1	..	preserved	4-98
■	D-EKDU	Albatros B.IIa	10019	preserved, marked as 'B.1302/17', ex 'NG-UR'	4-98
■	D-IRIK	Hawk II	H.81	preserved, unmarked, ex D-1515, ex D-3165	4-98
■	F-BGCQ	DH.82A	86514	preserved, marked as 'T8209/10', ex NM206	4-98
■	SP-AAG	LWD Szpak 4T	48-004	preserved	4-98
■	SP-AFO	Bü131B	13113	preserved	4-98
❏	SP-AJB	PWS 26	81.123	preserved, unmarked	4-98
■	SP-AKG	RWD21	331	preserved, marked as 'SP-PBE', ex YR-AEZ	4-98
❏	SP-AOP	Yak-18	EM005	preserved	4-98
■	SP-ARM	Zlin 26	640	preserved	4-98
■	SP-ASZ	Yak-12	5013	preserved	4-98
■	SP-AXT	CSS S13	9-3505	preserved, locall build Polikarpov Po-2	4-98
■	SP-BNU	RWD13	283	preserved, ex SP-ARL	4-98
■	SP-BPL	LWD Junak 3	13-9578	preserved	4-98
■	SP-FXA	L-60	150723	preserved	4-98
■	SP-GIL	BZ-1 Gil	1	preserved	4-98
■	SP-GLC	UC-78	112793	preserved, ex SP-LEM	4-98
■	SP-GLM	Yak-17W	3120132	preserved, marked as '02'	4-98
■	SP-LXH	Aero 145	172011	preserved	4-98
■	SP-NXA	L-200A	170409	preserved, ex SP-NAA	4-98
■	SP-PAK	PZL M4P	1/3	preserved	4-98
■	SP-PBB	PZL S4	02	preserved	4-98
■	SP-SAD	SM-1/300	S101003	preserved	4-98
■	SP-SAP	SM-2	S202016	preserved	4-98
■	..	Blériot XI	..	preserved, replica	4-98
■	..	Farman HF-4	..	preserved, replica, marked as '4'	4-98
■	..	Grigorovich M15	BnC.262	preserved	4-98
■	..	JK-1	2	preserved	4-98

The museum also has a number of aircraft in storage. Of these Lim-5R 506 has since gone to the District Military Museum in town. The stored Yak-18 SP-APR (EM019) went to Belgium and is in flying condition.

	Registration	Type	Serial	Status	Date
❏	42-98643	L-5 Sentinel	..	stored	6-97
❏	1909	Lim-5R	1C-1909	stored, outside	4-98
❏	23	SM-1	S1-15007	stored	4-98
❏	04	TS-11	11PR04	stored	4-98
❏	A.180/14	Stahltaube	76	stored	6-97
❏	C.17077/17	DFW C.V	473	stored	8-97
❏	D-EFRI	Etrich Taube	..	stored, replica	6-97
❏	D-EKYQ	Albatros L.101	245	stored	6-97
❏	D-INJR	Me209V-1	1185	stored, fuselage only	8-97
❏	D-OMIP	Heinkel HE5e	..	stored	8-97
❏	SP-AAA	LWD Szpak 2	01	stored, under restoration	6-97
❏	SP-AAB	LWD Szpak 3	02	stored	6-97
❏	SP-AAX	LWD Zak 3	10	stored	4-98
❏	SP-ADE	Polikarpov Po-2VS	9725	stored	8-97
❏	SP-ADM	LWD Junak 2	732	stored	8-97
❏	SP-API	CSS-13R	420-64	stored, local build Polikarpov Po-2	8-97
❏	SP-AFP	L-4A Cub	10524	stored, ex 43-29233	8-97
❏	SP-AOU	Yak-18	EM016	stored	8-97
❏	SP-BAD	LWD Zuch 1	17	stored	8-97
❏	SP-BAL	L-4H	11806	stored, ex SP-ALD, ex 43-30515	8-93
❏	SP-BAM	LWD Zuch 2	22	stored	8-97
❏	SP-BAO	LWD Zuch 2	24	stored	8-97
❏	SP-BAR	CSS-13	1	stored, cockpit only, local Polikarpov Po-2	6-97
❏	SP-BRI	Yak-18	9732	stored	8-97

POLAND - 364

☐ SP-GLA	LWD Junak 1	16		stored	8-97
☐ SP-GLB	LWD Zuraw	26		stored	6-97
☐ SP-GLF	TS-8	P1		stored	6-97
☐ SP-KFB	PZL 106A	07810131		stored	4-98
☐ SP-LXC	Aero Super 45	03002		stored, outside	4-98
☐ SP-TNA	Aero 145	172006		stored, outside, ex OK-NHH	4-98
☐ SP-590	HWL Pegaz	1		stored	8-97
☐ ..	BZ-4 Zuk	4		stored	6-97
☐ ..	Geest Moewe IV	4		stored	6-97
☐ ..	Levasseur Antoinette	..		stored	8-97
☐ ..	Lippisch Storch VII	..		stored	6-97
☐ ..	Wagner Eule E2	..		stored	8-97

KRZESINY
There are no recent reports from this airfield near Poznań and it is not known if the MiG-21 is still on the dump.

☐ 4010	MiG-21US	10685140	dumped	8-91

ŁASK
Town: In town two aircraft are preserved. Near the junctions of road 14 and 44 a Lim-2 is mounted on a pole.

☐ 708	Lim-2	1B-00708	preserved, marked as '1996', ex '1980'	10-97

Airfield: At the base three older MiG-21s should be preserved.

☐ 0703	MiG-21PF	760703	preserved, in barracks	2-93
☐ 1801	MiG-21PF	761801	preserved, in barracks	2-93
☐ ..	MiG-21PFM	..	preserved, marked as '8001'	8-95

LEGNICA
In the town centre a Soviet MiG-17 was noted in 1991.

☐ 30 red	MiG-17	..	preserved, ex Soviet	8-91

LEŹNICA WIELKA
Of the three stored Mi-6s 670 moved on to Łódź and 063 (741063) and 671 (10671) have been sold to the Ukraine. The preserved Mi-2 carries an incorrect serial as the real 5432 (535432127) is a Czech Air Force Mi-2.

☐ 3036	VEB-14P	14803036	dumped, ex preserved	4-98
☐ 2122	Mi-2FM	512122121	dumped, set in concrete	3-96
☐ ..	Mi-2	..	preserved, marked as '5432'	4-98
☐ 0314	Mi-8T	0314	dumped	2-93
☐ 1015	Mi-14PS	A1015	stored, in hangar, crashed, ex Navy	7-95
☐ ..	Lim-2	..	preserved, at gate, marked as '1973'	2-93

ŁOBEZ
Along the road to Wegorzyno an unknown Lim-2 is mounted on a pole.

☐ ..	Lim-2	..	preserved, marked as '1974'	8-97

ŁÓDŹ
Airfield: At the airfield, west of road number 14, maintenance is carried out on all types of helicopters. A large number of Mi-2s are stored here. Some may return to operational service, while others will be scrapped. Mi-24s, Mi-14s and Mi-8s can be seen here in outside storage, awaiting overhaul. Mi-2s removed for scrapping included

POLAND - 365

1245, 2124, 2125, 2127, 3104, 4602, 4720 and 6001. All went to Babimost.

❏ 0602	Mi-2RL	530602127	stored	4-96
❏ 0623	Mi-2RL	519623028	stored	4-98
❏ 2121	Mi-2FM	512121121	stored	4-96
❏ 2123	Mi-2URP	512123121	stored	4-98
❏ 2128	Mi-2URN	562128121	stored	8-95
❏ 2214	Mi-2TSz	542214012	stored	4-98
❏ 2639	Mi-2URN	562639112	stored	4-98
❏ 2641	Mi-2T	562641112	stored	9-94
❏ 2643	Mi-2	562643112	stored	9-94
❏ 2648	Mi-2URN	562648112	stored	12-97
❏ 3003	Mi-2P	533003063	stored	4-98
❏ 3107	Mi-2URN	563107093	stored	4-98
❏ 3223	Mi-2URN	563223113	stored	4-98
❏ 3225	Mi-2URN	563225113	stored	4-98
❏ 3228	Mi-2URP	563228123	stored	4-98
❏ 3621	Mi-2R	563621084	stored	5-93
❏ 3648	Mi-2R	563648094	stored	4-98
❏ 3720	Mi-2D	513720094	stored	4-98
❏ 3726	Mi-2D	513726084	stored	12-97
❏ 3728	Mi-2D	513728084	stored	4-98
❏ 3729	Mi-2D	513729084	stored	5-95
❏ 3802	Mi-2D	513802094	stored	4-98
❏ 3803	Mi-2D	513803094	stored	4-98
❏ 3804	Mi-2Ch	513804094	stored	4-98
❏ 3826	Mi-2Ch	513826104	stored	4-98
❏ 3828	Mi-2D	513828104	stored	10-97
❏ 3951	Mi-2TSz	543951035	stored	4-98
❏ 4038	Mi-2P	544038035	stored	4-98
❏ 4043	Mi-2TSz	544043035	stored	4-98
❏ 4314	Mi-2URP	564314105	stored	4-98
❏ 4402	Mi-2URP	564402105	stored	12-97
❏ 4403	Mi-2URP	564403115	stored	4-98
❏ 4437	Mi-2RL	554437115	stored	5-91
❏ 4546	Mi-2TSz	544546016	stored	8-96
❏ 4549	Mi-2TSz	544549016	stored	4-98
❏ 5245	Mi-2D	515245077	stored, ex navy	7-96
❏ 5338	Mi-2R	565338117	stored	8-96
❏ 5339	Mi-2R	565339117	stored	4-98
❏ 5340	Mi-2R	565340117	stored	4-98
❏ 5342	Mi-2R	565342117	stored	3-95
❏ 5345	Mi-2R	565345127	stored	4-98
❏ 5348	Mi-2R	565348127	stored	8-96
❏ 5825	Mi-2R	565825128	stored	4-98
❏ 6048	Mi-2Ch	516048049	stored	8-90
❏ 7801	Mi-2P	517801082	stored, at civil side	4-98
❏ 8221	Mi-2URP	568221063	stored	7-96
❏ 8222	Mi-2URP	568222063	stored	4-98
❏ 8223	Mi-2URP	568223063	stored	4-98
❏ 8825	Mi-2URPG	568825104	stored	12-97
❏ 8826	Mi-2URPG	568826104	stored	7-96
❏ 523	Mi-8T	0523	stored	12-97
❏ 611	Mi-8T	10611	stored	12-97
❏ 655	Mi-8T	10655	stored	4-98
❏ 014	Mi-24D	A1014	dumped, crashed 1990	2-92

POLAND - 366

❏ 213	Mi-24D	730213	stored, ex Cottbus, ex 9619/GAF, ex 543/NVA	9-97
❏ 269	Mi-24D	340269	stored, ex Cottbus, ex 9612/GAF, ex 524/NVA	9-97
❏ 271	Mi-24D	340271	stored, ex Basepohl, ex 9634/GAF, ex 529/NVA	9-97
❏ 277	Mi-24D	340277	stored, ex Basepohl, ex 9638/GAF, ex 544/NVA	9-97
❏ 458	Mi-24D	410458	dumped, crashed 11-7-90	2-92

The Wystawa Sprzętu Lotniczego I Wojskowego museum has no official opening times, but visiting the museum is not problem as the compiler discovered during a visit to the museum at 07:00 local time. The entry fee for the museum is only 3 zloty. In the museum's hangar are also some active An-2s and TS-8s. Su-7BKL 813 moved recently to Savigny lès Beaune.

❏ 36/10-RB	Mirage 3C	36	preserved, ex Savigny les Beaune	4-98
❏ 0916	Il-14P	146000916	preserved, unmarked	4-98
❏ ..	Il-28R	2402010	preserved, unmarked	4-98
❏ 5826	Mi-2R	565826128	preserved, inside	4-98
❏ 670	Mi-6A	10670	preserved, ex Leźnica Wielka, ex SP-ITB	4-98
❏ 606	Mi-8T	10606	stored, burnt out wreck	4-98
❏ 002	SBLim-2	1A-09002	preserved, inside	4-98
❏ 006	SBLim-2	1A-06006	preserved	4-98
❏ 109	Lim-2	1B-00109	preserved	4-98
❏ 193	MiG-15UTI	142693	preserved	4-98
❏ 307	Lim-2	1B-00307	preserved	4-98
❏ 311	Lim-2	1B-00311	preserved	4-98
❏ 316	Lim-2	1B-00316	preserved, inside	4-98
❏ 512	Lim-2	1B-00512	preserved	4-98
❏ 602	Lim-2	1B-00602	preserved	4-98
❏ 776	MiG-15UTI	712276	preserved	4-98
❏ 905	SBLim-2	1A-09005	preserved	4-98
❏ 1612	Lim-2	1B-01612	preserved	4-98
❏ 6008	SBLim-2M	1A-06008	preserved, ex navy	4-98
❏ 7039	SBLim-2M	1A-07039	preserved	4-98
❏ 305	Lim-6M	1D-0305	preserved	4-98
❏ 408	Lim-5	1C-0408	preserved, ex Bemowo	4-98
❏ 427	Lim-6bis	1J-0427	preserved	4-98
❏ 608	Lim-6R	1J-0608	preserved	4-98
❏ 637	Lim-6R	1J-0637	preserved	4-98
❏ 1730	Lim-5R	1C-1730	preserved	4-98
❏ 1913	Lim-5R	1C-1913	preserved, inside	4-98
❏ 06	MiG-21PFM	94ML06	preserved, inside, ex Oleśnica	4-98
❏ 09	MiG-21PFM	94ML09	preserved	4-98
❏ 10	MiG-21PFM	94ML10	preserved	4-98
❏ 1702	MiG-21PF	761702	preserved	4-98
❏ 1809	MiG-21PF	761809	preserved	4-98
❏ 2015	MiG-21F-13	742015	preserved, ex Bemowo	4-98
❏ 2720	MiG-21U-600	662720	preserved	4-98
❏ 5705	MiG-21PFM	94A5705	preserved, ex Mierzęcice	4-98
❏ 6509	MiG-21PFM	94A6509	preserved, ex Mierzęcice	4-98
❏ 6510	MiG-21MF	966510	preserved	4-98
❏ 3025	SM-2	S2-03025	preserved, ex Bemowo	4-98
❏ 12	Su-7BKL	6012	preserved, ex Mierzęcice	4-98
❏ 22	Su-7BKL	6022	preserved, ex Mierzęcice	4-98
❏ 702	Su-7UM	3702	preserved	4-98
❏ 6255	Su-20	76305	preserved	4-98
❏ 103	TS-11-100	1H-0103	preserved	4-98
❏ 219	TS-11-100	1H-0219	preserved, inside	4-98
❏ SP-DNG	An-2	1G86-09	stored, at civil side	4-98

POLAND - 367

☐ SP-PBH	PZL 106BR	04003	preserved	4-98
☐ SP-EFG	TS-8	1E-0401	preserved, inside	4-98
☐ SP-FBA	TS-8	1E-0426	preserved, inside, flyable	4-98
☐ SP-FID	An-2T	1G108-68	preserved, ex 0868	4-98
☐ SP-FNM	VEB-14P	14803010	preserved, ex Warszawa Okecie	4-98
☐ SP-FNP	TS-8	1E-0909	preserved, inside	4-98
☐ SP-LHE	Tu-134A	48405	stored	4-98
☐ SP-SPE	Mi-2	514520115	stored, civil side	4-98
☐ SP-TBA	An-2PF	1G159-01	stored, partly burnt	4-98
☐ SP-TVA	Mi-2	533443044	preserved, inside	4-98
☐ SP-WWE	An-2R	1G173-56	preserved, fuselage only	4-98

<u>Town</u>: One of the co-owners of the musuem has a Mercedes garage along road 71 on the north east side of Łódź. Here a MiG-21 is preserved in the front garden, while a number of other aircraft are stored at the back of the garage. Of these a TS-8 with the false markings '0309' went to <u>Cerbaiola</u>.

☐ 1703	MiG-21PF	761703	preserved	10-97
☐ 302	TS-8	1E-0302	stored	8-97
☐ 0309	TS-11-100 bisA	1H-0309	preserved	2-92
☐ ..	TS-11	..	stored, partly burned	8-97
☐ SP-CLZ	TS-8	1E-0925	preserved	2-92
☐ SP-CNO	TS-8	..	stored	8-97
☐ SP-CNS	TS-8	1E-0921	preserved	10-91
☐ SP-FBB	TS-8	1E-0419	dumped, crashed 18-8-91	2-92
☐ SP-FRL	PZL 110	19003	stored	2-92
☐ SP-SXD	SM-1W	W05019	stored	8-97

ŁOWICZ
Preserved within some military barracks is a Lim-1.

☐ 116	Lim-1	1A-10016	preserved	8-90

LUBLINIEC
The local military barracks have an An-2 which is used for para-training. The barracks are on the south side of town along road number 43 from Katowice.

☐ 9862	An-2T	1G98-62	instructional	10-97

MIASTKO
The Lim-5 here shows that it is always unwise to assume serials of aircraft from panels only. 409 has a number of panels and parts from other aircraft fitted. The aircraft is along the main road through town.

☐ 409	Lim-5	1C-0409	preserved	8-97

MIELEC
The local PZL factory here is building M-24s, An-2s and An-28s. A number of unsold aircraft are stored here.

☐ CCCP-28958	An-28	1AJ010-05	stored, not delivered	4-92
☐ CCCP-28959	An-28	1AJ010-06	stored, not delivered	9-96
☐ CCCP-	An-2	1G239-53	stored, not delivered	9-96
☐ CCCP-	An-2	1G239-54	stored, not delivered	9-96
☐ CCCP-	An-2	1G239-55	stored, not delivered, in hangar	9-96
☐ CCCP-	An-2	1G240-32	stored, not delivered	9-96
☐ ..	An-2	1G249-52	stored, not delivered	9-96
☐ SP-DLH	An-2P	1G238-51	stored	9-96

☐ SP-DNO	An-2P	1G119-70	stored		9-96
☐ SP-DNP	An-2R	1G148-60	stored		9-96
☐ SP-FBH	An-2	1G238-20	stored		9-96
☐ SP-FBO	An-2	1G238-18	stored		9-96
☐ SP-PDM	PZL M21	1ALP01-01	stored, derelict		9-96
☐ SP-PDN	PZL M21	1ALP01-02	stored, derelict		9-96
☐ SP-PFA	PZL M24W	1AKP01-02	stored, derelict		9-96
☐ SP-PFB	PZL M24W	1AKP01-03	stored, derelict		9-96
☐ SP-PFC	PZL M24W	1AKP01-04	stored, derelict		9-96
☐ SP-PFD	PZL M24W	1AKP01-05	stored, derelict		9-96

MIERZĘCICE
The airfield is signposted as Pyrzowice, which will bring you to the civil side. Preserved here is an unknown MiG-21. Another MiG-21 can be found at a gate near the (non restricted) military barracks.

☐ 1813	MiG-21PF	761813	preserved, at military gate	8-97
☐ ..	MiG-21MF	..	preserved, marked as '1994', at civil side	8-97

The airfield was home to a large number of stored aircraft. Between 27th September 1993 and 1st October 1993 54 aircraft were officialy scrapped here in accordance with the CFE treaty. These included five Lim-5s, 17 Lim-6bis's, 22 MiG-21PFMs, three MiG-21MFs and seven MiG-21Rs. All aircraft marked dismantled or sectioned were probably all scrapped in that period. The MiG-21Rs are all known and are listed in the next block. In 1997 only a handful of MiG-21s were still visible near the hangars. It is said that all the stored MiG-15s and MiG-17s have left. Known to have been here are:

☐ 0855	An-2T	1G108-55	stored, fuselage only	10-95
☐ 0860	An-2T	1G108-60	stored	2-95
☐ 1462	An-2	1G114-62	stored	9-95
☐ 5215	An-2T	1G52-15	stored	12-95
☐ 7355	An-2T	1G73-55	stored	2-95
☐ 9858	An-2T	1G98-58	stored, fuselage only	10-95
☐ 011	SBLim-2	1A-06011	stored	5-95
☐ 016	SBLim-2	1A-09016	stored	8-91
☐ 029	SBLim-2	1A-06029	stored	8-91
☐ 041	Lim-1 ?	..	stored	6-91
☐ 714	SBLim-2A	1A-07014	stored	8-91
☐ 2001	SBLim-2A	1A-02001	stored, ex Navy	8-91
☐ 3304	MiG-15UTI	3304	stored, ex Navy	8-91
☐ 33.8	MiG-15bis	..	stored, marked as '1963', dismantled on trailer	9-94
☐ 6021	SBLim-2	1A-06021	stored	8-91
☐ 7004	SBLim-2	1A-07004	stored	8-91
☐ ..	MiG-15UTI	522555	stored	8-91
☐ ..	SBLim-2	1A-01004	stored	8-91
☐ ..	SBLim-2	1A-01005	stored	8-91
☐ ..	SBLim-2A	1A-11014	stored	8-91
☐ 206	Lim-6bis	1F-0206	stored	8-91
☐ 209	Lim-5	1C-0209	stored	8-91
☐ 301	Lim-6bis	1F-0301	stored	8-91
☐ 306	Lim-5	1C-0306	stored	2-95
☐ 311	Lim-5	1C-0311	stored, sectioned (f/n 6-91, see next line)	5-93
☐ 311	Lim-6bis	1F-0311	stored	8-91
☐ 314	Lim-5	1C-0314	stored	6-91
☐ 317	Lim-6bis	1F-0317	stored	5-93
☐ 321	Lim-6bis	1F-0321	stored	9-93
☐ 404	Lim-5	1C-0404	stored	8-91
☐ 406	Lim-6bis	1J-0406	stored	8-91

POLAND - 369

☐ 410	Lim-6bis	1J-0410	stored, sectioned	5-93
☐ 422	Lim-6R	1J-0422	stored	5-93
☐ 426	Lim-6bis	1J-0426	stored	8-91
☐ 431	Lim-6bis	1J-0431	stored, sectioned	5-93
☐ 432	Lim-6bis	1J-0432	stored, sectioned	5-93
☐ 440	Lim-6bis	1J-0440	stored, sectioned	5-93
☐ 504	Lim-6R	1J-0504	stored, sectioned	5-93
☐ 509	Lim-5R	1C-0509	stored	8-91
☐ 512	Lim-6R	1J-0512	stored	5-93
☐ 520	Lim-5	1C-0520	stored	8-91
☐ 529	Lim-6R	1J-0529	stored, sectioned	5-93
☐ 609	Lim-6R	1J-0609	stored	9-94
☐ 615	Lim-6bis	1J-0615	stored	8-91
☐ 628	Lim-6R	1J-0628	stored	8-91
☐ 629	Lim-6R	1J-0629	stored	8-91
☐ 633	Lim-6R	1J-0633	stored	8-91
☐ 1008	Lim-5	1C-1008	stored	8-91
☐ 1011	Lim-5	1C-1011	stored	12-95
☐ 1101	Lim-5R	1C-1101	stored	2-95
☐ 1304	Lim-5	1C-1304	stored	8-91
☐ 1306	Lim-5	1C-1306	stored	12-95
☐ 1308	Lim-5	1C-1308	stored	9-94
☐ 1314	Lim-5	1C-1314	stored	8-91
☐ 1409	Lim-5	1C-1409	stored	2-95
☐ 1411	Lim-5	1C-1411	stored	8-91
☐ 1417	Lim-5	1C-1417	stored	12-95
☐ 1424	Lim-5R	1C-1424	stored	8-91
☐ 1425	Lim-5R	1C-1425	stored	8-91
☐ 1502	Lim-5	1C-1502	stored	12-95
☐ 1509	Lim-5	1C-1509	stored	8-91
☐ 1613	Lim-5R	1C-1613	stored	2-95
☐ 1615	Lim-5	1C-1615	stored	2-95
☐ 1620	Lim-5R	1C-1620	stored	8-91
☐ 1624	Lim-5R	1C-1624	stored	8-91
☐ 1629	Lim-5	1C-1629	stored	8-91
☐ 1706	Lim-5	1C-1706	stored	12-95
☐ 1710	Lim-5	1C-1710	stored	12-95
☐ 1711	Lim-5	1C-1711	stored	9-94
☐ 1725	Lim-5	1C-1725	stored	8-91
☐ 1728	Lim-5R	1C-1728	stored	12-95
☐ 1903	Lim-5R	1C-1903	stored	8-91
☐ 1906	Lim-5R	1C-1906	stored	12-95
☐ 1908	Lim-5	1C-1908	stored	8-95
☐ 022	MiG-21PFM	94MO22	stored, behind hangar	10-95
☐ 1506	MiG-21PF	761506	stored	8-91
☐ 1713	MiG-21PF	761713	stored	8-91
☐ 1714	MiG-21PF	761714	stored	8-93
☐ 1804	MiG-21PF	761804	stored, behind hangar	12-95
☐ 1808	MiG-21PF	761808	stored	8-91
☐ 1810	MiG-21PF	761810	stored	9-94
☐ 2005	MiG-21PF	762005	stored	8-91
☐ 4007	MiG-21US	07685140	stored	12-95
☐ 4010	MiG-21PFM	94A4010	stored, dismantled	5-93
☐ 4012	MiG-21PFM	94A4012	stored	5-93
☐ 4014	MiG-21PFM	94A4014	stored, dismantled	5-93

POLAND - 370

☐ 4015	MiG-21PFM	94A4015	stored		8-91
☐ 4102	MiG-21PFM	94A4102	stored		9-94
☐ 4108	MiG-21PFM	94A4108	stored		5-93
☐ 4201	MiG-21PFM	94A4201	stored		8-91
☐ 4203	MiG-21PFM	94A4203	stored		12-95
☐ 4206	MiG-21PFM	94A4206	stored, behind hangar		10-95
☐ 4701	MiG-21US	01685147	stored		5-93
☐ 4910	MiG-21PFM	94A4910	stored		5-93
☐ 4911	MiG-21PFM	94A4911	stored, hangared		5-93
☐ 4912	MiG-21PFM	94A4912	stored		12-95
☐ 4913	MiG-21PFM	94A4913	stored, sectioned		5-93
☐ 5001	MiG-21PFM	95A5001	stored		9-94
☐ 5003	MiG-21PFM	95A5003	stored		5-93
☐ 5004	MiG-21PFM	95A5004	stored, hangared		5-93
☐ 5311	MiG-21PFM	94A5311	stored, behind hangar		12-95
☐ 5312	MiG-21PFM	94A5312	stored, sectioned		5-93
☐ 5315	MiG-21PFM	94A5315	stored, sectioned		5-93
☐ 5401	MiG-21PFM	94A5401	stored, behind hangar		12-95
☐ 5403	MiG-21PFM	94A5403	stored		6-91
☐ 5405	MiG-21PFM	94A5405	stored, sectioned		5-93
☐ 5701	MiG-21PFM	94A5701	stored		4-96
☐ 5702	MiG-21PFM	94A5702	stored, hangared		5-93
☐ 5703	MiG-21PFM	94A5703	stored, hangared		5-93
☐ 6507	MiG-21PFM	94A6507	stored, dismantled, see next line		5-93
☐ 6507	MiG-21MF	966507	stored		8-95
☐ 6512	MiG-21PFM	94A6512	stored, behind hangar, see next line		2-95
☐ 6512	MiG-21MF	966512	stored		2-95
☐ 6513	MiG-21MF	966513	stored		12-95
☐ 6515	MiG-21PFM	94A6515	stored, sectioned		5-93
☐ 6601	MiG-21PFM	94A6601	stored, sectioned		5-93
☐ 6609	MiG-21PFM	94A6609	stored, sectioned		5-93
☐ 6613	MiG-21PFM	94A6613	stored, fuselage only		5-93
☐ 6615	MiG-21PFM	94A6615	stored, fuselage only		5-93
☐ 6701	MiG-21PFM	94A6701	stored		4-96
☐ 6804	MiG-21MF	966804	stored, hangared		5-93
☐ 6813	MiG-21MF	966813	stored		2-95
☐ 6913	MiG-21PFM	94A6913	stored		12-95
☐ 7013	MiG-21PFM	94A7013	stored		12-95
☐ 7015	MiG-21PFM	94A7015	stored		12-95
☐ 7103	MiG-21PFM	94A7103	stored		8-91
☐ 7675	MiG-21MF	96007675	stored, may go to Peutie		12-95
☐ 7788	MiG-21MF	96007788	stored		12-95
☐ 7810	MiG-21MF	967810	stored		12-95
☐ 7811	MiG-21PFM	94N7811	stored, behind hangar		12-95
☐ 7812	MiG-21PFM	94N7812	stored		12-95
☐ 7814	MiG-21PFM	94N7814	stored		12-95
☐ 7901	MiG-21PFM	94N7901	stored, dismantled, see next line		12-95
☐ 7901	MiG-21MF	967901	stored, dismantled		2-95
☐ 7902	MiG-21PFM	94N7902	stored		12-95
☐ 7903	MiG-21PFM	94N7903	stored, see next line		12-95
☐ 7903	MiG-21MF	967903	stored, behind hangar		12-95
☐ 7904	MiG-21PFM	94N7904	stored		12-95
☐ 7905	MiG-21PFM	94N7905	stored, hangared		10-95
☐ 7909	MiG-21MF	967909	stored		12-95
☐ 7914	MiG-21MF	967914	stored, sectioned		5-95

POLAND - 371

☐ 9311	MiG-21UM	516999311	stored		9-94
☐ 4246	Su-20	6606	stored		9-94
☐ 408	TS-11-100 bisA	1H-0408	stored		2-95
☐ 507	TS-11-100 bisB	1H-0507	stored		8-91
☐ 517	TS-11-100 bisB	1H-0517	stored		2-95
☐ 522	TS-11-100 bisB	1H-0522	stored		2-95
☐ 609	TS-11-100 bisB	1H-0609	stored		2-95
☐ 902	TS-11-200Art bisC	2H-0902	stored		2-95

The following aircraft are known to have left Mierzęcice.

4183	An-2P	1G141-83	stored, l/n 2-95	to SP-AOI
5222	An-2T	1G52-22	stored, l/n 12-95	to N2854G
9856	An-2T	1G98-56	stored, l/n 12-95	to N3576G
9863	An-2T	1G98-63	stored, l/n 12-95	to Poznań
9865	An-2T	1G98-65	stored, l/n 12-95	to N3576G
007	SBLim-2	1A-05007	stored, l/n 8-91	to N157GL
008	SBLim-2	1A-09008	stored, l/n 8-91	to Shoreham, UK
012	SBLim-2	1A-09012	stored, l/n 8-91	to Duiven
018	SBLim-2M	1A-06018	stored, ex navy, l/n 8-91	to Kraków
025	SBLim-2	1A-06025	stored, l/n 8-91	to N15MU
026	SBLim-2	1A-06026	stored, l/n 5-93	to Coolidge, USA
032	SBLim-2A	1A-07032	stored, ex navy, l/n 8-91	to N132DG
056	SBLim-2	1A-07056	stored, l/n 8-91	to N76584
116	SBLim-2	1A-10016	stored, l/n 6-91	to Łowicz
126	MiG-15UTI	10926	stored, l/n 5-93	to VH-EKI
203	MiG-15UTI	27003	stored, l/n 5-95	to N678
266	MiG-15UTI	242266	stored, l/n 5-95	to N41125
356	MiG-15UTI	3506	stored, l/n 8-91	to N15H
358	MiG-15UTI	3508	stored, l/n 5-95	to Santa Rosa, USA
604	MiG-15UTI	612304	stored, ex navy, l/n 8-91	to N14694
622	MiG-15UTI	622022	stored, l/n 5-93	to N151MG
627	SBLim-2	1A-06027	stored, l/n 8-91	to Scottsdale, USA
628	MiG-15UTI	622028	stored, l/n 8-91	to N115MG
710	SBLim-2M	1A-07010	stored, ex navy, l/n 8-91	to N710DW
773	MiG-15UTI	712273	stored, l/n 8-91	to Dordrecht
807	SBLim-2	1A-08007	stored, l/n 5-93	to VH-REH
6010	SBLim-2M	1A-06010	stored, ex navy, l/n 8-91	to Warszawa
6247	MiG-15UTI	622047	stored, l/n 8-91	to G-OMIG
7009	SBLim-2M	1A-07009	stored, ex navy, l/n 8-91	to Scottsdale, USA
7031	SBLim-2	1A-07031	stored, l/n 8-91	to N150MG
7048	SBLim-2	1A-07048	stored, l/n 8-91	to N78053
7050	SBLim-2	1A-07050	stored, l/n 8-91	to Australia
..	MiG-15UTI	26012	stored, l/n 8-91	to USA
..	MiG-15UTI	512032	stored, l/n 8-91	to N5557B
..	MiG-15UTI	512036	stored, l/n 8-91	to N115PW
..	MiG-15UTI	522546	stored, l/n 8-91	to N15UT
003	Lim-5	1C-1003	stored, l/n 8-91	to Duiven
305	Lim-5	1C-0305	stored, l/n 9-94	to Dęblin
319	Lim-6bis	1F-0319	stored, l/n 5-93	to Coolidge, USA
414	Lim-6bis	1J-0414	stored, l/n 5-93	to Śrem
418	Lim-6R	1J-0418	stored, l/n 5-93	to Driebruggen
438	Lim-6bis	1J-0438	stored, l/n 5-93	to N2153V, later to N438MG
505	Lim-6R	1J-0505	stored, l/n 5-93	to N2153K, later to N505MG
506	Lim-5R	1C-0506	stored, l/m 8-91	to Kraków
508	Lim-5	1C-0508	stored, l/n 5-93	to N1917M

510	Lim-6R	1J-0510	stored, l/n 7-94	to N17JL
511	Lim-6R	1J-0511	stored, l/n 5-93	to N6953X
514	Lim-6R	1J-0514	stored, l/n 5-93	to Coolidge, USA
522	Lim-6R	1J-0522	stored, l/n 5-93	to Coolidge, USA
523	Lim-6R	1J-0523	stored, l/n 5-93	to Coolidge, USA
528	Lim-6R	1J-0528	stored, l/n 5-93	to N17HQ
631	Lim-6R	1J-0631	stored, l/n 9-94	to N73568
1020	Lim-5	1C-1020	stored, l/n 5-93	to N117MG
1210	Lim-5	1C-1210	stored, l/n 12-95	to N1210
1211	Lim-5	1C-1211	stored, l/n 2-95	to G-BWUF
1217	Lim-5	1C-1217	stored, l/n 8-91	to Warszawa
1220	Lim-5	1C-1220	stored, l/n 8-91	to Dordrecht
1301	Lim-5	1C-1301	stored, l/n 5-93	to Coolidge, USA
1312	Lim-5R	1C-1312	stored, l/n 5-93	to USA
1319	Lim-5	1C-1319	stored, l/n 8-91	to Coolidge, USA
1330	Lim-5	1C-1330	stored, l/n 8-91	to Waganiec
1413	Lim-5R	1C-1413	stored, l/n 8-91	to USA
1414	Lim-5R	1C-1414	stored, l/n 8-91	to Kraków
1419	Lim-5	1C-1419	stored, l/n 5-95	to Poznań
1426	Lim-5	1C-1426	stored, l/n 2-95	to N1426D
1529	Lim-5R	1C-1529	stored, l/n 8-91	to N117BR
1603	Lim-5	1C-1603	stored, l/n 10-95	to USA
1604	Lim-5	1C-1604	stored, l/n 8-91	to Warszawa
1605	Lim-5	1C-1605	stored, l/n 8-91	to Coolidge, USA
1607	Lim-5	1C-1607	stored, l/n 5-93	to N606BM
1611	Lim-5	1C-1611	stored, l/n 8-91	to N217SH
1703	Lim-5R	1C-1703	stored, l/n 8-91	to Coolidge, USA
1705	Lim-5	1C-1705	stored, l/n 8-91	to N968
1707	Lim-5	1C-1707	stored, l/n 9-94	to N9143Z
1713	Lim-5	1C-1713	stored, l/n 8-91	to Coolidge, USA
1719	Lim-5R	1C-1719	stored, l/n 8-91	to N1719
1721	Lim-5R	1C-1721	stored, l/n 9-94	to Drzonów
1726	Lim-5	1C-1726	stored, l/n 8-91	to USA
1910	Lim-5R	1C-1910	stored, l/n 8-91	to Kraków
1139	MiG-21R	94R021139	stored, sectioned, l/n 5-93	scrapped
1375	MiG-21R	94R021375	stored, l/n 5-93	scrapped
1507	MiG-21R	94R021507	stored, l/n 5-93	scrapped
1802	MiG-21PF	761802	stored, l/n 8-91	to Warszawa
1811	MiG-21PF	761811	stored, l/n 7-92	to N21PF
2004	MiG-21PF	762004	stored, l/n 8-91	to Kraków
2009	MiG-21PF	762009	stored, l/n 8-91	to Kraków
2051	MiG-21R	94R022051	stored, l/n 5-93	scrapped
2355	MiG-21R	94R022355	stored, sectioned, l/n 5-93	scrapped
2426	MiG-21R	94R022426	stored, sectioned, l/n 5-93	scrapped
2657	MiG-21R	95R022657	stored, l/n 5-93	scrapped
3056	MiG-21UM	516913056	stored, l/n 8-91	sold as N317DM
4105	MiG-21PFM	94A4105	stored, l/n 8-91	to Ramstein, later to New York
4106	MiG-21PFM	94A4106	stored, l/n 2-95	to Poznań
4107	MiG-21PFM	94A4107	stored, l/n 9-94	to Kalamazoo, USA
4502	MiG-21US	02685145	stored, l/n 9-94	to N221MG
4608	MiG-21US	08685146	stored, l/n 2-95	to N9149F
5011	MiG-21UM	516905011	stored, l/n 8-91	to VH-XXI
5615	MiG-21PFM	94A5615	stored, l/n 12-95	to Gliwice
5705	MiG-21PFM	94A5705	stored, l/n 6-91	to Łódź
6006	MiG-21UM	06695160	stored, l/n 8-91	to Warszawa

POLAND - 373

6509	MiG-21PFM	94A6509	stored, l/n 8-91	to Łódź
6513	MiG-21PFM	94A6513	stored, l/n 8-91	to Kraków
6604	MiG-21PFM	94A6604	stored, l/n 8-91	to Warszawa
6604	MiG-21MF	966604	stored, l/n 10-95	to Balice
6614	MiG-21PFM	94A6614	stored, l/n 8-91	to Kraków
7436	MiG-21MF	96007436	stored, l/n 6-94	to Nowa Sól
7505	MiG-21UM	05695175	stored, l/n 8-91	to N711MG
7809	MiG-21PFM	94N7809	stored, l/n 8-91	to N221GL
7815	MiG-21PFM	94N7815	stored, l/n 9-94	to Drzonów
01	Su-7BM	5301	stored, l/n 8-91	to Kraków
02	Su-7BM	5302	stored, l/n 9-91	scrapped
12	Su-7BKL	6012	stored, l/n 7-90	to Łódź
13	Su-7BKL	6013	stored, l/n 9-91	to Warszawa
14	Su-7BKL	6014	stored, l/n 9-91	scrapped
16	Su-7BKL	6016	stored, l/n 9-91	scrapped
17	Su-7BKL	6017	stored, l/n 9-91	to Warszawa
18	Su-7BKL	6018	stored, l/n 9-91	scrapped
19	Su-7BKL	6019	stored, l/n 8-91	scrapped
20	Su-7BKL	6020	stored, l/n 9-91	scrapped
22	Su-7BKL	6022	stored, l/n 8-93	to Łódź
331	Su-7UM	3313	stored, l/n 8-91	to Warszawa
516	Su-7BKL	6516	stored, l/n 8-91	scrapped
517	Su-7UM	3517	stored, l/n 5-93	scrapped
706	Su-7UM	3706	stored, l/n 5-93	scrapped
804	Su-7BKL	8004	stored, l/n 10-91	scrapped
808	Su-7BKL	7808	stored, l/n 10-91	scrapped
807	Su-7BKL	7807	stored, l/n 8-93	to Kraków
809	Su-7BKL	7809	stored, l/n 8-93	to Skarżysko Kamienna
812	Su-7BKL	7812	stored, l/n 6-91	scrapped
813	Su-7BKL	7813	stored, l/n 5-93	to Łódź
815	Su-7BKL	7815	stored, l/n 8-93	to Warszawa
818	Su-7BKL	7818	stored, l/n 5-93	scrapped
819	Su-7BKL	7819	stored, l/n 5-93	scrapped
821	Su-7BKL	7821	stored, l/n 5-93	scrapped
911	Su-7BKL	7911	stored, l/n 8-91	scrapped
926	Su-7BKL	7926	stored, l/n 5-93	to Opole
4244	Su-20	6604	stored, l/n 2-95	to Bemowo
4245	Su-20	6605	stored, l/n 2-95	to Poznań
201	TS-11-100	1H-0201	stored, l/n 8-91	to N42GS
304	TS-11-100 bisA	1H-0304	stored, l/n 8-91	to N304WV
313	TS-11-100 bisA	1H-0313	stored, l/n 8-91	to N313TS
314	TS-11-100 bisA	1H-0314	stored, l/n 8-91	to N66EN
415	TS-11-100 bisA	1H-0415	stored, l/n 8-91	to N415J
501	TS-11-100 bisB	1H-0501	stored, l/n 8-91	to N501SH
502	TS-11-100 bisB	1H-0502	stored, l/n 8-91	to N24ZR
509	TS-11-100 bisB	1H-0509	stored, l/n 8-91	to N509J
518	TS-11-100 bisB	1H-0518	stored, l/n 2-95	to N2ZB
524	TS-11-100 bisB	1H-0524	stored, l/n 8-91	to N524SH
529	TS-11-100 bisB	1H-0529	stored, l/n 12-95	to N529J
707	TS-11-100 bisB	1H-0707	stored, l/n 2-95	to USA
710	TS-11-100 bisB	1H-0710	stored, l/n 2-95	to Poznań

MIROSŁAWIEC
Preserved outside the gate (in a housing area) at this Su-22 base is a Lim-6.

POLAND - 374

❏ 402	Lim-6M	1D-0402	reserved, marked as '1988/4499'		9-94

MODLIN
Modlin is the airfield where the Polish test unit is based. Of the stored aircraft Lim-5 1717 (1C-1717) became N1717M in 1996 and MiG-15UTI 3303 went to Warszawa.

❏ 6611	An-2	1G66-11	stored, ex instructional	9-94
❏ 7354	An-2T	1G73-54	stored, ex instructional	9-94
❏ 419	Lim-5	1C-0419	stored	9-94
❏ 543	MiG-15UTI	522543	stored	9-94
❏ 1704	Lim-5	1C-1704	stored	9-94

NOWA SÓL
Stored by a local car dealer are a former Mierzęcice MiG-21MF and an unknown Lim-5.

❏ ..	Lim-5	..	stored	97
❏ 7436	MiG-21MF	96007436	stored, ex Mierzęcice	97

NOWE MIASTO NAD PILICĄ
Nowe Miasto is one of Poland's helicopter bases and has two aircraft at its main gate.

❏ 1540	SM-1	..	preserved, at gate	10-97
❏ ..	Lim-5P	..	preserved, at gate, marked as '5058'	10-97

NOWY TOMYSL
In a park in this town, some 50km west of Poznań, is a TS-8.

❏ SP-CNR	TS-8	..	preserved	—

OLEŚNICA
The instructional school here (Centrum Szkolenia Inżynieryjno Lotniczego) has a large number of airframes. Some of these have their original serials removed and have received two figure code numbers, making their identification impossible. MiG-21PFM 06 has recently moved to Łódź. Lim-5 1103 went to East Germany and is now at Cottbus. Mi-24D 015 (A1015, l/n May 1995) has returned to service while Lim-5 413 and Su-7BM 09 went to Hermeskeil.

❏ 7356	An-2T	1G73-56	preserved	4-96
❏ 9868	An-2T	1G98-68	preserved	4-96
❏ 7	Il-28	..	instructional	4-96
❏ S3	Il-28U	67612	dumped	4-92
❏ 1243	Mi-2P(VIP)	531243119	instructional, soc 25-9-90	—
❏ 1244	Mi-2P(VIP)	531244119	instructional	5-95
❏ 3102	Mi-2	563102093	instructional	4-96
❏ 414	Mi-8T	0414	instructional	4-96
❏ 015	Mi-24D	A1015	instructional	5-95
❏ 460	Lim-2	1B-00460	dumped, ex instructional	4-96
❏ 408	Lim-5P	1D-0408	dumped	4-92
❏ 609	Lim-6M	1D-0609	preserved	4-92
❏ 03	MiG-21PFM	94ML03	instructional	4-94
❏ 05	MiG-21PFM	94ML05	instructional	4-94
❏ 430	MiG-21U-400	0430	instructional	4-92
❏ 1807	MiG-21PF	761807	instructional	5-94
❏ 1815	MiG-21M	961815	instructional	5-95
❏ 2005	MiG-21M	962005	instructional	4-92

POLAND - 375

☐ 4503	MiG-21US	03685145	instructional	4-94
☐ 4909	MiG-21PFM	94A4909	instructional	4-92
☐ 5609	MiG-21PFM	94A5609	instructional	5-94
☐ 6509	MiG-21MF	966509	instructional	4-92
☐ 7908	MiG-21MF	967908	instructional	4-92
☐ 8703	MiG-21MF	968703	instructional	4-92
☐ 8855	MiG-21bis	N75078855	instructional, marked as '55'	4-96
☐ ..	MiG-21PFM	..	instructional, marked as '01'	4-96
☐ ..	MiG-21PFM	..	instructional, marked as '05'	4-96
☐ ..	MiG-21PFM	..	instructional, marked as '06'	4-96
☐ ..	MiG-21	..	instructional, marked as '08'	5-95
☐ ..	MiG-21PFM	..	instructional, marked as '09'	4-96
☐ ..	MiG-21	..	instructional, marked as '10'	5-95
☐ ..	MiG-21	..	instructional, marked as '11'	5-95
☐ ..	MiG-21	..	instructional, marked as '12'	5-95
☐ 140	MiG-23MF	0390217140	instructional, marked as '40'	4-96
☐ ..	Su-7/Su-20	..	instructional, in covers	5-95
☐ 6137	Su-20	74827	instructional	4-92
☐ ..	Su-20	..	instructional, marked as '20'	5-95
☐ ..	Su-22M-4K	..	instructional, marked as '22'	5-95
☐ 706	TS-11-100 bisB	1H-0706	instructional	4-96
☐ 821	TS-11-100 bisB	1H-0821	instructional	4-96
☐ ..	TS-11	..	preserved, at gate, marked as '1978'	4-96

OLSZTYN
A large scrapyard in this town holds, beside a number of PZL106's, a large quantity of An-2s.

☐ SP-WCD	An-2R	1G105-49	stored	8-96
☐ SP-WCK	An-2R	1G126-25	dumped	4-96
☐ SP-WKA	An-2R	1G134-50	stored	8-96
☐ SP-WKD	An-2R	1G135-49	stored	8-96
☐ SP-WKK	An-2R	1G144-28	stored	8-96
☐ SP-WKM	An-2R	1G144-30	stored	8-96
☐ SP-WKN	An-2R	1G144-31	stored	8-96
☐ SP-WKY	An-2R	1G144-39	stored	8-96
☐ SP-WLG	An-2R	1G144-47	stored	8-96
☐ SP-WLK	An-2R	1G144-50	stored	8-96
☐ SP-WLW	An-2R	1G148-10	stored	8-96
☐ SP-WMD	An-2R	1G156-19	stored	8-96
☐ SP-WNR	An-2R	1G155-23	stored	8-96
☐ SP-WOF	An-2R	1G160-19	stored	8-96
☐ SP-WOO	An-2R	1G161-04	stored	8-96
☐ SP-WPB	An-2R	1G163-08	stored	8-96
☐ SP-WPI	An-2R	1G163-15	stored	8-96
☐ SP-WPK	An-2R	1G163-16	stored	8-96
☐ SP-WPM	An-2R	1G163-18	stored	8-96
☐ SP-WPW	An-2R	1G167-29	stored	8-96
☐ SP-WWA	An-2R	1G173-52	stored	8-96
☐ SP-WWD	An-2R	1G173-55	stored	8-96
☐ SP-WWH	An-2R	1G173-59	stored	8-96
☐ SP-WWM	An-2R	1G177-03	stored	8-96
☐ SP-WWN	An-2R	1G177-04	stored	8-96
☐ SP-WZL	An-2R	1G185-55	stored, ex SP-DNB	8-96
☐ SP-WNB	An-2R	1G156-39	dumped	4-96
☐ SP-ZEB	An-2R	1G187-10	stored	8-96

☐ SP-ZUM	An-2	..	stored		8-96
☐ SP-ZWI	An-2	..	stored		8-96
☐ SP-ZWY	An-2	..	stored		8-96

OPOLE
At the gate of this airfield an ex Mierzęcice Su-7 is preserved.

☐ 926	Su-7BKL	7926	preserved, ex Mierzęcice	9-96

OSTRÓW WIELKOPOLSKI
In the centre of this town an unknown MiG-17 is preserved.

☐ ..	Lim-6	..	preserved, marked as '1966'	9-96

OŚWIĘCIM
In an Army camp along the main road to the former Auschwitz concentration camp is an unidentified MiG-15.

☐ ..	MiG-15UTI	..	preserved	3-97

OTREBUSY
Along road number 719 from Warszawa, near the town of Grodzisk Maz, the village of Otrebusy can be found. Outside the Muzeum Motoryzacji a TS-8 with Polish markings is preserved.

☐ ..	TS-8	..	preserved, marked as 'B-53', ex 'SP-1'	8-97

PIŁA
Preserved on the former Su-22 airfield of Piła is a MiG-15. Where exactly is not known, but it is not at the gate.

☐ ..	Lim-2	..	preserved, marked as '1973'	4-96

PIOTRKÓW TRYBUNALSKI
The local aero club bought an An-2 in the mid 1990s. The aircraft is in good condition.

☐ 5155	An-2	1G15-155	preserved, ex 155	8-97

POTOK
The small village of Potok is just west of Krosno. Here an unknown TS-8 is mounted on a pole.

☐ ..	TS-8	..	preserved, on pole	..

POZNAŃ
Town: Just north of the town centre is a large park at the former citadel. Near the army museum and a large war monument is the Muzeum Cytadeli Poznańskiej. The museum is open daily between 09:00 and 17:00 (in summer) and 10:00 and 16:00 (in winter). All aircraft are on outside display.

☐ 9863	An-2T	1G98-63	preserved, ex Mierzęcice	4-98
☐ 4	Il-28R	1910	preserved	4-98
☐ ..	Lim-2 ?	..	preserved, marked as '01'	4-98
☐ 1419	Lim-5	1C-1419	preserved, ex Mierzęcice	4-98
☐ 4106	MiG-21PFM	94A4106	preserved, ex Mierzęcice	4-98
☐ 2018	SM-1WS	402018	preserved	4-98
☐ 4245	Su-20	6605	preserved, ex Mierzęcice	4-98
☐ 0306	TS-8	1E-0306	preserved	8-97

POLAND - 377

☐ 710	TS-11-100 bisB	1H-0710	preserved, ex Mierzęcice	4-98
☐ 04	Yak-11	17205	preserved	4-98
☐ 998	Yak-12M	210998	preserved	4-98

An unknown MiG-15 was reported as being preserved on a pole near one of the bridges over the river Warta. A ride along the river in 1997 did not reveal the aircraft.

☐ ..	Lim-2	..	preserved, marked as '1963'	8-93

Airfield - Ławica: Three MiGs can be found on this airfield. The two Lim-2s are preserved well inside the military part of the airfield. The MiG-15 is in fantasy colours (including shark's mouth) near some buildings on the civil side. An-2T 7357 (1G73-57) which was stored at the airfield has been sold as SP-AOH.

☐ 801	Lim-2	1B-00801	preserved, marked as '1988'	3-96
☐ 1920	Lim-2	1B-01920	preserved, at civil side, fantasy colours	4-98

RADOM
Amongst the many operational TS-11s and PZL-130s is an unknown TS-11 on the dump.

☐ ..	TS-11	..	dumped	9-94

RADOMYŚL WIELKOPOLSKI
The PZL M2 should be preserved in the town centre. Radomyśl Wielkopolski is just south west of Mielec.

☐ SP-PAC	PZL M2	..	preserved, poor condition	9-96

RZESZÓW
Stored near the flying school are a number of An-2s. These are in non-flying condition and many have fabric missing. Also stored here are 10 ex Soviet Air Force Mi-2s, these were impounded by the Polish customs for illegal import without proper documentation. Although these retain their military Soviet colours, all have RA markings applied.

☐ RA-23666	Mi-2	544122045	stored, ex 36 red/Soviet AF	9-96
☐ RA-23667	Mi-2	544119045	stored, ex 12 red/Soviet AF	9-96
☐ RA-23668	Mi-2	544115045	stored, ex 28 red/Soviet AF	9-96
☐ RA-23669	Mi-2	544120045	stored, ex 50 yellow/Soviet AF	9-96
☐ RA-23670	Mi-2	544131055	stored, ex 07 red/Soviet AF	9-96
☐ RA-23671	Mi-2	544439105	stored, ex 57 red/Soviet AF	9-96
☐ RA-23672	Mi-2	545008106	stored, ex 80 yellow/Soviet AF	9-96
☐ RA-23673	Mi-2	545429117	stored, ex 71 yellow/Soviet AF	9-96
☐ RA-23674	Mi-2	546101039	stored, ex 39 white/Soviet AF	9-96
☐ RA-23675	Mi-2	548446123	stored, ex 60 red/Soviet AF	9-96
☐ SP-TWA	An-2T	1G172-50	stored	9-96
☐ SP-TWB	An-2T	1G174-43	stored	9-96
☐ SP-TWC	An-2TP	1G177-37	stored	9-96
☐ SP-TWD	An-2TP	1G177-38	stored	9-96
☐ SP-TWL	An-2P	1G181-54	stored	9-96
☐ SP-TWM	An-2T	1G188-35	stored	9-96
☐ SP-TWN	An-2T	1G195-01	stored	9-96

SANOK
Along the road towards Rzeszów, in the extreme south east part of Poland, is an Il-14 in use as a restaurant. The serial is not yet known, but it may be the 3073 which was sold in the mid-1990s from Balice to serve as a restaurant.

☐ 3073 ?	VEB-14P	14803073	preserved, ex Balice?, see note	—

SIEMIROWICE
Inside the gate (visible from outside) of this Polish Navy TS-11 and Antonov airfield is a Lim-6. A second Lim-6 is preserved in the housing area in the village of Siemirowice, The village is to the south of the airfield. This airfield is also sometimes referred to as Cewice. This name is not correct and the name comes from western flight maps.

❏ 101	Lim-6MR	1D-0101	preserved, at gate		8-97
❏ 523	Lim-6MR	1D-0523	preserved, at housing area		8-97

SKARŻYSKO KAMIENNA
A number of aircraft are displayed outside the Museum Orła Białego. The museum is on the south side of town and signposted from the 7/E77 road.

❏ 3054	VEB-14P	14803054	preserved, ex 009, ex SP-LNR	4-98
❏ 611	Mi-4A	..	preserved	4-98
❏ 1526	Lim-2	1B-01526	preserved	4-98
❏ 635	Lim-6MR	1D-0635	preserved	4-98
❏ 1001	MiG-17PF	1001	preserved	4-98
❏ 2401	MiG-21PF	762401	preserved	4-98
❏ 6003	SM-1	S116003	preserved	4-98
❏ 809	Su-7BKL	7809	preserved, ex Mierzęciece	4-98
❏ 21	Yak-23	..	preserved	4-98
❏ SP-CLC	TS-8	1E-0207	preserved, ex 0207	4-98

SŁUPSK
The gate of Poland's only MiG-23 base is guarded by four aircraft. MiG-23M is ex Soviet and has never flown with the Polish Forces.

❏ ..	MiG-23M	4602	preserved, at gate, as '979', ex Soviet AF	8-97
❏ ..	Lim-5	..	preserved, at gate, marked as '1952'	8-97
❏ 908	MiG-19PM	650908	preserved, at gate	8-97
❏ 1815	MiG-21PF	761815	preserved, at gate, marked as '1974'	8-97

SMARDZOWE
On the road 8/E67 (Wrocław-Oleśnica) a gas station has a Lim-5. The aircraft has a coat of thick blue paint and is marked 'System Rolet Okiennych'. The village has also been reported as Zmardzow.

❏ ..	Lim-5	..	preserved	5-95

ŚREM
Along the perimeter fence of an Army signals unit, between some artillery pieces, a MiG is preserved. The barracks are along the main road leading south west towards Dalewo.

❏ 414	Lim-6bis	1J-0414	preserved, marked as '3486', ex Mierzęcice	8-97

STRUMIEŃ
Along route 93 an Il-18 is stored. The aircraft carries faded LOT colours. Its identity is unconfirmed. Strumień is north west of Bielsko, near the Slovak border.

❏ SP-LSH?	Il-18V	181002701	preserved, see note	9-96

ŚWIDNIK
Town: In the town centre an SM-1 is preserved. Its identity is not confirmed.

❏ SP-SAA	SM-1/300	100001	preserved, marked as 'SM-1', see note	9-96

Airfield: The PZL factory is currently building W-3 helicopters here. Beside the aircraft listed below some 80 civil Mi-2s with their rotors removed are also stored at this airfield.

❏ 02	Mi-2M-2	ZD0102113	preserved, unmarked, outside factory	9-96
❏ 306	Lim-2	1B-01306	preserved	4-92
❏ SP-PSB	W-3	300103	instructional, marked as 'W3-TR'	9-96
❏ SP-PSC	W-3	300104	stored	9-96
❏ SP-PSD	W-3	300105	instructional, unmarked, not airworthy	9-96
❏ ..	SM-1	..	preserved, marked as 'SM1', inside factory	9-96
❏ ..	SM-2	..	preserved, marked as 'SM2', inside factory	9-96

ŚWIDWIN

Świdwin is an Su-22 base were a Lim-5 is preserved on a pole in the town outside the station.

❏ ..	Lim-5P	..	preserved, at station, unmarked, ex '1976'	10-97
❏ 8001	Su-22M-4K	28001	dumped, wreck	5-91

SZYMAKI

Some 6km north of Płońsk, along the E77 from Warszawa to Gdańks, is the village of Szymaki. In use as a restaurant is a former Warszawa-Okecie Tu-134.

❏ SP-LHF	Tu-134A	3352005	preserved, ex Okecie, ex 101	4-98

TORUŃ

With the local fire brigade (the Państwowa Straż Pożarne) is some sort of museum. Noted from recent photos are a MiG-21, TS-8 and an other aircraft.

❏ 9015	MiG-21MF	969015	preserved	—
❏ SP-FFA	TS-8	..	preserved	—

WAGANIEC

In this village near Torun a Lim-5 should be preserved.

❏ 1330	Lim-5	1C-1330	preserved, ex Mierzęcice	—

WARSZAWA

Town: Besides the Novotel hotel near the airfield of Okecie is a Lim-5 preserved.

❏ 1501	Lim-5	1C-1501	preserved, near Novotel	4-98

The Museum Wojska Polskiego is along the river Wisła in the town centre. The address is Al Jerozolimski 3.

❏ 5928	An-2P	1G159-28	preserved	4-98
❏ 011	Avia B33	5523	preserved, unmarked, Czech built Il-10	4-98
❏ 21	Il-2m3	..	preserved, unmarked	4-98
❏ 1449	Mi-2	531449040	preserved	4-98
❏ 0614	Mi-8T	0614	preserved	4-98
❏ 0615	MiG-21PF	760615	preserved	4-98
❏ ..	Pe-2FT	..	preserved	4-98
❏ 101	TS-11-100	1H-0101	preserved	4-98
❏ 8	Tu-2S	..	preserved	4-97
❏ ..	Yak-9P	10107	preserved, marked as '23'	4-98
❏ 23	Yak-23	1017	preserved, marked as '12'	4-98

At Fort Sadyba the Katyn Museum is located. At the back of the fort, along road 7 near the Wilanow Palace, is a large storage area. Beside tanks and armoured vehicles a number of aircraft are stored.

POLAND - 380

☐ 22	Il-28	56729	stored, marked as '65'		4-98
☐ 873	MiG-15bis	231873	stored, marked as '365'		4-98
☐ 1312	Lim-2	1B-01312	stored, marked as '1132'		4-98
☐ 3303	MiG-15UTI	3303	stored, ex Modlin		4-98
☐ 6010	SBLim-2M	1A-06010	stored, ex Mierzęcice, ex navy		4-98
☐ 101	Lim-6bis	1F-0101	stored		4-98
☐ 1217	Lim-5	1C-1217	stored, ex Mierzęcice		4-98
☐ 1604	Lim-5	1C-1604	stored, ex Mierzęcice		4-98
☐ 1802	MiG-21PF	761802	stored, ex Mierzęcice		4-98
☐ 6006	MiG-21UM	06695160	stored, ex Mierzęcice		4-98
☐ 6604	MiG-21PFM	96A6604	stored, ex Mierzęcice		4-98
☐ 13	Su-7BKL	6013	stored, ex Mierzęcice		4-98
☐ 17	Su-7BKL	6017	stored, ex Mierzęcice		4-98
☐ 331	Su-7UM	3313	stored, ex Mierzęcice		4-98
☐ 815	Su-7BKL	7815	stored, ex Mierzęcice		4-98
☐ 6131	Su-20	76301	stored		8-97
☐ ..	TS-8	..	stored, fuselage only		3-97

Town – Zoliborz: In this part of town a Lim-6R should be preserved. It is not known if the aircraft is still current and where Zoliborz is located.

☐ 527	Lim-6R	1J-0527	preserved		—

Airfield – Bemowo: In the western suburbs of Warszawa is the airfield of Bemowo. The military side has an instructional school, who sometimes park their aircraft outside. A number of instructional airframes have moved on. SM-2 3025 went to Łódź as did Lim-5 408, MiG-21F-13 2015 and MiG21PFM 06. At the civil side an An-2 is stored.

☐ 7444	An-2T	1G74-44	instructional, at civil side, fuselage only		4-98
☐ 05	Mi-2M2	ZD0105094	instructional		8-97
☐ 0624	Mi-2	510624028	instructional		4-97
☐ 04	MiG-21PFM	94ML04	instructional		4-97
☐ 4205	MiG-21PFM	94A4205	instructional		4-97
☐ 6810	MiG-21MF	966810	instructional		4-97
☐ 4244	Su-20	6604	instructional, forward fuselage, ex Mierzęcice		4-97
☐ 0503	TS-11-100 bisB	1H-0503	instructional		4-97

Airfield – Okecie: Located at the international airfield is a PZL factory, a military area and it is also the main base of LOT. Stored here are some former LOT aircraft of which Il-14 SP-FNM and Tu-134A SP-LHE have moved on to Łódź. Stored Tu-134A SP-LHF went to Szymaki. The TS-8 is in the courtyard of the PZL buildings alongside the E67/E77 road.

☐ 002	SBLim-1	1A-05002	preserved, at barracks near airport		9-94
☐ 114	TS-8	1E-0114	preserved, at PZL factory		10-97
☐ SP-PCB	PZL 130	003	stored		9-94
☐ SP-LHA	Tu-134A	3351808	stored, ex 104		8-96
☐ SP-LHB	Tu-134A	3351809	stored, ex 103		8-96
☐ SP-LHC	Tu-134A	3351810	stored		2-98
☐ SP-LHD	Tu-134A	48400	stored		8-96
☐ SP-LHG	Tu-134A	3352008	stored		4-98
☐ SP-LNB	Il-14P	4340510	preserved		4-98
☐ SP-LNE	Il-14P	6341602	stored		4-98
☐ SP-LTA	An-24B	67302203	stored, near freight terminal		4-98

WIDEŁKA

Preserved in this town is the sole Piper PA35.

☐ N3535C	PA35	E1	preserved		10-95

WIERUSZÓW
Preserved in this town center is a former LOT Li-2
| ❏ SP-LKI | Li-2P | 23444804 | preserved | 8-90 |

WIKLON
In use as a restaurant along the E75 (Czestochowa - Piotrków Trybunalski) road is an Il-18. Wiklon is just north of Kruzyna and south of Radomsko. The registration has been checked in the cockpit.
| ❏ SP-LSD | Il-18V | 184007102 | preserved, unmarked | 4-98 |

WITKOWO
Preserved in the centre at a housing estate should be an Il-28. The TS-8 is preserved in a school yard.
| ❏ 8 | Il-28R | .. | preserved, marked as '10' | 8-96 |
| ❏ 0627 | TS-8 | 1E-0627 | preserved | 8-96 |

WOŁÓW
Preserved somewhere in town is an unknown MiG-17.
| ❏ ... | Lim-5P | .. | preserved, marked as '1985' | 9-96 |

WROCŁAW
At the airfield of Wrocław Strachowice a Lim-5 is preserved at the aero club. The military have an Il-28 and MiG-21. An unknown TS-8 is in a housing area, while preserved with the Lotnicze Zakłady Naukowe at the ul. Kielczcwska (street name) should be a Lim-1.
❏ 65	Il-28	..	preserved, inside base, see Dęblin	3-95
❏ 151	Lim-1	..	preserved	—
❏ 419	Lim-5P	1D-0419	preserved, at aero club, ex Zamość	9-96
❏ 8007	MiG-21MF	968007	preserved	4-96
❏ ..	TS-8	..	preserved, in housing area, marked as '1979'	3-95

WRZESNIA
Along the main road through this town, towards the railway station, an unknown Lim-5 was noted in 1995.
| ❏ .. | Lim-5 | .. | preserved | 8-95 |

ZAKOPANE
On the west side of this town are military barracks. Here two aircraft with false markings are preserved.
| ❏ ... | TS-11 | .. | preserved, marked as '1991' | 9-96 |
| ❏ ... | Yak-23 | .. | preserved, marked as '1957' | 9-96 |

ZAMOŚĆ
Zamość has a large instructional school. This school was closed in the mid 1990s and the fate of most of the airframes is unknown. Lim-5P 419 went to Wrocław. Lim-5P 521 (1D-0521) went to the USA and Lim-6R 404 went to Prešov. The TS-11 is preserved in flying attitude on a pole at the gate.
❏ 2616	An-2T	1G26-16	stored, fuselage only	9-96
❏ 0546	Mi-2	510546127	instructional, wfu 21-8-71	—
❏ 0616	Mi-2	510616018	preserved, no rotors, wfu 16-10-68	9-96
❏ 0627	Mi-2	510627038	instructional, wfu 16-9-70	—
❏ 4601	Mi-2Sz	544601016	preserved, ex instructional, wfu 5-3-91	—

❏ 404	Lim-5P	1D-0404	instructional		—
❏ 405	Lim-5P	1D-0405	instructional		—
❏ 407	Lim-5P	1D-0407	instructional		—
❏ 411	Lim-5P	1D-0411	instructional		—
❏ 509	Lim-5P	1D-0509	instructional		—
❏ 518	Lim-5P	1D-0518	instructional		—
❏ 604	Lim-6R	1J-0604	instructional		—
❏ 610	Lim-5P	1D-0610	instructional		—
❏ 1305	Lim-5	1C-1305	stored, behind hangars		9-96
❏ 1326	Lim-5	1C-1326	instructional		6-86
❏ 1912	Lim-5	1C-1912	instructional		6-86
❏ ..	TS-11	..	preserved, on pole, marked as '1995'		9-96

ZAWIERCIE
Zawiercie is north east of Katowice. Preserved here should be a Lim-6.

❏ 526	Lim-6R	1J-0526	preserved	—

ZGIERZ
In this small town just north of Łódź Mr Witold Wieczorek has rebuilt two Mi-2s as funfair attractions. Both travel throughout the country.

❏ 0101	Mi-2	520101086	preserved, marked 'Super Cobra', soc 18-2-93	4-98
❏ SP-FSF	Mi-2	563623084	preserved, marked 'Błekitnym Gromie'	10-94

PORTUGAL

Portugal is blessed with a large number of aircraft relevant to this survey. Besides the usual gate guards and inhabitants of scrap yards, playgrounds and dumps, the 'wrecks and relics scene' in this country is dominated by the Museu do Ar, a military aviation museum with a collection of over a hundred aircraft, among which are many interesting ones. Because of financial problems, the museum can only display about a dozen of small aircraft in the far too small building at Alverca. Since recently, some additional exhibits were placed in front of the museum. The majority of the collection, however, is stored at different places including an open storage compound at Alverca and storage hangars on the air bases of Sintra, Beja and Ota. Also several gate guards belong to the collection. The museum is campaigning for sufficient funding to begin the construction of the desired new museum hall on a site south of the runway of Sintra air base.

ALCOCHETE
The gate of Campo do Tiro Alcochete, a military target range south of the N119 and west of the N10, is guarded by an A-7P. The aircraft, previously stored at Monte Real as 15537, is preserved as 5537.

❑ 15537	A-7P	A-163	preserved, marked as '5537', ex 153254/USN	5-98

ALFEITE
Preserved at the Arsenal do Alfeite, the naval base near the village of Almada on the south bank of the River Tejo, is a G91 displayed on a pole. It is situated about 300m from the main gate to the left, just behind the high stone wall around the barracks. It can be seen from higher ground across the street or with one's face pressed against the emergency gate halfway between the G91 and the main gate.

❑ 5469	G91R/3	307	preserved, ex 3050/WGAF	2-97

ALVERCA
Various airframes can be found within the perimeters of Alverca airfield. On display in the Museu do Ar are a number of aircraft. During 1997 Tiger Moth 111 was successively noted at Lisboa airport, the Lisboa Zoo and Sintra air base, finally returning to Alverca. The Hurricane replica, noted together with the Tiger Moth in the Zoo, has also returned. The correct construction number of the G91R/3 5441 is 0064 and not 0065 as often wrongly quoted.

❑ 5563	Spitfire HF.IXc	..	preserved, marked as 'ML255/MR-Z', ex SAAF	1-98
❑ 248	Vampire FB.9	..	preserved, marked as '5801', ex SAAF	1-98
❑ 111	DH.82A	OGMA-P1	preserved, ex Sintra, ex Lisboa Zoo	1-98
❑ 129	G44 Widgeon	1251	preserved	1-98
❑ 2601	T-38A	N5209	preserved, outside, ex store OGMA, ex 61-0843	1-98
❑ 3212	L-21B	18-2534	preserved, ex 52-6234	1-98
❑ 3564	Auster D5/160	OGMA-70	preserved	1-98
❑ 5319	F-86F	191-958	preserved, outside, ex store, ex Ota, ex 52-5262	1-98
❑ 5420	G91R/4	0130	preserved, outside, nose only, ex BR-363/WGAF	1-98
❑ 5441	G91R/3	0064	preserved, outside, ex Montijo, ex 3011/WGAF	1-98
❑ PE-1	Grunau SG-38	..	preserved, glider	1-98
❑ CS-AXA	Jodel D.9	436	preserved	1-98
❑ CS-PAE	Grunau Baby IIb	3567	preserved, glider	1-98
❑ ..	Blériot XI	..	preserved, replica	1-98
❑ ..	Caudron G.3	..	preserved, replica	1-98
❑ ..	Fairey IIID	..	preserved, replica, marked as '17'	1-98
❑ ..	Hurricane II	..	preserved, replica, marked as '591/RV-J'	1-98
❑ ..	Farman MF-4	..	preserved, replica	1-98
❑ ..	Santos Dumont XX	..	preserved, replica	1-98

The museum compound is gradually losing its inhabitants. Many aircraft have moved to Ota, including Harvard IIA 1527 and T-6J 1737, RT-33A 1916, C-45 2513, Do27s 3487 and 3489, F-84G 5176 and F-86Fs 5337, 5338 and 5347. F-86F 5361 went to Sintra, while F-84G 5216 is now at Savigny les Beaune. Two Ju52s (6304 and 6306) have moved to Bodø for restoration of which the 6304 will return to the Museu do Ar. In exchange for the Hurricane replica, Harvard 1662 (88-16366) and Texan 1766 (CCF4-464) went to Sudbury, UK. The makers plate inside N2502 '6403' indicates construction number 6, which makes it 6405. The museum appears to confirm this identity and evidence available suggests a curious change of identity in 1977. Not noted since the 1980s and presumably scrapped are Auster 3553 (OGMA-50) and G91R/4 5412 (0149). The C-45 2517 is presumable an AT-11 or an SNB-5.

☐ 2517	C-45	CA-94	dumped, ex 2296/RCAF, LH tail from 2514	1-98
☐ 4620	PV-2C	15-1439	stored, ex 37473/US Navy	1-98
☐ 5333	F-86F	191-880	stored, ex 52-5184	1-98
☐ 5360	F-86F	202-119	stored, ex 53-1190	1-98
☐ 6300	Ju52/3mg3e	..	stored, no engines	1-98
☐ 6301	Ju52/3mg3e	893	stored, no nose	1-98
☐ 6310	Ju52/3mg7e	501219	stored, fuselage only, fin to Brussels for 6309	11-97
☐ 6311	AAC.1	205	stored, fuselage only	1-98
☐ 6315	AAC.1	005	stored, fuselage only, ex Sintra	1-98
☐ 6405	N2502A	6	stored, marked as '6403', ex F-BGZF	1-98
☐ 6412	N2502F	006F	stored, white c/s	1-98
☐ 6417	N2501D	059	stored, white fuselage only, ex GC-248/WGAF	1-98
☐ 6420	N2501D	032	stored, desert c/s	1-98
☐ 6606	C-54A	3069	stored, ex CS-TSC, ex 41-37279	1-98
☐ 6706	DC-6B	44116	stored, ex N6116C	1-98
☐ 7104	B-26B	28005	stored, fuselage, glass B-26C nose, ex 44-34726	1-98
☐ CS-ALQ	PA18-125	18-1290	stored, frame, ex 'CS-207', ex 3201	1-98

Adjacent to the museum storage compound is the compound of the Accident Investigation Laboratory which holds the wrecks of some accident victims and one or two Alouette 3s. SA316B 9337 needs further investigation as its construction number 1661 is also reported to have been sold to the Angola Air Force as H-205. Possibly only parts were dumped. All aircraft on this dump were noted in March 1998, but none were identified.

☐ 1322	Chipmunk Mk.20	OGMA-12	dumped, fuselage upside down	6-97
☐ 2418	T-37C	40743	dumped, crashed 5-11-86, ex 62-5943	6-97
☐ ..	T-37C	..	dumped	3-98
☐ 5433	G91R/4	0119	dumped, ex BR-245/WGAF	6-97
☐ 9337	Alouette 3	1661	dumped	5-95
☐ 9391	Alouette 3	1881	dumped, ex store OGMA	3-97

At the OGMA maintenance facility, a number of stored aircraft have been noted over the years. Of these the ex German Army Alouette 2 7625 (1612) became F-GLXL and 7700 (1847) became F-GLPI. Portuguese Alouette 2 9211 (1643) was sold as F-GMBG and 9218 went to Beja. Not noted since the 1980s are Alouette 3s 9358 (1750, a mix up with 9258?) and 9395 (1896). 9314 (1621) returned to flying status. Also SA330H 9501, which crashed on 13th June 1986, has not been seen since 1988. The stored Boeing 727-2M7 TN-AEB (21655, last noted May 1996) has flown out as VR-CDL.

☐ 7596	Alouette 2	1532	stored, ex WG Army, 9214 ntu	10-91
☐ 7632	Alouette 2	1632	stored, ex WG Army, 9215 ntu	1-91
☐ 1507	Harvard III	88-15790	dumped, ex 41-34099	5-95
☐ 19258	Alouette 3	1116	stored, exhibition a/c	7-95
☐ 9312	Alouette 3	1613	stored, shell only, rest on dump	1-91
☐ 9368	Alouette 3	1786	stored, poor condition	7-95
☐ 9387	Alouette 3	1858	stored, poor condition	8-95
☐ 19513	SA330L	1270	dumped, tail only	5-95
☐ N207GM	C-130A	3207	stored, ex EL-AJM, ex A97-207/RAAF	3-98
☐ N216CR	C-130A	3216	stored, ex N15FV, ex A97-216/RAAF	3-98
☐ ..	RC680FLP	1651-32	stored	3-98

The DGMFA (Depósito Geral de Material da Força Aérea) is at another part of Alverca. From here T-37C 2407 moved to Ota. Stored at DGMFA, have been several G91s, most of which have probably been scrapped by now. By 1997 all aircraft were gone, while in a scrap compound on the south (river) side of the museum storage area at least six G91s fuselages could be observed together with an unidentified Harvard. Only G91R/3 5462 was still upright and the only identifyable aircraft in this compound. The two Chadian AF C-130s, TT-PAD (3180) and TT-PAE, were partly scrapped in situ at DGMFA, but at least TT-PAE later turned up in a scrapyard in Arranhó. G91T/3 1801 (0003) went to Savigny lès Beaune, G91R/3 5452 moved to the Lisboa Zoo in Sete Rios, and two G91R/4s, 5428 and 5432, went to Montijo. Stored with DGMFA are:

❏ 1507	Harvard III	88-15790	stored, ex 7613/SAAF, ex EZ226, ex 41-34099	5-95
❏ 02411	T-37C	40735	preserved, on a pole at the gate, ex 62-5936	1-98
❏ 2406	T-37C	40730	stored, fuselage only, ex 62-5931	1-98
❏ 2422	T-37C	40747	stored, with wings of 2420, ex 62-5947	1-98
❏ 2429	T-37C	40783	stored, with wings of 2410, ex 62-12500	1-98
❏ 15507	A-7P	A-103	stored, ex Monte Real, ex 153194/US Navy	1-98
❏ 15508	A-7P	A-128	stored, ex Monte Real, ex 153219/US Navy	1-98
❏ 15515	A-7P	A-130	stored, ex 153221/US Navy	1-98
❏ 5517	A-7P	A-137	stored, ex store OGMA, ex 153228/US Navy	1-98
❏ 15522	A-7P	A-064	stored, ex store OGMA, ex 153155/US Navy	1-98
❏ 5526	A-7P	A-086	stored, ex store OGMA, ex 153177/US Navy	1-98
❏ 15527	A-7P	A-088	stored, ex store OGMA, ex 153179/US Navy	1-98
❏ 15529	A-7P	A-104	stored, ex store OGMA, ex 153195/US Navy	1-98
❏ 15532	A-7P	A-125	stored, ex store OGMA, ex 153216/US Navy	1-98
❏ 15534	A-7P	A-138	stored, ex store OGMA, ex 153229/US Navy	1-98
❏ 15536	A-7P	A-117	stored, ex store OGMA, ex 153208/US Navy	1-98
❏ 15545	TA-7P	A-110	stored, ex store OGMA, ex 153201/US Navy	1-98
❏ 15550	TA-7P	A-193	stored, ex 154354/US Navy	1-98

From the dump south of the museum storage area all G91s have now disappeared, probably including 5462. A recently noted C-45 (probably an AT-11 or SNB-5) is not identified yet. Although a likely candidate is 2514 (ex Pinhal Novo), it's identity needs confirmation. Besides the G91s, recent inhabitants are:

❏ 2514 ?	C-45		dumped, ex Pinhal Novo, see note	1-98
❏ 5462	G91R/3	301	dumped, ex store DGMFA, ex 3044/WGAF	3-97
❏ ..	Harvard	..	dumped	2-97

The aircraft which had gone by 1997 include:

3049	G91R/3	306	stored, l/n 1-91, ex Sintra, ex WGAF	gone
3265	G91R/3	535	stored, l/n 5-95, ex Sintra '5470', ex WGAF	gone
3277	G91R/3	547	stored, l/n 1-91, ex Sintra '5457', ex WGAF	gone
3279	G91R/3	549	stored, l/n 5-95, ex Sintra, ex WGAF	gone
3454	G91T/3	614	dumped, l/n 5-95, ex Sintra, ex WGAF	gone
3459	G91T/3	619	dumped, l/n 5-95, ex Sintra, ex WGAF	gone
1805	G91T/3	0019	stored, l/n 7-95, ex 3417/WGAF	gone
5403	G91R/4	0115	dumped, l/n 5-93, ex BR-241/WGAF	gone
5421	G91R/4	0135	stored, l/n 6-91, ex BR-367/WGAF	gone
5424	G91R/4	0152	stored, l/n 7-95, ex BR-384/WGAF	gone
5425	G91R/4	0100	stored, l/n 6-95, ex BR-233/WGAF	gone
5427	G91R/4	0111	stored, l/n 6-87, ex BR-237/WGAF	gone
5431	G91R/4	0136	stored, l/n 6-90 in crates, ex BR-368/WGAF	gone
5442	G91R/3	0077	stored, l/n 10-95, ex 3021/WGAF	gone
5450	G91R/3	428	stored, l/n 5-95 in crates, ex 3160/WGAF	gone
5456	G91R/3	449	dumped, l/n 7-88, ex 3181/WGAF	gone

Close to the museum storage compound, possibly in use by the fire department was an anonymous G91R/4 in February 1997. In June 1997, a G91R/4 with overpainted tail number was identified by its construction number 136, which makes it 5431, previously stored with DGMFA. In September 1997, a Fiat G91R/4 was noted at the runway-side of the hangars with overpainted but readable number 5420. If all sightings are of the same aircraft, it must be a composite. Notably the forward fuselage of 5420 is now preserved outside the Museum. The RF-10 was noted at the other side of the runway near the river Tejo.

❑ 1204	RF-10	13		dumped, glider, crashed 29-3-85	1-91
❑ ..	G91R/4	..		instructional	2-97

In the village of Alverca, in front of a technical school close to the airfield, another T-37 is displayed on a pole.

❑ 02412	T-37C	40736	preserved, ex 62-5937	11-97

ARRANHÓ
In this village about 20km north west of Vila Franca de Xira, a scrapyard was reported to contain about a dozen (13 were reported) unidentifiable G91R/3s and R/4s in September 1997. The aircraft probably came from the Alverca dump since the yard also contained the remains of one or two ex Chadian C-130s of which TT-PAE was identified.

❑ TT-PAE	C-130A	3037	scrapped, ex Alverca, ex Chadian AF	9-97

BARREIRO
In a children's playground at the far western end of the promenade along the river is an AT-11.

❑ 2501	AT-11	..	preserved, unmarked and overpainted	11-97

BEJA
On this former German Air Force base, many aircraft are stored in and around the hangars of Esq301, Esq552 and DGMFA, the latter referred to as the 'Factory'. Some belong to the Museu do Ar. The exact locations reported for each individual aircraft vary from report to report. In the overview below the latest reported locations are given. Although a substantial part of the Esq103/301 Alpha Jet fleet is normally stored in the units' hangars, most aircraft concerned frequently return to active service later on and, therefore, will not be listed here. The two Alpha Jets listed below are being canibalised for spares. T-38A 2611 moved to Ota, while T-38A 2605 and T-33A 1926 went to Sintra, T-38A 2610 to Vilar de Luz and 2603 to Lisboa. The tiger colourschemed G91R/3 5454 left for Sintra. In the Factory hangars were:

❑ 2419	T-37C	40744	stored, ex Sintra, ex 62-5944, special c/s	3-98
❑ 4803	P-3P	5404	stored, ex N4005X, ex A9-294/RAAF	11-97
❑ 5502	A-7P	A-109	stored, ex 153200/US Navy	11-97
❑ 5513	A-7P	A-097	stored, ex 153188/US Navy	11-97
❑ 5528	A-7P	A-096	stored, ex 153187/US Navy	11-97
❑ 5538	A-7P	A-169	stored, ex 153260/US Navy	11-97
❑ 5539	A-7P	A-135	stored, ex 153226/US Navy	11-97

Outside Factory hangars:

❑ 1905	T-33A	7476	stored, ex 51-17496	11-97
❑ 1907	T-33A	9352	stored, ex 54-1616	11-97
❑ 1909	T-33A	9624	stored, ex 55-3083	11-97
❑ 1911	T-33A	9629	stored, ex 55-3088	5-98
❑ 1914	T-33A	9626	stored, ex 55-3085	5-98
❑ 1918	T-33A	9806	stored, ex 55-4362	11-97
❑ 1919	T-33A	9952	stored, ex 56-1602	5-98
❑ 1922	T-33A	9974	stored, ex 56-1624	11-97
❑ 1923	T-33A	1189	stored, ex 57-0540	11-97
❑ 1924	T-33A	1190	stored, ex 57-0541	11-97
❑ 1927	T-33A	1344	stored, ex 57-0695	11-97
❑ 1928	T-33A	7573	stored, ex FT-21/BelgAF, ex 51-17513	11-97
❑ 2606	T-38A	N5281	stored, ex preserved, ex 61-0915	5-98

PORTUGAL - 387

❏ 2609	T-38A	N5206	stored, ex 61-0840	5-98
❏ 2612	T-38A	N5269	stored, ex 61-0903	5-98

Esq301 area:
❏ 1929	T-33A	9091	stored, ex FT-28/BelgAF, ex 53-5752	5-98
❏ 1930	T-33A	9093	stored, ex preserved, ex FT-30, ex 53-5874	11-97
❏ 15205	Alpha Jet	0021	stored, stripped for spares, ex 4021/GAF	11-97
❏ 15238	Alpha Jet	0110	stored, stripped for spares, ex 4110/GAF	11-97
❏ 5347	F-86F	202-12	stored, ex Alverca, ex 53-1083	3-98

Esq552 area (location not confirmed in latest report)
❏ 2604	T-38A	N5256	stored, ex 61-0890	11-97
❏ 2607	T-38A	N5181	stored, ex 61-0815	11-97
❏ 2608	T-38A	N5203	stored, ex 61-0837	11-97

Preserved and further recent non-flying residents include:
❏ 2099	F-104G	6620	preserved, gate, unmarked, ex '2381', ex WGAF	11-97
❏ 3238	G91R/3	507	preserved, officers' mess, as '3838', ex WGAF	3-97
❏ 1951	T-33AN	045	preserved, gate guard, ex Alverca, 21045/RCAF	11-97
❏ 2602	T-38A	N5219	preserved, gate guard, ex stored, ex 61-0853	11-97
❏ 9218	Alouette 2	1267	preserved, ex Alverca, ex 7515/WG Army	11-97
❏ 15533	A-7P	A-082	dumped, w/o 25-7-95, ex 153173/US Navy	7-96
❏ 19359	Alouette 3	1751	stored, in pieces	11-95

BENEDITA
Approximately halfway between Ota and Monte Real along the N1 highway are the premises of scrap and surplus dealers Amarino A. Mendes Lda. This yard is actually out in the middle of nowhere, but the next junction as you head south is to the village of Benedita. On top of some offices Chipmunk 1370 (OGMA-60, ex Ota) was displayed as an eyecatcher at least until May 1995. The aircraft was not found in November 1995.

CARVOEIRA
Some kilometers north west of Sintra, on the N247 road to Ericeira, the village of Carvoeira can be found. A G91 is positioned on concrete blocks in a small playground in front of a school. The aircraft is easily spotted from the main road through the village.
❏ 3083	G91R/3	346	preserved, ex Tancos, ex WGAF	11-97

CASCAIS
At the Cascais regional airport of Tires, three ex military aircraft used to be present, of which Chipmunk 1360 (OGMA-50) went to Spanhoe Lodge (UK) and the F-86F 5344 moved a few kilometers down the road to Trajouce. Cessna FTB337 3727 (028, ex Ota) did not move and after some time of open storage finally became CS-DBV. Newly arrived Texan 1635 (which was first noted here in May 1995) moved on to Vale de Lobos in late 1997.

CHAVES
The aero club at the airfield of this town, some 100km north east of Porto, has a T-33AN on display on a pole. A Texan was stored close to the airfield entrance, but has not been noted beyond 1995. The T-33AN caused some confusion some time ago, when its sistership 1953 in Lisboa Pedrouços was painted as '1952' as well in 1992. Both T-33s have been seen since.
❏ 1765	T-6J	CCF4-531	stored, ex Seixas, ex BF-070/WGAF, gone ?	5-95
❏ 1952	T-33AN	T33-228	preserved, on a pole, ex 21228/RCAF	7-96

COIMBRA
Town: Part of Parque Infantil Rainha Santa Izabel is 'Portugal dos Pequeninos'. Ju52/3mg3e 6304 that was displayed here at least until June 1985 was later replaced by a G91R/3 that, despite wearing full Portuguese Air Force colours, never saw active duty in Portugal.
☐ 5474 G91R/3 544 preserved, ex Ota, ex 3274/WGAF 10-96

Airfield: On the small civil airfield of Coimbra, a number of ex military Chipmunks have been stored on their way to the civil market. Among them were at least the following: 1304 (C1/0286) to CS-DAE, 1323 (OGMA-13) to CS-AZQ, 1334 (OGMA-24) to CS-AZX, 1340 (OGMA-30) to CS-DAI, 1342 (OGMA-32) to CS-DAJ, 1343 (OGMA-33) to CS-DAF, 1350 (OGMA-40) to CS-AZY and 1358 (OGMA-48) to CS-AZO.

EVORA
Another Ju52/3mg3e was preserved at a public swimming pool, just outside this town about 100km south east of Lisboa. The aircraft, 6303, was here at least until April 1985, but had gone by May 1992.

FARO
Along road N125 halfway between Faro and Albufeira the Constellation, previously in use as a bar at Faro Airport, was noted dismantled in 1997.
☐ CS-TLA L1049G 4616 stored, marked as '5N-83H', ex Faro Airport 97

FUNCHAL (Madeira)
Since many years a Noratlas (fuselage and one wing) lies on a dump near the civil airfield of the Portuguese Island of Madeira, close to Água de Pena. The aircraft was sabotaged in 1965 at Porto Santo.
☐ 6419 N2501D 044 dumped, ex GC-101/WGAF 7-92

GONDOMAR
In this small town, 12 kilometer east of Porto, G91T/3 1804 was offered for display near the townhall, where it was first noted in October 1989. It was not (yet) put on perminent display and is currently stored in a local municipal compound.
☐ 1804 G91T/3 0017 stored, ex 3415/WGAF 3-98

LAJES (Azores)
Upon the withdrawal of the G91 from active service with the Portuguese Air Force, the remaining operational aircraft at Lajes were initially put to open storage. They were not noted after July 1991 and some claim they were scrapped at Alverca as late as 1994. Can anyone confirm this with actual dates? Two are on display on the base: 5418 is guarding the gate at least until July 1996, while 5414 is reported hangared awaiting preservation, This G91R/4 carries construction number 4-101 belonging to 5426, which crashed on 14th September 1979. Is this a composite?

☐ 5401	G91R/4	0109	stored, ex BR-235/WGAF	7-91
☐ 5407	G91R/4	0132	stored, ex BR-250/WGAF	7-91
☐ 5410	G91R/4	0131	stored, ex BR-364/WGAF	7-91
☐ 5414	G91R/4	0133	preserved, in hangar, ex BR-365/WGAF	9-97
☐ 5415	G91R/4	0147	stored, ex BR-379/WGAF	7-91
☐ 5417	G91R/4	0110	stored, ex BR-236/WGAF	7-91
☐ 5418	G91R/4	0121	preserved, at gate, ex BR-247/WGAF	3-98
☐ 5422	G91R/4	0139	stored, ex BR-371/WGAF	7-91
☐ 5430	G91R/4	0134	stored, ex BR-366/WGAF	7-91
☐ 5434	G91R/4	0120	stored, ex BR-246/WGAF	7-91
☐ 5435	G91R/4	0127	stored, ex BR-253/WGAF	7-91

| PORTUGAL - 389 |

| ☐ 5439 | G91R/4 | 0143 | stored, ex BR-375/WGAF | 7-91 |
| ☐ 5440 | G91R/4 | 0151 | stored, ex BR-383/WGAF | 7-91 |

LEIRIA
Preserved in the municipal park along the river is an AT-11. The aircraft is situated next to a small snack bar and visible from outside the park.
| ☐ 2508 | AT-11 | .. | preserved | 3-97 |

LISBOA
Town – Alfragide: In and around Lisboa, various airframes are stored or preserved. In the western suburb Alfragide, along highway IC19, the Air Force have their headquarters. Preserved in front of the buildings are two G91s in flying position, while the Cadet is preserved inside.
☐ 501	Avro 631 Cadet	727	preserved, unmarked, ex Sintra	97
☐ 5444	G91R/3	0089	preserved, ex Montijo, ex 3031/WGAF	3-97
☐ 5467	G91R/3	517	preserved, ex Montijo, ex 3248/WGAF	3-97

Town – Belém: In this suburb along the N6 coastal highway on the north bank of the River Tejo the Museu de Marinha is located, west of the large bridge. This naval museum has three aircraft on display inside.
☐ 17	Fairey IIID	F402	preserved	5-95
☐ 120	G44 Widgeon	1242	preserved, marked as '128'	5-95
☐ 2	Schreck FBA	203	preserved	5-95

Town – Luz: The Colégio Militar has an anonymous G91R/4 on display on a pole. Despite the high wall around the barracks, the aircraft can be seen from the Estrada da Luz. It is unmarked, but still shows its construction plate underneath the cockpit, hopefully allowing for future identification.
| ☐ .. | G91R/4 | .. | preserved | 2-97 |

Town - Monsanto Park: In this large park in the north west of the city, an ex-Portuguese Navy G.44 Widgeon can be found in the 'Parque Infantil do Alvito', a children's playground in the eastern part of the park. Its empty shell is painted in fanciful colours consisting of green enginebays, a blue fuselage and yellow wings.
| ☐ 2401 | G44 Widgeon | .. | preserved | 2-97 |

Town – Pedrouços: A suburb immediately west of Belém, houses the Institute of Higher Military Studies. Preserved on a pole inside the barracks is T-33AN 1953. In the early 1990s the aircraft was restored and painted as '1952'. The real 1952 was still at Chaves at that time. Whether or not the Pedrouços aircraft was given the false serial on purpose or by mistake is not clear.
| ☐ 1953 | T-33AN | 317 | preserved, marked as '1952', ex 21317/RCAF | 3-97 |

Town - Sete Rios: In March 1997, some aircraft were noted displayed in the Lisboa Zoo, located in this part of town. This display was not permanent, because DH.82A 111 moved on to Sintra, G91R 5452 to Ota and the Hurricane replica back to Alverca.

Airfield – Portela de Sacavém: The fire department at the International Airport used to have an Aero Commander as an instructional airframe. It was parked next to the hulk of an ex Portuguese Air Force Texan, at least untill October 1995. The AC560F was recently dumped into a small compound against a building nearby, whereas the remains of the Texan have disappeared. Another Texan, 1635, belonging to Aero Fénix (see Vale de Lobos), was noted in a compound behind a brick wall close to the gate, until March 1996. It recently went to Cascais. The Texans have caused considerable confusion in recent years, because some reports incorrectly claimed the hulk to be 1635. Museu do Ar's Tiger Moth 111 was noted in the departure hall in February 1997 publicing for the Museum, but later moved on to the Lisboa Zoo in Sete Rios. Across the highway the Convair is still in use as a bar carrying no markings besides the title 'Avião'. West of the runway, B707 9T-MMS of the Republic of Zaire has been stored already for several years. The Talon will become a gate guard.

PORTUGAL - 390

❑ 1539	Harvard IIA	88-9261	stored, ex 7051/SAAF, ex EX182, ex 41-33155		98
❑ 2603	T-38A	N5234	stored, ex Ota, ex 61-0868		3-98
❑ 9T-MSS	B707-382B	19969	stored, ex CS-TBD		3-97
❑ CS-DGA	C-47A	19503	preserved, ex 4X-AOC, ex 43-15037		3-98
❑ G-ARDK	AC560F	992-6	dumped, ex fire department, CS-AJL ntu		2-97
❑ N8806E	Convair 880	21	preserved, unmarked		3-97

MONTE REAL
Of the two T-33As on the dump here in 1988, the remains of only one unidentified machine were noted in 1996. A report dated March 1997 claims this aircraft to be the 1912. Preserved at the main gate is an F-86. From here the ex German G91 and the tail of F-86F 5329 can readily be seen on the dump, which is located on the other side of the runway. The wreck of A-7P 15542 is in a compound along the fence, halfway in between the main gate and the temporary F-16 sheds.

❑ 3055	G91R/3	313	dumped, burnt fuselage, ex store, ex WGAF		3-98
❑ 1910	T-33A	9627	dumped, ex 55-3086		9-88
❑ 1912	T-33A	9631	dumped, crashed 1-82, ex 55-3090		3-97
❑ 5301	F-86F	191-864	preserved, at gate, ex 52-5168		5-98
❑ 5320	F-86F	191-964	stored, ex 52-5268		3-98
❑ 5329	F-86F	191-897	dumped, tail only, ex 52-5201		3-98
❑ 5535	A-7P	A-149	dumped, crashed 26-5-86, ex 153240/US Navy		6-87
❑ 15542	A-7P	A-190	dumped, crashed 15-3-94, ex 154351/US Navy		3-98

Several A-7 Corsairs were reported to have been withdrawn from use and put to storage here in 1996. Some of them moved on: 15503 went to Sintra, 15506 went to Ota, 15507, 15508 went to Alverca and 15537 is now preserved at Alcochete.

❑ 15544	A-7P	A-154	stored, ex 153245/US Navy		11-97
❑ 15547	TA-7P	A-133	stored, ex 153224/US Navy		11-97

MONTIJO
At the gate is an unknown G91R/3 on a pole which used to carry the false G91R/4 serial '5423' in 1985. It had changed identity to '5463' by June 1987, possibly as a tribute to the real 5463 which had crashed on 6th May 1986. The previous false identity '5423' was still visible under the new paint. In 1996, it underwent some maintenance work in the Esq751 hangar, after which it returned to the gate, still as '5463'. Near the air traffic control tower is G91R/4 preserved as '5404'. A number of aircraft have moved on: G91T 1806 and G91Rs 5436, 5438 and 5445 have gone to Ota. Gina 5441 went to Alverca and 5454 to Beja. The dumped ex WGAF G91R/3 3253 (522) has not been seen since 1985 and may have been scrapped.

❑ 3460	G91T/3	620	dumped, ex WGAF		3-96
❑ 1802	G91T/3	0006	dumped, on taxitrack, tail only, ex 3406/WGAF		7-95
❑ 5404	G91R/4	0117	preserved, on pole, ex BR-243/WGAF		5-98
❑ 5432	G91R/4	0145	stored, ex Alverca, ex BR-377/WGAF		7-93
❑ 5443	G91R/3	0080	dumped, on taxitrack, ex 3023/WGAF		11-95
❑ 5447	G91R/3	387	stored, in ammunition area, ex 3120/WGAF		3-97
❑ 5455	G91R/3	440	stored, ex 3172/WGAF		10-94
❑ 5471	G91R/3	530	stored, fuselage only, ex 3260/WGAF		7-95
❑ ..	G91R/3	..	preserved, at gate, marked as '5463', ex '5423'		5-98

Many of the phased-out G91s ended up stored here. Most of those not transported elsewhere were dismantled by the end of 1995 and were still present as such in March 1996, in three groups.

❑ 1803	G91T/3	0009	stored, in dispersal 2, ex 3409/WGAF		3-96
❑ 1807	G91T/3	0031	stored, on old runway, ex 3428/WGAF		3-96
❑ 1809	G91T/3	0034	stored, on old runway, ex 3430/WGAF		3-96
❑ 1810	G91T/3	0042	stored, in dispersal 1, ex 3438/WGAF		3-96

☐ 1811	G91T/3	613	stored, on old runway, ex 3453/WGAF	3-96
☐ 5428	G91R/4	0112	stored, old runway, ex Alverca, BR-238/WGAF	3-96
☐ 5446	G91R/3	370	stored, in dispersal 1, ex 3104/WGAF	3-96
☐ 5448	G91R/3	402	stored, in dispersal 2, ex 3134/WGAF	3-96
☐ 5451	G91R/3	433	stored, in dispersal 1, ex 3165/WGAF	3-96
☐ 5453	G91R/3	520	stored, on old runway, ex 3251/WGAF	3-96
☐ 5458	G91R/3	477	stored, in dispersal 1, ex 3209/WGAF	3-96
☐ 5459	G91R/3	551	stored, on old runway, ex 3281/WGAF	3-96
☐ 5460	G91R/3	582	stored, in dispersal 1, ex 3311/WGAF	3-96
☐ 5461	G91R/3	0097	stored, in dispersal 1, ex 3037/WGAF	3-96
☐ 5464	G91R/3	391	stored, in dispersal 2, ex 3124/WGAF	3-96
☐ 5465	G91R/3	450	stored, on old runway, ex 3182/WGAF	3-96
☐ 5466	G91R/3	533	stored, in dispersal 2, ex 3263/WGAF	3-96
☐ 5468	G91R/3	393	stored, fuselage on old runway, ex 3126/WGAF	3-96

For fire practice a Corsair and some G91s are or have been used.

☐ 3065	G91R/3	326	dumped, ex Sintra, ex WGAF	11-95
☐ 3241	G91R/3	510	dumped, remains, ex WGAF	3-96
☐ 3266	G91R/3	536	dumped, ex store, ex WGAF	3-96
☐ 3414	G91T/3	0016	dumped, remains, ex WGAF	11-95
☐ 5523	A-7P	A-068	dumped, remains, crashed 29-4-92, ex 153159	3-96

MURTEIRA
Along the main road (223) through this village, located between Ovar and Vila da Feira, is a scrap yard, or more accurately a house with a lot of junk around it, which has a Do27 sitting on top of a container. The tail of this aircarft is painted red and the tailnumber is written with a black marker. Since an aircraft with apparently the same identity has turned up in Souith Africa, the Murtiera Do27 needs further investigation.

☐ 3439 ?	Do27K-2	2133	dumped, fuselage only, see note	3-98

OTA
Ota air base, the former BA2 and home of the Chipmunks of Esq101 until the unit was disbanded in the early 1990s, still hosts the technical school 'Centro de Formação Militar e Técnica da Força Aérea'. Some of their instructional airframes are inside while others are lined-up in front of hangar 4, depending on the students' activities. Hangar 1 is used by the Museu do Ar for storage of a large part of the collection. In February 1997, German G91 3285 and Portuguese G91 5438 were outside in front of hangar 1, while G91T 1806 was in front of hangar 4, in the CFMTFA line. In the meantime there seems to have been a swop, because by March 1997, 1806 was noted in front of hangar 1 and may be a new item of the Museu do Ar, while 5438 and 3285 had moved to the technical school. By May 1997 5438 had moved on to Sintra. At the edge of the ramp in front of hangar 1 are the fuselages of a Do27A and a Texan. Museum aircraft that have moved on from here include F-86F 5319 (to Alverca) and 5347 (to Beja), Hurricane replica '591/RV-J' (to Alverca) and Harvard IIA 1527 and T-6J 1734 which both went to Savigny lès Beaune. The flyable Chipmunks that were stored in hangar 2 in May 1995 have not been noted since. Former instructional T-6J 1766 (CCF4-464) went to Sudbury, UK, while 1770 has not been noted since 1987. Previously stored FTB337G 3727 went to Cascais some time ago and became CS-DBV and 3731 (032) was sold as CS-DBD. G91R/3 5474 moved on to Coimbra. Ex German G91T/3 3427 (0030) may have gone to Barcelos but was later reported to have been scrapped at Alverca. Chipmunks 1315 and 1316, reported as stored in hangar 2 in May 1995, as well as 1306 and 1319 , previously stored in hangar 1, were upgraded with Lycoming 180hp engines at Alverca during the summer of 1997 and returned to active service with the Air Force Academy at Sintra shortly after. Also Chipmunk 1376 returned to service by late 1997. FTB337G 3717 (018) became CS-DBP in 1997. Alouette 2s 9216 has probably left for Beja, while 9217 is now the Museu do Ar at Sintra as is G91R 5445 in special markings. The mentioned three C-45s should be AT-11s or SNB-5s. Around and inside the Museu do Ar hangar 1 are:

☐ A13	Alouette 2	1566	stored, ex Belgian Army	1-98

PORTUGAL - 392

☐ 104750		CF-104	1050	stored, tail from 104648, ex Savigny, ex CAF	1-98
☐ 313	33-TU	Mirage 3R	313	stored, ex Savigny les Beaune, ex French AF	1-98
☐ 1305		Chipmunk Mk.20	C1/0292	stored	1-98
☐ 1517		Harvard III	88-14544	stored, ex São Jacinto, ex 7430/SAAF, ex EX873	1-98
☐ 1645		T-6G	188-38	stored, ex São Jacinto, ex 51-15175	1-98
☐ 1737		T-6J	CCF4-521	dumped, outside, ex Alverca, ex BF-077/WGAF	1-98
☐ 1806		G91T/3	0025	stored, ex Montijo, ex 3423/WGAF	1-98
☐ 1916		RT-33A	8813	stored, ex Alverca, ex 53-5474	1-98
☐ 2504		AT-11	1431	stored, ex Sintra, ex BC-4/Portugese Navy	1-98
☐ 2513		C-45	CA-76	stored, ex Alverca, ex 2278/RCAF	1-98
☐ 2515		C-45	..	stored, ex Sintra	1-98
☐ 2516		C-45	..	stored, ex instructional	1-98
☐ 3301		MH1521C	281/51C	stored, ex Sintra, ex F-WJSO	1-98
☐ 3303		MH1521C	282/53C	stored, ex Sintra, no tail, ex F-ZKAU	1-98
☐ 3304		MH1521C	283/54C	stored, ex Alverca, ex F-ZKAV	1-98
☐ 3487		Do27A-4	141	dumped, fuselage, ex Alverca, ex PB-104/WGAF	1-98
☐ 3489		Do27A-4	251	stored, ex Alverca, ex PA-105/WGAF	1-98
☐ 5176	FS-957	F-84G	..	stored, ex Alverca, ex FZ-41/Belgium AF	1-98
☐ 5187		F-84G	..	stored, ex Sintra, ex 19928/French AF	1-98
☐ 5337		F-86F	191-895	stored, ex Alverca, ex 52-5199	1-98
☐ 5338		F-86F	191-900	stored, ex Alverca, ex 52-5204	1-98
☐ 5436		G91R/4	0128	stored, ex Montijo, ex BR-254/WGAF	1-98
☐ 5452		G91R/3	469	stored, tiger c/s, ex 3201/WGAF, ex Lisboa Zoo	1-98
☐ 15506		A-7P	A-159	stored, ex Monte Real, outside hangar 4	1-98
☐ 9216		Alouette 2	1638	stored, ex Alverca, ex 7633/WGArmy	2-97
☐ CR-CAL		Dove 1B	04157	stored	1-98
☐ CS-AIB		Beagle A-109	B527	stored	1-98
☐ CS-PAI		Slingsby T.21	551	stored, ex BGA619	1-98

The Centro de Formação Militar e Técnica de Força Aérea has the following instructional airframes.

☐ 3004	G91R/3	0057	instructional, hangar 4, ex WGAF	1-98
☐ 3285	G91R/3	555	instructional, ouside hangar 4, ex WGAF	1-98
☐ 2401	T-37C	40737	instructional, hangar 4, ex Sintra, ex 62-5926	1-98
☐ 2407	T-37C	40731	instructional, wings of 2427, ex Alverca	3-96
☐ 2414	T-37C	40739	instructional, hangar 4, ex Alverca, ex 62-5939	1-98
☐ 2611	T-38A	N5238	instructional, outs hangar 4, ex Beja, ex 61-0872	1-98
☐ 3707	FTB337G	008	instructional, hangar 4	1-98
☐ 3718	FTB337G	019	stored, hangar 2	9-97
☐ 3719	FTB337G	020	stored, hangar 2	3-97
☐ 9379	Alouette 3	1835	instructional, hangar 4, c/n marked as '1612'	1-98

On a hill behind the control tower, the fire department uses ex German G91s for instruction. Due to their bad condition and missing c/n plates a G91T/3, an adjacent G91R/3 in the grass and another G91T/3 in the fire pit could not be identified during an inspection in January 1998.

☐ 3030	G91R/3	0088	instructional, on ramp, ex WGAF	5-92
☐ 3078	G91R/3	340	instructional, on ramp, ex WGAF	1-98
☐ 3220	G91R/3	488	instructional, on ramp, ex WGAF	1-98
☐ 3239	G91R/3	508	instructional, heavily burnt	87
☐ 3287	G91R/3	558	instructional, on ramp, ex WGAF	2-97
☐ 3442	G91T/3	602	instructional, heavily burnt	88
☐ 3443	G91T/3	603	instructional, on ramp, ex WGAF	1-98
☐ 3445	G91T/3	605	instructional, in grass, ex WGAF	1-98
☐ 3448	G91T/3	608	instructional, in grass, ex WGAF	1-98
☐ 3457	G91T/3	617	instructional, in grass, ex WGAF	1-98
☐ 3450	G91T/3	610	instructional, heavily burnt	5-95

Still guarding the gate is F-84G Thunderjet 5201 on a pole. Some ex German G91s have not been noted for a while and may have gone.

❑ 3038	G91R/3	0102	stored, dispersal area, ex WGAF	5-92
❑ 3207	G91R/3	475	stored, dispersal area, ex WGAF	5-92
❑ 3227	G91R/3	496	stored, dispersal area, ex WGAF	5-92
❑ 3458	G91T/3	618	stored, dispersal area, ex WGAF	88
❑ 5201	F-84G	..	preserved, at gate on a pole	1-98

OVAR
Stored in the south hangar at the airfield is a Corsair. This aircraft will be put on display in the near furure.

❑ 15504	A-7P	A-093	stored, ex 153184	3-98

PALHAIS
A recently discovered private scrap yard in this village, a few kilometers south of Barreiro, holds a large pile of aircraft remains, of which in some cases the identity can only be recognized by a piece of metal wearing a number. Most aircraft are not carefully dismantled, but simply torn into large sections. The chaos is completed by tall reed, overgrowing the scrap it is impossible to tell whether all aircraft are complete. The pieces by which the wrecks were identified are therefore indicated, if reported. The yard can be found just north of the village, on the left side of the road to Barreiro. To avoid unwanted contact with small but vicious watchdogs of dubious race, ask for permission from the owner before starting an archeological adventure!

❑ 141002	C-131F	..	dumped, nose/wings, ex US Navy	3-97
❑ 1530	Harvard IIA	..	dumped, wing	9-97
❑ 1558	Harvard IIA	88-9666	dumped, wing, ex 7103/SAAF, ex EX260	10-96
❑ 1610	AT-6A	..	dumped, tail	3-97
❑ 1630	T-6G	188-44	dumped, wing, ex 51-15181	3-97
❑ 1640	T-6G	188-55	dumped, wing, ex 51-15192	3-97
❑ 1665	Harvard III	88-17006	dumped, tail, ex EZ451, ex 42-85225	3-97
❑ 1692	T-6G	182-142	dumped, tail, ex 114455/FrenchAF, ex 51-14455	3-97
❑ 1742	T-6J	..	dumped, tail	3-97
❑ 3338	Do27A1	317	dumped, wing, ex PA-107/WG Army	3-97
❑ 3341	Do27A1	424	dumped, wing	10-96
❑ 3342	Do27A1	118	dumped, ex 5512/West Germany	5-96
❑ 3351	Do27A1	187	dumped, tail, ex 5555/West Germany	3-97
❑ 3370	Do27A1	356	dumped, ex 5662/West Germany	5-96
❑ 3443	Do27A1	134	dumped, wings, ex AC-934/WGAF	3-97
❑ 3458	Do27A1	213	dumped, ex AC-941/WGAF	5-96
❑ 3474	Do27A1	203	dumped, tail/wing, ex AC-948/WGAF	3-97
❑ 3491	Do27A1	266	dumped, tail, ex PB-107/WG Army	3-97
❑ 4701	SP-2E	426-5273	dumped, tail, ex Sintra, 086/RNethNavy	3-97
❑ 4704	SP-2E	426-5276	dumped, wing, ex 089/RNethNavy	3-97
❑ 4706	SP-2E	426-5278	dumped, wing, ex 091/RNethNavy	3-97
❑ 5310	F-86F	191-890	dumped, fuselage, ex 52-5194	3-97
❑ 5321	F-86F	202-19	dumped, ex 53-1090	5-96
❑ 5327	F-86F	191-877	dumped, fuselage, ex 52-5181	3-97
❑ 5335	F-86F	191-893	dumped, fuselage/wing, ex 52-5197	3-97
❑ 5339	F-86F	191-916	dumped, wing, ex 52-5220	3-97
❑ 5341	F-86F	191-948	dumped, fuselage/wing, ex 52-5252	3-97
❑ 5350	F-86F	202-39	dumped, fuselage, ex 53-1110	3-97
❑ 5354	F-86F	202-65	dumped, fuselage, ex 53-1136	3-97
❑ 5357	F-86F	202-93	dumped, fuselage, ex 53-1164	3-97
❑ 5364	F-86F	191-779	dumped, fuselage, ex 52-5083	3-97
❑ 7503	HC-54D	22155	dumped, wings, ex 43-17205	3-97

PINHAL NOVO
Along the road from Palhais to Montijo is the village of Pinhal Novo. Right in front of the railway station, in a children's playground, C-45 (or better AT-11 or SNB-5) 2514 was preserved. By November 1995 it had disappeared, and presumably went to Alverca where it was noted in January 1998. Only the concrete blocks marked its previous position. The aircraft had the starboard rudder of 2517 which is stored at Alverca.

SÃO JACINTO
In a children's playground next to a church in the southern end of this town close to Aveiro Texan 1734 used to be on display. This aircraft went to Ota. It was later replaced by an unknown ex Esq301 G91R/3.
❑ .. G91R/3 .. preserved, unmarked 97

SEIXAS
In 1984 T-6G Texan 1765 was noted here preserved on a pole in the town centre. It later moved on to Chaves.

SESIMBRA
Just east of the village of Sesimbra, near the sign of road N10 towards Setúbal and close to the restaurant 'a Fonte', is a contractor's yard with a rubbish skip which holds the mortal remains of an AT-11.
❑ 2502 AT-11 .. dumped 11-95

SETÚBAL
The local branch of the Associação de Especialistas da Força Aérea has a G91R/3 preserved as a monument.
❑ .. G91R/3 .. preserved —

SINTRA
Within Sintra air base several aircraft are on display on various locations. The Museu do Ar also has a storage facility, sharing their hangars with the operational glider unit Esq802. The number of T-37Cs in open storage is decreasing, as they gradually find their way to other places. On display outside are:

❑ 1926	T-33A	1343	preserved, Museu do Ar, ex Beja, 57-0694	5-98
❑ 2404	T-37C	40728	preserved, parade ground (parts of 2417)	11-97
❑ 2423	T-37C	40748	preserved, near hangar on flightline, ex 62-5948	5-95
❑ 2424	T-37C	40749	preserved, on pole at base Hq, ex 62-5949	5-98
❑ 2427	T-37C	40781	preserved, Museu do Ar, ex store, 62-12498	5-98
❑ 2430	T-37C	40784	preserved, on pole at Esq502, ex 62-12501	5-98
❑ 2605	T-38A	N5263	preserved, Museu do Ar, ex Beja, 61-0897	5-98
❑ 4711	SP-2E	5283	preserved, Museu do Ar, ex 096/RNethNavy	5-98
❑ 5361	F-86F	202-133	preserved, Museu do Ar, ex Alverca, 53-1204	5-98
❑ 5438	G91R/4	0142	preserved, Museu do Ar, ex Ota, Montijo,	5-98
❑ 5457	G91R/3	464	preserved, on pole with Academy, ex 3196	11-97
❑ 5472	G91R/3	559	preserved, on pole on parade ground, ex 3288	11-97
❑ 5473	G91R/3	590	preserved, on pole on parade ground, ex 3319	11-97
❑ 15503	A-7P	A-181	preserved, Museu do Ar, ex Monte Real	5-98
❑ 6157	C-47A	19755	preserved, Museu do Ar, ex 43-15289	5-98

The Esq802/Museu do Ar storage hangars hold a large part of the museum collection until the new museum buildings become available. Ota received MH1521s 3303 and 3304 from the museum store and Cadet 501 can now be found inside the Portuguese Air Force HQ building in Lisboa. Tiger Moth 111 returned to Alverca.

❑ MM57-5979	UH-19D	..	stored, marked as '9101', ex Italian AF	5-98
❑ NL928	DH.82A	..	stored, marked as '119'	11-97
❑ 1001	ASK-21	21394	stored, glider, wreck, w/o 1990	7-96

☐ 1201		RF10	6	stored, glider, dismantled	3-97
☐ 1202		RF10	7	stored, glider, dismantled	3-97
☐ 11203		RF10	12	stored, glider, dismantled	3-97
☐ 1769		T-6J	CCF4-517	stored, flyable, ex Alverca, ex BF-078/WGAF	5-98
☐ 1774		T-6J	CCF4-486	stored, flyable, ex Montijo, ex BF-079/WGAF	5-98
☐ 2307		DH.89A	6899	stored, ex 507, ex CS-ADI	5-98
☐ 2420		T-37C	40745	stored, ex open store, 62-5945	5-98
☐ 3218		L-21B	18-2573	stored, ex Alverca, 52-6237	5-98
☐ 3358		Do27A-1	235	stored, marked as '3357', ex 5585/WGAF	11-97
☐ 3548		Auster D5/160	OGMA-45	stored, flyable, ex Alverca	5-98
☐ 5445		G91R/3	0091	stored, ex store Ota, Montijo, ex 3032/WGAF	5-98
☐ 5454		G91R/3	532	stored, ex store Beja, Montijo, ex 3262/WGAF	5-98
☐ 9217		Alouette 2	1640	stored, ex store Ota, Alverca, ex 7635/WGArmy	5-98
☐ ..		DH.82A	DHTM.3A	stored, marked as '102'	11-97
☐ CR-AAC		DH.87B	8104	stored	11-97
☐ CS-ABY		J-3C Cub	17243	stored	5-95
☐ CS-AFF		DH.82A	DHTM.18A	stored	87
☐ CS-ALN		PA18-125	18-1297	stored, ex 3208	5-95
☐ CS-AQI		Do27A-3	350	stored, marked as '3480', ex 3339	5-98
☐ CS-AXB	9	Jurca MJ-2B	9	stored	11-97
☐ D-EERH ?		Do27H-2	514	stored, marked as '3422', ex V-606/Swiss AF	5-98

Of the aircraft previously in open storage, Alverca received ex German AF G91Rs 3049, 3265, 3277 and 3279 and G91Ts 3454 and 3459. AT-11 2504, C-45 2515, MH1521C 3301 and F-84G 5187 all went to Ota. T-37C 2419 moved to Beja and 2410 to Vilar de Luz. T-37C 2417 (40742) was loaded on a cradle in July 1995 and left to an unknown destination. The dump may still contain two aircraft.

☐ 1408		TB30	165	dumped, wreck, crashed 14-5-93	7-93
☐ 1512		Harvard III	88-15053	dumped, fuselage only, ex 7518/SAAF	6-85
☐ 1546		Harvard IIA	88-9687	stored, ex 7171/SAAF, ex EX271, ex 41-33244	5-98
☐ 2402		T-37C	40726	stored, ex 62-5927	5-98
☐ 2403		T-37C	40727	stored, ex 62-5928	5-98
☐ 2421		T-37C	40746	stored, ex 62-5946	5-98
☐ 2425		T-37C	40779	stored, ex 62-12496	5-98
☐ 2426		T-37C	40780	stored, ex 62-12497	5-98
☐ 2428		T-37C	40782	stored, ex 62-12499	5-98

TANCOS
Noted here in July 1984 were a number of ex German G91R/3 used as decoys. They were first parked near the hangars but by October 1990 six of them, 3088 (352), 3179 (447), 3250 (519), 3254 (524), 3276 (546) and 3298 (569) had moved to the eastern corner of the base whilst one other, 3082 (345), had gone to the western side. 3083 had left by the late 1980s and is now at Carvoeira. Also 3283 (553) had left by that time. In the summer of 1993 the airfield was closed and the remaining aircraft are believed to have been scrapped.

TRAJOUCE
The F-86 that guarded the gate of Cascais airfield was disposed of in the early 1990s. It was cut into major sections and dumped on a garbage area behind a gypsy camp in the village of Trajouce, about 7km from the airfield. It had no detectable markings left.

☐ 5344	F-86F	191-963	dumped, unmarked, ex Cascais, ex 52-5267	5-92

VALE DE LOBOS
In this village near Pero Pinheiro, about 13km north east of Sintra, the private association Aero Fénix have a T-6

stored in their president's backyard and another (1635) in the garage. The association intends to preserve historic aircraft in flying condition. They also own T-6 1539, reportedly at Lisboa Airport. Possibly 1716 will be used as 'currency' to pay for the overhaul of 1635 in the UK.

❑ 1635	T-6G	188-40	stored, ex Cascais, ex Lisboa, ex 51-15177	97
❑ 1716	T-6G	182-32	stored, ex 114345/French AF, ex 51-14345	97

VILA FRESCO DE AZEITÃO

A scrapyard (on the left side) along the E04/N-10 to Lisboa has the remains of a yet unknown Harpoon. It was last seen in May 1993 and not found during a search in April 1994.

❑ ..	PV-2	..	dumped	5-93

VILAR DE LUZ

This new municipal airport near Maia, 12km north of Porto, received two aircraft for preservation. They arrived in November 1997 to be placed on poles at the airport entrance.

❑ 2410	T-37C	40734	stored, ex Sintra, ex 62-5935	11-97
❑ 2610	T-38A	N5233	stored, ex Beja, ex 61-0867	11-97

VILA REAL

Preserved on a pole at the entrance to the civil airfield of Vila Real is an ex Portuguese Air Force G91.

❑ 5470	G91R/3	339	preserved, on pole, ex 3077/WGAF	10-97

VISEU

About 100km south of Vila Real, the local aerodrome has a G91 on a pole at the entrance.

❑ 5408	G91R/4	0153	preserved, ex BR-385/WGAF	11-95

SLOVAK REPUBLIC

Some 90% of the military *Wrecks & Relics* in the Slovak Republic still carry the roundel from the Czechoslovakian days. Only a handful of aircraft have received the new Slovak cross. The largest exception are the aircraft at Sliač. The ex Vodochody MiG-21s had their roundels removed, while the newer MiG-21s placed in storage after the separation from the Czech Republic had the new markings applied during their operational careers. For many years there was talk of an official military museum at Trenčín. Although there were many plans for building a new museum, all that has happened has been the gathering of some 'to be preserved' aircraft at the airfield.

BANSKÁ BYSTRICA
During 1944 Banská Bystrica was the centre of the Slovak National Uprising (SNP) against the German occupation of Slovakia. In memory of this, the town houses the SNP museum. Besides a memorial and some tanks, a former Czechoslovak Lisunov can be found in the museum.

| ☐ | 2105 | Li-2 | 23422105 | preserved, marked as '20 white', ex D-25 | 5-98 |

BRATISLAVA - IVANKA
At the international airport two former Aeroflot Turbolets are stored. Both are used for spare parts for local operators. An ex Slov Air aircraft is dumped near the fire station.

☐	CCCP-67365	L-410UVP	820921	stored, OK-NDN ntu, OM-NDN ntu	7-96
☐	CCCP-67367	L-410UVP	830935	stored	5-95
☐	OK-PDI	L-410UVP	851527	dumped	6-96

BUZICA
In a small village south west of Košice, along the Hungarian border, MiG-15bis 3933 (623933) was preserved at a school. It was removed and scrapped locally by late 1994.

ČERNÍK
A school one kilometer south of the village of Černík, has a MiG-19 preserved on a pole. Černík itself is some 8km north east of Šurany.

| ☐ | 1048 | MiG-19PM | 651048 | preserved | 5-95 |

DRIENOV
The airfield of Prešov used to have two Avia 14s. Both were relocated in 1994 and 3153 moved 20km to the south where it acts as a restaurant on the Drienov-Ličartovce exit of the highway.

| ☐ | 3153 | Avia 14T | 913153 | preserved | 8-97 |

HOLÍČ
Since the dismantling of the 'Iron Curtain', aircraft of former Warsaw Pact countries have become collectors items in the West. One of the early aircraft to move to the United States was L-29 1418 (591418) which used to be preserved with the aero club at Holíč. Its place has been taken during early 1993 by a MiG-15.

| ☐ | 3826 | MiG-15bis | 623826 | preserved | 5-95 |

KOŠICE
Town: In the northern part of the town, along the main road to Prešov, the Slovak Military Academy is housed.

SLOVAK REPUBLIC - 398

Just visible from the main road is the L-39 which is preserved in false markings on a pole on the parade ground.
☐ ..   L-39   ..   preserved, marked as '1973'   5-95

Airfield: Besides the operational aircraft of the VSL, the airfield holds a number of instructional airframes. Of these, MiG-21PF 1215 moved to Trenčín in early 1996. The local aero club has its own Delfin.

☐ 2402	L-29RS	792402	stored, in bad shape, with aero club	11-97
☐ 0315	MiG-21F-13	560315	instructional, wfu 1-90	10-94
☐ 4820	MiG-21U-600	664820	instructional, outside	8-95
☐ 9904	MiG-21F-13	269904	instructional, outside	5-98
☐ OK-188	L-39ZA	X10	instructional	10-94

KOSORÍN
Only a handful of MiG-15UTIs are known to be still around in the Republic. One of them is in Kosorín (10km north of Žiar nad Hronom).
☐ 2620   MiG-15UTI   142620   preserved   10-94

KRÁLOVSKÝ CHLMEC
Compared to the Czech Republic only a few MiG-15s are preserved in the countryside of Slovakia. Královský Chlmec is a small village located some 15km from the Ukrainian border in the south east of the country.
☐ 3010   MiG-15bisSB   713010   preserved   3-94

MALACKY
Town: MiG-19PM 1113 which was preserved at the entrance of a cable factory in town was moved to Prešov airfield during the late 1980s.

Airfield: Until the separation of Czechoslovakia only a handful of target-towing L-39Vs were based at Malacky (sometimes also referred to as Kuchyňa). After 1st January 1993 the Slovaks based a number of fighters here. The gate is still guarded by a target-towing version of the MiG-15bis and since 1996 a number of MiG-21MFs have been placed in storage here.

☐ 1928		L-29	691928	stored, wreck in compound, crashed 19-9-96	6-97
☐ 3806	023	MiG-15bisT	623806	preserved	5-98
☐ 1113		MiG-21MF	..	stored	5-98
☐ 1203		MiG-21MF	..	stored	5-98
☐ 1206		MiG-21MF	..	stored	5-98
☐ 1208		MiG-21MF	..	stored	5-98
☐ 1209		MiG-21MF	..	stored	5-98
☐		Su-22	..	stored, wreck in compound	6-97
☐ 8072		Su-25K	25508108072	instructional, forward fuselage only	6-97

Ranges: On the east and western side of the airfield army and air force firing ranges are located. Over the years a number of retired airframes have been used as targets. Most of them have since been scrapped, although two forward fuselages of Il-28s were moved to Druztová in 1991. On the range in the hills it is known that there should still be three MiG-15s in use as targets. Serials are not known.

MARTIN
The aero club of the local airfield is restoring an L-29, while the MiG-15 is temporarily stored and will be their next project. A civil L-200 Morava is on display on a pole at the airfield entrance.

☐ 2609	L-29R	792609	under restoration	6-94
☐ 3024	MiG-15bisR	713024	stored	6-94
☐ OK-RFR	L-200D	171115	preserved, on pole	6-94

SLOVAK REPUBLIC - 399

NEPORADZA
A local school at Neporadza (some 15km south west of Trenčín) holds an early production Delfin.
❏ 0914	L-29	290914	preserved	10-94

NOVÁ DUBNICA
The newly built town of Nová Dubnica has two relics. A MiG-15 is on a pole at the sports centre, while close by a unmarked Li-2 is in use as a restaurant. It is not known if this is a former military or civilian example.
❏ 1133	MiG-15SB	141133	preserved	6-94
❏ ..	Li-2F	..	preserved	9-95

OPATOVÁ
A local aviation collector at Opatová (just south of Nová Dubnica) has, besides a number of aero engines and the forward fuselage of a MiG-15, a Mi-1 in perfect shape.
❏ 6007	Mi-1	..	preserved	6-94

PETRŽALKA
The southern suburb of Bratislava is Petržalka. On the Gessayova (street name) a Delfin is preserved.
❏ 2826	L-29R	892826	preserved	6-97

PIEŠŤANY
The majority of the Slovak transport fleet are based on this airfield. The small civil terminal is guarded by an ex CSA Tu-134, which serves as a restaurant, while close by on the airfield a Mi-4 is used for parachute training.
❏ 1093	Mi-4	1093	instructional	5-98
❏ OK-AFB	Tu-134A	1351410	preserved, wfu 11-10-88	5-95

PODLIPNÍKY
Since mid-1994 the former Prešov airfield Avia 14 3156 can be found on a field some 15km east of Prešov.
❏ 3156	Avia 14T	013156	preserved, ex Prešov	8-97

PREŠOV
Town: For many years Ján Brehový's Dopravne Muzeum has constituted a number of aircraft in the owner's front- and backyard. Recent plans for a proper museum building have been approved to be erected on the same site. The small area on which the museum will be built means that a number of current airframes must be disposed of. After removal of the useful items, a number of airframes have been sold for scrap. These were Mi-24D 4012 (M34012), MiG-15s 1909 (231909) and 1974 (construction number unknown and serial is doubtfull), MiG-15bis 3257 (613257) and MiG-15bisSB 3214 (613214). MiG-19S 0316 went to Cerbaiola. The new museum at Trenčín recived MiG-19S 0409.

❏ 404	Lim-6	1J-0404	preserved, ex Zamość, ex Polish AF	11-97
❏ 6101	Avia 14FG	806101	preserved, cockpit only, ex Sliač	10-94
❏ 4369	Il-10	..	preserved, forward fuselage	11-97
❏ 0115	L-29	290115	preserved	11-97
❏ 0215	L-29	290215	preserved, cockpit only	8-97
❏ 1417	L-29	591417	preserved	8-97
❏ 2802	L-29RS	892802	preserved	11-97
❏ 3014	MiG-15bisSB	713014	preserved	11-97
❏ 3609	MiG-15bis	613609	preserved, composite with parts from 3257	8-97
❏ 3652	MiG-15bis	613652	preserved	8-97
❏ 3852	MiG-15bisR	623852	preserved	8-97

SLOVAK REPUBLIC - 400

❏ 0302	MiG-19S	150302	preserved	11-97
❏ 0005	MiG-21F-13	360005	preserved, marked as '0088', cockpit only	10-94
❏ 0714	MiG-21F-13	760714	preserved	11-97
❏ 1307	MiG-21PF	1307	preserved	11-97
❏ 5315	Su-7BM	5315	preserved	11-97
❏ OK-ADZ	Aero C-3A	..	preserved, fuselage only, local built Si-204D-1	8-97

In the centre of Prešov, close to the railway and road number 18, a preserved Avia 14 can be found in a park.

❏ 3132	Avia 14T	913132	preserved	8-97

Airfield: Besides a Slovak helicopter wing the base houses a large number of airframes. Two of them are preserved. The remainder (mostly ex instructional airframes) are held in storage for the museum at Trenčín. Some airframes, like the Mi-1 1043, L-29RS 2615, L-39 0004 and MiG-21F-13s 0104, 0109 and 0903 left for Trenčín in 1996. The aero club's Avia 14T 3153 is now at Drienov, while the former instructional 3156 is at the nearby village of Podlipniky.

❏ 0007	L-29	190007	preserved	5-98
❏ 0306	L-29	290306	stored	8-97
❏ 1514	Mi-4	11114	stored	5-98
❏ 5153	Mi-4	05153	stored	5-98
❏ ..	MiG-15	..	stored	11-97
❏ ..	MiG-15	..	stored	9-93
❏ 1113	MiG-19PM	651113	stored, ex Malacky	5-98
❏ 0515	MiG-21F-13	660515	stored	5-98
❏ ..	MiG-21F-13	..	stored	9-93
❏ 5608	Su-7BM	5608	stored	5-98
❏ 5617	Su-7BM	5617	preserved	5-98

SLIAČ

'Fightertown Slovakia' is without doubt Sliač. The large number of operational and stored MiGs makes it one of the bigger attractions in the Republic. Preserved Avia 14FG 6101 was very badly damaged by the departing Soviets in the late 1980s. It was finally scrapped in 1993, but the cockpit was saved and has moved to the museum at Prešov. A number of older MiG-21s arrived in 1993 from Vodochody. These are still stored with their Czech roundels removed. Two of these, MiG-21PFM 4415 and MiG-21U 2419 moved to Trenčín in 1996. Is has been reported that all the other MiG-21F-13s, MiG-21PFMs and the MiG-21U have been scrapped. After the separation of Czechoslovakia, the Slovaks received seven Su-7s from Přerov. Two of these Su-7BKLs (6504 and 6506) went to Trenčín. The remainder, Su-7BKLs 6430, 6501, 6503, 6505 and 6508, were scrapped here. The wreck of a crashed Police Mi-8S was noted in a hangar in 1993.

❏ 0368	MiG-21UM	03695168	stored, special c/s	6-97
❏ 0416	MiG-21F-13	660416	stored, ex Vodochody, arr 9-6-93 by train	8-95
❏ 0420	MiG-21F-13	660420	stored, ex Vodochody, arr 25-6-93 by train	8-95
❏ 0441	MiG-21US	..	stored	5-98
❏ 0475	MiG-21UM	..	stored, silver c/s	8-97
❏ 0502	MiG-21F-13	660502	stored, ex Vodochody, arr 9-6-93 by train	8-95
❏ 0516	MiG-21F-13	660516	stored, ex Vodochody, arr 25-6-93 by train	8-95
❏ 0518	MiG-21F-13	660518	stored, ex Vodochody, arr 9-6-93 by train	8-95
❏ 0604	MiG-21F-13	760604	stored, ex Vodochody, arr 9-6-93 by train	5-95
❏ 0605	MiG-21F-13	760605	stored, ex Vodochody, arr 25-6-93 by train	8-95
❏ 0609	MiG-21F-13	760609	stored, ex Vodochody, arr 25-6-93 by train	8-95
❏ 0614	MiG-21F-13	760614	stored, ex Vodochody, arr 9-6-93 by train	5-95
❏ 0617	MiG-21F-13	760617	stored, ex Vodochody, arr 9-6-93 by train	5-95
❏ 0646	MiG-21US	06685146	stored, silver c/s	5-98
❏ 0701	MiG-21F-13	760701	stored, ex Vodochody, arr 25-6-93 by train	8-95
❏ 0710	MiG-21F-13	760710	stored, ex Vodochody, arr 25-6-93 by train	11-97
❏ 1112	MiG-21MF	..	stored, silver c/s	11-97

SLOVAK REPUBLIC - 401

☐ 1114		MiG-21MF	..	stored	11-97
☐ 1201		MiG-21MF	..	stored, silver c/s	11-97
☐ 1502		MiG-21R	94R01502	stored, silver c/s	11-97
☐ 1702		MiG-21R	94R01702	stored, silver c/s	11-97
☐ 1703		MiG-21R	94R01703	stored, silver c/s	11-97
☐ 1918		MiG-21R	94R01918	stored, silver c/s	11-97
☐ 1920		MiG-21R	94R01920	stored	11-97
☐ 1922		MiG-21R	94R01922	stored, silver c/s	11-97
☐ 1923		MiG-21R	94R01923	stored, silver c/s	11-97
☐ 1924		MiG-21R	94R01924	stored	11-97
☐ 2615		MiG-21MF	..	stored	11-97
☐ 2705		MiG-21MF	..	stored	11-97
☐ 2707		MiG-21U-600	662707	stored, ex Letnany, ex Vodochody	11-97
☐ 3041		MiG-21UM	..	stored	11-97
☐ 3051		MiG-21UM	..	stored, silver c/s	11-97
☐ 3176		MiG-21UM	..	stored, silver c/s	6-97
☐ 4308		MiG-21MF	..	stored	11-97
☐ 4310		MiG-21MF	..	stored	11-97
☐ 4311		MiG-21MF	..	stored, silver c/s	11-97
☐ 4312		MiG-21MF	..	stored	11-97
☐ 4401		MiG-21MF	..	stored	11-97
☐ 4409		MiG-21PFM	94A4409	stored, ex Vodochody, arr 3-7-93 by train	8-95
☐ 4410		MiG-21MF	..	stored	9-96
☐ 4413		MiG-21PFM	94A4413	stored, ex Vodochody, arr 23-7-93 by train	8-95
☐ 4414		MiG-21PFM	94A4414	stored, ex Vodochody, arr 23-7-93 by train	8-95
☐ 5101		MiG-21UM	..	stored	11-97
☐ 5166		MiG-21UM	516921066	stored, silver c/s	11-97
☐ 5410		MiG-21PFM	94A5410	stored, ex Vodochody, arr 3-7-93 by train	8-95
☐ 7206		MiG-21PFM	94A7206	stored, ex Vodochody, arr 3-7-93 by train	8-95
☐ 7207		MiG-21PFM	94A7207	stored, ex Vodochody, arr 3-7-93 by train	8-95
☐ 7209		MiG-21PFM	94A7209	stored, ex Vodochody, arr 3-7-93 by train	9-96
☐ 7705 ?		MiG-21MF	967705	stored, should be at Kbely	6-97
☐ 7706		MiG-21MF	967706	stored, camo c/s	11-97
☐ 7707		MiG-21MF	967707	stored	8-97
☐ 7715		MiG-21MF	967715	stored, camo c/s	11-97
☐ 7908		MiG-21PFM	94N9808	stored, ex Vodochody, arr 23-7-93 by train	11-97
☐ 7910		MiG-21PFM	94N7910	stored, ex Vodochody, arr 23-7-93 by train	8-97
☐ 7913		MiG-21PFM	94N7913	stored, ex Vodochody, arr 3-7-93 by train	11-97
☐ 8207		MiG-21MF	..	stored	11-97
☐ 8208		MiG-21MF	..	stored	6-97
☐ 8209		MiG-21MF	..	stored	11-97
☐ 9404		MiG-21MF	..	stored	11-97
☐ 9406		MiG-21MF	..	stored, camo c/s	11-97
☐ 9501	SL	MiG-21MF	..	preserved	5-98
☐ 9714		MiG-21MF	..	stored, camo c/s	6-97
☐ 9813		MiG-21MF	..	stored	11-97
☐ 9814		MiG-21MF	..	stored, camo c/s	11-97
☐ 9815		MiG-21MF	..	stored	11-97
☐ B-8939		Mi-8S	10839	stored, wreck in hangar	7-93

SVIDNÍK
The Dukelske Muzeum in Svidník has had an Li-2 since 1969. The ex Czech aircraft was painted in Soviet markings during 1974 and its former identity is not 100% sure.

☐ 2107 ?   Li-2   23442107 ? preserved, in Soviet marks as '40 white'   4-93

TRENČÍN

The Letecke Opravovne Trenčín (LOT) is responsible for most of the heavy maintenance of the Slovak Air Force. In the past LOT was also contracted by the museum at Kbely for restoration of their aircraft. Some of the projects were the HC helicopters and the SARO Cloud OK-BAK, which went to Kbely in 1995. Although their civil registrations were reserved in the US register as long ago as 1993, the three former Iraqi L-39s are still stored at Trenčín. Stored L-29 0105 (290105) was sold as N82171, 2821 (marked as '8291') went to Líšeň and L-29RS 2603 (792603) became N29RZ. The Mi-1, MiG-21s and Su-7s are stored here on behalf of the new military museum which will be built at the airfield. The aero club had Mi-4 8518, but this was scrapped in 1993.

❏ 3105	L-39ZO	433105	stored, ex Iraq, to become N332BH	5-98
❏ 3108	L-39ZO	433108	stored, ex Iraq, to become N303BH	5-98
❏ 3110	L-39ZO	433110	stored, ex Iraq, to become N334BH	5-98
❏ 5101	Avia 14S	015101	preserved, with the aero club	5-98
❏ 2409	L-29RS	792409	stored	8-94
❏ 2615	L-29RS	792615	stored, ex Prešov	5-98
❏ 2619	L-29RS	792619	stored	8-94
❏ 2823	L-29R	892823	stored	8-93
❏ 0004	L-39	130004	preserved, ex Prešov	5-98
❏ 1043	Mi-1	401043	stored, ex Prešov	5-98
❏ 0104	MiG-21F-13	460104	stored, ex Prešov, wfu 6-73	5-98
❏ 0109	MiG-21F-13	460109	stored, ex Prešov	5-98
❏ 0306	MiG-21PF	760306	stored	5-98
❏ 0412	MiG-21F-13	660412	stored, camouflaged, wfu 3-91	5-98
❏ 1215	MiG-21PF	761215	preserved, ex Kosice	5-98
❏ 2419	MiG-21U-600	662419	preserved, ex Sliač	5-98
❏ 4415	MiG-21PFM	94A4415	stored, ex Sliač	5-98
❏ 5021	Su-7BM	5021	stored	7-97
❏ 5316	Su-7BM	5316	stored	5-98
❏ 6504	Su-7BKL	6504	stored, ex Sliač, ex Přerov	10-96
❏ 6506	Su-7BKL	6506	stored, ex Sliač, ex Přerov	7-97
❏ OK-06	HC-3A	3	stored, ex Kbely, ex Trenčín, ex OK-VZB	—
❏ OK-RVY	HC-102	0434	stored, ex Kbely, ex Trenčín	—

VYHNE

One of the relatively few MiG-15s (compared with the Czech republic) in this country can be found in the village of Vyhne, which is some 20km south of Žiar nad Hronom.

❏ 3717	MiG-15bis	623717	preserved	6-94

VYŠNY KOMÁRNIK

As a war memorial a 'Sturmovik' and some tanks are preserved along road number 73 from Svidník to the Polish border. The Il-10 is owned by the Dukelske Muzeum from Svidník.

❏ 5514 ?	Avia B33	..	preserved, marked as '40', locally built Il-10	4-93

ŽILINA

For many years the international civil airfield of Žilina has had a unmarked L-410 in storage.

❏ OK-ADN	L-410A	710004	stored, unmarked	9-94

ZLIECHOV

In the mountains east of Nová Dubnica the village of Zliechov has an L-29. The aircraft is parked at the back of a local school and is in a bad shape.

❏ 0109	L-29	290109	preserved	6-94

SPAIN

Although thousands of tourists flock to the Spanish beaches each year, the airfields of Spain are visited less often. This may be connected with the fact that aircraft spotting in Spain (from outside the fence) is uncommon and not always accepted by the local authorities. Pre-arranged visits to airbases however, are becoming more and more frequent and give a good opportunity to 'explore' the whole airfield.

ALBACETE - LOS LLANOS

All the aircraft mentioned in EWR-1 had left by the late 1980s and have been replaced by other types. Two of the CASA 1131Es auctioned on 16th September 1988 will remain in Spain, E.3B-321 becoming EC-ERP and E.3B-408 becoming EC-ERO. The other two, E.3B-153 (0157, to G-BPTS) and E.3B-429, left for Britain. A more modern aircraft is the wrecked F1CE (C.14-23) which was discovered on a dump on the base in May 1992. A year later another one was noted. It is not known if both were on the same dump and therefore if C.14-19 had left by October 1994. Ala Logística 53 (better known as Maestranza de Albacete) maintains a large number of Spanish Air Force aircraft. Near their hangars two concrete poles were noted in September 1991, one held a CASA 1131 while the other was still empty. The two dumped CASA 101s, E.25-70/74-24 and E.25-77/74-31, have both moved to San Javier.

❏ E.3B-174		CASA 1131E	..	preserved, with Ala 53	3-93
❏ E.3B-521		CASA 1131E	..	preserved, Maestranza gate	9-97
❏ C.14-10	14-10	Mirage F1CE	..	instructional	4-96
❏ C.14-19	14-19	Mirage F1CE	..	dumped, crashed 19-7-92	3-93
❏ C.14-23	14-23	Mirage F1CE	..	dumped, crashed 27-8-91	9-97
❏ C.14-65	462-14	Mirage F1EE	..	stored, fuselage only	11-95
❏ U.9-37	14-03	CASA 127	37	stored	4-96
❏ U.9-42		CASA 127	42	stored, ex Granada	3-97
❏ ..		CASA 101EB	..	preserved, Maestranza gate, as 'E.25-22/79-22'	9-97

During 1989 a large number of Mentors arrived from San Javier for storage. It was reported in the press that 23 Spanish Mentors had been sold to Uruguay, but of the 25 Spanish aircraft two have crashed some years before, one is preserved at San Javier and three at Cuatro Vientos. Therefore only 19 could be available for sale.

E.17-1	791-01	T-34A	G.21	stored, ex San Javier, ex 52-7640	gone
E.17-2	791-02	T-34A	G.33	stored, ex San Javier, ex 52-7652	to Uruguay AF
E.17-3	791-03	T-34A	G.35	stored, ex San Javier, ex 52-7654	to Uruguay AF
E.17-4	791-04	T-34A	G.36	stored, ex San Javier, ex 52-7655	to Uruguay AF
E.17-5	79-05	T-34A	G.42	stored, ex San Javier, ex 52-7661	to Uruguay AF
E.17-6	791-06	T-34A	G.46	stored, ex San Javier, ex 52-7665	to Uruguay AF
E.17-7	791-07	T-34A	G.47	stored, ex San Javier, ex 52-7666	to Uruguay AF
E.17-8	791-08	T-34A	G.50	stored, ex San Javier, ex 52-7669	gone
E.17-9	791-09	T-34A	G.52	stored, ex San Javier, ex 52-7671	to Uruguay AF
E.17-10	79-10	T-34A	G.56	stored, ex San Javier, ex 52-7675	to Uruguay AF
E.17-11	79-11	T-34A	G.61	stored, ex San Javier, ex 52-7680	to Uruguay AF
E.17-12	791-12	T-34A	G.66	stored, ex San Javier, ex 52-7685	to Uruguay AF
E.17-14	791-14	T-34A	G.81	stored, ex San Javier, ex 53-3320	to Uruguay AF
E.17-15	791-15	T-34A	G.111	stored, ex San Javier, ex 53-3350	to Uruguay AF
E.17-16	791-16	T-34A	G.786	stored, ex San Javier, ex 55-0229	to Cuatro Vientos
E.17-17	791-17	T-34A	G.732	stored, ex San Javier, ex 55-0175	to Uruguay AF
E.17-18	791-18	T-34A	G.107	stored, ex San Javier, ex 53-3346	to Uruguay AF
E17-20	791-20	T-34A	X100	stored, ex San Javier	to Cuatro Vientos
E.17-21	79-21	T-34A	X101	stored, ex San Javier	to Uruguay AF
E.17-22	791-22	T-34A	X102	stored, ex San Javier	to Uruguay AF
E.17-23	791-23	T-34A	X103	stored, ex San Javier	to Uruguay AF
E.17-24	79-24	T-34A	X104	stored, ex San Javier	to Cuatro Vientos

ALCALÁ

Disco 'Aeropuerto' received the T-33 and Stinson from Vicálvaro by October 1992. The nose of the Torrejón F-4C also came here, but regulations forbade its use in the disco.

☐ E.15-33	41-31	T-33A	1296	preserved, ex 57-0647, with tail from E.15-40	10-92
☐ ..		F-4C	..	preserved, nose only, ex Torrejón	..
☐ ..		Stinson 108-3	..	preserved, ex Vicálvaro	10-92

ALCANTARILLA

The CASA 352L is preserved at the military airfield, while the CASA 212 fuselage in use for paratroop training is a purpose built mock up

☐ T.2B-181	721-10	CASA 352L	72	preserved	3-97
☐ ..		CASA 212	..	instructional, mock-up	8-92

ALICANTE

No traces were found of the I-115s in 1994. They were probably cleared out of the airport as it has grown a lot in recent years. In use with the airport fire-fighters here is a metallic fuselage which looks very similar to that of a C-130. Any ideas?

☐ E.9-74	ST-189	AISA I-115	74	preserved	1-92
☐ E.9-93	221-31	AISA I-115	93	stored	1-92
☐ EC-AQB		C-47A	12844	stored, ex EC-WQB	4-96
☐ EC-BUG		TC-47B	32734	stored, ex T.3-47, ex N86441	4-96
☐ ..		C-130 ?	..	instructional	94

ALMAGRO

This FAMET base has the boom of Bo105 HE.15-7/ET-133 which was written of on 19th October 1980. It is planted vertically in the ground as a memorial.

ARANDA DE DUERO

Mounted on a pole at a roundabout (junction of roads N1 and N122) in this town is an Agusta Bell 47.

☐ HE.7-12	78-03	AB47G-2	277	preserved	4-96

ARANJUEZ

In this town, on road NIV at Km54 (near Madrid), is a 'well equipped scrapyard'. In 1990 the yard held, besides trucks, ambulances, etc, the wrecked fuselage of CASA 101 E.25-04/793-04 (EB01-04-004) which crashed on 4th February 1985. It was gone in 1992.

BARCELONA - SAINT CUGAT

The huge discotheque Saint Cugat Airport still has the two KC-97s. It can be found along the A7 motorway on the north side of the city.

☐ TK.1-1	123-01	KC-97L	16985	preserved, ex Albacete, ex 53-0172	3-96
☐ TK.1-2		KC-97L	17007	preserved, white c/s, ex Tarancon, ex 53-0225	3-96

BENIDORM

Displayed on top of the Mellow Yellow Beach Disco on the N332 road is a Whirlwind Srs.2. This chopper has been in Benidorm since the mid 1970s.

☐ ZD.1B-21	Whirlwind Srs.2	WA396	preserved, ex code 803-3	3-97

BÉTERA
A Bell arrived from storage at Colmenar at this Army airfield by 1987 for display on the gate with Unidad de Helicopteros II.
☐ HE.7B-27 ET-059-103 OH-13S 3903 stored, ex Colmenar Viejo, ex 65-13007 3-97

CANDÁS
The former Aviaco Douglas DC-8 arrived here (between Gijón and Avilés) by road from Madrid on 13th July 1987 to be converted into a bar for the Pub San Francisco.
☐ EC-AUM DC-8-52 45657 preserved —

COLMENAR VIEJO
Of the five Bell 47s which were stored here (as mentioned in the first issue), the last one also found a new home, HE.7B-29 became EC-DZJ. AB206A HR.12-3/ET-197 moved on to Cuatro Vientos.
☐ HE.7B-26 ET-102 OH-13S 3902 preserved, marked as 'OH.13-5', ex 65-13006 11-97
☐ HU.8B-10 ET-206 UH-1C 3110 preserved, ex 65-12764 11-97
☐ HR.12-1 ET-109 AB206A-1 8197 preserved, marked as 'HR.12B-15' 11-97
☐ HR.12-4 ET-198 AB206A-1 8140 instructional 5-93
☐ HR.12-5 ET-199 AB206A-1 8144 stored 12-91
☐ HR.15-25 ET-144 Bo105LOH S.465 stored, wreck 4-96

CÓRDOBA
Stored on this airfield in 1996 was UH-1H HE.10B-38/78-51 (13275, ex Granada, ex 72-21576). By 1997 it had become airworthy again and is now EC-GKZ.

CORRAL DE AYLLÓN
The aerodrome is owned by the Fundacion Vara de Rey. They used to be called Fundacion Milicia Aerea Universitaria. Now civil, the foundation still has close relations with the military. A few times a year tactical exercises (involving paratroopers, security police) are held here. Aircraft noted at this field are:
☐ U.9-7 78-92 CASA 127 7 stored, ex León 5-91
☐ U.9-12 CASA 127 12 stored, ex León 5-91
☐ U.9-75 407-55 Do27B-1 145 stored, ex León, ex 5528/WGAF 5-91

CORUÑA DEL CONDE
On a pole in this town, which is some 30km north east of Aranda, is an unknown 'T-bird'.
☐ .. T-33A .. preserved, with tail from E.15-15 95

DOS BARRIOS
Dos Barrios is 72km south of Madrid along the NIV/E25. The two CASA 207s which were stored next to a local hotel were no longer current here in 1995. Therefor the fate of T.7-13/405-13 and T.7-18/405-18 is unknown.

DOS HERMANAS
The Saeta is this village, next to the NIV road, some 10km south of Seville, is still unknown. It is preserved on a pole next to the hotel 'La Motilla'. The aircraft on the northern edge of town is certainly not, as has been suggested, A.10B-53 which is still preserved at Villanubla.
☐ .. HA200D .. preserved 11-97

ESCALONA
There have been no reports of this aircraft since the mid 1980s. It may still be current.
❏ N87805   C-47B   33558   stored   —

ESPARTINAS
Preserved here, near Sevilla, is T-33A E.15-56/'24-01' which is believed to be the example noted at Moron in 1996 as '41-13'.
❏ E.15-26 '24-01'   T-33A   ..   preserved   —

GIBRALTAR
Gibraltar is still British, but Vulcan XM571 here was scrapped in 1990.

GRANADA
Town: Preserved at the Parque de las Ciencas Granada, a museum which is in town and can be reached from the same orbital exit as Armilla airfield, is a 'Huey'.
❏ HE.10A-1 78-60   AB205A   4002   preserved, ex Armilla, ex EC-SSO   11-97

Airfield - Armilla: Of the stored helicopters AB205A HE.10A-1/78-60 went to Granada town, UH-1H HE.10B-38/78-51 went to Cordoba, while HE.10B-52/78-54 is now at Cuatro Vientos. The HE.10B-37/78-50 (13274, l/n 11-95) is now flying as EC-GKY. Stored CASA 127 U.9-42/78-91 went to Albecete.
❏ ..   CASA 1131   ..   preserved, marked as 'E.3B-001/781-01'   10-97
❏ HE.7B-16 782-6   AB-47G   1507   stored, code 782-16 on other side   10-97
❏ HE.7A-48 78-14   OH-13H   ..   preserved, ex 57-6223   11-97
❏ ..   AB47G-3B1   ..   stored, frame in workshop   11-95
❏ HE.10B-51 78-53   UH-1H   13551   stored, ex 73-22068   11-95
❏ HE.10B-53 78-55   UH-1H   13553   preserved, at gate, ex stored, ex 73-22070   1-98
❏ HE.10B-54 78-56   UH-1H   13554   stored, ex 73-22071   11-95
❏ HE.20-14 78-43   H269C   1280748   stored, wreck, crashed 2-2-96   4-96
❏ N5435X   S-76   760161   instructional   4-96
❏ PT-HRD   S-76A   760203   instructional, as flight simulator   11-95

Of the long list of CASA 1131Es from the EWR-1, some more fates are known: E.3B-312 to EC-FIK, E.3B-340 to EC-EYU, E.3B-397 to EC-ETT, E.3B-487 to G-BUVN (2092 ex EC-333), E.3B-489 to D-EBPP (2091), E.3B-508 to G-BUOR (2134), E.3B-509 to Cuatro Vientos, E.3B-532 to F-AZGI, E.3B-539 to G-BUVP (2139, ex EC-338), and E.3B-556 to EC-FTZ and E.3B-606 to D-EFCB (2225). The last CASA 1131s were noted in April 1996. These have also become civil: E.3B-542/781-51 became EC-GIO, E.3B-557/781-59 is now EC-GIP, E.3B-591/781-2 became EC-GIN, E.3B-610/781-27 became EC-GIS and E.3B-620/781-26 became EC-GIR.

JEREZ DE LA FRONTERA
The based Orion unit moved to Morón in the mid 1990s, taking their Albatross (ex Palma) with them. Their preserved P-3 will go to the museum at Cuatro Vientos. C-7A Caribou T.9-27 (47) moved to here for a restaurant/disco by late 1992. It had disappeared by 1994.
❏ P.3-7   22-26   P-3A   5042   stored, ex preserved, ex 150516   11-97

LAS BARDENAS REALES
In most countries the shooting- and gunnary ranges are not really accessible and Las Bardenas is no exception. Many F-86s, T-33s and Saetas should still be in use as targets.
❏ C.5-100   F-86F   227-169   instructional, ex 55-3984   —

SPAIN - 407

☐ C.5-172	F-86F	227-196	instructional, ex 55-4011	—
☐ C.5-239	F-86F	176-385	instructional, ex 51-13454	—
☐ C.5-251	F-86F	176-330	instructional, ex 51-13399, camo c/s	—
☐ E.15-1	T-33A	7682	instructional, ex 51-17537	—
☐ E.15-12	T-33A	8093	instructional, ex 52-9947	—
☐ E.15-31	T-33A	1614	instructional, ex 57-0645	—
☐ E.15-45	T-33A	1628	instructional, ex 57-0659	—

LAS PALMAS DE GRAN CANARIA
Airfield - Gando: The Falcon 20DC EC-ECB (120) which overran the runway on 9th September 1987 was no longer to be noted in 1989.

☐ EC-BSQ	DC-7C	45159	preserved	3-96

Airfield - San Agustín: The DC-7 has swopped its blue Blaupunkt colours for the white and yellow scheme of Rothmans. The aircraft is still at the local aero club.

☐ EC-BBT	DC-7C	45553	preserved, ex JA6306, ex SE-CCH, ex HB-IBP	3-96

LEÓN
Preserved at the Virgen Del Camino air base are a number of former instructional airframes.

☐ B.2I-39		CASA 2111B	..	preserved	9-90
☐ C.10B-70	462-70	HA200D	20-76	preserved	6-94
☐ C.12-01	12-01	F-4C	1289	preserved, ex 64-0884	6-94
☐ E.15-22		T-33A	9079	preserved, ex 53-5740	6-94

The technical school still has a number of instructional airframes. Of these HA220 A.10C-110/214-55 has been noted at Cuatro Vientos.

☐ E.15-9	79-3	T-33A	8087	instructional, ex 52-9941	12-91
☐ E.15-36	41-24	T-33A	1379	instructional, ex 57-0650	12-91
☐ E.15-52	41-5	T-33A	8332	instructional, ex 53-4993	12-91
☐ E.15-59	41-12	T-33A	8489	instructional, ex 53-5150	12-91
☐ E.16-70	793-41	T-6G	..	instructional	12-91
☐ E.16-89	793-32	T-6G	..	instructional	12-91
☐ E.16-107	793-36	T-6G	..	instructional	12-91
☐ E.16-119	793-40	T-6G	..	instructional	12-91

MADRID
Town: The Escuela Tecnica Superior de Ingenieros Aeronauticos on the edge of the city has a number of instructional airframes. The school is in the 'University Town' of Madrid.

☐ A.10B-64		HA200D	20-70	instructional	—
☐ C.6-124	421-45	SNJ-5	121-41805	instructional, ex 90974/USN, ex 44-81103	—
☐ C.6-128	421-45	SNJ-5	..	instructional, ex 43859/USN, ex 42-84250	—
☐ E.14-30	212-60	HA200R-1	..	instructional	—
☐ ZD.1B-19	803-1	Whirlwind Srs.2	WA394	instructional	—

With the departure of the Americans from Torrejón, the USAF instructional airframes also had to leave. Some were brought to a scrapyard close to the airfield of Barajas.

☐ 82-1015	F-16A	61-608	dumped, wreckage only, crashed 13-1-88	3-97
☐ 83-1067	F-16A	61-620	dumped, wreckage only, crashed 17-12-87	3-97
☐ 86-0316	F-16C	5C-422	dumped, wreckage only, ex Torrejón, ex USAF	3-97
☐ 63-8263	F-105G	..	dumped, ex Torrejón, ex USAF	3-97
☐ 63-8265	F-105G	..	dumped, ex Torrejón, ex USAF	3-97

SPAIN - 408

Airfield - Barajas: A few aircraft have been noted on the dump at this international airport, only the Dakota and Falcon 20 could be identified. Both stored DC-8s EC-DYA and EC-DYB have flown out in 1994 as N925BV and N7046H.

❑ EC-ASP		C-47B	26980	dumped, ex OO-SBK	2-97
❑ EC-EKK		Falcon 20C	106	dumped, ex N31V	3-96
❑ EC-FHB		SA226TC	TC-355	dumped	3-96

Airfield – Cuatro Vientos: Several aircraft can be found on the military side of the airfield. Of these SA319B Alouette 3 HD.16-4 (2212) has become EC-GAJ.

❑ E.24A-17	42-14	Beech F33C	CJ.115	stored, wreck, crashed 22-9-93	10-94
❑ HE.7-11	78-02	AB47G-2	276	preserved, at Maestranza, on pole	11-97
❑ HU.10A-6	78-62	AB205A	4011	stored	6-96
❑ HU.10A-7	78-63	AB205A	4012	stored	10-94
❑ HD.16-8	803-05	Alouette 3	2274	stored	4-93
❑ T.9-5		DHC-4A	262	stored, ex Villanubla, ex 37-05	11-97
❑ T.9-29		C-7A	70	preserved, ex 371-09, ex 61-2592	11-97
❑ U.9-8	407-70	CASA 127	8	stored	6-96
❑ U.9-28	407-28	CASA 127	28	stored	6-96
❑ U.9-48	407-31	CASA 127	48	stored	6-96
❑ EC-283		Ce421C	..	dumped, burned wreck	5-96

The civil side of the airfield has a flying museum with aircraft like the CASA 1131 and Saeta. Stored at this side are also some ex military aircraft. Of these PA23-250E E.19-4 (27-4810) has become EC-800 and to EC-GGF, E.19-5 (27-4811, to EC-516 and EC-GLA) and E.19-6 (27-4812, to EC-GLJ). Of the Caribous T.9-21 (22) has been scrapped and T.9-27 moved to San Martin de la Vega in the early 1990s. Stored T.9-2 (259) became N82NC, T.9-4 (261) became N51NC, T.9-6 (263) became N53NC, T.9-7 (284) became N84NC and T.9-8 (286) became N86NC. The preserved Do28A-1 EC-CPP (3002) near the tower was scrapped in the mid 1990s. Three stored DHC-4A Caribous entered the civil register in early 1998 for conversion to firebombers; T.9-1 (258) became EC-GQL, T.9-11 (289) became EC-GQM and T.9-12 (290) became EC-GQN. Al three are ex Villanubla and last seen here in November 1997.

❑ E.3B-509		CASA 1131E	..	stored, ex Granada	5-96
❑ E.17-20	791-20	T-34A	X.100	stored, with Fundación Infante de Orleans	11-97
❑ E.19-2	42-51	PA23-250E	27-4807	stored	5-96
❑ T.9-3		DHC-4A	260	stored	11-97
❑ T.9-9		DHC-4A	287	stored, ex Villanubla, ex 37-09	11-97
❑ T.9-10		DHC-4A	288	stored, ex Villanubla, ex 37-10	11-97
❑ EC-CTO		RC680F	1195-100	instructional, with Are Aviation , ex OE-FAI	11-97

Cuatro Vientos also houses the well known Museo del Aire. On outside display and in open storage are:

❑ 2012		MiG-23ML	0390324621	preserved, ex 331/NVA, ex Laage	11-97
❑ 2226		MiG-21SPS	944302	preserved, ex 740/NVA, ex Rothenburg	11-97
❑ 2518		Su-22M4	26205	preserved, ex 686/NVA, ex Laagc	11-97
❑ 2623		F-104G	9174	preserved, one side as 'C.8-15/104-15/32733'	11-97
❑ A.9-050	21-50	SF-5A	2050	preserved, ex Morón, ex C.9-050	11-97
❑ AR.9-062	23-62	SRF-5A	2062	preserved, ex Talavera, ex Morón, ex CR.9-062	11-97
❑ A.10C-91	214-91	HA220	22-96	preserved, ex Tablada, ex C.10C-91	11-97
❑ A.10C-106		HA220	22-111	stored, in bare metal, ex C.10C-106	1-97
❑ A.10C-110	214-55	HA220	22-115	stored, ex León, ex C.10C-110	3-97
❑ C.5-223		F-86F	176-381	preserved, as 'C.5-104/1-104', ex 51-13450	11-97
❑ C.11-9	11-09	Mirage 3EE	596	preserved	11-97
❑ C.12-37	12-29	F-4C	1151	preserved, ex 64-0820	5-97
❑ CR.12-42	12-51	RF-4C	1726	preserved, ex 65-0937	5-97
❑ AD.1B-8		HU-16B	186	preserved, ex 51-5304	11-97
❑ HD.5-2	58-2	Do24T-3	5341	preserved, ex HR.5-2, ex EC-DAB	11-97
❑ E.14A-9		HA200	20-16	preserved, on pole	11-97

SPAIN - 409

	Serial	c/n	Type	MSN	Notes	Date
☐	E.16-90	793-6	T-6G	168-462	preserved, ex 49-3348	11-97
☐	E.19-3	42-52	PA23-250E	27-4809	preserved, ex N14239	11-97
☐	XE.25-01		CASA 101	P1	preserved, ex EC-ZDF, ex EC-ZDY	11-97
☐	HE.7B-21	782-11	AB47G-3B	1512	preserved	11-97
☐	HE.7A-41	78-10	OH-13H	2386	stored, ex 58-5373	3-97
☐	HE.7A-52	78-18	OH-13H	2536	preserved, ex 59-4953	11-97
☐	HE.10B-39	78-52	UH-1H	13276	stored, in a compound, ex 72-21577	11-97
☐	HE.10B-52	78-54	UH-1H	13552	preserved, ex EC-STL, ex 73-22069	11-97
☐	HD.11-1		AB47J-3B-1	2094	preserved, unmarked, ex Z.11-1, ex EC-SSA	5-97
☐	TK.1-3	123-03	KC-97L	16971	preserved, ex 53-0189	11-97
☐	T.2B-211	911-16	CASA 352L	102	stored	11-97
☐	T.2B-254		CASA 352L	145	stored, marked as '4025', ex 'D-2521'	5-97
☐	T.3-36	721-9	C-47B	20600	preserved, ex N86444, ex 43-16134	11-97
☐	T.4-10	911-10	C-54A	10366	preserved, ex N88934, ex NC88934	11-97
☐	T.7-6	405-15	CASA 207A	6	preserved, ex Tablada	11-97
☐	T.7-17	405-17	CASA 207C	17	preserved	11-97
☐	T.9-25	371-05	C-7A	53	preserved, ex 61-2394	11-97
☐	U.9-10		CASA 127	10	preserved, ex L.9-10	3-97
☐	U.9-33	403-53	CASA 127	33	stored, ex L.9-33	11-97
☐	U.9-49	403-52	CASA 127	49	stored, ex L.9-49	5-97
☐	UD.13-1	43-01	CL-215	1010	preserved, ex EC-BXN, ex CF-TXD	11-97
☐	U.14-1	407-7	Do28A-1	3014	preserved, ex EC-AQD, ex EC-WQD	11-97
☐	ZD.1B-22	803-4	Whirlwind Srs.2	WA397	preserved	11-97
☐	Z.2-6	75-6	AC12	06 ?	preserved	11-97
☐	EC-AXE		AB47J-3B1	2077	preserved	11-97
☐	EC-CLV		Alouette 3	2373	preserved	11-97
☐	EC-693		PBY-5A	1960	preserved, marked as 'DR.1/74-21', ex EC-314	11-97
☐	F-BMMS		MS733	105	stored, in fake Aeronavale marks as '105'	11-97
☐	N86427		TB-25N	108-33440	preserved, marked as '74-17', ex Malaga	11-97
☐	VH-TRQ		Transavia T300A	G783	preserved	11-97

Displayed inside various museum hangars:

	Serial	c/n	Type	MSN	Notes	Date
☐	A.10C-104		HA220	22-109	preserved, ex C.10C-104	11-97
☐	B.2-82	25-82	He111E-3	2940	preserved, ex code 14-16	11-97
☐	C.1-328		HA132L	328 ?	preserved, marked as '3-52' and '262'	11-97
☐	C.4J-10	94-28	HA1112K-1L	46 ?	preserved	11-97
☐	C.4K-158	471-23	HA1112M-1L	226 ?	preserved	11-97
☐	C.5-58	102-4	F-86F	191-290	preserved, ex 52-4594	11-97
☐	C.6-155	421-35	SNJ-5	121-41833	stored, ex 90982/USN, ex 44-81111	11-97
☐	E.1-14	513-20	Bü133C	09	preserved, ex ES.1-14	11-97
☐	E.3B-198		CASA 1131	0203	preserved	11-97
☐	E.4-161		HM-1B	161 ?	preserved, marked as 'HM-1'	11-97
☐	E.9-119		AISA I-115	172	preserved	11-97
☐	XE.14-2		HA200R-1	20-02	preserved, ex 'X2-14-2', ex EC-ANN	11-97
☐	E.15-51	41-8	T-33A	8260	preserved, ex 53-4921	11-97
☐	E.17-16	791-16	T-34A	G.786	preserved, ex 55-0229	11-97
☐	HE.7-13	751-4	AB47G-2	278	preserved	11-97
☐	HR.12-3		AB206A-1	8136	preserved, SAR c/s one side, FAMET other	11-97
☐	HD.16-1	803-01	Alouette 3	1952	preserved, ex EC-STE	11-97
☐	L.2-21	90-53	Stinson 108-3	108-5162	preserved	11-97
☐	L.12-2	407-2	O-1A	22426	preserved, ex 51-12112	11-97
☐	T.4-5		C-54D	10824	preserved, nose section only, ex 42-72719	11-97
☐	T.8B-97	462-04	CASA 2111H	108	preserved	11-97
☐	D-ENAE		Klemm L-25	277	preserved, marked as '30-22', ex Schwenningen	11-97
☐	EC-AFJ		HS34	1	preserved, marked as '1E34'	11-97

SPAIN - 410

	Registration	Type	Serial	Status	Date
☐	EC-AFQ	DH.60G-3	..	preserved, marked as '30-89' and 'EM-016/P'	11-97
☐	EC-AIM	Cierva C.19 Mk.4P	5158	preserved, licence built Avro 620, ex EC-CAB	11-97
☐	EC-AKL	AISA I-11B	006	preserved	3-97
☐	EC-APR	Ka6CR	784	preserved, glider	11-97
☐	EC-AXK ?	Zlin 326A	..	preserved, marked as 'EC-AXL'	11-97
☐	EC-AZD ?	Stinson 108	108-4338 ?	preserved, unmarked frame	11-97
☐	EC-BLD	AISA I-11B	159	preserved, marked as 'L.8C-44/791-28'	5-97
☐	EC-MCQ	Grunau Baby II	..	preserved, glider	11-97
☐	EC-OBN	Kranich II	..	preserved, glider	11-97
☐	EC-ODK	Kranich III	..	preserved, glider	11-97
☐	EC-RAB	Weihe	..	preserved, glider	11-97
☐	EC-RAT	Slingsby T.34	..	preserved, glider	11-97
☐	F-AZET	N1002	97	preserved, marked as 'L.15-2/91-6'	11-97
☐	F-BFZJ	SV-4C	46	preserved	11-97
☐	G-ACYR	DH.89A	6261	preserved	11-97
☐	..	Avro 504K	..	preserved, replica, as 'A-28' and 'M-MABE'	11-97
☐	..	Blériox XI	1	preserved, Vilanova Acedo built	11-97
☐	..	Breguet 19GR	42	preserved, marked as '12-72', CASA built	11-97
☐	..	Breguet 19 Super TR	..	preserved, replica, marked as 'Cuatro Vientos'	3-97
☐	..	Bristol F.2b	..	preserved, replica, marked as 'B.21'	11-97
☐	..	Caudron C.272	..	preserved, replica, as '30-171' and 'EL-007'	11-97
☐	..	Caudron G.3	..	preserved, replica, marked as 'BC-6'	3-97
☐	..	Cierva C.6	..	preserved, replica, marked as 'C6.B'	11-97
☐	..	DH.82A	..	preserved, marked as '30-103' and 'EP-003'	11-97
☐	..	Dornier J Wal	..	preserved, replica, marked as 'M-MWAL'	11-97
☐	..	Eagle II	..	preserved, replica, marked as 'EC-CBB'	3-97
☐	..	Fi156C2	2027	preserved, marked as 'L.16-23/96-1'	11-97
☐	..	Fokker C.III	..	preserved, replica, marked as 'M-MOAB'	11-97
☐	..	Fokker Dr.I	..	preserved, replica, marked as '425/17'	11-97
☐	..	Hawk Major	..	preserved, marked as 'EN-002' and '30-145'	11-97
☐	..	MS181	..	preserved, marked as 'E-004'	11-97
☐	..	MS230	1066	preserved, marked as '005'	11-97
☐	..	Nieuport IVG	..	preserved, replica, marked as 'mN n5'	11-97
☐	..	Polikarpov I-15	..	preserved, replica, as 'CA-125' and 'A.4-103'	11-97
☐	..	Polikarpov I-16	..	preserved, replica, as 'CM-260' and 'C.8-25'	11-97

The museum should also have the following aircraft, although their current status is unknown. Stored HA1112M1L C.4K-162 has moved on to Chavenay, together with CASA 1131 E.3-574. Harvard E.16-97/793-2 is now at Vilanova, while T-34A E.17-24/79-24 (X104) has become EC-750. CASA 2111D BR.2I-10/462-06 became G-BFFS. SNJ-6 C.6-179/421-67 became F-AZJS at La Ferte Alais. C.6-168/421-59 (168-409) and C.6-135/421-46 also ended up in France. The fuselage of T-Bird E.15-48 was joined with the tail from E.15-19 and both went to Villanueve y la Geltrú. The stored T-6G C.6-188/421-68 (168-160) moved to La Ferté Alais by early 1998.

	Registration	Type	Serial	Status	Date
☐	FH426	Hudson IIIA	414-6716	stored, recovered from Med. Sea, ex 41-37227	10-92
☐	A.10A-12 793-46	HA200A	20-20	stored, ex C.10A-12, ex E.14A-12	8-96
☐	A.10C-111 214-111	HA220	22-116	stored, ex Tablada	10-94
☐	C.6-125 421-43	SNJ-5	121-42332	stored, ex 91036/USN, ex 44-81310, ex N5196V	8-96
☐	C.6-159 421-55	SNJ-4	88-13578	stored, ex 27842/US Navy	8-96
☐	E.3B-533	CASA 1131	..	stored	10-93
☐	E.3B-565 781-7	CASA 1131	2182	stored, no tail	6-96
☐	E.3B-605	CASA 1131	..	stored	10-93
☐	E.3B-625	CASA 1131	..	stored	10-93
☐	E.4-174	HM-1B	174 ?	stored, dismantled	10-94
☐	E.15-12	T-33A	8093	stored, tail only, ex 52-9947	10-94
☐	E.15-17 41-14	T-33A	8504	stored, ex 53-5165, tail at Murcia	6-94

☐ E.15-20	41-6	T-33A	8826	stored, forward fuselage only, ex 53-5487	4-96
☐ E.15-21	41-11	T-33A	8894	stored, forward fuselage only, ex 53-5555	4-96
☐ E.15-27	41-39	T-33A	9175	stored, tail only, ex 54-1544	3-96
☐ E.15-38	41-26	T-33A	1381	stored, fuselage with tail from E.15-12	10-94
☐ E.15-38	41-26	T-33A	1381	stored, tail only, ex 57-0732	3-96
☐ E.15-53	41-13	T-33A	8389	stored, fuselage with tail from E.15-56	3-96
☐ E.16-118	793-3	T-6G	168-584	stored, ex 49-3450	8-96
☐ U.9-76	407-72	Do27B-1	400	stored, ex L.9-76	8-96
☐ Z.2-7	75-7	AC12	07 ?	preserved	3-96
☐ Z.4-06		AC14	06 ?	preserved	3-96
☐ 60-3558	ET-E01	UH-1B	204	stored, ex Colmenar Viejo	8-92

Airfield – Getafe: The Madrid transport base still has two complete aircraft preserved. These have been joined by a number of aircraft tails. These are preserved inside the gate.

☐ E.24A-20	42-25	Beech F33C	CJ.118	dumped, collided 24-1-97	11-97
☐ E.24A-26	42-19	Beech F33C	CJ.124	dumped, collided 24-1-97	11-97
☐ T.4-8	35-08	DC-4	27323	preserved, ex 44-9087	11-97
☐ T.7-1	351-01	CASA 207	1	preserved, at gate	11-97

Parked between the Air Force and CASA side of the airfield are 14 Mirage 3s, which were brought here in the hope of reselling them. This never happened and all the Mirages were stored here. During 1992 a further batch of Mirages arrived from Valencia for storage, comprising of C.11-1/11-01, C.11-2/11-02, C.11-3/11-03, C.11-5/11-05, C.11- 9/11-09, C.11-16/11-16 and CE.11-26/11-71. Of these only Mirage 3EE C.11-9 has been seen after 1992, and it is now at the museum of <u>Cuatro Vientos</u>. The others may be in storage inside the CASA factory.

☐ C.11-4	11-04	Mirage 3EE	591	stored, ex Valencia	11-97
☐ C.11-6	11-06	Mirage 3EE	593	stored, ex Valencia	11-97
☐ C.11-8	11-08	Mirage 3EE	595	stored, ex Valencia	11-97
☐ C.11-10	11-10	Mirage 3EE	597	stored, ex Valencia	11-97
☐ C.11-17	11-17	Mirage 3EE	604	stored, ex Valencia	11-97
☐ C.11-18	11-18	Mirage 3EE	626	stored, ex Valencia	11-97
☐ C.11-21	11-21	Mirage 3EE	629	stored, ex Valencia	11-97
☐ C.11-22	11-22	Mirage 3EE	630	stored, ex Valencia	11-97
☐ C.11-23	11-23	Mirage 3EE	631	stored, ex Valencia	11-97
☐ C.11-24	11-24	Mirage 3EE	632	stored, ex Valencia	11-97
☐ CE.11-25	11-70	Mirage 3DE	424 ?	stored, ex Valencia	11-97
☐ CE.11-27	11-72	Mirage 3DE	542 ?	stored, ex Valencia	11-97
☐ CE.11-28	11-73	Mirage 3DE	547 ?	stored, ex Valencia	11-97
☐ CE.11-29	11-74	Mirage 3DE	562 ?	stored, ex Valencia	11-97

Displayed at the CASA factory are some of their earlier aircraft. These are visible from outside at one of the side gates.

☐ A.10B-49		HA200D	20-54	preserved	11-97
☐ T.7-19	351-02	CASA 207	19	preserved	5-97
☐ EC-DUJ		CASA 101	98	preserved	11-97

MÁLAGA

The eight American C-47s (see first edition) which were here in the late 1980s have all gone. Of the last two, N219G (9894) became YV-505C and N486F (20214) was cancelled from the register as destroyed. Of the Doves, one is unmarked between the terminal and coach station while the other is concrete blocks preserved at the fire station.

☐ N9888A	Dove 7	04534	stored, ex CN-MBA	3-97
☐ N9890A	Dove 7	04535	stored, ex CN-MBB	3-97
☐ N9886A	Beech D18	A-932	stored, ex CN-MAL	3-97

MARÍN
Preserved with the Spanish Navy officers school may still be a Agusta Bell.
❏ Z.8-1   003-1   AB204AS   3128   preserved, crashed 31-10-78   —

MORÓN DE LA FRONTERA
The F-5s have left the base and their place hase been taken by CASA 101s, EF-18s and the Orions from Jerez. Most of the F-5s in storage here have left the airfield. SF-5As A.9-037 and A.9-048 moved on to Talavera. Also going to Talavera were SRF-5As AR.9-045, AR.9-053, AR.9-055, AR.9-056, AR.9-058, AR.9-062, CR.9-063 and AR.9-064. SF-5A A.9-050 went to Cuatro Vientos. In 1996 and early 1997 a T-33A was noted with code 41-13, which is believed to be E.15-56, moved on to Espartinas.

❏ AR.9-060	21-55	SRF-5A	2060	preserved, ex instructional	1-98
❏ A.10B-52	214-52	HA200D	20-66	preserved, at gate	1-98
❏ C.5-231	151-21	F-86F	193-36	preserved, at gate, ex 52-5307	1-98
❏ AN.1B-13	221-13	HU-16B	..	preserved, ex Jerez, ex Palma	1-98
❏ U.9-46	21-91	CASA 127	46	stored	4-96
❏ U.9-72	407-54	Do27A-1	142	stored	4-96

MOTRIL
Motril is some 65km south of Granada near the sea, near the Hospital Santa Ana a Saeta is preserved.
❏ A.10B-89 'EVA.9-1' HA200D   20-95   preserved   11-97

MURCIA - SAN JAVIER
San Javier is still the home of Ala 79 with its T35s, CASA 101s and CASA 212s. Of their former T-34s only one has remained. The rest relocated to Albacete.

❏ E.3B-75	791-1	CASA 1131L	..	preserved	4-97
❏ E.15-27	41-39	T-33A	9175	preserved, ex 54-1544, with tail from E.15-17	4-97
❏ E.16-199	793-11	T-6G	..	preserved	4-97
❏ E.17-25	791-25	T-34A	X107	preserved	4-97
❏ U.9-56	11-92	Do27A-4	506	stored, ex 5756/WGArmy	4-96

PALMA DE MALLORCA
Town: In a shopping mall, close to the docks, an unmarked I-115 is hanging from the ceiling.
❏ ..   AISA I-115   ..   preserved, unmarked   97

Town – Malaguf la Porrasc: On the lawn of the 'Talbert Lawn Bowls' is a DC-6. The aircraft carries the titles and logo of the club and can be found just north of Magaluf (south west of Palma).
❏ EC-AVA   DC-6   43118   preserved, ex PH-DPP, ex PJ-DPP   3-96

Airfield - Son Ferriol: On small airfield some 6km north west of Palma international is a Skytrain in fairly good condition. It is in bare metal with only the rudder missing.
❏ EC-EJB   C-47A   4479   preserved, ex Malaga, ex EC-177   10-96

Airfield - Son San: Of the stored fleet of Convair 990s EC-BJC (22), EC- BJD (23), EC-BTE (21), EC-BXI (35), EC-CNF (8), EC-CNG (7), EC-CNH (17) and EC-CNJ (14) were all scrapped in 1991, followed by EC-BQA (36) and EC- BZP (18) in November 1995. Only two CV990s remain here. The two DC-8s noted here in the 1980s, EC-CDC (45567) and EC-CCN (45569) have also been scrapped. Also scrapped in the early 1990s were DC-3s EC-EIS (16066/32814) and EC-EQH (16310/33058). HU-16B AD.1B-13 moved to Jerez.

❏ EC-BQQ	Convair 990	34	stored, fuselage only, ex N5606	10-96
❏ EC-BZO	Convair 990	30	preserved, ex N7643	10-96
❏ EC-DSA	AC680T	1564-20	stored, ex I-ARBO	2-89

SPAIN - 413

REUS
There was no sign of the unknown AISA I-115 at the gate in 1994. Only a T-33A was present on a pole at the officers school here.

❑ E.15-43	41-15	T-33A	1306	preserved, ex 57-0657	9-94

ROTA
At Rota the United States and Spanish Navy have based aircraft, while other NATO countries also make use of the airfield.

❑ 146451	001	EA-3B	12403	dumped	5-93
❑ 146457	007	EA-3B	12409	preserved	11-97
❑ 156610	AC-604	RA-5C	316-5	dumped, SOC 31-10-78	5-93
❑ E.30-1	01-401	PA24-260	24-4071	stored, near hangars, wfu 15-12-88	11-97
❑ E.30-2	01-402	PA24-260	24-4075	stored, near hangars, wfu 11-9-89	11-97
❑ E.31-1	01-403	PA30-160	30-599	stored, near hangars, wfu 15-12-88	11-97
❑ E.31-2	01-404	PA30-160	30-653	preserved, wfu 10-10-92	4-96
❑ HE.7-2		Bell 47G	700	stored, ex code 01-102	8-92
❑ HE.7-3		Bell 47G	701	stored, ex code 01-103	8-92
❑ HE.7-4		Bell 47G	1519	stored, ex code 01-104	8-92
❑ HE.7-9	01-109	AB47G-2	272	stored	8-92
❑ HE.7-15	01-110	AB47G-2	281	stored	8-92
❑ HE.7-23		AB47G-2A	1523	stored, ex code 01-111	8-92
❑ HE.7A-24		AB47G-2A-1	1610	stored, ex code 01-112	8-92
❑ HE.7B-31		Bell 47G-5	7884	stored, ex code 01-114	4-96

SABADELL
On this airfield north west of Barcelona a DC-3 and AVD12 are stored.

❑ XL.10-1	AVD12C	1	stored, derelict	3-95
❑ EC-FDH	DC-3	11982	stored	95

SALAMANCA
Town: In town, on the road to Zamora (road Nr N630), next to the Moderno Hotel, near the railway station, is a scrapyard with a complete Skytrain. Stored at the Montero scrapyard is the fuselage of an Aviocar.

❑ T.3-28	744-28	C-47A	9914	preserved, ex N44V, ex 42-24052	11-97
❑ T.12B-38	37-10	CASA 212	74	stored, fuselage only	3-97

Airfield - Matacan: The Texan at the gate has been changing its code over the years. It has been seen marked as 793-104, 793-11 and 742-104. The T-33 at the gate may be in fake markings as by 1988 E.15-20/41-6, with the tail of E.15-15/41-43, had arrived here from Zaragoza to be mounted at the gate.

❑ C.16-122	74-103	SNJ-5	..	preserved, ex Albacete	11-97
❑ E.15-25	791-25	T-33A	..	preserved, see note	9-95
❑ E.25-70	74-24	CASA 101EB	EB01-70-076	dumped, crashed 10-5-91	5-93
❑ E.25-77	74-31	CASA 101EB	EB01-77-085	dumped, crashed 10-5-91	5-93

SANCHIDRIÁN - AVILA
The airfield was closed in the early 1990s and the stored AISA I-115s, E.9-131, E.9-143 and E.9-166, were all scrapped.

SAN FERNANDO
A museum at this marine base near Cádiz should have two helicopters on display.

☐ HA.14-8	01-708	AH-1G	21127	preserved	—
☐ ..		UH-1	..	preserved	—

SAN JUAN DE AZNALFARACHE
Noted in the cafe El Avion in this village near Gelves and Tablada is a CASA 207, at Km3 on the SE660 road.
☐ T.7-16   405-16   CASA 207C   16   preserved, ex Tablada   11-97

SAN MARTÍN DE LA VEGA
Arriving here on 26th April 1991 was ex Cuatro Vientos Caribou T.9-27. It departed to <u>Jerez</u> by late 1992.

SANTANDER - PARAYAS
An SNJ-6, which arrived in 1983, is still preserved near the tower.
☐ C.6-165   421-54   SNJ-6   ..   preserved, ex 112200/US Navy   11-95

SEVILLA
Airfield - San Pablo: The Saeta is preserved in the car park of the CASA factory. The stored CV440 T.14-1 (401, ex Tablada) was sold in early 1997 as 3C-JJO.

☐ A.10C-112	214-112	HA220	22-117	preserved, with CASA	1-98
☐ E.15-54	46-41	T-33A	8392	preserved, with CASA, ex Tablada, ex 53-5053	4-96
☐ TR.7-5	403-02	CASA 207	5	dumped	11-97
☐ T.7-8	351-8	CASA 207A	8	dumped	8-94
☐ T.7-10		CASA 207A	10	dumped, burned	4-96

Airfield – El Copero: The ex Colmenar Viejo Bell OH-13 is still displayed at the gate of this army airfield.
☐ HE.7B-28   ET-059-104   OH-13S   3904   preserved, at gate, ex 65-13008   3-97

Airfield - Tablada: The base was closed in the early 1990s. The preserved T-33 E.15-54 and stored Convair 440 T.14-1 moved to <u>San Pablo</u>. The HA200A E.14A-9 moved to Cuatro Vientos. CASA 2111B B.2I-103 (marked as 'B.2I-25') was sold to the OFMC at <u>Duxford</u>. The fate of the Beech F33C E.24A-4 is unknown.

TALAVERA LA REAL
The amount of *Wrecks & Relics* has grown rapidly over recent years with the arrival of the F-5s from Morón. All flyable F-5s are now based at Talavera la Real. F-5 AR.9-062/21-57 (ex Morón) went to <u>Cuatro Vientos</u>.

☐ E.15-60		T-33A	8764	preserved as 'E.15-1/E.73-1', ex 53-5425	11-97
☐ C.5-2		F-86F	176-170	preserved as 'C.5-199/732-1', ex 51-13239	11-97
☐ A.9-037	02	SF-5A	2037	dump, along platform, ex Morón, crashed 1-84	11-97
☐ A.9-038	21-01	SF-5A	2038	stored, fuselage only, camo c/s	4-96
☐ A.9-039		SF-5A	2039	stored, fuselage only, camo c/s	3-97
☐ A.9-041	23-27	SF-5A	2041	stored, in hangar	11-97
☐ A.9-046		SF-5A	2046	stored, cockpit only	4-96
☐ A.9-048	21-09	SF-5A	2048	stored, fuselage only, camo c/s, ex Morón	5-96
☐ A.9-052		SF-5A	2052	stored, fuselage only, camo c/s	3-97
☐ AR.9-053	21-50	SRF-5A	2053	stored, fuselage only, camo c/s, ex Morón	3-97
☐ AR.9-055	21-51	SRF-5A	2055	stored, fuselage only, camo c/s, ex Morón	4-96
☐ AR.9-056	212-56	SRF-5A	2056	stored, fuselage only, camo c/s, ex Morón	4-96
☐ AR.9-058	21-53	SRF-5A	2058	stored, fuselage only, camo c/s, ex Morón	4-96
☐ CR.9-063	211-63	SRF-5A	2063	stored, fuselage only, silver c/s, ex Morón	5-96
☐ AR.9-064	212-64	SRF-5A	2064	stored, fuselage only, camo c/s, ex Morón	3-97
☐ AR.9-066	23-31	SRF-5A	2066	stored, camo c/s	3-97

SPAIN - 415

❏ AR.9-069	23-32	SRF-5A	2069	stored	3-97
❏ U.9-17	23-90	CASA 127	17	stored, wreck, crashed 27-1-92	4-96
❏ U.9-36	31-06	CASA 127	36	stored, wreck, crashed 12-5-92	3-97

TENERIFE

Town - Santa Cruz: The second Viscount from Los Rodeos has moved here and is now in use as a discotheque/night club in the La Granja park.

❏ EC-DYC	Viscount 806	262	stored, ex G-AOYM	92

Airfield - Los Rodeos: Preserved at the Spanish Army base here is a Stinson with obviously incorrect markings. At the civil side two Viscounts were stored. Of these, EC-DYC moved on to <u>Santa Cruz</u>.

❏ ..	Stinson 108	..	preserved, marked as 'ET-1992'	1-93
❏ EC-DXU	Viscount 806	264	stored, ex G-AOYO	11-96

TORREJÓN DE ARDOZ

With the departure of the Americans the *Wrecks & Relics* were also disposed of. F-4C 63-7667 was scrapped locally, while F-16C 86-0316 and F-105G 63-8263 and 63-8265 appeared in a scrapyard near Barajas. The former instructional T-33 (ex French AF), was still in use for training in 1997. The stored R/F-4Cs are either stored on the ramp or in a compound near one of the hangars.

❏ 52-9975	TJ	T-33A	8206	dumped, ex French AF	3-97
❏ C.5-98		F-86F	227-167	preserved, as 'C.5-82/6-082', ex 55-3982	11-97
❏ C.12-02	12-01	F-4C	1331	preserved, at gate, ex 64-0900	11-97
❏ C.12-03	121-02	F-4C	1339	stored, in compound, ex 64-0903	3-97
❏ C.12-04	122-02	F-4C	1359	stored, in compound, ex 64-0909	3-97
❏ C.12-07	121-04	F-4C	1395	stored, in compound, ex 64-0920	4-96
❏ C.12-08	122-04	F-4C	1411	stored, in compound, ex 64-0925	3-97
❏ C.12-09	12-06	F-4C	1247	stored, in compound, ex 64-0867	3-97
❏ C.12-10	12-07	F-4C	1317	stored, in compound, ex 64-0895	3-97
❏ C.12-11	121-06	F-4C	1255	stored, ex 64-0871	3-96
❏ C.12-12	122-06	F-4C	1201	stored, ex 64-0846	4-93
❏ C.12-14	122-07	F-4C	1209	stored, on ramp, ex 64-0850	3-97
❏ C.12-15	12-11	F-4C	1244	stored, ex 64-0866	5-93
❏ C.12-16	12-12	F-4C	1249	stored, on ramp, ex 64-0868	3-97
❏ C.12-17	121-09	F-4C	1214	stored, in compound, ex 64-0853	3-97
❏ C.12-18	12-14	F-4C	1216	stored, on ramp, ex 64-0854	3-97
❏ C.12-19	12-15	F-4C	1257	preserved, near base museum, ex 64-0872	11-97
❏ C.12-20	122-10	F-4C	1220	stored, on ramp, ex 64-0855	3-97
❏ C.12-22	12-17	F-4C	1407	stored, in compound, ex 64-0924	4-96
❏ C.12-23	12-18	F-4C	1270	stored, in compound, ex 64-0877	3-97
❏ C.12-26	122-13	F-4C	1233	stored, on ramp, ex 64-0861	3-97
❏ C.12-27	121-14	F-4C	1235	stored, on ramp, ex 64-0862	3-97
❏ C.12-28	122-14	F-4C	1275	stored, on ramp, ex 64-0880	4-95
❏ C.12-29	12-22	F-4C	1225	stored, in compound, ex 64-0858	3-97
❏ C.12-30	122-15	F-4C	1272	stored, in compound, ex 64-0878	3-97
❏ C.12-31	121-16	F-4C	1229	stored, on ramp, ex 64-0859	3-97
❏ C.12-32	122-16	F-4C	1221	stored, in compound, ex 64-0856	3-97
❏ C.12-33	12-33	F-4C	1296	stored, on ramp, ex 64-0887	3-97
❏ C.12-34	122-17	F-4C	1353	stored, in compound, ex 64-0907	3-97
❏ C.12-36	122-18	F-4C	1240	stored, on ramp, ex 64-0864	4-96
❏ C.12-38		F-4C	1284	stored, in compound, ex 64-0882	3-97
❏ C.12-39	12-31	F-4C	1308	stored, on ramp, ex 64-0892	3-97
❏ C.12-40	12-32	F-4C	1320	stored, on ramp, ex 64-0896	3-97

SPAIN - 416

☐ CR.12-41		RF-4C	1718	stored, in compound, ex 65-0936	3-97
☐ CR.12-43		RF-4C	1736	stored, in compound, ex 65-0938	3-97
☐ CR.12-44		RF-4C	1782	stored, in compound, ex 65-0943	3-97
☐ E.15-13	6-47	T-33A	8419	preserved, ex 53-5080, with tail from E.15-47	3-97
☐ E.17-13	791-13	T-34A	CCF34-27	dumped, burned wreck, ex 52-8254	3-96
☐ T.2B-246	792-20	CASA 352L	137	preserved	3-97

UTRERA
Displayed outside the scrapyard of Desguaces Carmonais the fuselage of a HA200. Inside the yard is an HA220. Utrera is near Sevilla.

☐ A.10B-57	214-57	HA200D	20-63	stored, fuselage, on a container	11-97
☐ A.10C-105	214-105	HA220	22-110	stored, inside, compleet	11-97

VALENCIA - MANISES
Still preserved at the Ala11 base are two fighters. The Fairchild is stored next to the terminal on the civil side of the airfield.

☐ C.5-101		F-86F	227-185	preserved, marked as 'C.5-5/1-005', ex 55-4000	4-97
☐ C.11-7	11-07	Mirage 3EE	594	preserved, crashed 2-5-77	4-97
☐ N90713		F-27	82	stored, ex EC-CPO, ex TC-KOD	—

VICÁLVARO
Along the road to Coslada (south of Torrejón) a Spanish T-33 was preserved in American colours. It moved to Alcalá by October 1992, together with the unknown Stinson from the local restaurant.

VILANOVA I LA GELTRÚ
The town has changed its name and was formally known as Villanueve y la Geltrú. The museum also got a new name and is now called Centre Aeri. Not all the aircraft are really on display, some may be used as instructional airframes. The following were seen from outside:

☐ C.5-71	102-8	F-86F	191-414	preserved, ex 52-4718	4-96
☐ C.10B-45		HA200D	20-51	preserved	4-96
☐ E.15-48	41-16	T-33A	1491	preserved, tail from E.15-19, ex Cuatro Vientos	4-96
☐ E.16-97	793-2	T-6G	168-478	preserved, ex 49-3364	4-96
☐ HE.7A-56	78-20	OH-13H	1943	preserved, ex 56-2231	1-92
☐ Z.2-11	75-11	AC12	11 ?	preserved, ex Cuatro Vientos	4-96

VILA-SECA
On a Goodyear tyre shop the hulk of a Saeta was noted in 1992. The aircraft carries no serial.

☐ ..		HA200	..	preserved, coded 214-40 and 793-88	5-92

VILLAFRANCA DE CASTELLO
The fuselages of two former Iberia DC-3s are in use here as summer homes.

☐ ..		DC-3	..	preserved, unmarked, fuselage only	12-91
☐ ..		DC-3	..	preserved, unmarked, fuselage only	12-91

VILLANUBLA
The airfield of Villanubla-Valladolid is the home of Ala37 with its CASA 212s.

SPAIN – 417					

☐ A.10B-53	203-53	HA200D	20-59	preserved	11-97
☐ C.6-152	33-152	SNJ-5	..	preserved	11-97
☐ T.9-14	37-14	C-7A	17	stored, ex 60-3766	5-96
☐ T.9-23	37-23	C-7A	14	preserved, ex 60-3763	11-97

ZARAGOZA

The USAF has abandoned this base, leaving the Spanish with their Hornets and C-130s. The Americans used an F-4C and F-105G for instructional purposes. Both were last noted in November 1988 and have gone (scrapped?). Nearby are the ranges of Las Bardenas Reales.

☐ 55-3103		T-33A	9673	stored, ex instructional, ex French AF	11-95
☐ C.5-70	2-41	F-86F	191-379	preserved, ex 52-4683	4-96
☐ C.5-143		F-86F	227-156	preserved, marked as '25406/FU-406'	1-92
☐ C.15-18	15-05	EF-18A	533	stored, canopy/engines cocooned	6-96
☐ C.15-26	15-13	EF-18A	597	stored, canopy/engines cocooned	6-96
☐ C.15-28	15-15	EF-18A	608	stored, canopy/engines cocooned	5-96
☐ E.15-50	41-4	T-33A	8177	preserved, ex 52-9871	4-97
☐ EC-CPO		C-47B	34361	stored, ex T.3-50, ex N86442	5-96

SWEDEN

Sweden is one of those countries where they don't scrap their aircraft directly after their military service. The aircraft, mainly SAAB fighters, are used for training, decoy and other non-destructive uses after their flying career. This accounts for the numerous aircraft which are to be found on the various military airfields. Of the country's museums, the Kåremo Flugmuseum has recently closed its gates. Luckily the new museum at Visby has taken delivery of a number of former Kåremo aircraft. The Swedish military have used their own aircraft designation system. The more uncommon type designations will be explained in the text or in the columns.

ÄNGELHOLM
F10 is still flying with a mix of Drakens and Viggens. Some of the older Drakens have found new homes; J35B 35264 went to Halmstad, 35550 to Skavsta and J35F 35528 is now at Göteborg. The dumped Drakens 35248, 35354, 35391 and 35431 have all been scrapped. Tunnen 29589 moved to Hillerstorp.

❏ 22185	K	J22-1	22185	preserved, F10 markings, ex Linköping	97
❏ 35081	44	J35A	35081	dumped, F10 markings	6-90
❏ 35206	46	J35B	35206	dumped, F10 markings, ex Halmstad	8-96
❏ 35432	32	J35F	35432	preserved, F10 markings	8-96
❏ 35472	21	J35F	35472	dumped, F10 markings	6-90
❏ 35475	11	J35F	35475	dumped, fuselage only, red c/s, F10 markings	8-96
❏ 35495		J35F	35495	dumped, fuselage only, F10 markings	8-95
❏ 35509	39	J35F	35509	dumped, upside down, F10 marks, ex Skavsta	7-97
❏ 35538	48	J35F	35538	stored, F10 markings	8-94
❏ 35569	35	J35J	35569	dumped, F10 markings	7-97
❏ 35918	18	S35E	35918	dumped, F11 markings, wfu 9-6-78	8-90
❏ 35948	48	S35E	35948	dumped, F11 markings, ex 35279	8-90
❏ 37013		AJ37	37013	dumped, fuselage only, F13 markings	8-95
❏ 37017	12	AJ37	37017	stored, dismantled	8-95
❏ 37086		AJS37	37086	stored, on trailer, crashed 5-8-96	8-96

ARBOGA
Some 140km west of Stockholm is the town of Arboga. Preserved here outside a supermarket along the E20 (on the east side of Arboga) is a former Soviet MiG-21.

❏ 11 yel		MiG-21SMT	500AT01	preserved, ex Soviet	8-96

BERGA
Dumped on the Marinen Helikopterdivisionen base is an AB206.

❏ 06044		Hkp6B	8176	dumped, ex SE-HLY	8-95

BODEN
Boden houses one of the two Arméflygbataljons of the Swedish Army. Preserved is a Do27, while stored on the base is an Alouette 2.

❏ 02403	33	Hkp2	1269	stored	8-95
❏ 53271	81	Fpl53	2099	preserved, owned by Flygvapen Museum	8-95

ESKILSTUNA
The civil airfield houses a number of flyable vintage aircraft including some former Swiss Hunters and Vampires. The dual Draken has no intentions to fly again.

❏ 35810	87	Sk35C	35810	preserved, ex 35019	97

GÖTEBORG

Town: At the packhuskajen, near the new opera building, is the Maritima Centrum. Preserved on the deck of one of the ships (the *Småland*) is an Alouette 2.

| ❏ 02034 | 34 | Hkp2 | 1175 | preserved, on ship | 6-97 |

Airfield - Säve: Preserved on the ramp of the military side of the airport is a former Ängelholm Draken which received new markings here (it used to be 65 of F10). A large number of ex F10 J35Js should be in store at this airfield, but none have actually been seen here, except for one aircraft coded 20 with F10 markings in June 1997. The Alouette 2 cannot be seen from outside.

| ❏ 02036 | 36 | Hkp2 | 1246 | preserved | 8-93 |
| ❏ 35528 | 00 | J35F | 35528 | preserved, F9 markings, ex Ängelholm | 6-97 |

HALMSTAD

Of the instructional aircraft mentioned in edition 1, Draken 35206 is now at Ängelholm, 35448 went to Finland as DK-241 and 35483 as DK-255. 35502 returned to operational service as a J35J together with 35512, 35521, 35545 and 35546. 35609 is now at Skavsta. The fate of J35D 35366 is unknown. Viggen 37323 returned to service. A number of aircraft on the fire dump have expired: Lansens 32286 and 32579, Drakens 35407, 35418, 35488, 35560, 35574, 35806, 35813, 35939 and Pembroke 83015.

❏ 222		AB204B	3014	dumped, all green, ex Skavsta, ex RNethNavy	6-94
❏ 32094	P	A32A	32094	stored, F14 markings	8-94
❏ 35253	09	J35B	35253	instructional, fire dump, F10 markings	8-96
❏ 35264	68	J35B	35264	instructional, fire dump, F10 markings	8-96
❏ 35308	25	J35D	35308	instructional, fire dump, F4 markings	8-94
❏ 35367	39	J35D	35367	instructional. fire dump, F4 markings	8-96
❏ 35372	32	J35D	35372	instructional, fire dump, FC markings	8-94
❏ 35402	34	J35F	35402	instructional, fire dump, no tail, F10 markings	8-95
❏ 35434	22	J35F	35434	instructional, F10 markings	8-94
❏ 35439	01	J35F	35439	instructional, F10 markings	8-94
❏ 35440		J35F	35440	instructional, F10 markings, see also Skavsta	8-94
❏ 35461	59	J35F	35461	instructional, fire dump, no tail, F10 markings	8-94
❏ 35482	60	J35F	35482	instructional, F10 markings	8-94
❏ 35514	43	J35F	35514	instructional, fire dump, F10 markings	9-92
❏ 35516		J35F	35516	instructional, fire dump, upside down	9-92
❏ 35559	57	J35F	35559	instructional, fire dump, upside down, F10 mrks	8-96
❏ 35571	28	J35F	35571	instructional	8-94
❏ 35573	32	J35F	35573	instructional, fire dump, F10 markings	8-96
❏ 35579	67	J35F	35579	instructional, fire dump, F10 markings	8-96
❏ 35816	84	Sk35C	35816	instructional, fire dump, F16 markings, ex 35015	8-96
❏ 35825		Sk35C	35825	instructional, fire dump, F16 markings, ex 35034	8-96
❏ 37-3		AJ37	37-3	instructional, F14 markings	8-94
❏ 37006	06	AJ37	37006	instructional, fire dump	8-96
❏ 37007	07	AJ37	37007	instructional, fire dump	8-96
❏ 37016	16	AJ37	37016	instructional	8-94
❏ 37022	22	AJ37	37022	instructional	8-94
❏	22	SAAB 35	..	instructional, fire dump, F7 markings	8-96
❏	44	SAAB 35	..	instructional, fire dump, F7 markings	8-96

HILLERSTORP

Near the village of Hillerstorp, on the east side of road 152 is the 'Wild West' park called High Chapparal. In front of the shopping centre (köpcentrum) is a J29, on the roof a Viscount and inside the centre a Draken. The DC-3 and Norwegian Thunderjet are dumped on the west side of the park and are in bad condition. The aircraft mentioned in EWR-1 were damaged by a fire and have all been scrapped, except for the F-84G.

SWEDEN - 420

❏ 29978	MU-B	F-84G	..	stored, bad condition, ex RNoAF	6-97
❏ 29589	59	J29F	29589	preserved, F10 markings, ex Ängelholm	6-97
❏ 35220	20	J35B	35220	preserved, F18 markings, ex Malmslätt	6-97
❏ SE-EGR		DC-3D	42970	stored, bad condition, ex OH-VKC	8-96
❏ SE-IVY		Viscount 815	375	preserved, marked 'Big Airland', ex G-AVJB	5-96

KALMAR
The unmarked Draken at the terminal has been identified as 35404 with false F12 markings (really ex F17).

❏ 35404	01	J35F	35404	preserved, 'F12' markings	5-96

KÅREMO
The Kåremo Flygmuseum was located some 24km north west of Kalmar. The museum closed in late 1996 and their aircraft auctioned off. Some ten aircraft were brought to the new museum at Visby. The fate of the others is unknown. Only a few were seen during a visit in May 1996.

❏ ET-AGB	MFI-15-200	15006	stored, ex SE-FIM, ex LN-BIV	—
❏ LN-BDR	UC-64A	92	stored, ex RA-K/RNoAF, ex 492/RCAF	—
❏ N1021K	Pitts S1C	1021	stored	5-96
❏ SE-ANU	Taylorcraft	1227	stored, ex OH-KLB	—
❏ SE-AYS	Ercoupe 415C	3956	stored	—
❏ SE-CEA	P31C Proctor IV	H772	stored, ex G-ANVY, ex RM169/RAF	5-96
❏ SE-CPG	MFI-9	02	stored	—
❏ SE-CKM	PA22 Tri Pacer	22-6110	stored	—
❏ SE-CLC	Fw44J	..	stored	—
❏ SE-CNC	Aeronca 7EC	718	stored	—
❏ SE-CZL	PA22 Colt 108	22-9131	stored	—
❏ SE-FIT	MFI-15-200A	15832	stored	—
❏ SE-GEH	Ce402B	402B-1351	stored	5-96
❏ SE-GXO	DH.82A	82869	stored, ex G-APJP, ex N6670	—
❏ SE-HHN	Enstrom F.28A	295	stored, ex OY-HBL	—
❏ SE-HME	Bell 47G-3B1	WA583	stored, ex G-BBZL, ex XT404	—
❏ SE-HNS	AS350B	1508	stored, ex D-HLOO	—
❏ ..	MFI-15-200A	15847	stored, unfinished	—
❏ ..	MFI-15-200A	15848	stored, unfinished	—

KARLSBORG
The Lansen is preserved outside the main gate of this former Viggen base. Not much is visible at this base from outside and the current status of *Wrecks and Relics* here is unknown.

❏ 02412	91	Hkp2	1436	preserved (Alouette 2), ex 02212	6-89
❏ 32259	43	A32A	32259	preserved, outside gate, F6 markings	5-96
❏ 35504	13	J35F	35504	dumped, F10 markings	8-94
❏ 35535	01	J35F	35535	dumped, F10 markings	8-94

LINKÖPING
Town: Mounted high inside the Gränden Shopping Centre in the town centre is a SAAB 91, which is owned by the Flygvapenmuseum. This aircraft used to be displayed in the konseerthus.

❏ 50068		Sk50	91278	preserved, silver c/s	6-96

At each of the three highway exits a SAAB aircraft is preserved on a pole. The SF340 is at the exit to the SAAB airfield, while the is J29 at the exit to the Flygvapenmusuem. The SAAB 91 is at the exit in between these two.

❏ 29441	52	J29F	29441	preserved, F3 markings	6-97

SWEDEN - 421

☐ 50016		81	Sk50	91216	preserved, F3 markings	6-97
☐ SE-ISF			SF340	001	preserved, unmarked	6-97

Airfield - Malmslätt: The airfield at the western side of Linköping, also often referred to as Malmen, is used by the military and the Flygvapenmuseum. The unknown SAAB 35 with code 18 was first noted together with 35218/18 in June 1991.

☐ 05241			Hkp5B	S1258	instructional (Hughes 300C)	8-95
☐ 32255		25	A32A	32255	instructional, fire dump, FC markings	8-93
☐ 32945		45	S32C	32945	instructional, fire dump	8-93
☐ 32610		35	J32D	32610	instructional, fire dump	9-97
☐ 35090		52	J35A	35090	instructional, f16 markings	2-98
☐ 35218		18	J35B	35218	instructional, fire dump, F18 markings	9-97
☐ 35492		62	J35F	35492	instructional, fire dump, F10 markings	6-95
☐		23	SAAB 32	..	instructional, fire dump	8-93
☐		13	SAAB 35	..	dumped, upside down	9-97
☐		18	SAAB 35	..	stored	8-95
☐ 83010			Tp83	P66/56	instructional (Pembroke C.52), fire dump	9-97

The museum is located at the Carl Cederströms Gata at the northern side of the airfield. It is open daily from 1st June to 31st August between 10:00 and 17.00, from 1st September to 31st May between 12:00 and 16:00 except on Mondays, when it is closed. **Inside** are:

☐ 54			P-51D	122-31718	preserved, ex Israeli AF, ex 26020, ex 44-63992	9-97
☐ HS964			Spitfire PR.XIX	S6-683524	preserved, marked as '31051/51', ex Indian AF	9-97
☐ 04			Sk1	464	preserved (Albatros B.II)	9-97
☐ 80			Se101	..	preserved (Grunau SG-38), glider	9-97
☐ 155	A		B3C-2	0860412	preserved (Ju86K-4), F21 markings	9-97
☐ 278	H		J8	G5/59066	preserved (Gladiator), F19 markings	9-97
☐ 386		86	S6B	207	preserved (Fokker C.VE), F3 markings	9-97
☐ 515		19	Sk11A	47	preserved (DH.82A), F5 markings, ex SE-BYM	9-97
☐ 536		101	Sk10	20	preserved (Raab RK-26), F5 markings	9-97
☐ 558		58	Sk9	1720	preserved (DH.60T), as '5558', composite	9-97
☐ 714	M		B4A	52	preserved (Hawker Hart), F19 markings	9-97
☐ 814		61	P1	8	preserved (Sparmann S.1A), F8 markings	9-97
☐ 945			Macchi M7	..	preserved	9-97
☐ 947			J1	..	preserved (Phönix 122 DIII)	9-97
☐ 2134		53	J9	282-19	preserved (Seversky EP106), F3 markings	9-97
☐ 2340		40	J20	405	preserved (Re2000), F10 markings	9-97
☐ 2543			J11	921	preserved (Fiat CR42 Falco), F9 markings	9-97
☐ 3656			Ö1	147	preserved (CFM 01 Tummelisa), F3 markings	9-97
☐ 5075		116	Sk15	1596	preserved (Klemm KL.35B), F5 markings	9-97
☐ 8316			Se104	235	preserved (Weihe), glider	9-97
☐ 01001		01	Hkp1	497	preserved (Vertol 44) ex Marinen	9-97
☐ 02406		92	Hkp2	1279	preserved (Alouette 2), ex 02202/Armén	9-97
☐ 16109		82	Sk16A	14-366	preserved (AT-16), F10 markings, ex FE632	9-97
☐ 17005		5	S17B	005	preserved (SAAB 17B), F3 markings	9-97
☐ 18172	D		B18B	172	preserved (SAAB 18B), F14 markings	9-97
☐ 21364	R		J21A-3	21364	preserved (SAAB 21A-3), F6 markings	9-97
☐ 22280	L		J22-2	22280	preserved (FFVS 22), F3 markings	9-97
☐ 25114		314	Sk25	114	preserved (Bü181B), F5 markings, ex D-EBIH	9-97
☐ 28001	P		J28A	EEP42083	preserved (Vampire Mk.1), F3 markings	9-97
☐ 29398	F		J29B	29398	preserved (SAAB 29B), F22 markings	9-97
☐ 34016		06	J34	41H/680304	preserved (Hunter Mk.50), F9 markings	9-97
☐ 35410		3	J35F	35410	preserved (SAAB 35F), nose only	9-97
☐		18	J35		preserved, nose only	9-97
☐ 50046		46	Sk50	91258	preserved (SAAB 91B), ex Ljungbyhed	9-97

SWEDEN - 422

☐ 53273	83	Flp53	2110	preserved (Do27A-4), ex Armén	9-97
☐ SE-CLZ		Tp78	492	preserved (UC-64A), F2 markings, ex 78001	9-97
☐ SE-DCD		J33	12374	preserved (Venom Mk.51), ex 33025	9-97
☐ SE-EGB		Sk12	12	preserved (Fw44J), ex 670 and marked as such	9-97
☐ SE-GCT		Flp51B	18-6803	preserved (L-21B), ex 51256/56, ex SE-CKH	9-97
☐ SE-SAP		Se102	2152	preserved (Grunau Baby II), glider, F3 markings	9-97
☐ SE-SWN		Se103	072	preserved (Kranish II), glider, ex 8211	9-97
☐ SE-XBZ		SAAB 105XT	105-2	preserved, ex SE-502	9-97
☐ ..		M1	138	preserved, (Nieuport IV-G)	9-97
☐ ..		S14	..	preserved, (Fi156), composite as '3812/67'	9-97
☐ ..		SAAB 210	..	preserved	9-97
☐ ..		B1	..	preserved (Breguet U.III), replica, marked as '53'	9-97
☐ ..		S16	..	preserved (Caproni 313), replica, marked as '12'	9-97

Outside:

☐ 215		Lim-2	1B00215	preserved, marked as '24', ex PolishAF	9-97
☐ 29575	51	J29F	29575	preserved (SAAB 29F), F3 markings	9-97
☐ 32197	21	A32A	32197	preserved (SAAB 32A) F6 markings	9-97
☐ 35090	52	J35A	35090	preserved (SAAB 35B), F16 markings	9-97
☐ 35375	05	J35D	35375	preserved (SAAB 35D), F4 markings	9-97
☐ 35541	43	J35J	35541	preserved (SAAB 35J), F10 markings, yellow c/s	9-97
☐ 37-1	51	AJ37	37-1	preserved (SAAB 37)	9-97
☐ 37800	22	Sk37	37800	preserved, FC markings	9-97
☐ 47001	79	Tp47	CV-244	preserved (PBY-5A), ex 9810/RCAF	9-97
☐ 52002	02	Tp52	SH1648	preserved (Canberra B.2), F8 markings	9-97
☐ 79007	797	Tp79	13647	preserved (DC-3), F13 markings, ex SE-CFR	9-97
☐ 82001	80	Tp82	622	preserved (Varsity T.1), F8 markings, ex WJ900	9-97
☐ 83008	85	Tp83	P66/52	preserved (Pembroke C.52), F13 markings	9-97

The museum also has a large collection of stored aircraft. Most of them are on the military side of the airfield, although some may be under restoration off site. SAAB 91D SF-18 moved to Stockholm Arlanda. Stored J35B 35220 moved to Hillerstorp. Some of the museum aircraft have been placed on permanent loan to other museums, including Safir SE-AYC to Trollhäggen, J21 21311 to the F15 wing went to Söderhamn, J22 22185 to Ängelholm, S29C 29902 to Skavsta, 29970 to Luleå and 35906 to Skavsta. J28B 28317/A (V0604) went to Norway in parts and will not be rebuilt into a complete aircraft. Stored Hkp9B 09413/93 (S.722, not part of the museum) moved to the UK to act as a spares source for Bond Helicopters.

☐ 225		AB205B	3023	stored, ex Skavsta, ex RNN, to go to Soesterberg	97
☐ AT-160		TF-35	351160	stored, ex RDanAF	97
☐ 03422	93	Hkp3B	3004	stored (AB204B), F10 markings	97
☐ 04451	91	Hkp4A	401	stored (Vertol 107), F17 markings	97
☐ 05215		Hkp5A	22-0049	stored (Hughes 269A), ex Stockholm	97
☐ 21286		A21R	21286	stored (SAAB 21R), composite, ex J21A	97
☐ 25000	76	Sk25	181.5001	stored (Bül81B), F8 markings, ex D-EXWB	97
☐ 28311	17	J28B	V0590	stored (Vampire Mk.50), F15 markings	97
☐ 28451	81	Sk28C	15745	stored (Vampire Mk.55), F5 markings	97
☐ 29171	L	J29A	29171	stored (SAAB 29A), F13 markings	97
☐ 29507	53	J29F	29507	stored (SAAB 29F) F3 markings	97
☐ 29937	09	S29C	29937	stored (SAAB 29C)	97
☐ 32917	17	S32C	32917	stored (SAAB 32C), F11 markings	97
☐ 35-5		SAAB 35B-1	35-5	stored (SAAB 35B), ex instructional	97
☐ 35051	14	J35A	35051	stored (SAAB 35A), F16 markings	97
☐ 35902		S35E	35902	stored (SAAB 35E), ex Skavsta, FC markings	97
☐ 35959	35	S35E	35959	stored (SAAB 35E), FC markings	97
☐ 37108	55	AJS37	37108	stored, F10 markings	97
☐ 61030	30	Sk61	61030	stored (Bulldog), F5 markings	97

SWEDEN - 423

❏ N9887A		Beech D18S	A-932	stored, marked as '45003', ex CN-MAL	97
❏ N79901		JRF-5 Goose	B63	stored, marked as '81002', ex 37810/USN	97
❏ SE-BYH		B17A	17239	flyable (SAAB 17A), ex 17239/J	97
❏ SE-CAS		Meteor T.7	G5/1496	stored, ex WF833/RAF	97
❏ SE-CAW		Firefly F.1	F6121	stored, ex PP392/RNavy	97
❏ SE-CPI		Fpl54	03	stored (MFI-10B), ex 54382/Armén	97
❏ SE-DCA		J33	12364	stored (Venom Mk.51), ex 33015	97
❏ SE-DXB		J29F	29670	flyable (SAAB 29F), ex 29670	97
❏ SE-EBI		Skyraider AEW.1	7960	stored, ex G31-11, ex WT947/RNavy	97
❏ SE-EUK		MFI-9B	42	stored, marked as '801-42/01'	97
❏ SE-KEA		NC701	264	stored, French built Si204D	97
❏ SE-SAM		Sk5	235	stored (Heinkel HD35)	97
❏ SE-XCB		MFI-15	01	stored, ex SE-301	97
❏ VR-NAP		Dove Mk.1B	04082	stored, marked as '46002', ex G-ANVU	97
❏ ..		HM-14	..	stored	97
❏ ..		Sk14	..	stored, rebuild from Yale and Wirraway	97

Airfield - Vårdsberg: The airfield at the western side of Linköping is in use as a civil airfield and by SAAB. Both military and civilian aircraft are built here. The SAAB 2000 is stored wingless outside the production line. The very first SAAB 35 (non flying) is dumped outside the airfield on the eastern side on a small fire dump (on a dead end road along the airfield fence).

❏ 35-0		SAAB 35	35-0	instructional, fire dump	5-96
❏ SE-001		SAAB 2000	001	stored, last flight 8-9-95	6-96

LJUNGBYHED
The base of Ljungbyhed was closed in 1997 and its SAAB 105s (Sk60) and Bulldogs (Sk61) relocated. J35D 35352 moved on to Såtenäs, while 35307 has disappeared. The stored SAAB 91 50046 went to Malmslätt.

❏ 29666	64	J29F	29666	preserved, at gate, F4 markings	8-96
❏ 32932	32	S32C	32932	dumped, F11 markings	8-96
❏ 32934	34	S32C	32934	dumped, F11 markings	8-96
❏ 35046		J35A	35046	dumped, in small pieces, ex Halmstad	8-93
❏ 35059	23	J35A	35059	dumped, F16 markings, wfu 29-4-76	8-93
❏ 35235		J35B	35235	dumped, F10 markings, wfu 21-4-76	8-96
❏ 35312		J35D	35312	dumped, F4 markings, upside down	8-96
❏ 35562		J35F	35562	instructional, cockpit only, as 'SIM3882'	8-96
❏ 50039	39	Sk50	91251	stored, in hangar	8-96
❏ 50063		Sk50	91272	instructional, nose only	8-96
❏ 50064	64	Sk50	91273	preserved, at gate, on pole	8-96
❏ 60030	30	Sk60C	60030	stored, wreck in hangar, crashed 7-5-92	8-94
❏ 60090	90	Sk60B	60090	instructional, composite, F5 markings	8-96
❏ 60134	134	Sk60A	60134	stored, blue/yellow c/s, F5 markings	8-96
❏ 61019	19	Sk61E	122	stored, ex G-AZET, F5 markings	—
❏ 61022	22	Sk61E	126	stored, ex G-AZHX, F5 markings	8-94
❏ 61045	45	Sk61E	157	stored, F5 markings	8-93
❏ 61046	46	Sk61E	158	stored, F5 markings	—
❏ 61050	50	Sk61E	165	stored, F5 markings	8-93
❏ 61052	52	Sk61E	167	stored, F5 markings	8-93
❏ 61054	54	Sk61D	172	stored, F5 markings	8-93
❏ 61061	61	Sk61C	179	stored, F5 markings	—
❏ 61068	68	Sk61C	186	stored, F5 markings	—
❏ 61073	73	Sk61C	191	stored, F5 markings	—
❏ 61076	76	Sk61C	194	stored, F5 markings	8-93
❏ 61077	77	Sk61C	195	stored, F5 markings	—
❏ 61078	78	Sk61C	196	stored, F5 markings	8-93

LÖDDEKÖPINGE

Halfway between Malmö and Helsingborg is the village of Loddekopinge (some 20km north of Malmö). Preserved here is a fomer Soviet MiG-21.

❑ 09 yel		MiG-21SMT	50023098	preserved, ex Soviet	6-97

LULEÅ - KALLAX

The base of Luleå houses F21 wing with its three Viggen squadrons. As with all Swedish bases a number of preserved and instructional aircraft can be found on the airfield.

❑ 29929	21	S29C	29929	preserved, at military gate, F21 markings	6-97
❑ 29970	37	S29C	29970	preserved, ex Linköping	97
❑ 32529	05	J32E	32529	stored, F16M markings	—
❑ 35062	66	J35A	35062	stored, F16 markings, wfu 13-10-77	6-92
❑ 35357	32	J35D	35357	instructional, fire dump	6-91
❑ 35949	21	S35E	35949	preserved, at civil side, ex 35295, wfu 14-4-80	6-97
❑ 35952	68	S35E	35952	stored, ex 35290, wfu 14-4-80	6-96
❑ 50002	76	Sk50	91202	stored (SAAB Safir)	6-96
❑ 83005	82	Tp83	P66/47	instructional (Pembroke C.52), fire dump	6-91

MALMÖ

Town: The Tekniska Museum is open daily from Tuesday to Sunday between 12:00 and 16:00. During June, July and August the museum in the city centre park is also open on Mondays.

❑ 35484	55	J35F	35484	preserved, F10 markings	8-96
❑ OH-VKM		DH.60GIII	4/VK	preserved	—
❑ OY-DNU		Auster J/1	2102	preserved	—
❑ SE-BHG		Sk15A	1806	preserved (Klemm KL.35D), ex 5010/RsweAF	—
❑ SE-SHG		Meise	1	preserved, glider	—

Airfield - Sturup: Viscount SE-CNK at this international airport went, reunited together with its nose from the technical museum, to Stockholm Arlanda.

❑ 32915	15	S32C	32915	dumped, F11 markings	12-97
❑ 35612	12	J35J	35612	preserved, F10 markings	12-97
❑ LN-KLP		SE210-3	24	dumped, ex preserved	12-97

MARKARYD

This airfield has been closed for some years now and the owner of Sk16 16009 and S29C 29945 has also left, taking his aircraft with him.

MÄRSTA

Just west of Stockholm Arlanda airport is the small town of Märsta. Above the supermarket cash desks in a shopping mall, near the main road to Stockholm, is a Pembroke is preserved. The bar at the other end of the mall has a rare Avro 594.

❑ 83004	24	Tp83	P66/46	preserved, red/white c/s	5-96
❑ SE-ADT		Avro 594	318	preserved, ex G-AAHD	5-96

NORRKÖPING

Airfield - Bravalla: The northern airfield of Norrköping is still in use by the military as a reserve airfield. The only resident aircraft are two instructional airframes.

❑ 35-6	46	SAAB 35D	35-6	instructional, fire dump	8-96
❑ 35077	42	J35A	35077	instructional, fire dump	8-96

SWEDEN - 425

Airfield - Kungsängan: Stored at this civil airfield west of Norrköping in the Luftfartsverket compound is a Dove. The wingless aircraft is in bad condition and mounted on a lorry.

❏ SE-GRA	Dove 6		04437	stored, ex G-AMZN	5-96

ÖSTERSUND
Town: Located here is the Jämtlands Flyghistoriska Museum. In 1997 the museum was expecting to get Hkp3 03421/81 from F4.

❏ 16145	78	Sk16B	75-3497	preserved (AT-6A), F4 markings, ex 3223/RCAF	6-97
❏ 29373	12	J29F	29373	preserved, F4 markings	6-97
❏ 32601	15	J32E	32601	preserved, F16M markings	97
❏ 35392	52	J35D	35392	preserved, F4 markings, wfu 12-2-79	6-97
❏ 37097	29	AJS37	37097	preserved, F15 markings	97
❏ 50020	74	Sk50	91222	preserved, ex Västerås	6-97
❏ LN-BFN		UC-64A	649	preserved, as '78002', ex LN-PAD, ex 44-70384	6-97
❏ SE-AHU		GV-38	3006/12	preserved	6-95
❏ SE-SDE		Grunau Baby IIb	104	preserved, glider, marked as 'A'	—
❏ SE-SLL		Bergfalke II/55	218	preserved, glider	—
❏ SE-STZ		Pik-5	1	preserved, glider	—
❏ SE-XFC		Janowski J-1B	085	preserved	6-95

Airfield - Frösön: Preserved at the military barracks of F4 are two fighters. The Draken has been read off as 35388, while the 35392 mentioned in the first edition is not here but at the above mentioned museum.

❏ 29401	29	J29F	29401	preserved, F4 markings	6-97
❏ 35207		J35B	35207	dumped, F18 markings	6-96
❏ 35353		J35D	35353	dumped, F4 markings	6-96
❏ 35388	35	J35D	35388	preserved, F4 markings	6-96
❏ 35471	63	J35F	35471	dumped	6-96
❏ 35600	06	J35F	35600	dumped, FC markings	6-96
❏ 83014		Tp83	P66/64	dumped (Pembroke C.52)), F6 markings	6-96

RONNEBY - KALLINGE
F17 wing is still based here with its Viggens. A number of former *Wrecks & Relics* aircraft have not been seen for some time and may all have been scrapped; A32A Lansens 32051, 32206, 32240, 32285 and J35B Drakens 35225, 35241 and 35272.

❏ 32151	B	A32A	32151	preserved, at gate, F17 markings	7-95
❏ 32222	27	A32A	32222	dumped	6-94
❏ 32248	49	A32A	32248	dumped	7-95
❏ 32258	42	A32A	32258	dumped, F6 markings	7-95
❏ 32268		A32A	32268	dumped, F11 markings	6-94
❏ 35060	24	J35A	35060	dumped, F16 markings	8-93
❏ 35070	13	J35A	35070	dumped, F16 markings, wfu 11-12-75	6-94
❏ 35271		J35B	35271	dumped, F10 markings, wfu 21-4-71	6-94
❏ 35901	08	S35E	35901	dumped, ex 35278, wfu 24-6-74	7-95
❏ 83012	88	Tp83	P66/61	dumped, burned out	6-94

SÅTENÄS
All the Lansens mentioned in the first edition have been consumed as have most of the Drakens. The last report is from 1990 and the situation may well have changed since then.

❏ 02407	91	Hkp2	1291	preserved (Aloutte 2), ex SE-HDF	8-90
❏ 32085	33	A32A	32085	preserved, F7 markings	8-90
❏ 35073	37	J35A	35073	dumped, F16 markings, wfu 21-4-71	8-90

SWEDEN - 426					
☐ 35321	16	J35D	35321	dumped, F4 markings	8-90
☐ 35348		J35D	35348	dumped, F21 markings, no tail	8-90
☐ 35352		J35D	35352	dumped, ex Ljungbyhed, F21 markings, no tail	8-90
☐ 35808		Sk35C	35808	dumped, ex 35009, F16 markings, ex Halmstad	8-90
☐ 35818		Sk35C	35818	dumped, ex 35032, F16 markings, wfu 9-6-78	8-90
☐ 83016	86	Tp83	P66/69	dumped (Pembroke C.52), F13 markings	8-90

SKOKLOSTER
The car museum of Skokolster is on a peninsula some 20km south of Uppsala.

☐ 35945	35	S35E	35945	preserved, FC markings	6-97
☐ 50023	72	Sk50B	91225	preserved (SAAB 91), F1 markings, SE-IGI ntu	6-97
☐ 79002	792	Tp79	9103	preserved (DC-3), F13 markings, ex SE-APW	6-97

SÖDERHAMN
Town: On a pole at the junction of the E4 highway and road 304 a Viggen is preserved.

☐ 37031	54	AJ37	37031	preserved, F15 markings	8-96

Airfield: Not much has changed over the years at F15's base.

☐ 21311	A	J21A-3	21311	preserved, F15 markings, ex Linköping	97
☐ 32070	H	A32A	32070	preserved, F15 markings	8-96
☐ 32127	07	A32A	32175	dumped	6-91
☐ 32172	25	A32A	32172	instructional, fire dump, F15 markings	8-96
☐ 32175	27	A32A	32175	dumped, in woods	8-96
☐ 32505		J32B	32505	stored	8-93
☐ 32532	2	J32B	32532	dumped, F4 markings	8-93
☐ 35053	17	J35A	35053	instructional, F16 markings, wfu 18-7-75	5-96

STOCKHOLM
Town: The Sveriges Tekniska Museum still had their Junkers F13 SE-AAC (marked as S-AAAC), as well as some other smaller vintage aircraft on display in May 1996.

☐ S-AAAC		Junker F13	..	preserved, marked as 'SE-AAC'	5-96

Building work is being caried out at the Kungl Tekniska Hogskolan and the instructional airframes have been moved, Hkp5A 05215 went to **Malmslätt** and Saab 37 37-61 to **Västerås**.

Town - Agusta: On the site of the experimental nuclear power station, a Draken was noted in June 1991. This aircraft was only visible during an overflight.

Airfield - Arlanda: Sk25 Bestmann 25056 which was preserved inside terminal 5 has moved on to Ornskoldsvik. This is where the aircraft was originally built.

☐ 35515	49	J35F	35515	instructional, rescue trainer, ex Skavsta	5-96
☐ OY-KHO		DC-9-81	53003	fire dump, forward fuselage only, w/o 27-12-91	3-94
☐ SE-CNK		Viscount 745	227	instructional, ex Malmö	5-96
☐ SE-DAA		SE210-3	4	instructional, de-icing trainer, ex SAS	10-96
☐ SE-DAF		SE210-3	112	stored, museum owned	8-96
☐ SE-DEC		SE210-10B1R	263	instructional, fire and rescue trainer, ex EC-CYI	8-96
☐ SE-KAL		Si204D-1	159	stored, museum owned	5-96
☐ ..		HM-14	..	preserved, in terminal 4, museum owned	5-96

Located near the SAS Flight Academy (at the eastern side of the airfield) is the storage hangar of Stiftelsen Aerospace (formally known as Luftfartmuseet). The museum hopes to open a new site near one of the terminals before the year 2000. Meanwhile the storage site can normally be visited on working days between 10:00 and 15:00. Sk16A 16010 (14-565) was sold in the USA in 1993, while Hutter H17A LN-GBD moved to Denmark. The Thulin A has also gone and the Bu131B-3 SE-AGU (846) is now at Sundsvall-Harsosand airport.

❏ SF-18		SAAB 91D	91364	stored, ex Malmslätt, ex FinnAF	6-97
❏ 42-24049		C-47A	9911	stored, outside, ex USAF	6-97
❏ 01007		Hkp1	608	stored (Vertol 44), cockpit only	6-97
❏ 35937	21	S35E	35937	stored, cockpit only, wfu 20-11-69	6-97
❏ D-6680		Specht	808	stored, glider, ex D-4320	5-96
❏ LN-DBE		Beech G50	GH-98	stored, of site, ex N186AA	5-96
❏ S-AAR		Rieseler R.III	..	stored	5-96
❏ SE-ADR		Ju52/3mce	4017	stored, cockpit only, ex RSweAF, ex SE-ADK	6-97
❏ SE-AGF		DH.60GIII	5132	stored, ex Narrtalje	5-96
❏ SE-AGL		J-2 Cub	989	stored, incomplete	5-96
❏ SE-AHD		GV-38	5	stored, licence built Rearwin Sportster	5-96
❏ SE-BGA		Klemm KL.35D	1983	stored, ex 5054/RSweAF	6-97
❏ SE-BWX		Fw44J	2816	stored, ex 5773/RSweAF	6-97
❏ SE-BXE		UC-61K	..	stored, ex LN-MAD, ex HB625, ex 43-14898	6-97
❏ SE-BXU		L12A Electra	1313	stored, ex T-3, ex PJ-AKE	5-96
❏ SE-BYA		Junkers W34h	2835	stored, ex 6/RSweAF	5-96
❏ SE-BZE		L18-56	2593	stored, ex OH-VKP, ex G-AGIJ, ex 43-16433	6-97
❏ SE-CBT		Auster AOP.5	841	stored, ex LN-BDU, ex G-ANIU, ex MS977	6-97
❏ SE-CFI		MFI-10B	01	stored	5-96
❏ SE-CPB		UC-64A	89	stored, ex LN-AEN, ex RNoAF, ex 42-5050	5-96
❏ SE-EBB		Skyraider AEW.1	7962	stored, ex G-31-5, ex WT949/Royal Navy	6-97
❏ SE-ESE		CeF172F	0170	stored, marked as 'SE-KEU'	5-96
❏ SE-IGG		Bellanca 8KCAB	605-80	stored	5-96
❏ SE-SFA		Grunau Baby IIb	098	stored, glider	5-96
❏ SE-SGF		Meise	7	stored, glider	5-96
❏ SE-SUA		Bergfalke II	336	stored, glider	5-96
❏ 21		Grunau SG-38	210	stored, glider	5-96
❏ ..		Adams Wilson 101	..	stored, homebuilt helicopter	5-96
❏ ..		Bell 47D	..	stored, composite airframe, marked as 'SE-HAD'	5-96
❏ ..		MFH Junior	..	stored, marked as 'F-AHLE'	5-96
❏ ..		NAB 9	..	stored, locally built Albatros B.II, forward fusel.	5-96
❏ ..		Nilsson BEDA	..	stored, flying boat	5-96
❏ ..		Nyberg Flugan	..	stored	5-96
❏ ..		Persson Pappfokkers	..	stored	5-96
❏ ..		Thulin A	..	stored	5-96

Airfield - Barkarby: Four Hkp4s (Vertol 107s) were reworked here in 1993 for the civil market. These were 04452/92 (to N193CH), 04454/94 (to N194CH), 04456/96 (to N195CH) and 04457/97 (to N196CH). All had left by 1994. The remaining helicopter was robbed of all its spares. Stored out on the airfield is an unknown An-2. The wreck of Hkp6B (AB206B) 06052 and fuselage of Sk50 Safir 50057 were not found in May 1996.

❏ 04460	90	Hkp4A	410	stored, used for spares	5-96
❏ ..		An-2		stored	5-96

Airfield - Bromma: Stored on the platform at this Stockholm domestic airfield is a Convair of the Luftfartmuseet, which waits its turn to be relocated to the new museum site at Arlanda.

❏ SE-CCX	Convair 440	320	stored, Arlanda museum owned, ex LN-KLB	8-96

Airfield - Skavsta: The airfield of Nyköping Oxelösund has officially been renamed as Stockholm Skavsta although it is more than 100km south west of Stockholm. The Transport Teknik part of Gripenskolan here is a civil technical school with a number of military aircraft.

❏ 220	K	AB204B	3010	instructional, ex RNethNavy	6-97
❏ 02409	92	Hkp2	1303	instructional (Alouette 2), ex 02203	6-97

☐ 32507		01	J32E	32507	instructional, F16M markings	—
☐ 35550		40	J35F	35550	instructional, ex Ängelholm	5-96
☐ 35566		56	J35F	35566	instructional, F10 markings	6-96
☐ 35609		09	J35F	35609	instructional, F10 markings, ex Halmstad	6-96
☐ 37010			AJ37	37010	instructional, no tail and wings	8-95
☐		53	SAAB 37	..	instructional, grey c/s	6-97
☐ 50007		76	Sk50	91207	instructional	8-95
☐ 50051		72	Sk50	91260	instructional	8-95
☐ SE-HTZ			H369HS	102-0421S	instructional	5-96
☐ SE-ISA			SF340B	002	stored, arrived 11-5-96	5-96

The Swedish FMV, a civil firm which officially owns all Swedish military aircraft, also has some overhaul facilities here. Some of the former Dutch Navy AB204Bs have been relocated: 222 to Halmstad, 225 to Linköping, while 228 went to Västerås. Other former airframes which have left are Drakens 35221 to Uppsala and 35902 to Malmslätt, J35F 35509 to Ängelholm, 35461, 35514, 35559, 35579, 35816 and 35825 to Halmstad and 35515 to Stockholm.

☐ 4D-BI		AB204B	3118	stored, unmarked, ex Austrian AF	6-96
☐ 226		AB204B	3032	stored, ex RNethNavy	8-93
☐ 227		AB204B	3035	stored, ex RNethNavy	8-93
☐ 32935	35	S32C	32935	stored, on trailer near FMV hangar	8-95
☐ 35371	01	J35D	35371	stored, F10 markings	8-93
☐ 35409		J35F	35409	stored	6-94
☐ 35440		J35F	35440	stored, F10 markings, see also Halmstad	8-94
☐ 35506		J35F	35506	stored, F10 markings	6-94
☐ 35597		J35J	35597	dumped, fuselage only, near FMV hangar	8-95
☐ 35906	06	S35E	35906	stored, ex Linköping, FC markings	97

The airfield used to be the home base of F11 before the airfield went civil. Still located here is the F11 Museum. The serial and code of the S32C 32940 are correct. The main wheel doors are taken from 32513 and have a different code.

☐ 29902	B	S29C	29902	preserved, ex Västerås	6-97
☐ 32940	40	S32C	32940	preserved, F11 marks, code 13 on wheeldoors	6-97
☐ 35916	16	S35E	35916	preserved, F11 markings	6-97
☐ ..		J35B	..	preserved, nose only	6-97

The Nyge Aero company has a Mitsubishi which is used for ground running of overhauled engines.

☐ SE-FGO		Mu-2	102 ?	stored, no wings, use for engine runs	4-94

SUNDSVALL
The gate of this former military field was guarded by a Draken. This aircraft may have been removed as a new road was being built through the area.

☐ 35066	30	J35A	35066	stored, F16 markings, wfu 5-12-78	6-91

TROLLHAGGEN
The local SAAB car museum has a complete silver SAAB Safir mounted on a pole.

☐ SE-AYC		SAAB 91A	91104	preserved, marked as '91104/8', ex Linköping	9-97

TULLINGE
The Swedish Air Force left this base in 1991, leaving nothing of Drakens 35047/11 and 35955/69. The only relic here now is a stored An-2 with the local aero club.

☐ LY-AER		An-2R	1G191-15	stored, with aero club	8-95

UGGLARP

The Svedinos Bil Och Flygmuseum is run by volunteers and is only open from early June till late August between 11:00 and 16:00. In July it may sometimes be open between 10:00 and 18:00. Besides a collection of aircraft, the museum also has a large of number rare cars on display

☐ 16028	92	Sk16A	14-725	preserved (AT-16), F14 markings, ex FE991	8-96
☐ 16033	33	Sk16A	14-772	preserved (AT-16), F5 markings, ex FH138	8-96
☐ 22149	E	J22A	149	preserved (FFVS 22), F10 markings	8-96
☐ 28307	A	J28B	V0578	preserved (Vampire Mk.50), F10 markings	8-96
☐ 28444	74	Sk28C2	15378	preserved (Vampire T.55), F5 markings	8-96
☐ 29203	U	J29A	29203	preserved (SAAB 29A), F16 markings	8-96
☐ 32599	59	J32B	32599	preserved (SAAB 32B), F12 markings	8-96
☐ 34070	P	J34	B249	preserved (Hunter Mk.50), F10 markings	8-96
☐ 35-1		SAAB 35	35-1	preserved	8-96
☐ 52001	01	Tp52	EEP71174	preserved (Canberra B.2), unmarked, ex WH711	8-96
☐ 83007	84	Tp83	P66/51	preserved (Pembroke C.52), F8 markings	8-96
☐ N9012N		CASA 352L	..	preserved, unmarked, ex T.2B-142/Spanish AF	8-96
☐ SE-ABS		DH.60	261	preserved, marked as '5555', ex S-AABS	8-96
☐ SE-AHY		GV-38	15	preserved	5-96
☐ SE-AFT		Tipsy S.2	30	preserved, ex OO-ASC	5-96
☐ SE-AGP		Tipsy B	504	preserved, ex OO-DOT	5-96
☐ SE-BEW		Fw44J	2670	stored, ex OH-SZG (at Linköping)	5-96
☐ SE-BFY		Ercoupe 415D	4409	stored, of site, ex NC3784H	5-96
☐ SE-BGF		Klemm KL.35D	1899	preserved, marked as '5015/155'	8-96
☐ SE-BNA		Ercoupe 415D	4735	preserved, ex OY-FAC	6-96
☐ SE-BWR		Fw44J	..	preserved, ex 5787 and painted as such	8-96
☐ SE-BWZ		Fw44J	CVV.29	preserved, marked as '5-59', ex 647	8-96
☐ SE-BYY		M38 Messenger	6703	preserved, marked as 'L-H', ex G-AKAO	6-96
☐ SE-CGR		J/1 Autocrat	2230	preserved, ex G-AIZW	8-96
☐ SE-DCC		Meteor T.7	G5/1425	preserved, marked as 'WS774/4', ex G-ANSO	8-96
☐ SE-EBC		Skyraider AEW.1	7975	preserved, ex G-31-6, ex WT962/RNavy	8-96
☐ SE-FNA		DH.82A	82003	preserved, ex D-EMWT	8-96
☐ SE-HDM		Ka-26	7001307	preserved	6-96
☐ SE-SAI		Meise	210	preserved, glider	6-96
☐ SE-SNY		Grunau Baby 11b	033	preserved, glider, ex 8118	8-96
☐ 136		Grunau SG-38	136	preserved, glider	8-96
☐ ..		Grankvist Autogiro	..	preserved, marked as '001'	6-96

UPPSALA

The instructional J35F 35402 went to Halmstad. The fate of the J35F 35459 is unknown.

☐ 35056	20	J35A	35056	dumped, F16 markings, wfu 18-7-75	8-93
☐ 35221	21	J35B	35221	decoy, F18 markings, wfu 3-7-73, ex Skavsta	6-97
☐ 35269	69	J35B	35269	dumped, F18 markings, wfu 16-10-73	6-95
☐ 35427	67	J35F	35427	decoy, F16 markings	6-97
☐ 35490	35	J35F	35490	preserved, at gate, F16 markings	6-97

VÄSTERÅS - HÄSSLÖ

The Flygtecknikcentrum has replaced Harvard 16105, S29C 29902 (to Skavsta), J35F 35609 (to Halmstad), Saab 91 50020 (to Östersund) and Pembroke 83004 with some newer aircraft. Draken 35583 used to be preserved at the military part of the airfield, but was parked in front of the school's hangars in 1995.

☐ 228		AB204B	3017	instructional, ex Skavsta, RNethNavy	6-97
☐ 02201	01	Hkp2	1278	instructional (Alouette 2)	6-97
☐ 02404	92	Hkp2	1302	instructional (Alouette 2), ex SE-HDI, 02204	6-97

☐ 16073		Sk16	..	instructional (AT-16)	6-97
☐ 29969	16	S29C	29969	instructional, F3 markings	6-97
☐ 29974	20	S29C	29974	stored, ex instructional, F3 markings	6-97
☐ 35496	32	J35F	35496	stored, ex instructional, F16 markings	6-97
☐ 35555	22	J35F	35555	instructional, F10 markings	6-97
☐ 35583	44	J35F	35583	stored, ex preserved, F1 markings	6-97
☐ 35592	18	J35F	35592	instructional, F10 markings	6-97
☐ 37-61		AJ37	37-6	instructional, Paris code 57, ex Stockholm	6-97
☐ 50003	71	Sk50	91203	instructional, F6 markings	6-97
☐ 50012	71	Sk50	91212	instructional, F4 markings	6-97
☐ 50027	77	Sk50	91229	instructional, F13 markings	6-97
☐ 50029	73	Sk50	91231	instructional, F4 markings	6-97
☐ 50036	85	Sk50	91238	instructional, F6 markings	6-97
☐ 50054	74	Sk50	91263	instructional, F17 markings	6-97
☐ 50055	71	Sk50	91264	instructional, F21 markings	5-96
☐ 50065	75	Sk50	91274	instructional, F17 markings	6-97
☐ G-ANEF		DH.82A	83226	instructional, ex T5493 and marked as such	5-96
☐ N123VC		Mu-2B	20-214	instructional	5-96
☐ OY-SVC		PA31	..	instructional, ex LN-RAV ?	5-96
☐ SE-CKN		PA23 Apache	..	instructional	5-96
☐ SE-CZH		PA28-150	28-1188	instructional	5-96
☐ SE-FDO		PA28R-180	28R-30685	instructional	5-96
☐ SE-FHU		PA28-140	28-26068	instructional	5-96
☐ SE-FLG		PA31	31-361	instructional	5-96
☐ SE-GLB		PA31T	31T-7400002	instructional	5-96
☐ SE-GYB		CeF172	1552	instructional	5-96
☐ SE-HHF		H369D	116-0049D	instructional	5-96
☐ SE-IUO		Beech 95-B55	TC-1716	instructional	5-96

Located next to the school is a flying museum with some vintage aircraft (Pembroke, P-51 and others).

VIDSEL
Some 75km west north west of Luleå is a FMV test airfield, where a Draken is dumped.

☐ 35042	07	J35A	35042	fire dump, F16 markings	6-91

VISBY
The new Gotland museum opened here in June 1997 at the airfield of Visby. Part of the collection comes from the now closed museum at Kåremo. The museum is open daily from June to August between 12:00 and 20:00 and from September to May on Sundays only between 13:00 and 17:00.

☐ 01009	09	Hkp1	607	preserved (Vertol 44), ex Kåremo	97
☐ 02042	42	Hkp2	1830	preserved (Alouette 2), ex Kåremo, Marinen	97
☐ 29624	P	J29F	29624	preserved, F9 markings, ex Kåremo	97
☐ 32502	U	J32B	32502	preserved, fuselage only, ex Kåremo	97
☐ 32548	32	J32D	32548	preserved, arr 14-11-97	11-97
☐ 32612	12	J32E	32612	preserved, F16M markings	97
☐ 35429		J35F	35429	preserved	97
☐ 35624	24	J35J	35624	preserved, F10 markings	97
☐ 35824	82	Sk35C	35824	preserved, F10 markings	97
☐ 37972	56	SF37	37972	preserved, F10 markings	97
☐ CCCP-70501		An-2	1G83-34	preserved, ex Kåremo	97
☐ HB-EZE		L-60 Brigadyr	150911	preserved, ex Kåremo	97
☐ SE-AGA		Caudron C.510	7338/45	preserved, ex Kåremo, F-AOYC	97

SWEDEN - 431

☐ SE-EBP	MFI-9B	05	preserved, ex Kåremo, OH-MFA	97
☐ SE-HCP	H269C	38-0358	preserved, ex Kåremo	97
☐ SE-HHT	Enstrom F.28C	429	preserved, fuselage only	97
☐ SE-TNX	Pik-20	20009	preserved, glider, ex Kåremo	97
☐ SE-464	HM-14	..	preserved, ex Kåremo	97

SWITZERLAND

The end of the flying career of the Hunter was saddest thing that has happened in recent years in Switzerland. Luckily a large number of them can still be found in Europe, including a number of them flying with civilians. Also the era of the Alouette 2 and Vampire ended. Most aircraft of these two types have been brought by civil operators and continue to fly. Although the Swiss airfields are often visited, the *Wrecks & Relics* locations are not. A large number of sightings are for the early 1990s and require confirmation if they are still current.

AARBURG
Noted in this village at the rear of a shop in August 1994 was the tail of Vampire T.55 U-1232. Was the rest of the aircraft also here?

AIGLE
Some 60km east of Lausanne is the village of Aigle. A Vampire is preserved at the Zone Industrielle 1.
❏ J-1157   Vampire FB.6   666   preserved, on pole, ex Sion   9-96

ALPNACH
Stored here since 1992 were some 17 Alouette 2s. Sixteen of them were auctioned on 11th June 1994 and most found new home: V-45 (1229, to F-GJEB), V-46 (1230, to N3102U), V-48 (1236, to N92785), V-51 (1929, to HB-XBJ), V-53 (1931), V-54 (1897, to G-BVSD), V-56 (1902, to F-GJIE), V-57 (1903), V-59 (1908, to F-GNFJ), V-60 (1909, to France), V-61 (1910, to HB-XKJ), V-62 (1914, to F-GJLK), V-64 (1916), V-66 (1920, to F-GNVB), V-68 (to <u>Altenrhein</u>) and V-70 (1926, to TU-THM). The 17th chopper, V-43 (110), became HB-XBI and went later to <u>Luzern</u>.

ALTENRHEIN
Altenrhein is home of a number of flying vintage props and jets. Former Swiss AF Venoms, Vampires and Hunters have made here their home base. Most are in civil marking except for the following:

❏ ..	Sycamore HR.54	..	stored, marked as 'D-HFUM', ex WGArmy	7-96
❏ V-68	Alouette 2	1924	stored, ex F-WKQC	7-96
❏ C-509	C3605	289	stored	12-88
❏ J-1111	Vampire FB.6	620	stored, ex Sion	3-91
❏ J-4062	Hunter F.58	..	dumped	7-96

AU
Preserved near the Au/Waedenens railway station (at lake Zürich) is the cockpit section of a Tu-134.
❏ OK-CFC   Tu-134A   2351504   preserved, cockpit only, wfu 15-11-90   1-98

BASEL
Town: Preserved at a private location in the Fullingsdorf part of Basel is a Vampire, while a Venom should still be preserved at military barracks in town. Hunter J-4026 should be at a private museum in Basel.

❏ J-1150	Vampire FB.6	659	preserved, ex Sion	5-91
❏ J-1648	Venom FB.50	..	preserved	4-89
❏ J-4026	Hunter F.58	..	preserved, ex Emmen	—

Airfield: On the airfield two former Sunshine Friendships were broken up in mid the 1995s, by late 1997 only the fuselage of one aircraft was still visible. Hunter J-4086 is with Crossair Hunter Club. Stored Boeing 707-336

N14AZ (19498) flew out in the early 1990s.
❑ J-4086	Hunter F.58	..	preserved, ex Emmen	5-98
❑ HB-ISH	F27-200	10260	dumped, ex OH-LKC	8-94
❑ HB-ISQ	F27-500	10506	dumped, ex F-BSUM	8-95
❑ N617SE	DC-6B	44088	preserved, at firestation, 9Q-CVM ntu	8-95

BERN
Town: In the town of Bern is the Armeé Museum. Preserved inside should be a Venom.
❑ J-1649   Venom FB.50   ..   preserved   4-89

Airfield - Belp: An ex Army Air Coprs Sioux was noted dumped outside the Heli Swiss hangar. The serial quoted was XT441, but this belongs to a Predannack Wasp. Its fate is unknown. Passing through here in the early 1990s were some ex Sion Vampires; J-1138 went to Fulenbach, J-1161 went to Wohlen and J-1195 went to Pieterlen. The fate of J-1135 is unknown.
❑ J-1135   Vampire FB.6   644   stored, ex Sion   4-91

BEX
Town: There are no new reports of the C3605 at the Jordacier factory in the town.
❑ C-539   C3605   319   preserved   9-88

Airfield: Still preserved at the airfield is Venom J-1627. It is currently for sale.
❑ J-1627   Venom FB.50   837   preserved   5-98

BRAMOIS
Just outside Sion, in the direction of Turtmann, a Vampire is preserved at an investigation centre. This place has also been reported as Grone.
❑ J-1158   Vampire FB.6   667   preserved, ex Sion   11-92

BÜTZBERG
Between Langenthal and Herzogenburhsee is the village of Bützberg. Stored outside a surplus collector is an ex US Army UH-1 (short version, UH-1B or UH-1C). It was bought in Germany many years ago and was last noted in March 1995. The helicopter was sold to a film studio the next year and has gone.

CLARO
Mounted on a pole at Heli TV is the former Lodrino gate guard. Claro is near Lodrino, just north of Bellinzona.
❑ J-1580   Venom FB.50   790   preserved, ex Magadino   5-98

COLOMBIER
Their are two Colombiers in Switzerland, one near Neuchâtel and one near Lausanne. One of them has a Venom, but which one?
❑ J-1526   Venom FB.50   736   preserved, ex Gals   4-96

DIEPOLSAU
At Diepolsau, along the border with Austria, is the Gasser Autoabbruch und Handel. This car scrapyard has a Venom as gate guard.
❑ J-1766   Venom FB.54   ..   preserved   3-94

DÜBENDORF

Town: Preserved inside the town of Dübendorf, at the Hobbyrama garden centre, is a Venom. Its located on the Bettlistrasse near the railway station. It was noted noted in 1998.

❏ J-1641	Venom FB.50	..	preserved		8-96

Airfield: Preserveon at the airfield is Hunter J-4020. It is parked near some hangars.

❏ J-4020	Hunter F.58	..	preserved, Patrouille Suisse c/s		5-98

The airfield is well known due to its Museum der Fliegertruppen It is open on Tuesday to Friday between 13:30 and 17:00, on Saturday between 09:00 and 17:00 and on Sundays between 13:00 and 17:00.

❏ FR56	SO1221S	38	preserved, marked as 'V-23', ex French Army	5-98
❏ A-51	Bü131B	64	preserved	5-98
❏ A-57	Bü131B	70	preserved	5-98
❏ U-61	Bü133C	8	preserved	5-98
❏ A-100	Fi156C-3	1685	preserved	5-98
❏ U-134	P2-06	51	preserved, ex U-105	5-98
❏ A-209	Bf108B-2	..	preserved	5-98
❏ J-276	D-3801J	66	preserved, licence built MS406C-1	5-98
❏ U-328	Harvard T.2B	14-201	preserved, ex FE811, ex 42-12298	5-98
❏ J-355	Bf109E-3	31647	preserved	5-98
❏ C-534	C3603-1	314	preserved	5-98
❏ A-801	P3-02	318-1	preserved, ex HB-HOO	5-98
❏ A-803	P3-03	320-3	preserved, N321RD ntu	5-98
❏ J-1126	Vampire FB.6	635	preserved, on pole outside	5-98
❏ J-1153	Vampire FB.6	662	preserved	5-98
❏ U-1224	Vampire T.55	984	preserved, ex Sion	5-98
❏ J-1580	Venom FB.54	790	preserved, nose only	5-98
❏ J-1642	Venom FB.50	852	preserved	5-98
❏ J-1751	Venom FB.54	921	preserved, nose only	5-98
❏ J-1753	Venom FB.54	923	preserved	5-98
❏ J-2113	P-51D	122-39808	preserved, ex 44-73349	5-98
❏ J-4001	Hunter F.58	..	preserved, l/f 16-12-94, ex XE536	5-98
❏ KAB-202	UH-12B	387	preserved	5-98
❏ 180	EKW C35	395	preserved, licence built Fokker C.X	5-98
❏ 257	Dewoitine D27	..	preserved, ex Luzern	5-98
❏ 331	Fokker C.VE	5261	preserved, ex Luzern	5-98
❏ 607	Nieuport 28C-1	..	preserved	5-98
❏ 653	Hanriot HD.1	..	preserved	5-98
❏ D-EDOC	Bü181B	27	preserved, as 'A-251', ex 25057/RSweAF	5-98
❏ HB-GAC	Beech G18S	8343	preserved, ex G-8, ex SE-BTS, ex NC79848	5-98
❏ HB-HOI	N1203	122	preserved	5-98
❏ HB-USI	Comte AC4	33	preserved, ex CH-249	5-98
❏ X-HB-VAD	FFA P16	05	preserved, ex 'J-3004'	5-98
❏ ..	Alouette 2	..	preserved, frame only, parts from Alouette 3	5-98
❏ ..	Blériot XI	..	preserved, replica	5-98
❏ ..	Fokker D.VII	..	preserved, partial replica, marked as '640'	5-98
❏ ..	Hafeli DH-1	..	preserved, replica, marked as '245'	5-98
❏ ..	Hafeli DH-5	..	preserved, replica, marked as '459'	5-98
❏ ..	N20 Aiguillon	..	preserved	5-98

A number of museum aircraft should still be stored at the airfield. Of these Venom J-1742 went to Comignago and the pod of Venom J-1712 was noted at Hurn in the UK.

❏ A-16	Bü131B	25	stored, hanging inside hangar	7-94
❏ A-32	Bü131B	43	stored, ex HB-USP	98
❏ A-43	Bü131B	55	stored	98

SWITZERLAND - 435

☐ V-49	Alouette 2	1237	stored, for museum	98
☐ U-62	Bü133C	9	stored, hanging inside hangar	98
☐ A-204	C3603	..	stored	98
☐ C-497	C3605	277	stored	98
☐ A-713	Beech E50	EH-58	stored, in shelter, for museum, ex HB-HOW	98
☐ A-901	PC-7	509-58	stored, ex HB-HOZ, ex A-871 (P3-05)	98
☐ J-1049	Vampire FB.6	960	stored	5-97
☐ J-1629	Venom FB.50	839	stored	98
☐ J-1649	Venom FB.50	859	stored	98
☐ J-1717	Venom FB.54	..	stored	98
☐ J-2201	Mirage 3C	..	stored	98
☐ J-4152	Hunter F.58A	..	stored, l/f 11-4-90, ex G-9-385, ex WT716	98
☐ J-4204	Hunter T.68	..	stored, for museum, ex G-9-375, ex XE702	98
☐ HB-GCP	RC680FL	1633/19	stored	98

Passing through here were Vampires looking for new homes, J-1130 went to <u>Schwarzenbach</u>, J-1133 went to <u>Ramsen</u> and J-1144 (ex Emmen) is now at <u>Volketswill</u>. A number of Hunters were scrapped at the airfield.

J-4036	Hunter F.58	..	stored, l/f 27-6-91	scrapped
J-4112	Hunter F.58A	..	stored, l/f 30-4-92, ex G-9-329, ex XF976	scrapped
J-4120	Hunter F.58A	..	stored, l/f 6-7-92, ex G-9-336, ex WT713	scrapped
J-4127	Hunter F.58A	..	stored, l/f 2-9-92, ex G-9-212, ex N-212	scrapped
J-4141	Hunter F.58A	..	stored, l/f 31-1-92, ex G-9-408, ex XF990	scrapped
J-4142	Hunter F.58A	..	stored, l/f 7-10-92, ex G-9-393, ex XF994	scrapped

DULLIKEN
Preserved on poles at the Autowad garage in Dulliken are three aircraft. Dulliken is east of Olten.

☐ C-533	C3605	313	preserved	3-94
☐ J-1624	Venom FB.50	834	preserved	3-94
☐ HB-LCC	Ce310C	35856	preserved	3-94

DUSSNANG
Venom J-1639 was preserved here for many years. By the mid 1990s the aircraft moved to Langenthal.

EGERKINGEN
Outside a local factory in Egerkingen (west of Olten) is a Vampire mounted on a pole.

☐ J-1185	Vampire FB.6	694	preserved, ex Emmen	5-98

EMMEN
Emmen is mainly in use by the training aircraft of the Swiss Air Force. Also here are the FFW works, currently building the Hornets. Stored Hunter J-4057 is now at <u>Cerbaiola</u>, J-4026 and J-4086 went to <u>Basel</u>, J-4013 should be at <u>Langenthal</u>, J-4041 at <u>Nidau</u> and J-4095 at <u>Cuers</u>.

☐ J-1709	Venom FB.54	..	preserved, on pole at FFW gate, ex 'J-1700'	6-97
☐ J-1775	Venom FB.54	..	stored	6-89
☐ J-2336	Mirage 3S	..	stored, wreck, crashed 21-2-94	7-94
☐ J-4070	Hunter F.58	..	stored	9-97

In the late 1980s a number of Vampires were stored here. Of these J-1152, J-1154, J-1167, J-1173, J-1183, J-1184, J-1192, J-1196, U-1210, U-1212, U-1219, U-1225, U-1226 and U-1233 all went to <u>Sion</u> by 1991. Others were transported to their new owners from Emmen. Some Hunters were scrapped here:

J-1082	Vampire FB.6	993	stored, ex Sion, l/n 8-90 to Altenrhein as HB-RVE	

J-1127	Vampire FB.6	636	stored, l/n 8-90		to Melun as F-AZOO
J-1144	Vampire FB.6	653	stored, l/n 8-90		to Dübendorf
J-1145	Vampire FB.6	654	stored, ex Sion, l/n 8-90		to Nîmes Garons
J-1180	Vampire FB.6	689	stored, l/n 8-90		to Malters
J-1184	Vampire FB.6	693	stored, l/n 8-90		to Eskiltuna as SE-DXY
J-1185	Vampire FB.6	694	stored, l/n 8-91		to Egerkingen
J-1197	Vampire FB.6	706	stored, l/n 8-90		to Altenrhein as HB-RVN
J-4011	Hunter F.58	..	stored, l/f 10-10-93, ex XE555		scrapped
J-4033	Hunter F.58	..	stored, l/f 14-10-93		scrapped
J-4047	Hunter F.58	..	stored, l/f 21-7-93		scrapped

FREIENBACH
Freienbach is located on the south side of the Züricher See. Near the railway station is the factory of R Uiker where two aircraft were noted in 1991.

❏ C-464	C3605	244	preserved		8-91
❏ J-1105	Vampire FB.6	614	preserved, ex Sion		8-91

FRUTIGEN
No reports have been received from this place where P2-06 U-147 may have been preserved at a local museum.

FULENBACH
A private owner at Fulenbach, south west of Aarburg, has acquired a Vampire.

❏ J-1138	Vampire FB.6	647	stored, ex Bern, ex Sion		5-98

GALS
Venom J-1526, which used to be preserved here at a civil defence centre, moved on to Colombier.

GENEVE
Large parts of an ex French Navy Atlantic (from Nîmes Garons) are preserved inside the terminal. Stored C3605 C-552 (322) became civil as HB-RBJ.

❏ 27	Br1150	27	preserved, cut away frame, ex Nîmes		2-97
❏ J-4085	Hunter F.58	..	preserved, on pole, l/f 14-6-95		2-98
❏ N9498	DC-7C	45187	stored, ex instructional, marked as 'HB-SSA'		2-97

GRANDVILLARD
Venom J-1544, which used to be preserved at the place de tir, moved on to Payerne.

HESIRAU
Preserved along the railroad near the town centre are two Venoms.

❏ J-1559	Venom FB.50	769	preserved		3-94
❏ J-1778	Venom FB.54	..	preserved		3-94

HOCHSTETTEN
A Venom is preserved along the main road, the N1 from Bern to Zürich, through this village.

❏ J-1756	Venom FB.54	..	preserved		7-94

INTERLAKEN

The Vampire is a long term inmate at the base, while the Hunters here are stored on behalf of a preservation group named Hunterverein.

❑ J-1117	Vampire FB.6	626	preserved	8-95
❑ J-4007	Hunter F.58	..	preserved, l/f 16-12-93, ex XE545	5-96
❑ J-4018	Hunter F.58	..	stored, l/f 29-8-92	3-98
❑ J-4050	Hunter F.58	..	stored, l/f 15-12-93	—
❑ J-4093	Hunter F.58	..	stored	9-95

A large number of Hunters were stored here in the mid-1990s. All were scrapped locally, only one survived, J-4075 went to the England as G-BWKA.

J-4012	Hunter F.58	..	stored, l/f 18-11-93, ex XE553	scrapped
J-4019	Hunter F.58	..	stored, l/f 14-7-92	scrapped
J-4023	Hunter F.58	..	stored, l/f 4-5-94	scrapped
J-4028	Hunter F.58	..	stored, l/f 12-8-91	scrapped
J-4038	Hunter F.58	..	stored, l/f 7-5-92	scrapped
J-4044	Hunter F.58	..	stored, l/f 25-9-91	scrapped
J-4048	Hunter F.58	..	stored, l/f 21-4-92	scrapped
J-4052	Hunter F.58	..	stored, l/f 16-12-93	scrapped
J-4054	Hunter F.58	..	stored, l/f 5-12-91	scrapped
J-4108	Hunter F.58A	..	stored, l/f 11-4-90, ex G-9-314, ex WV411	scrapped
J-4115	Hunter F.58A	..	stored, l/f 23-9-92, ex G-9-321, ex XF436	scrapped
J-4116	Hunter F.58A	..	stored, l/f 27-11-91, ex G-9-256, ex XF937	scrapped
J-4118	Hunter F.58A	..	stored, l/f 20-2-92	scrapped, nose to Luzern
J-4119	Hunter F.58A	..	stored, l/f 21-2-92, ex G-9-335, ex WV380	scrapped
J-4122	Hunter F.58A	..	stored, l/f 2-11-89, ex G-9-337, ex XE717	scrapped
J-4123	Hunter F.58A	..	stored, l/f 21-6-90, ex G-9-339	scrapped
J-4124	Hunter F.58A	..	stored, l/f 28-9-90, ex G-9-340, ex XE674	scrapped
J-4125	Hunter F.58A	..	stored, l/f 5-7-90, ex G-9-341, ex WV257	scrapped
J-4126	Hunter F.58A	..	stored, l/f 5-11-91, ex G-9-342, ex WW659	scrapped
J-4128	Hunter F.58A	..	stored, l/f 12-12-89, ex G-9-194, ex N-318	scrapped
J-4132	Hunter F.58A	..	stored, l/f 14-2-92, ex G-9-392, ex XF933	scrapped
J-4136	Hunter F.58A	..	stored, l/f 15-6-90, ex G-9-387, ex XF370	scrapped
J-4137	Hunter F.58A	..	stored, l/f 19-2-92, ex G-9-384, ex WV404	scrapped
J-4139	Hunter F.58A	..	stored, l/f 2-11-89, ex G-9-374, ex XF941	scrapped
J-4148	Hunter F.58A	..	stored, l/f 13-3-90, ex G-9-388, ex WT797	scrapped
J-4149	Hunter F.58A	..	stored, l/f 14-2-92, ex G-9-401, ex XE659	scrapped
J-4150	Hunter F.58A	..	stored, l/f 3-12-91, ex G-9-405, ex XF312	scrapped

JEGENSDORF

Venom J-1739 should have been with the local Zivilschutzzentrum. The Venom was no longer noted in Jegensdorf in August 1991.

KUSSNACHT

Of the Venom at the local Ford Dealer/BP garage it has been said that it was sold to a private owner somewhere in the Bern area in 1991.

❑ J-1564	Venom FB.50	774	preserved, see note	5-90

LANGENTHAL

The C3605 should still be stored in the large building at the airfield near Bleienbach. Nothing is known about the Hunter which should have arrived in 1995. The Venom is under restoration at the Honda motor dealer which is along the main road through Langenthal.

☐ C-494	C3605	274		stored, ex Lodrino	8-91
☐ J-1639	Venom FB.50	..		under restoration, ex Dussnang	5-98
☐ J-4013	Hunter F.58	..		preserved, ex Emmen	—

LOCARNO
Both the Venoms mentioned in the first edition moved from here to Bournemouth, UK. J-1539 became G-DHUU and J-1611 became G-DHTT.

LODRINO
All the C3605s mentioned in the first edition have left, through not all their fates are known. C-547 (327) became HB-RBI. Some of the stored P3s are now civil, A-815 became HB-RCQ and A-839 is now N839A.

☐ A-812	P3-03	329-11	stored	2-91
☐ A-852	P3-05	490-39	stored	5-95

During April 1992 there was a sale of former Swiss Air Force P3-05s. Fates are known for most of the auctioned aircraft.

A-817	P3-05	455-4	stored, l/n 4-92	to C-FCBQ
A-819	P3-05	457-6	stored, l/n 4-92	gone
A-827	P3-05	465-14	stored, l/n 4-92	to HB-RCC
A-830	P3-05	468-17	stored, l/n 4-92	to HB-RBV
A-837	P3-05	475-24	stored, l/n 4-92	to N837A
A-853	P3-05	491-40	stored, l/n 4-92	to HB-RCA, to F-AZVB
A-855	P3-05	493-42	stored, l/n 4-92	to HB-RBX
A-857	P3-05	495-44	stored, l/n 4-92	to N857P
A-859	P3-05	497-46	stored, l/n 4-92	to HB-RCB
A-860	P3-05	498-47	stored, l/n 4-92	to HB-RBZ
A-863	P3-05	501-50	stored, l/n 4-92	to C-GPIL
A-866	P3-05	504-53	stored, l/n 4-92	to HB-RBT
A-867	P3-05	505-54	stored, l/n 4-92	to HB-RCD, to G-BUKM
A-868	P3-05	506-55	stored, l/n 4-92	to C-FNQB
A-869	P3-05	507-56	stored, l/n 4-92	to F-AZRF
A-872	P3-05	510-59	stored, l/n 4-92	to HB-RBY

LUZERN
Verkehrshaus der Schweiz. Vampire J-1068 (979) was last noted in the early 1990s and moved on to Schwenningen. The N20 Aiguillon moved to Dübendorf, as did Fokker C.VE 331. Vampire J-1120 (l/n May 1995) is no longer here. The museum is open daily between 09:00 and 18:00 (1st April to 31st October) and during the rest of the year between 10:00 and 17:00.

☐ U-60	Bü133C	7	preserved	5-97
☐ A-97	Fi156C	8063	preserved	5-97
☐ A-210	Bf108B	..	preserved	5-97
☐ C-537	C3603-1	317	preserved	5-97
☐ J-1200	Vampire FB.6	709	preserved, ex Sion	5-97
☐ J-1729	Venom FB.54	899	preserved	5-97
☐ J-4118	Hunter F.58A	..	preserved, nose only, ex Interlaken, ex G-9-334	5-97
☐ 23	Blériot XI	..	preserved	5-97
☐ 688	Nieuport 28C-1	..	preserved	5-97
☐ HB-HFB	FFA AS202/15	021	preserved, fuselage only	97
☐ HB-ICC	Convair 990	12	preserved	97
☐ HB-IRN	DC-3	16645/33593	preserved, ex KN683, ex 44-77061	97

SWITZERLAND - 439

☐ HB-KIL	Comte AC4	35	preserved, ex CH-264		97
☐ HB-LBO	Fokker F.VIIa	5005	preserved, ex CH-157		97
☐ HB-OPR	PA18-150	18-5786	preserved		97
☐ HB-RAE	Dewoitine D26	320	preserved, ex U-288 and marked as such		97
☐ HB-XAE	Bell 47G	689	preserved		97
☐ HB-XBI	Alouette 2	1120	preserved, arr 14-4-94, ex V-43, ex Alpnach		97
☐ HB-XDF	Alouette 3	1216	preserved, arr 27-3-80		97
☐ HB-112	Hug Spyr IIIb	..	preserved, glider		97
☐ HB-307	Spalinger S21H	..	preserved, glider		97
☐ HB-362	Z-12 Zögling	..	preserved, glider		97
☐ HB-935	Neukom S-3	..	preserved, glider		97
☐ NC12222	Lockheed L9C	180	preserved, marked as 'CH-167'		97
☐ ..	Farner WF-7	32	preserved, glider		97
☐ ..	HM-8	..	preserved		97
☐ ..	SE210-3	..	preserved, mock-up, cockpit only		97
☐ ..	Soldenhoff S-5	..	preserved, glider		97

MAGADINO
The former gate guarding Venom J-1580 moved on to Claro. Of the three C3605s mentioned in the first edition C-493 (273) became F-AZGC, C-550 (330) became F-AZGD and C-552 (332) became HB-RBJ.

☐ A-858	P3-05	496-45	preserved, at gate on pole	8-97

MALTERS
A private collector at Malters, west of Luzern, has his own Vampire.

☐ J-1180	Vampire FB.6	689	preserved, ex Emmen	—

MARTIGNY
Preserved at a school in this town between Payerne and Sion is an unmarked Vampire

☐ J-1176	Vampire FB.6	685	preserved, ex Sion	5-98

MATRAN
Matran is some 10km south west of Fribourg. A Venom may still be here.

☐ J-1584	Venom FB.50	794	preserved	8-91

MEIRINGEN
The scrapping of Hunters also took place at Meiringen. Of the aircraft marked * the nosewheel doors (with its serial) were still preserved on a hangar wall in August 1995. Hunter J-4107 was last noted here in June 1995 and moved to Nantes. The stored Hunter J-4015 moved on. It is now at Mollis, whilst J-4104 is now flying as G-PSST in the UK.

J-4009	Hunter F.58	..	stored, l/f 17-2-94, l/n 3-94, ex XE528	scrapped
J-4039 *	Hunter F.58	..	stored, l/f 25-9-91	scrapped
J-4049 *	Hunter F.58	..	stored, l/f 11-2-94, l/n 3-94	scrapped
J-4053	Hunter F.58	..	stored, l/f 17-12-93, l/n 3-94	scrapped
J-4106	Hunter F.58A	..	stored, l/f 21-10-93, ex G-9-316, ex WV405	scrapped
J-4109 *	Hunter F.58A	..	stored, ex G-9-297, ex XF365	scrapped
J-4111	Hunter F.58A	..	stored, l/f 15-10-90, ex G-9-310, ex XG272	scrapped
J-4114 *	Hunter F.58A	..	stored, l/f 27-8-90, ex G-9-254, ex XF981	scrapped
J-4117	Hunter F.58A	..	stored, l/f 21-3-90, ex G-9-260, ex XF361	scrapped
J-4129 *	Hunter F.58A	..	stored, l/f 16-8-90, ex G-9-269, ex XF992	scrapped

J-4131 *	Hunter F.58A	..	stored, l/f 13-2-90, ex G-9-373, ex XF429	scrapped
J-4135	Hunter F.58A	..	stored, l/f 3-4-90, ex G-9-381, ex XF308	scrapped
J-4138 *	Hunter F.58A	..	stored, l/f 18-5-93, ex G-9-386, ex WW590	scrapped
J-4143	Hunter F.58A	..	stored, l/f 26-2-92, ex G-9-394, ex XF973	scrapped
J-4146 *	Hunter F.58A	..	stored, l/f 25-10-93, ex G-9-399, ex WV266	scrapped
J-4147	Hunter F.58A	..	stored, l/f 27-8-90, ex G-9-400, ex WV261	scrapped

MELLINGEN
Another privately kept Vampire is at Melllingen, just south of Baden (east of Zürich).
❑ J-1164   Vampire FB.6   673   preserved, ex Sion   3-91

MOLLIS
Preserved at this reserve field is one of the famous Hunters.
❑ J-4015   Hunter F.58   .. preserved, ex Meiringen   11-96

MONTE TAMARO
Nothing is known abouth the location of this trainer Vampire.
❑ U-1218   Vampire T.55   978   preserved, ex Sion   7-91

MORGES
The Musée Militaire Vaudoise is located in an old castle. A Vampire has been reported as being preserved outside. Venom J-1545, which also used to be here, moved to St Julien de Cassagras.
❑ J-1055   Vampire FB.6   966   preserved   8-91

MUNSINGEN
Preserved in a private garden is a Venom. The more exact location is near the Industrie Strasse Sportzentrum.
❑ J-1646   Venom FB.50   ..   preserved   4-96

NEUHEIM
Stored Venom J-1545 was offered for sale in the late 1980s and moved on the Morges.

NIDAU
On the border with France, just south of Biel (Bienne), is a local museum with a Hunter.
❑ J-4041   Hunter F.58   .. preserved, ex Emmen   —

PAYERNE
Town: Preserved in town at a Renault garage is a Vampire. The garage can be found on the road from Payerne to Fribourg.
❑ J-1134   Vampire FB.6   643   preserved, ex Sion   9-97

Airfield: Both the instructional Vampire T.55s left in the late 1980s. U-1202 went to Sion, while the fate of U-1201 is unknown. Aircraft passing through here were Vampire J-1128 (637, l/n 3-91) to Bulle. Hunter F.58 J-4006 was stored here for a year before flying to Lyneham, on its way to Ontario, Canada. Other aircraft made their last flight from here. J-4065 went to Toulouse, J-4073 moved to Soesterberg and J-4099 is now preserved at Le Bourget.

SWITZERLAND - 441

☐ J-1051	Vampire FB.6	962	dumped, marked as 'J-1117'		88
☐ J-1142	Vampire FB.6	651	preserved, at viewing area		3-98
☐ J-1156	Vampire FB.6	665	instructional, ex preserved with barracks		3-98
☐ J-1544	Venom FB.50	754	stored, ex Grandvillard		8-95
☐ J-4003	Hunter F.58	..	preserved, ex stored, ex XE541		3-98
☐ J-4037	Hunter F.58	..	stored		3-94
☐ J-4045	Hunter F.58	..	preserved, at barracks		5-98
☐ J-4063	Hunter F.58	..	stored ?		5-95
☐ J-4078	Hunter F.58	..	stored, on trailer		9-95

PIETERLEN
On the Moosstrasse (along the Grienstrasse) in the Industrie West estate is the Seckler company, with a Vampire. Pieterlen is between Solothurn and Biel-Bienne.

☐ J-1195	Vampire FB.6	704	preserved, ex Bern, ex Sion	8-93

RAMSEN
Right on the border with Germany a private collector has a Vampire. Ramsen is some 40km west of Konstanz.

☐ J-1133	Vampire FB.6	642	preserved, ex Dübendorf, ex Sion	3-91

RARON
This reserve airfield has at least five Hunters stored in its mountain hangars.

☐ J-4032	Hunter F.58	..	stored, ex Stans	4-97
☐ J-4082	Hunter F.58	..	stored, ex Stans, arr 2-4-97	4-97
☐ J-4201	Hunter T.68	..	stored, ex Stans, arr 2-4-97, ex G-9-406	4-97
☐ J-4203	Hunter T.68	..	stored, ex Stans, arr 2-4-97, ex G-9-411	4-97
☐ J-4206	Hunter T.68	..	stored, ex Stans, arr 2-4-97, ex G-9-413	4-97

RICKENBACH (or Schlierbach)
Some 10km north of the Sursee is the village of Rickenbach. Preserved here with Mr Schmid, in a field behind his tractor repair shop, is a Venom.

☐ J-1578	Venom FB.50	788	preserved	8-91

SAMEDAN
At the small civil airfield of Samedan a Vampire is still preserved at the gate.

☐ J-1169	Vampire FB.6	678	preserved, at gate	1-98

SCHOTZ
At the Zivilschutzcentrum two aircraft should still be current. The Venom could clearly be identified but the Vampire gives some problems. It has been reported as J-1126, but this aircraft is confirmed as being at the Dübendorf museum.

☐ J-1126 ?	Vampire FB.6	..	instructional, see note	3-94
☐ J-1719	Venom FB.54	..	instructional	3-94

SCHWARZENBACH
There are several Schwarzenbachs in the country. It is not known in which one the Vampire is preserved.

☐ J-1130	Vampire FB.6	639	stored, ex Dübendorf, ex Sion	3-91

SWITZERLAND - 442

SION

Town: Preserved in the village of Sion, near the highway to Bramois, is a Vampire. It is at the Seat garage named Auto Pole.

❏ J-1154	Vampire FB.6	663	preserved	1-98

Airfield: A large gathering of stored Vampires took place at Sion in the late 1980s. Of these, only a handful still remain at the base.

❏ J-1080	Vampire FB.6	991	stored	5-98
❏ J-1103	Vampire FB.6	612	stored, to become HB-RVH	5-98
❏ J-1190	Vampire FB.6	699	preserved	5-98
❏ U-1211	Vampire T.55	971	stored, to become HB-RVL	3-98
❏ U-1222	Vampire T.55	982	stored	5-98
❏ U-1235	Vampire T.55	44352	stored, to become HB-RVI	5-98
❏ U-1239	Vampire T.55	DH13	stored	5-98
❏ ..	Vampire T.55	..	stored, on far side	5-98
❏ J-4100	Hunter F.58A	..	preserved	5-98
❏ HB-HOX	Twin Pioneer 3	570	stored	5-98

The Vampires have all been sold of and the following aircraft have all left the storage site of Sion.

J-1081	Vampire FB.6	992	stored, l/n 6-91	to Speyer
J-1082	Vampire FB.6	993	stored, l/n 10-89	to Emmen
J-1101	Vampire FB.6	610	stored, l/n 7-91	to Rennes as F-AZHY
J-1102	Vampire FB.6	611	stored, l/n 9-91	to Fort Lauderdale as N100VJ
J-1105	Vampire FB.6	614	stored, l/n 8-90	to Freienbach
J-1107	Vampire FB.6	616	stored, l/n 9-91	to Vigna di Valle
J-1111	Vampire FB.6	620	stored, l/n 8-90	to Altenrhein
J-1115	Vampire FB.6	624	stored, l/n 9-91	to Le Havre as F-AZHX ('VZ152')
J-1121	Vampire FB.6	630	stored, l/n 8-90	to Eskilstuna as SE-DXZ
J-1122	Vampire FB.6	631	stored, l/n 7-91	to Nîmes Garons
J-1124	Vampire FB.6	633	stored, l/n 8-93	for Air Vampire
J-1126	Vampire FB.6	635	stored, l/n 8-90	to Dübendorf
J-1129	Vampire FB.6	638	stored, l/n 7-91	to Cincinnati as N4024S ('WA235')
J-1130	Vampire FB.6	639	stored, l/n 8-90	to Dübendorf
J-1133	Vampire FB.6	642	stored, l/n 8-90	to Dübendorf
J-1134	Vampire FB.6	643	stored, l/n 3-91	to Payerne
J-1135	Vampire FB.6	644	stored, l/n 3-91	to Bern
J-1138	Vampire FB.6	647	stored, l/n 3-91	to Bern
J-1140	Vampire FB.6	649	stored, l/n 7-91	to Tamiami as N3160Y
J-1143	Vampire FB.6	652	stored, l/n 7-91	to Rennes as F-AZHI
J-1145	Vampire FB.6	654	stored, l/n 10-89	to Emmen
J-1146	Vampire FB.6	655	stored, l/n 6-91	to Gardermoen as LN-JET
J-1149	Vampire FB.6	658	stored, l/n 3-91	to Bournemouth as G-SWIS
J-1150	Vampire FB.6	659	stored, l/n 8-90	to Basel
J-1152	Vampire FB.6	661	stored, l/n 8-91	to Cincinnati as N152RD
J-1154	Vampire FB.6	663	stored, l/n 9-91	to Sion Town
J-1155	Vampire FB.6	664	stored, l/n 9-91	to Le Bourget as F-AZHZ
J-1157	Vampire FB.6	666	stored, l/n 9-91	to Aigle
J-1158	Vampire FB.6	667	stored, l/n 3-91	to Bramois
J-1159	Vampire FB.6	668	stored, l/n 6-91	to Cuers as F-AZHJ ('VZ221')
J-1161	Vampire FB.6	670	stored, l/n 8-90	to Bern
J-1164	Vampire FB.6	675	stored, l/n 8-90	to Mellingen
J-1167	Vampire FB.6	676	stored, l/n 3-91	to Cranfield as G-MKVI ('VZ304')
J-1170	Vampire FB.6	679	stored, l/n 9-90	to Rivolto
J-1173	Vampire FB.6	682	stored	to Southampton as G-DHXX ('LZ551')
J-1176	Vampire FB.6	685	stored, l/n 8-90	to Martigny

SWITZERLAND - 443

J-1178	Vampire FB.6	687	stored, l/n 6-91	to Savigny
J-1179	Vampire FB.6	688	stored, l/n 6-91	gone
J-1183	Vampire FB.6	692	stored, l/n 10-91	to Orange
J-1187	Vampire FB.6	696	stored, l/n 9-91	to Nîmes Garons
J-1191	Vampire FB.6	700	stored	to Nîmes Garons as F-AZIK
J-1192	Vampire FB.6	701	stored, l/n 5-92	to Melun as F-AZOP
J-1193	Vampire FB.6	702	stored, l/n 9-91	gone
J-1195	Vampire FB.6	704	stored, l/n 3-91	to Bern
J-1196	Vampire FB.6	705	stored	to Nîmes Garons, later to SE-DXS
J-1199	Vampire FB.6	708	stored, l/n 3-91	to Caen as F-AZHH
J-1200	Vampire FB.6	709	stored, l/n 8-90	to Luzern
U-1202	Vampire T.55	862	stored, l/n 7-91	scrapped
U-1203	Vampire T.55	863	stored, l/n 9-91	gone
U-1204	Vampire T.55	864	stored, l/n 3-91	to Nîmes Garons as F-AZHZ
U-1205	Vampire T.55	865	stored, l/n 9-91	gone
U-1206	Vampire T.55	866	stored, l/n 8-91	to Fort Lauderdale as N115DH
U-1208	Vampire T.55	868	stored, l/n 8-91	to Altenrhein as HB-RVF
U-1210	Vampire T.55	870	stored, l/n 7-91	to Rennes as F-AZHU
U-1212	Vampire T.55	972	stored, l/n 6-91	to Eskelstuna as SE-DXT
U-1213	Vampire T.55	973	stored, l/n 7-91	to Lancaster as N935HW
U-1214	Vampire T.55	974	stored, l/n 6-91	to Southampton as G-DHVV
U-1215	Vampire T.55	975	stored, l/n 8-91	to Bournemouth as G-HELV ('215')
U-1216	Vampire T.55	976	stored, l/n 5-90	to Boscombe Down as ZH653
U-1217	Vampire T.55	977	stored, l/n 6-91	to Sola
U-1218	Vampire T.55	978	stored, l/n 6-91	to Monte Tamaro
U-1219	Vampire T.55	979	stored, l/n 6-91	to Southampton as G-DHWW
U-1220	Vampire T.55	980	stored, l/n 7-91	to Miami as N391RH
U-1221	Vampire T.55	981	stored, l/n 5-90	to Ljungbyhed as SE-DXV
U-1223	Vampire T.55	983	stored, l/n 6-91	to Rennes as F-AZHV
U-1224	Vampire T.55	984	stored, l/n 8-89	to Dübendorf
U-1225	Vampire T.55	985	stored, l/n 9-91	to Altenrhein as HB-RVM
U-1226	Vampire T.55	986	stored, l/n 6-91	to Florence as N593RH
U-1227	Vampire T.55	987	stored, l/n 6-91	to Nîmes Garons
U-1228	Vampire T.55	988	stored, l/n 8-91	to Samedan as HB-RVJ
U-1229	Vampire T.55	989	stored, l/n 8-91	to Cannes as F-AZGU
U-1230	Vampire T.55	990	stored, l/n 7-91	to Southampton as G-DHZZ
U-1232	Vampire T.55	..	stored, l/n 10-89	to Aarburg
U-1233	Vampire T.55	DH37	stored, l/n 9-91	to Millau as HB-RVK
U-1234	Vampire T.55	84913	stored, l/n 9-91	to North Weald as G-DHAV
U-1236	Vampire T.55	40303	stored, l/n 6-91	to Västerås as SE-DXX
U-1237	Vampire T.55	22277	stored, l/n 8-91	to Charleroi as N4638F
U-1238	Vampire T.55	40279	stored, l/n 6-91	to Eskilstuna as SE-DXU

ST STEPHAN
Preserved at this reserve field is a Hunter which is painted in 'newspaper' colours. It was not noted outside in May 1998.

❑ J-4040	Hunter F58	..	stored, marked as 'J-4015', l/f 26-11-93		11-93

STANS
This base is also sometimes referred to as Buochs. Of the stored Hunters J-4010, should have moved to <u>Wangen</u>, although some parts of this aircraft were noted in a local camp site. Stored J-4032, J-4082, J-4201, J-4203 and J-4206 have all gone to <u>Raron</u>. The P2-05 should be inside the Pilatus factory. The exact location of

SWITZERLAND - 444

the Venom is unknown.
☐ U-131	P2-05	21	stored, ex A-101, ex U-101, ex HB-GAB	8-90
☐ J-1596	Venom FB.50	806	preserved, in town ?	4-89
☐ R-2105	Mirage 3RS	..	stored	6-97

ULRICHEN
The local aero club may still have their Venom FB.54.
| ☐ J-1776 | Venom FB.54 | .. | preserved | 4-89 |

VILLMERGEN
A Venom could still be preserved at Villmergen, some 16km south east of Lenzburg.
| ☐ J-1626 | Venom FB.50 | 836 | preserved | 4-89 |

VOLKETSWIL
Near Dübendorf is the village of Volketswill, preserved here on the parking lot of a place named Volkyland was Vampire J-1144 (653). It was last noted in 1994 and has since disappeared.

WANGEN AN DER AARE
A small military exercise area, along the river some kilometres west of the village, has the wingless Hunter J-4010. Some parts of this aircraft were noted at a campsite near Stans.
| ☐ J-4010 | Hunter F.58 | .. | preserved, ex Stans, ex XE554 | 5-98 |

WIEDLISBACH
Stored dismantled in the area near the village of Wiedlisbach are fife Venoms.
☐ J-1503	Venom FB.50	713	stored	99
☐ J-1535	Venom FB.50	745	stored	98
☐ J-1579	Venom FB.50	789	stored	98
☐ J-1640	Venom FB.50	850	stored	98
☐ J-1643	Venom FB.50	853	stored	98

WINTERTHUR
The Viking in false markings should still be preserved at the Technorama museum.
| ☐ G-AHPB | Viking 1 | 12 | preserved, marked as 'D-BABY' | — |

WOHLEN
A private collector here sold Vampire J-1161 to the museum of Kbely in the Czech republic. He still should have the J-1076.
| ☐ J-1076 | Vampire FB.6 | 987 | preserved | — |

ZÜRICH - KLOTEN
The Bücker and Bf108 inside the terminal of the International Airport are both on loan from the Dübendorf museum. The HM-14 is owned by the Luzern museum. The DC-8 and two dumped aircraft have not been seen for some time and may have been scrapped.
| ☐ A-67 | Bü131B | 80 | preserved, inside terminal | 9-96 |
| ☐ A-201 | Bf108B-2 | .. | preserved, inside terminal | 9-96 |

SWITZERLAND - 445

❏ 9H-ABN	Ce421B	0007	dumped, remains only	12-95
❏ HB-LAY	P166	359	dumped, remains only	12-95
❏ HB-SUR	HM-14	..	preserved, inside terminal	12-95
❏ TU-TCP	DC-8-53	45568	instructional, on fire dump, ex F-BJLB	12-95

TURKEY

Since the first edition of this book our knowledge of the Turkish *Wrecks & Relics* has expanded dramatically. Although Turkey remains one of the few European countries which is not often visited by aviation enthusiasts. Reports from this country are few and far between. That fact that most of the local authorities don't like people nosing about their airfields from outside does not help. Most of the *Wrecks & Relics* sightings are therefore during official base visits.

AKHISAR
At this base only a few *Wrecks & Relics* are present.

☐ 41756	3-756	F-100C	217-16	decoy, ex 54-1756	5-98
☐ 42022	3-022	F-100C	217-283	decoy, ex 54-2022	5-98
☐ 42100	3-100	F-100C	217-361	decoy, ex 54-2100	5-98
☐ 52724	3-724	F-100C	222-16	decoy, ex 55-2724	5-98
☐ 7164	9-164	F-104G	7164	preserved, at gate, ex 2282/WG Navy	5-98
☐ 12605	4-605	F-104G	4005	decoy, ex 61-2605	5-98

ANKARA
Airfield – Akinci: This airfield was formally known as Mürted. As well as a wing of F-16s, the Turkish Aircraft Industry facilities are also here, building F-16s and CN235s.

☐ 6840		AT-11	..	preserved, marked as 46-840	5-96
☐ 19103		CL-13 Sabre 2	3	preserved, special colours	5-96
☐ 0-53392		F-102A	..	preserved, ex 55-3392	5-96
☐ 0-54033		TF-102A	..	decoy, at range, ex 55-4033	3-95
☐ 2873	8-873	CF-104	1173	instructional, ex 104873/CAF	5-96
☐ 7051	4-051	F-104G	7051	preserved, ex 2182/WGAF	9-96
☐	260	F-104G	..	preserved, at officers mess	5-96
☐ 53106	4-106	T-33A	9676	preserved, ex 55-3106	5-96

Airfield – Etimesgut: Of the two Skytrains mentioned in the first edition, 6075 moved to Erkilet, while 6003 disappeared. A 'new' Skytrain is preserved at the airfield and is part of the new museum (see Stop Press).

☐ 6073	073	C-47A	19529	preserved, ex 43-15063	5-98
☐ 246		Viscount 794	246	dumped, ex TC-SEC	11-90

BALIKESIR
Town: Preserved in the centre (Kamel Ataturk park) of this town is an F-104G.

☐ 110011	9-011	F-84G	..	preserved, ex 51-10011	5-98
☐ 7122	9-122	F-104G	7122	preserved, ex 2244/WGAF	5-98
☐ 8090	9-090	F-104G	8090	preserved, ex D-8090/RNethAF	5-98

Airfield: That the 9th Wing used to fly the Starfighter is obvious, due to the large amount of stored 104s with the Wing's 9-codes. The two dual F-104s mentioned in the first edition, 5812 and 5814, both went to Eskisehir, while the third, 12263 (5508) has not been heard of since October 1988. Ex Dutch F-104G 6667//9-667 went to the new museum at Eskisehir (see Stop Press).

☐ 6865	9-865	AT-11	..	preserved, at gate	5-98
☐ 10531	5-531	F-5A	N6192	preserved, ex Vietnam AF, ex 1216/Taiwan AF	5-98
☐ 37214	9-214	F-84F	..	preserved, at gate, ex 53-7214	5-98
☐ ..966	9-966	F-84G	..	preserved	5-98
☐ 2045	9-045	F-104G	2045	preserved, ex 2038/WGAF	5-98

TURKEY - 447

☐ 2056	9-056	F-104G	2056	stored, ex 2048/WGAF	5-98
☐ 6622	9-622	F-104G	6622	preserved, at gate, ex 2101/WGAF	5-98
☐ 6652	9-652	F-104G	6652	dumped, tail from 6656, ex D-6652/RNethAF	4-94
☐ 6653	9-653	F-104G	6653	stored, ex D-6653/RNethAF	5-98
☐ 8049	9-049	F-104G	8049	dumped, ex D-8049/RNethAF	10-94
☐ 8058	9-058	RF-104G	8058	dumped, ex D-8058/RNethAF	4-94
☐ 8060	9-060	F-104G	8060	stored, ex D-8060/RNethAF	5-98
☐ 8089	9-089	F-104G	8089	stored, ex D-8089/RNethAF	4-94
☐ 8109	9-109	RF-104G	8109	stored, ex D-8109/RNethAF	4-94
☐ 8115	9-115	RF-104G	8115	dumped, ex D-8115/RNethAF	4-94
☐ 8273	9-273	RF-104G	8273	stored, ex D-8273/RNethAF	8-89
☐ 8324	9-324	F-104G	8324	dumped, ex D-8324/RNethAF	8-90
☐ 8342	9-342	F-104G	8342	dumped, ex D-8342/RNethAF	8-90
☐ 9039	9-039	F-104G	9039	dumped, ex preserved, tail from 8130, ex FX-17	4-94
☐ 9044	9-044	F-104G	9044	dumped, ex FX-19/Belgium AF	8-89
☐ 9051	9-051	F-104G	9051	dumped, ex FX-23/Belgium AF	8-89
☐ 9145 ?		F-104G	9145 ?	preserved, tiger colors	5-98
☐ 12233	9-233	F-104G	4035	dumped, ex 62-12233	8-90
☐ 21536	2-536	T-33AN	536	preserved, ex 21536/RCAF	5-98
☐	9-323	T-33A	..	dumped	4-94

BANDIRMA
Bandirma is home to a wing of F-16s. Two of the wing's instructional Starfighters have left, 7064 (with tail from 7188) went to Eskishir, while 8082 went to Yeni Foca for scrapping. The preserved AT-11 6833/6-833 left for Ankara (see Stop Press).

☐ 21158	6-158	F-5A	N6347	preserved, at gate, ex 1521/Taiwan, ex 67-21158	10-94
☐ 110001	6-001	F-84G	..	preserved, at gate, serial not 100%	2-96
☐ 37040	6-040	F-84F	..	preserved, at gate, unmarked, ex 53-7040	10-94
☐ 7186	6-186	F-104G	7186	preserved, at gate, unmarked, ex 2303/WGAF	10-96
☐ 8164	6-164	F-104G	8164	preserved	1-97
☐ 9050	4-050	F-104G	9050	preserved, ex FX-22/BAF	10-92
☐ 9064	6-064	F-104G	9064	instructional, ex FX-30/BAF	5-95
☐ 9082	6-082	F-104G	9082	preserved, ex FX-27/BAF	6-93
☐ 04432	6-432	T-33A	5528	preserved, white c/s, ex 55-4432/French AF	10-96

CEVEZILI
Along road D-100 (from Istanbul to Izmit) is the town of Cevezili, an unidentified Super Sabre is preserved.

☐ ..		F-100C	..	preserved	3-95

CURLU
Mounted at the gate of this airfield is a former Eskisehir Starfighter.

☐ 59415	9-415	TF-104G	5529	preserved, ex Eskisehir, ex CE.8-23/SpaAF	—

DIYARBAKIR
This most south eastern airfield is close to the Iraq border and is one of the country's five F-16 based. Beside the normal amount of preserved aircraft, a large number of Starfighters are still on the base. The preserved F-84F 28816 went to Istanbul.

☐ 6923	8-923	AT-11	..	preserved, near tower	4-94
☐ 111218	8-218	F-84G	..	preserved, ex 51-11218	4-94
☐ 52763	8-763	F-100D	224-30	preserved, ex 55-2763	4-94

TURKEY - 448

☐ 2066	8-066	F-104G	2066	stored, ex 2057/WGAF	4-94
☐ 2087	8-087	F-104G	2087	instructional, ex 2074/WGAF	4-94
☐ 2638	8-638	CF-104D	5308	stored, ex 104638/CAF	4-94
☐ 5703	8-703	TF-104G	5703	stored, ex 2702/WGAF, ex 61-3032	4-94
☐ 5704	8-704	TF-104G	5704	stored, ex 2703/WGAF, ex 61-3033	4-94
☐ 5722	8-722	TF-104G	5722	stored, ex 2721/WGAF, ex 61-3051	4-94
☐ 5723	8-723	TF-104G	5723	stored, ex 2722/WGAF, ex 61-3052	4-94
☐ 5911	8-911	TF-104G	5911	stored, ex 2782/WGAF	4-94
☐ 5945	8-945	TF-104G	5945	stored, ex 2815/WG Navy	4-94
☐ 7037	8-037	F-104G	7037	stored, ex 2168/WGAF	4-94
☐ 7081	8-081	F-104G	7081	stored, ex 2211/WG Navy	4-94
☐ 7108	8-108	F-104G	7108	preserved, ex 2230/WG Navy	4-94
☐ 7432	8-432	F-104G	7432	stored, ex 2686/WG Navy	4-94
☐ 8128	8-128	F-104G	8128	preserved, ex Eskisehir, ex 2399/WGAF	4-94
☐ 8302	8-302	F-104G	8302	stored, ex 2526/WGAF	4-94
☐ 8303	8-303	F-104G	8303	stored, ex 2527/WGAF	4-94
☐ 9135	8-135	F-104G	9135	stored, ex 2607/WGAF	4-94
☐ 9185	8-185	F-104G	9185	preserved, ex 2633/WGAF	8-93
☐ 62-713	8-713	CF-104	1013	preserved, ex 104713/CAF	4-94
☐ 62-865	8-865	CF-104	1165	stored, ex 104865/CAF	4-94
☐ 62-883	8-883	CF-104	1183	stored, ex 104883/CAF	4-94
☐ 65-661	8-661	CF-104D	5331	stored, ex 104661/CAF	4-94
☐ 104788	8-788	CF-104	1088	stored, with tail from 2796, ex 104788/CAF	4-94
☐ 104847	8-847	CF-104	1147	stored, ex 104847/CAF	4-94

ERHAC
At the Phantom base of Erhac relatively few *Wrecks & Relics* are known to exist.

☐ 6904	7-904	AT-11	..	preserved	4-94
☐ 73-1016	7-016	F-4E	4525	instructional, wreck	4-94
☐ ..	7-808	F-84F	..	preserved, maybe 52-8808 ?	4-94
☐ 52782	7-782	F-100D	224-49	preserved, at gate, ex G-782//RDanAF	4-94
☐ 7073	7-073	F-104G	7073	instructional, ex 2203/WGAF	4-94
☐ 62-739	8-739	CF-104	1039	instructional, ex 104739/CAF	4-94

ERKILET
The airfield of Erkilet (8km south of Kayseri) houses a large number of stored Skytrains. Only two are known to have left; 6025 to <u>Kayseri</u> and 6032/032 to <u>Eskisehir</u>.

☐ 6023	023	C-47A	19595	stored, ex 43-15129	4-94
☐ 6037	037	C-47A	12910	stored, ex 42-93042	4-94
☐ 6038	038	C-47A	19565	stored, ex 43-15099	4-94
☐ 6042	042	C-47A	12949	stored, ex 42-93077	4-94
☐ 6046	12-046	C-47A	20117	stored, ex 43-15651	2-96
☐ 6048	048	C-47A	19516	stored, ex 43-15050	4-94
☐ 6054	054	C-47A	12713	stored, ex 42-92865	4-94
☐ 6056	056	C-47A	19341	stored, ex 42-100878	4-94
☐ 6059	12-059	C-47A	12757	stored, ex 42-92905	2-96
☐ 6068	H-068	C-47A	13880	stored, ex 43-30729	1-96
☐ 6075	12-075	C-47A	19357	stored, ex Ankara, ex 42-100894	2-96
☐ 6076	E-076	C-47A	12771	stored, ex 42-92917	2-96
☐ 6079	H-079	C-47A	19644	dumped, ex 43-15178	4-94
☐ 6096	ETI-96	C-47A	12955	stored, ex 42-93083	4-94
☐ 691		C-47B	26952	stored, ex Libya AF, ex 43-49691	4-94

TURKEY - 449

☐ 149850	850	S-2E	125C	stored	2-96
☐ 151663	663	S-2E	196C	stored	2-96
☐ 151668	668	S-2E	201C	stored	2-96

ESKISEHIR
Town: The university in town used a Caravelle as instructional airframe.

☐ TC-ASA	SE210-10B1R	222	instructional, ex HB-ICQ, ex F-BNRB	12-92

Airfield: Of the large number of stored Starfighters, some are known to have found new 'homes'. F-104s 2072, 5801, 7165, 7302, 8290 and 6906 went to <u>Yeni Foca</u> for scrapping, 8128 has gone to <u>Diyarbakir</u> and 6895 and 62-733 both went to <u>Istanbul</u>. RF-4E 69-7465/1-7465, RF-84F 8733/733, F-104G 7190/9-190 and T-33AN 21321/2-321 went to the new museum in town, while RF-4E 69-7503/1-7503, RF-84F 11924/1-924 and F-104G 8205/9-205 have gone to the mew museum at <u>Ankara</u> (see Stop Press).

☐ 6933	1-933	AT-11	..	preserved, ex 42-37585	5-98
☐ 68-0348	7-348	F-4E	3402	stored, wreck	3-95
☐ 69-7448		F-4E	3861	stored, for spares, ex 9801/WGAF	5-96
☐ 69-7452	1-7452	RF-4E	3905	stored, for spares, ex 3505/WGAF	5-96
☐ 69-7467	1-7467	RF-4E	4048	stored, for spares, ex 3520/WGAF	5-96
☐ 69-7478	1-7478	RF-4E	4081	preserved, ex 3531/WGAF	5-98
☐ 69-7483	1-7483	RF-4E	4094	stored, for spares, ex 3536/WGAF	5-96
☐ 69-7497	1-7497	RF-4E	4121	stored, for spares, ex 3550/WGAF	5-96
☐ 69-7510	1-7510	RF-4E	4145	stored, for spares, ex 3563/WGAF	6-95
☐ 69-7513	1-7513	RF-4E	4152	stored, for spares, ex 3566/WGAF	5-96
☐ 69-7515	1-7515	RF-4E	4156	stored, for spares, ex 3568/WGAF	5-98
☐	5-571	F-5A	..	preserved	5-98
☐ 7234	1-7234	RF-84F	..	preserved, ex 52-7234/FAF	4-97
☐ 8758	8758	RF-84F	..	stored, ex 52-8758	3-95
☐ 8759	759	RF-84F	..	stored, ex 52-8759	3-95
☐ 8764	764	RF-84F	..	stored, ex 52-8764	5-98
☐ 110987	1-10987	F-84G	..	preserved, ex 51-10987	5-98
☐ 111019	11019	F-84G	..	preserved, ex 51-11019	4-95
☐ 27123 ?		F-84F	..	preserved, in Thunderbirds c/s as 7123	2-96
☐ 27142		F-84F	..	preserved, in Thunderbirds c/s	2-96
☐ 27153 ?		F-84F	..	preserved, in Thunderbirds c/s as 7153	10-93
☐ 27301	301	RF-84F	..	stored, ex 52-7301/FAF	3-95
☐ 28722	8722	RF-84F	..	stored, ex 52-8722/FAF	3-95
☐ 41946	3-946	F-100C	217-207	preserved, ex 54-1946	5-96
☐ 52775	1-775	F-100D	224-42	preserved, ex G-775/RDanAF, ex 55-2775	5-98
☐ 52779	1-779	F-100D	224-46	preserved, ex G-779/RDanAF, ex 55-2779	7-92
☐ 55-2795		F-100D	224-62	preserved	5-96
☐ 2001	9-001	F-104G	2001	stored, CFE a/c, ex 2001/WGAF	6-95
☐ 2080	9-080	F-104G	2080	stored, CFE a/c, ex 2068/WGAF	6-95
☐ 2089	9-089	F-104G	2089	stored, ex 2076/WGAF	8-93
☐ 2837	8-837	CF-104	1137	stored, near hangars, ex 104837/CAF	3-95
☐ 5702	9-702	TF-104G	5702	stored, ex D-5702/RNethAF	5-90
☐ 5728	9-728	TF-104G	5728	stored, CFE a/c, ex 2727/WGAF, ex 61-3057	6-95
☐ 5740	8-740	TF-104G	5740	stored, near hangars, ex 2738/WGAF	3-95
☐ 5741	8-741	TF-104G	5741	stored, ex 2739/WGAF	3-95
☐ 5812	9-812	TF-104G	5812	stored, ex Balikesir, ex D-5812/RNethAF	10-89
☐ 5814	9-814	TF-104G	5814	stored, ex Balikesir, ex D-5814/RNethAF	8-89
☐ 5817	9-817	TF-104G	5817	stored, CFE a/c, ex D-5817/RNethAF	6-95
☐ 5914	9-914	TF-104G	5914	stored, CFE a/c, ex 2785/WGAF	6-95
☐ 6859	9-859	F-104S	1159	stored, near hangars	6-95
☐ 6862	9-862	F-104S	1162	stored, near hangars	2-96

TURKEY - 450

☐ 6868	9-868	F-104S	1168	stored, near hangars	5-96
☐ 6882	9-882	F-104S	1182	stored, CFE a/c	6-95
☐ 6885	9-885	F-104S	1185	stored, CFE a/c	6-95
☐ 6893	9-893	F-104S	1193	stored, CFE a/c	6-95
☐ 6897	9-897	F-104S	1197	stored, CFE a/c	6-95
☐ 6900	9-900	F-104S	1200	stored, CFE a/c	6-95
☐ 7005	8-005	F-104G	7005	stored, near hangars, ex 2137/WGAF	6-95
☐ 7012	4-012	F-104G	7012	stored, CFE a/c, ex 2144/WGAF	6-95
☐ 7034	9-034	F-104G	7034	stored, CFE a/c, ex 2165/WGAF	6-95
☐ 7047	6-047	F-104G	7047	stored, ex 2178/WGAF	8-93
☐ 7050	4-050	F-104G	7050	stored, CFE a/c, ex 2181/WGAF	6-95
☐ 7064	9-064	F-104G	7064	stored, CFE a/c, tail from 7188, ex 2195/WGAF	6-95
☐ 7068	9-068	F-104G	7068	stored, ex 2199/WGAF	8-93
☐ 7083	6-083	F-104G	7083	stored, CFE a/c, ex 2213/WG Navy	6-95
☐ 7089	6-089	F-104G	7089	stored, CFE a/c, ex 2218/WG Navy	6-95
☐ 7125	9-125	F-104G	7125	stored, ex 2247/WGAF	8-93
☐ 7130	9-130	F-104G	7130	stored, ex 2250/WGAF	8-93
☐ 7154	9-154	F-104G	7154	stored, ex 2273/WG Navy	8-93
☐ 7171	9-171	F-104G	7171	stored, CFE a/c, ex 2288/WG Navy	6-95
☐ 7178	9-178	F-104G	7178	stored, CFE a/c, ex 2295/WG Navy	6-95
☐ 7179	9-179	F-104G	7179	stored, CFE a/c, ex 2296/WG Navy	6-95
☐ 7185	6-185	F-104G	7185	stored, CFE a/c, ex 2302/WG Navy	6-95
☐ 7188	6-188	F-104G	7188	stored, CFE a/c, ex 2305/WG Navy	6-95
☐ 7189	9-189	F-104G	7189	stored, ex 2306/WG Navy	3-95
☐ 7209	9-209	F-104G	7209	stored, CFE a/c, ex 2325/WG Navy	6-95
☐ 7210	6-210	F-104G	7210	stored, CFE a/c, ex 2326/WG Navy	6-95
☐ 7304	8-304	F-104G	7304	stored, near hangars, ex 2644/WG Navy	5-96
☐ 7431	6-431	F-104G	7431	stored, ex 2685/WG Navy	8-93
☐ 7436	4-436	F-104G	7436	stored, ex 2690/WGAF	8-93
☐ 8105	4-105	RF-104G	8105	stored, ex D-8105/RNethAF	8-93
☐ 8116	8-116	F-104G	8116	stored, near hangars, ex 2395/WGAF	5-96
☐ 8144	9-144	F-104G	8144	stored, CFE a/c, ex 2406/WGAF	6-95
☐ 8163	9-163	F-104G	8163	stored, CFE a/c, ex 2421/WGAF	6-95
☐ 8185	9-185	F-104G	8185	stored, CFE a/c, ex 2442/WGAF	6-95
☐ 8223	9-223	F-104G	8223	stored, CFE a/c, ex 2474/WGAF	6-95
☐ 8289	8-289	F-104G	8289	stored, CFE a/c, ex 2518/WGAF	6-95
☐ 8310	9-310	F-104G	8310	stored, CFE a/c, ex 2532/WGAF	6-95
☐ 8316	9-316	F-104G	8316	stored, CFE a/c, ex 2535/WGAF	6-95
☐ 8339	8-339	F-104G	8339	stored, CFE a/c, ex 2547/WGAF	6-95
☐ 8340	9-340	F-104G	8340	stored, CFE a/c, ex 2548/WGAF	6-95
☐ 9071	9-071	F-104G	9071	stored, ex FX-34/Belgium AF	8-93
☐ 9128	9-128	F-104G	9128	stored, CFE a/c, ex 2603/WGAF	6-95
☐ 9130	9-130	F-104G	9130	stored, CFE a/c, ex 2605/WGAF	6-95
☐ 9150	6-150	F-104G	9150	stored, CFE a/c, ex 2612/WGAF	6-95
☐ 9156	9-156	F-104G	9156	stored, CFE a/c, ex 2615/WGAF	6-95
☐ 9186	9-186	F-104G	9186	stored, ex 2634/WGAF	8-93
☐ 13027		TF-104G	5503	stored, ex 61-3027	8-89
☐ 13028	9-028	TF-104G	5504	stored, ex 61-3028	8-89
☐ 13029	9-029	TF-104G	5505	stored, ex 61-3029	5-90
☐ 212279	9-279	TF-104G	5524	stored, CFE a/c, ex CE.8-22, ex CE.8-2/SpaAF	6-95
☐ 62-737	8-737	CF-104	1037	stored, CFE a/c, ex 104737/CAF	6-95
☐ 62-747	8-747	CF-104	1047	stored, CFE a/c, ex 104747/CAF	6-95
☐ 62-3751	8-751	CF-104	1051	preserved, ex 104751/CAF	5-98
☐ 14228	4-228	T-33A	5522	stored, ex 51-4228	5-98
☐ 14511	8-511	T-33A	5806	stored, ex M-1/RNethAF, ex 51-4511	5-96

TURKEY - 451

☐ 16764	8-764	T-33A	6096	stored, ex 51-6764	5-96
☐ 16772	4-772	T-33A	6104	stored, ex 51-6772	5-98
☐ 16773	3-773	T-33A	6105	stored, ex 51-6773	5-96
☐ 17491	4-491	T-33A	7471	stored, ex M-21/RNethAF, ex 51-17491	5-96
☐ 17517	7-517	T-33A	7577	stored, ex 51-17517	5-96
☐ 17541	5-541	T-33A	7686	stored, ex M-28/RNethAF, ex 51-17541	5-96
☐ 18931	9-931	T-33A	6715	stored, ex 51-8931	5-96
☐ 19250	4-250	T-33A	7034	stored, ex 51-9250	5-96
☐ 21030	2-030	T-33AN	030	stored, ex 21030/RCAF	4-95
☐ 21309	2-309	T-33AN	309	stored, ex 21309/RCAF	4-95
☐ 21321	2-321	T-33AN	321	stored, ex 21321/RCAF	5-96
☐ 21407	2-407	T-33AN	407	stored, ex 21407/RCAF	5-96
☐ 21410	2-410	T-33AN	410	preserved, ex 21410/RCAF	5-98
☐ 21461	2-461	T-33AN	461	stored, ex 21461/RCAF	6-95
☐ 21493		T-33AN	493	stored, ex 21493/RCAF	4-95
☐ 29857	3-857	T-33A	8163	stored, ex 52-9857	5-96
☐ 35742	6-742	T-33A	9081	stored, ex 53-5742	5-96
☐ 35747	6-747	T-33A	9086	stored, ex 53-5747	5-96
☐ 41543	8-543	RT-33A	9174	stored, ex 54-1543/FAF	5-96
☐ 41548	1-548	RT-33A	9179	preserved, ex 54-1548/FAF	5-98
☐ 53095	7-095	T-33A	9636	stored, ex 55-3095	5-96
☐ 53096	5-096	T-33A	9637	stored, ex 55-3096	5-96
☐ 53108	7-108	T-33A	9678	stored, ex 55-3108	5-96
☐ 54961	5-961	T-33A	9921	stored, ex M-39/RNethAF, ex 55-4961	4-95
☐ 5327		T-33A	8666	stored, ex 53-5327	4-95
☐ 80454	1-454	T-33A	1423	stored, ex 9484/WGAF, ex 58-0454	5-96
☐ 80640	1-640	T-33A	1609	stored, ex 9507/WGAF, ex 58-0640	5-96
☐ 9866	4-866	T-33A	8172	stored, ex 52-9866	4-95

Airfield - Anadolu: The Alouette 2 is from the Polis Havacilik Daire Baskanligi, not to be confused with the other police force, the Jandarma.

☐ 6091	TK-91	C-47A	19602	dumped, ex 43-15136	4-94
☐ 31732	3-732	F-100C	214-24	preserved, ex 53-1732	4-94
☐ 12733	4-733	F-104G	6085	preserved, ex 63-12733	4-94
☐ 8298	8-298	F-104G	8298	preserved, ex 2523/WGAF	4-94
☐ E-2120		Alouette 2	2120	stored, ex 7765/WG Army	4-94
☐ 431 ?		Viscount 794D	..	instructional	4-94
☐ 40x		F-104G	..	stored	4-94

GÜVERCINLIK

Güvercinlik is the Turkish army main base, with a regiment of AH-1 Cobras and a large number of training and communication aircraft and helicopters. O-1G 10160 is marked as such on it's construction plate.

☐ ..		AB204	..	preserved, marked as '901'	5-98
☐ 10016		Do28D-2	4081	preserved, base museum, ex 5806/WGAF	5-98
☐ 10101		Citabria	..	preserved, base museum	5-98
☐ 10160		O-1G	..	preserved, base museum	5-98
☐ 10395		OH-13S	3946	preserved, base museum, ex 67-15897	5-98
☐ 10437		TH-13T	3478	preserved, ex 65-8045	5-95
☐ 11032		AB204B	3179	preserved, base museum	5-98
☐ 11364		O-1E	..	stored	4-95
☐ 12262		O-1E	..	stored	4-95
☐ 13168		O-1E	..	stored	4-95
☐ 14064		O-1E	..	stored	4-95
☐ ..		U-17	A185E-01699	instructional, unmarked	5-98

INCIRLIK
This airfield is mainly used by the Americans and NATO. Both dumped aircraft were used by the USAF for battle damage repair training and may both have gone by now.

❏ 63-7474		F-4C	465	dumped, ex instructional, ex USAF	9-95
❏ 62-4423 ?		F-105G	..	dumped, ex instructional, ex USAF	12-92

ISTANBUL
Town: In the town is the Askeri Muzeri Istanbul or the Istanbul Army Museum. The Starfighter came from the nearby museum at Yesilköy (and before that from Eskisehir).

❏ 62-733	8-733	CF-104	1033	preserved, ex Yesilköy, ex 104733/CAF	5-98
❏ 10397		OH-13S	3959	preserved	5-98

An industrial museum in town, the Rahmi Koc Museum, has yet another F-104.

❏ 6895	9-895	F-104S	1195	preserved, ex Eskisehir	10-96

The Turkish Air Force academy, the Hava Harp Okulu, is located along the Sahil Yolu road. It seems that the two former Belgium F-104s have been replaced by two ex German Starfighters. Preserved Sabre 2 19190/190 move to Ankara.

❏ 28816	816	F-84F	..	preserved, ex 52-8816, ex Diyarbakir	5-98
❏ 42059	E-059	F-100C	217-330	preserved, ex 54-2059	5-98
❏ 078		F-104G	9078	preserved, ex FX-38/Belgium AF	5-98
❏ 083		F-104G	9083	preserved, ex FX-40/Belgium AF	5-98
❏ 8277	8-277	F-104G	8277	preserved, ex 2513/WGAF	5-98
❏ 8299	8-299	F-104G	8299	preserved, ex 2524/WGAF	5-98

Airfield – Yesilköy: The main aviation museum, the Havacilik Muzezi, is at the military side of the airfield of Istanbul. The museum is open Wednesday to Sunday, between 09:00 and 12:00 and 13:00 and 17:00. TF-102A went to the new museum at Ankara.

❏ 6930	9-930	AT-11	4561	preserved, ex 42-37565	5-98
❏ 6008	H-008	C-47B	15011/26456	preserved, ex 43-49195	5-98
❏ 6052	YSL-52	C-47A	13877	preserved, ex 43-30726	5-98
❏ 10683	ETI-683	C-54D	10788	preserved, ex 42-72683	5-98
❏ 19207	207	CL-13 Sabre 2	107	preserved	5-98
❏ 19268	268	CL-13 Sabre 2	168	preserved	9-97
❏ 14460		F-5A	N6314	preserved, ex Vietnam, ex 1242/Taiwan AF	5-98
❏ 3070	3-070	NF-5A	3070	preserved, ex K-3070/RNethAF	5-98
❏ 97147	5-147	RF-5A	RF1027	preserved, ex 69-7147	5-98
❏ 19953	953	F-84G	..	preserved, ex 51-9953	5-98
❏ 110572		F-84G	..	preserved, special c/s, ex 51-10572	5-98
❏ 28941	941	F-84F	..	preserved, ex 52-8941	5-98
❏ 11901	901	RF-84F	..	preserved, ex 51-1901	5-98
❏ 11917	1917	RF-84F	..	preserved, ex 51-1917	5-98
❏ 42089	3-089	F-100C	217-350	preserved, ex 54-2089	5-98
❏ 42245	E-245	F-100D	223-125	preserved, ex 54-2245	5-98
❏ 63788	3-788	F-100F	243-64	preserved, ex Konya, ex 56-3788	5-98
❏ 53386	386	F-102A	..	preserved, ex 55-3386	5-98
❏ 12344	4-344	F-104G	6043	preserved, on pole, ex 62-12344	5-98
❏ 12619	4-619	F-104G	4019	preserved, tail from 8223, ex 61-2619	5-98
❏ 5725	9-725	TF-104G	5725	preserved, ex 2724/WGAF, ex 61-3043	5-98
❏ 52-7577		UH-19B	..	preserved	9-97
❏ TC-21		P-47D	..	preserved, ex 44-33712	9-97
❏ 7504	04	T-6G	182-266	preserved, ex 51-14579	7-96
❏ 35744	5744	T-33A	9083	preserved, ex 53-5744	9-97
❏ 24220	2-220	T-34A	34-24	preserved, ex 54-5220	9-97

TURKEY - 453

☐ 430	Viscount 794D	430	preserved, ex TC-SEL		5-98
☐ 10012	Do28B-1	3079	stored, ex D-IBON		5-98
☐ 10013	Do28B-2	3083	stored		5-98
☐ 10020	Do28D-1	4021	stored, ex D-IBBA		5-98
☐ 10022	Do28D-2	4106	preserved, ex 5831/WGAF		5-98
☐ 10293	Do27H-2	2142	preserved, ex D-EFCI		7-96
☐ 10294	Do27H-2	2114	preserved, ex D-ECPI		7-96
☐ 10306	L-18C	..	preserved		5-98
☐ 60	M14 Magister 1	..	preserved		7-96
☐ 77	M14 Magister 1	..	preserved		5-98
☐ 2015	PZL P24G	..	preserved		5-98
☐ 5824	Bergfalcke II	..	preserved, glider		5-98
☐ TC-TK-15	CW.22B Falcon	..	preserved, ex 15-13571		5-98
☐ TC-ABA	SE210-10B1R	253	preserved, ex HB-ICN		5-98
☐ TC-EAF	SIAT 223K-1	026	preserved, ex 023/EgyptAF, ex D-EABU		5-98
☐ TC-KUJ	MKEK 4 Ugur	5144	preserved, marked as '5144', ex 'TC-KUS'		5-98
☐ ..	DH.89A	..	preserved, marked as 'TC-ERK'		5-98
☐ ..	Grigorovich M5	..	preserved		5-98
☐ ..	Mavi Isik G	..	preserved		5-98

At the civil side, stored A300B4-203 TC-ALP (153) was burned out on 17th May 1996 and was scrapped by the end of the year. Scrapped in 1996 was B727-51 TC-AJZ (18802). It used to be stored by the museum, but was in too bad shape to be restored.

☐ TC-ALI		C-47A	12830	stored, ex 71288/Yugoslav AF, ex N62DN	11-97

IZMIR
Town: In a public park at the town of Izmir a Starfighter and T-33 are mounted on poles.

☐ 12316	4-316	F-104G	6015	preserved, ex 62-12316, with tail from 104733	11-97
☐ 4952	2-952	T-33A	9912	preserved, ex M-41/RNethAF, ex 55-4952	11-97

Along the main road through town towards Cigli is a Starfighter preserved.

☐ 62-786	8-786	CF-104	1086	preserved, on pole, ex 104786/CAF	5-98

Airfield - Çigli: Izmir has a number of airfields. At Çigli jet training is carried out by the 2nci AJU. Of the 'T-Birds' mentioned in the first edition none has survived into the 1990s. The UH-1 is unmarked and a plate in the cockpit reeds '12667'. In May 1996 two F-5s were noted on the dump.

☐ 6033	E-033	C-47A	12950	instructional, ex 42-93078	5-96
☐ 97145	.-145	RF-5A	RF1025	dumped, ex 69-7145, serial not 100% sure	4-94
☐ ..		UH-1D	..	instructional, see note	5-96
☐ 14087 ?		T-33A	..	dumped	4-94
☐ 21058	2-058	T-33AN	058	preserved, ex 21058/RCAF	5-98
☐ 21066	2-066	T-33AN	066	preserved, ex 21066/RCAF	4-94
☐ 21436	2-436	T-33AN	436	preserved, marked as '21221', ex 21436/RCAF	5-96
☐ 54952	2-952	T-33A	9912	preserved, ex M-41/RNethAF, ex 55-4952	4-94
☐ 24201	2-201	T-34A	34-4	preserved, ex 54-5201	2-96
☐ 13221		T-38A	N5650	stored, ex 64-13221	5-96
☐ 3823.		T-38A	..	dumped, maybe 38233 (N5580)	5-96

Airfield – Gaziemir: On the southern side of Izmir is the airfield of Gaziemir. Beside based CN235s, there is a large military technical school. Also from here an aircraft left for one of the new museums, F-100C 41766/E-766 went to Ankara.

☐ 6057	E-057	C-47A	25711	instructional, ex 43-48450	4-94
☐ 9211	E-211	F-5A	N7034	instructional, ex 66-9211	8-90
☐ 6876	867	F-84F	..	instructional, ex 52-6876	8-90

TURKEY - 454

❑ 7196	196	F-84F	..	preserved, ex 52-7196	8-90
❑ 11185	FS-185	F-84G	..	instructional, ex 51-1185	8-90
❑ 27299		RF-84F	..	decoy, ex 52-7299/FAF	8-90
❑ 54-2059	E-059	F-100C	..	instructional, to Istanbul	8-90
❑ 54-2172	E-172	F-100D	223-52	preserved	8-90
❑ 42202	E-202	F-100D	223-82	instructional, ex 54-2202	8-90
❑ 54-2245	E-245	F-100D	223-125	instructional, to Istanbul	8-90
❑ 53413		F-102A	..	dumped, ex 53-3413	8-90
❑ 7161	E-161	F-104G	7161	instructional, ex 2279/WG Navy	8-90
❑ 7758	E-758	F-104G	6130	instructional, ex 758/RNoAF, ex 64-17758	8-90
❑ 12620	E-620	F-104G	4020	instructional, ex 61-2620	8-90
❑ 12633	E-633	F-104G	4033	instructional, ex 633/RNoAF, ex 61-2633	8-90
❑ 22324	E-324	F-104G	6023	instructional, ex 62-12324	8-90
❑ 51-17551	E-551	T-33A	7696	instructional, ex 51-17551	8-90

KAYSERI
Noted at the local airfield at the north side of this town, which itself is near the airfield of Erkilet, is a former Erkilet Skytrain.

❑ 6025	H-025	C-47A	19531	preserved, ex Erkilet, ex 43-15065	8-93

KONYA
Konya is nowadays home base for F-4 Phantoms and F-5 Freedom Fighters.

❑ 6833	3-833	AT-11	..	preserved, see Bandirma	4-97
❑ 68-0313	3-313	F-4E	3335	preserved	4-97
❑ ..	'3-133'	F-84G	..	preserved	4-95
❑ ..	'3-131'	F-84F	..	preserved, ex 52-7023 or 52-7025 ?	4-97
❑ 42013	'3-132'	F-100C	217-274	preserved, ex 54-2013	4-95
❑ 52910	3-910	F-100D	224-177	instructional, ex 55-2910	4-97
❑ 63921	3-921	F-100F	243-197	preserved, ex 56-3921	4-97
❑ 29922	3-922	T-33A	7893	preserved, ex M-27/RNethAF, ex 52-9922	8-97

In April 1994 still some 40 Supe Sabres (half of those listed below) were noted at the airfield. Rumours have it that all of the aircraft have gone by now. Also 52778/3-778 was reported here in August 1989, but this aircraft crashed way back in 1970 while in service with the Danish Air Force. F-100F 63788 went to <u>Istanbul</u> and F-100C 41732 to <u>Eskisehir</u>.

❑ 41741	3-741	F-100C	217-2	dumped, ex 54-1741	8-89
❑ 41749	3-749	F-100C	217-10	dumped, ex 54-1749	1-89
❑ 41782	3-782	F-100C	217-43	dumped, ex 54-1782	7-89
❑ 41788	3-788	F-100C	217-49	dumped, ex 54-1788	7-89
❑ 41794	3-794	F-100C	217-55	dumped, ex 54-1794	7-89
❑ 41798	3-798	F-100C	217-59	dumped, ex 54-1798	8-89
❑ 41800	3-800	F-100C	217-61	dumped, ex 54-1800	1-89
❑ 41805	3-805	F-100C	217-66	dumped, ex 54-1805	1-89
❑ 41818	3-818	F-100C	217-79	dumped, ex 54-1818	8-89
❑ 41835	3-835	F-100C	217-96	dumped, ex 54-1835	1-89
❑ 41845	3-845	F-100C	217-106	dumped, ex 54-1845	8-89
❑ 41872	3-872	F-100C	217-133	dumped, ex 54-1872	8-89
❑ 41878	3-878	F-100C	217-139	dumped, ex 54-1878	8-89
❑ 41883	3-883	F-100C	217-144	dumped, ex 54-1883	8-89
❑ 41942	3-942	F-100C	217-203	dumped, ex 54-1942	1-89
❑ 41944	3-944	F-100C	217-205	dumped, ex 54-1944	8-89
❑ 41945	3-945	F-100C	217-206	dumped, ex 54-1945	8-89

TURKEY - 455

☐ 41949		F-100C	217-210	dumped, ex 54-1949	8-89
☐ 41972	3-972	F-100C	217-233	dumped, ex 54-1972	1-89
☐ 41975	3-975	F-100C	217-236	dumped, ex 54-1975	8-89
☐ 41978	3-978	F-100C	217-239	dumped, ex 54-1978	8-89
☐ 42048	3-048	F-100C	217-309	dumped, ex 54-2048	8-89
☐ 42056	3-056	F-100C	217-317	dumped, ex 54-2056	1-89
☐ 42057	3-057	F-100C	217-318	dumped, ex 54-2057	1-89
☐ 42058	3-058	F-100C	217-319	dumped, ex 54-2058	8-89
☐ 42060	3-060	F-100C	217-321	dumped, ex 54-2060	1-89
☐ 42066	3-066	F-100C	217-327	dumped, ex 54-2066	8-89
☐ 42068	3-068	F-100C	217-329	dumped, ex 54-2068	8-89
☐ 42092	3-092	F-100C	217-353	dumped, ex 54-2092	1-89
☐ 42115	3-115	F-100C	217-376	dumped, ex 54-2115	8-89
☐ 42177	3-177	F-100D	223-57	dumped, ex G-177/RDanAF, ex 54-2177	1-92
☐ 42206		F-100D	223-86	dumped, ex G-206/RDanAF, ex 54-2206	1-92
☐ 42222	3-222	F-100D	223-102	dumped, ex G-222/RDanAF, ex 54-2222	1-92
☐ 42242	3-242	F-100D	223-122	dumped, ex 54-2242	8-89
☐ 42261		F-100D	223-141	dumped, ex G-261/RDanAF, ex 54-2261	1-92
☐ 42266	3-266	F-100D	223-146	dumped, ex G-266/RDanAF, ex 54-2266	1-92
☐ 42275	3-275	F-100D	223-155	dumped, ex 54-2275	1-92
☐ 42276	3-276	F-100D	223-156	dumped, ex 54-2276	8-89
☐ 42303	3-303	F-100D	223-183	dumped, ex G-303/RDanAF, ex 54-2303	1-92
☐ 52744	3-744	F-100D	224-11	dumped, ex 55-2744	8-89
☐ 52748	3-748	F-100D	224-15	dumped, ex G-748/RDanAF, ex 55-2748	1-92
☐ 52752	3-752	F-100D	224-19	dumped, ex 55-2752	8-89
☐ 52768	3-768	F-100D	224-35	dumped, ex G-768/RDanAF, ex 55-2768	1-92
☐ 52769	3-769	F-100D	224-36	dumped, ex G-769/RDanAF, ex 55-2769	1-92
☐ 52771	3-771	F-100D	224-38	dumped, ex G-771/RDanAF, ex 55-2771	1-92
☐ 52798	3-798	F-100D	224-65	dumped, ex 55-2798	8-89
☐ 52874	3-874	F-100D	224-141	dumped, ex 55-2874	8-89
☐ 52916	3-916	F-100D	224-183	dumped, ex 55-2916	89
☐ 52919	3-919	F-100D	224-186	dumped, ex 55-2919	1-89
☐ 53596	3-596	F-100D	223-278	dumped, ex 55-3596	8-89
☐ 53700	3-700	F-100D	223-382	dumped, ex 55-3700	1-89
☐ 63355	3-355	F-100D	245-5	dumped, ex 56-3355	8-89
☐ 63396	3-396	F-100D	245-46	dumped, ex 56-3396	8-89
☐ 63732	3-732	F-100F	243-8	dumped, ex 56-3732	8-89
☐ 63752	3-752	F-100F	243-28	dumped, ex 56-3752	8-89
☐ 63766	3-766	F-100F	243-42	dumped, ex 56-3766	1-89
☐ 63783	3-783	F-100F	243-59	dumped, ex 56-3783	1-89
☐ 63793	3-793	F-100F	243-69	dumped, ex 56-3793	1-89
☐ 63831	3-831	F-100F	243-107	dumped, ex 56-3831	1-89
☐ 63843	3-843	F-100F	243-109	dumped, ex 56-3843	1-89
☐ 63852	3-852	F-100F	243-118	dumped, ex 56-3852	1-89
☐ 63854	3-854	F-100F	243-120	dumped, ex 56-3854	8-89
☐ 63867	3-867	F-100F	243-133	dumped, ex 56-3867	1-89
☐ 63890	3-890	F-100F	243-166	dumped, ex 56-3890	1-89
☐ 63896	3-896	F-100F	243-172	dumped, ex 56-3896	1-89
☐ 63909	3-909	F-100F	243-185	dumped, ex 56-3909	1-89
☐ 63946	3-946	F-100F	243-222	dumped, ex 56-3946	1-89
☐ 63950	3-950	F-100F	243-226	dumped, ex 56-3950	1-89
☐ 63952	3-952	F-100F	243-228	dumped, ex 56-3952	7-89
☐ 63967	3-967	F-100F	243-243	dumped, ex 56-3967	8-89
☐ 63976	3-976	F-100F	243-252	dumped, ex 56-3976	8-89
☐ 86976		F-100F	261-2	dumped, ex GT-976/RDanAF, ex 58-6976	1-92

TURKEY - 456

MERZIFON
Only a few *Wrecks & Relics* aircraft are known to be at this F-16 airfield in the mid north of the country.

☐ 6815	5-815	AT-11	..	preserved	3-95
☐		F-5A	..	dumped, fire dump	3-95
☐	5-419	Sabre 2	..	preserved, at gate in red/white c/s	2-96

SILA
In a small village, some 70km from Istanbul, along the Black Sea coast, a Starfighter is preserved

☐ 7019	4-109	F-104G	7109	preserved, ex 2150/WGAF	9-97

SIVRIHISAR
This reserve airfield is some 10km west of the town along road number 200 towards Eskisehir. The two preserved Super Sabre's may be the same aircraft, while F-100C 41877/3-877 went to Ankara (see Stop Press).

☐ 41803	803	F-100C	217-64	preserved, ex 54-1803	4-94
☐ 41950	3-950	F-100C	217-211	dumped, ex 54-1950	4-94
☐		F-100	..	preserved, marked as 'TC2469/1-903'	5-98

TOPEL
Topel is the Turkish Navy base. The Trackers are stored as they have run out of airframe hours and spares. The AB212ASW was dumped after it had crashed.

☐ ..	163	S-2A	..	preserved, at gate	5-96
☐ 147895	895	S-2E	35C	stored	5-96
☐ 149263	263	S-2E	107C	preserved	5-96
☐ 149275	275	S-2E	119C	stored	5-96
☐ 149849	849	S-2E	124C	stored	5-96
☐ 149858	858	S-2E	133C	stored	5-96
☐ 149874	874	S-2E	149C	stored	5-96
☐ 149877	877	S-2E	152C	stored	5-96
☐ 149883	883	S-2E	158C	stored	5-96
☐ 149887	887	S-2E	162C	stored	5-96
☐ 150610	610	S-2E	168C	stored	5-96
☐ 151639	639	S-2E	172C	stored	5-96
☐ 151651	651	S-2E	184C	stored	5-96
☐ 151652	652	S-2E	185C	stored	5-96
☐ 151655	655	S-2E	188C	stored	5-96
☐ 151673	673	S-2E	206C	stored, crashed	5-96
☐ 151679	679	S-2E	212C	stored	5-96
☐ 152368	368	S-2E	255C	stored	5-96
☐ TCB-40		AB212ASW	..	dumped	10-93

YENI FOCA
Some 2km north of this town, which itself is some 50km south of Izmir, a large scrapyard was discovered in 1996. At least 40 Starfighters were seen, though only just over a dozen could be identified. By August 1997 still some 50 plus aircraft were present.

☐ 2072	9-072	F-104G	2072	dumped, ex Eskisehir, ex 2062/WGAF	9-96
☐ 2082	9-082	F-104G	2082	dumped, ex 2070/WGAF	9-96
☐ 5801	9-801	TF-104G	5801	dumped, ex Eskisehir, ex D-5801/RNethAF	9-96
☐ 6906	9-906	F-104S	1206	dumped, ex Eskisehir	8-97
☐ 7032	6-032	F-104G	7032	dumped, ex 2163/WGAF	9-96
☐ 7101	6-101	F-104G	7101	dumped, ex 2226/WG Navy	9-96

TURKEY 457

☐ 7119	9-119	F-104G	7119	dumped, ex 2241/WGAF	8-97
☐ 7165	6-165	F-104G	7165	dumped, ex Eskisehir, ex 2283/WG Navy	8-97
☐ 7302	9-302	F-104G	7302	dumped, ex Eskisehir, ex 2642/WGAF	9-96
☐ 8029	9-029	F-104G	8029	dumped, ex 2350/WG Navy	9-96
☐ 8082	9-082	F-104G	8082	dumped, ex Bandirma, ex D-8082/RNethAF	9-96
☐ 8290	6-290	F-104G	8290	dumped, ex Eskisehir, ex 2519/WGAF	9-96
☐ 8333	6-333	F-104G	8333	dumped, ex 2544/WGAF	9-96
☐ 9136	6-136	F-104G	9136	dumped, ex 2608/WGAF	8-97
☐ 9151	9-151	F-104G	9151	dumped, ex 2613/WGAF	9-96
☐ 9167	9-167	F-104G	9167	dumped, ex 2619/WGAF	8-97
☐ 104842	8-842	CF-104	1142	dumped, ex 104842/CAF	9-96
☐	6-002	F-104G	..	dumped	8-97
☐	9-191	F-104G	..	dumped	8-97
☐ 21108	2-108	T-33AN	108	dumped, ex 21108/RCAF	8-97
☐ 21237	2-237	T-33AN	237	dumped, ex 21237/RCAF	9-96
☐ 21289	2-289	T-33AN	289	dumped, ex 21289/RCAF	9-96
☐ 21602	2-602	T-33AN	602	dumped, ex 21602/RCAF	9-96
☐ 33649	2-649	T-33AN	649	dumped, ex 133649/CAF	9-96
☐ 90310	TE-310	T-37B	40472	dumped, ex 59-0310	9-96

STOP PRESS

ANKARA – ETIMESGUT (Turkey)
A new museum has opened on this airfield.

☐ 9308		MiG-21MF	969308	stored, ex Hungarian AF	5-98
☐ 6833	3-833	AT-11		preserved, ex Bandirma	5-98
☐ 69-7503	1-7503	RF-4E	4133	preserved, ex Eskisehir, ex 3556/WGAF	5-98
☐ 10575	5-575	F-5A	N7023	preserved, ex 575/RNoAF, ex 65-10575	5-98
☐ 21208	3-208	RF-5A	N6395	stored, ex 67-21208	5-98
☐ 7186	186	F-84F		preserved	5-98
☐ 11924	1-924	RF-84F		preserved, ex Eskisehir, ex P-24/RNethAF	5-98
☐ 19190	1-190	CL-13 Sabre 2	90	preserved, ex Istanbul	5-98
☐ 41766	3-766	F-100C	217-27	preserved, ex Izmir, ex 54-1766	5-98
☐ 41877	3-877	F-100C	217-138	preserved, ex Sivrihisar, ex 54-1877	5-98
☐ 62368		TF-102A		stored, ex Istanbul, ex 56-2368	5-98
☐ 62-642	8-642	CF-104D	5312	preserved, ex 104642/CAF	5-98
☐ 62-711	8-711	CF-104	1011	preserved, ex 104711/CAF	5-98
☐ 62-770	8-770	CF-104	1070	preserved, ex 104770/CAF	5-98
☐ 8205	8-205	F-104G	8205	preserved, ex Eskisehir, ex 2457/WGAF	5-98
☐ 14284	4-284	T-33A	5579	preserved, ex 51-4284	5-98
☐ 01410		T-41D		preserved	5-98
☐ TC-CCA		T-34A		preserved	5-98

BÜCKEBURG (Germany)
Airfield: Additional helicopters have been noted at the airfield.

☐ ..		H-34	..	under restoration, ex French	3-98
☐ 5616		Do27A-1	284	under restoration, marked as 'BAW', ex Celle	4-98
☐ 7182		UH-1D	8302	stored	4-98
☐ 7186		UH-1D	8306	stored	4-98
☐ 7210		UH-1D	8330	stored	4-98
☐ 7240		UH-1D	8360	stored	4-98
☐ 7278		UH-1D	8398	stored	4-98
☐ 7366		UH-1D	8486	stored	4-98
☐ 7501		Alouette 2	1178	stored	4-98
☐ 7539		Alouette 2	1340	stored	4-98
☐ 7542		Alouette 2	1351	stored	4-98
☐ 7582		Alouette 2	1495	stored	4-98

CUERS – PIERREFEU (France)
Eleven F-8Js were seen at Davis Monthan in April 1990 being prepared for departure to France. These should be the aircraft at Cuers and should include : 149145, 149149, 149201, 149204, 149210, 149215, 150658, 150680, 150683, 150845 and 150871.

ESKISEHIR (Turkey)
In new museum in the town of Eskisehir the following aircraft were seen:

☐ 6032		C-47A	18973	preserved, as '052', ex Erkilet, ex 42-100510	5-98
☐ 69212	5-212	F-5A	N7035	preserved, ex 212/RNoAF, ex 66-9212	5-98
☐ 69-7465	1-7465	RF-4E	4043	preserved, ex 3513/WGAF	5-98
☐ 8733	1-733	RF-84F		preserved, ex 52-8733	5-98
☐ 31732	1-732	F-100C	214-24	preserved, ex Konya, ex 54-1732	5-98

STOP PRESS – 459

☐ 6667	9-667	F-104G	6667	preserved, ex Balikasir, ex D-6667/RNethAF	5-98
☐ 7190	8-190	F-104G	7190	preserved, ex 2307/WG Navy	5-98
☐ 21321	2-321	T-33AN	321	preserved, ex 21321/RCAF	5-98
☐ 29919	4-919	T-33A	7890	preserved, ex 52-9919	5-98
☐ E-2077		Alouette 2	2077	preserved, ex Polis, ex 7745/WG Army	5-98
☐ TC-AUA		L-21B		preserved, ex 10325	5-98
☐ TC-YSB		L-18C		preserved	5-98

MILANO – BRESSO (Italy)
Four rotorless AB204Bs were noted in outside store at this airfield.

☐ MM80439	EI-249	AB204B	4005	stored	5-98
☐ MM80452	EI-262	AB204B	4067	stored	5-98
☐ MM80687	EI-316	AB204B	4213	stored	5-98
☐ MM80721	EI-350	AB204B	4382	stored	5-98

POWIDZ (Poland)
On this Su-22 airfield a number of An-2s were noted in storage, all may be offered for onward sale. One has left; An-2TD 5709 (1G157-09) was sold as SP-AOR.

☐ 0863	An-2T	1G108-63	stored, no engine	9-97
☐ 4723	An-2T	1G147-23	stored, no engine	9-97
☐ 5705	An-2T	1G157-05	stored ?	9-97
☐ 5706	An-2T	1G157-06	stored, wfu 5-11-96	9-97
☐ 7353	An-2TD	1G73-53	stored	9-97
☐ 8551	An-2T	1G85-51	stored ?	9-97
☐ 9869	An-2T	1G98-69	stored ?	9-97

PRZYLEP (Poland)
Stored on this civil airfield near Zielona Góra is an An-2.

☐ UR-70289	An-2P	1G139-48	stored	4-98

SAN POSSIDONIO (Italy)
Some new arrivals have been noted in museum.

☐ MM57259	SM1019	1-069	stored	5-98
☐ MM57263	SM1019	1-073	stored	5-98
☐ MM80568	AB206A-2	8211	stored, wreck	5-98
☐ MM80796	OH-13H	2139	stored, hulk, ex Frosinone, ex 57-6210	5-98
☐ 5x	AB47	..	stored	5-98

TRENČÍN (Slovak Republic)
'New' aircraft have arrived for the proposed musuem.

☐ 1901	L-29	691901	stored	5-98
☐ 0716	Mi-2	5110716098	stored	5-98
☐ 0409	MiG-19S	150409	stored, ex Prešov	5-98
☐ 0903	MiG-21F-13	960903	stored, ex Prešov	5-98

VESTONE (Italy)
Noted at the aero club of the local airfield is an MB326.

☐ MM54208	4-08	MB326	..	preserved	5-98

TYPES INDEX

A19 Cloud (SARO)	40
A101	288
A102	326
A106	259, 266
A109 Hirundo	265, 281
A129 Mangusta	264, 269
A-1 Skyraider	111
	423, 427, 429
A-3 Skywarrior	413
A-5 Vigilante	413
A-7 Corsair	243
	267, 383, 385, 386, 387
	390, 391, 392, 393, 394
A-10 Thunderbolt	226
A-26 Invader	18, 74, 111
A-37	96, 361
AA-1 Trainer	14
AAC.1	see Junkers 52
AB47	see H-13
AB204	see H-1
AB205	see H-1
AB206	see H-58
AB212	see H-1
AB412	see Bell 412
AC12	409, 411, 416
AC14	411
Adams Wilson 101	427
Adaridi	66
Aerfer Ariete	325
Aerfer Sagittario II	325
Aero (Super) 45	37, 40, 42
	43, 65, 112, 179, 252, 364
Aero 145	41, 208, 363, 364
Aero A-10	39
Aero A-11	64
Aero A-12	39
Aero A-18	39
Aero A-32	70
Aero Ab-11	39
Aero Ae-10	39
Aero Ap-32	39
Aero C-3	see Siebel 204
Aero C-103	see Siebel 204
Aero C-104	see Bücker 131
Aero Commander	55, 102
	112, 163, 214, 227, 240
	242, 329, 333, 354, 355
	384, 390, 408, 412, 435
Aeronca 7 Champion	23, 420
Air & Space 18	174

Airbus A300	19, 102
Airbus A310	19
Airbus A340	107, 205
AISA I-11	410
AISA I-115	404, 409, 412
Akaflieg Hannover	209
Albatros B.I	13
Albatros B.II	363, 421
Albatros C.I	362
Albatros L.13	228
Albatros L.101	363
Alouette 2	15, 16
	17, 20, 27, 73, 75, 89, 93
	96, 102, 104, 111, 113, 115
	116, 123, 124, 126, 128, 163
	165, 168, 170, 174, 175, 177
	178, 184, 186, 188, 197, 198
	199, 202, 203, 207, 211, 213
	217, 222, 226, 341, 384, 387
	391, 392, 395, 418, 419, 420
	421, 425, 427, 429, 430, 432
	434, 435, 439, 451, 458, 459
Alouette 3	10, 54, 56, 78
	79, 89, 93, 98, 99, 105, 107
	109, 113, 124, 126, 239, 335
	336, 339, 341, 342, 345, 346
	384, 387, 392, 408, 409, 439
Alpha Jet	96, 125
	166, 168, 185, 186, 188
	189, 190, 193, 199, 206
	212, 213, 226, 229, 387
AM-3	322
Ambrosini S2	321
Ambrosini S7	278, 325
An-2	10, 11, 12
	40, 42, 68, 162, 163
	167, 178, 179, 194,198
	199, 202, 208, 213, 215, 217
	222, 224, 227, 230, 247, 248
	249, 255, 256, 267, 336, 348
	356, 357, 360, 366, 367, 368
	371, 374, 375, 376, 377, 379
	380, 381, 427, 428, 430, 459
An-12	226
An-14	167, 176
An-24	38, 40
	252, 254, 360, 380
An-26	167, 195, 223, 227, 253
An-28	367
Anatra Anasalja	46

Ansaldo A-1 Balilla	260, 321
Ansaldo AC2	324
Ansaldo SVA5	263, 288, 324
Anson (Avro)	19, 339
Aquilon 203	see Venom
Arado 66	70, 209
	350, 351, 354
Arado 96	354
Arado 196	354
Arsenal Air 100	110
AS350	125, 420
AS565	73
ASK-21	394
Astir CS77 (Grob)	55
Astra Wright	107
AT-11 Kansan	see C-45
AT-16	see T-6
Atlantique NG	105
Auster	16, 18, 22, 23, 39
	61, 342, 383, 395, 424, 427
AVD12	413
Avia 14	see Il-14
Avia B-10	39
Avia B-33	see Il-10
Avia B-534	39
Avia Ba-122	40
Avia BH-9	46
Avia BH-10	46
Avia BH-11	39
Avia CB-33	see Il-10
Avia CS-199	see Bf109
Avia S-92	see Me262
Avia S-199	see Bf109
Avia FL3	263, 275, 321, 325
Aviatik C.I	17
Aviatik C.III	362
Avro 504	55, 68, 166
	349, 351, 410
Avro 594 Avian	424
Avro 631 Cadet	389
B-17 Flying Fortress	112
B-25	335, 338, 341, 409
B-26 Marauder	384
Ba349 Natter	209
Baaden 152	179, 221
BAC111	237
BAK-01	40
Bataille Triplane	18
Battle (Fairey)	16

TYPES INDEX – 461

Bazzochi EB-4 323
Beagle A-109 392
Beagle B-121 Pup 333, 340
Beech 18 77, 334, 337
　　411, 423, 434
Beech 33 408, 411
Beech 45 11
Beech 50 226, 427, 435
Beech 65 65, 73, 269
Beech 95 Baron 65, 430
Beech 99 126
Beech T7 342
Bell 47 see H-13
Bell 412 239, 282, 352
Bellanca 8 Decathlon 427
Bensen B-8 40, 174, 275
Berg & Storm BS.III 55
Bergfalke II 61, 194, 223
　　224, 354, 425, 427, 453
Bernard 191 Grand Raid 109
Bf108 18, 194, 209, 213
　　410. 434, 438, 444
Bf109 39, 69, 70, 108, 166
　　202, 209, 213, 222, 224
　　230, 350, 354, 409, 434
BK117 174, 224
Blackburn Ripon 70
Blackburn Skua 350
Blenheim (Bristol) 65, 69, 234
Blériot XI 18, 46, 107, 194
　　209, 278, 324, 326, 353
　　363, 383, 410, 434, 438
Blomqvist & Nyberg 66
BN-2 Islander 15, 17, 256, 354
Bo46 174
Bo102 174
Bo103 174
Bo105 168, 174, 178
　　184, 185, 208, 209, 217
　　222, 229, 333, 339, 405
Bo108 174
Boeing 377SG 125, 192
Boeing 707 18, 19, 29, 31
　　100, 107, 110, 169, 193
　　209, 224, 338, 390, 433
Boeing 720 232, 234, 329
Boeing 727 72, 100, 112, 237
Boeing 737 102
Bolingbroke (Bristol) 16
Brantly B2 194
Breda Ba15 278
Breda Ba19 288
Breguet 14 70, 107
Breguet 19 108, 410
Breguet 763 Provence 92

Breguet 765 Sahara 92, 125
Breguet 901 Mouette 110
Breguet 904 Nympahle 104
Breguet 905 Fauvette 23, 112
Breguet 941 73, 111
Breguet 1001 Taon 111
Breguet 1050 Alize 89, 93, 94
　　97, 105, 111, 115, 116, 119
Breguet 1150 Atlantic 89, 97
　　105, 110, 114, 212, 230, 436
Breguet G111 109
Breguet UIII 422
Bristol F.2B Fighter 18, 410
Bristol Coanda 321
Bryan HP-16 66
Bücker 131 Jungmann 42
　　209, 222, 321, 363, 403, 406
　　408, 409, 410, 412, 434, 444
Bücker 133 Jungmeister 409
　　434, 435, 438
Bücker 181 Bestmann 18
22 U.　22, 61, 104, 112, 166
　　222, 226, 421, 422, 434
Bulldog (Bristol) 64
Bulldog (Scottish Av) 422, 423
BZ-1 Gil 363
BZ-4 Zuk 364

C3603 98, 434, 435, 438
C3605 193, 224, 432
　　433, 435, 436, 438
C.5 Polar 350
C-7 Caribou 329, 408, 409, 417
C-45 Expeditor 76, 92
　　97, 112, 116, 125, 264
　　265, 275, 280, 281, 284, 285
　　288, 322, 325, 330, 384, 385
　　386, 389, 392, 394, 446, 447
　　448, 449, 452, 454, 456, 458
C-47 Skytrain 12, 14, 17, 26
　　27, 29, 41, 55, 57, 65, 66
　　68, 70, 91, 95, 103, 105, 109
　　111, 115, 122, 125, 166, 168
　　187, 195, 210, 213, 223, 225
　　227, 234, 235, 236, 237, 238
　　243, 244, 256, 265, 267, 281
　　285, 288, 329, 330, 331, 332
　　340, 341, 351, 390, 394, 404
　　406, 408, 409, 412, 413, 416
　　417, 419, 422, 426, 427, 438
　　446, 448, 451, 452, 453, 458
C-54 Skymaster 26, 169, 187
　　384, 393, 409, 452
C-60 see Lockheed L18
C-61 Argus 18, 22, 55, 62

　　104, 263, 321, 326, 425, 427
C-64 Norseman 69, 95
　　420, 422, 427
C-78 Bobcat 363
C-97 103, 404, 409
C-117 Skytrooper 256
C-119 Flying Boxcar 17, 19
　　24, 265, 280, 282
　　283, 323, 326, 327
C-130 Hercules 19, 75, 99
　　237, 384, 386, 404
C-131 Samaritan 393
C-160 92, 106, 161, 186
Camel F.1 (Sopwith) 17, 363
Canberra 110, 167, 194, 206
　　222, 225, 350, 422, 429
Cap 10 (Mudry) 73
Cap 20 (Mudry) 109
Caproni Ca6 321
Caproni Ca9 321
Caproni Ca33 324
Caproni Ca53 321
Caproni Ca100 321, 326
Caproni Ca113 321
Caproni Ca163 321
Caproni Ca193 321
Caprino Ca313 422
Caproni Campini CC1 324
Caproni Campini CC2 278
Caproni TM2 277
Caproni Trento F5 321
Caproni Vizzola C22 321
CASA 101 403, 409, 411, 413
CASA 127 see Do27
CASA 207 Azor 409, 411, 414
CASA 212 107, 329, 404, 413
CASA 352 see Junkers 52
CASA 1131 see Bücker 131
CASA 2111 see Heinkel 111
Castel C25 104, 112
Castel C242 110
Castel C301 110
Castel C310 112
Castel C311 104, 112
Caudron C.59 70
Caudron C.60 65, 109
Caudron C.109 108
Caudron C.272 Luciole 410
Caudron C.277 Luciole 109
Caudron C.282 Phalene 125
Caudron C.366 Atalante 108
Caudron C.510 Pelican 430
Caudron C.635 Simoun 109
Caudron C.714 70, 108, 112
Caudron C.800 18, 109, 116

TYPES INDEX – 462

Caudron G.3	17, 64	
	108, 383, 410	
Caudron G.4	108	
CCS-13	see Po-2	
Cessna 140	343	
Cessna 150	11, 62, 65	
	126, 333, 336	
Cessna 170	330	
Cessna 171	333	
Cessna 172	67, 69, 427, 430	
Cessna 182	20	
Cessna 195	214	
Cessna 206	98	
Cessna 208	61	
Cessna 310	23, 55, 435	
Cessna 337	55, 350, 354, 392	
Cessna 402	420	
Cessna 411	76	
Cessna 421	26, 61, 408, 445	
Cessna 500	67	
CF-100 Canuck	16	
CFM01 Tummelisa	421	
CG-4 Hadrian	122	
Chandelon Helicopter	23	
Chipmunk	see DHC-1	
Cierva C.6	410	
Cierva C.8	110	
Cierva C.19	410	
Cierva C.30	277, 341	
Citabria	451	
CL-13 Sabre	see F-86	
CL-215	99, 409	
CM170	15, 17, 20, 21, 26, 29	
	65, 66, 67, 68, 70, 71, 72	
	73, 74, 75, 76, 77, 78, 79	
	80, 84, 85, 87, 88, 90, 92	
	93, 94, 95, 96, 97, 98, 101	
	102, 103, 104, 105, 106, 107	
	109, 111, 113, 114, 115, 116	
	117, 118, 119, 121, 122, 123	
	124, 125, 126, 127, 128, 168	
	177, 195, 203, 206, 212, 226	
CM170M Esquif	116	
CM173	103	
CM175 Zephyr	94, 110, 111	
	115, 116, 119	
CM191	226	
Colomban MC.10	110	
Comet (DH.106)	195	
Comte AC4	434, 439	
Concorde	110, 112, 125, 195	
Convair 440	65, 273	
	353, 354, 427	
Convair 880	390	
Convair 990 Coronado	412, 438	

Croses EC.6	110	
Curtiss A1	326	
CVT M200	275	
CVV6	263, 285, 325, 326	
CVV8 Bonaventura	275, 325	
CW.22 Falcon	453	
D-3801	see MS406	
DC-2	65, 69, 340	
DC-3	see C-47	
DC-4	411	
DC-6	31, 59, 86, 216, 259	
	263, 265, 285, 384, 412, 433	
DC-7	55, 58, 59, 111, 407, 434	
DC-8	59, 111, 187	
	338, 405, 445	
DC-9	426	
Deperdussin B	108	
Deltaviex	125	
Devon (Sea)	208, 340	
Dewoitine D26	439	
Dewoitine D27	434	
Dewoitine D520	108, 116	
Dewoitine D530	108, 111	
DFS230	166	
DFW C.V	363	
DH.9	108	
DH.60 Moth Major	69, 410	
	421, 424, 427, 429	
DH.80 Puss Moth	277	
DH.82 Tiger Moth	18, 19, 22	
	23, 39, 56, 61, 224, 234, 340	
	341, 349, 353, 363, 383, 394	
	395, 410, 420, 421, 429, 430	
DH.87 Hornet Moth	61, 395	
DH.89 Dragon Rapide	18, 57	
	109, 340, 341, 395, 410, 453	
DHC-1 Chipmunk	16, 55, 56	
	58, 61, 62, 166, 384, 392	
DHC-2 Beaver	see U-6	
DHC-3 Otter	18, 350	
DHC-4 Caribou	see C-7	
DHC-5 Buffalo	90	
DHC-6 Twin Otter	350	
Dijkhaster FB25	343	
Dittmar Condor	214	
Doppelraab IV	61, 168, 213	
Donnet Leveque A	108	
Dominie	see DH.89	
Dornier A Libelle	209	
Dornier Do24	212, 213,	
	227, 341, 408	
Dornier Do27	15, 17, 163, 165	
	168, 169, 171, 184, 185, 188	
	194, 206, 207, 209, 214, 222	

	223, 228, 333, 391, 392, 393	
	395, 403, 405, 408, 409, 411	
	412, 415, 418, 422, 453, 458	
Dornier Do28 Skyservant	25	
	26, 112, 168, 186	
	195, 199, 202, 203, 204	
	205, 206, 210, 212, 213, 231	
	236, 237, 333, 409, 451, 453	
Dornier Do29	168	
Dornier Do31	213, 214	
Dornier Do32	209	
Dornier Do34 Kiebitz	174, 199	
Dornier Do228	107, 213	
Dornier Do328	213	
Dornier DS10	168	
Dornier J Wal	410	
Dove	13, 55, 62, 126, 224	
	321, 340, 392, 411, 423, 425	
DOWA 81	209	
Dragonfly	344	
Druine Turbulent	61	
E-3 Sentry	241	
EC38-56 Uribel	263	
Eagle II (British Aircraft)	410	
Eklund TE-1	66	
EKW C35	434	
Ellehammer	55, 57, 61	
Elliott AP8	354	
EMB110 Bandeirante	67	
EMB121 Xingu	25	
Enstrom F.28	10, 67, 420, 431	
Ercoupe 415	23, 420, 429	
Etendard 4	74, 76	
	89, 94, 95, 96, 97, 102	
	110, 114, 116, 119, 126	
Etrich Taube	363	
Europa Jet	222	
Evans VP-1	222, 341	
F4U Corsair	96	
F24 Forwarder	see C-61	
F27 Friendship	59, 90, 91	
	107, 170, 184, 333, 334, 339	
	340, 342, 345, 354, 416, 433	
F28 Fellowship	91	
F-4 Phantom	17, 31, 38	
	114, 168, 173, 185, 194	
	196, 198, 202, 211, 213, 216	
	226, 227, 233, 239, 242, 243	
	256, 259, 267, 331, 333, 349	
	350, 404, 407, 408, 415, 416	
	448, 449, 452, 454, 455, 458	
F-5 Freedom Fighter/Tiger	38	
	234, 235, 238, 239	

TYPES INDEX – 463

240, 241, 242, 243, 244, 245
280, 332, 333, 334, 335, 336
341, 342, 343, 345, 346, 347
348, 350, 351, 352, 353, 354
361, 408, 412, 414, 415, 446
447, 449, 452, 453, 456, 458
F-8 Crusader   89, 95, 116, 126
F-15 Eagle   216, 226, 227, 341
F-16 Fighting Falcon   15
   17, 20, 27, 28, 29, 226
   259, 333, 337, 339, 343
   344, 345, 352, 353, 407
F/A-18 Hornet   287, 417
F-27 (Fairchild)   see F27
F-84   14, 15, 17, 19, 20, 21, 22
   23, 25, 26, 27, 29, 54, 55
   56, 57, 58, 59, 60, 61, 62, 63
   79, 100, 101, 103, 109, 113
   115, 119, 120, 121, 122, 124
   125, 167, 170, 173, 183, 188
   193, 195, 196, 197, 199, 204
   207, 208, 212, 213, 217, 223
   227, 229, 233, 234, 235, 236
   238, 239, 240, 241, 242, 243
   244, 245, 259, 263, 265, 267
   271, 272, 274, 275, 277, 278
   279, 280, 282, 283, 286, 287
   325, 327, 333, 334, 335, 336
   338, 339, 340, 341, 343, 345
   349, 350, 351, 352, 353, 354
   355, 392, 393, 420, 446, 447
   448, 449, 452, 453, 454, 458
F-86 Sabre   17, 54
   57, 60, 109, 119
   166, 167, 168, 170, 179, 185
   191, 192, 195, 197, 199, 200
   203, 211, 213, 217, 222, 224
   225, 226, 229, 230, 232, 233
   234, 235, 240, 241, 242, 244
   257, 259, 262, 263, 265, 271
   272, 273, 274, 277, 280, 282
   283, 288, 321, 322, 325, 327
   332, 341, 342, 343, 349, 350
   351, 352, 353, 354, 383, 383
   387, 390, 392, 393, 394, 395
   406, 407, 408, 409, 412, 414
   414, 415, 416, 417, 446, 456
F-100 Super Sabre   55
   57, 58, 59, 60, 79, 103, 109
   119, 120, 124, 125, 193, 223
   259, 342, 446, 447, 448, 449
   451, 452, 454, 455, 456, 458
F-101   23, 125, 216, 223, 227
F-102 Delta Dagger   194
   235, 237, 240, 242, 243

244, 341, 446, 452, 454, 458
F-104 Starfighter   14, 15, 17, 20
   21, 22, 23, 24, 25, 26, 27
   29, 54, 55, 56, 57, 59, 60
   61, 62, 103, 106, 111, 119
   120, 121, 124, 161, 162, 163
   166, 167, 168, 169, 173, 175
   178, 183, 184, 185, 187, 188
   191, 192, 193, 194, 196, 197
   198, 199, 200, 202, 203, 206
   207, 208, 209, 210, 211, 212
   213, 217, 222, 224, 225, 226
   227, 228, 229, 230, 232, 233
   234, 235, 236, 238, 239, 240
   242, 244, 258, 261, 262, 264
   266, 267, 269, 270, 271, 272
   273, 274, 275, 279, 281, 282
   284, 286, 287, 288, 324, 326
   327, 331, 332, 333, 336, 337
   339, 340, 341, 342, 343, 344
   345, 347, 348, 349, 350, 351
   352, 353, 354, 387, 392, 408
   446, 447, 448, 449, 450, 451
   452, 454, 456, 457, 458, 459
F-105 Thunderchief   110, 119
   194, 226, 407, 452
Fa330 Bachstelze   56, 174, 209
Fairey III   383, 389
Falcon 20   75, 112, 408
Farman F.11   18
Farman F.46   351
Farman F.60 Goliath   110
Farman F.192   108
Farman F.455   109
Farman HF-4   278, 363
Farman HF-20   108, 341
Farman MF-4   383
Farman MF-7   108, 234
Farman-Voisin   18
Farner WF-7   439
Fauvel AV36   75, 110, 175
   214, 222, 263
Fauvel AV45   112
Ferber 6   108
FES-530 Lehrmeister   350
FF7 Hauk   353
FF9 Kaie   350
FFA AS202 Bravo   438
FFA P16   434
FFVS 22   418, 421, 429
FH-227 (Fairchild)   see F27
Fiat C29   324
Fiat CR42   421
Fiat G5   325
Fibera KK-1 Utu   66

Fieseler 156 Storch   13, 16
   18, 22, 40, 65, 90, 104, 108
   166, 209, 224, 226, 231, 326
   332, 354, 410, 422, 434, 438
Finsterwalder Bergfex   209
Firebird M1 (Schweiger)   199
Firefly (Fairey)   423
Flying Jeep (Bölkow)   174
FN305   325
FN333 Riviera   278
Fokker 50   340
Fokker 70   340
Fokker 100   340
Fokker B.IV   340
Fokker C.III   410
Fokker C.V   70, 340
   350, 421, 434
Fokker C.X   69, 341
Fokker D.VI   208
Fokker D.VII   108, 208
   209, 341, 434
Fokker D.VIII   288
Fokker D.X   69
Fokker D.XXI   55, 68, 341
Fokker Dr.I   18
   166, 194, 202, 208, 209
   226, 230, 267, 341, 410
Fokker E.III   166, 202
   222, 224
Fokker F.II   341
Fokker F.VII   166, 340, 439
Fokker F.VIII   340
Fokker G.1   341
Fokker Spin   340, 341
Fouga 90   122
Frebel F5 Aeolus   224
Fs24 Phönix   209
Fs26 Moseppl   222
Fw44 Stieglitz   69, 70, 202
   209, 213, 420, 422, 427, 429
Fw61   174
Fw190   108, 222, 224, 352, 361

G44 Widgeon   117, 353
   383, 389
G46 (Fiat)   13, 16, 267, 270
   278, 283, 287, 322, 325, 326
G49 (Fiat)   325
G51 (Fiat)   321
G59 (Fiat)   275, 276, 325
G80 (Fiat)   281
G82 (Fiat)   324, 325, 326
G91 (Fiat)   17
   101, 103, 110, 119, 162
   166, 167, 168, 171, 173, 174

TYPES INDEX – 464

175, 178, 185, 188, 191, 192
193, 195, 196, 197, 199, 203
204, 206, 207, 211, 212, 213
216, 218, 222, 224, 225, 227
229, 230, 231, 258, 260, 261
262, 263, 265, 267, 268, 269
271, 272, 273, 274, 275, 276
277, 279, 280, 281, 282, 283
284, 286, 287, 288, 321, 322
324, 325, 326, 327, 330, 383
384, 385, 387, 388, 389, 390
391, 392, 393, 394, 395, 396

G159 Gulfstream 1	99, 235
G164 Ag-Cat	236, 333
G212 (Fiat)	325
G222 (Alenia)	280, 281, 324
Gamecock (Gloster)	65, 70
Gannet	168, 194, 212, 227
Gardan GY80 Horizon	125
Gary R.01 Gyrocoptère	110
Geest Moewe IV	364
Georges G1 Papillon	174
Georges G2	174
Gladiator (Sea)	330, 350, 421
Glenten (Svendsen)	57
Gnat (Folland)	64, 65
	67, 68, 69, 70
Göppingen Gö IV	18, 213
Gourdou Lesseurre B7	108
Gourdou Lesseurre GL22	68
GP2 Asiago	321
Grade A	209
Grade Monoplane	222
Grandvist Autogiro	429
Grasshopper (Slingsby)	213
Grigorovich M5	453
Grigorovich M15	363
Grunau Baby II	17, 18, 55, 57
	61, 65, 69, 125, 166, 194
	202, 213, 214, 222, 223, 224
	383, 410, 422, 425, 427, 429
Grunau Baby III	17
Grunau ESG-9	65, 350
Grunau SG-38	17, 61, 65, 110
	127, 166, 179, 209, 222
	350, 383, 421, 427, 429
GV-38	425, 427, 429
Gull Four (Percival)	18
Gyrocopter VS17	55
H369 Courier	325
H-1 Iroquois	11, 12, 162, 166
	175, 178, 184, 193, 203, 211
	217, 229, 235, 236, 260, 261
	263, 264, 266, 269, 270, 271

273, 276, 279, 281, 282, 284
285, 287, 323, 324, 327, 328
342, 349, 350, 351, 352, 353
354, 405, 406, 408, 409, 411
412, 414, 419, 422, 427, 428
429, 451, 453, 456, 458, 459

H-3 (Sea King)	230, 276
	281, 352
H-12	90, 434
H-13	12, 13, 65, 89, 103, 110
	115, 173, 208, 209, 235, 236
	242, 243, 257, 259, 260, 261
	262, 263, 264, 265, 266, 267
	269, 270, 271, 273, 275, 276
	277, 278, 279, 281, 284, 285
	287, 288, 322, 325, 327, 328
	350, 352, 353, 354, 404, 405
	406, 408, 409, 413, 414, 416
	420, 427, 439, 451, 452, 459
H-19 Chickasaw	55, 59, 89
	91, 93, 98, 103, 119, 125
	209, 235, 340, 351, 394, 452
H-21 Shawnee	72, 89, 90
	93, 96, 103, 116, 125, 127
	168, 174, 193, 195, 207, 223
H-23 Raven	173, 341
H-34 Choctaw	17, 23
	73, 76, 77, 78, 89, 90, 92
	93, 95, 96, 97, 99, 100, 102
	103, 105, 111, 114, 116, 119
	123, 124, 125, 126, 127, 165
	168, 174, 187, 199, 203, 213
	223, 224, 225, 276, 341, 458
H-43 Huskie	174
H-53 Sea Stallion	185
H-55 Osage	174
H-58 Kiowa	11, 193
	328, 409, 418, 458
HA132 (Hispano)	324, 409
HA200 Saeta	405, 407, 408
	409, 410, 411, 412, 416, 417
HA220 Saeta	408, 409
	410, 414, 416
HA300 (Helwan)	213
HA1112 (Hispano)	see Bf109
Habicht	110
Hafeli DH-1	434
Hafeli DH-5	434
Halberstadt C.V	17
Halberstadt CL.II	362
Halberstadt CL.IV	213
Hanriot HD.1	17, 324, 434
Harakka I	65, 67
Harakka II	65, 67
Harakka III (Pik-7)	67
Harakka H-17	70
Harakka H-54	70
Harrier (Hawker Siddeley)	166
	173, 194, 202, 269
Harvard	see T-6
Hastings (Handley Page)	166
Haukka I	70
Haukka II	64
Havertz HZ-5	174
Hawk II (Curtiss)	363
Hawk Major	410
Hawker Dankock	55
Hawker Hart	421
HC-2 (VZLU)	40
HC-3 (VZLU)	402
HC-4 (VZLU)	40
HC-102 (Zlin)	40, 42, 46, 402
HD-10 (Hurel Dubois)	109
HD-34 (Hurel Dubois)	89, 112
HD-321 (Hurel Dubois)	78
Heinkel HD35	423
Heinkel HE5	363
Heinkel He46	112
Heinkel He111	111, 168, 194
	213, 224, 351, 407, 409
Heinkel He162	110
Heinonen HK-1	65
Heron (Sea)	195, 354
HF24 Marut	213
HFB320 Hansa	111, 167
	186, 192, 202, 206, 209
Hirsch C100	109
HKS3	209
HM-1 (Huarte)	409, 410
HM-8 (Mignet)	110, 125, 439
HM-14 Pou du Ciel	39, 46
	56, 66, 69, 103, 108
	120, 125, 224, 227, 330
	341, 423, 426, 431, 445
HM-290 (Mignet)	18
HM-293 (Mignet)	18
HM-320 (Mignet)	112
HM-504 (Mignet)	202
Holländer HT.1	61
Horvath III	202
HP137	see Jetstream
HS34 (Hispano Suiza)	409
HS748 (Hawker Siddeley)	123
HSS-1	see H-34
HSX-2 Cierva	174
H/T-O 2G	55, 61
Hudson (Lockheed)	410
Hughes 269	59, 406, 422, 431
Hughes 300	421
Hughes 369	428, 430

TYPES INDEX – 465

Hughes 500	281	
Hug Spyr III	439	
Hunter (Hawker)	11	
	14, 15, 17, 18, 23	
	28, 55, 56, 89, 101, 103	
	110, 119, 125, 167, 193, 194	
	202, 222, 227, 252, 259, 280	
	336, 337, 339, 340, 341, 342	
	344, 351, 354, 421, 429, 432	
	433, 434, 435, 436, 437, 438	
	439, 440, 441, 442, 443, 444	
HUP Retriever	111, 337, 339	
Hurricane	17, 69, 330, 350, 383	
Hütter H17	56, 194, 214	
HWL Pegaz	364	
Hydravion Fabre	108	
I-22 Iryda (PZL)	359	
Il-2	39, 70, 171, 231, 352, 379	
Il-10	39, 40, 41	
	359, 362, 379, 399, 402	
Il-12	53	
Il-14	40, 42, 43, 45	
	47, 51, 52, 53, 168, 175	
	179, 184, 186, 194, 203, 216	
	225, 226, 247, 252, 356, 360	
	361, 364, 366, 367, 377, 378	
	380, 397, 398, 399, 400, 402	
Il-18	33, 38, 42, 48	
	170, 178, 184, 193, 195, 205	
	225, 247, 248, 252, 378, 381	
Il-28	34, 38, 40, 69, 167, 217	
	249, 252, 267, 358, 359, 360	
	361, 366, 374, 376, 380, 381	
Il-62	161, 191, 205, 227	
IMAM Ro43	325	
IVL A.22 Hansa	65	
IVL C.24	64	
IVL K.1 Kurki	70	
J-2 Cub	427	
J-3 Cub	see L-4	
J/1 Autocrat	429	
Jaguar (SEPECAT)	78, 79	
	80, 81, 85, 111, 116	
	119, 120, 124, 128, 173	
Janowski J-1	425	
Javelin (Gloster)	267	
Jet Commander	see Aero Commander	
Jet Provost	25, 32, 335, 340	
Jetstream	56, 59, 335, 336	
JK-1 Trzmiel	363	
Job 15-150	13	
Jodel D.9 Bebe	18, 110, 383	

Jodel D.119	112, 125	
JRB Expeditor	see C-45	
JRF Goose	423	
Junkers A50 Junior	65, 209	
Junkers F13	108, 202	
	209, 426, 228	
Junkers J4	278	
Junkers J9	108, 166	
Junkers Ju52	19, 111, 168	
	177, 194, 195, 209, 210, 224	
	225, 226, 231, 349, 350, 351	
	384, 404, 409, 416, 427, 429	
Junkers Ju86	421	
Junkers Ju87	222, 224	
Junkers Ju88	163, 224, 349, 351	
Junkers W33	171	
Junkers W34	427	
Jurca MJ-2 Tempete	395	
Jurca MJ-5 Sirocco	226, 321	
K8 (Schleicher)	67	
K-65 Cap	see Fieseler 156	
Ka2 (Schleicher)	22	
Ka4 (Schleicher)	194, 224, 226	
Ka6	18, 103, 209, 410	
Ka-26	67, 162, 170, 174	
	175, 178, 186, 187, 194, 207	
	213, 215, 224, 247, 252, 429	
Kaiser Ka1	214, 224	
Karhu 48	64, 66	
Kassel 12	18, 66	
Kjølseth PK X-1	352	
Klemm L-25	65, 169	
	202, 209, 409	
Klemm KL.35	56, 421	
	424, 427, 429	
Knechtel KN-1	208	
Kneller Bechereau E60	108	
Knoller C II	46	
Kokkola Ko-04	66	
Kranich II	214, 410, 422	
Kranich III	410	
Kreit & Lambrickx KL2	18	
Kurir L	see Fieseler 156	
KZ I (SAI)	61	
KZ II (SAI)	55, 61	
KZ III (SAI)	18, 55, 61	
KZ IV (SAI)	55, 61	
KZ VII (SAI)	59, 61	
KZ VIII (SAI)	61	
KZ X (SAI)	61	
KZ G1 (SAI)	61	
L-4 Cub	22, 40, 61, 125	
	194, 213, 332, 354, 363, 395	

L-5 Sentinel	41, 61, 275, 325	
	326, 363, 404, 409, 410, 415	
L-13 Blanik	56, 66, 224, 350	
L-18 Super Cub	17, 22, 61, 90	
	111, 166, 235, 284, 328, 342	
	350, 351, 395, 439, 453, 459	
L-19 Bird Dog	10, 12, 13	
	75, 76, 89, 90, 95, 100	
	103, 409, 166, 261, 328	
	329, 350, 351, 352, 451	
L-21 Super Cub	193	
	235, 242, 243, 284, 328	
	337, 383, 384, 422, 459	
L-29 Delfin	10, 16, 33, 36	
	37, 38, 39, 40, 42, 43, 44	
	45, 46, 48, 50, 53, 167, 179	
	192, 197, 218, 221, 226, 247	
	252, 398, 399, 400, 402, 459	
L-39 Albatros	10, 38, 39	
	40, 42, 43, 47, 51, 162, 163	
	167, 168, 177, 186, 191, 193	
	195, 215, 218, 220, 221, 223	
	225, 227, 249, 253, 398, 402	
L-40 Meta Sokol	40	
L-60 Brigadyr	40, 228, 247	
	252, 363, 430	
L-160 Brigadyr	41	
L-200 Morava	39, 40, 43	
	52, 187, 363, 398	
L-410 Turbolet	38, 42, 43	
	53, 267, 329, 357, 397, 402	
La-7 (Lavochkin)	39	
Lancaster	111	
Landmann La11	194	
Larsen	57	
Larsen Special	352	
Leduc 0.16	108	
Leduc 0.22	108	
LeO C.302	109	
LET C11	see Yak-11	
Letov S-20	39	
Letov S-21	68	
Letov S-218 Smolik	39, 66	
Levasseur Antoinette	108, 364	
LF-1 Zaunkönig	213, 222	
LFG Roland C.VI	362	
Li-2	41, 246, 247, 251, 254	
	360, 361, 381, 397, 399, 401	
Libellula I (Manzolini)	321	
Lightning (BAC)	31, 119	
	167, 194, 208	
Lim-1	see MiG-15	
Lim-2	see MiG-15	
Lim-5	see MiG-17	
Lim-6	see MiG-17	

Lippisch Storch VII	364	
Lockheed L9 Orion	439	
Lockheed L12 Electra	342, 427	
Lockheed L18	66, 351, 427	
Lockheed L100	see C-130	
Lockheed L749	111	
Lockheed L1049	104, 114, 195, 210, 388	
Lockheed L1101 Tristar	205	
Loening C2 Air Yacht	353	
Lohner L-1	324	
Lom 57/I Libelle	202	
L-Spatz (Schiebe)	224	
Lund HL-1	56	
Luscombe 8 Silvaire	11	
LVG B.II	362	
LVG C.VI	16	
LWD Januk 1	364	
LWD Januk 2	363	
LWD Januk 3	358, 363	
LWD Szpak 2	363	
LWD Szpak 3	363	
LWD Szpak 4	363	
LWD Zak 3	363	
LWD Zuch 1	363	
LWD Zuch 2	363	
LWD Zuraw	364	
LWF V Scout	46	
Lynx (Westland)	54, 56, 62, 92, 116, 352	
M-1 Sokol	40, 46, 226	
M14 Magister (Miles)	17, 453	
M28 Mercury (Miles)	61	
M38 Messenger (Miles)	22, 429	
M65 Gemini (Miles)	55, 354	
Maagen	57	
Macchi AL60	265, 323	
Macchi M7	421	
Macchi M20	288	
Macchi M39	324	
Macchi M67	324	
Macchi M416	266, 325, 326, 327	
Macchi MB308	263, 321, 323, 326	
Macchi MB323	325	
Macchi MB326	257, 262, 264, 265, 267, 268, 269, 271, 272, 273, 274, 275, 276, 278, 279, 280, 281, 282, 283, 285, 286, 287, 288, 322, 323, 324, 325, 326, 327, 459	
Macchi MC72	324	
Macchi MC200 Saetta	274	
Macchi MC202 Folgore	325	
Macchi MC205	277, 323, 325	
Macchi Nieuport Ni10	277, 286	
Magni PM3/4 Vale	277	
Mantelli AM10	263	
Martinsyde F.4 Buzzard	68	
Mauboussin M.121	75	
Mauboussin M.123	111	
Mavi Isik G	453	
MBB223	see SIAT 223	
MD12 (Misztal Duleba)	362	
MD311 Flamant	71, 77, 103, 105, 111, 118	
MD312 Flamant	32, 73, 74, 76, 78, 87, 90, 92, 98, 101, 102, 103, 105, 112, 113, 115, 116, 117, 125, 126, 127	
MD315 Flamant	79, 92, 111, 113, 122	
MD450 Ouragan	16, 73, 79, 101, 103, 105, 109, 118, 119, 120, 126	
Me163	166, 208, 209	
Me209	363	
Me262	39, 41, 209, 222, 213, 226, 354, 424, 427, 429	
Meise	66, 69, 110, 127, 202	
Me M17	209	
Merckle SM.67	174	
Mercure 100	75, 99, 102, 110, 112, 227	
Meteor	15, 17, 19, 20, 23, 38, 54, 57, 59, 62, 63, 76, 103, 111, 119, 120, 125, 199, 337, 341, 342, 343, 423, 429	
Meteor FL55	11	
Meteor MS-30	see L-Spatz	
MFI-9	420, 423, 431	
MFI-10 Vipan	423, 427	
MFI-15 Safari	352, 420, 423	
MFH Junior	427	
MH1521 Broussard	71, 72, 74, 76, 87, 89, 90, 92, 94, 95, 100, 101, 102, 103, 104, 105, 106, 111, 113, 115, 116, 118, 125, 126, 128, 169, 392	
Mi-1	34, 38, 41, 42, 44, 45, 51, 53, 65, 69, 168, 174, 178, 193, 207, 221, 247, 248, 252, 253, 255, 359, 360, 363, 367, 374, 376, 378, 379, 399, 402	
Mi-2	34, 38, 39, 41, 53, 67, 127, 163, 168, 171, 172, 174, 175, 176, 177, 178, 183, 184, 192, 194, 195, 197, 207, 215, 218, 221, 223, 226, 231, 251, 252, 253, 254, 267, 332, 337, 339, 348, 356, 357, 359, 362, 364, 365, 366, 367, 374, 377, 379, 380, 381, 382, 459	
Mi-4	10, 36, 37, 38, 41, 43, 45, 51, 53, 65, 68, 167, 176, 179, 194, 252, 359, 360, 361, 378, 399, 400	
Mi-6	195, 366	
Mi-8	10, 12, 38, 41, 52, 53, 65, 67, 70, 89, 163, 164, 167, 169, 170, 171, 172, 174, 176, 177, 178, 182, 183, 185, 186, 187, 188, 195, 207, 212, 214, 215, 218, 223, 225, 227, 231, 247, 251, 252, 253, 254, 364, 365, 366, 374, 379, 401	
Mi-9	164, 167, 176, 195, 228, 266	
Mi-14	195, 214, 215, 218, 221, 227, 229, 364	
Mi-17	252	
Mi-24	17, 28, 38, 164, 165, 167, 174, 176, 177, 195, 196, 199, 206, 251, 252, 365, 366, 374	
MiG-3	70	
MiG-15	10, 11, 12, 16, 26, 33, 34, 35, 36, 37, 38, 39, 41, 42, 43, 44, 45, 46, 47, 48, 49, 51, 52, 53, 64, 69, 70, 119, 161, 166, 168, 171, 176, 177, 184, 186, 193, 202, 207, 211, 213, 221, 222, 223, 224, 226, 234, 246, 247, 248, 249, 250, 251, 252, 254, 255, 266, 350, 356, 358, 359, 360, 361, 362, 364, 366, 367, 368, 371, 374, 376, 377, 378, 379, 380, 381, 397, 398, 399, 400, 402, 422	
MiG-17	26, 38, 39, 41, 66, 67, 119, 167, 168, 172, 175, 176, 179, 186, 191, 194, 210, 215, 221, 223, 227, 246, 248, 249, 251, 252, 255, 267, 338, 339, 356, 357, 358, 359, 360, 361, 362, 363, 364, 366, 367, 368, 369, 371, 372, 374, 376, 378, 379, 380, 381, 382, 399	
MiG-19	10, 33, 35, 38, 39, 41, 43, 44, 48, 52, 53, 167, 200, 246, 248, 249, 250, 252, 254, 255, 266, 361, 362, 378, 397, 398, 400	
MiG-21	10, 11, 17, 20, 25, 33

TYPES INDEX – 467

34, 35, 37, 38, 39, 41, 42	Mitka Mi-1 222	74, 89, 92, 94, 95
43, 44, 45, 46, 47, 48, 50	MKEK 4 Ugar 453	103, 106, 111, 125, 168
51, 52, 53, 64, 65, 67, 68	Monospar ST-25 56	Nord N1203 Norecrin 14
70, 101, 103, 105, 106, 119	Mooney M.20 22, 222	103, 109, 114, 434
124, 162, 163, 164, 165, 166	Morane H 108	Nord N1300 see Grunau Baby
167, 168, 170, 173, 175, 176	Mosquito (DH.98) 17, 349	Nord N1500 Griffon II 109
177, 178, 179, 180, 181, 182	MS30 108, 326	Nord N2000 see Meise
185, 186, 191, 192, 193, 194	MS50 68	Nord N2200 126
196, 199, 200, 202, 206, 208	MS181 410	Nord N2501 Noratlas 72, 76
210, 211, 212, 213, 215,216	MS230 18, 109, 410	77, 78, 79, 84, 89, 92, 96, 97
217, 218, 219, 220, 221, 222	MS315 17	100, 103, 104, 105, 106, 111
223, 225, 226, 227, 228, 229	MS317 18, 109	112, 113, 114, 117, 118, 120
230, 231, 246, 247, 248, 249	MS406 111, 434	122, 125, 126, 127, 128, 161
250, 252, 253, 254, 255, 266	MS472 Vanneau 109	169, 192, 195, 195, 199, 203
288, 331, 334, 335, 341, 342	MS500 Criquet see Fieseler 156	208, 221, 223, 227, 229, 231
343, 347, 348, 350, 356, 357	MS502 Criquet see Fieseler 156	235, 237, 243, 244, 384, 388
358, 359, 360, 361, 362, 364	MS505 Criquet see Fieseler 156	Nord N2502 Noratlas 384
366, 367, 368, 369, 370, 371	MS733 Alcyon 16, 98	Nord N2504 Noratlas 71
372, 373, 374, 375, 376, 378	103, 113, 114, 125, 409	Nord N3202 90, 92
379, 380, 381, 398, 400, 401	MS760 Paris 78	100, 104, 109, 111
402, 408, 418, 424, 458, 459	83, 86, 96, 116, 123	Nord N3400 71, 74, 90
MiG-23 10, 16	MS880 Rallye 22, 103, 112	95, 100, 103, 111, 125
28, 35, 36, 38, 39, 42	MS892 Rallye 18	Nord NC701 see Siebel 204
48, 52, 53, 64, 101, 162, 163	MS893 Rallye 125	Nord NC702 see Siebel 204
165, 167, 168, 170, 175, 177	Mu-2 (Mitsubishi) 428, 430	Nord NC856 Norvigie 19, 71
179, 180, 184, 185, 186, 193	Mü10 Milan 213	74, 76, 90, 104, 111
194, 199, 200, 206, 210, 213	Mü13 Bergfalke I 56, 213	Nord NC900 see Fw190
215, 217, 218, 219, 223, 225	Muegyetemi M.24 278	Norseman (Noorduyn) 351, 352
227, 230, 231, 250, 266, 331	Mystère 2 94, 105	Northrop N-3PB 351
336, 357, 362, 375, 378, 408	106, 119, 124	Nyberg Flugan 427
MiG-25 68	Mystère 4 23, 28, 71, 74, 76	
MiG-29 36, 200	78, 79, 90, 97, 101, 103, 104	O-1 Bird Dog see L-19
Milita MB3 Leonardo 326	106, 109, 111, 113, 118, 119	Oxford (Airspeed) 17
Mirage 3 72, 74, 76	120, 121, 123, 124, 125, 126	
77, 78, 79, 80, 85, 86, 87		P2 (Pilatus) 222, 434, 444
88, 89, 90, 92, 94, 96, 97	N20 Aiguillon 434	P3 (Pilatus) 222, 226
98, 99, 101, 102, 103, 104	N24 Nomad 335	434, 438, 439
106, 109, 110, 111, 112, 115	NAB 9 (Albatros B.II) 427	P31 Proctor IV 17, 420
117, 118, 119, 120, 121, 122	Nagler Rolz NR54 174	P34 Proctor III 58
123, 124, 125, 126, 127, 128	Neukom S-3 Elfe 439	P34 Proctor T.1 22
194, 199, 226, 350, 358, 366	NH500 (Nardi) see Hughes 500	P40 Prentice 23
392, 408, 411, 416, 435, 444	NHI H-3 336, 340	P44 Proctor V 22
Mirage 4 73, 76, 78, 80	Nielsen & Winther Aa 55	P66 Pembroke 13, 17, 19, 69
86, 101, 106, 110	Nieuport IV 410, 422	167, 195, 204, 208, 212, 225
111, 117, 119, 122	Nieuport 11 42, 108, 339	421, 422, 424, 425, 426, 429
Mirage 5 17, 20, 21, 23, 24, 25	Nieuport 17 202	P66 President 55
27, 28, 78, 80, 81, 86, 88	Nieuport 23 17	P-2 Neptune 76, 93, 97
90, 113, 119, 120, 121, 194	Nieuport 28 434, 438	103, 110, 114, 116, 223, 333
Mirage 2000 74, 81, 90	Nieuport Delage 29 108	340, 341, 342, 344, 393, 394
98, 109, 117, 123, 242	Nilsson BEDA 427	P-3 238, 243, 256, 386, 406
Mirage F1 72, 77	Nord N260 89, 116	P-39 Airacobra 69
78, 79, 81, 86, 96, 101	Nord N262 91, 94	P-47 Thunderbolt 110
113, 116, 120, 128, 403	102, 111, 169	163, 325, 452
Mirage F2 124	Nord N1002 see Bf108	P-51 Mustang 110, 325
Mirage G8 101, 109	Nord N1101 16, 71	341, 421, 434

TYPES INDEX – 468

PA16 Clipper	61	
PA18 Super Cub	see L-18	
PA22 Tri-Pacer	90, 350, 420	
PA23 Apache	55, 408, 409, 430	
PA24 Comanche	413	
PA27 Aztec	336	
PA28	66, 68, 350, 430	
PA30 Twin Comanche	413	
PA31	116, 126, 247, 430	
PA35 Pocono	380	
Päätalo Tiira	69	
Partenavia P53	321, 326	
Partenavia P64 Oscar	276, 282	
Partenavia P66	263	
Pasotti F9 Sparviero	278	
Paumier Biplan	108	
Payen Pa49 Katy	109	
PBV Catalina	116	
PBY Catalina	57, 344, 350, 354, 409, 422	
PC-7 (Pilatus)	435	
PD808 (Piaggio Douglas)	281	
Pe-2 (Petyalakov)	350, 379	
Persson Pappfokkers	427	
Pfalz D.XII	108	
Phantom	see F-4	
Phönix 122	421	
Piaggio P136	326	
Piaggio P148	263, 264, 265, 266, 267, 268, 274, 275, 280, 284, 285, 323	
Piaggio P149	162, 163, 165, 166, 167, 168, 173, 174, 185, 186, 188, 199, 202, 203, 206, 207, 211, 212, 222, 228, 230, 231, 269, 348, 354	
Piaggio P150	325	
Piaggio P166	260, 262, 263, 264, 265, 267, 269, 271, 272, 275, 280, 285, 286, 324, 325, 328, 445	
Piel CP301 Emeraude	16, 354	
Piel CP1310	109	
Pik-3	66, 69	
Pik-5	66, 69, 425	
Pik-10	66	
Pik-11	66	
Pik-12	66	
Pik-16	66	
Pik-20	66, 431	
Pitts S1 Special	420	
PL-12 Airtruk (Transavia)	61	
Polikarpov Po-2	39, 179, 253, 357, 361, 362, 363	
Polikarpov I-15	410	

Polikarpov I-16	65, 410	
Polikarpov I-153	110	
Polikarpov U-2	69	
Polyt I (Polyteknisk)	55	
Polyt II (Polyteknisk)	57	
Polyt III (Polyteknisk)	54	
Potez 36	108	
Potez 43	109	
Potez 53	87, 110	
Potez 94	103	
Potez 842	112	
Praga E-114 Air Baby	40	
Praha PB.3	46	
PT-17 Kaydet	110, 112	
PT-18 Kaydet	194	
PT-19 Cornell	351	
PT-26 Cornell	55, 350, 353	
Pützer Motorraab	213	
PV Ventura	324, 384, 396	
PWS 26	363	
PZL 101 Gawron	246, 247, 255	
PZL 104 Wilga	42, 162, 252	
PZL 106 Kruk	162, 176, 177, 187, 223, 226, 364, 367	
PZL 110 Koliber	367	
PZL 130 Orlik	359, 380	
PZL P11	362	
PZL P24	453	
PZL M2	377	
PZL M4 Tarpan	363	
PZL M18	37, 187, 236	
PZL M21	368	
PZL M24	368	
PZL S4 Kania	363	
Quickie	66, 209	
R4D Skytrain	see C-47	
Raab Krahe 2	226	
Raab Katzenstein RK-26	421	
RAF BE.2	351	
RAF RE.8	17	
Rallye 235 (SOCATA)	107	
RC-3 Seabee	321	
Re2000 Falco	321, 421	
Re2001	285	
Re2002 Ariete	97	
Rearwin 9000L Sportster	61	
REP type D	112	
REP type K	108	
RF2 (Fournier)	109	
RF3 (Fournier)	125	
RF10 (Fournier)	384, 395	
Ricci 6	278	
Rieseler R.III	427	

Robin HR200 Club	336	
Rockwell Commander	see Aero Commander	
Rumpler 6	64	
Rumpler C.IV	18, 209	
Rumpler Taube	166, 209	
RW3	222	
RWD13	363	
RWD21	363	
Ryan NYP	202	
S.1A Cadet (Interstate)	351	
S-2 Tracker	262, 263, 279, 325, 336, 340, 341, 344, 449, 456	
S-11 Instructor	331, 333, 339, 340, 341, 342	
S-12 (Fokker)	348	
S-13 (Fokker)	343	
S-14 Machtrainer	333, 339, 340, 342	
S-55	see H-19	
S-58	see H-34	
S-61 (Sikorsky)	269	
S-64 (Sikorsky)	194	
S-76 (Sikorsky)	406	
SA226 Metro	11, 15, 59, 67, 408	
SA321 Super Frelon	94, 95, 99, 109, 124	
SA330 Puma	93, 102, 386	
SA341 Gazelle	89, 93, 99, 126	
SA349	90	
SA360	116	
SA361	90, 99	
SA365	128	
SA610 Ludion	109	
SA-103 Emouchet	112	
SA-104 Emouchet	110	
SAAB 17	55, 421, 423	
SAAB 18	421	
SAAB 21	421, 422, 426	
SAAB 29	10, 11, 12, 13, 55, 111, 324, 420, 421, 422, 423, 424, 425, 428, 429, 430	
SAAB 32 Lansen	11, 57, 111, 419, 420, 421, 422, 423, 424, 425, 426, 428, 429, 430	
SAAB 35 Draken	10, 11, 17, 54, 55, 56, 57, 58, 59, 60, 62, 64, 65, 67, 68, 110, 213, 350, 354, 418, 419, 420, 421, 422, 423, 424, 425, 426, 427, 428, 429, 430	
SAAB 37 Viggen	227, 418	

TYPES INDEX – 469

419, 422, 425, 426, 428, 430
SAAB 91 Safir   10, 11, 12, 13
   65, 67, 68, 69, 70, 107, 333
   336, 345, 348, 349, 350, 351
   352, 354, 355, 420, 421, 423
   424, 425, 426, 427, 428, 430
SAAB 105   11, 422, 423
SAAB 210   422
SAAB 2000   423
SAAB SF340   421, 428
SABCA Junior   18
SABCA Vivette   18
SAIMAN 202   267
   277, 321, 326
Sandringham (Short)   112
Santos Dumont XX   108, 383
SB2C Helldriver   235
SC7 Skyvan   67, 237, 343
Schreck FBA   17, 108, 389
Scout   178
SD330 (Short)   237
SE210 Caravelle   18, 25, 55
   59, 73, 74, 75, 76, 87
   88, 91, 98, 99, 100, 103
   104, 107, 109, 110, 111, 112
   113, 115, 125, 126, 244, 266
   267, 276, 278, 286, 327, 340
   353, 424, 426, 439, 449, 453
SE535 Mistral   see Vampire
SE3101   110
SE5003 Baroudeur   111
Sea Fury   341
Sea Hawk   167, 183
   212, 222, 333, 341
Secan SUC10 Courlis   109
Seversky EP106   421
SF-260   15, 19, 20, 21, 25, 323
Shackleton   32
SIAI 3 Eolo   325
SIAI S205   106, 275
SIAI S211   269, 453
SIAT 223 Flamingo   213
Siebel 204   40, 42, 46, 53, 95
   118, 168, 353, 400, 423, 426
Sioux   192
Siren Bertin C34   110
Skeeter (SARO)   168, 174, 222
Skyraider   see A-1
Slingsby T.21   392
Slingsby T.34 Sky   410
SM30 (Merville)   125
SM56   325
SM79   288, 325
SM80   321
SM82   325

SM102   278, 321
SM1019   260, 261
   262, 270, 286, 328, 459
SM-1 (WSK)   see Mi-1
SM-2 (WSK)   42, 358
   359, 361, 363, 366, 379
SN601 Corvette   107
SNB Expeditor   see C-45
SNECMA C.400   109
SNJ Texan   see T-6
SO30 Bretagne   122
SO1110 Ariel   109
SO1220   109
SO1221 Djinn   90, 94, 99
   100, 104, 123, 125, 174, 434
SO6000 Triton   109
SO6025 Espadon   125
SO9000 Trident   109
Soldenhoff S-5   439
Sopwith 1½ Strutter   17, 108
Sopwith F.1 Camel   202
SPAD VII   39, 108, 326
SPAD XIII   17, 108
SPAD 52   108
SPAD 54   109
Spalinger S21   439
Sparmann S.1   421
Spatz B   61
Specht   427
Spitfire   17, 20, 39, 55, 73, 110
   202, 235, 325, 330, 332, 335
   337, 338, 340, 341, 342, 343
   350, 351, 353, 362, 383, 421
Sportavia C1   168
SR-7   18
Stahltaube (Jeannin)   363
Stinson 108 Voyager   see L-5
Su-7   34, 39, 41, 43, 45
   46, 47, 52, 53, 119, 165
   191, 194, 228, 231, 266
   359, 362, 366, 373, 375
   376, 378, 380, 400, 402
Su-20   206, 337, 357, 358, 366
   371, 373, 375, 376, 380
Su-22   10, 163, 167, 175, 177
   186, 187, 194, 200, 201, 206
   212, 215, 217, 218, 220, 223
   225, 227, 229, 250, 254, 266
   357, 362, 375, 379, 398, 408
Su-25   51, 398
Super Etendard   95, 116
Super Mirage 4000   110
Super Mystère B2   23
   28, 72, 74, 75, 77, 78, 79
   89, 97, 98, 100, 101, 103

   104, 106, 109, 111, 118, 120
   121, 125, 126, 127, 166, 194
SV-4   16, 17
   18, 19, 61, 90, 109, 116
   202, 222, 227, 339, 410
Sycamore   13, 163, 168
   174, 193, 203, 207, 432
SZD-8 Jaskolka   18
SZD-9 Bocian   66
SZD-10 Czapla   65, 69
SZD-22 Mucha   69
SZD-24 Foka   110
SZD-30 Pirat   55

T-6 Texan   17, 39, 55, 59
   61, 72, 74, 75, 87, 94, 95, 97
   100, 102, 107, 109, 116, 119
   121, 125, 166, 194, 203, 224
   228, 231, 234, 235, 236, 257
   259, 262, 263, 264, 267, 269
   271, 272, 273, 275, 276, 277
   278, 279, 282, 283, 284, 285
   286, 287, 321, 325, 326, 327
   328, 335, 336, 337, 338, 340
   341, 342, 345, 349, 354, 362
   385, 386, 387, 389, 392, 393
   395, 396, 407, 409, 410, 411
   412, 413, 414, 416, 417, 421
   423, 425, 429, 430, 434, 452
T-28 Trojan   16, 111
T-33   15, 17, 30, 55, 57, 59
   62, 74, 75, 76, 89, 101, 103
   107, 112, 115, 120, 122, 125
   126, 127, 162, 167, 168, 173
   188, 191, 192, 193, 195, 197
   203, 213, 222, 223, 225, 227
   233, 234, 235, 237, 238, 239
   240, 242, 243, 244, 258, 262
   263, 267, 269, 272, 274, 275
   276, 279, 282, 283, 324, 325
   327, 333, 336, 342, 344, 345
   351, 354, 384, 387, 389, 390
   392, 394, 404, 405, 406, 407
   409, 410, 411, 412, 413, 414
   415, 416, 417, 447, 450, 451
   452, 453, 454, 457, 458, 459
T-34 Mentor   403, 408
   409, 412, 416, 452, 453, 458
T-37   351, 384, 385, 386
   392, 394, 395, 396, 457
T-38 Talon   383, 386, 387
   389, 392, 394, 396, 453
T-39 Sabreliner   287
T-41 Mescalero   458
Taylorcraft   61, 354, 420

Taylor E2 Cub	39	
TB30 Epsilon	88, 395	
Ternen	57	
Texan	see T-6	
Thubulent D (Stark)	209	
Thulin A	427	
Thulin D	69	
Tipsy B	429	
Tipsy Nipper	18	
Tipsy S.2	18, 429	
Tipsy Trainer	19	
TOM-8	39	
Tornado (Panavia)	173	
	183, 184, 185, 198, 206	
	229, 262, 264, 271, 325	
Transavia T300	409	
Trident	31	
TRS-1 Hummel	174	
TS-8 Bies	267, 356, 358, 359	
	361, 362, 364, 367, 374	
	376, 378, 379, 380, 381	
TS-11 Iskra	11, 28, 357	
	358, 359, 360, 362, 363	
	366, 367, 371, 373, 375	
	377, 379, 380, 381, 382	
Tu-2	362, 379	
Tu-104	31, 42, 45, 49	
Tu-134	33, 40, 48	
	169, 178, 187, 195, 205	
	214, 225, 247, 252, 276	
	367, 379, 380, 399, 432	
Tu-154	179, 228, 247	
Tu-A3	247	
Tu ANT-40	70	
Tu SB-2	70	
Twin Pioneer (Scottish Av)	442	
U-2	349	
U-6 Beaver	65, 69	
	169, 340, 342	
U-16 Albatross	235, 238, 243	
	262, 263, 265, 272, 274	
	275, 284, 324, 408, 412	
U-17	240, 451	
UTI-4	see Polikarpov I-16	
V-10 Bronco	101, 166	
	168, 174, 179, 185	
	193, 199, 222, 226	
VAK-191	199, 206, 214	
Valmet L.90	69	
Valmet LEKO 70	70	
Valmet Myrsky	69	
Vampire	11, 17, 40, 65, 69, 71	
	101, 106, 109, 111, 113, 115	
	119, 125, 193, 222, 226, 277	
	283, 325, 330, 350, 351, 353	
	354, 383, 421, 422, 429, 432	
	433, 434, 435, 436, 437, 438	
	439, 440, 441, 442, 443, 444	
Vanguard	113	
Varsity (Vickers)	168, 422	
Vautour 2	71, 74, 75	
	76, 79, 88, 91, 94, 98	
	99, 101, 103, 106, 112	
	115, 120, 121, 122, 125	
VBS-1 Kunkadlo	46	
VC10 (Vickers)	195	
VEB-14	see Il-14	
Venom	11, 111, 115, 119, 194	
	224, 269, 422, 423, 432	
	433, 434, 435, 436, 437	
	438, 439, 440, 441, 444	
Vertol 44	421, 427, 430	
Vertol 107	422, 427	
VFW614	171, 227	
VFW H2	174	
VFW H3 Sprinter	174	
VJ101	209	
Vi-22 Autogyro	194, 199	
Viberti Musca	321	
Viking	13, 102, 444	
Viscount	27, 96, 100, 184, 187	
	195, 225, 228, 283, 415	
	420, 426, 446, 451, 453	
Vizzola II	321	
VL Humu	68	
VL Kotha	70	
VL Pyörremyrsky	69	
VL Pyry	65, 69, 70	
VL Sääski	64, 65, 69, 351	
VL Tuisku	65, 70	
VL Tuuli	66	
VL Vihuri	65, 69	
VL Viima	66, 68	
Voisin LA5	18, 108	
Voisin Farman 1	108	
Vollmoller	214	
Vuia 1	108	
W-3 Sokol (PZL)	379	
WA30 Bijave	112	
WA51 Pacific	112	
WA54 Atlantic	66	
Wagner Eule	364	
Wagner Rotocar 3	174	
Wasp (Westland)	178, 333	
Weihe	22, 56, 69	
	110, 112, 410, 421	
Wessex	178, 192	
WGM-21 (Aerotechnik)	174	
Whirlwind	31, 178, 193	
	194, 404, 407, 409	
Wiegel Harz Fink	222	
WSK M15 Belphegor	252, 362	
Wright Flyer	54, 108, 341	
Wright Type A	209	
WW1123	349	
WWS-1 Salamandra	66	
X-113	175	
X-114	175	
XA-66 Aeron	33	
XM-12 Makrol	41	
Yak-3	108	
Yak-9	360, 379	
Yak-11	11, 16, 39, 41, 42, 43	
	166, 176, 252, 359, 362, 377	
Yak-12	42, 253, 360, 363, 377	
Yak-17	39, 363	
Yak-18	168, 179, 202	
	253, 358, 360, 363	
Yak-23	39, 358, 360, 361	
	362, 378, 379, 381	
Yak-27	177, 178	
Yak-28	246	
Yak-40	38, 41	
Yak-50	213	
YS-11	237, 238	
Z506 (CANT)	325	
Z-12 Zögling	18, 439	
Zlin XII	39	
Zlin XIII	46	
Zlin 22 Junak	40	
Zlin 26 Trener	40, 363	
Zlin 37 Cmelak	13, 42, 43, 52	
	53, 65, 164, 170, 176	
	177, 178, 186, 187, 207	
	213, 215, 217, 222, 230	
Zlin 42	42, 164	
Zlin 43	42	
Zlin 50	40, 46	
Zlin 126	42	
Zlin 135	39	
Zlin 226	33, 40, 42, 43, 247	
Zlin 326	40, 43	
	109, 247, 253, 410	
Zlin 381	40, 224	
Zlin 526	192, 222, 247	

WRECKS & RELICS
16th Edition
Ken Ellis

AVIATION MUSEUMS OF BRITAIN
Aviation Pocket Guide 2 (2nd edition)
Ken Ellis

We hope you enjoyed this book...

Aerofax and Midland Publishing titles are edited and designed by an experienced and enthusiastic trans-Atlantic team of specialists.

Further titles are in preparation but we always welcome ideas from authors or readers for books they would like to see published.

In addition, our associate company, Midland Counties Publications, offers an exceptionally wide range of aviation, spaceflight, astronomy, military, naval and transport books and videos for sale by mail-order around the world.

For a copy of the appropriate catalogue, or to order further copies of this book, and any of the titles mentioned elsewhere on this or the next page, please write, telephone, fax or e-mail to:

Midland Counties Publications
Unit 3 Maizefield,
Hinckley, Leics, LE10 1YF,
England

Tel: (+44) 01455 233 747
Fax: (+44) 01455 233 737
E-mail: midlandbooks@compuserve.com

Wrecks & Relics is an institution. Each edition is eagerly awaited by enthusiasts, historians, owners and operators of historic aircraft and curators of aviation collections as the most trusted and hard-working of references. Now in its 37th year of publication, the 16th edition marks the author's 25th year at the helm and to celebrate, takes an occasional look back at the scene in 1974.

The book takes the reader on a geographical journey through the fascinating world of museums, military stores and dumps, 'geriatric' airliners awaiting the axe, restoration workshops, technical schools, treasures in garages and barns and much more. Fully revised and updated, the 16th edition has an array of appendices and the usual extensive indexing and cross-referencing. The comprehensive photographic section is full of out-of-the-way subjects.

Previous editions still in print:
12th edition (1990) 252pp £9.95
14th edition (1994) 336pp £12.95
15th edition (1996) 350pp £14.95

Hardback
210 x 148 mm, 352 pages
191 b/w and 3 colour photographs
1 85780 079 6
£14.95

Aviation Museums of Britain has established itself as a valued on-the-spot guide to the nation's aviation museums. Much thought was given to making the guide really work hard for its reader with particular attention being paid to what facilities are available and what else can be seen of non-aviation interest in the area – vital information for family outings.

Here is the answer to questions that most people need to know before setting out: When is it open? Which ones are open off-season? Is there a cafe? Is there somewhere for the youngsters to play? Is there a shop to browse around? This book answers all of these questions and more while providing an easy-to-read review of aircraft exhibits plus a breakdown of other displays, features and themes within Britain's 93 aviation museums.

All of these features combine to make this a unique guide which opens up many ideas and possibilities. A constant companion in the glove box or the pocket.

Softback
150 x 105 mm, 128 pages
86 b/w and 3 colour photographs
1 85780 078 8
£6.95

Advertisement

Other titles from Midland Publishing

McDonnell Douglas F-15 Eagle
Sbk, 280 x 216 mm, 112pp
c.200 photos. £14.95

McDonnell Douglas DC-10 and KC-10 Extender
Sbk, 280 x 216 mm, 128pp
290 photos. £19.95

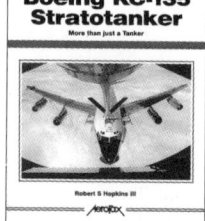

Boeing KC-135 Stratotanker
Sbk, 280 x 216 mm, 224pp
256 photos. £24.95

MiG-21 'Fishbed'
Sbk, 280 x 216 mm, 144pp
381 photos. £16.95

MiG-25 'Foxbat' MiG-31 'Foxhound'
Sbk, 280 x 216 mm, 96pp
110 photos. £12.95

Tupolev Tu-95/-142 'Bear'
Sbk, 280 x 216 mm, 128pp
260 photos. £14.95

Tupolev Tu-22 'Blinder' Tu-22M 'Backfire'
Sbk, 280 x 216 mm, 96pp
c.140 photos. £14.95

Convair B-58 Hustler
Sbk, 280 x 216 mm, 152pp
477 photos. £16.95

Eurofighter 2000
Sbk, 280 x 216 mm, 48pp
58 colour, 4 b/w photos. £8.95

AIR FORCES UK
Sbk, 148 x 105mm, 160pp
142 photos, 33 3-views. £6.95

DISCOVER AVIATION TRAILS
Sbk, 148 x 105mm, 128pp
97 b/w, 3 colour photos. £5.95

OKB SUKHOI
Hbk, 280 x 216 mm, 296pp
678 photographs. £29.95

AIRLINE TAIL COLOURS
Sbk, 150 x 105mm, 128pp
485 colour photos. £6.95

U.S. Military Aircraft Designations and Serials
Sbk, 210 x 148 mm, 252pp
£12.95

U-BOAT FACT FILE
Sbk, 234 x 156mm, 224pp
12 b/w photos. £16.95

LOCATIONS INDEX

Aalsmeerderbrug	331	Annecy	72	Basel	432
Aarburg	432	Annemasse	72	Baschutz	164
Abbenes	331	Ansbach	162	Basepohl	164
Abbeville	71	Antwerpen	14	Basiliano	260
Acharnae	232	Apt	72	Bautzen	165
Acqui Terme	257	Arlon	14	Bayeux	73
Agrinion	232	Aranda de Duero	404	Bayreuth	165
Ahaus	161	Aranjuez	404	Beaune	73
Ahlhorn	161	Aráxos	232	Beauvais	74
Ähtäri	64	Arboga	418	Bechyně	33
Aigen in Ennstall	10	Arcen	331	Beelitz	165
Aigle	432	Arranhó	386	Beervelde	14
Air sur L'Adour	71	Aschersleben	162	Beja	386
Aix	71	Assen	331	Békéscsaba	246
Ajka	246	Ath	14	Benedita	387
Akhisar	446	Athíne	232	Benidorm	404
Akrotiri	31	Atina	258	Bentlage	165
Alach	161	Au	432	Berga	418
Albano Terme	257	Aubenas	73	Bergamo	260
Albacete	403	Auenhausen	162	Bergen	349
Albert	71	Augsburg	162	Bergerac	74
Albertville	72	Augusta	259	Berlin	166
Ålborg	54	Aulnay sous Bois	73	Bern	433
Alcalá	404	Aurich	163	Bernsdorf	169
Alcantarilla	404	Aviano	259	Best	332
Alcochete	383	Avignon	73	Bétera	405
Alès	72	Avord	73	Bevekom	14
Alesssandria	257			Beveren aan de IJzer	15
Alicante	404	Baak	331	Beverlo	15
Alfeite	383	Baarlo	331	Bevingen	15
Alghero	257	Babie Doły	356	Bex	433
Ålholm	54	Babimost	356	Beynes	74
Allstedt	161	Bachero di Cíngolia	259	Beziers	74
Almagro	404	Bad Düden	163	Biała Błota	357
Alpnach	432	Badhoevedorp	332	Biała Podlaska	357
Alsónémedi	246	Bad Ischl	10	Biarritz	74
Altenburg	161	Bad Oeynhausen	163	Bibano	260
Altenrhein	432	Bad Sooden - Allendorf	163	Biberach	169
Altenstadt	161	Bakov	33	Biella	260
Altes Lager	162	Balen	14	Bierset	15
Alverca	383	Balice	356	Bilé Poličany	33
Ambérieu en Bugey	72	Balikesir	446	Billund	54
Améndola	257	Banak Lakselv	349	Birkenfeld	169
Andøya	349	Bandirma	447	Birzebuggia	329
Andravida	232	Banská Bystrica	397	Bitburg	169
Andrychów	356	Barcelona	404	Blomberg	170
Ängelholm	418	Bardufoss	349	Bochum	170
Angers	72	Barcelonnette	73	Bócsa	246
Ankara	446, 458	Bareggio	259	Boden	418
Anklam	162	Bari	259	Bodø	349
Ankum	162	Barreiro	386	Bologna	260

LOCATIONS INDEX – 474

Bonn	170	Candás	405	Coimbra	388
Bordeaux	74	Cannes	77	Cognac	88
Börgönd	246	Capelle aan de IJssel	332	Coleman Barracks	175
Borkheide	170	Captieux	77	Colmar	88
Borkum	171	Capua	263	Colmenar Viejo	405
Bouilly	75	Carpi	263	Colombier	433
Bourg en Bresse	75	Carrara San Pelagio	263	Comignago	269
Bourges	75	Carvoeira	387	Compiègne	88
Bořetice	32	Casale	263	Conegliano D'Otranto	269
Borgo Piave	261	Cascais	387	Contrexéville	88
Bracciano	261	Caserta	263	Córdoba	405
Bramois	433	Čáslav	34	Corral de Ayllón	405
Brandis	171	Cassino	264	Coruña del Conde	405
Brasschaat	16	Castel del Rio	264	Cottbus	175
Bratislava	397	Castello di Annone	264	Coulanges sur L'Autize	88
Braunschweig	171	Castelnau Magnoac	78	Crailo	332
Breda	332	Castel Volturno	264	Creil	88
Bree	16	Catania	266	Cressanges	89
Bremen	171	Cavalcaselle	266	Cuers	89, 458
Bremerhaven	171	Cazaux	78	Curlu	447
Bremgarten	171	Cecchina	266	Czaplinek	358
Breno	261	Celle	175		
Brétigny sur Orge	76	Ceolini	266	Darłówko	358
Brezová	33	Cerbaiola	266	Darmstadt	177
Briare	76	Cergnago	267	Dax	89
Brienne le Château	76	Černík	397	Decimomannu	269
Briest	171	Červene Janovice	34	Dęblin	358
Brignoles	76	Cérvia	267	Deelen	332
Brindisi	261	Cesena	268	Dehtín	36
Brive	76	Česká Olešná	34	De Kooy	332
Brno	33	Česká Skalice	35	Delft	333
Broumov Bylnice	34	Česká Třebová	35	Den Dolder	333
Brüggen	173	České Budějovice	35	Den Haag	333
Brussels	16	Cevezili	447	Den Ham	334
Brustem	19	Chambon sur Voueize	78	Den Helder	334
Bruz	77	Charleroi	19	De Peel	334
Bubovice	34	Chartes	78	Dermsdorf	177
Büchel	173	Châteaudun	78	Dessau	177
Bückeburg	173	Châteauroux	87	Detmold	178
Budaörs	246, 458	Châtellerault	87	Dhekeli	31
Budapest	247	Châtillon en Diois	87	Diepensee	178
Budel	332	Cherance	87	Diepholz	178
Burgau	175	Cherbourg	87	Diepolsau	433
Bützberg	433	Chièvres	19	Dijon	90
Butzweilerhof	175	Cisterna di Latina	268	Diyarbakir	447
Buzica	397	Chaves	387	Dinan	90
Bydgoszcz	357	Chorzów	358	Dinant	20
		Chotusice	36	Dinard	90
Cadibona	262	Chrudim	36	Doetinchem	334
Caen	77	Claro	433	Dolni Poustevna	36
Cagliari	262	Clères	87	Domme	91
Calais	77	Clermont Ferrand	87	Dordrecht	334
Cambrai	77	Cloyes sur le Loir	88	Dos Barrios	405
Cámeri	262	Coazze	268	Dos Hermanas	405
Cämmerswalde	175	Codróipo	269	Dresden	178

LOCATIONS INDEX – 475

Drewitz	179	Fiorano Modenese	270	Granada	406
Driebruggen	334	Firenze	270	's Graveland	335
Drienov	397	Flensburg	187	Graz	10
Drzonów	359	Florennes	20	Grazzanise	272
Druztová	36	Fonte	270	Grenå	56
Dübendorf	434	Fontenay	92	Grenoble	93
Duiven	334	Forlì	270	Grimstad	352
Dulliken	435	Fossano	270	Grodzisko	360
Dunaújváros	247	Frankfurt	187	Grontardo	272
Düren	183	Fredensborg	56	Gross Dolln	191
Düsseldorf	183	Frederikshavn	56	Grossenbrode	191
Dussnang	435	Freienbach	436	Grossenhain	191
		Fréjus	92	Grosser Weserbogen	191
Ecrouves	91	Friedrichshafen	187	Grosseto	272
Egerkingen	435	Friedrichsthal	188	Grossmachnow	191
Egeskov Castle	56	Frosinone	270	Grossrohrsdorf	192
Eggebeck	183	Fritzlar	188	Grottáglie	273
Eilenburg	183	Frutigen	436	Gudsø	57
Eindhoven	334	Fulenbach	436	Guiscriff	93
Elbląg	360	Funchal	388	Gütersloh	192
Elefsís	237	Fürstenfeldbruck	188	Gut Kummersdorf	192
Emmen	435			Güvercinlik	451
Épinal	91	Gaanderen	335	Györ	248
Ercan	31	Gap	92		
Erding	183	Gallarate	271	Haarlem	335
Erfurt	184	Gals	436	Hadsund	57
Erhac	448	Gambéttola	271	Hahn	192
Erkilet	448	Ganderkesee	190	Hal Far	329
Esbjerg	56	Gardermoen	351	Halle	192
Escalona	406	Gavres	93	Halli	64
Escherhausen	184	Gdynia	360	Halmstad	419
Eskilstuna	418	Geel	21	Hamburg	192
Eskisehir	448, 458	Geiselwind	191	Hammelburg	193
Esneux	20	Geldenaken	21	Hanau	193
Espartinas	406	Geldrop	335	Harbke	193
Essen	184	Geneve	436	Havlíčkuv Brod	37
Esztergom	248	Gerasdorf	10	Heiloo	336
Étain	91	Germersheim	191	Heino	336
Étampes	91	Ghedi	271	Helchteren	21
Eu sur Mer	92	Gibraltar	406	Helsinge	57
Evere	20	Gilching	191	Helsingki	64
Evora	388	Gilze Rijen	335	Helsingør	57
Fagioli di San Ilario	269	Gioia del Colle	271	Hénin Beaumont	93
Évreux	92	Gits	21	Heráklion	238
		Gliwice	360	Hermeskeil	193
Falaise	92	Glons	21	Herning	58
Falconara	270	Goetsenhoven	21	Herstal	21
Falkenberg	184	Goleniów	360	Hesirau	436
Faro	388	Gondomar	388	Hildesheim	195
Farum	56	Gorizia	272	Hillerstorp	419
Fassberg	184	Goslar	191	Hnjotur	256
Féniers	92	Gosselies	21	Hochstetten	436
Fichtelberg	186	Goszczyń	360	Hoek van Holland	336
Finow	186	Göteborg	419	Hoeven	336
Finsterwalde	187	Grandvillard	436	Hohn	195

LOCATIONS INDEX – 476

Holíč	397	Kalmhout	22	Kuopio Rissala	67		
Holstebro	58	Kalo Lakatamia	31	Kussnach	437		
Holzdorf	195	Kaltwasser	197	Kymi	67		
Hoofddorp	336	Kamenz	197				
Hoornsterzwaag	336	Kampenhout	22	Laage	199		
Hopsten	196	Kapellen	22	Laarbruch	201		
Hořice	37	Kaposvár	248	Laatzen	202		
Hourtin	93	Kåremo	420	La Baule	94		
Hoyerswerda	196	Karlsborg	420	La Ferté Alais	94		
Hrabyně	37	Karlsruhe	198	Lahr	202		
Hradec Králové	37	Karstula	66	L'Aigle	95		
Hrbov	37	Karup	58	Lajes	388		
Hubhof	11	Kassel	198	Lagny sur Marne	95		
Husum	196	Kastéllion	238	Lampedusa	274		
Hvalsmoen	352	Katahas	238	Lancenigo	274		
Hyères	93	Kato Achia	238	Landivisiau	95		
Hyrylä	66	Kaufbeuren	198	Landsberg	203		
		Kauhava	66	Landshut	203		
Imatra	66	Kbely	38	Laneuveville en Saulnois	95		
Imola	273	Kecel	248	Langenbernsdorf	203		
Incirlik	452	Kecskemét	248	Langenthal	437		
Ingolstadt	196	Keiheuvel	23	Lannion	95		
Inowrocław	360	Keflavík	256	Lanvéoc	95		
Interlaken	437	Kerpen	198	L'Aquila	274		
Ioánnina	238	Kiel	199	Lárissa	239		
Iserlohn	196	Kirkenes	352	Larnaca	31		
Isola del Liri	273	Kirkop	329	Las Bardenas Reales	406		
Isola Sacre	273	Kjeller	352	Lasclaveries	95		
Istanbul	452	Kjevik	352	Łask	364		
Istrana	273	Klatovy	42	Las Palmas de Gran Can.	407		
Istres	94	Klecany	42	Latina	274		
Iraklion	238	Kleine Brogel	23	Latresne	96		
Itzehoe	196	København	59	Lariano	275		
Izmir	453	Koblenz	199	Lauda	203		
		Kolbenmoor	199	Laupheim	203		
Jacou	94	Kolín	42	La Valette	96		
Jagel	197	Köln	199	Le Castellet	96		
Jagerspris	58	Kołobrzeg	361	Lecce	276		
Jahnsdorf	197	Kongelunden	59	Lechfeld	203		
Jakabszállás	248	Kongsberg	353	Leck	204		
Jegensdorf	437	Konya	454	Ledeberg	25		
Jerez de la Frontera	406	Koksijde	23	Leeuwarden	337		
Jever	197	Kopřivnice	42	Legnica	364		
Jihlava	37	Kormu	67	Le Havre	96		
Jilem	37	Kortrijk – Wevelgem	25	Leini	276		
Jimlín Zeměchy	37	Košice	397	Leipheim	204		
Jonstrup	58	Kosorín	398	Leipzig	205		
Josefodol	37	Kotroni	238	Leiria	389		
Józefów	361	Kozienice	361	Le Luc	96		
		Kraków	361	Lelystad	337		
Kalamáta	238	Královský Chlmec	398	Lempdes	96		
Kalisz	361	Kralupy nad Vltava	42	Lemwerder	205		
Kalken	21	Krzesiny	364	León	407		
Kalker	197	Kunčina Ves	43	Leopoldsburg	25		
Kalmar	420	Kunovice	43	Les Andelys	96		

LOCATIONS INDEX – 477

Les Mureaux	97	Marín	412	Morgenröthe Rautenkranz	208
Letňany	43	Markaryd	424	Morges	440
Levaldigi	275	Marl	206	Mořkov	43
Leźnica Wielka	364	Marmende	98	Morlaix	102
Libin	25	Marseille	98	Morón de la Frontera	412
Lido di Iésolo	275	Märsta	424	Mortagne au Perche	102
Limnos	240	Martigny	439	Mosbach	208
Limoges	97	Martin	398	Mošnov	43
Líně	44	Marxzell	207	Motril	412
Linköping	420	Masera	277	Mourmelon le Grand	102
Linz	11	Matran	439	Mulhouse	102
Lisboa	389	Meaux	99	München	208
Líšeň	44	Meerhout	25	Musingen	440
Livorno	276	Meiringen	439	Munster	210
Ljungbyhed	423	Mégara	240	Murcia	412
Łobez	364	Megliadino San Fidénzio	277	Murteira	391
Locarno	438	Mellingen	440		
Löddeköpinge	424	Melun	99	Náměšť nad Oslavou	45
Lodrino	438	Memmingen	207	Nancy	103
Łódź	364	Mendig	207	Nantes	104
Loheac	97	Mengam	99	Napoli	278
Lommel	25	Mengen	207	Narbonne	104
Lonate Pozzolo	276	Merseburg	207	Néa Ankhíalos	240
Lons le Saunier	97	Merville	99	Nechvalin	45
Loreto	276	Merzifon	456	Nedašov	45
Lorient	97	Messolongi	240	Neporadza	399
Łowicz	367	Messtetten	208	Neubiberg	210
Lubliniec	367	Metz	100	Neubrandenburg	210
Lucheux	97	Miastko	367	Neuburg	211
Lugo di Romagna	276	Midden Zeeland	337	Neubuz	45
Luleå	424	Mielec	367	Neuhardenberg	211
Lunéville	97	Mierzęcice	367	Neuhausen ob Eck	211
Luni	276	Milano	277, 459	Neuheim	440
Luqa	329	Minden	208	Neu Ulm	211
Luxeuil	98	Mirosławiec	373	Neuville	26
Lužany	44	Modena	278	Nice	104
Luzern	438	Modlin	374	Nicosia	31
Lyon	98	Moissac	100	Nidau	440
		Moisselles	100	Niederalteich	211
Maasbracht	337	Molenheide	26	Niederstetten	211
Macerata	276	Mollis	440	Nieuw Loosdrecht	337
Madrid	407	Mönchengladbach	208	Nieuw Millingen	337
Magadino	439	Montauban	100	Nijmegen	337
Mahlwinkel	206	Montceau les Mines	100	Nîmes	104
Malacky	398	Mont de Marsan	100	Nordhausen	212
Málaga	411	Montegaldella	278	Nordholz	212
Malmö	424	Montelimar	101	Norrköping	424
Malters	439	Monte Real	390	Norvenich	212
Manching	206	Monte San Savino	278	Nová Dubnica	399
Manerbio	277	Monte Tamaro	440	Nové Město nad Metuje	45
Maniago	277	Montijo	390	Novi Ligure	279
Mannheim	206	Montlaur en Diois	101	Nowa Sól	374
Mantagrava	277	Montluçon	101	Nowe Miasto nad Pilicą	374
Mantova	277	Montmirault	102	Nowy Tomysl	374
Marche	25	Montpellier	102	Noyant sur Allier	105

LOCATIONS INDEX – 478

Oberpfaffenhofen	212	Pescara	279	Raná	47
Oberschleissheim	213	Petříkovice	46	Rangsdorf	216
Odense	59	Petržalka	399	Raron	441
Oirschot	338	Peutie	26	Raufoss	353
Oksbøl	59	Pézenas	113	Ravenna	282
Oldenburg	214	Pferdsfeld	215	Rechlin	216
Oleśnica	374	Phalsbourg	113	Reek	338
Olomouc	45	Piacenza	280	Reggio Nell'Emilia	282
Olsztyn	375	Piamio	67	Reims	115
Opatová	399	Piešťany	399	Reinsdorf	217
Opole	376	Pieterlen	441	Rennes	115
Oostende	26	Piła	376	Reus	413
Orange	105	Pinhal Novo	394	Reykjavík	256
Orte	279	Pinneberg	216	Rezzato	282
Ørland	353	Piotrków Trybunalski	376	Ribnitz Damgarten	217
Orléans	106	Pisa	280	Rickenbach	441
Oroszlány	249	Plaisir	113	Rieti	282
Orzinuovi	279	Plassac	114	Rijsen	338
Oschersleben	214	Plejerup	60	Rimini	282
Osnabruck	214	Plobannalec	114	Rivolto	282
Oslo	353	Ploneis	114	Roanne	115
Oss	338	Plzeň	46	Rocamadour	115
Östersund	425	Podlipníky	399	Rochefort	115
Ostrów Wielkopolski	376	Poitiers	114	Rodez	117
Oświęcim	376	Pontarleir	114	Roma	282
Ota	391	Pont D'Ain	114	Romans sur Isère	118
Otrebusy	376	Pontedera	280	Romorantin	118
Oude Tongen	338	Pordenone	280	Ronneby	425
Oud Loosdrecht	338	Pori	68	Ronov nad Doubravka	48
Ovar	393	Potok	376	Rosmalen	338
Overboelare	26	Pourville sur Mer	114	Rota	413
Overloon	338	Poznań	376	Rotenburg	217
Ozigetvar	249	Pozzuoli	280	Roth	217
		Praha	46	Rothenburg	217
Paderborn	214	Prática di Mare	280	Rotte	339
Palermo	279	Prato	282	Rottenburg an der Laaber	221
Palhais	393	Předměřice nad Labem	47	Rotterdam	339
Palma de Mallorca	412	Přerov	47	Roudnice	48
Pápa	249	Preschen	216	Rouen	118
Paphos	31	Prešov	399	Rovaniemi	68
Pardubice	45	Preussisch Oldendorf	216	Rovereto	286
Paris	106	Préveza	241	Rygge	353
Parma	279	Přibor	47	Ryvangen	60
Parow	214	Prinsenbosch	338	Rzeszów	377
Pätz	215	Prostějov	47		
Pau	112	Pulsforde	216	Sabadell	413
Payerne	440	Punitz	12	Sabáudia	286
Pease	279	Purmerend	338	Saintes	118
Pécs	251			Saffraanberg	27
Peenemünde	215	Radom	377	Salamanca	413
Peer	26	Radomyśl Wielkopolski	377	Salerno	286
Perdasdefogu	279	Randers	60	Salon de Provence	118
Pergau	215	Ramsau am Dachstein	12	Salto di Quirra	286
Perpignan	113	Ramsen	441	Salzburg	12
Persan Beaumont	113	Ramstein	216	Samedan	441

LOCATIONS INDEX – 479

San Angelo in Villa	286	Söderhamn	426	Stuttgart	228
Sanchidrián	413	Soest	225	Sundsvall	427
Sanok	377	Soesterberg	341	Sulmona	287
San Possionio	286, 459	Sola	354	Svidník	401
San Fernando	413	Solbiate Olona	287	Świdnik	378
San Juan de Aznalfarache	414	Solenzara	121	Świdwin	379
San Martín de la Vega	414	Söllingen	225	Szeged	251
Santander	414	Solt	251	Szentkirályszabadja	251
Sao Jacinto	394	Soltau	225	Szolnok	252
Sarre Union	118	Sontheim	225	Szymaki	379
Sassuolo	287	Sonthofen	225		
Såtenäs	425	Sora	287	Tábor	49
Savigny en Septaine	118	Soúda	241	Talavera la Real	414
Savigny lès Beaune	118	Spa	27	Tampere	68
Savona	287	Spangdahlem	226	Tanágra	242
Schaarsbergen	339	Sperenberg	226	Tancos	395
Schaffen – Diest	27	Speyer	226	Tannheim	228
Schaijk	339	Śrem	378	Ta'Qali	329
Schiedam	339	St Augustin	227	Táranto	287
Schiphol	340	St Brieuc	121	Tarbes	123
Schönhagen	221	St Cyr	121	Tarnos	123
Schotz	441	St Dalmas de Tende	121	Taszár	254
Schwarzenbach	441	St Dizier	121	Taverny	123
Schwaz	12	St Gatien des Bois	122	Temmes	68
Schwelm	221	St Geoirs	122	Templin	228
Schwenningen	222	St Isidorushoeve	342	Temploux	27
Sec	48	St Jacques de Grasse	122	Tenerife	415
Seckenhausen	222	St Johann im Pongau	12	Texel	343
See	223	St Julian de Cassagnas	122	Tielen	27
Seifertshofen	223	St Malo	122	Tiendeveen	343
Seixas	394	St Martin	12	Tilburg	343
Sembach	223	St Mère Èglise	122	Til Châtel	124
Sesimbra	394	St Nazaire	122	Thessaloníki	243
Setúbal	394	St Péravy la Colombe	122	Thisted	62
Sevilla	414	St Philibert des Champs	123	Tikkakoski	68
Sézanne	121	St Rambert D'Albon	123	Tirstrup	62
Sezimove Ústí	48	St Stephan	443	Tököl	254
Siegburg	224	St Valery en Caux	123	Tønder	62
Siemirowice	378	Stade	227	Topel	456
Sigonella	287	Stans	443	Torino	287
Sila	456	Stara Boleslav	48	Torp	355
Sinsheim	224	Stauning	60	Torrejón de Ardoz	415
Sintra	394	Stefanavíkion	242	Toruń	379
Sion	442	Štěpánov	48	Touchay	124
Sivrihisar	456	Stilisa	242	Toul	124
Skarżysko Kamienna	378	Stod	48	Toulon	124
Skedsmo	354	Stockholm	426	Toulouse	124
Skokloster	426	Stölln	227	Tours	126
Skrydstrup	60	Stolzenau	228	Toussous le Noble	126
Slangerup	60	Stračov	49	Toužim	49
Sliač	400	Strasbourg	123	Trajouce	395
Słupsk	378	Strausberg	228	Trapani	288
Slušovice	48	Strumień	378	Trenčín	402
Smardzowe	378	Studenec	49	Trento	288
Soběslav	48	Studenká	49	Treviso	321

LOCATIONS INDEX – 480

Trino Vercellese	322	Vilanova I la Geltrú	416	Wildenrath	230
Tripolos	244	Vilar de Luz	396	Wilhelmshafen	230
Trollhaggen	428	Vila Real	396	Winterthur	444
Tullinge	428	Vila-Seca	416	Witkowo	381
Tulln – Langenlebarn	12	Villacoublay	127	Wittmund	230
Twello	343	Villanubla	416	Woensdrecht	345
Twenthe	343	Villafranca de Castello	416	Wohlen	444
		Villafranca di Verona	326	Wolgast	230
Überlingen	228	Villefranche sur Saône	127	Wolls Petersdorf	230
Udine	322	Villa Fresco de Azeitão	396	Wołów	381
Uetersen	228	Villebon sur Yvette	127	Wrocław	381
Ugglarp	429	Villeperdue	127	Wrzesnia	381
Uithuizen	343	Villmergen	444	Wünsdorf	230
Ulft	344	Villorba Castrette	327	Wunstorf	231
Ulm	229	Visan	127	Wuppertal	231
Ulrichen	444	Visby	430	Würzburg	231
Ummendorf	229	Viseu	396		
Uppsala	429	Viterbo	328	Yeni Foca	456
Ústí nad Ladem	49	Vitrolles	128	Ypenburg	347
Utrera	416	Viuz en Sallez	128		
Utti	70	Vizzola Ticino	328	Zábřeh	52
Uusikaupunki	70	Vlachovice	49	Zakopane	381
		Vlissingen	344	Zamość	381
Værløse	62	Vodochody	49	Zaragoza	417
Værnes	355	Vojens	63	Zawiercie	382
Vale de Lobos	395	Vojkovice	51	Zbraslavice	52
Valence	126	Volkel	344	Zele	30
Valencia	416	Volketswil	444	Zellik	30
Valetta	330	Volkmarsen	229	Zeltweg	13
Valkenburg	344	Vólos	245	Zepfenhain	231
Vandel	62	Vonitsa	245	Zgierz	382
Vannes	126	Vyhne	402	Žilina	402
Varennes sur Allier	127	Vyškov	51	Zliechov	402
Varpalota	255	Vyšny Komárnik	402	Zlin	52
Västerås	429			Zorgvliet	347
Vecsés	255	Waarschot	27	Zruč	52
Vedelago	323	Wageniec	379	Zürich	444
Veghel	344	Wangen an der Aare	444	Zwannenburg	348
Vejle	63	Warszawa	379	Zweibrücken	231
Veľká Losenise	49	Weelde	27	Zwijndrecht	348
Velke Chvalkovice	49	Weesp	344	Zwolle	348
Venegono	323	Well	345		
Venezia	323	Werneuchen	229		
Vergiate	323	Westerland	229		
Verona	323	Westerloo	29		
Vesivehmaa	70	Westerschouwen	345		
Vestone	324	Westouter	29		
Vicálvaro	416	Wetteren	29		
Vicenza	324	Widełka	379		
Videbæk	63	Wiedlisbach	444		
Vidsel	430	Wien	13	**Stop Press:**	
Vierzon	127	Wiener Neudorf	13		
Vigna di Valle	324	Wiener Neustadt Ost	13	Powidz	459
Vignale	326	Wieruszów	381	Przylep	459
Vijfeiken	344	Wiklon	381	Vestone	459